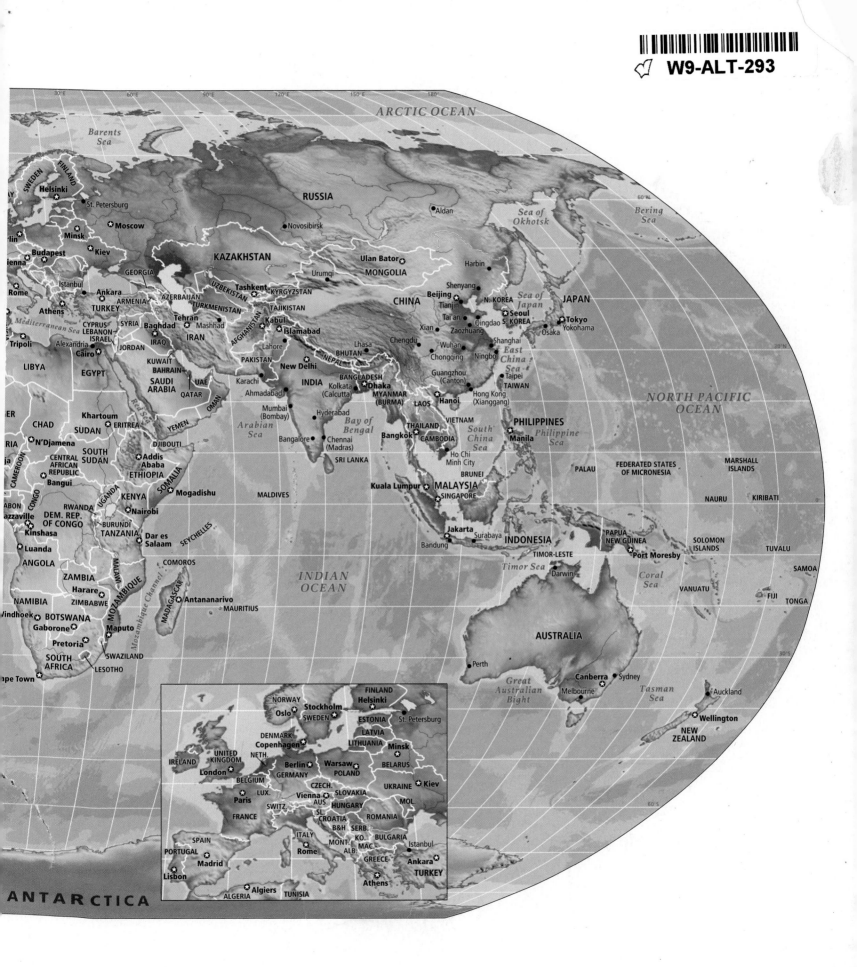

WORLD REGIONAL GEOGRAPHY CONCEPTS

SECOND EDITION

Lydia Mihelič Pulsipher
University of Tennessee

Alex A. Pulsipher
Geographer and Independent Scholar

with the assistance of
Conrad "Mac" Goodwin
Anthropologist/Archaeologist and Independent Scholar

W. H. Freeman and Company
New York

Dedication

To John Mihelič, who has helped envision and guide this project since its inception, and to Allison Pulsipher, who has kept us healthy, fed, cleansing, and ascending.

EXECUTIVE EDITOR: **Steven Rigolosi**

DEVELOPMENTAL EDITOR: **Elaine Epstein**

PROJECT EDITOR: **Laura McGinn**

DIRECTOR OF MARKETING: **John Britch**

COVER AND TEXT DESIGNER: **Vicki Tomaselli**

SENIOR ILLUSTRATION COORDINATOR: **Bill Page**

ASSISTANT AND SUPPLEMENTS EDITOR: **Kerri Russini**

EDITORIAL ASSISTANT: **Stephanie Ellis**

MAPS: **University of Tennessee, Cartographic Services Laboratory, Will Fontanez, Director; Maps.com**

PHOTO EDITOR: **Ted Szczepanski**

PHOTO RESEARCHER: **Alex Pulsipher**

PRODUCTION MANAGER: **Julia De Rosa**

COMPOSITION: **MPS Limited, a Macmillan Company**

PRINTING AND BINDING: **RR Donnelly**

Library of Congress Control Number: 2011934105

ISBN-13: 978-1-4292-5366-6
ISBN-10: 1-4292-5366-5

Printed in the United States of America
First printing

W.H. Freeman and Company
41 Madison Avenue
New York, NY 10010
Houndmills, Basingstoke RG21 6XS, England
www.whfreeman.com/geography

ABOUT THE AUTHORS

Lydia Mihelič Pulsipher is a cultural-historical geographer who studies the landscapes of ordinary people through the lenses of archaeology, geography, and ethnography. She has contributed to several geography-related exhibits at the Smithsonian Museum of Natural History in Washington, D.C., including "Seeds of Change," which featured the research she and Conrad Goodwin did in the eastern Caribbean. Lydia Pulsipher has ongoing research projects in the eastern Caribbean (historical archaeology) and in central Europe, where she is interested in various aspects of the post-Communist transition. Her graduate students have studied human ecology issues in the Caribbean and border issues and issues of national identity in several central European countries. She has taught cultural, gender, European, North American, and Mesoamerican geography at the University of Tennessee at Knoxville since 1980; through her research, she has given many students their first experience in fieldwork abroad. Previously she taught at Hunter College and Dartmouth College. She received her B.A. from Macalester College, her M.A. from Tulane University, and her Ph.D. from Southern Illinois University.

Alex A. Pulsipher is an independent scholar in Knoxville, Tennessee, who has conducted research on vulnerability to climate change, sustainable communities, and the diffusion of green technologies in the United States. In the early 1990s, while a student at Wesleyan University in Connecticut, Alex spent time in South Asia working for a sustainable development research

center. He then completed his B.A. at Wesleyan, writing his undergraduate thesis on the history of Hindu nationalism. Beginning in 1995, Alex contributed to the research and writing of the first edition of *World Regional Geography* with Lydia Pulsipher. In 1999 and 2000, he traveled to South America, Southeast Asia, and South Asia, where he collected information for the second edition of the text and for the Web site. In 2000 and 2001, he wrote and designed maps for the second edition. He participated in the writing of the fourth edition and in restructuring the content of, and creating photo essays and maps for, the fifth edition. Alex worked extensively on the first and second editions of *World Regional Geography Concepts*. He has a master's degree in geography from Clark University in Worcester, Massachusetts.

Conrad McCall Goodwin In the writing of *World Regional Geography Concepts*, Lydia and Alex Pulsipher were assisted in many ways by Lydia's husband, **Conrad "Mac" Goodwin**, an anthropologist and historical archaeologist with a B.A. in anthropology from the University of California at Santa Barbara, an M.A. in historical archaeology from the College of William and Mary, and a Ph.D. in archaeology from Boston University. He specializes in sites created during the European colonial era in North America, the Caribbean, and the Pacific. He has particular expertise in the archaeology of agricultural systems, gardens, domestic landscapes, and urban spaces. When not working on the textbook, Mac is a master organic gardener and slow food chef.

BRIEF CONTENTS

Preface xii

1 **Geography: An Exploration of Connections** 1

2 **North America** 56

3 **Middle and South America** 98

4 **Europe** 136

5 **Russia and the Post-Soviet States** 170

6 **North Africa and Southwest Asia** 206

7 **Sub-Saharan Africa** 246

8 **South Asia** 286

9 **East Asia** 326

10 **Southeast Asia** 368

11 **Oceania: Australia, New Zealand, and the Pacific** 406

Epilogue: Antarctica 439

Glossary G-1

Photo Credits PC-1

Figure Credits FC-1

Index I-1

CONTENTS

1 Geography: An Exploration of Connections 1

Where Is It? Why Is It There? Why Does It Matter?	2
What Is Geography?	3
Geographers' Visual Tools	3

Understanding Maps 3 • Longitude and Latitude 3 • Map Projections 6 • Geographic Information Science (GISc) 7 • The Detective Work of Photo Interpretation 7

The Region as a Concept	7
Thematic Concepts and Their Role in This Book	9

POPULATION — 9

Global Patterns of Population Growth	9
Local Variations in Population Density and Growth	14
Age and Sex Structures	15
Population Growth Rates and Wealth	16

GENDER — 17

Gender Roles	18

Gender Issues 18

DEVELOPMENT — 19

Measures of Economic Development	19
Geographic Patterns of Human Well-Being	20
Sustainable Development and Political Ecology	22
Human Impact on the Biosphere	22

FOOD — 22

Agriculture: Early Human Impacts on the Physical Environment	23

Agriculture and Its Consequences 23

Modern Food Production and Food Security	23

URBANIZATION — 26

Why Are Cities Growing?	27
Patterns of Urban Growth	27

GLOBALIZATION — 27

What Is the Global Economy?	28
Workers in the Global Economy	30
The Debate over Globalization and Free Trade	31

DEMOCRATIZATION — 33

The Expansion of Democracy	33
Democratization and Geopolitics	34
International Cooperation	36

WATER — 37

Calculating Water Use Per Capita	37
Who Owns Water? Who Gets Access to It?	37

Water Quality 38 • Water and Urbanization 38

GLOBAL CLIMATE CHANGE — 39

Drivers of Global Climate Change	39
Climate Change Impacts	41
Vulnerability to Climate Change	41
Responding to Climate Change	42

PHYSICAL GEOGRAPHY PERSPECTIVES — 44

Landforms: The Sculpting of the Earth	44

Plate Tectonics 44 • Landscape Processes 44

Climate	45

Temperature and Air Pressure 45 • Precipitation 46 • Climate Regions 47

HUMAN AND CULTURAL GEOGRAPHY PERSPECTIVES — 50

Ethnicity and Culture: Slippery Concepts	50
Values	50
Religion and Belief Systems	51
Language	51
Race	53

PHOTO ESSAYS

Understanding Maps	4
The Detective Work of Photo Interpretation	8
Thematic Concepts	10
1.1 Human Impacts on the Biosphere	24
1.2 Urbanization	28
1.3 Democratization and Conflict	34
1.4 Vulnerability to Climate Change	42
1.5 Climate Regions of the World	48
1.6 Major Religions around the World	52

2 North America 56

I THE GEOGRAPHIC SETTING — 58

Physical Patterns	58

Landforms 58 • Climate 59

Environmental Issues	59

Loss of Habitat for Plants and Animals 59 • Oil Drilling 62 • Logging 62 • Coal Mining 65 • Urbanization and Habitat Loss 65 • Climate Change and Air Pollution 65 • Water Resource Depletion and Pollution 66 • Green Living at a Global Scale 68

Human Patterns over Time   69

The Peopling of North America 69 • The European
Transformation 70 • Expansion West of the Mississippi 72 •
European Settlement and Native Americans 73 • The Changing
Regional Composition of North America 73

II CURRENT GEOGRAPHIC ISSUES   74

Political Issues   74

The Wars in Afghanistan and Iraq
74 • The United States Abroad 74 •
Dependence on Oil Questioned 74

**Relationships between Canada
and the United States**   75

Asymmetries 75 • Similarities 75 •
Interdependencies 76 • Democratic
Systems of Government: Shared
Ideals, Different Trajectories 76 •
The Social Safety Net: Canadian
and U.S. Approaches 79 • Gender
in National Politics 80

Economic Issues   80

North America's Changing Food-Production Systems 80 •
Changing Transportation Networks and the North American
Economy 81 • The New Service and Technology Economy 82 •
Globalization and the Economy 83 • Repercussions of the 2008
Global Economic Downturn 85 • Women in the Economy • 85

Sociocultural Issues   86

Urbanization and Sprawl 86 • Immigration and Diversity 88 •
Race and Ethnicity in North America 90 • Religion 91 • Gender
and the American Family 92

Population Patterns   94

Mobility in North America 95 • Aging in North America 95

PHOTO ESSAYS

Thematic Overview of North America   57
2.1 Climates of North America   63
2.2 Human Impact on the Biosphere   64
2.3 Vulnerability to Climate Change in North America   67
Visual Timeline of North America   70
2.4 Democratization and Conflict in North America   76
2.5 Urbanization in North America   87
Reasons for Optimism in North America   97

3 Middle and South America   98

I THE GEOGRAPHIC SETTING   101

Physical Patterns   101

Landforms 101 • Climate 104

Environmental Issues   107

Tropical Forests, Climate Change, and Globalization 107 •
Environmental Protection and Economic Development 107 •
The Water Crisis 109

Human Patterns over Time   109

The Peopling of Middle and South America 111 • European
Conquest 111 • A Global Exchange of Crops and Animals 112 •
The Legacy of Underdevelopment 112

II CURRENT GEOGRAPHIC ISSUES   113

Economic and Political Issues   113

Economic Inequality and Income Disparity 113 • Phases of
Economic Development 114 • The Role of Remittances 119 •
The Informal Economy 119 • Regional Trade and Trade
Agreements 119

Food Production and Contested Space   120

Political Reform and Democratization   121

The Drug Trade, Conflict, and Democracy 123 • Foreign
Involvement in the Region's Politics 124 • The Political and
Economic Impacts of Information Technology 124

Sociocultural Issues   126

Population Patterns 126 • Cultural Diversity 127 • Race and the
Social Significance of Skin Color 128 • The Family and Gender
Roles 128 • Migration and Urbanization 129 • Religion in
Contemporary Life 132

PHOTO ESSAYS

Thematic Overview of Middle and South America   99
3.1 Climates of Middle and South America   105
3.2 Human Impact on the Biosphere   108
3.3 Vulnerability to Climate Change   110
Visual Timeline of Middle and South America   114
3.4 Democratization and Conflict   122
3.5 Urbanization in Middle and South America   130
Reasons for Optimism in Middle and South America   134

4 Europe   136

I THE GEOGRAPHIC SETTING   139

Physical Patterns   139

Landforms 139 • Vegetation 142

Environmental Issues   142

Europe's Impact on the Biosphere 142

Human Patterns over Time   148

Sources of European Culture 148 • The Inequalities of
Feudalism 149 • The Role of Urbanization in the Transformation
of Europe 149 • European Colonialism: An Acceleration
of Globalization 149 • Urban Revolutions in Industry and
Democracy 150 • Two World Wars and Their Aftermath 151

II CURRENT GEOGRAPHIC ISSUES   153

Economic and Political Issues   154

The European Union: A Rising Superpower 154 • The Euro and
Debt Crisis 156 • Food Production and the European Union 159
• Europe's Growing Service Economies 160

Sociocultural Issues   160

Population Patterns 160 • Immigration and Migration: Needs and
Fears 163 • Changing Gender Roles 164 • Social Welfare Systems
and Their Outcomes 167

PHOTO ESSAYS

Thematic Overview of Europe   137
4.1 Climates of Europe   143
4.2 Human Impact on the Biosphere in Europe   144

4.3 Vulnerability to Climate Change in Europe   147

Visual Timeline of Europe   152

4.4 Democratization and Conflict   154

4.5 Urbanization in Europe   162

Reasons for Optimism in Europe   169

5 Russia and the Post-Soviet States   170

I THE GEOGRAPHIC SETTING   173

Physical Patterns   173

Landforms 173 • Climate and Vegetation 176

Environmental Issues   176

Urban and Industrial Pollution 178 • The Globalization
of Nuclear Pollution 178 • The Globalization of Resource
Extraction and Environmental Degradation 178 • Water
Issues 180 • Climate Change 181 • Better Times Ahead? 181

Human Patterns over Time   183

The Rise of the Russian Empire 183 • The Communist
Revolution and Its Aftermath 185 • World War II and the Cold
War 186

II CURRENT GEOGRAPHIC ISSUES   188

Economic and Political Issues   188

Oil and Gas Development: Fueling Globalization 188 •
Economic Reforms in the Post-Soviet Era 189 • Food Production
in the Post-Soviet Era 192 • Democratization in the Post-Soviet
Years 194 • Corruption and Organized Crime 197

Sociocultural Issues   198

Population Patterns 198 • Gender: Challenges and Opportunities
in the Post-Soviet Era 201 • Religious Revival in the Post-Soviet
Era 203

PHOTO ESSAYS

Thematic Overview of Russia and the Post-Soviet States   171

5.1 Climates of Russia and the Post-Soviet States   177

5.2 Human Impact on the Biosphere in Russia and
the Post-Soviet States   179

5.3 Vulnerability to Climate Change in Russia
and the Post-Soviet States   182

Visual Timeline of Russia and the Post-Soviet States   184

5.4 Democratization and Conflict in Russia and
the Post-Soviet States   195

5.5 Urbanization in Russia and the Post-Soviet States   200

Reasons for Optimism in Russia and the Post-Soviet States   205

6 North Africa and Southwest Asia   206

I THE GEOGRAPHIC SETTING   209

Physical Patterns   209

Climate 209 • Landforms and Vegetation 209

Environmental Issues   213

An Ancient Heritage of Water Conservation 213 • Water and
Food Production 214 • Vulnerability to Climate Change 216 •
Climate Change and Desertification 216

Human Patterns over Time   218

Agriculture and Development of Civilization 218 • Agriculture
and Gender Roles 219 • The Coming of Monotheism: Judaism,
Christianity, and Islam 220 • The Spread of Islam 220 • Western
Domination and State Formation 221

II CURRENT GEOGRAPHIC ISSUES   222

Sociocultural Issues   222

Religion in Daily Life 222 • Family
Values and Gender 223 • Gender
Roles and Gendered Spaces 225 •
The Rights of Women in Islam 226 •
Changing Population Patterns 227 •
Population Growth and Gender
Status 227 • Urbanization,
Globalization, and Migration 227

Economic and Political Issues   231

Globalization, Development, and
Fossil Fuel Exports 232 • Economic
Diversification and Growth 234 •
Democratization and Islamism in a Globalizing World 235 •
The Role of Press and Media 236 • Democratization and
Women 238 • The Iraq War (2003–2010) 238 • The State of
Israel and the "Question of Palestine" 239

PHOTO ESSAYS

Thematic Overview of North Africa and Southwest Asia   207

6.1 Climates of North Africa and Southwest Asia   212

6.2 Human Impact on the Biosphere in North Africa
and Southwest Asia   215

6.3 Vulnerability to Climate Change in North Africa
and Southwest Asia   217

Visual Timeline of North Africa and Southwest Asia   222

6.4 Urbanization in North Africa and Southwest Asia   230

6.5 Democratization and Conflict in North Africa
and Southwest Asia   236

Reasons for Optimism in North Africa and Southwest Asia   244

7 Sub-Saharan Africa   246

I THE GEOGRAPHIC SETTING   249

Physical Patterns   249

Landforms 249 • Climate and Vegetation 252

Environmental Issues   252

Deforestation and Climate Change 252 • Food Production,
Water Resources, and Vulnerability to Climate Change 254 •
Wildlife and Climate Change 258

Human Patterns over Time 259
The Peopling of Africa and
Beyond 259 • Early Agriculture,
Industry, and Trade in Africa
259 • Europeans and the African
Slave Trade 259 • The Scramble
to Colonize Africa 261 • The
Aftermath of Independence 263

**II CURRENT
GEOGRAPHIC ISSUES** 264

Economic and Political Issues 264
Commodity-Based Economic
Development and Globalization
in Africa 264 • Succeeding Eras of
Globalization 264 • Regional and
Local Economic Development 269
• Democratization and Conflict 271

Sociocultural Issues 273
Population Patterns 275 • Gender Issues 280 • Religion 281

PHOTO ESSAYS

Thematic Overview of Sub-Saharan Africa 247
7.1 Climates of Sub-Saharan Africa 253
7.2 Human Impact on the Biosphere in Sub-Saharan Africa 255
7.3 Vulnerability to Climate Change in Sub-Saharan Africa 256
Visual Timeline of Sub-Saharan Africa 260
7.4 Democratization and Conflict in Sub-Saharan Africa 274
7.5 Urbanization in Sub-Saharan Africa 278
Reasons for Optimism in Sub-Saharan Africa 284

8 South Asia 286

I THE GEOGRAPHIC SETTING 289

Physical Patterns 289
Landforms 292 • Climate and Vegetation 292

Environmental Issues 295
South Asia's Vulnerability to Climate Change 295 •
Deforestation 296 • Water 298 • Industrial Air Pollution 300

Human Patterns over Time 300
The Indus Valley Civilization 300 • A Series of Invasions 300 •
Globalization and the Legacies of British Colonial Rule 301

II CURRENT GEOGRAPHIC ISSUES 304

Sociocultural Issues 304
Language and Ethnicity 305 •
Religion, Caste, and Conflict
305 • Geographic and Social
Patterns in the Status of
Women 309 • Population
Patterns 311

Economic Issues 315
Economic Trends 316 • Food
Production and the Green
Revolution 316 • Microcredit: A South Asian Innovation for the
Poor 317 • Industry over Agriculture: A Vision of Self-Sufficiency
318 • Economic Reform: Globalization and Competitiveness 318

Political Issues 320
Democratization and Conflict 320 • Religious Nationalism 323

PHOTO ESSAYS

Thematic Overview of South Asia 287
8.1 Climates of South Asia 294
8.2 Vulnerability to Climate Change in South Asia 297
8.3 Human Impact on the Biosphere in South Asia 299
Visual Timeline of South Asia 304
8.4 Urbanization in South Asia 314
8.5 Democratization and Conflict in South Asia 321
Reasons for Optimism in South Asia 324

9 East Asia 326

I THE GEOGRAPHIC SETTING 332

Physical Patterns 332
Landforms 332 • Climate 332

Environmental Issues 334
Climate Change: Emissions and Vulnerability in East Asia 334 •
Food Security and Sustainability in East Asia 337 • Three Gorges
Dam: The Power of Water 338 • Air Pollution: Choking on
Success 340

Human Patterns over Time 341
Bureaucracy and Imperial
China 341 • Confucianism
Molds East Asia's Cultural
Attitudes 342 • Why Did China
Not Colonize an Overseas
Empire? 342 • European and
Japanese Imperialism 343 •
China's Turbulent Twentieth
Century 343 • Japan Becomes a
World Leader 344 • Chinese and
Japanese Influences on Korea,
Taiwan, and Mongolia 345

II CURRENT GEOGRAPHIC ISSUES 347

Economic and Political Issues 347
The Japanese Miracle 348 • Mainland Economies: Communists
in Command 349 • Market Reforms in China 349 •
Urbanization, Development, and Globalization in East Asia 350 •
Democratization in East Asia 355

Sociocultural Issues 359
Population Patterns 359 • Responding to an Aging Population 359 •
China's One-Child Policy 360 • Population Distribution 361 •
Cultural Diversity in East Asia 361 • East Asia's Largest Cultural
Export: The Overseas Chinese 364

PHOTO ESSAYS

Thematic Overview of East Asia 327
9.1 Climates of East Asia 333

9.2 Vulnerability to Climate Change
in East Asia 335

9.3 Human Impact on the Biosphere
in East Asia 339

Visual Timeline of East Asia 346

9.4 Urbanization in East Asia 351

9.5 Democratization and Conflict in East Asia 356

Reasons for Optimism in East Asia 365

10 Southeast Asia 368

I THE GEOGRAPHIC SETTING 371

Physical Patterns 374
Landforms 374 • Climate and Vegetation 374

Environmental Issues 375
Climate Change and Deforestation 379 • Climate
Change and Food Production 379 • Climate Change
and Water 381 • Responses to Climate Change 381

Human Patterns over Time 383
The Peopling of Southeast Asia 383 • Diverse Cultural
Influences 383 • European Colonization 384 • Struggles
for Independence 384

II CURRENT GEOGRAPHIC ISSUES 387

Economic and Political Issues 387
Strategic Globalization: State Aid and Export-Led Economic
Development 387 • Economic Crisis and Recovery: The Perils of
Globalization 388 • Regional Trade and ASEAN 389 • Pressures
For and Against Democracy 391

Sociocultural Issues 395
Population Patterns 395 • Urbanization 396 • Migration Related
to Globalization 399 • Cultural and Religious Pluralism 400 •
Gender Patterns in Southeast Asia 401 • Globalization and
Gender: The Sex Industry 402

PHOTO ESSAYS
Thematic Overview of Southeast Asia 369

10.1 Climates of Southeast Asia 376

10.2 Human Impact on the Biosphere in
Southeast Asia 378

10.3 Vulnerability to Climate Change in
Southeast Asia 382

Visual Timeline of Southeast Asia 386

10.4 Democratization and Conflict in
Southeast Asia 394

10.5 Urbanization in Southeast Asia 398

Reasons for Optimism in Southeast Asia 404

11 Oceania: Australia, New Zealand, and the Pacific 406

I THE GEOGRAPHIC SETTING 409

Physical Patterns 409
Continent Formation 409 • Island Formation 412 •
Climate 412 • Fauna and Flora 414

Environmental Issues 416
Global Climate Change 416 • Invasive Species and Food
Production 419 • Globalization and the Environment in the
Pacific Islands 420

Human Patterns over Time 421
The Peopling of Oceania 422 • Arrival of the Europeans 422 •
The Colonization of Australia and New Zealand 422 • Oceania's
Shifting Global Relationships 424

II CURRENT GEOGRAPHIC ISSUES 425

Economic and Political Issues 425
Globalization, Development, and Oceania's New Asian
Orientation 425 • The Stresses of Asia's Economic Development
"Miracle" on Australia and New Zealand 426 • The Advantages
and Stresses of Tourism Development 427 • The Future: Diverse
Orientations? 428

Population Patterns 428
Disparate Population Patterns in Oceania 428 • Urbanization
in Oceania 429

Sociocultural Issues 431
Ethnic Roots Reexamined 431 • Politics and Culture: Different
Definitions of Democracy 433 • Gender Roles in Oceania 435 •
Gender, Democracy, and Economic Empowerment 436

PHOTO ESSAYS
Thematic Overview of Oceania 407

11.1 Climates of Oceania 413

11.2 Human Impact on the Biosphere in Oceania 417

11.3 Vulnerability to Climate Change in Oceania 418

Visual Timeline of Oceania 424

11.4 Urbanization in Oceania 430

11.5 Democratization and Conflict in Oceania 434

Reasons for Optimism in Oceania 437

Epilogue: Antarctica 439

Glossary G-1

Photo Credits PC-1

Figure Credits FC-1

Index I-1

In this text, we portray the rich diversity of human life across the world and humanize geographic issues by representing the daily lives of women, men, and children in various regions of the globe. Our goal is to make global patterns of trade and consumption meaningful for students by showing how these patterns affect not only world regions but also ordinary people at the local level. This second edition of *World Regional Geography Concepts* strives to reach this goal with improvements to make the text as current, instructive, and visually appealing as possible. At just over 450 pages, the text is designed to allow instructors to cover all world regions in a single semester.

NEW TO THE SECOND EDITION

Thematic Concepts

Teaching world regional geography is never easy. Many instructors have found that focusing their courses on a few key ideas makes their teaching more effective and helps students retain information. With that goal in mind, we have identified nine thematic concepts that provide a few basic hooks on which students can hang their growing knowledge of the world and each of its regions. These thematic concepts are, in alphabetical order:

- **Climate Change**: What are the indications that climate change is underway? How are places, people, and ecosystems in a particular region vulnerable to the shifts that climate change may bring? How are people and governments in the region responding to the threats posed by global warming? Which human activities contribute significant amounts of greenhouse gases?

- **Democratization**: What is the history and current status of democracy in a region? Which factors are working for or against the political freedoms that make democracy possible? What other viable political systems have emerged?

- **Development**: How do shifts in economic, social, and other dimensions of development affect human well-being? What paths have been charted by the so-called "developed" world and how are they relevant, or irrelevant, to the rest of the world? What new "homegrown" solutions are emerging from the so-called "less developed" countries?

- **Food**: How do food production systems impact environments and societies in a region? How has the use of new agricultural technologies impacted farmers? How have changes in food production created pressure to urbanize?

- **Gender**: How do the lives and livelihoods of men and women differ and how do gender roles influence societies in a region? To what extent do men and women differ in their contributions to family and community well-being? What is the nature of the disparities in income, education, and rights that persist between the genders?

- **Globalization**: How has a particular region been impacted by globalization, historically and currently? How are lives changing as flows of people, ideas, products, and resources become more global?

- **Population**: What are the major forces driving population growth or decline in a region? How have changes in gender roles influenced population growth? How are changes in life expectancy, family size, and the age of the population influencing population change?

- **Urbanization**: Which forces are driving urbanization in a particular region? How have cities responded to growth? How is the region affected by the changes that accompany urbanization—for example, changes in employment, education, and access to health care?

- **Water**: How do issues of water scarcity, water pollution, and water management affect people and environments in a particular region? Water issues are often discussed in the context of global warming or food production.

Thematic Overview and Learning Goals: Alone and in combination, the thematic concepts form the basis of five to six **Learning Goals** for each chapter. These goals are stated at the beginning of the chapter and discussed at the relevant point in the text. They also are reviewed in the "Things to Remember," sections found throughout the chapter and in the "Learning Goals Review" questions, which appear at the end of each chapter.

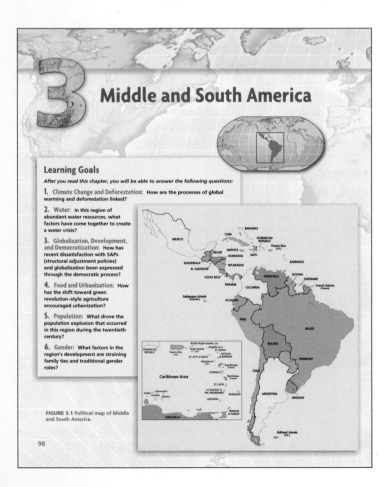

FIGURE 3.1 Political map of Middle and South America.

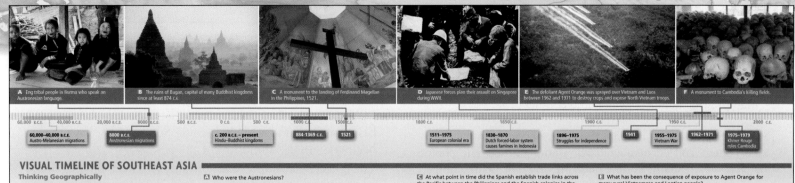

A Eng tribal people in Burma who speak an Austronesian language.

B The ruins of Bagan, capital of many Buddhist kingdoms since at least 874 C.E.

C A monument to the landing of Ferdinand Magellan in the Philippines, 1521.

D Japanese forces plan their assault on Singapore during WWII.

E The defoliant Agent Orange was sprayed over Vietnam and Laos between 1962 and 1971 to destroy crops and expose North Vietnam troops.

F A monument to Cambodia's killing fields.

60,000–40,000 B.C.E. Austro-Melanesian migrations

8000 B.C.E. Austronesian migrations

c. 200 B.C.E. – present Hindu–Buddhist kingdoms

884-1369 C.E.

1521

1511–1975 European colonial era

1830–1870 Dutch forced-labor system causes famines in Indonesia

1896–1975 Struggles for independence

1941

1955–1975 Vietnam War

1962–1971

1975–1979 Khmer Rouge rules Cambodia

VISUAL TIMELINE OF SOUTHEAST ASIA
Thinking Geographically

After you have read about the human history of Southeast Asia, you will be able to answer the following questions:

A Who were the Austronesians?

B How did Hinduism and Buddhism arrive in Southeast Asia?

C At what point in time did the Spanish establish trade links across the Pacific between the Philippines and the Spanish colonies in the Americas?

D During World War II, Japan conquered which parts of Southeast Asia?

E What has been the consequence of exposure to Agent Orange for many rural Vietnamese and Laotian people?

F How many Cambodians were executed or starved in the killing fields?

Reasons for Optimism in Sub-Saharan Africa

Democratization: Authoritarian political structures, many of which date to the colonial era, are giving way to more democratic systems of government. **A** Nigeria during recent elections.

Development: Recent years have seen growth in the manufacturing industries that could help Africa lift itself out of poverty. **B** The manufacturing of mosquito nets in Congo (Kinshasa).

Gender: The contributions of women to society, and the importance of gender issues in slowing population growth, are increasingly being recognized. **C** A girl's school in Zanzibar, Tanzania.

Thinking Geographically: What shortcoming of democratization in Nigeria does this photo suggest?

Thinking Geographically: How might manufacturing help lift Africa out of poverty?

Thinking Geographically: How might more girls' schools contribute to slower population growth in Tanzania?

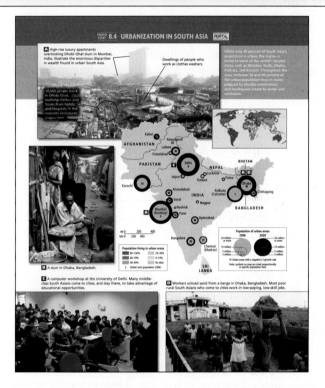

PHOTO ESSAY **8.4** URBANIZATION IN SOUTH ASIA

A High-rise luxury apartments overlooking Dhobi Ghat slum in Mumbai, India, illustrate the enormous disparities in wealth found in urban South Asia.

A A slum in Dhaka, Bangladesh.

C A computer workshop at the University of Delhi. Many middle-class South Asians come to cities, and stay there, to take advantage of educational opportunities.

D Workers unload sand from a barge in Dhaka, Bangladesh. Most poor rural South Asians who come to cities work in low-paying, low-skill jobs.

Photo Essays, Visual Timelines, Reasons for Optimism, and Thinking Geographically

Photos are a rich source of geographic information, and the number of photos in the second edition has been greatly expanded. Early in each regional chapter, a **Thematic Overview Photo Essay** provides examples of how each of the nine thematic concepts applies to the region. Subsequent **Map Photo Essays** illustrate particular concepts. For example, a photo essay about urbanization might include a map of urban patterns in that region as well as photos that illustrate various aspects of current urban life in that part of the world.

Also new to this edition are **Visual Timelines** for each region that use images to illustrate key points in the region's history. Finally, in light of the often overwhelming and, at times, depressing nature of the information presented in any world regional geography course, each chapter concludes with a new and unabashedly optimistic photo feature called **Reasons for Optimism.** Here we explore some of the more hopeful patterns and opportunities within each region.

To help instructors make use of all these new photo features in their teaching, the text offers **Thinking Geographically** questions with many photos. The answers can be found on this book's Web site, where they form the basis of computer-graded exercises that can be assigned and automatically graded and entered into each instructor's grade book.

Restructured Chapters

Each chapter includes a variety of features to support the teaching and learning of world regional geography.

Things to Remember At the close of every main section, a few concise statements review the section's important points. The statements emphasize some key themes while encouraging students to think through the ways in which the material illustrates these points. They also review the Learning Goals stated at the beginning of each chapter.

Marginal Glossary of Key Terms Terms important to the chapter content are boldfaced on first usage and defined in the center of each page on which they appear. These terms are listed at the end of the chapter with the page numbers on which they are defined. They are also listed alphabetically in the glossary at the end of the book.

Learning Goals Review At the end of each chapter a series of questions, many tied to the chapter's Learning Goals, encourages broader analysis of the chapter content. These questions could be used for assignments, group projects, or class discussion.

Consistent Base Maps This edition focuses on improving further what has often been cited as a principal strength of this text: high-quality, relevant, and consistent maps. To help students make conceptual connections and to compare regions, every chapter contains the following:

• Political map at the beginning of each chapter

• Regional map

• Climate map

- Map of the human impact on the Biosphere
- Map of the region's vulnerability to climate change
- Urbanization map
- Map showing democratization and conflict

New to the second edition are the maps on vulnerability to climate change and on democratization and conflict. The photos that accompany many of these maps are also new to this edition.

New Photos

An ongoing aim of this text has been to awaken the student to the circumstances of people around the world, and photos are a powerful way to accomplish this objective. This second edition promotes careful attention to photos by including in Chapter 1 a short lesson on photo interpretation. Students are encouraged to use these skills as they look at every photo in the text, and instructors are encouraged to use the photos as lecture themes and to help generate analytical class discussions.

The second edition has more than twice as many photos as the previous edition, and each new photo was chosen to complement a Thematic Concept or situation described in the text. All photos are numbered and referenced in the text, making it easier for students to integrate the text with the visuals as they read. Moreover, the photos—like all of the book's graphics, including the maps—have been given significantly more space and prominence in the page layout. The result is a more visually engaging, dynamic, and instructive text. All photo credits are at the end of the book, listed by according to figure numbers.

Videos

Over 300 videos clips (an average of 27 per chapter) are available with the second edition. Most videos are 2 to 6 minutes long and cover key issues discussed in the text. They can help instructors gain further expertise or can be used to generate class discussion. Each video is keyed to the text with the icon ❚❚ ▶ at the point in the discussion where it is most relevant. These videos, along with a related multiple-choice quiz, can be accessed at www.whfreeman.com/geographyvideos. Questions can be automatically graded and entered into a grade book. To access the videos, students need a password that can be bundled free with this textbook.

Up-to-Date Content

Because the world is constantly changing, it is essential that a world regional geography text be as current as possible. To that end, the second edition discusses the revolutions in the North Africa and Southwest Asia; the 2011 earthquake in Japan; the BP oil spill in the Gulf of Mexico; the earthquakes in Chile and Haiti; the varying effects of the global recession on regions, countries, and individuals; civil unrest in once-stable Thailand; the new influence of Arabic media outlets, such as Al Jazeera, which now affect thinking around the world; and the consequences of global climate change on Pacific Island nations.

Some of the major content areas of the book that have been updated include:

- Political revolutions and conflict in North Africa and Southwest Asia (the Middle East)
- Maps reflecting the new country of South Sudan (data remain based on Sudan as a whole because South Sudan has not yet begun reporting statistics)
- The global economic recession and its effect on migrants, labor outsourcing, and job security in importing and exporting countries
- Domestic and global implications of the U.S. political, economic, and military stances
- The role of terrorism in the realignment of power globally and locally
- Immigration and the ways it is changing countries economically and culturally
- Recent economic crises in the European Union which may bring about significant reorganization that has consequences for the original EU members, new and potential member states, and the global community
- Changing gender roles, particularly in developing countries
- The increasing role and influence of Islam around the world
- Climate change and its environmental, political, and economic implications

THE ENDURING VISION: GLOBAL AND LOCAL PERSPECTIVES

The Global View

In addition to the new features and enhancements to the text, we retain the hallmark features that have made the first edition of this text successful for instructors and students. For the second edition, we continue to emphasize global trends and the interregional linkages that are changing lives throughout the world, including those trends related to changing gender roles. The following linkages are explored in every chapter, as appropriate:

- The **multifaceted economic linkages** among world regions. These include (1) the effects of colonialism; (2) trade; (3) the role in the world economy of transnational corporations, such as Walmart, BP, and Cisco; (4) the influence of regional trade organizations, such as ASEAN and NAFTA; and (5) the changing roles of the World Bank and the International Monetary Fund as the negative consequences of structural adjustment programs become better understood.
- **Migration.** Migrants are changing economic and social relationships in virtually every part of the globe. The societies they leave are changed radically by their absence, just as the host societies are changed by their presence. The text explores the local

and global effects of foreign workers in places such as Japan, Europe, Africa, the Americas, and Southwest Asia, and the increasing number of refugees resulting from conflicts around the world. Also discussed are long-standing migrant groups, such as the Overseas Chinese and the Indian diasporas.

- **Mass communications and marketing techniques** are promoting **world popular culture** across regions. The text integrates coverage of popular culture and its effects in discussions of topics such as tourism in the Caribbean and Southeast Asia, and the blending of Western and traditional culture in places such as South Africa, Malawi, Japan, and China.

- **Gender issues** are covered in every chapter with the aim of covering more completely the lives of ordinary people. Gender is intimately connected to other patterns, such as internal and global migration, and these connections and other region-wide gender patterns are illustrated in a variety of maps and photos and in vignettes that illustrate gender roles as played out in the lives of individuals. The lives of children, especially their roles in families, are also covered, often in concert with the treatment of gender issues.

The Local Level

Our approach pays special attention to the local scale—a town, a village, a household, an individual. Our hope is, first, that stories of individual people and families will make geography interesting and real to students and, second, that seeing the effects of abstract processes and trends on ordinary lives will dramatize the effects of these developments for students. Reviewers have mentioned that students particularly appreciate the personal vignettes, which are often stories of real people (with names disguised). For each region, we examine the following local responses:

- **Cultural change:** We look closely at changes in the family, gender roles, and social organization in response to urbanization, modernization, and the global economy.

- **Impacts on well-being:** Ideas of what constitutes "well-being" differ from culture to culture, yet broadly speaking, people everywhere try to provide a healthful life for themselves in a community of their choosing. Their success in doing so is affected by local conditions, global forces, and their own ingenuity.

- **Issues of identity:** Paradoxically, as the world becomes more tightly knit through global communications and media, ethnic and regional identities often become stronger. The text examines how modern developments such as the Internet are used to reinforce particular cultural identities.

- **Local attitudes toward globalization:** People often have ambivalent reactions to global forces: They are repelled by the seeming power of these forces, fearing effects on their own lives and livelihoods and on local traditional cultural values, but they are also attracted by the economic opportunities that may emerge from greater global integration. The text looks at how the people of a region react to cultural or economic globalization.

ALSO AVAILABLE

To better serve the different needs of diverse faculty and curricula, two other versions of this textbook are available.

World Regional Geography with Subregions, Fifth Edition (1-4292-3241-2)

The fifth edition continues to employ a consistent structure for each chapter. Each chapter beyond the first is divided into three parts: I: The Geographic Setting; II: Current Geographic Issues; III: Subregions.

The subregion coverage provides a descriptive characterization of particular countries and places within the region that expands on coverage in the main part of the chapter. For example, the sub-Saharan Africa chapter considers the West, Central, East, and Southern Africa subregions, providing additional insights into differences in well-being and in social and economic issues across the African continent.

World Regional Geography without Subregions, Fifth Edition (1-4292-3244-7)

The briefer version provides essentially the same main text coverage as the version described above, omitting only the subregional sections. This version contains all the types of pedagogy found in the main version.

FOR THE INSTRUCTOR

The authors have taught world regional geography many times and understand the need for quick, accessible aids to instruction. Many of the new features were designed to streamline the job of organizing the content of each class session with the goal of increasing student involvement through interactive discussions.

Ease of instruction and active student involvement were the principal motivations behind the book's key features–the thematic concepts, the photo essays, the content maps that facilitate region-to-region comparisons, and the wide selection of videos.

Map Builder: Custom Mapping Program

Instructors use maps in the classroom in many different ways, and we have tried to make the book's maps easy to use for instructors and students by creating a flexible, friendly, online mapping tool. W.H. Freeman and Maps.com have built a revolutionary, first-of-its-kind teaching tool for instructors. It is a program that allows instructors to create, display, and print *custom maps*. The program permits in-class display of thousands of different custom maps and allows instructors to zoom in on any portion of a map they want to emphasize. With only a few mouse clicks, instructors can choose the data from selected maps that they would like to show on a single map. For example, instructors can compare climate and agriculture in North America by choosing to display together the continental climate zone and dairying. These maps can be printed in full color or displayed as slides in the classroom. In addition, the book's companion Web site offers a map-building exercise for each chapter. For a demonstration of the Map Builder program and its accompanying exercises, please ask your W.H. Freeman sales representative for a guided tour, or visit www.whfreeman.com/pulsipherconcepts2e.

Instructor's Resource CD-ROM (1-4292-8252-5)

To help instructors create their own Web sites and orchestrate dynamic lectures, the CD contains:

- **All text images** in PowerPoint and JPEG formats with enlarged labels for better projection quality.
- **PowerPoint lecture outlines**, by Rebecca Johns, University of South Florida. The main themes of each chapter are outlined and enhanced with images from the book, providing a pedagogically sound foundation on which to build personalized lecture presentations.
- **Instructor's Resource Manual**, by Jennifer Rogalsky, State University of New York, Geneseo, and Helen Ruth Aspaas, Virginia Commonwealth University, contains suggested lecture outlines, points to ponder for class discussion, and ideas for exercises and class projects. It is offered as chapter-by-chapter Word files to facilitate editing and printing.
- **Test Bank**, by Rebecca Johns (University of South Florida), expanded from the original test bank created by Jason Dittmer, University College London, and Andy Walter, West Georgia University. The Test Bank is designed to match the pedagogical intent of the text and offers more than 2500 test questions (multiple choice, short answer, matching, true/false, and essay) in a Word format that is easy to edit, add, and resequence questions. A **computerized test bank** (powered by Diploma) with the same contents is also available.
- **Clicker Questions**, by Rebecca Johns, University of South Florida. Written in PowerPoint for easy integration into lecture presentations, Clicker questions allow instructors to jump-start

discussions, illuminate important points, and promote better conceptual understanding during lectures.

Instructor's Web Site (password-protected)
www.whfreeman.com/pulsipherconcepts2e
In addition to all the resources available on the *Instructor's CD-ROM*, this password-protected Web site also includes:

- **Syllabus posting** functionality
- An integrated **grade book** that records student performance on online and video quizzes
- The W.H. Freeman **World Regional Geography DVD**, with 35 projection-quality videos and an accompanying instructor's video manual
- A link to the **Geography Faculty Lounge**, an online community of geographers offering shared resources

Course Management
All instructor and student resources are also available via **BlackBoard, WebCT, Angel, Moodle, Sakai,** and **Desire2Learn**. W.H. Freeman and Company offers a course cartridge that populates your site with content tied directly to the book.

W.H. Freeman World Regional Geography DVD

This DVD, available free to adopters of the second edition, builds on the book's purpose of putting a face on geography by giving students and instructors access to the fascinating personal stories of people from all over the world. The DVD contains 35 projection-quality video clips from 3 to 7 minutes in length. An **instructor's video manual** is also included on the DVD.

FOR THE STUDENT

World Regional Geography Online
www.whfreeman.com/pulsipherconcepts2e
This free companion Web site offers a wealth of resources to support the textbook, including:

- **Chapter Quizzes**—These multiple-choice quizzes help students assess their mastery of each chapter.
- **Thinking Geographically Questions**—These multiple-choice questions relate to select photos found throughout the book. The question sets form the basis of computer-graded exercises that can be assigned and automatically graded and entered into the instructor's online grade book.
- **Thinking Critically About Geography**—These activities, fully updated for the second edition, allow students to explore a set of current issues, such as deforestation, human rights, or free

trade, and see how geography helps clarify our understanding of them. Linked Web sites are matched with a series of questions or with brief activities that help students think about the ways in which they themselves are connected to the places and people they read about in the text.

- **Map Builder Software and Map Builder Exercises**—The Map Builder program allows students to create layered thematic maps on their own, while Map Builder Exercises offer a specific activity for each chapter in the second edition.

- **Map Learning Exercises**—Students can use this interactive feature to identify and locate countries, cities, and the major geographic features of each region.

- **Blank Outline Maps**—Printable maps of the world, and of each region, for use in note-taking or exam review or both, as well as for preparing assigned exercises.

- **Flashcards**—Matching exercises teach vocabulary and definitions.

- **Audio Pronunciation Guide**—Spoken guide of place names, regional terms, and names of historical figures.

- **World Recipes and Cuisines**—From *International Home Cooking*, the United Nations International School cookbook.

GEOGRAPHY PORTAL www.yourgeographyportal.com
This comprehensive resource is designed to offer a complete solution for today's classroom. Geography Portal offers all the instructor and student resources listed above, as well as premium resources available only on the portal:

- An eBook of *World Regional Geography Concepts* 2e, complete and customizable. Students can quickly search the text and personalize it just as they would the printed version, with highlighting, bookmarking, and note-taking features

- **A Guide to Using Google Earth** for the novice, plus step-by-step **Google Earth** exercises for each chapter

- Selected articles from *Focus on Geography* magazine (one for each chapter in the textbook) and accompanying quizzes for each article

- **Animated photo essays** with sound, easily downloadable as .mp3 files

- **Physical geography videos**, for instructors who want to cover physical geography topics in more detail

- Online **news feeds** for highly respected magazines such as the *Economist*

For more information or to schedule a demo, please contact your W.H. Freeman sales representative.

Rand McNally's Atlas of World Geography, 176 pages
This atlas contains:

- 52 physical, political, and thematic maps of the world and continents; 49 regional, physical, political, and thematic maps; and dozens of metro-area inset maps

- Geographic facts and comparisons, covering topics such as population, climate, and weather

- A section on common geographic questions, a glossary of terms, and a comprehensive 25-page index

ACKNOWLEDGMENTS

The authors wish to acknowledge the many geographers whose insights and suggestions have informed this book.

Second Edition, *World Regional Geography Concepts*

Heike C. Alberts
University of Wisconsin, Oshkosh

John All
Western Kentucky University

Robert G. Atkinson
Tarleton State University

Christopher A. Badurek
Appalachian State University

Bradley H. Baltensperger
Michigan Technological University

Denis A. Bekaert
Middle Tennessee State University

Mikhail Blinnikov
St. Cloud State University

Mark Bonta
Delta State University

Patricia Boudinot
George Mason University

Lara M.P. Bryant
Keene State College

Craig S. Campbell
Youngstown State University

Bruce E. Davis
Eastern Kentucky University

L. Scott Deaner
Owens Community College

James V. Ebrecht
Georgia Perimeter College

Kenneth W. Engelbrecht
Metropolitan State College of Denver

Natalia Fath
Towson University

Alison E. Feeney
Shippensburg University

John H. Fohn II
Missouri State University, West Plains

Stephen Franklin
Coconino Community College

Joy Fritschle
West Chester University

Matthew Gerike
University of Missouri

Carol L. Hanchette
University of Louisville

Ellen R. Hansen
Emporia State University

Katie Haselwood-Weichelt
University of Kansas

Kari Jensen
Hofstra University

Cub Kahn
Oregon State University

Curtis A. Keim
Moravian College

Jeannine Koshear
Fresno City College

Hsiang-te Kung
University of Memphis

Chris Laingen
Eastern Illinois University

Leonard E. Lancette
Mercer University

Heidi Lannon
Santa Fe College

James Leonard
Marshall University

John Lindberg
Scott Community College

Max Lu
Kansas State University

Peter G. Odour
North Dakota State University

Kefa M. Otiso
Bowling Green State University

Lynn M. Patterson
Kennesaw State University

James Penn
Grand Valley State University

Paul E. Phillips
Fort Hays State University

Gabriel Popescu,
Indiana University, South Bend

Jennifer Rahn
Samford University

Amanda Rees
Columbus State University

Robert F. Ritchie IV
Liberty University

Ginger L. Schmid
Minnesota State University, Mankato

Cynthia L. Sorrensen
Texas Tech University

Jennifer Speights-Binet
Samford University

Sam Sweitz
Michigan Technological University

Michael W. Tripp
Vancouver Island University

Julie L. Urbanik
University of Missouri, Kansas City

Irina Vakulenko
University of Texas at Dallas

Jean Vincent
Santa Fe College

Linda Wang
University of South Carolina, Aiken

Kelly Watson
Florida State University

Laura A. Zeeman
Red Rocks Community College

Sandra Zupan
University of Kentucky

First Edition, *World Regional Geography Concepts*

Gillian Acheson
Southern Illinois University, Edwardsville

Tanya Allison
Montgomery College

Keshav Bhattarai
Indiana University, Bloomington

Leonhard Blesius
San Francisco State University

Jeffrey Brauer
Keystone College

Donald Buckwalter
Indiana University of Pennsylvania

Craig Campbell
Youngstown State University

John Comer
Oklahoma State University

Kevin Curtin
George Mason University

Ron Davidson
California State University, Northridge

Tina Delahunty
Texas Tech University

Dean Fairbanks
California State University, Chico

Allison Feeney
Shippensburg University of Pennsylvania

Eric Fournier
Samford University

Qian Guo
San Francisco State University

Carole Huber
University of Colorado at Colorado Springs

Paul Hudak
University of North Texas

Christine Jocoy
California State University, Long Beach

Ron Kalafsky
University of Tennessee, Knoxville

David Keefe
University of the Pacific

Mary Klein
Saddleback College

Max Lu
Kansas State University

Donald Lyons
University of North Texas

Barbara McDade
University of Florida

Victor Mote
University of Houston

Darrell Norris
State University of New York, Geneseo

Gabriel Popescu
Indiana University, South Bend

Claudia Radel
Utah State University

Donald Rallis
University of Mary Washington

Pamela Riddick
University of Memphis

Jennifer Rogalsky
State University of New York, Geneseo

Tobie Saad
University of Toledo

Charles Schmitz
Towson University

Sindi Sheers
George Mason University

Ira Sheskin
University of Miami

Dmitri Siderov
California State University, Long Beach

Steven Silvern
Salem State College

Ray Sumner
Long Beach City College

Stan Toops
Miami University

Karen Trifonoff
Bloomsburg University of Pennsylvania

Jim Tyner
Kent State University

Michael Walegur
University of Delaware

Scott Walker
Northwest Vista College

Mark Welford
Georgia Southern University

Fifth Edition, *World Regional Geography*

Gillian Acheson
Southern Illinois University, Edwardsville

Greg Atkinson
Tarleton State University

Robert Begg
Indiana University of Pennsylvania

Richard Benfield
Central Connecticut State University

Fred Brumbaugh
University of Houston, Downtown

Deborah Corcoran
Missouri State University

Kevin Curtin
George Mason University

Lincoln DeBunce
Blue Mountain Community College

Scott Dobler
Western Kentucky University

Catherine Doenges
University of Connecticut at Stamford

Jean Eichhorst
University of Nebraska, Kearney

Brian Farmer
Amarillo College

Eveily Freeman
Ohio State University

Hari Garbharran
Middle Tennessee State University

Abe Goldman
University of Florida

Angela Gray
University of Wisconsin, Oshkosh

Ellen Hansen
Emporia State University

Nick Hill
Greenville Technical College

Johanna Hume
Alvin Community College

Edward Jackiewicz
California State University, Northridge

Rebecca Johns
University of South Florida, St. Petersburg

Suzanna Klaf
Ohio State University

Jeannine Koshear
Fresno City College

Brennan Kraxberger
Christopher Newport University

Heidi Lannon
Santa Fe College

Angelia Mance
Florida Community College, Jacksonville

Meredith Marsh
Lindenwood University

Linda Murphy
Blinn Community College

Monica Nyamwange
William Paterson University

Adam Pine
University of Minnesota, Duluth

Amanda Rees
Columbus State University

Benjamin Richason
St. Cloud State University

Amy Rock
Kent State University

Betty Shimshak
Towson University

Michael Siola
Chicago State University

Steve Smith
Missouri Southern State University

Jennifer Speights-Binet
Samford University

Emily Sturgess Cleek
Drury University

Gregory Taff
University of Memphis

Catherine Veninga
College of Charleston

Mark Welford
Georgia Southern University

Donald Williams
Western New England College

Peggy Robinson Wright
Arkansas State University, Jonesboro

Fourth Edition, *World Regional Geography*

Robert Acker
University of California, Berkeley

Joy Adams
Humboldt State University

John All
Western Kentucky University

Jeff Allender
University of Central Arkansas

David L. Anderson
Louisiana State University, Shreveport

Donna Arkowski
Pikes Peak Community College

Jeff Arnold
Southwestern Illinois College

Richard W. Benfield
Central Connecticut University

Sarah A. Blue
Northern Illinois University

Patricia Boudinot
George Mason University

Michael R. Busby
Murray State College

Norman Carter
California State University, Long Beach

Gabe Cherem
Eastern Michigan University

Brian L. Crawford
West Liberty State College

Phil Crossley
Western State College of Colorado

Gary Cummisk
Dickinson State University

Kevin M. Curtin
University of Texas at Dallas

Kenneth Dagel
Missouri Western State University

Jason Dittmer
University College London

Rupert Dobbin
University of West Georgia

James Doerner
University of Northern Colorado

Ralph Feese
Elmhurst College

Richard Grant
University of Miami

Ellen R. Hansen
Emporia State University

Holly Hapke
Eastern Carolina University

Mark L. Healy
Harper College

David Harms Holt
Miami University

Douglas A. Hurt
University of Central Oklahoma

Edward L. Jackiewicz
California State University, Northridge

Marti L. Klein
Saddleback College

Debra D. Kreitzer
Western Kentucky University

Jeff Lash
University of Houston, Clear Lake

Unna Lassiter
California State University, Long Beach

Max Lu
Kansas State University

Donald Lyons
University of North Texas

Shari L. MacLachlan
Palm Beach Community College

Chris Mayda
Eastern Michigan University

Armando V. Mendoza
Cypress College

Katherine Nashleanas
University of Nebraska, Lincoln

Joseph A. Naumann
University of Missouri at St. Louis

Jerry Nelson
Casper College

Michael G. Noll
Valdosta State University

Virginia Ochoa-Winemiller
Auburn University

Karl Offen
University of Oklahoma

Eileen O'Halloran
Foothill College

Ken Orvis
University of Tennessee

Manju Parikh
College of Saint Benedict and Saint John's University

Mark W. Patterson
Kennesaw State University

Paul E. Phillips
Fort Hays State University

Rosann T. Poltrone
Arapahoe Community College, Littleton, Colorado

Waverly Ray
MiraCosta College

Jennifer Rogalsky
State University of New York, Geneseo

Gil Schmidt
University of Northern Colorado

Yda Schreuder
University of Delaware

Tim Schultz
Green River Community College, Auburn, Washington

Sinclair A. Sheers
George Mason University

D. James Siebert
North Harris Montgomery Community College, Kingwood

Dean Sinclair
Northwestern State University

Bonnie R. Sines
University of Northern Iowa

Vanessa Slinger-Friedman
Kennesaw State University

Andrew Sluyter
Louisiana State University

Kris Runberg Smith
Lindenwood University

Herschel Stern
MiraCosta College

William R. Strong
University of North Alabama

Ray Sumner
Long Beach City College

Rozemarijn Tarhule-Lips
University of Oklahoma

Alice L. Tym
University of Tennessee, Chattanooga

James A. Tyner
Kent State University

Robert Ulack
University of Kentucky

Jialing Wang
Slippery Rock University of Pennsylvania

Linda Q. Wang
University of South Carolina, Aiken

Keith Yearman
College of DuPage

Laura A. Zeeman
Red Rocks Community College

Third Edition, *World Regional Geography*

Kathryn Alftine
California State University, Monterey Bay

Donna Arkowski
Pikes Peak Community College

Tim Bailey
Pittsburg State University

Brad Baltensperger
Michigan Technological University

Michele Barnaby
Pittsburg State University

Daniel Bedford
Weber State University

Richard Benfield
Central Connecticut State University

Sarah Brooks
University of Illinois at Chicago

Jeffrey Bury
University of Colorado, Boulder

Michael Busby
Murray State University

Norman Carter
California State University, Long Beach

Gary Cummisk
Dickinson State University

Cyrus Dawsey
Auburn University

Elizabeth Dunn
University of Colorado, Boulder

Margaret Foraker
Salisbury University

Robert Goodrich
University of Idaho

Steve Graves
California State University, Northridge

Ellen Hansen
Emporia State University

Sophia Harmes
Towson University

Mary Hayden
Pikes Peak Community College

R. D. K. Herman
Towson University

Samantha Kadar
California State University, Northridge

James Keese
California Polytechnic State University, San Luis Obispo

Phil Klein
University of Northern Colorado

Debra D. Kreitzer
Western Kentucky University

Soren Larsen
Georgia Southern University

Unna Lassiter
California State University, Long Beach

David Lee
Florida Atlantic University

Anthony Paul Mannion
Kansas State University

Leah Manos
Northwest Missouri State University

Susan Martin
Michigan Technological University

Luke Marzen
Auburn University

Chris Mayda
Eastern Michigan University

Michael Modica
San Jacinto College

Heather Nicol
State University of West Georgia

Ken Orvis
University of Tennessee

Thomas Paradis
Northern Arizona University

Amanda Rees
University of Wyoming

Arlene Rengert
West Chester University of Pennsylvania

B. F. Richason
St. Cloud State University

Deborah Salazar
Texas Tech University

Steven Schnell
Kutztown University

Kathleen Schroeder
Appalachian State University

Roger Selya
University of Cincinnati

Dean Sinclair
Northwestern State University

Garrett Smith
Kennesaw State University

Jeffrey Smith
Kansas State University

Dean Stone
Scott Community College

Selima Sultana
Auburn University

Ray Sumner
Long Beach City College

Christopher Sutton
Western Illinois University

Harry Trendell
Kennesaw State University

Karen Trifonoff
Bloomsburg University

David Truly
Central Connecticut State University

Kelly Victor
Eastern Michigan University

Mark Welford
Georgia Southern University

Wendy Wolford
University of North Carolina, Chapel Hill

Laura Zeeman
Red Rocks Community College

Second Edition, *World Regional Geography*

Helen Ruth Aspaas
Virginia Commonwealth University

Cynthia F. Atkins
Hopkinsville Community College

Timothy Bailey
Pittsburg State University

Robert Maxwell Beavers
University of Northern Colorado

James E. Bell
University of Colorado, Boulder

Richard W. Benfield
Central Connecticut State University

John T. Bowen, Jr.
University of Wisconsin, Oshkosh

Stanley Brunn
University of Kentucky

Donald W. Buckwalter
Indiana University of Pennsylvania

Gary Cummisk
Dickinson State University

Roman Cybriwsky
Temple University

Cary W. de Wit
University of Alaska, Fairbanks

Ramesh Dhussa
Drake University

David M. Diggs
University of Northern Colorado

Jane H. Ehemann
Shippensburg University

Kim Elmore
University of North Carolina, Chapel Hill

Thomas Fogarty
University of Northern Iowa

James F. Fryman
University of Northern Iowa

Heidi Glaesel
Elon College

Ellen R. Hansen
Emporia State University

John E. Harmon
Central Connecticut State University

Michael Harrison
University of Southern Mississippi

Douglas Heffington
Middle Tennessee State University

Robert Hoffpauir
California State University, Northridge

Catherine Hooey
Pittsburg State University

Doc Horsley
Southern Illinois University, Carbondale

David J. Keeling
Western Kentucky University

James Keese
California Polytechnic State University

Debra D. Kreitzer
Western Kentucky University

Jim LeBeau
Southern Illinois University, Carbondale

Howell C. Lloyd
Miami University of Ohio

Judith L. Meyer
Southwest Missouri State University

Judith C. Mimbs
University of Tennessee, Chattanooga

Monica Nyamwange
William Paterson University

Thomas Paradis
Northern Arizona University

Firooza Pavri
Emporia State University

Timothy C. Pitts
Edinboro University of Pennsylvania

William Preston
California Polytechnic State University

Gordon M. Riedesel
Syracuse University

Joella Robinson
Houston Community College

Steven M. Schnell
Northwest Missouri State University

Kathleen Schroeder
Appalachian State University

Dean Sinclair
Northwestern State University

Robert A. Sirk
Austin Peay State University

William D. Solecki
Montclair State University

Wei Song
University of Wisconsin, Parkside

William Reese Strong
University of North Alabama

Selima Sultana
Auburn University

Suzanne Traub-Metlay
Front Range Community College

David J. Truly
Central Connecticut State University

Alice L. Tym
University of Tennessee, Chattanooga

First Edition, *World Regional Geography*

Helen Ruth Aspaas
Virginia Commonwealth University

Brad Bays
Oklahoma State University

Stanley Brunn
University of Kentucky

Altha Cravey
University of North Carolina, Chapel Hill

David Daniels
Central Missouri State University

Dydia DeLyser
Louisiana State University

James Doerner
University of Northern Colorado

Bryan Dorsey
Weber State University

Lorraine Dowler
Penn State University

Hari Garbharran
Middle Tennessee State University

Baher Ghosheh
Edinboro University of Pennsylvania

Janet Halpin
Chicago State University

Peter Halvorson
University of Connecticut

Michael Handley
Emporia State University

Robert Hoffpauir
California State University, Northridge

Glenn G. Hyman
International Center for Tropical Agriculture

David Keeling
Western Kentucky University

Thomas Klak
Miami University of Ohio

Darrell Kruger
Northeast Louisiana University

David Lanegran
Macalester College

David Lee
Florida Atlantic University

Calvin Masilela
West Virginia University

Janice Monk
University of Arizona

Heidi Nast
DePaul University

Katherine Nashleanas
University of Nebraska, Lincoln

Tim Oakes
University of Colorado, Boulder

Darren Purcell
Florida State University

Susan Roberts
University of Kentucky

Dennis Satterlee
Northeast Louisiana University

Kathleen Schroeder
Appalachian State University

Dona Stewart
Georgia State University

Ingolf Vogeler
University of Wisconsin, Eau Claire

Susan Walcott
Georgia State University

These world regional geography textbooks have been a family project many years in the making. Lydia Pulsipher came to the discipline of geography at the age of 5, when her immigrant father, Joe Mihelič, hung a world map over the breakfast table in their home in Coal City, Illinois, where he was pastor of the New Hope Presbyterian Church. They soon moved to the Mississippi Valley of eastern Iowa, where Lydia's father, then a professor at the Presbyterian theological seminary in Dubuque, continued his geography lessons on the passing landscapes whenever Lydia accompanied him on Sunday trips to small country churches. Lydia's sons, Anthony and Alex, got their first doses of geography in the bedtime stories she told them. For plots and settings, she drew on Caribbean colonial documents she was then reading for her dissertation. They first traveled abroad and learned about the hard labor of field geography when, as 12- and 8-year-olds, they were expected to help with the archaeological and ethnographic research conducted by Lydia and her colleagues on the eastern Caribbean island of Montserrat. It was Lydia's brother John Mihelič who first suggested that Lydia, Alex, and Mac write a book like this one, after he too came to appreciate geography. He has been a loyal cheerleader during the process, as have our extended family and friends in Knoxville, Montserrat, San Francisco, Slovenia, and beyond.

Graduate students and faculty colleagues in the geography department at the University of Tennessee have been generous in their support, serving as helpful impromptu sounding boards for ideas. Ken Orvis, especially, has advised us on the physical geography sections of all editions. Yingkui (Philippe) Li provided information on glaciers and climate change; Russell Kirby wrote one of the vignettes based on his research in Vietnam; Toby Applegate, Alex Pulsipher (in his capacity as an instructor), Michelle Brym, and Sara Beth Keough have helped us understand how to better assist instructors; and Ron Kalafsky, Tom Bell, Margaret Gripshover and Micheline Van Riemsdijk have chatted with the authors many times on specific and broad issues related to this textbook.

Maps for this edition were conceived by Mac Goodwin and Alex Pulsipher and produced by Will Fontanez and the University of Tennessee cartography shop staff and by Maps.com under the direction of Mike Powers. Alex Pulsipher chose the photos and produced the photo essays and the map photo essays.

Liz Widdicombe and Sara Tenney at W.H. Freeman and Company were the first to persuade us that together we could develop a new direction for *World Regional Geography*, one that included the latest thinking in geography written in an accessible style. In accomplishing this goal, we are especially indebted to our first developmental editor, Susan Moran, and to the W.H. Freeman staff for all they have done to ensure that this book is well written, beautifully designed, and well presented to the public.

We would also like to gratefully acknowledge the efforts of the following people at W.H. Freeman: Steven Rigolosi, executive editor for this second edition; Elaine Epstein, developmental editor; Laura McGinn, project editor; John Britch, director of marketing; Anna Paganelli, copyeditor; Diana Blume, design manager; Bill Page, senior illustration coordinator; Julia De Rosa, production manager; and Kerri Russini, assistant editor.

Given our ambitious new photo program, we are especially grateful for Vicki Tomaselli's brilliant work and responsiveness as designer for the second edition, as well as for Ted Szczepanski's guidance and direction as our photo editor for the second edition. We are also grateful to the supplements authors, who have created what we think are unusually useful, up-to-date, and labor-saving materials for instructors who use our book.

1 Geography: An Exploration of Connections

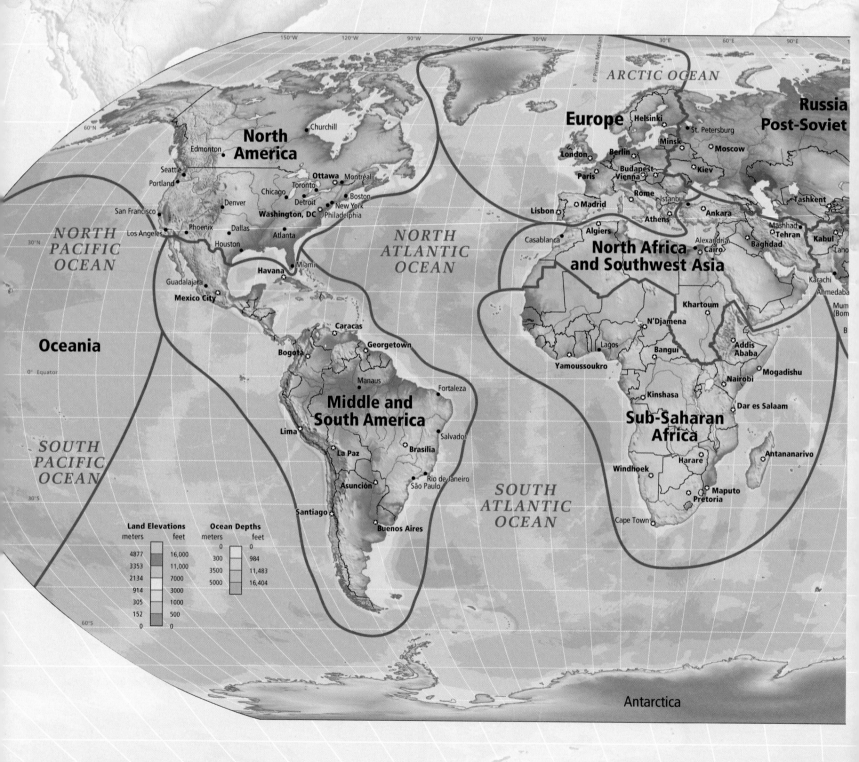

150°W   120°W   90°W   60°W   30°W   0° Prime Meridian   30°E   60°E   90°E

ARCTIC OCEAN

North America

Churchill
Edmonton
Seattle
Portland
Chicago   Toronto   Ottawa   Montréal
Denver   Detroit   Boston
San Francisco   Washington, DC   New York   Philadelphia
Phoenix   Dallas   Atlanta
Los Angeles   Houston

NORTH PACIFIC OCEAN

Guadalajara
Mexico City
Havana   Miami

Oceania

NORTH ATLANTIC OCEAN

Europe   Helsinki
St. Petersburg
London   Berlin   Minsk   Moscow
Paris   Budapest   Kiev
Madrid   Vienna
Lisbon   Rome   Istanbul   Ankara
Athens

Russia Post-Soviet

Tashkent

Mashhad   Tehran   Kabul
Casablanca   Algiers   Alexandria   Cairo   Baghdad   Laho

North Africa and Southwest Asia

Khartoum   Karachi
Ahmedaba
N'Djamena   Mum (Bom)

Lagos   Bangui   Addis Ababa
Yamoussoukro   Mogadishu
Kinshasa   Nairobi
Dar es Salaam

Caracas
Georgetown
Bogotá

Manaus   Fortaleza

Middle and South America

Lima   Salvador
La Paz   Brasília
Asunción   Rio de Janeiro   São Paulo
Santiago
Buenos Aires

SOUTH PACIFIC OCEAN

0° Equator

30°S

60°S

Sub-Saharan Africa

Windhoek   Harare   Antananarivo

Maputo
Pretoria
Cape Town

SOUTH ATLANTIC OCEAN

60°N
30°N

Land Elevations		Ocean Depths	
meters	feet	meters	feet
4877	16,000	0	0
3353	11,000	300	984
2134	7000	3500	11,483
914	3000	5000	16,404
305	1000		
152	500		
0	0		

Antarctica

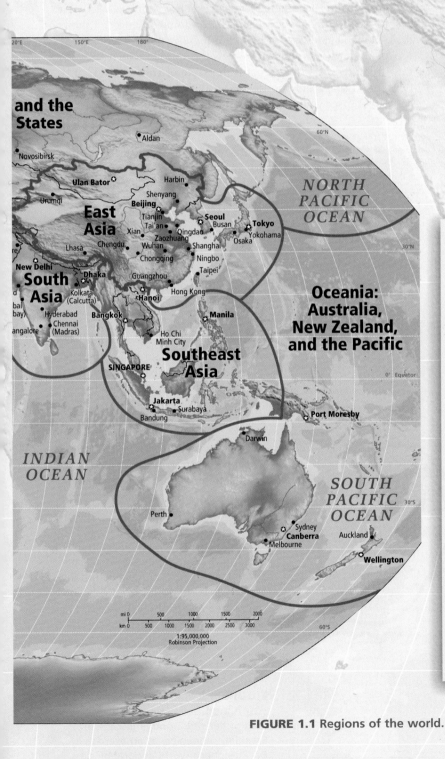

and the
States

· Novosibirsk

Ulan Bator ⊛

· Aldan

Harbin
Shenyang

East
Asia

Urumqi

Beijing
Tianjin
Xian   Tai'an   ⊛ Seoul
Zaozhuang   Busan   · Tokyo
Chengdu   Qingdao   Yokohama
Wuhan   Shanghai   Osaka
Chongqing   Ningbo

Lhasa

NORTH
PACIFIC
OCEAN

60°N

30°N

New Delhi
South
Asia

Dhaka

Guangzhou   Taipei

Kolkata
(Calcutta)

Hanoi   Hong Kong

bay)

Hyderabad
Bangalore   · Chennai
(Madras)

Bangkok

Manila

Oceania:
Australia,
New Zealand,
and the Pacific

SINGAPORE

Ho Chi
Minh City

Southeast
Asia

0° Equator

INDIAN
OCEAN

Jakarta
⊛ · Surabaya
Bandung

⊛ Port Moresby

· Darwin

SOUTH
PACIFIC
OCEAN

30°S

· Perth

⊛ Sydney
Canberra
· Melbourne

Auckland ·

⊛ Wellington

mi 0   500   1000   1500   2000
km 0   500   1000   1500   2000   2500   3000

60°S

1:95,000,000
Robinson Projection

FIGURE 1.1 Regions of the world.

Learning Goals

After you read this chapter, you will be able to answer the following questions:

1. Regions: What is a region?

2. Population and Gender: How is the shift toward greater equality between the genders influencing population growth rates, patterns of economic development, and politics?

3. Food and Urbanization: How are changes in food production pushing people out of rural areas and pulling them into urban ones?

4. Globalization and Development: How are increased global flows of information, goods, and people transforming patterns of economic development?

5. Democratization: What factors are related to the shift from authoritarian modes of government—based on the power of the state or community leaders—to more democratic systems in which individuals have a greater say in how governments are run?

6. Climate Change and Water: In what ways does water influence how different places are vulnerable to climate change?

Where Is It? Why Is It There? Why Does It Matter?

Where are you? You may be in a house or a library or sitting under a tree on a fine fall afternoon. You are probably in a community (perhaps a college or university), and you are in a country (perhaps the United States) and a region of the world (perhaps North America, Southeast Asia, or the Pacific). Why are you where you are? Some answers are immediate, such as "I have an assignment to read." Other explanations are more complex, such as your belief in the value of an education, your career plans, and your or someone's willingness to sacrifice to pay your tuition. Even past social movements that opened up higher education to more than a fortunate few may help explain why you are where you are.

The questions *where* and *why* are central to geography. Think about a time you had to find the site of a party on a Saturday night, the location of the best grocery store, or the fastest and safest route home. You were interested in location, spatial relationships, and connections between the environment and people. Those are among the interests of geographers.

Geographers seek to understand why different places have different sights, sounds, smells, and arrangements of features. They study what has contributed to the look and feel of a place, to the standard of living and customs of the people, and to the way people in one place relate to people in other places. Furthermore, geographers often think on several scales, from the local to the global. For example, when choosing the best location for a new grocery store, a geographer might consider the physical characteristics of potential sites, socioeconomic circumstances of the neighborhood, traffic

> **scale (of a map)** the proportion that relates the dimensions of the map to the dimensions of the area it represents; also, variable-sized units of geographical analysis from the local scale to the regional scale to the global scale

patterns for the broader area in which the neighborhood is located, as well as the store's location relative to the main population concentrations for the whole city. She could also consider national or even international transportation routes, possibly to determine cost-efficient connections to suppliers.

To make it easier to understand a geographer's many interests, try this exercise. Draw a map of your most familiar childhood landscape. Relax, and recall the objects and experiences that were most important to you in that place. If the place was your neighborhood, you might start by drawing and labeling your home. Then fill in other places you encountered regularly, such as your backyard, your best friend's home, or your school. For example, Figure 1.2 shows the childhood landscape of Julia Stump in Franklin, Tennessee.

Consider how your map reveals the ways in which your life was structured by space. What is the **scale** of your map? That is, how much space did you decide to illustrate on the map? The amount of space your map covers may represent the degree of freedom you had as a child, or how aware you were of the world around you. Were there places you were not supposed to go? Does your map reveal, perhaps subtly, such emotions as fear, pleasure, or longing? Does it indicate your sex, your ethnicity, or the makeup of your family? Did you use symbols to show certain features? In making your map and analyzing it, you have engaged in several aspects of geography:

- Landscape observation
- Description of the earth's surface and consideration of the natural environment
- Spatial analysis (the study of how people, objects, or ideas are related to one another across space)
- The use of different scales of analysis (your map probably shows the spatial features of your childhood at a detailed *local scale*)
- Cartography (the making of maps)

As you progress through this book and this course, you will acquire geographic information and skills. Perhaps you are planning to travel to other lands or thinking about investing in East Asian timber stocks. Maybe you are searching for a good place to market an idea or are trying to understand current events in your town within the context of world events. Knowing how to practice geography will make your task easier and more engaging.

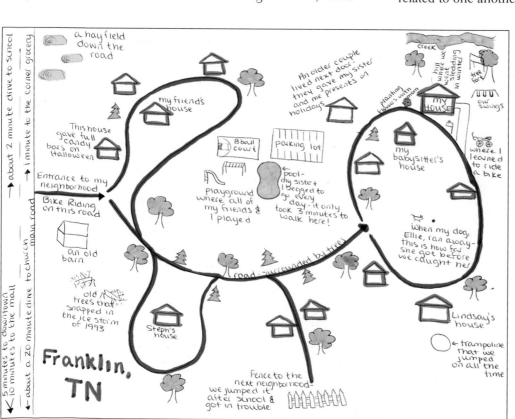

FIGURE 1.2 A childhood landscape map. Julia Stump drew this map of her childhood landscape in Franklin, Tennessee, as an exercise in Dr. Pulsipher's world geography class.

What Is Geography?

Geography is the study of our planet's surface and the processes that shape it. Yet this definition does not begin to convey the fascinating interactions of human and environmental forces that have given the earth its diverse landscapes and ways of life.

Geography, as an academic discipline, is unique in that it links the physical sciences—such as geology, physics, chemistry, biology, and botany—with the social sciences—such as anthropology, sociology, history, economics, and political science. Physical geographers have generally focused on how the earth's physical processes work independently of humans, but increasingly many are interested in how physical processes may affect humans, and how humans affect these processes in return. Human geography is the study of the various aspects of human life that create the distinctive landscapes and regions of the world. Physical and human geography are often tightly linked. For example, geographers might aim to understand:

- How and why people came to occupy a particular place.
- How people use the physical aspects of that place (climate, landforms, and resources) and then modify them to suit their particular needs.
- How people may create environmental problems.
- How people interact with other places, far and near.

Geographers usually specialize in one or more fields of study, or subdisciplines. Some of these particular types of geography are mentioned over the course of the book. Despite their individual specialties, geographers often cooperate in studying **spatial interaction** between people and places and the **spatial distribution** of relevant phenomena. For example, in the face of increasing global warming, climatologists, cultural geographers, and economic geographers work together to understand the spatial distribution of carbon dioxide emissions, and the cultural and economic practices that might be changed to limit such emissions. For example, an increasing number of geographers are interested in efforts to reduce transportation-related carbon dioxide emissions by changing patterns of human spatial interaction. This could take the form of redesigning urban areas so that people can live closer to where they work, or encouraging food production in locations closer to where the food will be consumed.

Many geographers specialize in a particular region of the world, or even in one small part of a region. Regional geography is the analysis of the geographic characteristics of a particular place, the size and scale of which can vary radically. The study of a region can reveal connections among physical features and ways of life, as well as connections to other places. These links are key to understanding the present and the past, and are essential in planning for the future. This book follows a "world regional" approach, focusing on general knowledge about specific regions of the world. We will see just what geographers mean by *region* a little later in this chapter.

spatial interaction the flow of goods, people, services, or information across space and among places

spatial distribution the arrangement of a phenomenon across the Earth's surface

longitude the distance in degrees east and west of Greenwich, England; lines of longitude, also called meridians, run from pole to pole (the line of longitude at Greenwich is 0° and is known as the prime meridian)

latitude the distance in degrees north or south of the equator; lines of latitude run parallel to the equator, and are also called parallels

Geographers' Visual Tools

Among geographers' most important tools are maps, which they use to record, analyze, and explain spatial relationships, as you did on your childhood landscape map. Geographers who specialize in depicting geographic information on maps are called cartographers.

Understanding Maps

Figure 1.3 on pages 4–5 explains the various features of maps. Different *scales of imagery* are demonstrated using maps, photographs, or a satellite image. Read the captions carefully to understand the scale being depicted in each image. Throughout this book, you will encounter different kinds of maps at different scales. Some will show physical features, such as landforms or climate patterns at the regional or global scale. Others will show aspects of human activities at these same regional or global scales—for example, the routes taken by drug traders. Yet other maps will show settlement or cultural features at the scale of countries or regions, or cities or even local neighborhoods.

Longitude and Latitude

Most maps contain lines of latitude and longitude, which enable a person to establish a position on the map relative to other points on the globe. Lines of **longitude** (also called meridians) run from pole to pole; lines of **latitude** (also called parallels) run around the earth parallel to the equator (Figure 1.4 on page 6).

Both latitude and longitude lines describe circles, so there are 360 degrees (designated with the symbol °) in each circle of latitude and 180 degrees in each pole-to-pole semicircle of longitude. Each degree spans 60 minutes (designated with the symbol '), and each minute has 60 seconds (designated with the symbol "). Keep in mind that these are measures of relative space on a circle, not measures of time. They do not even represent real distance because the circles of latitude get successively smaller to the north and south of the equator, until they become a virtual dot at the poles.

The globe is also divided into hemispheres. The prime meridian, 0° longitude, runs from the North Pole through Greenwich, England, to the South Pole. The half of the globe's surface west of the prime meridian is called the Western Hemisphere; the half to the east is called the Eastern Hemisphere. The longitude lines both east and west of the prime meridian are labeled from 1° to 180° by their direction and distance in degrees from the prime meridian. For example, 20 degrees east longitude would be written as 20° E. The longitude line at 180° runs through the Pacific Ocean and is used roughly as the International Date Line; the calendar day officially begins when midnight falls at this line.

The equator divides the globe into the Northern and Southern hemispheres. Latitude is measured from 0° at the equator to 90° at the North or South Pole.

FIGURE 1.3 Understanding Maps: The Legend

Being able to read a map legend is crucial to understanding the maps in this book. The colors in the legend convey information about different areas on the map. In the population density map below, the lowest density (0–3 persons per square mile), is colored pale yellow. A part of North America that has this density is shown in the map inset to the right of the legend. On the far right is a picture of this area. Related maps and pictures also appear for two other densities (27–260 and above 2000 people per square mile).

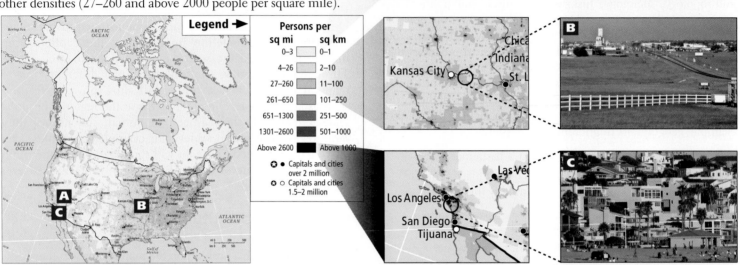

Legend →	Persons per	
	sq mi	**sq km**
	0–3	0–1
	4–26	2–10
	27–260	11–100
	261–650	101–250
	651–1300	251–500
	1301–2600	501–1000
	Above 2600	Above 1000

✪ ● Capitals and cities over 2 million
✪ ○ Capitals and cities 1.5–2 million

Understanding Maps: Choropleth versus Isorhythmic

There are two main types of maps presented in this book. In choropleth maps, the shapes visible on the map correspond to predefined areas (such as countries, counties, and census tracts) that are shaded according to the measurement of a statistical variable (for example, population density). On map **D** you can see examples of predefined areas, the blocky yellowish shapes around Las Vegas, which in this case are census tracts that the U.S. government uses to gather data on the population. The mapmaker assigned each census tract a color based on the density of the population in that tract. Another way to represent data spatially is to make an isorhythmic map, in which the shapes visible on the map correspond to areas defined by data patterns. Instead of fitting the data into predefined areas, the data define the areas. If map **D** had been done as an isorhythmic map it would look something like map **E**, an isorhythmic population density map. The maker of this map created shapes that correspond to each level of population density represented in the legend above. In this particular case, making the map involved using other data, such as satellite images that show where people really are living and a good deal of estimation about where they live. In this book we create maps using the method that best reflects the data we are trying to display. Since population density data are collected according to census tracts or other predefined boundaries, we use choropleth maps here. However, when the data used to make a map are not collected according to predefined boundaries (for example, the elevation data used to create the maps found at the beginning of each chapter), we use isorhythmic maps.

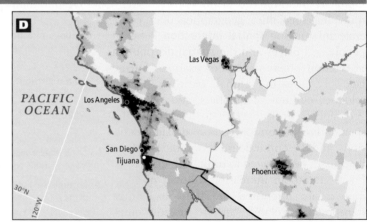

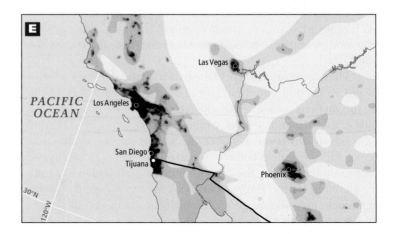

Understanding Maps: Scale

Maps often display information at different spatial scales, which means that lengths, areas, distances, and sizes can appear dramatically different on otherwise similar maps. This book often combines maps at several different scales with photographs taken by people on the surface of the earth and photos taken by satellites or astronauts in space. All of these visual tools convey information at a spatial scale. Here are some of the scales you might encounter in this book. The scale is visible in the corners of the map.

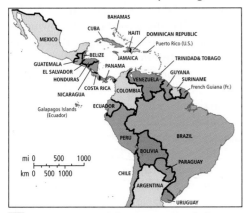

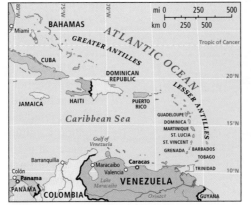

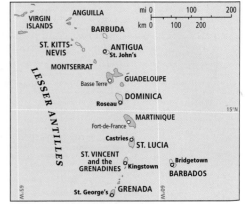

F Here is most of the region we define as Middle and South America, with just the tip of South America missing, at a scale of 1:200,000,000, meaning that 1 inch on the map equals 200,000,000 inches in reality. This is roughly the scale of the world maps, such as Figure 1.1, that appear throughout this book.

G Here is a close-up from the previous map, focusing on the Caribbean Sea, at a scale of 1:45,000,000. While the larger islands are clearly shown, some smaller islands are too small to identify clearly. This is roughly the scale of the full-page maps for Chapter 3, such as Figure 3.12 (population).

H The eastern Caribbean at a scale of 1:15,000,000. The shape of the islands and the locations of some capital cities are clearly visible.

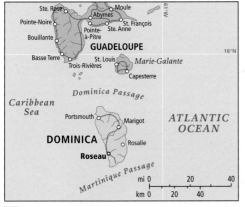

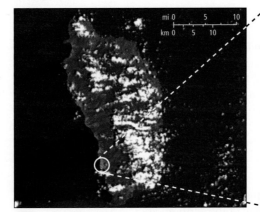

I Zooming in further, we arrive at a map of Guadeloupe and Dominica, in the eastern Caribbean, at a scale of 1:3,000,000. This scale makes it possible to show towns and a few roads and rivers.

J A satellite image of Dominica at a scale of 1:950,000. The town of Roseau, shown in the following picture, is circled. The dashed lines indicate that the following picture was taken within this circle. The white puffs are clouds.

K A photo overlooking part of the town of Roseau, capital of Dominica. It's hard to give a precise scale for this picture because the houses at the bottom are about 2000 feet closer to the photographer than the shoreline. The cruise boat is at a scale of roughly 1:3200.

Understanding Maps: Conventions in This Textbook

Rivers are darker blue. Lakes and oceans are lighter blue.

Major roads are in red.

Railroads are in black.

Cities are given an icon and print size indicating how large they are.

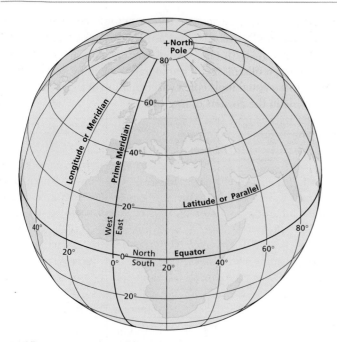

FIGURE 1.4 Summary of longitude and latitude. Lines of longitude (meridians) extend from pole to pole. The distance between them on the globe decreases steadily toward the poles, where they all meet. Lines of latitude (parallels) are equally spaced north and south of the equator and intersect the longitude lines at right angles. The only line of latitude that spans the complete circumference of the earth is the equator; all other lines of latitude describe ever-smaller circles heading away from the equator.

Lines of longitude and latitude form a grid that can be used to designate the location of a place. In Figure 1.3I, you can see that the island of Marie-Galante lies just south of the parallel (line of latitude) at 16° N and just about 16' west of the 61st west meridian. Hence, the position of the northernmost point on Marie-Galante's north coast is approximately 16° N by 61°16' W. Google Earth gives a more precise location of 16° 0'26" N by 61°16'33" W.

Map Projections

Printed maps must solve the problem of showing the spherical earth on a flat piece of paper. Imagine drawing a map of the earth on an orange, peeling the orange, and then trying to flatten out the orange-peel map and transfer it exactly to a flat piece of paper. The various ways of showing the spherical surface of the earth on flat paper are called **map projections**. All projections create some distortion. For maps of small parts of the earth's surface, the distortion is minimal.

> **map projections** the various ways of showing the spherical earth on a flat surface

Developing a projection for the whole surface of the earth that minimizes distortion is much more challenging.

The *Mercator projection* (**Figure 1.5A**) is popular, but geographers rarely use it because of its gross distortion near the poles. To make his flat map, the Flemish cartographer Gerhardus Mercator (1512–1594) stretched out the poles, depicting them as lines equal in length to the equator! As a result, Greenland, for example, appears about as large as Africa, even though it is only about one-fourteenth Africa's size. Nevertheless, Mercator's projection is still useful for navigation because it portrays the shapes of landmasses

more or less accurately, and because a straight line between two points on this map gives the compass direction between them.

Goode's interrupted homolosine projection (Figure 1.5B) flattens the earth rather like an orange peel, thus preserving some of the size and shape of the landmasses. In this projection, the oceans are split. The *Robinson projection* (Figure 1.5C) shows the longitude lines curving toward the poles to give an impression of the earth's curvature, and it has the advantage of showing an uninterrupted view of land and ocean; however, as a result, the shapes of landmasses are slightly distorted. In this book we often use the Robinson projection for world maps.

Maps are not unbiased. Most currently popular world map projections reflect the European origins of modern cartography. Europe or North America is usually placed near the center of the map, where distortion is minimal; other population centers, such as East Asia, are placed at the highly distorted periphery. For a less-biased study of the modern world, we need world maps that center on different parts of the globe. For example, much of the world's economic activity is now taking place in and around Japan, Korea, China, Taiwan, and Southeast Asia. Discussions of the world economy require maps that focus on these regions and include other parts of the world in the periphery. Another source of bias in maps is the convention that north is always at the top of the map. Some cartographers think that this can lead to a subconscious assumption that the northern hemisphere is somehow superior to the southern hemisphere.

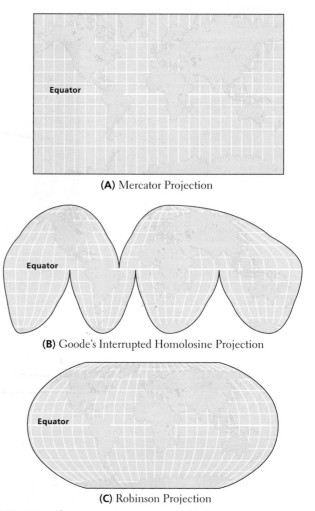

(A) Mercator Projection

(B) Goode's Interrupted Homolosine Projection

(C) Robinson Projection

FIGURE 1.5 Three common map projections.

Geographic Information Science (GISc)

The acronym GISc is now widespread and usually refers to **Geographic Information Science**, meaning the body of science that supports spatial analysis technologies. GISc is multidisciplinary, using techniques from cartography (mapmaking), geodesy (measuring the earth's surface), and photogrammetry (the science of making reliable measurements, especially by using aerial photography). But additional sciences, such as cognitive psychology and spatial statistics (geomatics or geoinformatics) are increasingly being used to give greater depth and breadth to three-dimensional spatial analysis. GISc, then, can be used in medicine to analyze the human body, in engineering to analyze mechanical devices, in architecture to analyze buildings, in archaeology to analyze sites above and below ground, and in geography to analyze the earth's surface and the space above and below the earth's surface.

GISc is a burgeoning field in geography, with wide practical applications in government and business and in efforts to assess and improve human and environmental conditions. GISc can also be applied to the computerized analytical systems that are the tools of this newest of spatial sciences.

The widespread use of GISc, particularly by governments and corporations, has dramatically increased the amount of information that is collected and stored, and changed the way it is analyzed and distributed. These changes create many new opportunities for solving problems, for example, by increasing the ability of local governments to plan future urban growth. However, these technologies also raise serious questions. What rights do people have over the storage, analysis, and distribution of information on their location and movements, which can now be gathered from their cell phones? Should this information reside in the public domain? Should individuals have the right to have their location-based information suppressed from public view? Should a government or corporation have the right to sell information to anyone, without special permission, about where people spend their time and how frequently they go to particular places? Progress on these societal questions has not kept pace with the technological advances in GISc.

The Detective Work of Photo Interpretation

Geographers use photographs to help them understand or explain a geographic issue or depict the character of a place. Interpreting a photo to extract its geographical information can sometimes be like detective work. In Figure 1.6 on page 8 are some points to keep in mind as you look at the pictures throughout this book; try them out first with the photo in Figure 1.6.

The Region as a Concept

Learning Goal 1
Regions: What is a region?

A **region** is a unit of the earth's surface with distinct physical and human features.

It could be a desert region, a region that produces rice, or a region experiencing ethnic violence. In this book, it is rare for any two regions to be defined by the same set of indicators. For example, the region of the southern United States might be defined by its distinctive vegetation, architecture, foods, and historical experience. Meanwhile Siberia, in eastern Russia, could be defined primarily by its climate, vegetation, remoteness, and sparse settlement.

Another issue in defining regions is that their boundaries are rarely crisp. The more closely we look at the border zones, the less distinct the divisions appear. Consider the case of the boundary between the United States and Mexico. The clearly delineated political border does not mark a separation between cultures or economies. In a wide band extending over both sides of the border, there is a blend of Native American, Spanish colonial, Mexican, and Anglo-American cultural features. Languages, local and regional economies, place-names, food customs, music, and family organization are only a few examples of this blend. And the economy of the border zone depends on interactions across a broad swath of territory.

Why, then, does this book place Mexico in the chapter that includes Middle America? Although northern Mexico has much in common with parts of the southwestern United States, overall Mexico still has more in common with Middle America than with North America. The use of Spanish as its official language ties Mexico to Middle and South America and separates it from the United States and Canada, where English dominates (except in Québec, where French is the dominant language). These language patterns are symbolic of the larger cultural and historical differences between the two regions, which will be discussed in Chapters 2 and 3.

If regions are so difficult to define and describe, why do geographers use them? To discuss the whole world at once would be impossible, and so geographers seek a reasonable way to divide the world into manageable parts. There is nothing sacred about the criteria or the boundaries we use. They are just practical aids to learning. In defining the world regions for this book, we have considered such factors as physical features, political boundaries, cultural characteristics, history, how the places now define themselves, and what the future may hold.

This book organizes the material into three *scales of analysis*: the global scale, the world regional scale, and the **local scale**. The word *scale*, whether on maps or in text, refers to the relative size of an area. At the **global scale**, explored in this chapter, the entire world is treated as a single area—a unity that is more and more relevant as our planet operates as a global system. We use the term **world region** for the largest divisions of the globe, such as East Asia and North America (Figure 1.1). We have defined ten world regions, each of which is covered in a separate chapter. Each regional chapter considers the interaction of human and physical geography in relation to cultural, social, economic, population, environmental, and political topics.

Geographic Information Science (GISc) the body of science that underwrites multiple spatial analysis technologies and keeps them at the cutting edge

region a unit of the earth's surface that contains distinct patterns of physical features and/or of human development

local scale the level of geography that describes the space where an individual lives or works; a city, town, or rural area

global scale the level of geography that encompasses the entire world as a single unified area

world region a part of the globe delineated according to criteria selected to facilitate the study of patterns particular to the area

(A) Landforms: Notice the lay of the land and the landform features. Are there any indications about the ways the landforms and humans have influenced each other? Is environmental stress visible?

(B) Vegetation: Notice whether the vegetation indicates a wet or dry, or warm or cold environment. Can you recognize specific species? Does the vegetation appear to be natural or is it influenced by human use?

(C) Material culture: Are there buildings, tools, clothing, foods, or vehicles that give clues about the cultural background, wealth, values, or aesthetics of the people who live where the picture was taken?

(D) What do the people in the photo suggest about the situation pictured?

(E) Can you see evidence of the global economy (for example, goods that probably were not produced locally)?

(F) Location: Can you tell where the picture was taken or narrow down the possible locations that might be being depicted?

You can use this system to analyze any of the photos in this book or elsewhere. Practice by analyzing the photos in this book before you read their captions. Here is an example of how you could do this with the photo below.

(A) Landforms:

1. The flat horizon suggests a plain or river delta. Environmental stress is visible in several places.
2. This fire could be burning out of control. No one seems to be tending it.
3. This looks like the charred remains of a tree trunk.
4. This black oily-looking liquid doesn't look natural. Could it be crude oil? That would explain the burning.

5. What are these lumps on the ground?
6. This could be a tree trunk coming out of one of the lumps. Could the lumps be what is left of trees that have burned?

   What would have caused all this landscape transformation? Maybe an oil spill? Maybe the oil caught on fire?

(B) Vegetation:

7. This looks like a palm tree. There are quite a few palm trees and other trees.

   Must be in the tropics and be fairly wet and warm.

(C) Material culture: See (D).

   There is nothing here but a single person. The whole area might be abandoned.

(D) People:

8. The clothing on this person doesn't look like he made it. It looks mass produced.

   This suggests that he has access to goods produced some distance away, maybe in a nearby city. Or he could buy things in a market where imported goods are sold.

9. His worn flip-flops and callused feet suggest he walks a lot.

   He might not have a car.

(E) Global economy: See (D).

(F) Location: Somewhere tropical where there could have been an oil spill. Hint: Use this book! See Figure 6.1 to see where the member countries of OPEC are. This suggests that the photo could be of Venezuela, Nigeria, or Indonesia. Suggestion: Read Chapter 7!

FIGURE 1.6 Oil development and the environment. On December 3, 2003, part of an oil pipeline burst in Rukpokwu state in Nigeria. No action was taken by the owner of the pipeline, Shell Oil, or the Nigerian government for a week. Fires burned for more than 6 weeks, and no cleanup was attempted afterward. About 740 acres of once-fertile farmland and fish ponds were destroyed. This picture was taken in 2004.

In summary, regions have the following traits:

- A region is a unit of the earth's surface that contains distinct environmental or cultural patterns.
- No two regions are necessarily defined by the same set of indicators.
- Regions can vary greatly in size (scale).
- The boundaries of regions are usually indistinct and hard to agree upon.

Thematic Concepts and Their Role in This Book

Within the world regional framework, this book is also organized around nine thematic concepts of special significance in the modern world—concepts that operate in every world region and interact in numerous ways:

- Population
- Gender
- Development
- Food
- Urbanization
- Globalization
- Democratization
- Water
- Climate change

The following sections explain each of these nine thematic concepts, how they tie in with the main concerns of geographers, and how they are related to each other. This information has a special relevance in each of the regional chapters, where the thematic concepts provide a useful framework for your rapidly accumulating knowledge of the world. The nine concepts are not presented in a particular order in each chapter, but rather are discussed as they naturally come up in the coverage of issues. For example, in some regions, water is chiefly a scarcity issue; in others, managing water resources equitably under conditions of climate change is the main concern.

THINGS TO REMEMBER

1. Geography is the study of the earth's surface and the processes that shape it. There is a continual interaction between physical and human processes.

2. Learning Goal 1: Regions A region is a unit of the earth's surface that has a combination of distinct physical and/or human features; the complex of features can vary from region to region, and regional boundaries are rarely clear and crisp.

3. This book consistently pursues nine thematic concepts—population, gender, development, food, urbanization, globalization, democratization, water, and climate change.

POPULATION

To study population is to study the growth and decline of numbers of people on earth, their distribution across the earth's surface, age and sex distributions, migration patterns and what makes people move (see Thematic Concepts A, B, C on pages 10–13; see also Figure 1.8 on page 14). Over the last several hundred years, the global human population has boomed, but growth rates are now slowing in most societies and, in a few, have even moved into negative growth. Much of this pattern has to do with changes in economic development and gender roles that have reduced incentives for large families.

Global Patterns of Population Growth

It took between 1 million and 2 million years (at least 40,000 generations) for humans to evolve and to reach a global population of 2 billion, which happened around 1945. Then, remarkably, in just 66 years—by the year 2011—the world's population more than tripled to 7 billion (Figure 1.7). What happened to make the population grow so quickly in such a short time?

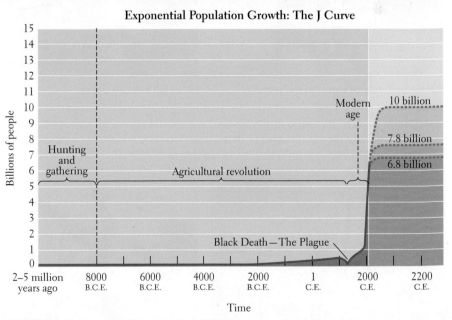

FIGURE 1.7 Exponential growth of the human population. The curve's J shape is a result of successive doublings of the population: it starts out nearly flat, but as doubling time shortens, the curve bends ever more sharply upward. Note that B.C.E. (before the common era) is equivalent to B.C. (before Christ); C.E. (common era) is equivalent to A.D. (anno domini).

Thematic Concepts: Population • Gender • Development • Food • Urbanization • Globalization • Democratization • Water • Climate Change

Population: Rapid global population growth has occurred over the last several hundred years, but growth rates are now slowing and some societies are even shrinking. Changes in economic development patterns, government policy, access to health care, and gender roles have reduced incentives for large families. These and other factors also shape the distribution and movement of human populations.

A As growth slows, many populations are rapidly aging. These retired men are in Portugal.

B Families with only one child are now common in Japan and elsewhere in East Asia.

C The most rapidly growing populations are poor and rural. These women are in Congo (Kinshasa).

Gender: Many places are moving toward greater equality between the genders. As more women pursue educational and employment opportunities outside the home, birth rates are declining. Meanwhile, economic development and politics are becoming transformed by the increasing participation of women.

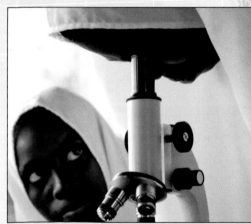

D Female high school students in Tanzania.

E Israeli military police.

F A woman votes in Congo (Kinshasa).

Development: Parts of the world (often labeled "the developing world") are shifting from lower-value and labor-intensive raw materials–based economies to higher-value and higher-skill–based manufacturing and service economies. This shift depends in part on the availability of social services, such as education and health care, that enable people to contribute to economic growth.

G A young boy cultivating by hand in Uganda.

H Students in a science class in Dakar, Senegal.

I A nurse delivers prenatal care to a pregnant woman in South Africa.

Food: So far, food production systems are keeping pace with global population growth, in part by shifting away from labor-intensive, small-scale, subsistence agriculture toward mechanized, chemically intensive, large-scale, commercial agriculture. This process increases productivity, but at the cost of environmental degradation that threatens further growth in food production. Moreover, many farmers are unable to afford the chemicals and machinery required for commercial agriculture and have to give up farming as a result.

J A crop duster sprays pesticides in Texas.

K A tractor is used for harvesting papayas in Brazil.

L Mechanized farming, as practiced above in Oregon, often exposes soils to wind and rain, resulting in erosion over time.

Urbanization: Changes in food production are pushing people out of rural areas, while the development of manufacturing and service economies is pulling them into cities. Living standards increase for some rural migrants, as access to jobs, health care, and education often improves. However, many are forced into vast slums with poor housing and inadequate access to water or social services.

M Many migrants to Dhaka, Bangladesh, work as bicycle rickshaw drivers.

N Kibera slum in Nairobi, Kenya, is home to more than 1 million people.

O The pressures of life in urban slums break up many families. These orphans are in Kibera slum in Nairobi, Kenya.

Globalization: Local self-sufficiency is giving way to global interdependence as goods, money, and people move across vast distances faster and on a larger scale than ever before. Influences from afar are transforming even seemingly isolated societies.

P In Guangdong, China, a woman manufactures laptop computer cases destined for the United States.

Q An Indian construction worker next to an English-language advertisement for the building he works on in Dubai, United Arab Emirates.

R A Masai herder in Tanzania displays his new cell phone.

Democratization: Authoritarianism, based on the authority of the state or community leaders, is giving way to more democratic systems in which each individual is given a greater voice in how governments are run. This shift is strongly linked to the growth of political freedoms, such as the right to protest and take action against injustice, especially through the media and the legal system.

S Women line up to vote in Yemen.

T Riot police disperse a peaceful protest for better access to water in Uganda.

U A journalist in Gambia protests censorship of the media.

Water: Fresh water is becoming scarce as human impacts on the environment increase. Pressure to reduce water use and pollution are rising as conflicts intensify over access to water for drinking and irrigation, and over the resources of aquatic ecosystems.

V For many urban dwellers, such as this boy in Kenya, a communal spigot is the only source of water.

W Dams can provide electricity and water for irrigation, but they flood potentially vast areas, and downstream river flows are forever altered.

X Irrigated rice terraces, such as these in China, feed billions of people worldwide.

Climate Change: Human activities that emit large amounts of carbon dioxide, methane, and other greenhouse gases are trapping heat in the atmosphere. The industrialized and rapidly industrializing countries, who are responsible for most of these emissions, are attempting to reduce their output of greenhouse gases. Meanwhile, the poorest countries of the world are highly vulnerable to the changes in climate brought about by global warming.

Y This coal-fired power plant in Russia spews tons of carbon dioxide into the atmosphere each day.

Z Global warming is bringing higher temperatures and greater acidity to the oceans, threatening coral reefs and the fishing industries they support.

AA Global warming is likely to increase flooding, especially in low-lying coastal areas such as Bangladesh (shown above).

Thematic Concepts Are Often Used in Combination to Explain Regional Issues Such as:

Globalization and Development in Madagascar, Sub-Saharan Africa: Highly globalized patterns of economic development dominate most of Africa. Originally set up by European colonial powers but still in place today, these patterns keep Africa relatively impoverished. Countries rely on exports of low-value raw materials to wealthier regions. They then have to import expensive manufactured goods needed to support their raw materials industries.

BB Women sort vanilla beans, a major export of Madagascar. Prices for vanilla are low on global markets.

CC Madagascar imports the engines for its trains from China at considerable expense.

DD Madagascar's largely agricultural economy leaves many in poverty.

Climate Change, Food, and Water in South Asia: South Asia's largest rivers are fed by glaciers high in the Himalayan Mountains that have been shrinking due to faster-than-normal melting. Most could be gone in 50 years. While flooding is the result for now, water shortages will be severe once the glaciers are gone, especially during the dry winter season when rivers are fed mostly by meltwater.

EE During the dry winter months, many rivers are fed by glacial meltwater. This valley bottom was once covered by a glacier.

FF An Indian farmer irrigates his rice field with age-old technology. Irrigation water will be less available during the winter if glaciers melt completely.

GG Faster-than-normal glacial melting causes flooding. Farmers in Bangladesh try to salvage their harvest after a flood.

Population and Gender in East Asia: China's attempt to control population growth with a one-child policy has had the unintended effect of producing a shortage of women. Cultural preferences for sons lead many couples to abort female fetuses or commit infanticide. China's government estimates that by 2020 there will be 30 million fewer women than men.

HH Most Chinese families would prefer that the one child they are allowed be a boy. A man and his son play in a park in Xian, China.

II With so many fewer women than men, and with many women choosing careers instead of child raising, brides are increasingly scarce in China.

JJ Some studies suggest that a large population of bachelors could lead to higher crime rates.

13

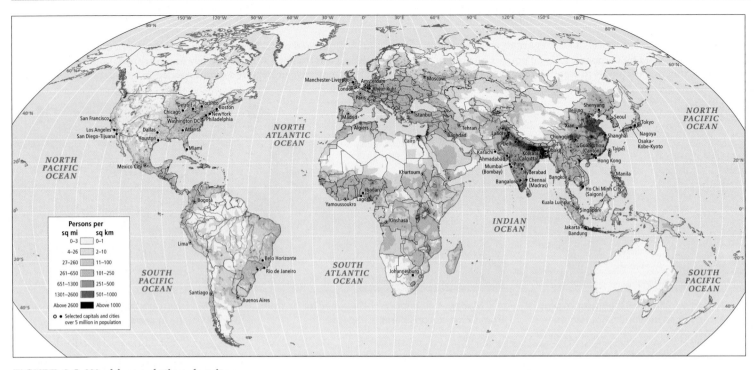

FIGURE 1.8 World population density.

The explanation lies in changing relationships between humans and the environment. For most of human history, fluctuating food availability, natural hazards, and disease kept human death rates high, especially for infants. Periodically, there were even crashes in human population—for example, the Black Death, a deadly pandemic throughout Europe and Asia in the 1300s.

An astonishing upsurge in human population began about 1500, at a time when the technological, industrial, and scientific revolutions were beginning in some parts of the world. Human life expectancy increased dramatically, and more and more people lived long enough to reproduce successfully, often many times over. The result was an exponential pattern of growth (see Figure 1.7). This pattern is often called a *J curve* because the ever-shorter periods between doubling and redoubling of the population cause an abrupt upward swing in the growth line when depicted on a graph.

Today, the human population is growing in virtually all regions of the world, but more rapidly in some places than in others. The reason for this growth is that currently a very large group of young people has reached the age of reproduction, and more will shortly join them. Nevertheless, the rate of global population growth is slowing. Since 1993 it has dropped from 1.7 percent per year to roughly 1.1 percent per year, which is where it stands today. If present slower growth trends continue, the world population may level off at between 7.8 billion and 10 billion before 2050. However, this projection is contingent on couples in less-developed countries having the education and economic security to choose to have smaller families and to practice birth control, using the latest information and technology.

In a few countries, especially in Central Europe, the population is actually declining and aging, due to low birth and death rates. This situation could prove problematic as younger workers become scarce and those who are elderly and dependent become more numerous, posing a financial burden to working-age people. HIV/AIDS is affecting population patterns to varying extents in all world regions. In Africa, the epidemic is severe; several African countries are experiencing sharply lowered life expectancies among young and middle-aged adults.

‖▶ 11. WORLD POPULATION TO BE CONCENTRATED IN DEVELOPING NATIONS, AS TOTAL EXPECTED TO REACH 9 BILLION BY 2050

Local Variations in Population Density and Growth

If the more than 7 billion people on earth today were evenly distributed across the land surface, they would produce an *average population density* of about 121 people per square mile (47 per square kilometer). But people are not evenly distributed. Nearly 90 percent of all people live north of the equator, and most of them live between 20° N and 60° N latitude. Even within that limited territory, people are concentrated on about 20 percent of the available land. They live mainly in zones that have climates warm and wet enough to support agriculture along rivers, in lowland regions, or fairly close to the sea. In general, people are located where resources are available.

Usually, the variable that is most important for understanding population growth in a region is the rate of natural increase (often called the growth rate). The **rate of natural increase (RNI)** is the relationship in a given population between the number of people being born, the **birth rate**, and the number dying, the **death rate**, without regard to the effects of migration.

> **rate of natural increase (RNI)** the rate of population growth measured as the excess of births over deaths per 1000 individuals per year without regard for the effects of migration
>
> **birth rate** the number of births per 1000 people in a given population, per unit of time, usually per year
>
> **death rate** the ratio of total deaths to total population in a specified community, usually expressed in numbers per 1000 or in percentages

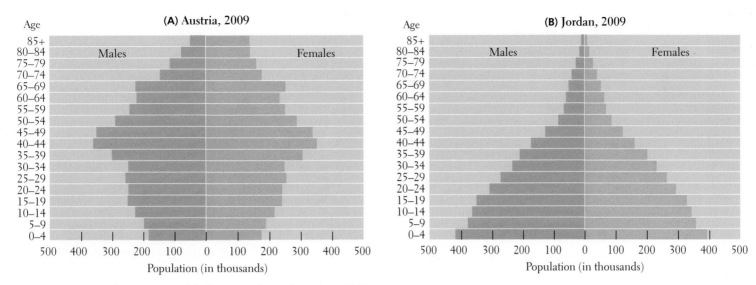

FIGURE 1.9 Population pyramids for Austria and Jordan, 2009. (A) Population pyramid for Austria. **(B)** Population pyramid for Jordan.

The rate of natural increase is expressed as a percentage per year. For example, in 2008, the annual birth rate in Austria (in Europe) was 9 per 1000 people, and the death rate was 9 per 1000 people. Therefore, the annual rate of natural increase was 0 per 1000 (9 − 9 = 0), or 0 percent (Figure 1.9).

For comparison, consider Jordan (in Southwest Asia). In 2008, Jordan's birth rate was 28 per 1000, and the death rate was 4 per 1000. Thus the annual rate of natural increase was 24 per 1000 (28 − 4 = 24), or 2.4 percent per year (Figure 1.9). At this rate, Jordan's population will double in just 29 years.

Total fertility rate (TFR) is another term used to indicate trends in population. The TFR is the average number of children a woman in a country is likely to have during her reproductive years (15–49). The TFR for Austrian women is 1.4; for Jordanian women, it is 3.6. As education rates for women increase, and as they postpone childbearing into their late 20s, total fertility rates tend to decline.

Another powerful contributor to population growth is **migration**. In Europe, for example, the rate of natural increase is quite low, but the region's economic power attracts immigrants from throughout the world. In 2005, international migration accounted for 85 percent of the European Union's population growth. All across the world people are on the move, seeking to improve their circumstances; often they are fleeing war or natural disasters.

Age and Sex Structures

Age and sex structures reflect past and present social conditions, and can help predict future population trends. The age distribution, or age structure, of a population is the proportion of the total population in each age-group. The sex structure is the proportion of males and females in each age-group.

The **population pyramid** is a graph that depicts age and sex structures. Consider the population pyramids for Austria and

Jordan (Figure 1.9). Notice toward the bottom of the Jordan pyramid that the largest groups are in the age categories 0 through 14.

In contrast, Austria's pyramid has an irregular vertical shape that tapers toward the bottom. The narrow base indicates that there are now fewer people in the younger age categories than in young adulthood or middle age, and that those over 70 greatly outnumber the youngest (ages 0 to 4). This age distribution tells us that many Austrians now live to an old age; also, in the last several decades, Austrian couples have been choosing to have only one child, or none. If this trend continues, Austrians will need to support and care for large numbers of elderly people, and those responsible will be an ever-declining group of working-age people.

Population pyramids also reveal *sex imbalance* within populations. Look closely at the right (female) and left (male) halves of the pyramids in Figure 1.9. In several age categories the sexes are not evenly balanced on both sides of the line. In the Austrian pyramid, there are more women than men near the top (especially age 80 and older). In Jordan, there are more males than females near the bottom (especially ages 20 to 24).

Demographic research on the reasons behind statistical sex imbalance is relatively new, and many different explanations are being proposed. In Austria, the predominance of elderly women reflects the trend in countries with long life expectancies for women to live about 5 years longer than men (a trend still poorly understood). But the sex imbalance in younger populations has different explanations. The normal ratio worldwide is for about 95 females to be born for every 100 males. Because baby boys are somewhat weaker than girls, the ratio normally evens out naturally within the first 5 years. However, in many places the ratio is as low as 80 females to 100 males (see Thematic Concepts HH, II, JJ) and continues throughout the life cycle. The widespread cultural preference for boys over girls (discussed in later chapters) becomes more observable as couples choose to have fewer children. Some female fetuses are aborted; girls and women are

> **total fertility rate (TFR)** the average number of children that women in a country are likely to have at the present rate of natural increase
>
> **migration** movement of people from a place or country to another, often for safety or economic reasons
>
> **population pyramid** a graph that depicts the age and gender structures of a country

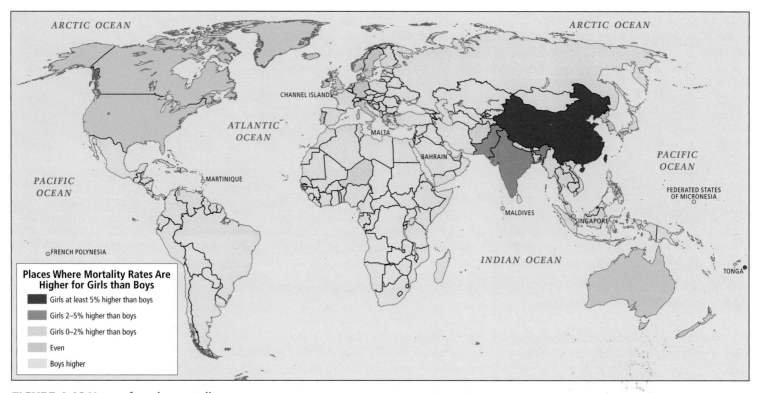

FIGURE 1.10 Young female mortality rates. In countries shown in the darkest shades (color), the mortality rate for girls is abnormally high.

sometimes fed less and receive less health care than males, especially in poverty-stricken areas. Hence, females are more likely to die, especially in early childhood (Figure 1.10).

Population Growth Rates and Wealth

Despite a wide range of variation, regions with slow population growth rates usually tend to be affluent, and regions with fast growth rates tend to have widespread poverty. The reasons for this difference are complicated; again, Austria and Jordan are useful examples.

Austria has an annual **gross domestic product (GDP)** per capita of $33,700. This figure represents the total production of goods and services in a country in a given year divided by the population. (All GDP per capita figures cited in this book are adjusted for purchasing power parity (PPP) so they represent comparable purchasing power.) Austria has a very low infant mortality rate of 3.7 per 1000. Its highly educated population is 100 percent literate and is employed largely in technologically sophisticated industries and services. Large amounts of time, effort, and money are required to educate a child to compete in this economy. Moreover, it is highly likely that a child will survive to adulthood and hence be available to care for aging parents. No surprise, then, that many Austrian couples choose to have only one or two children.

By contrast, Jordan has a GDP per capita of $5,530 and a high infant mortality rate of 24 per 1000. Much everyday work is still

> **gross domestic product (GDP)** the market value of all goods and services produced by workers and capital within a particular country's borders and within a given year
>
> **demographic transition** the change from high birth and death rates to low birth and death rates that usually accompanies a cluster of other changes, such as change from a subsistence to a cash economy, increased education rates, and urbanization
>
> **subsistence economy** an economy in which families produce most of their own food, clothing, and shelter

done by hand, so each new child is a potential contributor to the family income at a young age. There is also a much greater risk of children not surviving into adulthood, so having more children ensures that someone will be there to provide care for aging parents.

Circumstances in Jordan have changed rapidly over the last 25 years. In 1985, Jordan's per capita GDP was $993, infant mortality was 77 per 1000, and women had an average of 8 children. Now Jordanian women have an average of only 3.7 children. Geographers would say that Jordan is going through a **demographic transition**, meaning that a period of high birth and death rates is giving way to a period of much lower birth and death rates. In the middle phases, when death rates decline and birth rates have not yet followed suit, population numbers can increase rapidly as attitudes toward optimal family size adjust (see the green line in Figure 1.11). Eventually, populations may stabilize but at a much higher level than was the case at the beginning of the transition.

Much of the demographic transition has to do with the shift from subsistence to cash economies and from rural to urban ways of life. In a **subsistence economy**, a family, usually in a rural setting, produces most of its own food, clothing, and shelter, so there is little need for cash. Many children are needed to help perform the work that supports the family, so birth rates are high. Most needed skills are learned around the home, farm, and surrounding lands, so expensive educations at schools and colleges are not needed. Today, subsistence economies are

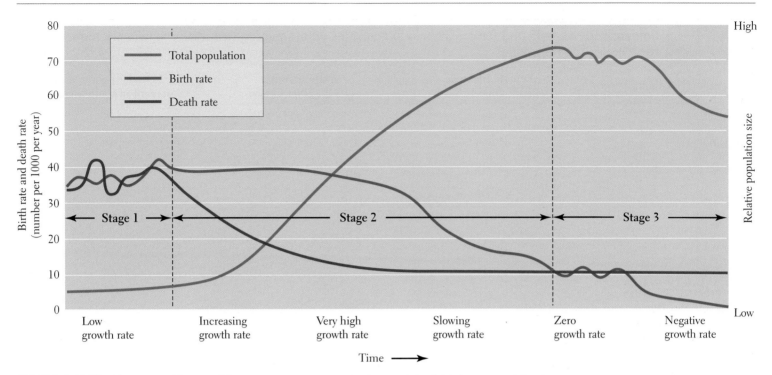

FIGURE 1.11 The demographic transition. In traditional societies (Stage 1), both birth rates and death rates are usually high (left vertical axis), and population numbers (right vertical axis) remain low and stable. With advances in food production, education, and health care (Stage 2), death rates usually drop rapidly, but strong cultural values regarding reproduction remain, often for generations, so birth rates drop much more slowly, with the result that for decades or longer, the population continues to grow significantly. When changed social and economic circumstances enable most children to survive to adulthood and it is no longer necessary to produce a cadre of family labor (Stage 3), population growth rates slow and may eventually drift into negative growth. At this point, demographers say that the society has gone through the demographic transition.

disappearing as people seek cash with which to buy food and goods, such as television sets and bicycles. In a **cash economy**, which tends to be urban but may be rural, skilled workers, well-trained specialists, and even farm laborers are paid in money. Each child needs years of education to qualify for a good cash-paying job and does not contribute to the family budget while in school. Having many children, therefore, is a drain on the family's resources. Perhaps most important, cash economies are more likely to have better health care, increasing the likelihood that each child will survive to adulthood.

THINGS TO REMEMBER

1. Rapid global population growth has occurred over the last several hundred years. Although growth will continue for many years, rates are now slowing in most places and in a few places the population is even shrinking.

2. As the population growth rate slows, many populations are rapidly aging.

GENDER

Note the difference between the terms *gender* and *sex*. **Gender** indicates how a particular social group defines the differences between the sexes. **Sex** refers to the biological category of male or female but does not indicate how males or females may behave or identify themselves. Gender definitions and accepted behavior for the sexes can vary greatly from one social group to another. Here we consider gender.

For women, the historical and modern global gender picture is puzzlingly negative. In nearly every culture, in every region of the world, and for a great deal of recorded history, women

cash economy an economic system in which the necessities of life are purchased with monetary currency

gender the sexual category of a person

sex the biological category of male or female

have had (and still have) an inferior status. Exceptions are rare, although the intensity of this second-class designation varies considerably. On average, females have less access to education, medical care, and even food. They start work at a younger age and work longer hours than males. Around the world, people of both sexes still routinely accept the idea that males are more productive and intelligent than females. The puzzling question of how and why women became subordinate to men has not yet been well explored because, oddly enough, few thought the question significant until recently.

TABLE 1.1	Comparisons of male and female income in countries where average education levels are higher for females than for males		
Country (HDI rank)	Female income (PPP[a], 2005 U.S.$)	Male income (PPP[a], 2005 U.S.$)	Female income as a percent of male income
Austria (15)	18,397	40,000	46
Barbados (31)	12,868	20,309	63
Canada (4)	25,448	40,000	64
Japan (8)[b]	17,802	40,000	45
Jordan (86)	2566	8270	31
Kuwait (33)	12,623	36,403	35
Poland (37)	10,414	17,493	60
Russia (67)	8476	13,581	62
Saudi Arabia (61)[c]	4031	25,678	16
Sweden (6)	29,044	36,059	81
United Kingdom (16)	26,242	40,000	66
United States (12)	25,005	40,000	63

[a] PPP, purchasing power parity, is the amount that the local currency equivalent of U.S.$1 will purchase in a given country.
[b] The education level for males in 2005 was 2% higher than for females.
[c] The education level was reported as equal in 2005.

In nearly all cultures, families prefer boys over girls because, as adults, boys will have greater earning capacity (Table 1.1) and more power in society. This preference for boys has some unexpected side effects. When families are limited to one child, fewer girls may be born and eventually there will be a shortage of marriageable women, leaving many men without the hope of forming a family (see Thematic Concepts HH, II, JJ).

II ▶ 6. WOMEN STILL LAG BEHIND MEN IN TOP BOARDROOM JOBS

Gender Roles

Geographers have begun to pay more attention to **gender roles**—the socially assigned roles for males and females—in different culture groups. In virtually all parts of the world, and for at least tens of thousands of years, the biological fact of maleness and femaleness has been translated into specific roles for each sex. The activities assigned to men and to women can vary greatly from culture to culture and from era to era, but they remain central to the ways societies function. Indeed, geographers' increasing attention to gender roles has been driven largely by interest in the shifts that occur when traditional ways are transformed by modernization.

There are some striking consistencies regarding gender roles around the globe and over time. Men are expected to fulfill public roles, and women are expected to fulfill private roles. Certainly there are exceptions in every culture, and customs are changing, especially in wealthier countries. But generally, men work

gender roles the socially assigned roles of males and females

outside the home in positions such as executives, animal herders, hunters, farmers, warriors, or government leaders. Women keep house, bear and rear children, care for the elderly, grow and preserve food and prepare the meals, among many other tasks. In nearly all cultures, women are defined as dependent on men—their fathers, husbands, brothers, or adult sons—even when the women may produce most of the family sustenance.

Gender Issues

Because their activities are focused on the home, women tend to marry early. One quarter of the girls in developing countries are mothers before they are 18. This is crucial in that pregnancy is the leading cause of death among girls 15 to 19 worldwide, primarily because immature female bodies are not ready for the stress of pregnancy and birth. Globally, babies born to women under 18 have a 60 percent greater chance of dying in infancy than do those born to women over 18.

Typically, women also have less access to education than men (globally, 70 percent of youth who leave school early are girls). They are less likely to have access to information and paid employment, and hence have less access to wealth and political power. When they do work outside the home (as is the case increasingly in every world region), women tend to fill lower-paid positions, such as laborers, service workers, or lower-level professionals. And even when they work outside the home, most women retain their household duties, so they work a *double day*.

Cultural ideas about masculinity and femininity, proper gender roles, and sexual orientation vary widely among groups. Yet within groups these ideas, handed down from generation to generation, have enormous effects on the everyday lives of men, women, and children. Perhaps more than for any other culturally defined human characteristic, significant agreement exists that gender is important, but just how gender roles are defined varies greatly across places and over time.

Traditional notions of gender roles are now being challenged everywhere. In many countries, including conservative Muslim

> **Learning Goal 2**
> **Population and Gender:** How is the shift toward greater equality between the genders influencing population growth patterns, patterns of economic development, and politics?

countries, females are acquiring education at higher rates than males. (see Thematic Concepts D). Eventually, this fact should make women competitive with men for jobs and roles in public life as policy makers and government officials, not just as voters (see Thematic Concepts F, S). Unless discrimination persists, women should also begin to earn pay equal to that of men (see Table 1.1 on page 18). Research data suggest that there is a ripple effect that benefits the whole group when developing countries pay attention to girls:

- Girls in developing countries, who get 7 or more years of education, marry 4 years later than average and have 2.2 fewer children.

- An extra year of secondary schooling over average boosts a girl's lifetime income by 15 to 25 percent.

- The children of educated mothers are healthier and more likely to finish secondary school.

- When women and girls earn income, 90 percent of their earnings are invested in the family, as compared to just 40 percent of males' earnings.

Considering only women's perspectives on gender, however, misses half the story. Men are also affected by strict gender expectations, often negatively. For most of human history, young men have borne a disproportionate share of burdensome physical tasks and dangerous undertakings. Until recently, only young men left home to migrate to distant, low-paying jobs. Overwhelmingly, it has been young men who die in wars or suffer physical and psychological injuries from combat.

This book will return repeatedly to the question of gender disparities because they play such a central role in changes occurring in every country on earth.

THINGS TO REMEMBER

1. Gender—the sexual category of a person—is both a biological and a cultural phenomenon. Gender indicates how a particular social group defines the differences between the sexes.

2. Sex is the biological category of male or female. Sex does not indicate how males or females may behave or identify themselves.

3. **Learning Goal 2: Population and Gender** There are significant social gender disparities throughout the world, yet many places are moving toward greater equality between the sexes. This shift is influencing population growth rates, among other things.

DEVELOPMENT

The economy is the forum in which people make their living, and resources are what they use to do so. Extraction (mining and agriculture), industrial production, and services (such as teaching or engineering) are three types of economic activities often described as *sectors of the economy*. Generally speaking, as people in a society shift from extractive activities, such as farming and mining, to industrial and service activities, their material standards of living rise—a process typically labeled **development**. *Extractive resources* are resources that must be mined from the earth's surface (mineral ores) or grown from its soil (timber and plants). There are also *human resources*, such as skills and brainpower, which are used to transform extractive resources into useful products (such as refrigerators or bread) or bodies of knowledge (such as books or computer software).

The development process has several facets, one of which is a shift from economies based on extractive resources to those based on human resources. In many parts of the world, especially in poorer societies (often referred to as "underdeveloped" or "developing"), there are shifts away from labor-intensive and low-wage,

> **development** usually used to describe economic changes like greater productivity of agriculture and industry that lead to better standards of living or simply to increased mass consumption
>
> **gross domestic product (GDP) per capita** the market value of all goods and services produced by workers and capital within a particular country's borders and within a given year, divided by the number of people in the country

often agricultural economies (see Thematic Concepts G) toward higher-wage but still labor-intensive manufacturing economies. Meanwhile, the richest countries (often labeled "developed") are lessening their dependence on labor-intensive manufacturing and shifting toward more highly skilled mechanized production or knowledge-based service and technology industries. As these changes occur, societies must provide adequate education, health care, and other social services to help their people contribute to economic development (see Thematic Concepts H, I).

Measures of Economic Development

The most popular economic measure of development is **gross domestic product (GDP) per capita**. Recall that GDP is the total value of all goods and services produced in a country in a given year. When GDP is divided by the number of people in the country, the result is GDP per capita.

Using GDP per capita as a measure of how well people are living has several disadvantages. First is the matter of wealth distribution. Because GDP per capita is an average, it can hide the fact that a country has a few fabulously rich people and a mass of abjectly poor people. For example, a GDP per capita of U.S.$50,000 would be meaningless if a few lived on millions per year and most lived on less than $10,000 per year.

Second, the purchasing power of currency varies widely around the globe. A GDP of U.S.$18,000 per capita in Barbados might represent a middle-class standard of living, whereas that same amount in New York City could not buy even basic food and shelter. Because of these purchasing power variations, in this book GDP per capita figures have been adjusted for **purchasing power parity (PPP)**. PPP is the amount that the local currency equivalent of U.S.$1 will purchase in a given country. For example, according to *The Economist*, on July 21, 2010, a Big Mac at McDonald's in the United States cost U.S.$3.73. In Norway it cost the equivalent of U.S.$7.20, and in China it cost U.S.$1.95. Of course, for the consumer in China, where annual per capita income averages about $5400, this would be a rather expensive meal.

A third disadvantage of using GDP per capita is that it measures only what goes on in the **formal economy**—all the activities that are officially recorded as part of a country's production. Many goods and services are produced outside formal markets, in the **informal economy**. Here, work is often traded for *in-kind* payments (food or housing, for example) or for cash payment that is not reported to the government as taxable income. It is estimated that one-third or more of the world's work takes place in the informal economy. Examples of workers in this category include anyone who contributes to her/his own or someone else's well-being through unpaid services such as housework, gardening, herding, animal care, or elder and child care. Hence, much of the work done by women is in the informal economy. *Remittances*, or pay sent home by migrants, become part of the formal economy of the receiving society if they are sent through banks or similar financial institutions—because records are kept and taxes levied. If they are transmitted in "off the books" (perhaps illegal) ways, such as via mail or cash, they are part of the informal economy.

Geographic Patterns of Human Well-Being

Some development experts, such as the Nobel Prize–winning economist Amartya Sen, advocate a broader definition of development that includes measures of **human well-being**. This term generally means a healthy and socially rewarding standard of living in an environment that is safe. The following section explores the three measures of human well-being that are used in this book: GDP per capita (calculated at PPP), the United Nations Human Development Index (HDI), and a comparison

of **female earned income as a percent of male earned income (F/MEI)**.

Global GDP per capita PPP is mapped in Figure 1.12A. Comparisons between regions and countries are possible, but as discussed above, GDP per capita figures ignore all aspects of development other than economic ones. For example, there is no way to tell from GDP per capita figures how fast a country is consuming its natural resources, or how well it is educating its young or maintaining its environment. Therefore, along with the traditional GDP per capita figure, geographers increasingly use several other measures of development. This book uses three maps: **gross domestic product (GDP) per capita PPP** (Figure 1.12A); the **United Nations Human Development Index (HDI)**, which calculates a country's level of well-being by considering income adjusted to PPP, data on life expectancy at birth, and data on educational attainment (Figure 1.12B); and to assess how well a country is doing in ensuring gender equality, we map female earned income as a percent of male earned income (Figure 1.12C). Together, these measures reveal some of the subtleties and nuances of well-being and make comparisons between countries somewhat more valid. Because these more sensitive indices (HDI and F/MEI) are also more complex than the purely economic GDP per capita, they are all still being refined by the United Nations.

A geographer looking at these maps might make the following observations:

- The GDP per capita figures (Figure 1.12A) show a wide range of difference across the globe with very obvious concentrations of high and low GDP per capita. The most populous parts of the world—China and India—rank rather low, but sub-Saharan Africa ranks the lowest.

- The HDI rank map (Figure 1.12B) shows a similar pattern, but look closely. Middle and South America, Southeast Asia, and Italy and Spain rank a bit higher on HDI than they do on GDP, and several countries in Africa as well as Iran rank lower on HDI than on GDP, thus illustrating the disconnect between GDP per capita and human well-being.

- The map of female earned income as a percent of male earned income (Figure 1.12C) shows, first of all, that nowhere on earth are average female earnings equal to average male earnings. Some poor parts of the world (East Africa; Vietnam; Romania; and Ghana in West Africa) have closer gender equity in pay than some of the richest, but pay scales for both genders are low. A few parts of Europe, notably Scandinavia, are in the category closest to equity—70 percent to 84 percent. Pay inequality is most extreme in North Africa and Southwest Asia, South Asia, and parts of Southeast Asia.

purchasing power parity (PPP) the amount that the local currency equivalent of U.S.$1 will purchase in a given country

formal economy all aspects of the economy that take place in official channels

informal economy all aspects of the economy that take place outside official channels

human well-being various measures of the extent to which people are able to obtain a healthy life in a community of their choosing

female earned income as a percent of male earned income (F/MEI) a measure of pay equity that shows average female earned income as a percent of average male earned income

gross domestic product (GDP) per capita PPP the market value of all goods and services produced by modern workers and capital within a particular country's borders and within a given year, divided by the number of people in the country and adjusted for purchasing power parity

United Nations Human Development Index (HDI) the ranking of countries based on three indicators of well-being: life expectancy at birth, educational attainment, and income adjusted to purchasing power parity

FIGURE 1.12 Global maps of human well-being. (A) Gross domestic product (GDP) per capita; **(B)** human development index (HDI); **(C)** female earned income as a percent of male earned income (F/MEI).

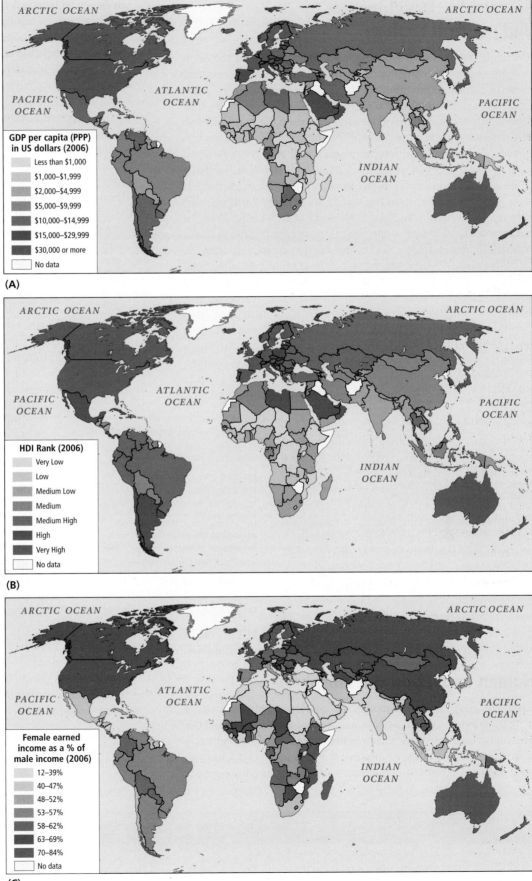

(A)

(B)

(C)

Sustainable Development and Political Ecology

The United Nations defines **sustainable development** as the effort to improve present living standards in ways that will not jeopardize those of future generations. Destroying ecosystems (as in deforestation) or poisoning them (as in pollution of water and air), may deprive future generations of needed resources if the impacted ecosystems can't recover. Sustainability has only recently gained widespread recognition as an important goal, well after the developed parts of the world had already achieved high standards of living based on mass consumption of resources. However, sustainability is particularly important for the vast majority of the earth's people who do not yet enjoy an acceptable level of well-being. Without sustainable development strategies, efforts to improve living standards will increasingly be foiled by degraded or scarce resources.

Geographers who study the interactions among development, politics, human well-being, and the environment are called **political ecologists**. They are known for asking the "Development for whom?" question, meaning, "Who is actually benefiting from so-called development projects?" Political ecologists examine how the power relationships in a society affect the ways in which development proceeds. For instance, in a Southeast Asian country, the clearing of forests to grow oil palm trees might at first seem to benefit many people. It would earn profits for the growers and raise tax revenue for the government through the sale of palm oil. However, these gains must be balanced against the loss of highly biodiverse forest ecosystems and the human cultures that depend on them. Indeed, valuable knowledge that could aid in more sustainable uses of forest ecosystems may be lost if forest dwellers are forced to migrate to crowded cities, where their woodland skills are useless and therefore soon forgotten.

Political ecologists often argue that development should be measured by the improvements it brings to overall human well-being and long-term environmental quality. By these standards, converting forests to oil palm plantations might appear less attractive, since only a few will benefit at the cost of widespread and often irreversible ecological and social disruption.

Human Impact on the Biosphere

Concerns over the sustainability of development grow out of increasing awareness that we humans are profoundly impacting the ecosystems we depend on. This is nothing new. From the beginning of human life, in seeking to improve our own living conditions we have overused resources, sometimes with disastrous consequences for our own well-being. What is new is the scale of human impacts on the planet, which can now be found virtually everywhere.

Mounting awareness of environmental impacts has prompted numerous proposals to limit damage to the biosphere. Societies have become so transformed by intensive use of the earth's resources that going back is enormously difficult. For example, how possible would it be for you and your entire family to live for just one day without using any fossil fuels for transportation or climate control? Intensive resource consumption is now deeply built into many societies, especially in the rich countries of the world, where 20 percent of the world's population consumes more than 80 percent of the available world resources.

Increasingly, human consumption of natural resources is being examined through the concept of the **ecological footprint**. This is a method of estimating the amount of biologically productive land and sea area needed to sustain a human population at its current standard of living. It is particularly useful for drawing comparisons. For example, the worldwide average biologically productive area per person—in other words, one individual's ecological footprint—is about 4.5 acres. However, in the United States, ecological footprints average about 24 acres, and in China about 4 acres. You can calculate your own footprint at http://www.earthday.net/Footprint/index.asp. A similar concept more closely related to global warming is the *carbon footprint*, which measures the greenhouse gas emissions a person's activities produce; for a calculation of your family's carbon footprint, see http://www.nature.org/ initiatives/climate-change/calculator/.

Photo Essay 1.1 on pages 24–25 shows a global map of the relative intensity of human impact on the **biosphere**, the global ecological system that integrates all living things and their relationships. The map includes insets that show particular trouble spots in South America, Europe, Southwest Asia, and Southeast Asia. Of the nine thematic concepts stressed in this book, this figure focuses on the interactions of urbanization, population, water, development, and food.

> **sustainable development** improvement of standards of living in ways that will not jeopardize those of future generations
>
> **political ecologist** a geographer who studies power allocations in the interactions among development, human well-being, and the environment
>
> **ecological footprint** the biologically productive area per person needed to sustain a particular standard of living
>
> **biosphere** the global ecological system that integrates all living things and their relationships

THINGS TO REMEMBER

1. Development refers to the rise in material standards of living that usually accompanies the shift from extractive activities, such as farming and mining, to industrial and service activities.

2. For development to happen, social services, such as education and health care, are necessary to enable people to contribute to economic growth.

FOOD

Over time, food production has undergone many changes. It started with hunting and gathering and changed to labor intensive, small-scale subsistence agriculture. It then became large scale in what we consider modern and mechanized commercial agriculture, involving an intensive use of chemicals. Modern processes of food production, distribution, and consumption have greatly increased the supply and, to some extent, the security of food systems. However, this has come at the cost of environmental pollution that may strain future food production.

Agriculture: Early Human Impacts on the Physical Environment

Agriculture includes animal husbandry, or the raising of animals, as well as the cultivation of plants. The ability to produce food, as opposed to the dependence on hunting and gathering, has had long-term effects on human population growth, rates of natural resource use, the development of towns and cities, and ultimately, the development of civilization.

Very early humans hunted animals and gathered plants and plant products (seeds, fruits, roots, and fibers) for their food, shelter, and clothing. To successfully use these wild resources, humans developed an extensive folk knowledge of the needs of the plants and animals they favored. The transition from hunting to tending animals in pens and pastures and from gathering in the wild to planting in gardens, orchards, and fields probably took place gradually over thousands of years.

Where and when did plant cultivation and animal husbandry first develop? Genetic studies support the view that at varying times between 8000 and 20,000 years ago, people in many different places around the globe independently learned to develop especially useful plants and animals through selective breeding, a process known as **domestication**.

Why did agriculture and animal husbandry develop in the first place? Certainly the desire for more secure food resources played a role, but the opportunity to trade may have been just as important. Many of the known locations of agricultural innovation lie near early trade centers. There, people would have had access to new information and new plants and animals brought by traders.

agriculture the practice of producing food through animal husbandry, or the raising of animals, and the cultivation of plants

domestication the process of developing plants and animals through selective breeding to live with and be of use to humans

food security the ability of a state to consistently supply a sufficient amount of basic food to the entire population

green revolution increases in food production brought about through the use of new seeds, fertilizers, mechanized equipment, irrigation, pesticides, and herbicides

Agriculture and Its Consequences

Agriculture made amassing of surplus stores of food for lean times possible, and allowed some people to specialize in activities other than food procurement. It also may have led to several developments now regarded as problems: rapid population growth, environmental degradation, and occasionally, famine.

Agriculture could support more people on a given piece of land than hunting and gathering. As populations expanded and as more land was turned over to agriculture, natural habitats were destroyed, reducing opportunities for hunting and gathering.

Through the study of human remains, archaeologists have learned of a previously unrecognized consequence of the development of agriculture. At some times and in some places, the nutritional quality of human diets may actually have declined as people stopped eating diverse wild plants and animals and began to eat primarily one or two species of domesticated plants.

Another consequence was that the storage of food surpluses not only made it possible to trade food, but also made it possible for people to live together in larger concentrations, which then facilitated the spread of disease. Moreover, land clearing increased vulnerability to drought and other natural disasters that could wipe out an entire harvest. Thus, as ever-larger populations depended solely on cultivated food crops, episodic famine actually became more common.

Modern Food Production and Food Security

For most of human history, people produced most of what they needed for food, clothing, and shelter. They subsisted. However, over the past five centuries of increasing global interaction and trade, people have become ever more removed from their sources of food. Today, occupational specialization means that food is increasingly mass-produced. Far fewer people work in agriculture, and food is usually purchased with money. Now most humans work for cash to buy food and other necessities.

A side effect of this dependence on money is that the **food security** of individuals and families can be threatened by economic disruptions even in distant places. As countries get more involved with the global economy, their food markets can become vulnerable to price swings beyond their control. For example, a crisis in food security began to develop in 2007 when the world price of corn spiked. Speculators in alternative energy thought that corn would be an ideal raw material to make ethanol, a substitute for gasoline, so they invested heavily in this commodity. As a result, global corn prices rose. Many families in developing countries that depend on corn as a dietary staple could not afford to send their children to school and even went without food. Figure 1.13 on page 26 identifies countries where undernourishment is an ongoing problem and periodic food security crises are especially intense.

Another way that modernized agriculture impacts food security is through its reliance on machines, chemical fertilizers, and pesticides. The shift to this kind of agriculture is called the **green revolution**. When successfully implemented, the results of green revolution agriculture, at least in the beginning, are soaring production levels and high profits for those farmers who can afford the additional investment. But many poorer farmers often can't afford the machinery and chemicals. As increased production leads to lower crop prices, these farmers often can't compete. Often they are forced to sell their land and move to crowded cities. Here they join masses of urban poor people whose access to food is often precarious.

Green revolution agriculture can also impact food security by damaging the environment. As rains wash fertilizers and pesticides into streams, rivers, and lakes, these bodies of water often become polluted. Over time, the pollution destroys fish and other aquatic animals vital to food security. Soil degradation can also increase as green revolution techniques often leave soils exposed to rains, and soil fertility may decline over time as natural nutrients are washed away. Indeed, many of the most agriculturally productive parts of

North America, Europe, and Asia have already suffered moderate to serious loss of soil through **erosion**. Globally, soil erosion and other problems related to food production affect about 7 million square miles (2000 million hectares), putting at risk the livelihoods of a billion people (see Thematic Concepts J, K, L).

Green revolution agriculture has also changed most places' **carrying capacity**, which refers to the maximum number of people that can be supported sustainably on a given piece of land. While green revolution technology has increased food production remarkably, it is unclear how sustainable these gains are. In the 25 years between 1965 and 1990, depending on the region, total global food production rose between 70 and 135 percent. In response, populations also rose quickly during this period. But the environmental impacts of green revolution technologies mean that the carrying capacity of certain places may have already been exceeded as the population rose due to gains in productivity that cannot be sustained over the long term.

Estimations of carrying capacity are increasingly involved with questions of agricultural technology. Scientists from many disciplines estimate that within the next 50 years the world will reach the limit of its carrying capacity because of growing environmental problems such as water scarcity and global climate change. It is also estimated that to feed the population projected for 2050, global food output must increase by 70 percent. Some technological advances that could make present agricultural systems more productive are increasingly controversial. In North America, **genetic modification (GM)**, the practice of splicing together the genes from widely divergent species to achieve particular characteristics, is increasingly being used to boost productivity. However, elsewhere many worry about the side effects of such agricultural manipulation. Europeans have tried (unsuccessfully) to keep GM food products out of Europe, fearing that they could lead to unforeseen ecological consequences or catastrophic crop failures. They point out that the main advance in GM agriculture has been the production of seeds that can tolerate high levels of environmentally damaging pesticides. Hence, the use of GM crops could lead to more, not less, environmental degradation. Many developing countries are also concerned that genetically modified seeds are much more expensive than traditional seeds, further adding to the burdens placed on poor farmers. As a result of these uncertainties of GM crops, many are returning to a much older idea of

Thinking Geographically

After you have read about Human Impacts on the Biosphere, you will be able to answer the following questions:

A What sector of the economy is mining in?

B What kind of pollution does this photo show most directly?

D What kind of agriculture are these women likely practicing?

E What is often lost when forests are cut down?

---

erosion the process by which fragmented rock and soil are moved over a distance, primarily by wind and water

carrying capacity the maximum number of people that a given territory can support sustainably with food, water, and other essential resources

genetic modification (GM) in agriculture, the practice of splicing together the genes from widely divergent species to achieve particular desirable characteristics

---

Humans have had enormous impacts on the biosphere, the global ecological system integrating all living things and their relationships. The map and insets show varying levels of human impact on the biosphere as of 2002. The impacts depicted here are derived from a synthesis of hundreds of studies. High-impact areas are associated with intense urbanization. Medium-high impact areas are associated with roads, railways, agriculture, or other intensive land uses. Low-medium impact areas are experiencing biodiversity loss and other disturbances related to human activity. For more details, go to www.whfreeman/pulsipher.com.

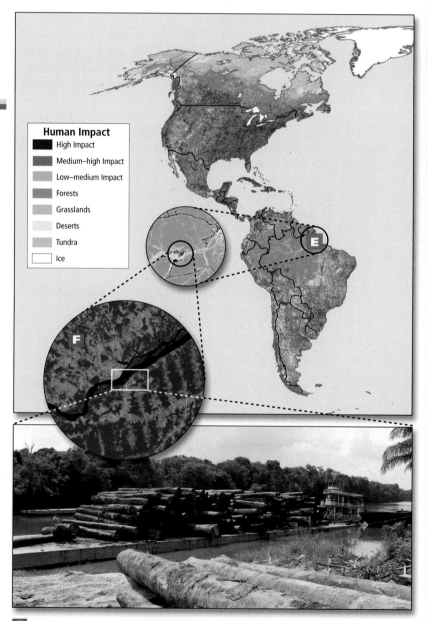

Human Impact
- High Impact
- Medium–high Impact
- Low–medium Impact
- Forests
- Grasslands
- Deserts
- Tundra
- Ice

E Development and Deforestation: In the Brazilian Amazon, deforestation often occurs in regularized spatial patterns, such as the "fishbone" pattern (see sattellite image inset **F**). This pattern results from regulations guiding the location of roads used for settlement and logging. Whole logs are brought by road to rivers where they are put on barges and taken to a port for export.

A Development and Mining: Perhaps no other human activity has as striking an impact on the landscape as mining. This open-pit coal mine is located in one of the most industrialized and densely inhabited parts of Germany.

B War and political conflict can have a devastating effect on the environment. Iraqi fire fighters work to contain a fire at an oil facility set off by rocket fire.

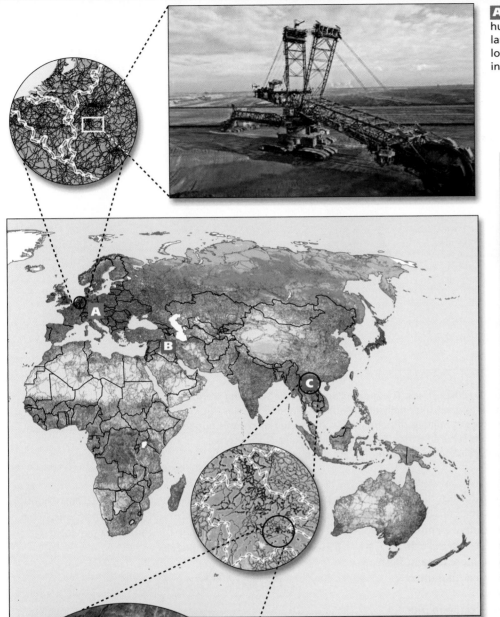

C Food and Deforestation: Farmers practicing shifting cultivation (see Chapter 3, page 111) plant "hill rice" in Laos. Shifting cultivation is an ancient technique that can be sustained indefinitely given sufficient land and fallow periods long enough for forest to regrow—20 years or more. Today, more and more forest is being turned over to short-fallow cultivation—3 to 6 years cultivation (see inset **D**) resulting in loss of habitat and biodiversity.

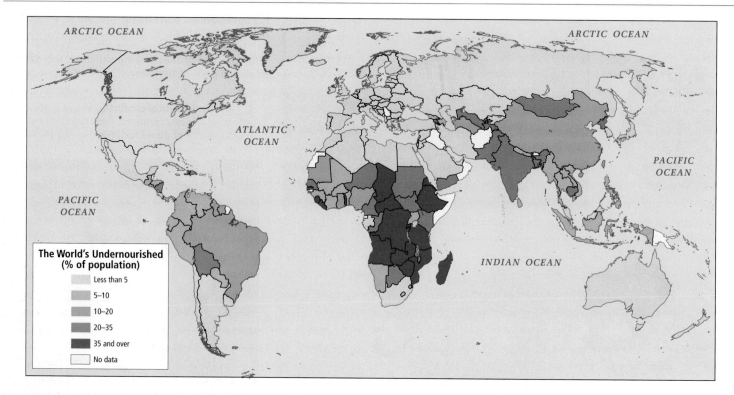

FIGURE 1.13 Undernourishment in the world, 2003–2005. The proportion of people suffering from undernourishment—the lack of adequate nutrition to meet their daily needs—has declined in the developing world over the past several years. However, hundreds of millions of people are still affected by chronic hunger. As you can see from the map, people in much of sub-Saharan Africa, South Asia, Mongolia, North Korea, Cambodia, Nicaragua, Bolivia, and Haiti and the Dominican Republic in the Caribbean suffer the most.

sustainable agriculture, which meets human needs with less damage to the environment and to soils. Often these systems avoid chemical inputs entirely, as in the case of increasingly popular methods of *organic agriculture*. However, while these systems can be highly productive, they often require significantly more human labor, resulting in higher food prices. Hence, many fear that these systems could reduce poor people's access to food.

Although global food production systems can support the global population, hunger is still widespread for a variety of reasons. According to the United Nations Food and Agriculture Organization, one-fifth of humanity subsists on a diet too low in total calories and vital nutrients to sustain adequate health and normal physical and mental development (Figure 1.13). As we will see in later chapters, this massive

> **sustainable agriculture** farming that meets human needs without poisoning the environment or using up water and soil resources
>
> **urbanization** the movement of people from rural areas to cities

hunger problem is partly due to political instability, corruption, and poor distribution systems. Within all these constraints, what food there is goes to those who have the money to pay for it.

❚❚▶ 3. DEFORESTATION: WORLDWIDES CONCERNS

THINGS TO REMEMBER

1. While modern processes of food production and distribution have greatly increased the supply of food, environmental damage could strain future food production.

2. Many farmers are unable to afford the chemicals and machinery required for commercial agriculture or new genetically modified seeds; as a result, they are forced to give up farming.

3. Sustainable agriculture is farming that meets human needs without harming the environment or depleting water and soil resources.

URBANIZATION

Learning Goal 3
Food and Urbanization: How are changes in food production pushing people out of rural areas and pulling them into urban ones?

In 1700, fewer than 7 million people, or just 10 percent of the world's total population, lived in cities, and only five cities had populations of several hundred thousand people

or more. The world we live in today has been transformed by **urbanization**, the process whereby cities, towns, and suburbs grow as populations shift from rural to urban livelihoods. Now about half of the world's population lives in cities, and there are more than 400 cities of more than 1 million (see Thematic Concepts M, N, O; see also the map in Photo Essay 1.2 on pages 28–29).

Why Are Cities Growing?

For some time, changes in food production have been pushing people out of rural areas, while the development of manufacturing and service economies and the possibility of earning cash incomes is pulling them into cities. This process is called the **push/pull phenomenon of urbanization**. Numerous cities, especially in poorer parts of the world, have been unprepared for the massive inflow of rural migrants, many of whom now live in **slum** areas plagued by poor housing and inadequate access to food, water, education, and social services. Often a substantial portion of the migrants' cash income goes to support their still-rural families.

> **push/pull phenomenon of urbanization** conditions, such as political instability, that encourage (push) people to leave rural areas, and urban factors, such as job opportunities, that encourage (pull) people to move to the urban area
>
> **slum** densely populated area characterized by crowding, run-down housing, and poverty

Patterns of Urban Growth

The most rapidly growing cities are in developing countries in Asia, Africa, and Middle and South America. The settlement pattern of these cities bears witness to their rapid and often unplanned growth, fueled in part by the steady arrival of masses of poor rural people looking for work. Cities like Mumbai (in India), Cairo (in Egypt), Nairobi (in Kenya), and Rio de Janeiro (in Brazil), sprawl out from a small affluent core, often the oldest part, where there are upscale businesses, fine old buildings, banks, shopping centers, and residences for wealthy people. Surrounding these elite landscapes are sprawling commercial, industrial, and middle-class residential areas, interspersed with pockets of extremely dense slums that provide housing for the poorest of the poor. Also known as *barrios, favelas, hutments, shantytowns, ghettos,* and *tent villages,* housing in slums is often built out of any materials the residents can find: cardboard, corrugated metal, masonry, scraps of wood and plastic. There are usually no building codes, sewers, or clean water; electricity is often obtained from illegal and dangerous connections to nearby power lines. Schools are few and overcrowded, and transport is provided only by informal, nonscheduled transportation services. Because these slums can pop up on any vacant land overnight, people may be sleeping on the street just a few blocks from soaring modern skyscrapers (see Photo Essay 1.2 A, B; see also Thematic Concepts M, N, O).

The UN estimates that currently over a billion people are living in slums, with as many as 2 billion predicted to live in slums by 2030. Life in these areas can be insecure and chaotic as criminal gangs often assert control, with violence, rioting, and looting—especially likely during periods of political instability. And yet remarkably, in many slums there are examples of self-initiated community development efforts. In some instances, even recently arrived migrants have successfully lobbied local governments for social support services, such as job training, daycare centers, and medical care.

Slums are only part of the story of urbanization today. Those who are financially able to come to urban areas for education, once they complete their studies, tend to find employment in modern industries and business services. They constitute the new middle class and leave their imprint on urban landscapes via the high-rise apartments they occupy and the shops and entertainment facilities they frequent (see Photo Essay 1.2 A, D). Cities such as Mumbai in India, São Paulo in Brazil, Cape Town in South Africa, and Shanghai in China are now home to this more educated group of new urban residents, many of whom may have started life on farms and in villages.

In the past, most migrants in cities were young males, but increasingly they are young females. Cities offer women more than better paying jobs. They also provide access to education, better health care, and more personal freedom. Nonetheless, young women are particularly vulnerable in harsh urban environments. Raised in sheltered conditions and initially possessing little education and few skills, they can be unwittingly pressed into the sex trade or involuntary servitude.

▌▌▶ 331. UN HABITAT AGENCY SAYS HALF THE WORLD POPULATION LIVES IN CITIES

THINGS TO REMEMBER

1. Today about half of the world's population live in cities; there are over 400 cities with more than 1 million people and about 25 cities of more than 10 million people.

2. **Learning Goal 3:** **Food and Urbanization** Changes in food production are pushing people out of rural areas, while the development of manufacturing and service economies is pulling them into cities.

3. For some, urbanization means improved living standards, while for others it means being forced into slums with inadequate food, water, and social services.

GLOBALIZATION

Learning Goal 4
Globalization and Development: How are increased global flows of information, goods, and people transforming patterns of economic development?

Throughout the world, globalization is transforming patterns of economic development, as local self-sufficiency is giving way to global interdependence and international trade. The economic recession that began in 2008 cast a spotlight on this phenomenon. Economic disruptions in the United States and Europe resulted in powerful ripple effects reaching around the world. Foreclosures in the U.S. housing market meant that European banks failed and Chinese factory workers lost their jobs. These connections between distant regions are known as **interregional linkages**. The term **globalization** encompasses the worldwide

> **interregional linkages** economic, political, or social connections between regions, whether contiguous or widely separated
>
> **globalization** the growth of interregional and worldwide linkages and the changes these linkages are bringing about

changes brought about by many types of these interregional linkages and flows that reach well beyond economics. Attitudes and values are modified as a result of global connections; ethnic identity may be reinforced or erased; personal or collegial relationships established between people who will never actually meet. Globalization is the most complex and far-reaching of the thematic concepts described in this book (see Thematic Concepts P, Q, R, Y, Z, AA, BB, CC, DD, EE, FF, GG).

What Is the Global Economy?

The **global economy** includes the parts of any country's economy that are involved in global flows of resources—mined minerals, agricultural commodities, manufactured products, money, and people and their ideas. Most of us participate in the global economy every day. For example, books like this can be made from trees

> **global economy** the worldwide system in which goods, services, and labor are exchanged

cut down in Southeast Asia or Siberia and shipped to a paper mill in Oregon. Many books are now printed in Asia because labor costs are lower there. Such long-distance movement of resources and products has grown tremendously in the past 500 years and remains possible because fuel costs for water transport are low. Globalization is not new. It existed at least 2500 years ago, when silk and other goods were traded along Central Asian land routes that connected Greece and then Rome in the Mediterranean with China.

European colonization was an early expansion of globalization. Starting in about 1500, European countries began extracting resources from distant parts of the world they had conquered. The colonizers organized systems to process those resources into higher-value goods to be traded wherever there was a market. Sugarcane, for example, was grown on Caribbean and Brazilian plantations with slave labor from Africa (see Figure 1.14 on page 30) and made locally into crude sugar, molasses, and rum, and then shipped to Europe and North America, where it was further refined and sold at considerable profit. The global economy grew as each region not only produced goods for export, rather than just for local consumption, but at the same time it also became increasingly dependent on imported food, clothing, machinery, energy, and knowledge.

The new wealth derived from the colonies, and the ready access to global resources led to Europe's *Industrial Revolution*, a series of innovations and ideas that changed the way goods were produced. No longer was one woman producing the cotton or wool for cloth, spinning thread, weaving the thread into cloth, and sewing

Urbanization and Urban Areas. The color of the country indicates the percentage of the population living in urban areas. The circles represent the populations of the world's largest urban areas in 2006 (blue circle) and projected in 2020 (black circle).

Thinking Geographically

After you have read about Urbanization, you will be able to answer the following questions:

A What group of urban migrants are these young men most likely a part of?

B What may have pushed these boys into the city?

C What trend do these female workers exemplify?

D Which agent of globalization is Tokyo Disneyland most related to?

E What problem faced by many growing cities is exemplified by this photo?

A Cities have always been centers of innovation, entertainment, and culture, in large part because they attract both money and talented people. These young residents of Shanghai are engaging in a new fad, parachuting off one of the new skyscrapers that now dominate the city's skyline.

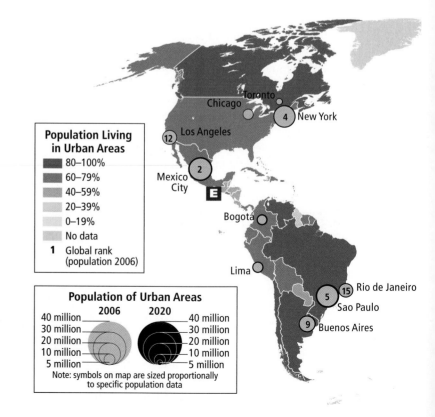

Population Living in Urban Areas
- 80–100%
- 60–79%
- 40–59%
- 20–39%
- 0–19%
- No data
- **1** Global rank (population 2006)

Population of Urban Areas

2006	2020
40 million	40 million
30 million	30 million
20 million	20 million
10 million	10 million
5 million	5 million

Note: symbols on map are sized proportionally to specific population data

B Recent migrants sleep on a sidewalk in Dhaka. Many of the world's fastest-growing cities are attracting more people than they can support with their existing housing and infrastructure.

C Employment is the greatest "pull" factor drawing in people. Most recent migrants work at physically demanding, low-paying, and often hazardous jobs. These women are handing buckets full of cement up a scaffolding on a construction site in Delhi.

D Tokyo Disneyland, Cinderella's Castle. Like most of the world's great cities, Tokyo has long been a major conduit for globalization. Tokyo has been the world's largest city since 1970 and is projected to retain that distinction for the foreseeable future.

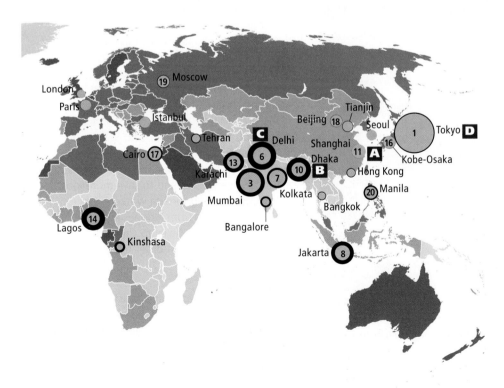

E Some cities struggle with major environmental problems. Mexico City, currently the world's second largest city, occasionally suffers from severe flooding due to its location on an old, now-sinking, lake bed and to its antiquated drainage and sewage infrastructure.

a garment. Instead, these separate tasks were spread out among many workers, often in distant places, with some people specializing in producing the fiber and others in spinning, weaving, or sewing. These innovations were followed by labor-saving improvements such as mechanized reaping, spinning, weaving, and sewing.

This larger-scale mechanized production accelerated globalization as it created a demand for raw materials and a need for markets in which to sell finished goods. European colonies in the Americas, Africa, and Asia provided both. For example, in the British Caribbean colonies, hundreds of thousands of African slaves wore garments made of cloth woven in England from cotton grown in British India. The sugar they produced on British-owned plantations with iron equipment from British foundries was transported to European markets in ships made in the British Isles of trees and resources from various parts of the world.

Until the early twentieth century, much of the activity of the global economy took place within the colonial empires of a few European nations. By the 1960s, global economic and political changes brought an end to these empires, and now almost all colonial territories are independent countries. Nevertheless, the global economy persists in the form of banks and **multinational corporations** such as Shell, Walmart, Bechtel, and Cisco that operate across international borders. These corporations extract resources from many places, make products in factories located where they can take advantage of cheap labor and transportation facilities, and market their products wherever they can make the most profit. Their global influence, wealth, and importance to local economies enable the multinationals to influence the economic and political affairs of the countries in which they operate.

> **multinational corporation** a business organization that operates extraction, production, and/or distribution facilities in multiple countries

FIGURE 1.14 European use of colonial resources. Among the first global economic institutions were Caribbean plantations like Old North Sound on Antigua, shown here in an old painting. In the eighteenth century, thousands of sugar plantations in the British West Indies, subsidized by the labor of slaves, provided huge sums of money for England and helped fund the Industrial Revolution.

Workers in the Global Economy

Vignette Sixty-year-old Olivia lives near Soufrière on St. Lucia, an island in the Caribbean (**Figure 1.15A**). She, her daughter Anna, and her three grandchildren live in a wooden house surrounded by a leafy green garden dotted with fruit trees. Anna has a tiny shop at the side of the house, from which she sells various small everyday items and preserves that she and her mother make from the garden fruits.

On days when the cruise ships dock, Olivia heads to the market shed on the beach with a basket of homegrown goods. She calls out to the passengers as they near the shore, offering her spices and snacks for sale. In a good week she makes U.S.$50. Her daughter makes about U.S.$100 per week in the shop, and is constantly looking for other ways to earn a few dollars.

Olivia and Anna support their family of five on about U.S.$170 a week (U.S.$8840 per year). From this income they take care of their bills and other purchases, including school fees for the granddaughter who will go to high school in the capital next year and perhaps college if she succeeds. Their livelihood puts them at or above the standard of living of most of their neighbors.

In Malacca, Malaysia, 30-year-old Setiya, an illegal immigrant from Tegal, Indonesia, is laying concrete blocks for a new tourist hotel (Figure 1.15B). Like a million other Indonesians attracted by the booming economy, he snuck into Malaysia, risking arrest, because in Malaysia average wages are four times higher than at home.

This is Setiya's second trip to Malaysia. His first trip was to do work legally on a Malaysian oil palm plantation. Upon arrival, however, Setiya found that he would have to work for 3 months just to pay off his boat fare. Not one to give in easily, he quietly caught a bus to another city and found a construction job earning U.S.$10 a day (about U.S.$2600 a year), which allowed him to send money home to his family in Indonesia.

In 2008, Malaysia announced it was expelling 500,000 foreign workers, and Setiya was one of them. The country was suffering from growing unemployment due to the global recession, and its leaders wanted to save more jobs for locals by expelling foreign workers. Once again, young fathers from Indonesia (like Setiya) may risk trips on leaky boats to illegally enter Malaysia and Singapore in order to support their families.

Fifty-year-old Tanya works at a fast-food restaurant outside of Charleston, South Carolina, making less than U.S.$7.50 an hour. She had been earning U.S.$8 an hour sewing shirts at a textile plant until it closed and moved to Indonesia. Her husband is a delivery truck driver for a snack-food company.

Between them, Tanya and her husband make $27,000 a year, but from this income they must cover all their expenses, including mortgage and car payments. In addition, they help their daughter, Rayna, who quit school after 11th grade and married a man who is now out of work. They and their baby live at the back of the lot in an old mobile home (Figure 1.15C).

With Tanya's now lower wage ($4000 less a year), there will not be enough money to pay the college tuition for her son, who is in high school. He had hoped to become an engineer, and would have been the first in the family to go to college. For now, he is working at the local gas station.

These people, living worlds apart, are all part of the global economy. Workers around the world are paid startlingly different rates for jobs that require about the same skill level. Varying costs of living and varying local standards of wealth make a difference in how people live and regard their own situation. Though Tanya's family has the highest income by far, compared to their neighbors they live in poverty, and their hopes for the future are dim. Olivia's family, on the other hand, are not well off, but they do not think of themselves as poor because they have what they need, others around them live in similar circumstances, and their children seem to have a future. They can subsist on local resources, and the tourist trade promises continued cash income. But their subsistence depends on circumstances beyond their control; in an instant, the cruise-line companies can choose another port of call. Setiya, by far the poorest, seems trapped by his status as an illegal worker, which robs him of many of his rights. Still, the higher pay that he can earn in Malaysia offers him a possible way out of poverty.

[Adapted from Lydia Pulsipher's field notes, 1992–2000 (Olivia and Tanya), and Alex Pulsipher's field notes, 1999–2008 (Setiya).] ∎

The Debate over Globalization and Free Trade

The term **free trade** refers to the unrestricted international exchange of goods, services, and capital. Free trade is an ideal that has not been achieved and possibly never will be. Currently, all governments impose some restrictions on

> **free trade** the movement of goods and capital without government restrictions

trade to protect their own national economies from foreign competition, although such restrictions are far fewer than in the 1980s. Restrictions take two main forms: *tariffs* and *import quotas*. Tariffs are taxes imposed on imported goods that increase the cost of those goods to the consumer, thus giving price advantage to locally made competing goods. Import quotas set limits on the amount of a given good that may be imported over a set period of time, thus curtailing supply.

These and other forms of trade protection are subjects of contention. Proponents of free trade argue that the removal of all tariffs and quotas encourages efficiency, lowers prices, and gives consumers more choices. Companies can sell to larger markets and take advantage of mass-production systems that lower costs further. As a result, they can grow faster, thereby providing people with jobs and opportunities to raise their standard of living. These pro-free trade arguments have been quite successful,

FIGURE 1.15 Workers in the global economy. (A) Soufrière, St. Lucia, as seen from a cruise ship. **(B)** Setiya working at the construction job in Malacca, Malaysia, before being expelled from the country, along with 500,000 other foreign workers, during the recession that began in 2008. **(C)** The trailer at the back of Tanya's lot where her daughter lives.

Thinking Geographically: (A) What about this photo might suggest that Olivia and her neighbors share a similar standard of living? **(B)** What about this photo suggests that Setiya is a low-paid worker? **(C)** How does this photo suggest that Rayna, Tanya's daughter, has a low income?

and in recent decades restrictions on trade imposed by individual countries were greatly reduced. Several *regional trade blocs* have been formed; these are associations of neighboring countries that agree to lower trade barriers for one another. The main ones are the North American Free Trade Agreement (NAFTA), the European Union (EU), the Southern Common Market (Mercosur) in South America, and the Association of Southeast Asian Nations (ASEAN).

A main global institution supporting the ideal of free trade is the **World Trade Organization (WTO)**, whose stated mission is to lower trade barriers and to establish ground rules for international trade. Related institutions, the *World Bank* (officially named the International Bank for Reconstruction and Development) and the *International Monetary Fund* (IMF), both make loans to countries that need money to pay for economic development projects. Before approving a loan, the World Bank or the IMF may require a borrowing country to reduce and eventually remove tariffs and import quotas. These requirements are part of larger *structural adjustment policies* (SAPs) that the IMF imposes on countries seeking loans,

> **World Trade Organization (WTO)** a global institution made up of member countries whose stated mission is the lowering of trade barriers and the establishment of ground rules for international trade
>
> **living wages** minimum wages high enough to support a healthy life
>
> **fair trade** trade that values equity throughout the international trade system; now proposed as an alternative to free trade

such as the requirement to close government enterprises and to reduce government services, mostly to the detriment of the poor. SAPs have become highly influential and controversial in virtually every region of world. Chapter 3 and Chapter 7 provide a detailed explanation of SAPs and their effects in specific regions.

Those opposed to free trade, and the SAPs that have long promoted it, argue that a less-regulated global economy can lead to rapid cycles of growth and decline that only increase global wealth disparity and can wreak havoc on smaller national economies (Figure 1.16). Labor unions point out that as corporations relocate factories and services to poorer countries where wages are lower, jobs are lost in richer countries. In the poorer countries, multinational corporations often work with governments to prevent workers from organizing labor unions that could bargain for **living wages**, wages that support a minimum healthy life. Environmentalists argue that in newly industrializing countries, which often lack effective environmental protection laws, multinational corporations tend to use highly polluting and unsafe production methods to lower costs. Many fear that a "race to the bottom" in wages, working conditions, government services, and environmental quality is underway as countries compete for profits and potential investors.

In response to the now widely recognized failures of SAPs, and the overemphasis on the power of markets to guide development, the IMF and the World Bank replaced SAPs with "Poverty Reduction Strategy Papers," or PRSPs. Each needy country works with World Bank and IMF personnel to design a broad-based plan for both economic growth and poverty reduction. PRSPs still push market-based solutions and aim toward reducing the role of government in the economy. These programs are highly bureaucratic, but they do focus on poverty reduction rather than just "development" per se. They also promote broader participation in civil society and include the possibility that all or some of a country's debt be "forgiven" (written off by the IMF and the World Bank), thus alleviating one of the worst side effects—bankrupting debt—that stopped progress in the poorest countries.

Fair trade, proposed as an alternative to free trade, seeks to provide a fair price to producers and to uphold environmental and safety standards in the workplace. Economic relationships surrounding trade are rearranged in order to provide better prices for producers from developing countries. For example, "fair trade" coffee and chocolate is now sold widely in North America and Europe. Prices are somewhat higher for consumers, but the extreme profits of middlemen are eliminated, and growers of coffee and cocoa beans receive living wages and improved working conditions.

In evaluating free trade and globalization, consider how many of the things you own or consume were produced in the global economy—computer, clothes, furniture, appliances, car, and foods. These products are cheaper for you to buy, and your standard of living is higher as a result of lower production costs and competition among many global producers (see Thematic Concepts P). However, you or someone you know may have lost a job because a company moved to another location where labor

FIGURE 1.16 Anti-WTO demonstration by Korean farmers and students (2005). The demonstrators were protesting WTO regulations that reduced farmers' freedom and income.

and resources are cheaper. Cheap products may be made by underpaid workers (even children) working under harsh conditions that possibly generate high levels of pollution. Given all these factors, consider the advantages and drawbacks of both free trade and fair trade.

THINGS TO REMEMBER

1. Globalization encompasses many types of worldwide and interregional flows and linkages, especially the ways in which goods, capital, labor, and resources are exchanged among distant and very different places.

2. **Learning Goal 4: Globalization and Development** Throughout the world, globalization is transforming patterns of economic development as local self-sufficiency is giving way to global interdependence and international trade.

3. Fair trade, a proposed alternative to free trade, seeks to provide a fair price to producers and to uphold environmental and safety standards in the workplace.

DEMOCRATIZATION

Learning Goal 5
Democratization: What factors are related to the shift from authoritarian modes of government—based on the power of the state or community leaders—to more democratic systems in which individuals are given a greater say in how governments are run?

For geographers studying globalization, an area of increasing interest is **democratization**, the transition toward political systems guided by competitive elections. Democratization runs counter to **authoritarianism**, a form of government that subordinates individual freedom to the power of the state or elite regional and local leaders (see Thematic Concepts T). In more democratic systems of government, individuals have more economic and political freedoms, such as the right to be an entrepreneur, the right to protest government policies, the right to access information of all types, the right to marshal public support through the media for particular programs, and the right to take action against injustice, especially through legal systems.

Geographers do not necessarily think that democracy is the "best" system of government. Indeed many geographers are critical of the imposition of democracy, usually by foreign governments or organizations, in places where long-standing cultural traditions support other political arrangements. Nevertheless, few would deny that the shift toward more democratic systems of government over the past century is extremely significant. In this regard, geographers and other scholars are particularly interested in the role of social movements, international organizations, and the role that a free media plays in democratization (see Thematic Concepts S, T, U). These phenomena, which sometimes accompany globalization, may give ordinary people a greater voice in how societies are run.

▮▶ 23. PROMOTING DEMOCRACY A CONTROVERSIAL THEME

The Expansion of Democracy

The twentieth century saw a steady expansion of democracy throughout the world, with more and more countries holding elections of their leaders—at least at the national level (see Thematic Concepts S). While democracy has become an ideal to which most countries aspire, there is little agreement on just what constitutes democratic institutions. Can the principles of a specific religion be part of democratic constitutions? Is it possible to have democratic government at the national level and yet quite authoritarian rule at the local level? What about forms of democratic participation other than voting, such as the ability to speak openly to leaders, to protest, to lobby for or against particular laws? Are these also essential components of democracy? The map in Photo Essay 1.3 on pages 34–35 is an attempt by *The Economist* magazine to depict the global pattern of democratization and to link that pattern with the occurrence of violent political conflict. Most violent conflicts happen in places that are authoritarian, not democratic.

What factors encourage democratization? Here are some of the most widely agreed on factors that support the public in seeking a more active role in their own governance:

- Peace: Elections held during times of war, particularly civil war, can rarely be considered "free and fair." Peace upholds many other factors supporting democratization as well.

- Broad prosperity: As access to more than the bare essentials of life is shared by a broad segment of the population (usually called the "middle class"), there is generally a shift toward giving more power to citizens through free elections of leaders. Still widely debated is whether general prosperity must occur before truly stable democracy can be established. And will prosperity necessarily lead to democracy?

- Education: Better-educated people tend to want a stronger voice in how they are governed. Although democracy has spread to countries with relatively undereducated populations, leaders in such places sometimes become more authoritarian once elected.

- **Civil society**: Institutions that encourage a sense of unity and informed common purpose among the general population are widely seen as supportive of democracy. Such institutions can include the media (see Thematic Concepts U), *nongovernmental organizations* (NGOs; discussed below), political parties, academia, unions, political parties, community service organizations such as Rotary and Lions Club, and in some cases, religious organizations.

democratization the transition toward political systems guided by competitive elections

authoritarianism a political system based on the power of the state or of elitist regional and local leaders

civil society the social groups and traditions that function independently of the state and its institutions

Democratization and Geopolitics

As the map in Photo Essay 1.3 shows, democratization at the global level has not yet been achieved (in fact, it is a goal that is not shared by all). A possible explanation is that democratization is often at odds with **geopolitics**, the strategies that countries use to ensure that their own interests are served in relations with other countries. Geopolitics was perhaps most obvious during the *Cold War era*, the period from 1946 to the early 1990s when the United States and its allies in Western Europe faced off against the Union of Soviet Socialist Republics (USSR) and its allies in Eastern Europe and Central Asia. Ideologically, the United States promoted a version of free market **capitalism**—an economic system based on the private ownership of the means of production and distribution of goods, driven by the profit motive and characterized by a competitive marketplace. By contrast, the USSR and its allies favored what was called **communism,** but what was actually a state-controlled economy—a socialized system of public services and a centralized government in which citizens participated indirectly through the Communist Party.

The Cold War became a race to attract the loyalties of unallied countries and to arm them. Sometimes the result was that unsavory dictators were embraced as allies by one side or the other. Eventually, the Cold War influenced the internal and external policies of virtually every country on earth, often oversimplifying complex local issues into a contest of democracy versus communism.

geopolitics the use of strategies by countries to ensure that their best interests are served

capitalism an economic system based on the private ownership of the means of production and distribution of goods, driven by the profit motive and characterized by a competitive marketplace

communism an ideology, based largely on the writings of the German revolutionary Karl Marx, that calls on workers to unite to overthrow capitalism and establish an egalitarian society in which workers share what they produce

In the post–Cold War period of the 1990s, geopolitics shifted. The Soviet Union dissolved, creating many independent states, nearly all of which began to implement some democratic and free market reforms. Globally, countries jockeyed for position in what looked like it might become a new era of trade and amicable prosperity, rather than war. But throughout the 1990s, while the developed countries enjoyed unprecedented prosperity, many unresolved political conflicts emerged in Asia, Africa, Southwest Asia, and southeastern Europe (Photo Essay 1.3 B, E). Too often these disputes erupted

Thinking Geographically

After you have read about Democratization and Conflict, you will be able to answer the following questions:

A What about this photo suggests that Álvaro Uribe has done some controversial things?

B Are there any clues in this photo that suggest these men might not be members of a government-backed army?

C How can you tell that these men are addressing foreigners?

(D & E) Of the factors mentioned in the text as necessary for democracy to flourish, which is most evident in each of these photos?

The map and accompanying photo essay show how countries with lower levels of democratization also suffer the most from violent conflict. Countries are colored according to their score on a "democracy index" created by *The Economist* magazine, which uses a combination of statistical indicators to capture elements crucial to the process of democratization. These elements include the ability of a country to hold peaceful elections that are accepted as fair and legitimate; the ability of people to participate in elections and other democratic processes; the strength of civil liberties (such as a free media and the right to hold political gatherings and peaceful protests); and the ability of governments to enact the will of their citizens free of corruption. Also displayed on the map are major conflicts initiated or in progress since 1990 that have resulted in at least 10,000 casualties. Aspects of the connections between democratization and armed conflict are explored in the photo captions.

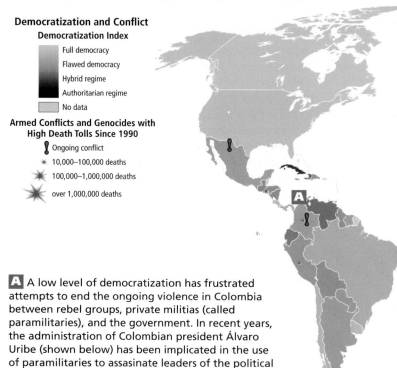

Democratization and Conflict
Democratization Index

- Full democracy
- Flawed democracy
- Hybrid regime
- Authoritarian regime
- No data

Armed Conflicts and Genocides with High Death Tolls Since 1990

- Ongoing conflict
- 10,000–100,000 deaths
- 100,000–1,000,000 deaths
- over 1,000,000 deaths

A A low level of democratization has frustrated attempts to end the ongoing violence in Colombia between rebel groups, private militias (called paramilitaries), and the government. In recent years, the administration of Colombian president Álvaro Uribe (shown below) has been implicated in the use of paramilitaries to assasinate leaders of the political opposition and labor union organizers. Repressive tactics like these diminish the potential to resolve disputes through peaceful democratic processes and help maintain support for violent confrontation.

B Violence in Sudan's Darfur region relates directly to the country's very low level of democratization. Disputes over control of resources, whether they be land, water, or Sudan's oil resources, often turn violent because there are no functioning democratic institutions capable of resolving conflicts. Shown on the left are members of the Sudan Liberation Movement Army, which opposes government-backed attacks on civilians in Darfur.

C Bringing democracy to Iraq was a main justification for the U.S. invasion and subsequent occupation of the country. Iraq's newly empowered parliament responded to widespread opposition to the continuing U.S. military presence by setting a deadline of 2011 for the withdrawal of U.S. forces, which the U.S. has agreed to honor. These men in Baghdad are protesting the presence of U.S. troops in Iraq.

D New Zealand is among the world's most democratized nations. Shown above are members of the Service and Food Workers Union, making use of their well-protected civil liberties by staging a rally for better wages and working conditions for its largely immigrant and female members. The protection of civil liberties is a crucial aspect of democratization.

E Democratic processes often become compromised during times of war. Shown here is a woman and her child running from the burning compound of presidential candidate Jean-Pierre Bemba's bodyguards in Kinshasa, Congo. The compound was attacked by demonstrators who wanted Congo's recent elections, the first in 40 years, postponed because of the country's ongoing civil war and accusations that the vote counting was being rigged. The civil war in Congo has resulted in 5.4 million deaths so far, mostly due to disease and malnutrition among people displaced by the violence.

FIGURE 1.17 Help from a nongovernmental organization (NGO). Members of the Kenyan Red Cross move an injured person to safety during postelection violence in Nairobi, Kenya, in January of 2008.

into bloodshed and the systematic attempt to remove (**ethnic cleansing**) or kill (**genocide**) all members of a particular ethnic or religious group.

The terrorist attacks on the United States on September 11, 2001, ushered in a new geopolitical era that is still evolving. Because of the size and the geopolitical power of the United States, the attacks and the U.S. reactions to them affected virtually every international relationship, public and private. The ensuing adjustments are directly or indirectly affecting the daily lives of billions of people around the world.

International Cooperation

So far there has been no serious effort to engage the principles of democratization at the global scale. However, many aspects of globalization favor international cooperation over national self-interest. Globalization also increases the need to enforce laws governing business, trade, and human rights at the international level.

The prime example today of international cooperation is the **United Nations (UN)**, an assembly of 192 member states. The member states sponsor programs and agencies, focusing on economic development, general health and well-being, democratization, peacekeeping assistance in "hot spots" around the world, humanitarian aid, and scientific research. Thus far, countries have been unwilling to relinquish *sovereignty*, the right of a country to conduct its internal affairs as it sees fit without interference from outside. Consequently, the United Nations has limited

ethnic cleansing the deliberate removal of an ethnic group from a particular area by forced migration

genocide the deliberate destruction of an ethnic, racial, or political group

United Nations (UN) an assembly of 192 member states that sponsors programs and agencies that focus on scientific research, humanitarian aid, planning for development, fostering general health, and peacekeeping assistance

nongovernmental organization (NGO) an association outside the formal institutions of government in which individuals, often from widely differing backgrounds and locations, share views and activism on political, social, economic, or environmental issues

legal authority and often can enforce its rulings only through persuasion. Even in its peacekeeping mission, there are no true UN forces. Rather, troops from member states wear UN designations on their uniforms and take orders from temporary UN commanders.

Nongovernmental organizations (NGOs) are an increasingly important embodiment of globalization. In such associations, individuals, often from widely differing backgrounds and locations, agree on political, economic, social, or environmental goals. For example, some NGOs work to protect the environment (as does, for instance, the World Wildlife Fund). Others, such as Doctors Without Borders, provide medical care to those who need it most. The Red Cross and Red Crescent provide emergency relief after disasters (Figure 1.17). The educational efforts of an NGO such as Rotary International can raise awareness among the global public about important issues, such as childhood vaccinations.

NGOs can be an important component of civil society, yet there is some concern that the power of huge international NGOs might undermine democratic processes, especially in small countries. Some critics feel that NGO officials are a powerful, do-gooder elite that does not interact sufficiently well with local people. A frequent target of such criticism is OXFAM International, a group of 13 NGOs that is the world leader in emergency famine relief. OXFAM was a major provider of relief after the Indian Ocean tsunami of 2004 and the Haiti earthquake of 2010. It has now expanded to cover long-term efforts to reduce poverty and injustice, which OXFAM sees as the root causes of famine. This more politically active role has brought OXFAM into conflict with local officials and with WTO policies on trade. At the local level, the best NGOs solicit input from a wide range of individuals—a feature that political scientists consider essential to building civil society.

❚❚▶ 22. NGOs PLAY LARGER ROLE IN WORLD AFFAIRS

THINGS TO REMEMBER

1. Democratization is the transition toward political systems that are guided by competitive elections in which individuals have a greater voice in how their governments are run.

2. Democratic systems are gradually replacing many authoritarian regimes worldwide.

3. Learning Goal 5: Democratization Among the most widely agreed upon factors necessary for the flourishing of democracy are peace, broad prosperity, education, and civil society.

WATER

Water is emerging as the major resource issue of the twenty-first century. Demand for clean water skyrockets as people move out of poverty. A huge increase in water consumption occurs with the production of foods, the manufacture of goods, and the development of services (like cleaning or even entertainment). This production inevitably creates *water pollution*. With clean, fresh water becoming scarce in so many parts of the world, water disputes are proliferating. This is especially true where rivers cross international borders and upstream users use more than their perceived fair share. Controversy also surrounds the sale of clean water. When water becomes a commodity rather than a free good, scarcity can result in water that is priced too high for many to afford.

> **virtual water** the volume of water used to produce all that a person consumes in a year
>
> **water footprint** the water used to meet a person's basic needs for a year, added to the person's annual virtual water

Calculating Water Use Per Capita

Humans require an average of 5 to 13 gallons (20 to 50 liters) of clean water per day for basic domestic needs: drinking, cooking, and bathing/cleaning. This domestic water consumption tends to increase as incomes rise, with the average person in a wealthy country consuming as much as 20 times the amount of water,

per capita, as the average person in a very poor country. However, this domestic water consumption is only a fraction of a person's actual water consumption, which must account for **virtual water**—the volume of water required to process all the goods and services a person consumes. Every apple eaten or cup of coffee drunk requires many liters of water for production and distribution. When we add up an individual's domestic water consumption with her virtual water consumption for an entire year, we have that person's total annual **water footprint**. The more one consumes, the larger one's virtual water footprint. Table 1.2 shows the amounts of water used to produce some commonly consumed products. (As you look at Table 1.2, note that there are 1000 liters, or 263 gallons, in a cubic meter (m^3).)

Like domestic consumption, personal water footprints vary widely according to standards of living and rates of consumption (see Figure 1.18 on page 38). Moreover, the amount of virtual water used to produce 1 ton of a specific product varies widely from country to country due to climate conditions along with agricultural and industrial technology. For example, on average, to produce 1 ton of corn in the United States requires 489 m^3 of virtual water, whereas in India the same amount of corn requires 1935 m^3 of virtual water; in Mexico 1744 m^3; and in the Netherlands, just 408 m^3. In the case of corn, water is lost to evapotranspiration in the field, to the evaporation of standing irrigation water, and to evaporation as water flows to and from the field. A further component of virtual water is that the water that becomes polluted in the production process is also lost to use.

For help calculating your individual water footprint, go to the water footprint Web site: http://www.waterfootprint.org/?page=cal/waterfootprintcalculator_indv.

Who Owns Water? Who Gets Access to It?

Though many people consider water a human right that should not cost anything to access, water has become the third most valuable commodity, after oil and electricity. Water in wells and running in streams and rivers is increasingly being *privatized*. This means that its ownership is being transferred from governments—which can be held accountable for protecting the rights of all citizens to access water—to individuals, multinational corporations, and other private entities that manage the water primarily for profit. Water is often privatized under the rationale that private enterprise will make needed investments that will boost the efficiency of water distribution systems and increase the overall quality of the water supply. Regardless of whether or not these potential gains are actually realized, privatization usually brings higher water costs that can become quite controversial. For example, in the city of Cochabamba, Bolivia, in 1999, water costs rose beyond what the urban poor could afford, following the sale of the Cochabamba public water company to a group of multinational corporations, led by Bechtel of

TABLE 1.2	**The global average virtual water content of everyday products is the volume of water used to produce that product**

Product[a]	Virtual water content (in liters)
1 potato	25
1 cup tea	35
1 slice of bread	40
1 apple	70
1 glass of beer	75
1 glass of wine	120
1 egg	135
1 cup of coffee	140
1 glass of orange juice	170
1 lb of chicken meat	2000
1 hamburger	2400
1 lb of cheese	2500
1 pair of bovine leather shoes	8000

[a] To see the water footprint of additional products, go to http://www.waterfootprint.org/?page=files/productgallery

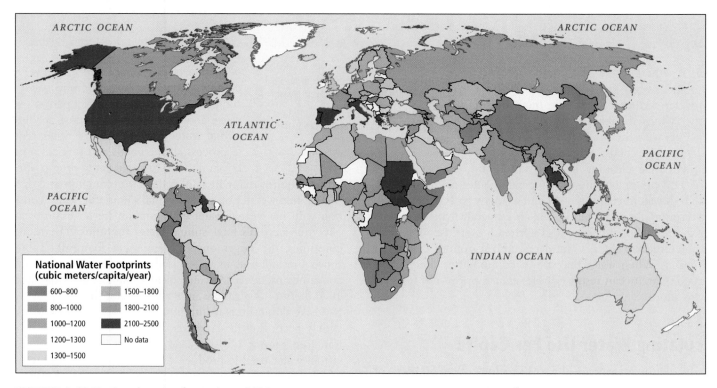

FIGURE 1.18 National water footprints, 2004. Average national water footprint per capita (m³ per capita per year). The color green indicates that the nation's water footprint is equal to or smaller than global average. Countries in red have a water footprint beyond the global average.

San Francisco, USA. The result was a nationwide series of riots that led to the abandonment of privatization.

Water Quality

About one-sixth of the world's population does not have access to clean drinking water, and dirty water kills over 6 million people each year. More people die this way than in all of the world's armed conflicts for an average year. In Europe and the United States, demand for higher water quality has resulted in a $100 billion bottled water industry. Meanwhile, in the poorer parts of cities in the developing world, many people draw water with a pail from a communal spigot (see Thematic Concepts Part V; see also Figure 1.19A). Usually this water should be boiled before use, even for bathing. Truly safe drinking water must be purchased in sealed containers, at a prohibitively high cost. These water quality problems help explain why so many people are chronically ill and why 24,000 children under the age of 5 die every day from water-borne diseases.

▌▶ 245. CLEAN WATER PROJECT IMPROVES LIVES IN SENEGAL

Water and Urbanization

Urban development patterns dramatically affect the management of water. In most urban slum areas in poor countries, and even in some relatively wealthy cities in places like the United States and Canada, crucial water management technologies such as sewage treatment systems are entirely absent. Germ-laden human waste from toilets and kitchens is often deposited in urban gutters that drain into creeks, rivers,

and bays. And yet, building adequate wastewater collection and treatment systems in cities, which already house several million inhabitants, is often deemed prohibitively costly, especially in poor countries.

Independent of sewage, water becomes polluted in cities. As cities grow, parking lots and rooftops replace areas that were once covered with natural vegetation. Rainwater quickly runs off these hard surfaces, collects in low places, and becomes stagnant instead of being absorbed into the ground. In urban slums, flooding can spread polluted water over wide areas, carrying it into homes and places where children play (Figure 1.19B). Diseases such as malaria and cholera, carried in this water, can spread rapidly as a result. Fortunately, new technologies and urban planning methods are being developed that can help cities avoid these problems, but they are not yet in widespread use.

THINGS TO REMEMBER

1. Water scarcity is emerging as the major resource issue of the twenty-first century, with population growth, skyrocketing per capita demand for clean water, and water pollution all straining water supplies.

2. Much of the water that humans use is virtual water.

3. Urbanization often leads to water pollution through untreated sewage, industrial production, and mismanaged storm water.

(A)

(B)

FIGURE 1.19 Access to water. (A) A young man carries water to his home in the hills outside Kabul, Afghanistan. The city is home to 3.4 million people, but has no central sewage system, and only 18 percent of its people have access to city water piped into their homes. **(B)** A boy drinks water from a temporary pool that formed outside the slum where he lives in Kenya. Rains often bring flooding to urban slums in Africa, which rarely have any planned drainage system.

Thinking Geographically: (A) What pressures to conserve water are evident in this photo? **(B)** Why might this boy be endangering his health by drinking this water?

GLOBAL CLIMATE CHANGE

Planet Earth is continually undergoing **climate change**, a slow shifting of climate patterns due to general cooling or warming of the atmosphere. The present trend of **global warming**—which refers to the observed warming of the earth's surface and climates during the twentieth century—is extraordinary because it appears to be due primarily to human agency and may be happening more quickly than climate changes in the past. **Greenhouse gases** (carbon dioxide, methane, and other gases) that trap heat in the atmosphere are being produced at accelerating rates by the burning of fossil fuels, and by deforestation (see Thematic Concepts Y, Z, AA).

Most scientists now agree that there is an urgent need to reduce greenhouse gas emissions to avoid catastrophic climate change in coming years. Climatologists, biogeographers, and other scientists are documenting long-term global warming and cooling trends by examining evidence in tree rings, fossilized pollen and marine creatures, and glacial ice. These data indicate that the twentieth century was the warmest century in 600 years and the decade of the 1990s was the hottest since the late nineteenth century. Evidence is mounting that these are not normal fluctuations. It is estimated that, at present rates of emissions, by 2100 average global temperatures could rise between 2.5°F and 10°F (about 2°C to 5°C).

> **climate change** a slow shifting of climate patterns due to the general cooling or warming of the atmosphere
>
> **global warming** the warming of the earth's climate as atmospheric levels of greenhouse gases increase
>
> **greenhouse gases** harmful gases, such as carbon dioxide and methane, released into the atmosphere by human activities

However, a key problem is that those most responsible for global warming (the world's most wealthy and industrialized countries) have the least incentive to reduce emissions because they are the least vulnerable to the changes global warming causes in the physical environment. Meanwhile, the most vulnerable to these changes (poor countries with low human development) are the least responsible for the growth in greenhouse gases, and hence the least able to effect any reduction in emissions (see Figure 1.20 on page 40).

▮▶ 330. U.S. GOVERNMENT SCIENTISTS CALL FOR URGENT ACTION ON GLOBAL WARMING

Drivers of Global Climate Change

Greenhouse gases—carbon dioxide, methane, nitrous oxide and other gases—exist naturally in the atmosphere. It is their heat-trapping ability that makes the earth warm enough for life to exist. Increase their levels, as humans are doing now, and the earth becomes warmer still.

Over the last several hundred years, humans have greatly intensified the release of greenhouse gases. Electricity generation, vehicles, industrial processes, and the heating of homes and

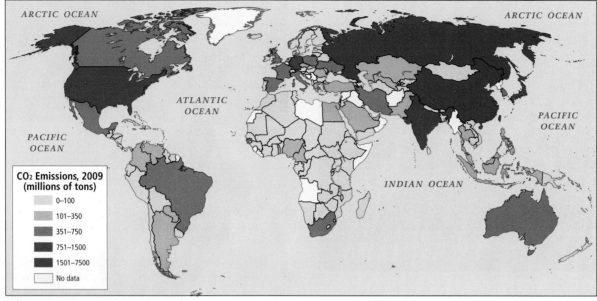

(A)

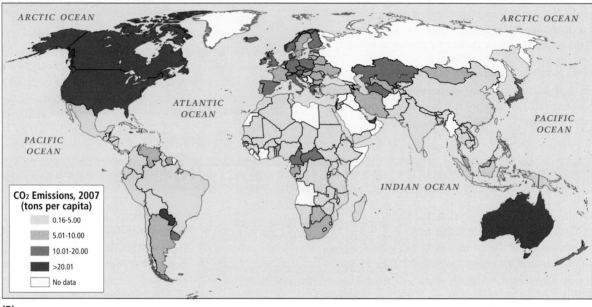

(B)

FIGURE 1.20 Greenhouse gas emissions around the world, 2006. **(A)** Total emissions by millions of tons, 2006. The United States leads the entire world (22.85 percent of the world total), with China second (13.22 percent), Russia third (7.13 percent), Japan fourth (4.36 percent), and India fifth (3.95 percent). These top five countries contribute 51.5 percent of the world's greenhouse gas emissions. **(B)** Tons of emissions per capita. The United Arab Emirates is the top per capita emitter at 56.71 tons of CO_2 per capita, followed by Jamaica, Bahrain, and Paraguay. The U.S. is eighth at 22.96 tons per capita. China is far down on the list at 3.39 tons per capita.

businesses all burn large amounts of CO_2-producing fossil fuels such as coal, natural gas, and oil. Even the large-scale raising of grazing animals contributes methane through the animals' flatulence. Unusually large quantities of greenhouse gases from these sources are accumulating in the earth's atmosphere, and their presence has already led to significant warming of the planet's climate.

Widespread deforestation worsens the situation. Living forests take in CO_2 from the atmosphere, release the oxygen, and store the carbon in their biomass. As more trees are cut down and their wood is used for fuel, more carbon enters the atmosphere, less is taken out, and less is stored. The loss of trees and other forest organisms produces as much as 30 percent of the buildup of CO_2 in the atmosphere. The use of fossil fuels accounts for the remaining 70 percent.

The largest producers of total greenhouse gas emissions (Figure 1.20) are the industrialized countries and large, rapidly industrializing countries. The United States produces the most emissions (22.85 percent); China is second (13.22 percent); Russia is third (7.13 percent); Japan is fourth (4.36 percent), and India fifth

(3.95 percent). In emissions per capita, the United Arab Emirates lead with 56.71 tons of CO_2 per capita, followed by Jamaica, Bahrain, and Paraguay. The United States is eighth at 22.96 tons per capita. China is far down the list at 3.39 tons per capita.

For the period of 1859 to 1995, developed countries produced roughly 80 percent of the greenhouse gases from industrial sources, and developing countries produced 20 percent. But by 2007, the developing countries were catching up, accounting for nearly 30 percent of total emissions. As developing nations industrialize over the next century and continue to cut down their forests, they will release more and more greenhouse gases every year. If present patterns hold, greenhouse gas contributions by the developing countries will exceed those of the developed world by 2040.

Climate Change Impacts

While it is not clear just what the impacts of rising global temperatures will be, it is clear that they will not be uniform across the globe. One prediction is that the glaciers and polar ice caps will melt, causing a corresponding rise in sea level. In fact, this phenomenon is already observable. Satellite imagery analyzed by scientists at the National Aeronautics and Space Administration (NASA) shows that between 1979 and 2005—just 26 years—the polar ice caps shrank by about 23 percent. The melting released thousands of trillions of gallons of meltwater into the oceans. If this trend continues, at least 60 million people in coastal areas and on low-lying islands could be displaced by rising sea levels. Another issue is the melting of high mountain glaciers, which are a major source of water for many of the world's large rivers. Scientists have monitored mountain glaciers across the globe for more than 30 years; while some are growing, the majority are melting rapidly. Over the short term, melting mountain glaciers could result in flooding in many rivers, but eventually flows would reduce as mountain glaciers shrank or disappeared entirely.

Over time, higher temperatures will shift with warmer climate zones northward in the Northern Hemisphere and southward in the Southern Hemisphere. Such climate shifts might lead to the displacement of huge numbers of people, because the zones where specific crops can grow would change. Animal and plant species that cannot adapt rapidly to the changes will likely disappear. Higher temperatures also will lead to stronger hurricanes, as these storms are powered by warm, rising air (see Photo Essay 1.4A, on page 42). In some areas, drought and water scarcity may also become more common since higher temperatures would increase water evaporation rates from soils, vegetation, and bodies of water. Another effect of global warming could be a shift in ocean currents. The result would be more chaotic and severe weather, especially for places where climates are strongly influenced by ocean currents, such as Western Europe.

Vulnerability to Climate Change

Learning Goal 6
Climate Change and Water: In what ways does water influence how different places are vulnerable to climate change?

A place's vulnerability to climate change can be thought of as its risk of suffering damage to human or natural systems as a result of such impacts as (among others) sea level rise, drought, flooding, or increased storm intensity. Many of these vulnerabilities are water related, though many are not. Understanding the full range of vulnerability in a place is crucial if humans are to adapt to climate change there. This is because the rising global temperatures that drive climate-change impacts are unavoidable at this point. Scientists who study climate change agree that while we can minimize temperature increases, we can't stop them entirely, much less reverse those that have already occurred. Hence, we are going to have to live with and adapt to the impacts of climate change for quite some time. The first step in doing this is to understand how and where humans and ecosystems are especially vulnerable to climate change.

Three concepts are important in understanding a place's vulnerability to climate change: exposure, sensitivity, and resilience. Here we explore them in the context of the vulnerability to water-related climate-change impacts in Mumbai, India. *Exposure* refers to the extent to which a place is exposed to climate-change impacts. For example, a low-lying coastal city like Mumbai, India, is highly exposed to sea level rise. *Sensitivity* refers to how sensitive a place is to those impacts. For example, many of Mumbai's inhabitants are very sensitive to sea level rise because they are extremely poor and can only afford to live in low-lying slums that have no sanitation and hence pollute nearby waterways. Flooding of these waterways would bring epidemics of waterborne illness that could kill many. *Resilience* refers to a place's ability to "bounce back" from the disturbances that climate-change impacts create. Mumbai's resilience to sea level rise is bolstered by its status as the wealthiest city in South Asia, which enables it to afford relief and recovery systems that could help it deal with sea level rise over the short and long term. Over the short term, Mumbai benefits from more and better hospitals and emergency response teams than any other city in South Asia. Over the long term, Mumbai's well-trained municipal planning staff can create and execute plans to help sensitive populations, like poor slum dwellers, adapt to sea level rise. This could be, for example, through planned relocation to areas less exposed to sea level rise, or by building sea walls and dikes that could keep sea waters out of low-lying slum areas. Of course Mumbai's overall vulnerability is more complicated than these examples suggest because the city faces many more climate-change impacts than just sea level rise. However, these examples help us understand vulnerability to climate change as a combination of exposure, sensitivity, and resilience.

One effort at understanding the global pattern of vulnerability to climate change can be seen in Photo Essay 1.4 on pages 42–43, which features a map of vulnerability to climate change and photos of the types of problems that are already observable. The pattern of greatest vulnerability covers the world regions that have the lowest human development. This reflects the reality that both sensitivity and resilience are related to the many factors that influence human development. For example, many of the qualities that make Mumbai more sensitive to sea level rise are less present in urban areas located in highly developed countries. For example, New York City has virtually none of the large unplanned slums found in Mumbai. In addition, numerous world-class hospitals, emergency response teams, plus large and well-trained municipal planning staffs boost New York's resilience.

Responding to Climate Change

In 1992, an agreement known as the **Kyoto Protocol** was drafted. The protocol called for scheduled reductions in CO_2 emissions by the highly industrialized countries of North America, Europe, East Asia, and Oceania. The agreement also encouraged, though it did not require, developing countries to curtail their emissions. One hundred eighty-three countries had signed the agreement by 2009. The only developed country that had not signed was the United States, the world's biggest per capita producer of CO_2.

In December 2009 in Copenhagen, 181 countries attempted to halt the rising concentrations of CO_2 in the atmosphere by 2020. However, wealthy nations did not offer to curb emissions sufficiently to make a difference. In the end, reaching agreement on specific emissions reductions proved impossible, although the roughly 30 countries that emit 90 percent of greenhouse gases agreed that they would begin to curb their own emissions. They also agreed to help developing countries achieve clean-energy economies and otherwise adapt to climate change.

> **Kyoto Protocol** an amendment to a United Nations treaty on global warming, the Protocol is an international agreement, adopted in 1997 and in force in 2005, that sets binding targets for industrialized countries for the reduction of emissions of greenhouse gases

THINGS TO REMEMBER

1. Planet Earth is continually undergoing climate change—a slow shifting of climate patterns resulting from general cooling or warming of the atmosphere.

2. Human activities that emit large amounts of carbon dioxide, methane, and other greenhouse gases are trapping heat in the atmosphere, causing what is now widely recognized as global warming.

3. **Learning Goal 6: Climate Change and Water** Climate-change impacts are unavoidable at this point in time. Hence, it is crucial to understand how places are vulnerable to climate impacts. Many of these are water related, such as sea level rise, drought, and flooding.

Thinking Geographically

After you have read about Vulnerability to Climate Change, you will be able to answer the following questions:

A Does this photo relate most to short-term or long-term resilience?

B Of the three factors shown in this graphic, which is discussed the most in the text?

D What sign of an orderly response to disaster is visible in this picture?

E How can you tell that food supplies are low in this camp?

This map shows overall human vulnerability to climate change based on a combination of human and environmental factors. Areas shown in darkest red are vulnerable to floods, hurricanes, droughts, sea level rise, or other impacts related to climate change. When a population is exposed to an impact that it is sensitive to and has little resilience towards, vulnerability results. For example, many populations are exposed to drought, but generally speaking, the poorest populations are the most sensitive. However, sensitivity to drought can be compensated for if adequate relief and recovery systems, such as emergency water and food distribution systems, are in place. These systems lend an area a level of resilience that can reduce its overall vulnerability to climate change. Hence, a place's vulnerability to climate change can be thought of as a result of its sensitivity, exposure, and resilience in the face of multiple climate impacts.

F **Climate change and hurricanes.** Climatologists predict that hurricanes will increase in intensity as the planet warms. Indeed recent decades have seen an increase in powerful storms. Poverty (high sensitivity) and inadequate recovery systems (low resilience) make much of Central America particularly vulnerable to many hurricane-related impacts. Shown here is flood damage along the Choluteca River, Honduras, caused by Hurricane Mitch in 1999. Over 9000 people died in the storm, making Mitch the second most deadly hurricane in history.

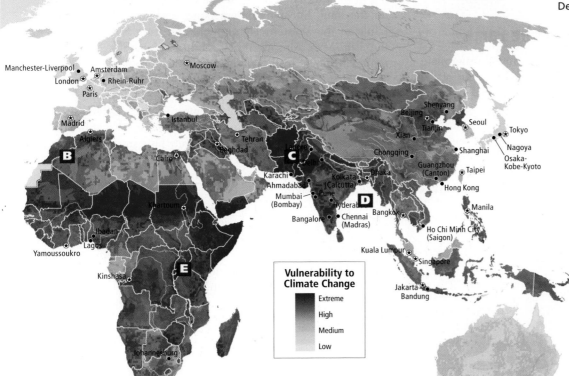

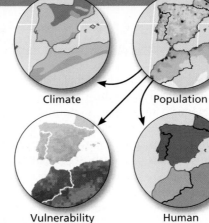

A **High resilience in the United States.** Effective and well-funded recovery and relief systems give the United States high resilience to climate impacts. This contributes to generally low vulnerability. Shown here are ambulances lined up to respond as Hurricane Ike approaches the Texas coast.

Climate

Population

Vulnerability

Human Development

B **Spain and Morocco: The multiple dimensions of vulnerability.** A wide variety of information is used to make the global map of vulnerability shown below. For example, the contrast in vulnerability between Spain and Morocco relates to (among other things) differences in climate, population density and distribution, and human development (which is itself based on many factors).

Manchester-Liverpool
Amsterdam
London
Rhein-Ruhr
Paris
Moscow
Madrid
Istanbul
Algiers
Tehran
Cairo
Baghdad
Lahore
Shenyang
Beijing
Tianjin
Seoul
Xian
Tokyo
Nagoya
Osaka-Kobe-Kyoto
Chongqing
Shanghai
Karachi
Kolkata (Calcutta)
Dhaka
Guangzhou (Canton)
Taipei
Ahmadabad
Pune
Mumbai (Bombay)
Hyderabad
Hong Kong
Khartoum
Bangalore
Chennai (Madras)
Bangkok
Manila
Ibadan
Lagos
Ho Chi Minh City (Saigon)
Yamoussoukro
Kuala Lumpur
Singapore
Kinshasa
Jakarta
Bandung
Johannesburg

Vulnerability to Climate Change

Extreme
High
Medium
Low

C **Afghanistan's extreme vulnerability.** A boy in a Kabul refugee camp brings home food donated by the UN. Afghanistan's poor population is highly exposed and sensitive to drought and flooding. Meanwhile, decades of civil war have eroded the country's resilience as the government is unable to mount effective recovery and relief efforts. Many refugees from war are dependent on foreign aid, which is often lacking.

E **Northern Uganda and southern Sudan.** The situation here is very similar to Afghanistan; however, sensitivity to drought is somewhat lower as access to water is better and poverty is less widespread. Left: Rebels in southern Sudan's civil war have complicated relief efforts. Below: Refugees in northern Uganda pick up bits of donated grain that have been dropped.

D **Bangladesh: Moderate resilience.** Rural Bangladeshis line up for food and water after a hurricane. Advances in government-led disaster recovery have increased Bangladesh's resilience. This has reduced the vulnerability of its poor population, which is highly exposed and sensitive to sea level rise, flooding, hurricanes (cyclones), drought, and other disturbances that climate change could create or intensify.

43

PHYSICAL GEOGRAPHY PERSPECTIVES

Physical geography is concerned with the processes that shape the earth's landforms, climate, and vegetation. In this sense, physical geographers are similar to scientists from other disciplines who focus on these phenomena, though as geographers they often look at problems spatially. In this book, physical geography provides a backdrop for the many aspects of **human geography** we discuss. However, physical geography is a fascinating, large, and growing field of study that you should consider taking an introductory course in. What follows are just the basics of landforms and climate.

Landforms: The Sculpting of the Earth

The processes that create the world's varied **landforms**—mountain ranges, continents, and the deep ocean floor—are some of the most powerful and slow-moving forces on earth. Originating deep beneath the earth's surface, these internal processes can move entire continents, often taking hundreds of millions of years to do their work. However, external processes form many of the earth's features, such as a beautiful waterfall or a rolling plain. These more rapid and delicate processes take place on the surface of the earth. The processes that constantly shape and reshape the earth's surface are studied by geomorphologists.

Plate Tectonics

Two key ideas related to internal processes in physical geography are the *Pangaea hypothesis* and *plate tectonics*. The geophysicist Alfred Wenger first suggested the Pangaea hypothesis in 1912. This hypothesis proposes that all the continents were once joined in a single vast continent called Pangaea (meaning "all lands"), which fragmented over time into the continents we know today (**Figure 1.21**). As one piece of evidence for his theory, Wegener pointed to the neat fit between the west coast of Africa and the east coast of South America.

For decades, most scientists rejected Wegner's hypothesis. We now know, however, that the earth's continents have been assembled into supercontinents at least three different times, only to break apart again. All of this activity is made possible by plate tectonics, a process of continental motion discovered in the 1960s, long after Wegener's time.

According to **plate tectonics**, the earth's surface is composed of large plates that float on top of an underlying layer of molten rock. The plates are of two types. Oceanic plates are dense and relatively thin, and they form the floor beneath the oceans. Continental plates are thicker and less dense. Much of their surface rises above the oceans, forming continents.

These massive plates drift slowly, driven by the circulation of the underlying molten rock flowing from hot regions deep inside the earth to cooler surface regions and back. The creeping movement of tectonic plates fragmented and separated Pangaea and created the continents we know today (Figure 1.21E).

Plate movements influence the shapes of major landforms, such as continental shorelines and mountain ranges. Huge mountains have piled up on the leading edges of the continents as the plates carrying them collided with other plates, folding and warping in the process. Hence, the theory of plate tectonics accounts for the long, linear mountain ranges extending from Alaska to Chile in the Western Hemisphere and from Southeast Asia to the European Alps in the Eastern Hemisphere. The highest mountain range in the world, the Himalayas of South Asia, was created when what is now India, at the northern end of the Indian-Australian Plate, ground into Eurasia. The only continent that lacks these long, linear mountain ranges is Africa. Often called the "plateau continent," Africa is believed to have been at the center of Pangaea and to have moved relatively little since the breakup.

Humans encounter tectonic forces most directly as earthquakes and volcanoes. Plates slipping past each other create the catastrophic shaking of the landscape we know as an earthquake. When plates collide and one slips under the other, this is known as *subduction*. Volcanoes arise at zones of subduction or sometimes in the middle of a plate, where gases and molten rock (called magma) can rise to the earth's surface through fissures and holes in the plate. Volcanoes and earthquakes are particularly common around the edges of the Pacific Ocean, an area known as the **Ring of Fire** (see Figure 1.22 on page 46).

Landscape Processes

The landforms created by plate tectonics have been further shaped by external processes, which are more familiar to us because we can observe them daily. One such process is **weathering**. Rock, exposed to the onslaught of sun, wind, rain, snow, ice, and the effects of life-forms, fractures and decomposes into tiny pieces. These particles then become subject to another external process, erosion. During erosion, wind and water carry away rock particles and any associated decayed organic matter and deposit them in new locations. The deposition of eroded material can raise and flatten the land around a river, where periodic flooding spreads huge quantities of silt. As small valleys between hills are filled in by silt, a **floodplain** is created. Where rivers meet the sea, floodplains often fan out roughly in the shape of a triangle, creating a **delta**. External processes tend to smooth out the dramatic mountains and valleys created by internal processes.

---

physical geography the study of the earth's physical processes: how they work, how they affect humans, and how they are affected by humans

human geography the study of patterns and processes that have shaped human understanding, use, and alteration of the earth's surface

landforms physical features of the earth's surface, such as mountain ranges, river valleys, basins, and cliffs

plate tectonics the scientific theory that the earth's surface is composed of large plates that float on top of an underlying layer of molten rock; the movement and interaction of the plates create many of the large features of the earth's surface, particularly mountains

Ring of Fire the tectonic plate junctures around the edges of the Pacific Ocean; characterized by volcanoes and earthquakes

weathering the physical or chemical decomposition of rocks by sun, rain, snow, ice, and the effects of life-forms

floodplain that flat land around a river where sediment is deposited during flooding

delta the triangular-shaped plain of sediment that forms where a river meets the sea

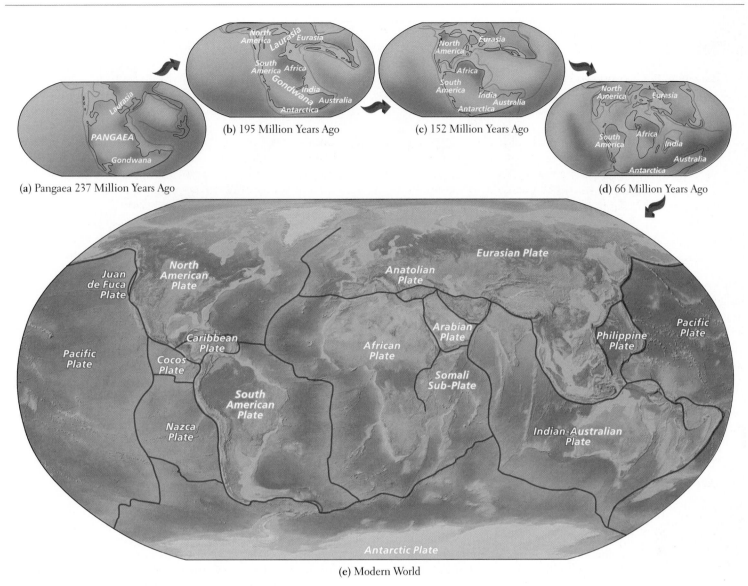

(a) Pangaea 237 Million Years Ago

(b) 195 Million Years Ago

(c) 152 Million Years Ago

(d) 66 Million Years Ago

(e) Modern World

FIGURE 1.21 The breakup of Pangaea. The modern world map **(E)** depicts the current boundaries of the major tectonic plates. Pangaea is only the latest of several global configurations that have coalesced and then fragmented over the last billion years.

Human activity often contributes to external landscape processes. By altering the vegetative cover, agriculture and forestry expose the earth's surface to sunlight, wind, and rain. These agents in turn increase weathering and erosion. Flooding becomes more common because the removal of vegetation limits the ability of the earth's surface to absorb rainwater. As erosion increases, rivers may fill with silt, and deltas may extend into the oceans.

Climate

The processes associated with climate are generally more rapid than those that shape landforms. Weather, the short-term and spatially limited expression of climate, can change in a matter of minutes. **Climate** is the long-term balance of temperature and precipitation that keeps weather patterns fairly consistent from year to year. By this definition, the last major global climate change took place about 15,000 years ago, when the glaciers of

climate the long-term balance of temperature and precipitation that characteristically prevails in a particular region

the last ice age began to melt. As we have seen, human activity is producing a new global climate change in our own time.

Solar energy is the engine of climate. The earth's atmosphere, oceans, and land surfaces absorb huge amounts of solar energy, and the differences in the amounts they absorb account for part of the variations in climate we observe. The most intense direct solar energy strikes the earth more or less head-on in a broad band stretching about 30 degrees north and south of the equator. The highest average temperatures on earth occur within this band. Moving away from the equator, solar energy strikes the earth's surface at an angle. This reduces its heating effect and results in drops in average temperatures.

Temperature and Air Pressure

The wind and weather patterns we experience daily are largely a result of variations in solar energy absorption that create complex

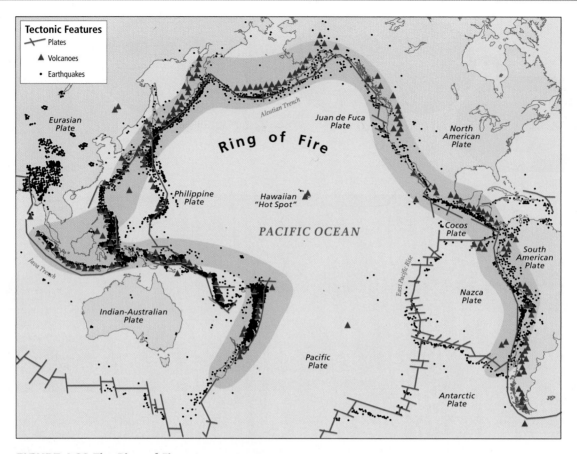

FIGURE 1.22 The Ring of Fire. Volcanic formations encircling the Pacific Basin form the Ring of Fire, a zone of frequent earthquakes and volcanic eruptions.

patterns of air temperature and *air pressure*. To understand air pressure, think of air as existing in a particular unit of space—for example, a column of air above a square foot of the earth's surface. Air pressure is the amount of force (due to the pull of gravity) exerted by that column on that square foot of surface. Air pressure and temperature are related: the gas molecules in warm air are relatively far apart and are associated with low air pressure. In cool air, the gas molecules are relatively close together (dense) and are associated with high air pressure.

As a unit of cool air is warmed by the sun, the molecules move farther apart. The air becomes less dense and exerts less pressure. Air tends to move from areas of higher pressure to areas of lower pressure, creating wind. If you have been to the beach on a hot day, you may have noticed a cool breeze blowing in off the water. This happens because land heats up (and cools down) faster than water. So on a hot day, the air over the land warms, rises, and becomes less dense than the air over the water. This causes the cooler, denser air to flow inland. At night the breeze often reverses direction, blowing from the now cooling land onto the now relatively warmer water.

These air movements have a continuous and important influence on global weather patterns. Over the course of a year, continents heat up and cool off much more rapidly than the oceans that surround them. Hence, the wind tends to blow from the ocean to the land during summer and from the land to the ocean during winter. It is almost as if the continents were breathing once a year, inhaling in summer and exhaling in winter.

> **orographic rainfall** rainfall produced when a moving moist air mass encounters a mountain range, rises, cools, and releases condensed moisture that falls as rain

Precipitation

Perhaps the most tangible way we experience changes in air temperature and density is through rain or snow. Precipitation occurs primarily because warm air holds more moisture than cool air. When this moist air rises to a higher altitude, its temperature drops, which reduces its ability to hold moisture. The moisture condenses into drops to form clouds and may eventually fall as rain or snow.

Several conditions that encourage moisture-laden air to rise influence the pattern of precipitation observed around the globe. When moisture-bearing air is forced to rise as it passes over mountain ranges, the air cools, and the moisture condenses to produce rainfall (Figure 1.23). This process, known as **orographic rainfall**, is most common in coastal areas where wind blows moist air from above the ocean onto the land and up the side of a coastal mountain range. Most of the moisture falls as rain as the cooling air rises along the coastal side of the range. On the inland side, the descending air warms and ceases to drop its moisture. The drier side of a mountain range is said to be in the *rain shadow*. Rain shadows may extend for hundreds of miles across the interiors of continents, as they do on the Mexican Plateau, east of California's Pacific coastal ranges, or north of the Himalayas of Eurasia.

A central aspect of earth's climate is the "rain belt" that exists in equatorial areas. Near the equator, moisture-laden tropical

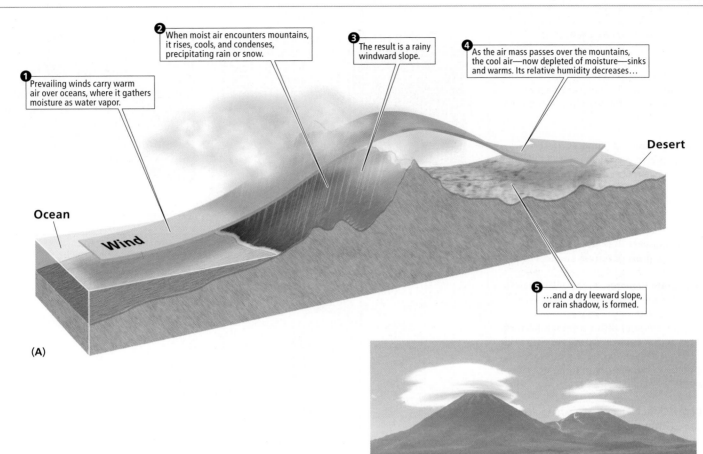

① Prevailing winds carry warm air over oceans, where it gathers moisture as water vapor.

② When moist air encounters mountains, it rises, cools, and condenses, precipitating rain or snow.

③ The result is a rainy windward slope.

④ As the air mass passes over the mountains, the cool air—now depleted of moisture—sinks and warms. Its relative humidity decreases…

⑤ …and a dry leeward slope, or rain shadow, is formed.

Ocean

Wind

Desert

(A)

FIGURE 1.23 Orographic rainfall (and rain shadow). (B) As moist air is blown ashore at Alaska's Cook Inlet, it hits a mountain, which forces the air to rise, resulting in the formation of a cloud.

(B)

air is heated by the strong sunlight and rises to the point where it releases its moisture as rain. Neighboring nonequatorial areas also receive some of this moisture when seasonally shifting winds move the rain belt north and south of the equator. The huge downpours of the Asian summer monsoon are an example.

In the **monsoon** season, the Eurasian continental landmass heats up during the summer, causing the overlying air to expand, become less dense, and rise. The somewhat cooler, yet moist, air of the Indian Ocean is drawn inland. The effect is so powerful that the equatorial rain belt is sucked onto the land (see Figure 8.4 [winter and summer monsoons] on page 293). The result is tremendous, sometimes catastrophic, rains throughout virtually all of South and Southeast Asia, as in Pakistan during the summer of 2010, and much of coastal and interior East Asia. Similar forces pull the equatorial rain belt south during the Southern Hemisphere's summer.

Much of the moisture that falls on North America and Eurasia is *frontal precipitation* caused by the interaction of large air masses of different temperatures and densities. These masses develop when air stays over a particular area long enough to take on the temperature of the land or sea beneath it. Often when we listen to a weather forecast, we hear about warm fronts or cold fronts. A *front* is the zone where warm

monsoon a wind pattern in which in summer months, warm, wet air coming from the ocean brings copious rainfall, and in winter, cool, dry air moves from the continental interior toward the ocean

and cold air masses come into contact, and it is always named after the air mass whose leading edge is moving into an area. At a front, the warm air tends to rise over the cold air, carrying warm clouds to a higher altitude. Rain or snow may follow. Much of the rain that falls along the outer edges of a hurricane is the result of frontal precipitation.

Climate Regions

Geographers have several systems for classifying the world's climates that are based on the patterns of temperature and precipitation just described. This book uses a modification of the widely known Köppen classification system, which divides the world into several types of climate regions, labeled A, B, C, D, and E on the climate map in Photo Essay 1.5 on pages 48–49. As you look at the regions on this map, examine the photos, and read the accompanying climate descriptions, the importance of climate to vegetation becomes evident. Each regional chapter will include a climate map; when reading these maps, refer to the verbal descriptions in Photo Essay 1.5 as necessary. Keep in mind that the sharp boundaries shown on climate maps are in reality much more gradual transitions.

(A) Tropical Humid Climates. In *tropical wet climates,* rain falls predictably every afternoon and usually just before dawn. The *tropical wet/dry climate,* also called a *tropical savanna,* has a wider range of temperatures and a wider range of rainfall fluctuation than the tropical wet climate.

(B) Arid and Semiarid Climates. *Deserts* generally receive very little rainfall (2 inches or less per year). Most of that rainfall comes in downpours that are extremely rare and unpredictable. *Steppes* have climates similar to those of deserts, but that are more moderate. They usually receive about 10 inches more rain per year than deserts and are covered with grass or scrub.

(C) Temperate Climates. Areas with temperate climates are moist all year and have short, mild winters and long, hot summers. *Subtropical climates* differ from midlatitude climates in that subtropical winters are dry. *Mediterranean climates* have moderate temperatures but are dry in summer and wet in winter.

(D) Cool Humid Climates. Stretching across the broad interiors of Eurasia and North America are *continental climates,* with either dry winters (northeastern Eurasia) or moist all year (North America and north-central Eurasia). Summers in cool humid climates are short but can have very warm days.

(E) Coldest Climates. *Arctic* and *high-altitude climates* are by far the coldest and are also among the driest. Although moisture is present, there is little evaporation because of the low temperatures. The Arctic climate is often called *tundra,* after the low-lying vegetation that covers the ground. The high-altitude version of this climate, which may occur far from the Arctic, is more widespread and subject to greater daily fluctuations in temperature. High-altitude microclimates, such as those in the Andes and the Himalayas, can vary tremendously depending on factors such as available moisture, orientation to the sun, and vegetation cover. As one ascends in altitude, the climate changes loosely mimic those found as one moves from lower to higher latitudes. These changes are known as temperature-altitude zones (see Figure 3.7 on page 117).

Note to reader: Red letter designations on photos (A–J) correspond to the red letters on the map, not with black climate classification letters on the figure legend (A–E).

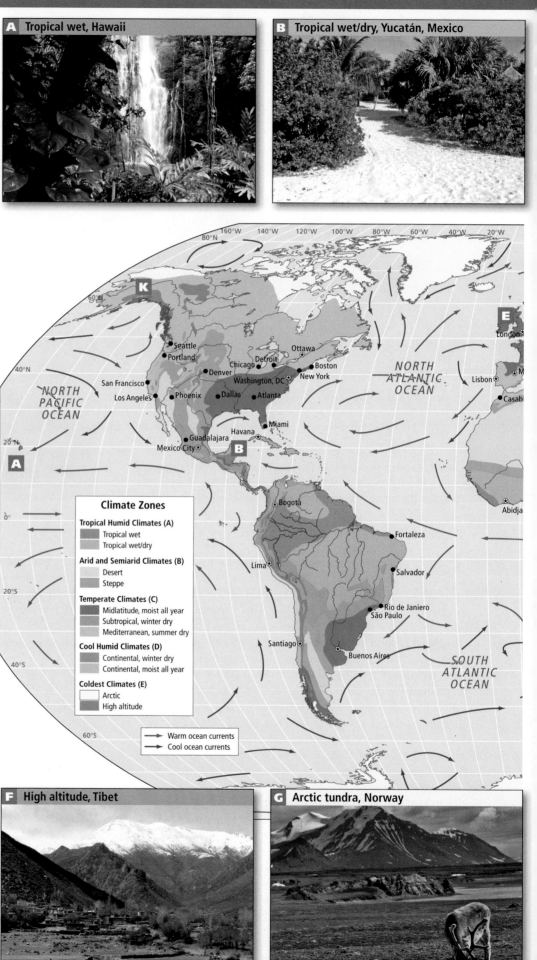

A Tropical wet, Hawaii

B Tropical wet/dry, Yucatán, Mexico

Climate Zones

Tropical Humid Climates (A)
Tropical wet
Tropical wet/dry

Arid and Semiarid Climates (B)
Desert
Steppe

Temperate Climates (C)
Midlatitude, moist all year
Subtropical, winter dry
Mediterranean, summer dry

Cool Humid Climates (D)
Continental, winter dry
Continental, moist all year

Coldest Climates (E)
Arctic
High altitude

Warm ocean currents
Cool ocean currents

F High altitude, Tibet

G Arctic tundra, Norway

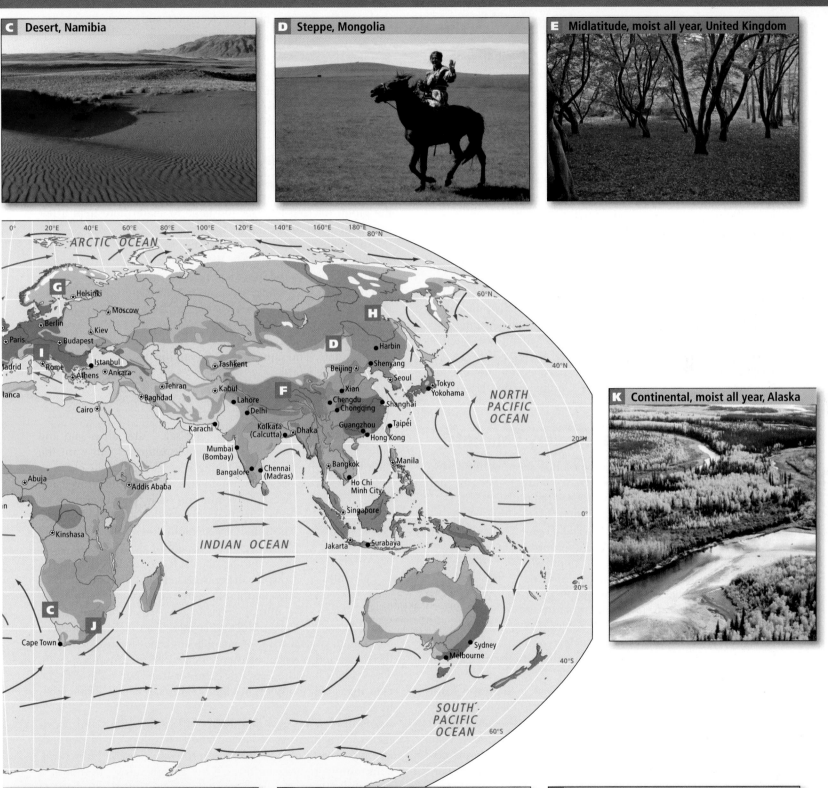

C Desert, Namibia

D Steppe, Mongolia

E Midlatitude, moist all year, United Kingdom

K Continental, moist all year, Alaska

H Continental, winter dry, Russia

I Mediterranean, summer dry, Italy

J Subtropical, winter dry, South Africa

1. Physical geography focuses on the processes that shape the earth's landforms and its climate, and on how human practices interact with physical patterns.

2. Internal processes, like plate tectonics, and external processes, such as weathering, shape landforms.

3. Climate is the long-term balance of temperature and precipitation that keeps weather patterns fairly consistent from year to year. Weather is the short-term and spatially limited expression of climate that can change in minutes.

HUMAN AND CULTURAL GEOGRAPHY PERSPECTIVES

Human geographers are interested in the economic, social, and cultural practices of a people, and in the spatial patterns these factors create. An important component of human geography is cultural geography, which focuses on culture as an important distinguishing characteristic of human societies. It comprises everything people use to live on earth that is not directly part of biological inheritance. **Culture** is represented by the ideas, materials, methods, and social arrangements that people have invented and passed on to subsequent generations, such as methods of producing food and shelter. Culture includes language, music, gender roles, belief systems, and moral codes (for example, those prescribed in Confucianism, Islam, and Christianity).

Ethnicity and Culture: Slippery Concepts

A group of people who share a location, a set of beliefs, a way of life, a technology, and usually a common ancestry and sense of common history form an **ethnic group**. The term *culture group* is often used interchangeably with ethnic group. The concepts of culture and ethnicity are imprecise, however, especially as they are popularly used. For instance, as part of the modern globalization process, migrating people often move well beyond their customary cultural or ethnic boundaries to cities or even distant countries. In these new places they take on many new ways of life and beliefs, yet they still may identify with their culture of origin.

For example, long before the U.S. war in Iraq, the Kurds in Southwest Asia were asserting their right to create their own country in the territory where they traditionally lived as nomadic herders. (Syria, Iraq, Iran, and Turkey now claim this area). Many Kurds who actively support the cause of the herders are now urban dwellers, living and working in modern settings in Turkey, Iraq, or even London. Although these people think of themselves as ethnic Kurds and are so regarded in the larger society, they do not follow the traditional Kurdish way of life (**Figure 1.24**). Hence, we could argue that these urban Kurds have a new identity within the Kurdish culture or ethnic group.

Another problem with the concept of culture is that it is often applied to a very large group that shares only the most general of characteristics. For example, one often hears the terms American culture, African-American culture, or Asian culture. In each case,

> **culture** all the ideas, materials, and institutions that people have invented to use to live on earth that are not directly part of our biological inheritance
>
> **ethnic group** a group of people who share a set of beliefs, a way of life, a technology, and usually a geographic location

the group referred to is far too large to share more than a few broad characteristics.

It might fairly be said, for example, that U.S. culture is characterized by beliefs that promote individual rights, autonomy, and individual responsibility. But when we look at specifics, just what constitutes the rights and responsibilities of the individual are quite debatable. In fact, American culture encompasses many subcultures that share some of the core set of beliefs, but disagree over parts of the core and over a host of other matters. The same is true, in varying degrees, for all other regions of the world.

Values

Occasionally you will hear someone say, "After all is said and done, people are all alike," or "People ultimately all want the same thing." It is a heartwarming sentiment, but an oversimplification. True, we all want food, shelter, health, love, and acceptance; but culturally, people are not all alike, and that is one of the qualities that make the study of geography interesting. We would be wise not to expect or even to want other people to be like us. It is often more fruitful to look for the reasons behind differences among people than to search hungrily for similarities. Cultural diversity has helped humans to be successful and adaptable animals. The various cultures serve as a bank of possible strategies for responding to the social and physical challenges faced by the human species. The reasons for differences in behavior from one culture to the next are usually complex, but they are often related to differences in values.

Consider this example that contrasts the values and *norms* (accepted patterns of behavior based on values) held by modern and urban individualistic cultures with those held by rural, community-oriented cultures. One recent rainy afternoon, a beautiful 40-something Asian woman walked alone down a fashionable street in Honolulu, Hawaii. She wore high-heeled sandals, a flared skirt that showed off her long legs, and a cropped blouse that allowed a glimpse of her slim waistline. She carried a laptop case and a large fashionable handbag. Her long, shiny black hair was tied back. Everyone noticed and admired her because she exemplified an ideal Honolulu businesswoman: beautiful, self-assured, and rich enough to keep herself well-dressed.

(A)

(B)

FIGURE 1.24 What does it mean to be Kurdish? (A) In rural areas away from the war zone in Iraq, being Kurdish often means leading a fairly peaceful, agriculturally based life. Here a young Kurdish woman in Semalka, northern Iraq, brings home fuel for cooking. **(B)** In urban areas, Kurdish identity tends to be more politicized. Here Kurds in Istanbul, Turkey, celebrate their new year holiday of Newroz. According to Kurdish legend, the day marks a deliverance of the Kurds from a tyrant. Now it is a day when Kurds show support for their main political party, the DTP or "Democratic Society Party," which works for greater recognition of the Kurdish language and Kurdish autonomy in southeastern Turkey.

Thinking Geographically: (A) What suggests that this photo depicts a semi-rural area? **(B)** What suggests that this photo was taken in an urban area?

In the village of this woman's grandmother—whether it be in Japan, Korea, Taiwan, or rural Hawaii—the dress that exposed her body to open assessment and admiration by strangers of both sexes would signal that she lacked modesty. The fact that she walked alone down a public street—unaccompanied by her father, husband, or female relatives—might even indicate that she was not a respectable woman. Thus a particular behavior may be admired when judged by one set of values and norms, but may be considered questionable or even disreputable when judged by another.

If culture groups have different sets of values and standards, does that mean that there are no overarching human values or standards? This question increasingly worries geographers, who try to be sensitive both to the particularities of place and to larger issues of human rights. Those who lean too far toward appreciating difference could end up tacitly accepting inhumane behavior, such as the oppression of minorities or violence against women.

Religion and Belief Systems

The religions of the world are formal and informal institutions that embody value systems. Most have roots deep in history, and many include a spiritual belief in a higher power (such as God, Yahweh, or Allah) as the underpinning for their value systems. Today religions often focus on reinterpreting age-old values for the modern world. Some formal religious institutions—such as Islam, Buddhism, and Christianity—proselytize; that is, they try to extend their influence by seeking converts. Others, such as Judaism and Hinduism, accept converts only reluctantly. Informal religions, often called belief systems, have no formal central doctrine and no firm policy on who may or may not be a practitioner.

Religious beliefs are often reflected in the landscape. For example, settlement patterns often demonstrate the central role of religion in community life: village buildings may be grouped around a mosque or synagogue, or an urban neighborhood may be organized around a Catholic church. In some places, religious rivalry is a major feature of the landscape. Certain spaces may be clearly delineated for the use of one group or another, as in Northern Ireland's Protestant and Catholic neighborhoods.

Religion has also been used to wield power. For example, during the era of European colonization, religion was a way to impose a change of attitude on conquered people. And the influence lingers. Photo Essay 1.6 on page 52 shows the distribution of the major religious traditions on earth today; it demonstrates some of the religious consequences of colonization. Note, for instance, the distribution of Roman Catholicism in the parts of the Americas, Africa, and Southeast Asia, all places colonized by European Catholic countries.

Religion can also spread through trade contacts. In the seventh and eighth centuries, Islamic people used a combination of trade and political power (and less often, actual conquest) to extend their influence across North Africa, throughout Central Asia, and eventually into South and Southeast Asia.

Language

Language is one of the most important criteria used in delineating cultural regions. The modern global pattern of languages reflects the complexities of human interaction and isolation over several hundred thousand years. Between 2500 and 3500 languages are spoken on earth today, some by only a few dozen people in

The small symbols on the map indicate a localized concentration of a particular religion within an area where another religion is predominant.

A Islam, Egypt

B Hinduism, India

C Indigenous religion, New Guinea

Predominant Religions and Belief Systems

Buddhism	Indigenous religions	Protestantism	Mixed Christian	▲ Roman Catholicism	■ Shintoism
Hinduism	Roman Catholicism	Sunni Islam	Mormon	● Protestantism	◆ Sikhism
Confucianism	Orthodox and other Eastern churches	Shi`ite Islam	No listing	✴ Judaism	

D Indigenous religion, Mexico

E Christianity, Ethiopia

F Buddhism, Tibet

isolated places. Many languages have several *dialects*—regional variations in grammar, pronunciation, and vocabulary.

The geographic pattern of languages has continually shifted over time as people have interacted through trade and migration. The pattern changed most dramatically around 1500, when the languages of European colonists began to replace the languages of people they conquered. For this reason, English, Spanish, Portuguese, or French are spoken in large patches of the Americas, Africa, Asia, and Oceania. Today, with increasing trade and instantaneous global communication, a few languages have become dominant. English is now the most important language of international trade, but Arabic, Spanish, Chinese, Hindi, and French are also widely used. At the same time, other languages are becoming extinct because children no longer learn them.

▐▌▶ 250. NEW DOCUMENTARY FILM TRACKS LANGUAGES

Race

Like ideas about gender roles, ideas about race affect human relationships everywhere on earth. However, according to the science of biology, all people now alive on earth are members of one species, *Homo sapiens sapiens*. Biologically, race is a meaningless concept; the characteristics we popularly identify as **race** markers—skin color, hair texture, face, and body shapes—have no significance as biological categories. For any supposed *racial trait*, such as skin color, there are wide variations within human groups. In addition, many invisible biological characteristics, such as blood type and DNA patterns, cut across skin color distributions and other so-called *racial attributes* and are shared across what are commonly viewed as different races. In fact, over the last several thousand years there has been such massive gene flow among human populations that no modern group presents a discrete set of biological characteristics. Although we may look

> **race** a social or political construct that is based on apparent characteristics such as skin color, hair texture, and face and body shape, but that is of no biological significance

quite different, from the biological point of view we are all closely related.

It is likely that some of the easily visible features of particular human groups evolved to help them adapt to environmental conditions. For example, biologists have shown that darker skin evolved in regions close to the equator, where sunlight is most intense (Figure 1.25). All humans need the nutrient vitamin D, and sunlight striking the skin helps the body absorb vitamin D. Too much of the vitamin, however, can result in improper kidney functioning. Dark skin absorbs less vitamin D than light skin and thus would be a protective adaptation in equatorial zones. In higher latitudes, where the sun's rays are more dispersed, light skin facilitates the sufficient absorption of vitamin D. Similar correlations have been observed between skin color, sunlight, and another essential vitamin, folate, which if deficient can result in birth defects.

Over time, race has acquired enormous social and political significance as humans from different parts of the world have encountered each other in situations of unequal power. *Racism*—the belief that genetic factors, usually visually apparent ones such as skin color, are a primary determinant of human abilities, has often been invoked to justify the enslavement of particular groups, or confiscation of their land and resources. Race and its implications in North America will be covered in Chapter 2, and the topic will be discussed in several other world regions as well.

Recognizing all the ills that have emerged from racism and similar prejudices, we need not infer that human history has been marked primarily by conflict and exploitation. Actually, humans have probably been so successful because of a strong inclination toward *altruism*, the willingness to sacrifice one's own well-being for the sake of others. It is probably our capacity for altruism that causes us such deep distress over the relatively infrequent occurrences of inhumane behavior.

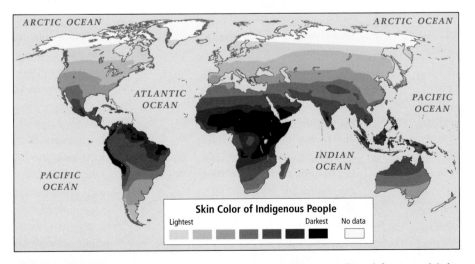

FIGURE 1.25 Skin color map for indigenous people as predicted from multiple environmental factors. Skin plays a twin role with respect to the sun: protection from excessive UV radiation and absorption of enough sunlight to trigger the production of vitamin D.

1. Human geographers are interested in the economic, social, and cultural practices of a people and in the spatial patterns these factors create.

2. While race is biologically meaningless, it has acquired enormous social and political significance.

Learning Goals Review

1. Regions: What is a region?

How might you understand the place where you live as part of a region? What combination of distinct physical and/or human features describes this region?

2. Gender and Population: How is the shift toward greater equality between the genders influencing population growth rates, patterns of economic development, and politics?

How do your plans for a career and a family compare to those of your grandparents and great-grandparents? Ask your parents who the first woman in your family was who had a career. Did this impact the number of children she had?

3. Urbanization and Food: How are changes in food production pushing people out of rural areas and pulling them into urban ones?

Would you say that the food you eat is more a product of "green revolution" agriculture or other food production systems, like organic agriculture? Do you grow any of your own food? How far back in your family history would you need to go to find ancestors who lived in a rural area and grew almost all their own food? Assuming you are not still in a rural area, what factors drew you and your family, or your ancestors, into the city?

4. Globalization and Development: How are increased global flows of information, goods, and people transforming patterns of economic development?

How is globalization evident in your life? Where were the clothes you wear made? From where do your favorite foods originate? How are your career plans influenced by globalization and the patterns of economic development it has brought to the place where you live or wish to live?

5. Democratization: What factors are related to the shift from authoritarian modes of government—based on the power of the state or community leaders—to more democratic systems in which individuals are given a greater say in how governments are run?

What aspects of democratization in your native country influence your day-to-day life? Are the factors necessary for the flourishing of democracy present? Are there peace, broad prosperity, education, and civil society? How do the political freedoms you enjoy shape the way you live?

6. Climate Change and Water: In what ways does water influence how different places are vulnerable to climate change?

How is the place where you live vulnerable to climate change? What water-related climate impacts are you exposed to? How are you sensitive to these impacts, and what kinds of resilience do you have that would enable you to "bounce back" from a disturbance?

Key Terms

agriculture 23
authoritarianism 33
biosphere 22
birth rate 14
capitalism 34
carrying capacity 24
cash economy 17
civil society 33
climate change 39
climate 45
communism 34
culture 50
death rate 14
delta 44
democratization 33
demographic transition 16
development 19
domestication 23
ecological footprint 22
erosion 24
ethnic cleansing 36

ethnic group 50
fair trade 32
female earned income as a percent
  of male earned income (F/MEI) 20
floodplain 44
food security 23
formal economy 20
free trade 31
gender 17
gender roles 18
genetic modification (GM) 24
genocide 36
Geographic Information Science (GISc) 7
geopolitics 34
global economy 28
global scale 7
global warming 39
globalization 27
green revolution 23
greenhouse gases 39
gross domestic product (GDP) 16

gross domestic product (GDP) per capita 19
gross domestic product (GDP)
  per capita PPP 20
human geography 44
human well-being 20
informal economy 20
interregional linkages 27
Kyoto Protocol 42
landforms 44
latitude 3
living wages 32
local scale 7
longitude 3
map projections 6
migration 15
monsoon 47
multinational corporation 30
nongovernmental organization (NGO) 36
orographic rainfall 46
physical geography 44
plate tectonics 44

political ecologist 22

population pyramid 15

purchasing power parity (PPP) 20

push/pull phenomenon of urbanization 27

race 53

rate of natural increase (RNI) 14

region 7

Ring of Fire 44

scale (of a map) 2

sex 17

slum 27

spatial distribution 3

spatial interaction 3

subsistence economy 16

sustainable agriculture 26

sustainable development 22

total fertility rate (TFR) 15

United Nations (UN) 36

United Nations Human Development Index (HDI) 20

urbanization 26

virtual water 37

water footprint 37

weathering 44

world region 7

World Trade Organization (WTO) 32

2 North America

Learning Goals

After you read this chapter, you will be able to answer the following questions:

1. Climate Change and Urbanization: How is North America's large per capita production of greenhouse gases related to its dominant pattern of urbanization?

2. Water and Food: How have North American food production systems contributed to water pollution?

3. Democratization: How has the ideal of democracy influenced North America's involvement in the affairs of foreign countries?

4. Globalization and Development: How has globalization transformed economic development in North America, thereby changing the kinds of jobs available in this region?

5. Population and Gender: What changes have contributed to the aging of North America's population?

FIGURE 2.1 Political map of North America.

Thematic Overview of North America

Climate Change: North America produces 25 percent of the world's greenhouse gases with only 5 percent of the world's population. **A** *A natural gas power plant in California.* ▼

Urbanization: Since World War II, North America's urban populations have increased by 150%, but the amount of land they occupy has increased by 300 percent. **B** *A suburb outside Edmonton, Alberta, Canada.* ▼

Water: Water pollution is a major problem, especially in the U.S. where 40 percent of the rivers are too polluted for fishing or swimming. **C** *Oil pollution on a river near Boston, Massachusetts.* ▼

Food: Most North American farms need few workers but require huge investments in land, machinery, fertilizers, and pesticides to be profitable. **D** *Pesticides being sprayed on lettuce in Arizona.* ▼

Democracy: While Canada and the U.S. have similar democratic systems of governments, they take different approaches to major issues such as how to best provide health care. **E** *A Canadian receives publicly funded health care.* ▼

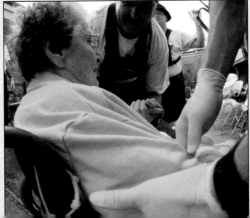

Globalization: North America is a major promoter of globalization and free trade throughout the world. **F** *Security guards outside a Kentucky Fried Chicken stand in China.* ▼

Development: North America's economy is shifting toward more knowledge-intensive industries that require education and specialized training. **G** *Programmers design social networking software in San Francisco.* ▼

Population: As more of North America's population reaches retirement age (65 and over), there are fewer young people to take care of them. **H** *An elderly woman is evacuated from a nursing home during Hurricane Katrina.* ▼

Gender: North American women are getting closer to achieving equal pay with men. In the 1940s women earned around 57 percent of what men did. Now they earn 80 percent. **I** *A factory worker in 1940s Texas.* ▼

Global Patterns, Local Lives Javier Aguilar, a 39-year-old father of three, has worked as an agricultural laborer in California's Central Valley for 20 years. He and hundreds of thousands like him tend the fields of crops (**Figure 2.2**) that feed the nation, especially during the winter months. Aguilar used to make $8.00 an hour, but now he is unemployed and standing in a church-sponsored food line. "If I don't work, [we] don't live. And all the work is gone here," he says, grimly.

In Mendota, also in California's Central Valley, young Hispanic men wait on street corners to catch a van to the fields. None come. In the winter of 2009, the unemployment rate in Mendota was 35 percent and rising. Mayor Robert Silva says his community is dying on the vine, and he sees the trouble spreading. Many small businesses are closing. Silva worries about the drug use, alcohol abuse, family violence, and malnutrition that can accompany severe unemployment in any community.

This level of unemployment in the heart of the nation's biggest producer of fruit and vegetables is partly due to drought and partly to a global economic recession. The drought is related to natural dry cycles as well as global climate change, which is worsening drought conditions globally. Meanwhile, the global economic recession has reduced overall demand for many of the fresh fruits and vegetables that California produces, as families turn to cheaper foods. All of these stresses have forced farmers to remove from production as much as 1 million of the 4.7 million acres once cultivated and irrigated in the Central Valley. Eventually, as many as 80,000 jobs and as much as $2.2 billion may be lost in California agriculture and related industries.

The confluence of troubles in California's Central Valley has been particularly devastating for low-wage Hispanic male agricultural workers like Javier Aguilar, because they have so little to fall back on in terms of savings, education, or skills. Hispanic women are more likely to remain employed because they are in domestic and caregiving jobs and have somewhat higher levels of education. One potential silver lining in this tale is the fact that economic hardship often encourages unskilled workers like Aguilar to take advantage of government-sponsored adult education programs. His children may

FIGURE 2.2 Grape harvest in California. Soter Mineral Springs Vineyard harvest in 2008.

Thinking Geographically: How might grape production for wine decline during a recession?

also have a greater incentive to graduate from high school. During the Great Depression of the 1930s, similar programs encouraged people to go back to school, causing the nation's high school graduation rate to jump from 20 percent to 60 percent. *[Adapted from: Jesse McKinley, "Drought Adds to Hardships in California," New York Times, February 21, 2009; David Leonhardt, "Job Losses Show Breadth of Recession," New York Times, March 3, 2009; and Catherine Rampell, "As Layoffs Surge, Women May Pass Men in Job Force," New York Times, February 5, 2009.]* ∎

THINGS TO REMEMBER

1. The story of Javier Aguilar in the opening vignette illustrates the complexity of several thematic concepts—water, food, development, gender, and global warming. Reviewing how these concepts are linked here will help you throughout the book.

I THE GEOGRAPHIC SETTING

Terms in This Chapter

The world region discussed in this chapter consists of Canada and the United States. The term *North America* is used to refer to both countries. Other terms relate to the growing cultural diversity in this region. The text uses the term **Hispanic** to refer to all Spanish-speaking people from Middle and South America, although their ancestors may have been African, European, Asian, or Native American.

> **Hispanic** term used to refer to all Spanish-speaking people from Middle and South America, although their ancestors may have been black, white, Asian, or Native American

Physical Patterns

The continent of North America is a huge expanse of mountain peaks, ridges and valleys, expansive plains, long winding rivers, myriad lakes, and extraordinarily long coastlines. Here the focus is on a few of the most significant landforms.

Landforms

A wide mass of mountains and basins, known as the Rocky Mountain zone, dominates western North America (see Figure 2.3D on page 60). It stretches down from the Bering Strait in the far north, through Alaska, and into Mexico. This zone formed about 200 million years ago when, as part of the breakup of the supercontinent Pangaea (see Figure 1.21 on page 45), the Pacific Plate pushed against the North American Plate, thrusting up mountains. These plates still rub against each other, causing earthquakes along the Pacific coast of North America.

The much older, and hence more eroded, Appalachian Mountains stretch along the eastern edge of North America from New Brunswick and Maine to Georgia. This range resulted from very ancient collisions between the North American Plate and the African Plate.

Between these two mountain ranges lies the huge central lowland of undulating plains that stretches from the Arctic to the Gulf of Mexico. This landform was created by the deposition of deep layers of material eroded from the mountains and carried to this central North American region by wind and rain and by the rivers flowing east and west into what is now the Mississippi drainage basin.

> **aquifers** natural underground reservoirs of water

During periodic ice ages over the last 2 million years, glaciers have covered the northern portion of North America. In the most recent ice age (between 25,000 and 10,000 years ago), the glaciers, sometimes as much as 2 miles (about 3 kilometers) thick, moved south from the Arctic, picking up soil and scouring depressions in the land surface. When the glaciers later melted, these depressions filled with water, forming the Great Lakes. Thousands of smaller lakes, ponds, and wetlands that stretch from Minnesota and Manitoba to the Atlantic were formed in the same way. Melting glaciers also dumped huge quantities of soil throughout the central United States. This soil, often many meters deep, provides the basis for large-scale agriculture, but remains susceptible to wind and water erosion.

East of the Appalachians, the coastal lowland stretches from New Brunswick to Florida. This lowland then sweeps west to the southern reaches of the central lowland along the Gulf of Mexico. In Louisiana and Mississippi, much of this lowland is filled in by the Mississippi river delta—a low, flat, swampy transition zone between land and sea. The delta was formed by massive loads of silt deposited during floods over the past 150-plus million years by the Mississippi, North America's largest river system. The delta deposit originally began at what is now the junction of the Mississippi and Ohio rivers at Cairo, Illinois; slowly, as ever more sediment was deposited, the delta advanced 1000 miles (1600 kilometers) into the Gulf of Mexico.

Over the centuries, human activities such as deforestation, deep plowing, and heavy grazing have led to erosion and added to the silt load of the rivers (**Figure 2.4** on page 62). At the same time, the construction of levees along riverbanks has drastically reduced flooding. Because of this flood control, much of the silt that used to be spread widely across the lowlands during floods is being carried to the southern part of the Mississippi delta—a low, flat zone characterized by swamps, lagoons, and sandbars. The silt is destroying wetlands and the extra weight is causing the delta to sink into the Gulf of Mexico as the silt also extends further out into deep waters.

Climate

The landforms over this continental expanse influence the movement and interaction of air masses and contribute to its enormous climate variety (**Photo Essay 2.1** on page 63). Along the southern west coast of North America, the climate is generally mild (Mediterranean)—dry and warm in summer, cool and moist in winter. North of San Francisco, the coast receives moderate to heavy rainfall. East of the Pacific coastal mountains, climates are much drier because as the moist air sinks into the warmer interior lowlands, it tends to hold its moisture (see Figure 1.23

on page 47). This interior region becomes increasingly arid as it stretches east across the Great Basin (see Photo Essay 2.1C) and Rocky Mountains. Many dams and reservoirs for irrigation projects have been built to make agriculture and urbanization possible. Because of the low level of rainfall, however, efforts to extract water for agriculture and urban settlements are exceeding the capacity of ancient underground water basins (**aquifers**) to replenish themselves.

On the eastern side of the Rocky Mountains, the main source of moisture is the Gulf of Mexico. When the continent is warming in the spring and summer, the air masses above it rise, sucking in warm, moist, buoyant air from the Gulf. This air interacts with cooler, drier, heavier air masses moving into the central lowland from the north (Photo Essay 2.1A) and west, often creating violent thunderstorms and tornadoes. Generally, central North America is wettest in the eastern (Photo Essay 2.1B) and southern parts and driest in the north and west (Photo Essay 2.1C). Along the Atlantic coast, moisture is supplied by warm, wet air above the Gulf Stream—a warm ocean current that flows north from Florida, and follows the coastline of the eastern United States and Canada before crossing the Atlantic Ocean.

The large size of the North American continent creates wide temperature variations. Because land heats up and cools off more rapidly than water, temperatures in the continental interior are hotter in the summer and colder in the winter than in coastal areas where temperatures are moderated by the oceans.

THINGS TO REMEMBER

1. North America has two main mountain ranges, the Rockies and the Appalachian, separated by long, winding rivers and expansive plains.

2. The size and variety of landforms influence the movement and interaction of air masses, creating enormous climatic variation. Because land heats up and cools off more rapidly than water, temperatures in the continental interior are higher in the summer and colder in the winter than in coastal areas.

Environmental Issues

North America's wide range of resources and seemingly limitless stretches of forest and grasslands long diverted attention from the environmental impacts of settlement and development. Increasingly, however, it is impossible to ignore the many environmental consequences of the North American lifestyle. This section focuses on a few of those consequences: habitat loss, climate change and air pollution, depletion and pollution of water resources and fisheries, and the accumulation of hazardous waste.

Loss of Habitat for Plants and Animals

Before the European colonization of North America, which began soon after 1500, the environmental impact of humans in the region was relatively low. Though North America was by no

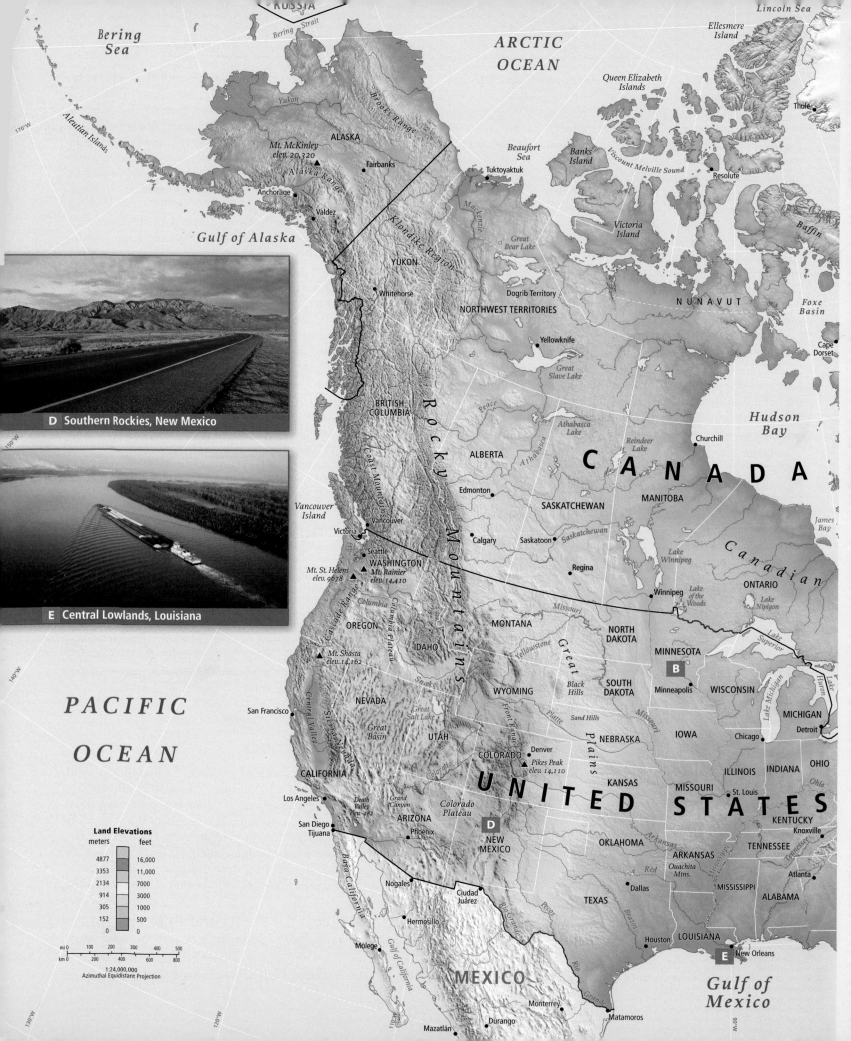

Bering Sea

RUSSIA

Bering Strait

Yukon

ARCTIC OCEAN

Lincoln Sea

Ellesmere Island

Brooks Range

ALASKA

▲ Mt. McKinley
elev. 20,320
Alaska Range

Fairbanks

Anchorage

Valdez

Beaufort Sea

Tuktoyaktuk

Queen Elizabeth Islands

Thule

Resolute

Gulf of Alaska

Klondike Region

Mackenzie

Great Bear Lake

Banks Island

Viscount Melville Sound

Victoria Island

Baffin

Aleutian Islands

170°W

YUKON

Whitehorse

Dogrib Territory

NORTHWEST TERRITORIES

Great Slave Lake

Yellowknife

N U N A V U T

Foxe Basin

Cape Dorset

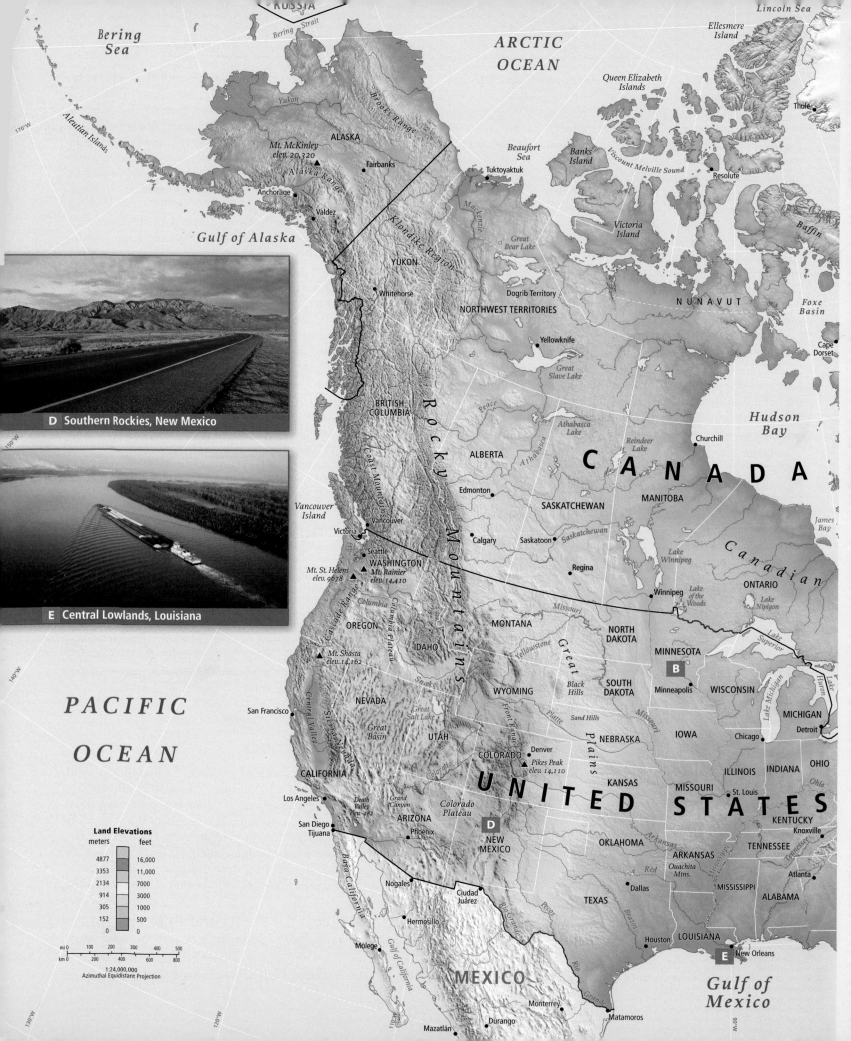

D Southern Rockies, New Mexico

BRITISH COLUMBIA

R o c k y

Peace

Athabasca

Athabasca Lake

Reindeer Lake

ALBERTA

Edmonton

C A N A D A

Churchill

Hudson Bay

150°W

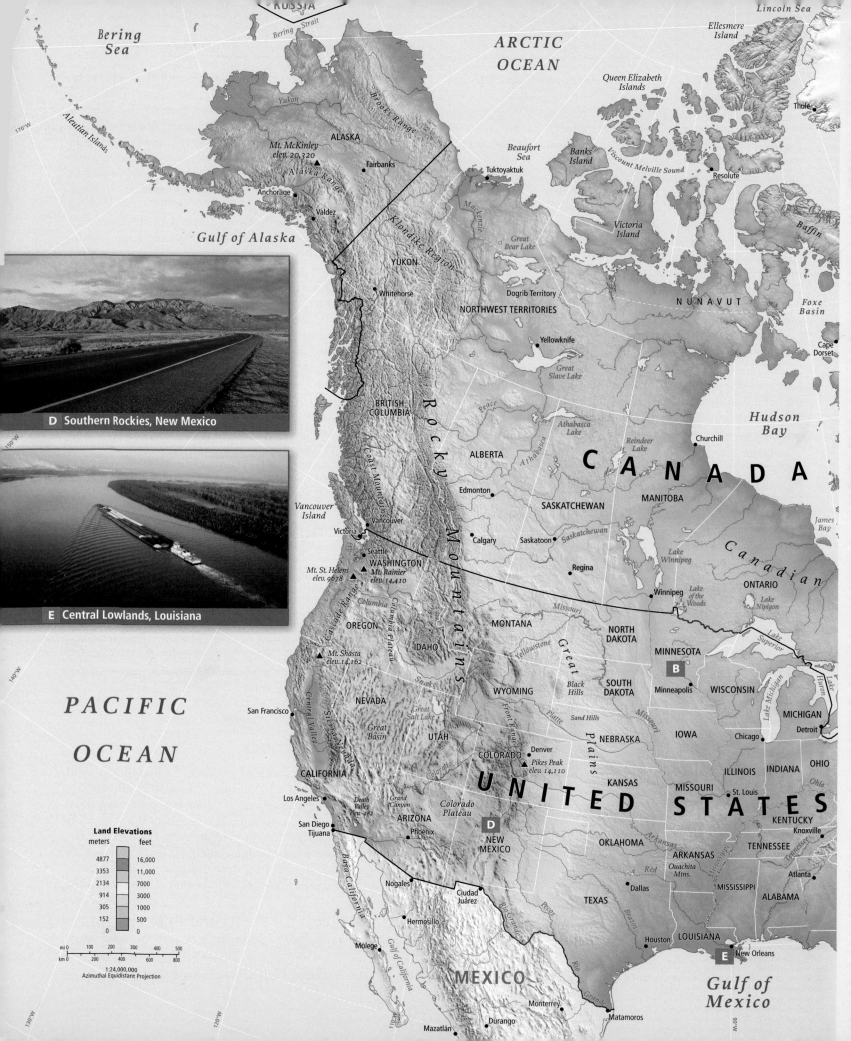

E Central Lowlands, Louisiana

Vancouver Island

Victoria

Vancouver

Seattle

WASHINGTON

Coast Mountains

M o u n t a i n s

SASKATCHEWAN

Calgary

Saskatoon

Saskatchewan

Regina

MANITOBA

Lake Winnipeg

James Bay

C a n a d i a n

ONTARIO

140°W

▲ Mt. St. Helens
elev. 9678

▲ Mt. Rainier
elev. 14,410

OREGON

Cascade Range

Columbia

Columbia Plateau

IDAHO

Snake

MONTANA

Yellowstone

Missouri

G r e a t

Winnipeg

Lake of the Woods

Lake Nipigon

Lake Superior

Lake Huron

PACIFIC OCEAN

▲ Mt. Shasta
elev. 14,162

San Francisco

Central Valley

Sierra Nevada

NEVADA

Great Basin

Great Salt Lake

UTAH

WYOMING

Black Hills

SOUTH DAKOTA

Front Range

Platte

Sand Hills

Missouri

P l a i n s

Minneapolis

NEBRASKA

IOWA

MINNESOTA

WISCONSIN

MICHIGAN

Lake Michigan

Chicago

Detroit

130°W

Los Angeles

*Death Valley
elev. −282*

Grand Canyon

Colorado Plateau

COLORADO

Denver

▲ Pikes Peak
elev. 14,110

KANSAS

U N I T E D

Arkansas

MISSOURI

St. Louis

ILLINOIS

INDIANA

OHIO

Ohio

CALIFORNIA

San Diego

Tijuana

ARIZONA

Phoenix

D

NEW MEXICO

Colorado

S T A T E S

OKLAHOMA

ARKANSAS

Ouachita Mtns.

Red

Mississippi

KENTUCKY

Knoxville

TENNESSEE

Tennessee

Land Elevations

meters		feet
4877		16,000
3353		11,000
2134		7000
914		3000
305		1000
152		500
0		0

Baja California

Nogales

Ciudad Juárez

Hermosillo

TEXAS

Dallas

Brazos

Pecos

MISSISSIPPI

ALABAMA

Atlanta

mi 0   100   200   300   400   500
km 0   200   400   600   800
1:24,000,000
Azimuthal Equidistant Projection

Mulegé

Gulf of California

MEXICO

Rio Grande

Houston

LOUISIANA

E

New Orleans

Gulf of Mexico

Monterrey

Durango

Mazatlán

Matamoros

A Appalachian Mountains, Virginia

B Central Lowlands, Minnesota

C Coastal Lowlands, Charleston, South Carolina

FIGURE 2.3 Regional map of North America.

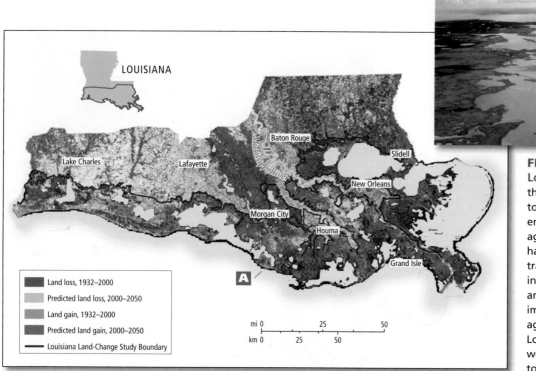

FIGURE 2.4 Wetland loss in Louisiana. The Louisiana coastline and the lower Mississippi River basin are vital to the nation's interests. They are the end point for the vast Mississippi drainage basin. They provide coastal wildlife habitats, recreational opportunities, transportation lanes that connect the vast interior of the country with the ocean, and access to offshore oil and gas. Most important, the wetlands provide a buffer against damage from hurricanes. However, Louisiana has lost one-quarter of its total wetlands over the last century, largely due to human impacts on natural systems. The remaining 3.67 million acres constitute 14 percent of the total wetland area in the lower 48 states. **(A)** A section of vanishing wetland in Terrebonne Parish, Louisiana. Map: USGS/National Wetlands Research Center.

means a pristine paradise when Europeans arrived, subsequently millions of acres of forests and grasslands that had served as habitats for native plants and animals were cleared to make way for European-style farms, cities, and industries (Photo Essay 2.2 on page 64). This was particularly true in the area that became the United States.

Oil Drilling

In many coastal and interior areas of North America, oil drilling is a large and potentially environmentally devastating industry. This often-overlooked reality was made clear in the spring and summer of 2010, when U.S. waters in the Gulf of Mexico became the site of the largest accidental marine oil spill in history. In April of 2010, an explosion aboard the Deepwater Horizon, an offshore oil-drilling rig, caused it to sink. One result was a massive leak from the rig's wellhead, located on the seafloor. Due to its location more than a mile beneath the sea surface, the wellhead could not be capped for almost 4 months, during which time it spewed out at least 200 million gallons of oil. While some of this oil made its way to the surface, where it has damaged shorelines in all the Gulf Coast states, most of the oil remains beneath the surface due to the use of chemical dispersants by BP, the company responsible for the oil spill. While the masses of oil that remain below the surface are less damaging to beaches and coastal marshes, there is mounting concern that this oil poses an ongoing threat to the Gulf's many fish, shrimp, and other aquatic species. Moreover, the dispersants used by BP are a health concern because of their high toxicity, which has already led to their being banned in Europe. Controversy surrounded the U.S.

government's reopening of many Gulf fisheries in the summer of 2010, and independent studies finding that both oil and dispersants in Gulf seafood have clashed with government reports. Clearly, the effects of the Deepwater Horizon spill will be felt for years to come (see Photo Essay 2.2A).

Elsewhere oil-drilling operations also have a dramatic effect on the environment. Along the northern coast of Alaska, the Trans-Alaska Pipeline runs southward for 800 miles to the port of Valdez. Often running above ground to avoid shifting as the earth freezes and thaws, the pipeline poses a constant risk of rupture, which potentially could result in devastating oil spills. Moreover, the pipeline interferes with migrations of caribou and other animals that Alaska's indigenous people have depended on for food in the past. However, today all Alaskan citizens receive a yearly rebate of several thousand dollars from oil revenues, which tends to quiet protests about threats to the environment from oil extraction.

Logging

Though widespread forest clearing for agriculture is now rare, logging occurs throughout North America, where it remains especially important along the northern Pacific coast and in the southeastern United States. Logging in these areas provides most of the construction lumber and an increasing amount of the paper used in Canada, the United States, and increasingly, in parts of Asia.

Wet Pacific air blows into coastal mountains, bringing rain and moderate temperatures. Continental effect makes winters colder and summers hotter.

Climate Zones

Tropical humid climates (A)
- Tropical wet/dry

Arid and semiarid climates (B)
- Steppe
- Desert

Temperate climates (C)
- Midlatitude, moist all year
- Subtropical, winter dry
- Mediterranean, summer dry

Cool humid climates (D)
- Continental, moist all year

Coldest climates (E)
- Arctic
- High altitude

→ Winds
→ Ocean currents

Wet Pacific air blows into coastal mountains, bringing rain and moderate temperatures.

Continental effect makes winters colder and summers hotter.

Gulf Stream current brings warm tropical water, warming the air over the Atlantic coast.

Moist Gulf of Mexico air brings rain and moderate temperatures.

mi 0 250 500
km 0 250 500

A Arctic, Alaska

B Temperate, midlatitude, Maryland

C Desert, Utah

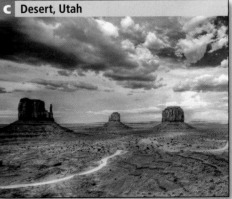

While parts of North America remain relatively less impacted by humans, much of this region has seen low-to-medium impacts, and the parts where most people live are highly impacted.

Photo (**A**) shows an old-growth cedar in the Olympic National Park. Clear-cutting, in which all trees on a plot of land are cut down (**B**), has become an increasingly controversial method of logging. Since 1971, over 30 percent of the forests on Washington State's Olympic peninsula (**C** and **D**) have been clear-cut. Even more controversial is the logging of old-growth forests that have never been cut. This practice has been declining in recent years, but still continues in some places.

E In remote areas, such as Alaska, mining and oil industries have the largest impacts. Photo shows the Alaskan pipeline, which runs for 800 miles.

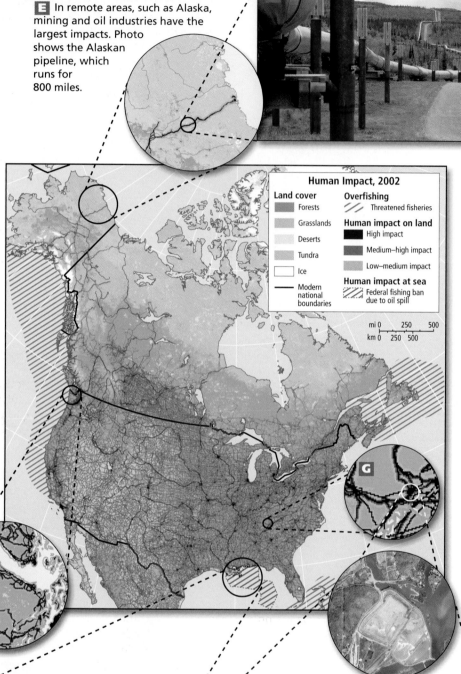

Human Impact, 2002

Land cover		Overfishing
	Forests	Threatened fisheries
	Grasslands	**Human impact on land**
	Deserts	High impact
	Tundra	Medium–high impact
	Ice	Low–medium impact
	Modern national boundaries	**Human impact at sea**
		Federal fishing ban due to oil spill

mi 0   250   500
km 0   250 500

In December of 2008 (**G–I**) a billon-and-a-half gallons of coal ash from a power plant spilled over 300 acres of river and shoreline near Kingston, Tennessee. A dike (**H**) that had been holding in the ash since the 1960s burst after a heavy rain, releasing heavy metals and radioctivity (a natural property of much coal) into the river. Thousands of fish were killed and drinking water supplies degraded. Coal generates about half of the electric power used in the U.S.

F A pelican is cleaned of oil after the sinking of the Deepwater Horizon oil rig in April of 2010 off the coast of Louisiana, resulted in the largest oil spill in U.S. history.

H Before

I After

Thinking Geographically

After you have read about the Human Impact on the Biosphere, you will be able to answer the following questions:

A Why might it make economic sense to keep old-growth trees, like the one shown here, standing?

E What are some threats to the environment posed by the Alaska oil pipeline?

F Even if this bird never again comes into contact with an oil slick, how might it still be endangered as a result of the Deepwater Horizon spill?

(G–I) Why is coal ash toxic?

Coal Mining

In many remote interior areas of North America, coal mining is also a large and environmentally damaging industry. Strip mining, where vast quantities of earth and rock are removed in order to extract underlying coal, can result in huge piles of mining waste called "tailings" that pollute waterways and threaten communities that depend on well water. Particularly damaging is a form of mining known as *mountaintop removal*, where the entire top of a mountain may be leveled and the tailings pushed into surrounding valleys, resulting in the pollution of whole watersheds. But the U.S. dependence on coal to generate electricity creates environmental concerns beyond the harm to landscapes and habitats done by strip mining. When coal is burned, finding a safe place to store the huge volumes of toxic ash that result is difficult. Left to dry out, the material becomes airborne and contributes to air pollution, so it is stored wet. In December 2008, a large earthen-walled pond of wet coal ash burst after a heavy rainstorm, spilling a billion-and-a-half gallons of toxic sludge (the largest industrial spill in U.S. history) over 300 acres of beautiful lakeshore and forestland in rural Tennessee (see Photo Essay 2.2 I). The sludge, containing heavy metals and harmful chemicals, ruined the ecology of the immediate area and threatened hundreds of square miles with water and air pollution. Cleanup will cost well over $1 billion.

However, as forests have been depleted, environmentalists have focused on the damage created by the logging industry. Much attention is focused on **clear-cutting**, the dominant logging method used throughout North America. In this method, all trees on a given plot of land are cut down, regardless of age, health, or species (Photo Essay 2.2 B, C). Clear-cutting destroys wild animal and plant habitats, thereby reducing species diversity; it leaves forest soils uncovered and highly susceptible to erosion. Concerns about the environmental impacts of logging are boosted by the dominance of service-sector jobs in the major logging states and provinces. For example, in the Pacific Northwest, even in many remote areas where logging was once the backbone of the economy, people are now dependent on tourism and other occupations that rely on the beauty of intact forest ecosystems (Photo Essay 2.2A).

clear-cutting the cutting down of all trees on a given plot of land, regardless of age, health, or species

urban sprawl the encroachment of suburbs on agricultural land

Urbanization and Habitat Loss

An important aspect of urbanization is **urban sprawl** (discussed more fully on pages 86–88). In many areas, and for several decades, middle- and upper-income urbanites have sought lower-density suburban neighborhoods. Farms that were once highly productive are giving way to expansive, low-density urban and suburban residential developments where pavement, golf courses, office complexes, and shopping centers cover the landscape. In the process, natural habitats are being degraded even more intensely than they were by farming. The loss of farmland and natural habitat in the urban fringe affects recreational land and our ability to produce local, affordable food for urban populations.

As North American plants and animals have been forced into ever smaller territories, many have died out entirely and been replaced by nonnative species (the domestic cat, for example) brought in by humans either inadvertently or for gardens and as pets. Estimates vary, but at least 4000 nonnative species have invaded North America. An example is the Asian snakehead fish, which is rapidly invading the Potomac River, where it eats baby bass and fiercely competes with native fish for food.

▐▌▶ 29. SNAKEHEAD REPORT

Climate Change and Air Pollution

Learning Goal 1
Climate Change and Urbanization: How is North America's large per capita production of greenhouse gases related to its dominant pattern of urbanization?

On a per capita basis, North Americans contribute more to climate change than any other people on the planet. With only 5 percent of the world's population, this region produces 26 percent of the greenhouse gases released globally by human activity. This large share can be traced to North America's high consumption of fossil fuels, which in turn is related to several factors. One of these is North America's dominant pattern of urbanization, characterized by vast and still-growing suburbs where people depend on their automobiles for almost all of their transportation needs (see Thematic Overview B on page 57). Spread-out, free-standing dwellings and businesses also require more energy to heat and cool than do the densely packed, high-rise buildings typical of cities in most other world regions (see Thematic Overview A). Industrial and agricultural production is also highly dependent on fossil fuels in North America.

Canada's government was one of the first in the world to commit to reducing the consumption of fossil fuels. Until recently, the United States resisted such moves, fearing damage to its economy. Both countries are now exploring alternative sources of energy, such as solar, wind, geothermal, and nuclear power. So far, neither country has been able to reduce its levels of greenhouse gas emissions, or even the rate at which these emissions are growing. However, both Canada and the United States possess the technological capabilities needed to lead the world in shifting over to cleaner sources of energy.

▐▌▶ 31. ENERGY REPORT
▐▌▶ 32. GREEN BUILDING

Vulnerability to Climate Change Both Canada and the United States are vulnerable to the effects of climate change. Dense population centers on the Gulf of Mexico and Atlantic coasts are highly exposed to hurricanes, which may become more violent as oceans warm (Photo Essay 2.3 B, D). Sea level rise and coastal erosion, due to the thermal expansion of the oceans, is already affecting many coastal areas along the Arctic coast of North America (Photo Essay 2.3A). Here at least 26 coastal villages are being forced to relocate inland at an estimated cost of $130 million per village. Meanwhile, many arid farming zones will dry further as higher temperatures reduce soil moisture, making irrigation crucial (Photo Essay 2.3C). In all of these areas, resilience to climate change is bolstered over the short term by excellent emergency response and recovery systems. Long-term resilience is boosted by careful planning as well as North America's large and diverse economy, which can provide alternative livelihoods to people highly exposed to climate impacts.

▐▐ ▶ 30. GLACIER NATIONAL PARK

Air Pollution In addition to climate change, most greenhouse gases contribute to various forms of air pollution, such as smog and acid rain. *Smog* is a combination of industrial emissions and car exhaust that frequently hovers as a yellow-brown haze over many North American cities, causing a variety of health problems. These same emissions also result in **acid rain**, which is created when pollutants dissolve in falling precipitation and make the rain acidic. Acid rain can kill trees and, when concentrated in lakes and streams, poison fish and wildlife.

The United States, with its large population and extensive industry, is responsible for the vast majority of acid rain in North America. Due to continental weather and wind patterns, however, the area most affected by acid rain encompasses a wide area on both sides of the eastern U.S.–Canada border (see Figure 2.5 map on page 68). The eastern half of the continent, which includes the entire eastern seaboard from the Gulf Coast to Newfoundland, is significantly affected by acid rain.

Water Resource Depletion and Pollution

People who live in the humid eastern part of North America find it difficult to believe that water is becoming scarce even there. Consider the case of Ipswich, Massachusetts, where the watershed is drying up as a result of overuse. There, innovators are saving precious water through conservation strategies in their homes and businesses (see the *Rainwater Cash* video). Elsewhere, as populations grow and per capita water usage increases, conflicts over water are more and more common.. In the Southeast, Atlanta, Georgia, with over 5 million residents, absorbs water so insatiably that downstream users in Alabama and Florida have sued Atlanta for depriving them of water.

▐▐ ▶ 34. RAINWATER CASH

Depletion In North America, water becomes increasingly precious the farther west one goes. On the Great Plains, rainfall is highly variable year to year, and in any given year may be too sparse to support healthy crops and animals. To make farming more secure and predictable, taxpayers across the continent have subsidized the building of pumps and stock tanks for farm animals and aqueducts and reservoirs for crop irrigation.

Irrigation is increasingly based on "fossil water" that has been stored over the millennia in natural underground reservoirs called *aquifers*. The *Ogallala aquifer* (Figure 2.6 on page 69) underlying the Great Plains is the largest in North America. In parts of the Ogallala, water is being pumped out at rates that exceed natural replenishment by 10 to 40 times.

Irrigation of the fruit and vegetable crops in California accounts for some of the water that goes into the *virtual water footprint* of U.S. consumers, as discussed in Chapter 1 (page 37 and Table 1.2). Billions of dollars of federal and state funds have paid for massive engineering projects that bring water from Washington, Oregon, Colorado, and northern California to farms in central and southern California. This water also supplies the cities of southern California, which are built on land that was once desert. Water is pumped from hundreds of miles away and over mountain ranges. California uses more energy to move water than some states use for all purposes. Moreover, irrigation in Southern California deprives Mexico of this much-needed resource. The mouth of the Colorado River (which is in Mexico) used to be navigable; now it is dry and sandy and only a mere trickle of water gets to Mexico.

Increasingly, citizens in western North America are recognizing that the use of scarce water for irrigated agriculture or for keeping lawns and golf courses green in desert environments is unsustainable. Conflicts over transporting water from wet regions to dry ones, or from sparsely inhabited to urban areas, are ongoing and have halted some new water projects. However, government subsidies have kept water artificially cheap, and in the past, new water supplies had always been found and harnessed. Hence, there is little incentive to change.

Pollution In the United States, 40 percent of rivers are too polluted for fishing and swimming, and over 90 percent of *riparian areas* (the interface between land and flowing surface water) have been lost or degraded. Pollution in the rivers of North America comes mainly as storm-water runoff from agricultural areas, urban and suburban developments, and industrial sites (see Thematic Overview C).

> **acid rain** falling precipitation that has formed through the interaction of rainwater or moisture in the air with sulfur dioxide and nitrogen oxides emitted during the burning of fossil fuels, making it acidic

> **Learning Goal 2**
> **Water and Food:** How have North American food production systems contributed to water pollution?

Thinking Geographically

After you have read about the Vulnerability to Climate Change in North America, you will be able to answer the following questions:

A What climate-change impact does this photo illustrate?

B What element of this community's emergency response system has clearly been compromised?

C How might irrigation reduce vulnerability to climate change, but also result in higher greenhouse gas emissions?

D Why are stronger hurricanes more likely as the climate warms?

North America's wealth and its well-developed emergency response systems give it high resilience that reduces its overall vulnerability. However, certain regions are highly exposed to temperature increases, drought, hurricanes, and sea level rise.

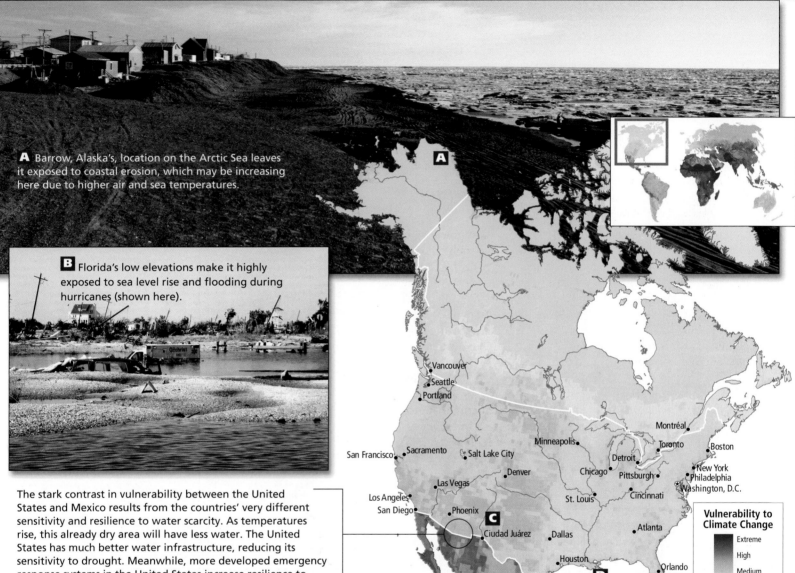

A Barrow, Alaska's, location on the Arctic Sea leaves it exposed to coastal erosion, which may be increasing here due to higher air and sea temperatures.

B Florida's low elevations make it highly exposed to sea level rise and flooding during hurricanes (shown here).

Vulnerability to Climate Change

Extreme
High
Medium
Low

The stark contrast in vulnerability between the United States and Mexico results from the countries' very different sensitivity and resilience to water scarcity. As temperatures rise, this already dry area will have less water. The United States has much better water infrastructure, reducing its sensitivity to drought. Meanwhile, more developed emergency response systems in the United States increase resilience to water shortage. With such drastic differences, the U.S.-Mexico border could become an even more contentious zone as the climate changes.

C Irrigation. Higher temperatures raise the rate at which plants lose water, increasing the need for irrigation. Farms that get their water from shrinking aquifers are thus less able to increase irrigation, and hence are highly sensitive to the higher temperatures and drought that climate change is bringing.

D Hurricanes gain strength with warmer temperatures. Many low-lying coastal cities are highly exposed and sensitive, though their resilience varies. Shown here are evacuees from Hurricane Katrina, an event which showed the weaknesses in local, state, and federal emergency response systems.

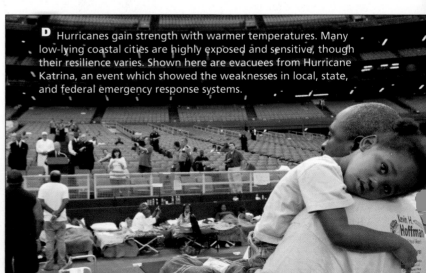

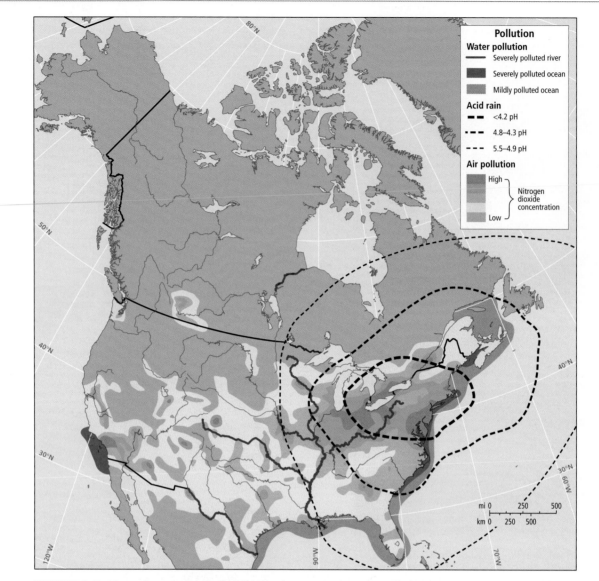

FIGURE 2.5 Air and water pollution in North America. This map shows two aspects of air pollution, as well as polluted rivers and coastal areas. Red and yellow indicate concentrations of nitrogen dioxide (NO$_2$), a toxic gas that comes primarily from the combustion of fossil fuels by motor vehicles and power plants. This gas interacts with rain to produce nitric acid, a major component of acid rain, as well as toxic organic nitrates that contribute to urban smog. The map also shows polluted coastlines (including all of the coastline from Texas to New Brunswick) as well as severely polluted rivers, which include much of the Mississippi River and its tributaries.

In the 1970s, scientists studying coastal areas began noticing *dead zones* where water is so polluted that it supports almost no life. Dead zones occur near the mouths of major river systems that have been polluted by fertilizers and pesticides washed from farms and lawns when it rains. A large dead zone is in the Gulf of Mexico near the mouth of the Mississippi, but similar zones have been found in all U.S. coastal areas. Even Canada, where much lower population density means that rivers are generally cleaner, has dead zones on its western coast.

Green Living at a Global Scale

As awareness of the environmental issues (some discussed above) has increased in North America, citizens often ask what they can do in daily life to ameliorate the situation. Simple solutions, such as

greener living—which involves recycling, driving less, and improving home energy efficiency—are important and can collectively make an impact. However, global climate change and other human impacts on the biosphere now take place on such a vast scale that collective action at every level of human interaction, from local to global, is more crucial than ever.

THINGS TO REMEMBER

1. **Learning Goal 1: Climate Change and Urbanization** North America produces 26 percent of the world's greenhouse gases, even though it has only 5 percent of the world's population. This is related to the region's pattern of urban development, as well as its fossil fuel–dependent industrial and agricultural production methods.

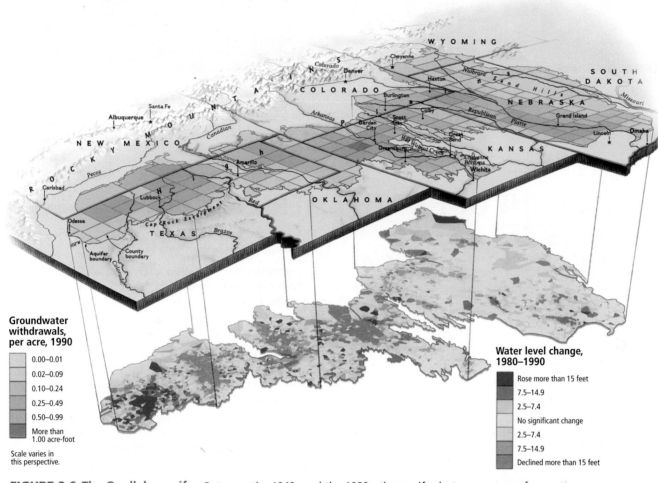

Groundwater withdrawals, per acre, 1990

- 0.00–0.01
- 0.02–0.09
- 0.10–0.24
- 0.25–0.49
- 0.50–0.99
- More than 1.00 acre-foot

Scale varies in this perspective.

Water level change, 1980–1990

- Rose more than 15 feet
- 7.5–14.9
- 2.5–7.4
- No significant change
- 2.5–7.4
- 7.5–14.9
- Declined more than 15 feet

FIGURE 2.6 The Ogallala aquifer. Between the 1940s and the 1980s, the aquifer lost an average of 10 feet (3 meters) of water overall, and more than 100 feet (30 meters) of water in some parts of Texas. During the 1980s, abundant rain and snow meant less decline in the aquifer. However, in this area the climate fluctuates from moderately moist to very dry. When a drought began in mid-1992 (continuing until late 1996), and large agribusiness firms pumped water for irrigation to supplement precipitation, water levels in the aquifer declined an average of 1.35 feet per year during the mid-1990s.

2. Water pollution is a major problem, especially in the United States, where 40 percent of the rivers are too polluted for fishing or swimming.

3. While many parts of North America are exposed to multiple climate-change impacts, this region is also highly resilient. Hence, overall vulnerability to climate change is generally low compared to other world regions.

Human Patterns over Time

The human history of North America is a series of arrivals and dispersals of people across the vast continent. In prehistoric times, humans came from Eurasia via Alaska. Beginning in the 1600s, waves of European immigrants, enslaved Africans, and their descendants, spread over the continent, primarily from east to west. Today, immigrants are coming mainly from all of Asia and from Middle and South America. They are arriving mainly in the Southwest and West. In addition, internal migration is still a

defining characteristic of life for most North Americans, who are among the world's most mobile people.

The Peopling of North America

Recent evidence suggests that humans first came to North America from northeastern Asia by at least 25,000 years ago and perhaps earlier, most arriving during an ice age. At that time, the global climate was cooler, polar ice caps were thicker, and sea levels were lower. The Bering land bridge, a huge low landmass more than 1000 miles (1600 kilometers) wide, connected Siberia to Alaska. Bands of hunters crossed by foot or small boats into Alaska and traveled down the west coast of North America.

By 15,000 years ago, humans had reached nearly to the tip of South America and had spread deep into that continent. By 10,000 years ago, global temperatures began to rise. As the ice caps melted, sea levels rose and the Bering land bridge was submerged beneath the sea.

Over thousands of years, the people settling in the Americas domesticated plants, created paths and roads, cleared

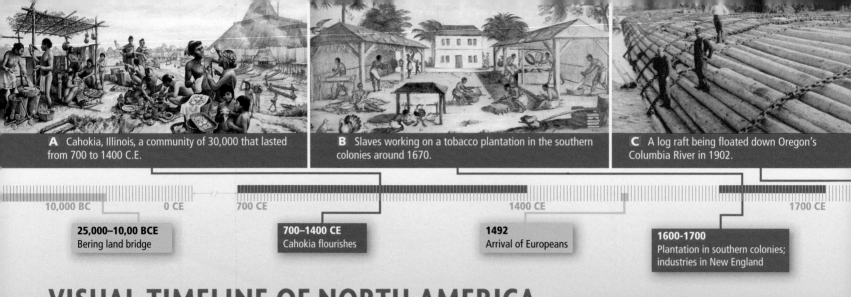

A Cahokia, Illinois, a community of 30,000 that lasted from 700 to 1400 C.E.

B Slaves working on a tobacco plantation in the southern colonies around 1670.

C A log raft being floated down Oregon's Columbia River in 1902.

10,000 BC   0 CE   700 CE   1400 CE   1700 CE

25,000–10,00 BCE
Bering land bridge

700–1400 CE
Cahokia flourishes

1492
Arrival of Europeans

1600-1700
Plantation in southern colonies; industries in New England

VISUAL TIMELINE OF NORTH AMERICA

Thinking Geographically

After you have read about the human history of North America, you will beable to answer the following questions:

A How did food crops like corn, beans, and squash influence the development of settlements like Cahokia?

B Why did the plantation system not lead to the establishment of better roads, communication networks, and infrastructure that could have boosted other forms of economic activity in the southern colonies?

forests, built permanent shelters, and sometimes created elaborate social systems. The domestication of corn is thought to be closely linked to settled life and the resulting population growth in North America. About 3000 years ago, corn was introduced from Mexico (into what is now the U.S. southwestern desert) along with other Mexican domesticated crops, particularly squash and beans.

These foods provided surpluses that allowed some community members to engage in activities other than agriculture, hunting, and gathering, making possible large, citylike regional settlements. For example, by 1000 years ago, the urban settlement of Cahokia (in what is now central Illinois) covered 5 square miles (12 square kilometers) and was home to an estimated 30,000 people (Timeline A). Here people could specialize in crafts, trade, or other activities beyond the production of basic necessities.

The Arrival of the Europeans North America was completely transformed by the sweeping occupation of the continent by Europeans. In the sixteenth century, Italian, Portuguese, and English explorers came ashore along the eastern seaboard of North America, and the Spanish penetrated from Florida deep into the heartland of the continent. In the early seventeenth century, the British established colonies along the Atlantic coast in what is now Virginia (1607) and Massachusetts (1620). Over the next two centuries, colonists from northern Europe built villages, towns, port cities, and plantations along the eastern coast. By the mid-1800s, they had occupied most Native American lands throughout the Appalachian Mountains and into the central part of the continent.

Disease, Technology, and the Native Americans The rapid expansion of European settlement was facilitated by the vulnerability of Native American populations to European diseases. Having long been isolated from the rest of the world, Native Americans had no immunity to diseases such as measles and smallpox. Transmitted by Europeans and Africans who had built up immunity to them, these diseases killed around 90 percent of Native Americans within the first 100 years of contact.

The European Transformation

European settlement erased many of the landscapes familiar to Native Americans and imposed new ones that fit the varied physical and cultural needs of the new occupants.

The Southern Settlements European settlement of eastern North America began with the Spanish in Florida in the mid-1500s and the establishment of the British colony of Jamestown in Virginia in 1607. By the late 1600s, the colonies of Virginia, the Carolinas, and Georgia were cultivating cash crops, such as tobacco and rice, on large plantations.

To secure a large, stable labor force, Europeans brought enslaved African workers into North America beginning in 1619. Within 50 years, enslaved Africans were the dominant labor force on some of the larger southern plantations (see Timeline B). By the start of the Civil War in 1861, slaves made up about one-third of the population in the southern states and were often a majority in the plantation regions. North America's largest concentrations of African-Americans are still in the southeastern states (Figure 2.7).

The plantation system concentrated wealth in the hands of a small class of landowners, who made up just 12 percent of southerners in 1860. Planter elites kept taxes low and invested their money in Europe or the more prosperous northern colonies, instead of in **infrastructure** at home. As a result, the road, rail, communication networks, and other facilities necessary for economic growth were not built.

> **infrastructure** road, rail, and communication networks and other facilities necessary for economic activity

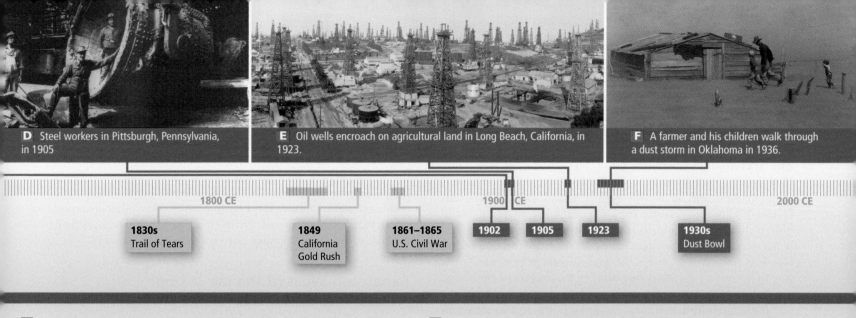

1800 CE   1900 CE   2000 CE

1830s
Trail of Tears

1849
California
Gold Rush

1861–1865
U.S. Civil War

1902

1905

1923

1930s
Dust Bowl

C Why might these men be surprised at the state of logging in the Pacific Northwest today?

D How did the steel industry stimulate development in the "economic core" of North America?

E Is there any evidence of economic diversification in this photo?

F What caused the dust bowl?

More than half of southerners were poor white farmers. Both they and the slaves lived simply, so their meager consumption did not provide a demand for goods. Hence, there were few market towns and almost no industries. Plantations tended to be self-sufficient and generated little *multiplier effect*. Enterprises like shops, garment making, small restaurants and bars, small manufacturing, and transport and repair services that normally "spin off" from, or serve, main industries simply did not develop in the South. Competition between the weak southern economy and the stronger, more diversified northern economies was a main cause of the Civil War (1861–1865), perhaps equal to the abolition movement to free enslaved Africans and their descendents. After the war, the victorious North returned to promoting its own industrial development, the plantation economy declined, and the South sank deeply into poverty. The South remained economically and socially underdeveloped well into the 1970s.

The Northern Settlements Throughout the seventeenth century, relatively poor subsistence farming communities dominated the colonies of New England and southeastern Canada. There were no plantations and few slaves, and not many cash crops were exported. What exports there were consisted of raw materials like timber, animal pelts, and fish from the Grand Banks off Newfoundland and the coast of Maine. Generally, farmers lived in interdependent communities that prized education, ingenuity, self-sufficiency, and thrift.

By the late 1600s, New England was implementing ideas and technology from Europe that led to the first industries. By the 1700s, diverse industries were supplying markets in North America and the Caribbean with metal products, pottery, glass, and textiles.

By the early 1800s, women made up a substantial part of the labor force. In 1822, "factory girls" were working in the textile

Percent of Population

70.0–100
50.0–69.9
25.0–49.9
12.9–24.9
5.0–12.8
1.0–4.9
0.0–0.9

FIGURE 2.7 African-American population, 2007. Many African-Americans live in cities across the United States, but the majority continue to live in the southeastern parts of the country. More African-Americans are returning to the South than are leaving, many retiring on pensions earned elsewhere.

mills of Lowell, Massachusetts, and living as single women in company-owned housing. Southern New England, especially the region around Boston, became the center of manufacturing in North America. It drew largely on young male and female immigrant labor from French Canada and Europe.

The Mid-Atlantic Economic Core The colonies of New York, New Jersey, Pennsylvania, and Maryland eventually surpassed in population and in wealth New England and southeastern Canada. This mid-Atlantic region benefited from more fertile soils, a slightly warmer climate, multiple deepwater harbors, and better access to the resources of the interior. By the end of the Revolutionary War in 1783, the mid-Atlantic region was on its way to becoming the **economic core**, or the dominant economic region, of North America.

Both agriculture and manufacturing in the region grew and diversified in the early nineteenth century, drawing immigrants from much of northwestern Europe. As farmers prospered, they bought mechanized equipment, appliances, and consumer goods. Many of these items were made in nearby cities. Port cities such as New York, Philadelphia, and Baltimore prospered as the intermediary for trade between Europe and the vast American continental interior.

> **economic core** the dominant economic region within a larger region

By the mid-nineteenth century, the economy of the core was increasingly based on the steel industry, which diffused westward to Pittsburgh and the Great Lakes industrial cities of Cleveland, Detroit, and Chicago. The steel industry further stimulated the mining of deposits of coal and iron ore throughout the region and beyond. Steel became the basis for mechanization, and the region was soon producing heavy farm and railroad equipment, including steam engines.

By the early twentieth century, the economic core stretched from the Atlantic to St. Louis, Chicago, and Milwaukee, and from Ottawa to Washington, D.C. It dominated North America economically and politically well into the mid-twentieth century. Most other areas produced food and raw materials for the core's markets and depended on the core's factories for manufactured goods (see Timeline D).

Expansion West of the Mississippi

The east-to-west trend of settlement continued as land in the densely settled eastern parts of the continent became too expensive for new immigrants. By the 1840s, immigrant farmers from central and northern Europe, as well as European descendants born in eastern North America, were pushing their way west across the Mississippi River (Figure 2.8).

The Great Plains Much of the land west of the Mississippi River was dry grassland or prairie. The soil usually proved very productive in wet years, and the area became known as the region's breadbasket. But the naturally arid character of this land eventually created an ecological disaster for Great Plains farmers. In the 1930s,

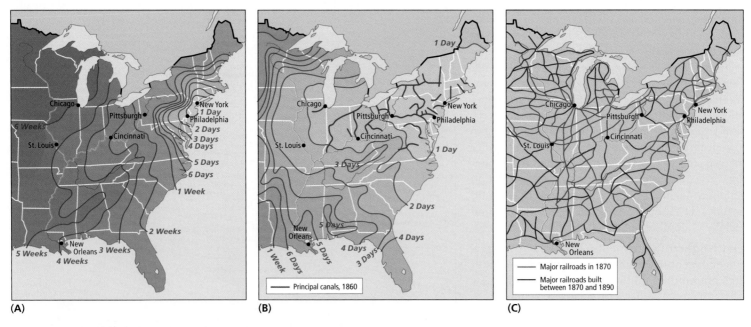

(A) **(B)** **(C)**

FIGURE 2.8 Nineteenth-century transportation. (A) Travel times from New York City, 1800. It took a day to travel by wagon from New York City to Philadelphia and a week to go to Pittsburgh. **(B)** Travel times from New York City, 1857. The travel time from New York to Philadelphia was now only 2 or 3 hours and to Pittsburgh less than a day because people could go part of the way via canals (dark blue). Via the canals, the Great Lakes, and rivers, they could easily reach principal cities along the Mississippi River, and travel was less expensive and onerous. **(C)** Railroad expansion by 1890. With the building of railroads, which began in the decade before the Civil War, the mobility of people and goods increased dramatically. By 1890, railroads crossed the continent, though the network was most dense in the eastern half.

after 10 especially dry years, a series of devastating dust storms blew away topsoil by the ton. This hardship was made worse by the widespread economic depression of the 1930s. Many Great Plains farm families packed up what they could and left what became known as the Dust Bowl (see Timeline F), heading west to California and other states on the Pacific Coast.

The Mountain West and Pacific Coast Some Europeans skipped over the Great Plains entirely, alerted to the possibilities farther west. By the 1840s, they were coming to the valleys of the Rocky Mountains, to the Great Basin, and to the well-watered and fertile coastal zones of what was then known as the Oregon Territory and California. News of the discovery of gold in California in 1849 created the *Gold Rush*, which drew thousands with the prospect of getting rich quickly. The vast majority of gold seekers were unsuccessful, however, and by 1852 they had to look for employment elsewhere. Farther north, logging eventually became a major industry (see Timeline C).

The extension of railroads across the continent in the nineteenth century facilitated the transportation of manufactured goods to the West and raw materials to the East. Today, the coastal areas of this region, often called the Pacific Northwest, have booming, diverse, high-tech economies and growing populations. Perhaps in response to their history, residents of the Pacific Northwest are on the forefront of so many efforts to reduce human impacts on the environment that the region has been nicknamed "Ecotopia."

The Southwest People from the Spanish colony of Mexico first colonized the Southwest at the end of the sixteenth century. Their settlements were sparse, and as immigrants from the United States expanded into the region, Mexico found it increasingly difficult to maintain control. By 1850, nearly the entire Southwest was under U.S. control.

By the twentieth century, a vibrant agricultural economy had developed in central and southern California, made possible by massive government-sponsored irrigation schemes. There, the mild Mediterranean climate made it possible to grow vegetables almost year-round. With the advent of refrigerated railroad cars, fresh California vegetables could be sent to the major population centers of the East. Southern California's economy rapidly diversified to include oil (see Timeline E), entertainment, and a variety of engineering- and technology-based industries.

European Settlement and Native Americans

As settlement relentlessly expanded west, Native Americans living in the eastern part of the continent, who had survived early encounters with Europeans, were occupying land that European newcomers wished to use. During the 1800s, almost all the surviving Native Americans were either killed off in innumerable skirmishes with European newcomers, absorbed into the societies of the Europeans (through intermarriage), or else forcibly relocated west to relatively small reservations with few resources. The largest relocation, in the 1830s, involved the Choctaw, Seminole, Creek, Chickasaw, and Cherokee of the southeastern states. These people had already adopted many European methods of farming, building, education, government, and religion. Nevertheless, they were rounded up by the U.S. Army and marched to Oklahoma, along a

route that became known as the Trail of Tears because of the more than 4000 who died along the way.

As Europeans occupied the Great Plains and prairies, many of the reservations were further shrunk or relocated onto ever less desirable land. Today, reservations cover just over 2 percent of the land area of the United States.

In Canada the picture is somewhat different. Reservations now cover 20 percent of Canada, mostly due to the creation of the Nunavut territory in 1999 and the Dogrib territory in 2003, both in the far north of Canada (see Figure 2.3). The Nunavut and Dogrib stand out as having won the right to legal control of these lands. In contrast to the United States, it had been unusual for native groups in Canada to have legal control of their territories.

After centuries of mistreatment, many Native Americans still live in poverty, and as in all communities under severe stress, rates of addiction and violence are high, especially in the United States. However, in recent decades some tribes have found avenues to greater affluence on the reservations by establishing manufacturing industries; developing fossil fuel, uranium, and other mineral deposits underneath reservations; or opening gambling casinos. One measure of this economic resurgence is population growth. Expanding from a low of 400,000 in 1907, the Native American population now stands at almost 4 million in 2011, the majority of which lives in the United States.

The Changing Regional Composition of North America

The areas of European-led settlement still remain in North America, but they are now less distinctive. The economic core region is less dominant in industry, which has spread to other parts of the continent. Some regions that were once dependent on agriculture, logging, or mineral extraction now have high-tech industries as well. The West Coast in particular has boomed with a high-tech economy and a rapidly growing population. The West Coast also benefits from trade with Asia, which now surpasses trade with Europe in volume and value.

THINGS TO REMEMBER

1. In 1492, roughly 18 million Native Americans lived in North America. By 1542, after only a few Spanish expeditions, there were half that many. By 1907, slightly more than 400,000, or just 2 percent of the original population, remained. By 2006, there were about 4 million.

2. By the late 1600s, New England was implementing ideas and technology from Europe that led to the first industries.

3. By the early twentieth century, North America's economic core stretched from the Atlantic Ocean to St. Louis, Chicago, and Milwaukee, and from Ottawa to Washington, D.C. It dominated North America economically and politically well into the mid-twentieth century.

II CURRENT GEOGRAPHIC ISSUES

A huge regional economy and still-plentiful resources privilege North America, but the region faces complex challenges posed by globalization, increasingly diverse populations, and rising energy and environmental concerns. These long-standing issues must be understood now in the context of a globalized world in which economic downturns impact life everywhere. Meanwhile, North America—especially the United States, because of its leadership in economic development and political affairs—faces increasing pressure to help resolve a wide range of global conflicts and disaffections arising from widening disparities of wealth and opportunity.

Political Issues

Politics in North America are characterized by a relatively high level of democratization. Many political issues in this region are also linked to national identity and the image North Americans would like their countries to have. Here we will look at just a few defining issues of the present day and especially at the ways in which the two countries are similar and different in terms of political culture.

The Wars in Afghanistan and Iraq

Immediately after the attacks in the United States on September 11, 2001 (9/11), the international community extended warm sympathy to the country and generally supported the strategies of President George W. Bush. In the fall of 2001, his administration, with the advice and consent of Congress, launched what was then called the War on Terror. The first target was Afghanistan, which was then thought to be host to Osama bin Laden and the elusive Al Qaeda network that openly claimed credit for masterminding the 9/11 attacks. The aim was to capture bin Laden—an aim still unaccomplished (as of this writing)—and to remake Afghanistan into a stable ally. Although U.S. forces in Afghanistan were joined by NATO (North Atlantic Treaty Organization) forces (including Canadian troops), the war has proved difficult to resolve because of heavy resistance from tribal leaders within the country and from militant Muslim insurgents in adjacent Pakistan. When, in the spring of 2003, President Bush brought the War on Terror to Iraq, U.S. troop support was diverted, and the momentum in Afghanistan was lost.

In 2009, the Obama administration announced a phased withdrawal from Iraq, which appeared to be stabilizing politically, and a temporary intensification of the war in Afghanistan, which was to be followed by a similar phased withdrawal. By the summer of 2010, all official U.S. combat troops had been withdrawn from Iraq, and a "surge" of U.S. forces was well underway in Afghanistan. The rationale was that increasing political instability in Afghanistan would empower the most radical elements there to recruit new terrorists that would prevent Afghanistan from ever becoming a stable U.S. ally, and possibly even pose a threat to the U.S. homeland.

Regardless of the official withdrawals, most analysts agree that a large U.S. military presence will remain in both Iraq and Afghanistan for years to come. High-ranking officials in the U.S. Department of Defense have publicly and repeatedly stated that the United States will have an ongoing strategic and economic interest in securing access to the oil and mineral resources of both Afghanistan and Iraq, thereby keeping these resources out of the hands of groups hostile to the United States and its allies. The history of U.S. involvement in conflicts abroad confirms the likelihood of substantial numbers of troops remaining in both countries for the time being (Photo Essay 2.4 on pages 76–77).

The United States Abroad

> **Learning Goal 3**
> **Democratization:** How has the ideal of democracy influenced North America's involvement in the affairs of foreign countries?

While it is an official policy of the U.S. government to promote democracy abroad, the global distribution of U.S. military bases as well as U.S. spending on aid and military assistance to foreign countries suggest that other influences determine U.S. involvement in the affairs of foreign countries. Bases and spending are both concentrated in places where the United States has for years had strong strategic and economic interests. Three main concentrations of bases and spending can be seen on the map in Photo Essay 2.4: Europe, Island and Peninsular East Asia, and South/Southwest Asia. The first two relate to strategic and economic interests dating from World War II that have remained relevant ever since, between the Cold War (see Chapter 1 on page 34), the subsequent collapse of the Soviet Union, and the rise of major trading partners in Europe and East Asia. The third relates directly to the U.S. interests in Iraq, Afghanistan, and the oil and mineral resources of South and Southwest Asia in general. Notice also the many places where the United States does not have significant bases and where spending on foreign aid is at low or moderate levels. These are generally places where the United States has fewer strategic and economic interests. For example, sub-Saharan Africa's poverty and ongoing political instability has kept U.S. economic interests here few, relative to other wealthier regions. The same could be said of Haiti. If U.S. policies to support democracy abroad were more of a factor, one might expect that the focus of U.S. spending on foreign aid and military assistance would be the parts of sub-Saharan Africa that have low levels of democratization. And yet this part of the world has almost no U.S. bases and only average levels of U.S. spending on foreign aid and military assistance.

II ▶ 39. CONSEQUENCES REPORT

II ▶ 40. SECURITY AND PERSONAL LIBERTY REPORT

II ▶ 41. U.S. IMAGE REPORT

Dependence on Oil Questioned

The 9/11 attacks in 2001 alerted the U.S. public to the fact that roughly 25 percent of the oil imported into the United States (which amounts to roughly 14 percent of the total oil used in the United States) came from countries along the Persian Gulf, where the terrorists had originated. By 2008, this figure was down slightly (Figure 2.9). This rate of dependency makes many in the United States uncomfortable, especially given the ongoing and expensive U.S. military presence in the Persian Gulf. Despite the fact that few, if any, oil-producing countries would participate in disrupting the flow of oil to the United States (they are too dependent on the income), finding more domestic sources of energy has become a major U.S. issue. This has boosted interest in renewable energy, as well as increased pressure to drill for oil in the coastal waters of the United States—though this last strategy was dealt a blow by the massive spill at the Deepwater Horizon.

Over the long term, incentives to find alternatives to fossil fuels, imported or not, are likely to rise. Oil analysts now agree that although oil prices are likely to fluctuate, the general trend will be upward, partly because producers will keep the supply tight, but primarily because demand from the fast-growing Asian economies of India and China will push prices higher. Political pressure to reduce greenhouse gas emissions may also intensify, especially if climate-change impacts increase.

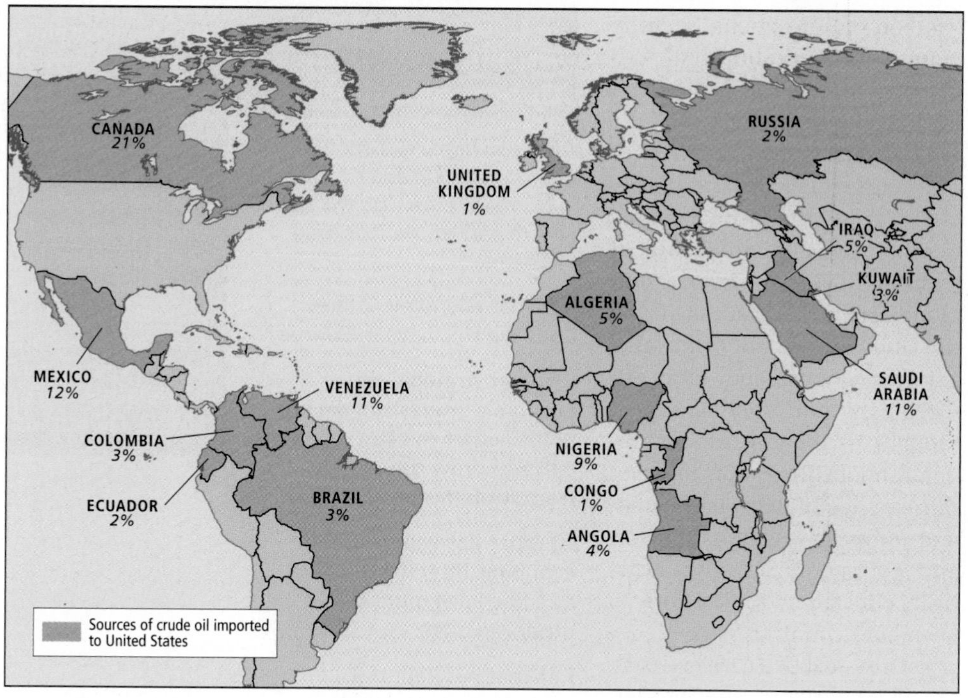

FIGURE 2.9 Sources of average daily crude oil imported to the United States in September, 2009. The United States is dependent on crude oil from many locations around the world. This map shows the percentage of average daily crude oil (millions of barrels) imported to the United States from specific nations in September 2009. In 2008, Canada supplied 22 percent of U.S. petroleum imports. The country sources and percentages have remained roughly the same for the past 5 years.

Relationships between Canada and the United States

Citizens of Canada and the United States share many characteristics and concerns. Indeed, in the minds of many people—especially those in the United States—the two countries are one. Yet that is hardly the case. Three key factors characterize the interaction between Canada and the United States: *asymmetries*, *similarities*, and *interdependencies*.

Asymmetries

Asymmetry means "lack of balance." Although the United States and Canada occupy about the same amount of space, much of Canada's territory is cold and sparsely inhabited. The U.S. population is about ten times the Canadian population. While Canada's economy is one of the largest and most productive in the world, producing U.S. $1.3 trillion purchasing power parity (PPP) in goods and services in 2010, it is dwarfed by the U.S. economy, which is more than ten times larger, at $14.8 trillion PPP in 2010.

There is also asymmetry in international affairs. The United States is an economic, military, and political superpower preoccupied with maintaining a world leadership role. Canada is only an afterthought in U.S. foreign policy, in part because

the country is so secure an ally. But for Canada, managing its relationship with the United States is the top foreign policy priority. As former Canadian Prime Minister Pierre Trudeau once told the U.S. Congress, "Living next to you is in some ways like sleeping with an elephant: No matter how friendly and even-tempered the beast, one is affected by every twitch and grunt." In 2008, for example, the robust debt-free Canadian economy slowed abruptly, due in large part to the slump in U.S. consumer spending, which hurt Canadian exports such as energy and cars.

Similarities

Notwithstanding the asymmetries, the United States and Canada have much in common. Both are former British colonies that also experienced settlement and exploration by the French. From their common British colonial experience, they developed comparable democratic political traditions. Both are federations (of states or provinces), and both are representative democracies. Their legal systems are also similar.

Not the least of the features they share is a 4200-mile (6720-kilometer) border, which until 2009 contained the longest sections of unfortified border in the world. For years, the Canadian border had just 1000 U.S. border guards, while the Mexican border, which is half as long, had nearly 10,000 agents. In 2009, the Obama administration decided to equalize surveillance of the two borders for national security reasons. Where some rural residents used to pass unobserved in and out of Canada and the United States many times in the course of a day simply by going about their usual activities, now there are drone aircraft with night-vision cameras and cloud-piercing radar scanning the landscape for smugglers, illegal immigrants, and terrorists. Canadians and Americans, who thought of their common border as nonexistent, now discover intrusive technology, barricades, and crossing delays. Tourism and commerce are also negatively affected by these efforts to keep both countries safer.

▌▌▶ 42. U.S. BORDER SECURITY REPORT

Well beyond the border, Canada and the United States share many other landscape similarities. Their cities and suburbs look much the same. The billboards that line their highways and freeways advertise the same brand names. Shopping malls and satellite business districts have followed suburbia into the countryside, encouraging similar patterns of mass consumption and urban sprawl. The two countries also share similar patterns of ethnic diversity that developed in nearly identical stages of immigration from abroad.

Interdependencies

Canada and the United States are perhaps most intimately connected by their long-standing economic relationship. The two countries engage in mutual tourism, direct investment, migration, and most of all, trade. By 2005, that trade relationship had evolved into a two-way flow of U.S.$1 trillion annually (**Figure 2.10** on page 78). Each country is the other's largest trading partner. Canada sells 80 percent of its exports to the United States and buys 54 percent of its imports from the United States. The United States, in turn, sells 21.4 percent of its exports to Canada and buys 15.7 percent of its imports from Canada.

Notice, however, that asymmetry exists even in the realm of interdependencies: Canada's smaller economy is much more dependent on the United States than the reverse. Nonetheless, if Canada were to disappear tomorrow, as many as 1 million American jobs would be threatened.

Democratic Systems of Government: Shared Ideals, Different Trajectories

Canada and the United States have similar democratic systems of government, but there are differences in the way power is divided between the federal government and provincial or state governments.

Thinking Geographically

After you have read about Democratization and Conflict in North America, you will be able to answer the following questions:

A What economic interest does the U.S. have in Europe?

B What strategic and economic interests make it unlikely that the U.S. military will withdraw completely from Afghanistan and Iraq?

D What has kept U.S. economic interests in Africa at a low level relative to other regions?

E What has kept U.S. economic interests in Haiti at a low level relative to other countries?

It is sometimes argued that the United States uses its power abroad primarily to promote democracy. There is some truth to this. For example, the largest recipient of U.S. foreign aid since 2002 has been Iraq, where the United States strongly promoted democratization during the Iraq War. However, officials who helped plan the war point out that U.S. strategic and economic interests played a much larger role than democratization as a motive of U.S. involvement. The same is true elsewhere. The strategic and economic interests of the United States generally determine where U.S. foreign aid goes and where the most military bases are. For example, if promoting democracy were the primary goal of the United States abroad, sub-Saharan Africa would be the major region of focus for the United States. While the United States has been increasing its presence in this region in recent years, the map below shows that sub-Saharan Africa receives relatively little U.S. aid and has few U.S. military bases.

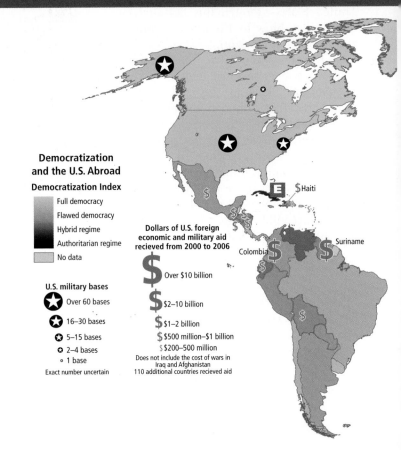

Democratization and the U.S. Abroad

Democratization Index
- Full democracy
- Flawed democracy
- Hybrid regime
- Authoritarian regime
- No data

U.S. military bases
- ★ Over 60 bases
- ★ 16–30 bases
- ✦ 5–15 bases
- ◉ 2–4 bases
- ○ 1 base

Exact number uncertain

Dollars of U.S. foreign economic and military aid recieved from 2000 to 2006
- $ Over $10 billion
- $ $2–10 billion
- $ $1–2 billion
- $ $500 million–$1 billion
- $ $200–500 million

Does not include the cost of wars in Iraq and Afghanistan
110 additional countries recieved aid

E Haitian refugees off the coast of Florida being detained by the U.S. Coast Guard. Haiti is far from being a functional democracy and is plagued by economic and political instability, but compared to Israel, Egypt, or Afghanistan, it receives relatively little U.S. foreign or military aid.

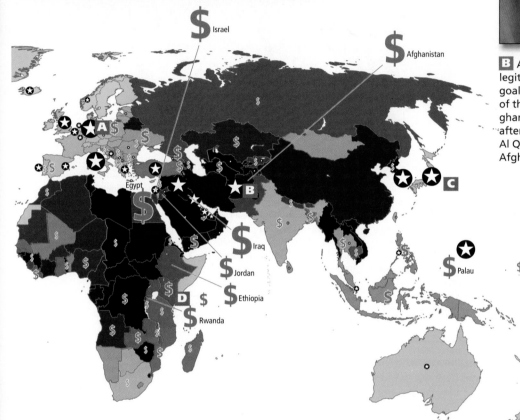

$ Israel

$ Afghanistan

Egypt

$ Iraq

$ Jordan

$ Ethiopia

$ Rwanda

$ Palau

A U.S. forces destined for Iraq wait to board aircraft at Ramstein Air Base, one of 260 U.S. bases in Germany. The large U.S. military presence in Europe is a legacy of World War II and the Cold War era that followed. Their purpose was to discourage potential aggression from the (now-defunct) Soviet Union against U.S. allies in Western Europe.

B Afghan women line up to vote in the country's first legitimate elections in decades. Democratization is a major goal of U.S. foreign aid in Afghanistan, but in the context of the larger strategic interests of the "War on Terror," Afghanistan became a focus for the United States only after the attacks of September 11th masterminded by the Al Qaeda terrorist network, which was partly based in Afghanistan.

C Japanese officials surrender aboard a U.S. battleship at the end of World War II. After the war both Japan and its occupied territories in South Korea were reorganized as democracies by the United States, and numerous military bases were established. However the bases are first and foremost a projection of U.S. power designed to counter any future agression by China, North Korea, the former Soviet Union, or by Japan and South Korea themselves.

D A Kenyan throws teargas back at riot police during several days of violence that followed the presidential elections in 2008. Democracy is fragile throughout much of sub-Saharan Africa, and elections are often plagued by violence. While the United States is giving increasing support to democracy in sub-Saharan Africa, U.S. foreign aid to the area is relatively small and historically has done little to promote democracy in the region. Military bases are relatively few.

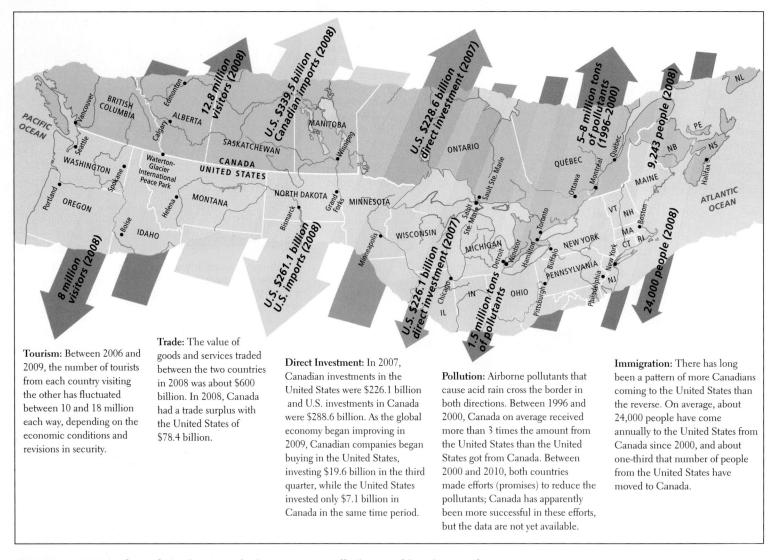

Tourism: Between 2006 and 2009, the number of tourists from each country visiting the other has fluctuated between 10 and 18 million each way, depending on the economic conditions and revisions in security.

Trade: The value of goods and services traded between the two countries in 2008 was about $600 billion. In 2008, Canada had a trade surplus with the United States of $78.4 billion.

Direct Investment: In 2007, Canadian investments in the United States were $226.1 billion and U.S. investments in Canada were $288.6 billion. As the global economy began improving in 2009, Canadian companies began buying in the United States, investing $19.6 billion in the third quarter, while the United States invested only $7.1 billion in Canada in the same time period.

Pollution: Airborne pollutants that cause acid rain cross the border in both directions. Between 1996 and 2000, Canada on average received more than 3 times the amount from the United States than the United States got from Canada. Between 2000 and 2010, both countries made efforts (promises) to reduce the pollutants; Canada has apparently been more successful in these efforts, but the data are not yet available.

Immigration: There has long been a pattern of more Canadians coming to the United States than the reverse. On average, about 24,000 people have come annually to the United States from Canada since 2000, and about one-third that number of people from the United States have moved to Canada.

FIGURE 2.10 Transfers of tourists, goods, investment, pollution, and immigrants between the United States and Canada. Canada and the United States have the world's largest trading relationship. The flows of goods, money, and people across the long Canada–U.S. border are essential to both countries. However, because of its relatively small population and economy, Canada is more reliant on the United States than the U.S. is on Canada. All amounts shown are in U.S. dollars.

There are also differences in the way the division of power has changed since each country achieved independence.

Both countries have a federal government, in which a union of states (provinces, in Canada) recognizes the sovereignty of a central authority while many governing powers are retained by state/provincial or local governments. In both Canada and the United States, the federal government has an elected executive branch, elected legislatures, and an appointed judiciary. In Canada, the executive branch is more closely bound to follow the will of the legislature. At the same time, the Canadian federal government has more and stronger powers (at least constitutionally) than does the U.S. federal government.

Over the years, both the Canadian and U.S. federal governments have moved away from the original intentions of their constitutions. Canada's originally strong federal government has become somewhat weaker. This change is largely in response to

demands by provinces, such as the French-speaking province of Québec, for greater autonomy over local affairs.

Meanwhile, the initially more limited federal government in the United States has expanded its powers. The U.S. federal government's original source of power was its mandate to regulate trade between states. Over time, this mandate has been interpreted ever more broadly. Now the U.S. federal government powerfully affects life even at the local level. This power is exercised primarily through its ability to dispense federal tax monies in programs such as grants for school systems, federally assisted housing, military bases, and the rebuilding of interstate highways. Money for these programs is withheld if state and local governments do not conform to federal standards. This practice has made some poorer states dependent on the federal government. However, it has also encouraged some state and local governments to enact more enlightened laws than they might have done otherwise. For example, in the 1960s the federal government promoted civil rights for African-American citizens by requiring states to end racial segregation in schools in order to receive federal support for their school systems.

The Social Safety Net: Canadian and U.S. Approaches

The Canadian and U.S. governments have responded differently to the displacement of workers by economic change. Ultimately, these differences derive from prevailing political positions and widely held notions each country has about what the government's role in society should be. In Canada there is broad political support for a robust **social safety net**, the services provided by the government—such as welfare, unemployment benefits, and health care—that prevent people from falling into extreme poverty (see Thematic Overview E). In the United States there is much less support for these programs, though recent years have seen the U.S. social safety net start to resemble Canada's a bit more.

For many decades, Canada has spent more per capita than the United States on social programs that lighten the financial burdens of working people in times of economic crisis. These policies have generally made the financial lives of working people more secure. They also reflect the workings of Canada's democracy, where voters have supported several major expansions of the social safety net during the twentieth century. Recently, more Canadian voters have supported weakening the social safety net, with opponents of these policies arguing that they place too much of an economic burden on Canadians. Canada's unemployment rate, which by some measures seems to be slightly higher than that of the United States, is often brought up as justification for weakening Canada's social safety net. However, there is evidence that the Canadian and U.S. governments measure their unemployment rates differently—for when the same methods are used for both countries, Canada actually has slightly lower unemployment.

Compared to Canada, the United States provides much less of a social safety net. During the twentieth century, there were several major expansions of U.S. social programs meant to benefit poor working people, but the influence of business interests in U.S. democracy has worked to keep this expansion at a much lower level than in Canada. For years the prevailing political argument against expansion of the social safety net has been that the U.S. system offers lower taxes for businesses, which in turn attract new investment and new jobs, the

> **social safety net** the services provided by the government—such as welfare, unemployment benefits, and health care—that prevent people from falling into extreme poverty

benefits of which *trickle down* to those most in need. The fit of this position with the reality of what actually happens has long been challenged, especially during times of economic recession. Many U.S. workers who have lost jobs during the recession have experienced increases in ill health, violent crime, drug abuse, family disintegration, and homelessness. The costs imposed on society from these multiple problems have grown high enough that some U.S. cities (for example Miami, Florida; Dallas, Texas; Portland, Oregon; and New York City, New York) have calculated that it actually costs less to make certain key social services more robust, such as subsidized or even free housing for the homeless.

Two Health-Care Systems In the mid-2000s, some U.S. firms began eyeing Canada as a desirable place to relocate, precisely because the country's social safety net included a government-sponsored health-care system. The contrasts between the two health-care systems are indeed striking. The Canadian system is heavily subsidized and covers 100 percent of the population. In the United States, health care is largely private and relatively expensive. The government subsidizes care for the elderly, disabled, children, veterans, and the poor, though for these groups the quality of care is often much lower than in Canada. About 37 million, or 15.3 percent of the U.S. population, are left with no health-care coverage. At the same time, the United States spends 15.4 percent of its GDP on health care, while Canada spends only 11.8 percent and gets better outcomes, outranking the United States on most indicators of overall health (Table 2.1).

The differences between these two health-care systems narrowed considerably in 2010 when the U.S. Congress passed a health-care reform law. The new law attempted to address many of the shortcomings of the private health-care system, particularly the large number of people with no health-care insurance. While the new law falls far short of Canada's universal health-care system, its passage marks a shift towards more robust social safety nets in the United States. Again, much of the shift is driven by an effort to reduce the overall costs to society of major problems. Indeed, throughout the intense political debates that ultimately led to the U.S. law's passage, the cheaper Canadian system was held up as a prime example of the potential savings that could be realized with a more robust social safety net for health care.

❚❚▶ 48. SICKO REPORT

TABLE 2.1	Health-related indexes for Canada and the United States						
Country	Health-care cost as a percentage of GDP	Percentage of population with no insurance	Deaths per 1000	Infant mortality per 1000 live births	Maternal mortality per 100,000 live births	Life expectancy at birth (years)	Annual health expenditures per capita (PPP* U.S.$)
Canada	11.8	0	7.4	5	7	79.8	$3173
United States	15.4	15.3	8.2	6.6	11	77.4	$6096

*PPP: Purchasing power parity.

Gender in National Politics

North America has some powerful political contradictions with regard to gender. While women voters are an ever more potent political force, successful female politicians are not as numerous as one might expect. Women cast the deciding votes in the U.S. presidential election of 1996, voting overwhelmingly for Bill Clinton. In the 2006 interim elections, 55 percent of women voted for the Democratic Party candidates, registering strong anti-war sentiment and making it possible for the Democratic Party to gain many seats in Congress. In 2008, Barack Obama won 56 percent of the female vote. Deciding factors in the election included such women's issues as family leave, health care, equal pay, day care, reproductive rights, and ending the war in Iraq.

❚❚▶ 49. WOMEN VOTING REPORT

And yet, women have not had the success as political leaders their powerful role as voters might warrant. Of the 538 people in the U.S. Congress as of 2008, only 91 members, or 17 percent, were women. Gender equity was somewhat closer at the state level, where 24.2 percent of legislators were women as of the 2008 elections. As of 2010, Canadian women in the House of Parliament and provincial government held, respectively, 21 and 20 percent of the legislative seats. By comparison, at least 30 percent of lawmakers were female in Argentina, Costa Rica, Denmark, Finland, Germany, Iceland, Mozambique, the Netherlands, Norway, South Africa, and Sweden. Things are no more equitable for women at the executive level. In 2010, the United States had only six female governors, and only one Canadian province (Nunavut) had a female executive. At the national level, Canada briefly had a female prime minister in 1993, while the United States has never had a female president. Nevertheless, in 2008, Hillary Clinton was the first woman to have a serious chance at becoming the U.S. president.

> **agribusiness** the business of farming conducted by large-scale operations that purchase, produce, finance, package, and distribute agricultural products

---

THINGS TO REMEMBER

1. **Learning Goal 3: Democratization** While it is an official policy of the U.S. government to promote democracy abroad, the global distribution of U.S. spending on foreign aid and military assistance is more influenced by strategic and economic interests.

2. Canada has a much more robust social safety net than does the United States, though recent reforms in the U.S. health-care system may reduce differences between the two countries in coming years.

3. While women voters are an ever more potent political force in the United States and Canada, the number of successful female politicians has not been commensurate with the percentage of women in the general population of each country.

Economic Issues

The economic systems of Canada and the United States, like their political systems, have much in common. Both countries evolved from societies based mainly on family farms. Then came an era of industrialization, followed by a move to primarily service-based economies. Both countries have important technology sectors and economic influence that reach worldwide.

North America's Changing Food-Production Systems

North America benefits from an abundant supply of food, and it is an important producer of food for foreign as well as domestic consumers (Figure 2.11). At one time, exports of agricultural products were the backbone of the North American economy. However, because of growth in other sectors, agriculture now accounts for less than 1.2 percent of the region's GDP. Moreover, because both countries are involved in the global economy, food is increasingly imported.

❚❚▶ 43. FOOD GLOBALIZATION

The shift to mechanized agriculture in North America has brought about sweeping changes in employment and farm management. In 1790, agriculture employed 90 percent of the American workforce; in 1890, it employed 50 percent. Until 1910, thousands of highly productive family-owned farms, spread over much of the United States and southern Canada, provided for most domestic consumption and the majority of all exports. Today, agriculture employs less than 1 percent of the U.S. workforce and less than 2 percent of the Canadian.

Farms have become highly mechanized operations that need few workers but require huge investments in land, machinery, fertilizers, and pesticides in order to be profitable (see Thematic Overview D). While the wealthier farmers have been able to invest in these "green revolution" methods (see chapter 1) and have prospered, poorer farmers have not been able to afford the ever-increasing costs. Most have long since sold their land, primarily to wealthier farmers, and moved to the cities to find other work.

Large **agribusiness** corporations that sell machinery, seeds, and chemicals or that buy the crops that farmers produce have facilitated the transition to green revolution methods. Their large financial resources have enabled them to invest in research, develop new products, and provide loans and cash to individual farmers. Over time, most farmers have been forced to enter into contracts and other relationships with agribusiness corporations that leave them with little actual control over what crops are grown or what farming methods are used.

Green revolution–style agriculture provides a wide variety of food at low prices for North Americans. However, in many rural areas, the shift to these production methods has depressed local economies and created social problems. Communities in places such as rural Alberta and Saskatchewan in Canada and Iowa, Nebraska, and Kansas in the United States were once made up of farming families with similar middle-class incomes, social standing, and commitment to the region. Today, farm communities are increasingly composed of a few wealthy farmer-managers amid a majority of poor, often migrant Hispanic or Asian laborers, who work on large farms and in food-processing plants for wages that are too low to provide a decent standard of living.

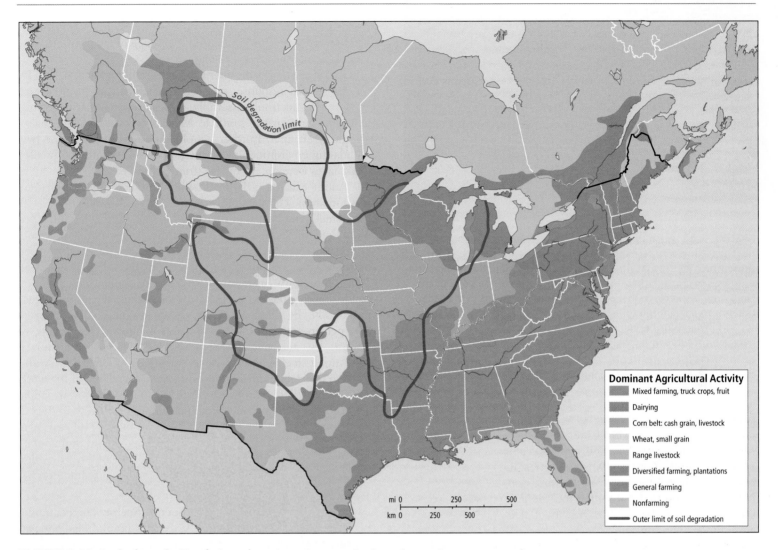

FIGURE 2.11 Agriculture in North America. Throughout much of North America, some type of agriculture is possible. The major exceptions are the northern reaches of Canada and Alaska and the dry mountain and basin region (the continental interior) lying between the Great Plains and the Pacific coastal zone. However, in some marginal areas, such as southern California, southern Arizona, and the Utah Valley, irrigation is needed for cultivation. (Hawaii is not included here because it is covered in Oceania, Chapter 11.)

Food Production and Sustainability Green revolution agriculture relies heavily on the use of highly modified seeds and chemical fertilizers, pesticides, and herbicides to increase crop yields. These methods can contaminate food, pollute nearby rivers, and even affect distant coastal areas (see page 68). In addition, many North American farming areas have lost as much as one-third of their topsoil due to farming methods that create soil erosion.

Growing concerns about sustainability involve changing patterns of food production. Throughout North America there is a revival of small family farms that supply organic vegetables and fruits (grown without chemical fertilizers and pesticides) to restaurants and groceries (see **Figure 2.12** on page 82; see also **Reasons for Optimism C** on page 97). Consumers are now buying organic foods in ever greater numbers, which is inspiring greater interest from large corporate farms that see the potential for high profits in sustainable food production.

Changing Transportation Networks and the North American Economy

An extensive network of road and air transportation, enabling high-speed delivery of people and goods, is central to the productivity of North America's economy. This system emerged along with the development of inexpensive mass-produced automobiles in the 1920s. Where railroads and shipping had once dominated transportation, trucks could now deliver cargo more quickly and conveniently. Beginning in the 1950s, automobile- and truck-based transport was boosted by the Interstate Highway System—a 45,000-mile (72,000-kilometer) network of high-speed, multilane roads. Because this network was connected to the vast system of local roads, it made the delivery of manufactured products quicker and more flexible than rail. Thus the highway system made possible the dispersal of industry and related services into suburban and

FIGURE 2.12 A small family farm in Allamakee County in northeast Iowa practicing a number of techniques that make agriculture more sustainable. To curb soil erosion, the farmers contour rows of crops so that they are perpendicular to the slope of the land, resulting in curve-shaped fields. Trees are also left in the valley bottoms to prevent the formation of gullies when it rains. To maintain soil fertility and further minimize soil erosion, crops are alternated between corn (brown in the photo) and alfalfa (green in the photo).

Thinking Geographically: Why have small farms like the one shown here historically taken a greater interest in sustainable food production?

rural locales across the country, where labor, land, and living costs were lower.

After World War II, air transportation also served economic growth in North America. Its primary niche is business travel, because face-to-face contact remains essential to American business culture despite the growth of telecommunications and the Internet. Because many industries are widely dispersed in numerous medium-size cities, air service is organized as a *hub-and-spoke network*. Hubs are strategically located airports, such as those in Atlanta, Chicago, Dallas, and Los Angeles. These airports are used as collection and transfer points for passengers and cargo continuing on to smaller locales. Most airports are also located near major highways, which provide an essential link for high-speed travel and cargo shipping.

The New Service and Technology Economy

> **Learning Goal 4**
> **Globalization and Development:** How has globalization transformed economic development in North America, thereby changing the kinds of jobs available in this region?

North America's job market has become more oriented toward knowledge-intensive jobs that require education and specialized professional training, often in technology and management (see Thematic Overview G). Meanwhile, low-skill, mass-production industrial jobs are increasingly being moved abroad.

▐▐ ▶ 46. U.S. GEOGRAPHY REPORT

Decline in Manufacturing Employment By the 1960s, the geography of manufacturing was changing. In the old economic core, higher pay and benefits and better working conditions won by labor unions led to increased production costs. Many companies

began moving their factories to the southeastern United States, where wages were lower due to the absence of labor unions.

▐▐ ▶ 44. U.S. LABOR TRANSITION REPORT

In 1994, the **North American Free Trade Agreement (NAFTA)** was passed. In response, many manufacturing industries, such as clothing, electronic assembly, and auto parts manufacturing, began moving farther south to Mexico or overseas. Here labor was vastly cheaper. Further, employers could save because laws mandating environmental protection as well as safe and healthy workplaces were absent or less strictly enforced.

Another factor in the decline of manufacturing employment is automation. The steel industry provides an illustration. In 1980, huge steel plants, most of them in the economic core, employed more than 500,000 workers. At that time, it took about 10 person-hours and cost about $1000 to produce 1 ton of steel. Spurred by more efficient foreign competitors in the 1980s, 1990s, and 2000s, the North American steel industry applied new technology to lower production costs, improve efficiency, and increase production. By 2006, steel was being produced at the rate of 0.44 person-hour per ton and at a cost of about $165 per ton. As a result, the steel industry in the United States has reorganized with much steel produced in small, highly efficient mini-mills. In total, the steel industry now employs fewer than half the workers it did in 1980. Throughout North America, far fewer people are now producing more of a given product at a far lower cost than was the case 30 years ago. Hence, even though employment in manufacturing has declined over the last three decades, the actual amount of industrial production has steadily increased.

> **North American Free Trade Agreement (NAFTA)** a free trade agreement made in 1994 that added Mexico to the 1989 economic arrangement between the United States and Canada
>
> **service sector** economic activity that involves the sale of services

Growth of the Service Sector The economic base of North America is now a broad and varied **service sector** in which people are engaged in the sale of services such as transportation, utilities, wholesale and retail trade, health, leisure, maintenance, finance, government, information, and education.

As of 2011, in both Canada and the United States, about three-fourths of jobs and a similar proportion of the GDP were in the service sector. High-paying jobs exist in all the service categories, but low-paying jobs are more common. The largest private employer in the United States is Walmart (1.4 million employees), where the average wage is $12 an hour or $24,000 a year, full time. This is just barely above the poverty level for a family of four in the United States.

Service jobs are often connected in some way to international trade. They involve the processing, transport, and trading of agricultural and manufactured products and information that are either imported to or exported from North America. Hence, international events can shrink or expand the numbers of these jobs.

An important subcategory of the service sector involves the creation, processing, and communication of information—what is often labeled the "knowledge economy." The knowledge economy includes workers who manage information, such as those employed in finance, journalism, higher education, research and development, and many aspects of health care.

Industries that rely on the use of computers and the Internet to process and transport information are freer to locate where they wish than were the manufacturing industries of the old economic core, which depended on locally available steel and energy, especially coal. These newer industries are more dependent on highly skilled managers, communicators, thinkers, and technicians, and often locate near major universities and research institutions. They may also locate in clusters of companies involved with the subset of the knowledge economy known as the "**information technology**" or **IT** sector, which deals with computer software and hardware and the management of digital data.

The Internet Economy The Internet is emerging as an economic force more rapidly in North America than in any other region in the world. It was here that the Internet was first widely available. Though North America has only 5 percent of the world's population, it accounts for 22 percent of the world's Internet users. As of 2010, roughly 77 percent of the population of the United States used the Internet, as did 85 percent of the Canadian population, compared to 58 percent in the European Union and 28.7 percent for the world as a whole. The total economic impact of the Internet in North America is hard to assess, but retail Internet sales increase every year. Indeed, during the recession of 2008 to 2009, in both the United States and Canada, while overall purchases were down, online purchases increased steadily.

Internet-based social networking (Facebook, MySpace, Twitter, YouTube, and others) has now moved well beyond mere personal communication to play a rapidly increasing role in marketing, not only for large retail firms, but also especially for small businesses. Social networking has also entered the political sphere, having played a very large role in recruiting volunteers and cash contributions during the last several national election cycles.

All the growth of Internet-based activity in North America makes access to the Internet increasingly crucial. Unfortunately, a **digital divide** has developed, as important portions of the population are not yet able to afford computers and Internet connections. For the poor, the elderly, and many women and minorities, the public library or county courthouse may be their only access to computers and the Internet.

Globalization and the Economy

North America is wealthy, technologically advanced, and hugely influential in the global economy. The economy of the United States, in particular, has an impact on the world economy, for it is almost as large as the economy of the entire European Union. North America's advantageous position in the global economy is reflected in the pro-globalization policies of free trade promoted by major corporations and the governments of Canada and the United States (see Thematic Overview F).

This was not always the case. Before its rise to prosperity and global dominance, trade barriers were important aids to North American development. For example, when it achieved independence from Britain in 1776, the new U.S. government imposed tariffs and quotas on imports and gave subsidies to domestic producers. This protected fledgling domestic industries and commercial agriculture, allowing its economic core region to flourish.

Now, as wealthy and globally competitive exporters, both Canada and the United States see tariffs and quotas in other countries as obstacles to North America's economic expansion abroad. Hence, they are usually powerful advocates for the reduction of trade barriers worldwide. Critics of free trade policies point out a number of fallacies in the present North American position on free trade. First, just as North America once did, many poorer countries still need tariffs and quotas. Furthermore, both countries, contrary to their own free trade precepts, still give significant subsidies to their farmers. These subsidies make it possible for North American farmers to sell their crops on the world market at such low prices that farmers elsewhere are hurt or even driven out of business. For example, many Mexican small farmers have been driven out of business by competition from large U.S. corporate farms, many of which receive subsidies from the U.S. government. These farms can now sell their produce in Mexico or even relocate there under NAFTA agreements. Beyond agriculture, the critics add, in North America the benefits of free trade go mostly to large manufacturers and businesses and their managers, while many workers end up losing jobs to cheaper labor overseas, or see their income stagnate.

NAFTA Trade between the United States and Canada has long been relatively unrestricted. The process of trade barrier reduction formally began with the Canada–U.S. Free Trade Agreement of 1989. The creation of NAFTA in 1994 brought in Mexico as well. The major long-term goal of NAFTA is to increase the amount of trade among Canada, the United States, and Mexico. Today, it is the world's largest trading bloc in terms of the GDP of its member states.

The effects of NAFTA are hard to assess because it is difficult to tell whether many changes are due to the actual agreement or to other changes in regional and global economies. However, a few things are clear. NAFTA has increased trade, and many companies are making higher profits because they have larger markets. Since 1990, exports among the three countries have increased by more than 300 percent in value. NAFTA exports to the world economy, by value, have increased by about 300 percent for the United States and Canada and by 600 percent for Mexico. Some U.S. companies, such as Walmart, expanded aggressively into Mexico after NAFTA was passed. Mexico now has more Walmarts (1262) than any country except the United States (**Figure 2.13** on page 84).

NAFTA seems to have worsened the long-standing tendency for the United States to spend more money on imports than it earns from exports. This imbalance is called a **trade deficit**. Before NAFTA, the United States had much smaller trade deficits with Mexico and Canada. After the agreement was signed, these deficits rose dramatically, especially with Mexico. For example, between 1994 and 2009, the value of U.S. exports to Mexico increased by about 153 percent, while the value of imports increased 265 percent.

NAFTA has also resulted in a net loss of around 1 million jobs in the United States. Increased imports from Mexico and Canada have displaced about 2 million U.S. jobs, while increased exports to these countries have created only about 1 million jobs. Some new

---

information technology (IT) the part of the service sector that deals with the design and distribution of computer software and hardware and the management of digital data

digital divide the discrepancy in access to information technology between small, rural, and poor areas and large, wealthy cities that contain major government research laboratories and universities

trade deficit the extent to which the money earned by exports is exceeded by the money spent on imports

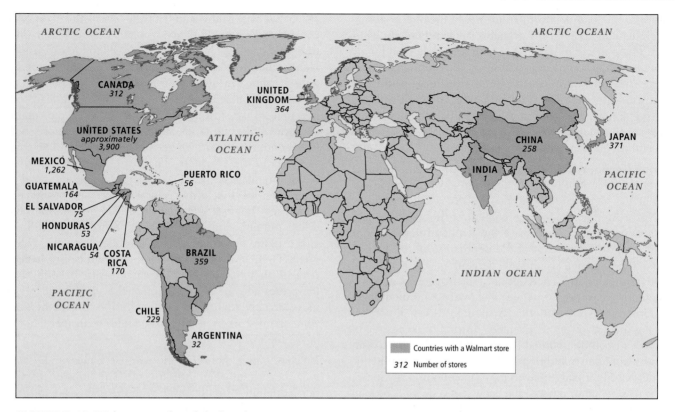

FIGURE 2.13 Walmart on the global scale. At the end of July 2009, Walmart had nearly 4400 store operations in the United States and over 3760 in 15 other countries. Walmart draws it products from over 70 countries, deals with 61,000 U.S. businesses, and is instrumental in generating 3 million U.S. jobs. Walmart itself employs over 1.4 million workers in the United States and 694,000 workers in other countries.

NAFTA-related jobs do pay up to 18 percent more than the average North American wage. However, those jobs are usually in different locations from the ones that were lost, and the people who take them tend to be younger and more highly skilled than those who lost jobs. Former factory workers often end up with short-term contract jobs or low-skill jobs that pay the minimum wage and carry no benefits.

NAFTA and Immigration from Mexico In Mexico, NAFTA appears to have increased exports and levels of foreign investment. However, these gains have been concentrated in only a few firms along the U.S. border and have not increased growth in the Mexican economy. Moreover, stiff competition from U.S. agribusiness has resulted in about 1.3 million job losses in the Mexican agricultural sector. These job losses, in turn, fuel legal and illegal immigration from Mexico to the United States, which increased from 350,000 people per year in 1992 to 1.2 million people (arriving either legally or illegally) in 2006. From 2007 through 2009, immigration from Mexico slowed by an estimated 42 percent, due first to a crackdown on illegal entry and then to the global economic recession.

As the drawbacks and benefits of NAFTA are being assessed, talk of extending it to the entire Western Hemisphere has stalled. Such an agreement, which would be called the Free Trade Area of the Americas (FTAA), would have its own drawbacks and benefits. A number of countries, such as Brazil, Bolivia, Ecuador, and Venezuela, are wary of being overwhelmed by the U.S. economy. Even in the absence of such an agreement, trade between North America and Middle and South America is growing faster than trade with Asia and Europe. This emerging trade is discussed in Chapter 3.

The Asian Link to Globalization NAFTA is only one way in which the North American economy is becoming globalized. The lowering of trade barriers has encouraged the growth of trade between North America and Asia. One huge category of trade with Asia is the seemingly endless variety of goods imported from China—everything from underwear to the chemicals used to make prescription drugs. China's lower wages make its goods cheaper than similar products imported from Mexico, despite Mexico's proximity and membership in NAFTA. Indeed, many factories that first relocated to Mexico from the southern United States have now moved to China to take advantage of its enormous supply of cheap labor. While recent concerns about lead and other toxins in Chinese imports have made many North Americans wary of these goods, trade with China promises to remain quite robust for some time.

North American investment in Asia, especially in China, is also driven by U.S. and Canadian companies that want to take advantage of China's vast domestic markets. For example, the United States fast-food chain Kentucky Fried Chicken (KFC) now has more than 2100 locations in 450 cities in China, and its business is growing rapidly there. The U.S. government has a powerful incentive to encourage such overseas expansion by companies like KFC that are headquartered in the United States and whose profits are taxable by the federal government.

Asian investment in North America is also growing. For example, Japanese and Korean automotive companies located plants in North America to be near their most important pool of car buyers—commuting North Americans. They often located their plants in the rural mid-South of the United States or in southern Canada close to arteries of the Interstate Highway System. Here the Asian companies found a ready, inexpensive labor force. These workers could access high-quality housing in rural settings within a commute of 20 miles (32 kilometers) or so from secure automotive jobs that paid reasonably well and included health and retirement benefit packages.

Japanese and Korean carmakers succeeded in North America even as U.S. auto manufacturers such as General Motors were struggling. This occurred primarily because more-advanced Japanese automated production systems—requiring fewer but better educated workers—produce higher-quality cars, which sell better both in North America and globally. Some analysts predict that foreign carmakers will eventually take over the entire North American market, while others predict that the old American car companies will be restructured to turn out more competitive, smaller, better-built, and more fuel-efficient cars.

IT Jobs Face New Competition from Developing Countries By the early 2000s, globalization was resulting in the *offshore outsourcing* of information technology (IT) jobs. A range of jobs—from software programming to telephone-based, customer-support services—were shifted to lower-cost areas outside North America. By the middle of 2003, an estimated 500,000 IT jobs had been outsourced, and forecasts are that 3.3 million more will follow by 2020. By 2009, IT jobs were still being created at a faster rate than they were ending due to the recession, but the overall trend pointed to fewer and fewer of the world's total IT jobs in North America. New IT centers are developing in India, China, Southeast Asia, the Baltic states in North Europe, Central Europe, and Russia. In these areas, large pools of highly trained, English-speaking young people work for wages that are less than 20 percent of their American counterparts. Some argue that rather than depleting jobs, outsourcing will actually help job creation in North America by saving corporations money, which will then be reinvested in new ventures. The viability of this argument remains to be seen.

▌▌▶ 45. U.S. COMPETITIVENESS

Repercussions of the 2008 Global Economic Downturn

The severe worldwide economic downturn that began in 2008 came on the heels of a long global upward trajectory of economic expansion. A booming housing industry and related growth in banks that finance home mortgages fueled this growth in the United States. In 2008, the housing industry collapsed as it became clear that much of the growth in previous years had been based on banks allowing millions of buyers to purchase homes with mortgages that were well beyond their means. When too many homebuyers could no longer afford their mortgage payments, the banks that had lent them money started to fail. The bank failures produced worldwide ripple effects because so many foreign banks were involved in the U.S. housing market. Between September 2008 and March 2009, the U.S. stock market fell by nearly half, wiping out the savings and pensions for millions of Americans. Similar plunges followed in foreign stock markets, ultimately resulting in a worldwide economic downturn as businesses could no longer find money to fund expansion.

As the recession intensified, job losses in the United States caused a sharp drop in consumption, which further affected world markets. Canada did not experience bank failures, but because so much of Canada's economy is linked to exports and imports from the United States (see Figure 2.10), Canada also suffered a high rate of job losses.

Women in the Economy

While women have made steady gains in achieving equal pay and overall participation in the labor force, there are still important ways in which they lag behind their male counterparts. On average, U.S. and Canadian female workers earn about 80 cents for every dollar that male workers earn for doing the same job. This means that, for example, a female architect will earn approximately 80 percent of what a male architect earns for performing comparable work. This situation is actually an improvement over previous decades. During World War II, when large numbers of women first started working in once male-dominated jobs, North American female workers earned, on average, only 57 percent of what male workers earned (see Thematic Overview I). The advances made by this older generation of women and the ones that followed have transformed North American workplaces. For the first time in history, women now represent more than half of the North American labor force, though most still work for male managers.

Throughout North America, women entrepreneurs are increasingly active, starting close to half of all new businesses. Women-owned businesses tend to be small, however, and less financially secure than those in which men have dominant control. This is true in part because it is harder for women business owners to obtain loans and large contracts—this despite special government credit programs and contract opportunities created for women business owners.

In secondary and higher education, North American women have equaled or exceeded the level of men in most categories. In 2008 in the United States, 33 percent of women age 25 and over held an undergraduate degree, compared to just 26 percent of men. This imbalance is likely to increase because U.S. women age 25 to 29 were receiving 7 percent more undergraduate degrees than men.

THINGS TO REMEMBER

1. North American farms have become highly mechanized operations that need few workers. To be profitable, however, they require huge investments in land and machinery, as well as the use of fertilizers and pesticides that contribute to water pollution.

2. In the twentieth century, the mass production of inexpensive automobiles and trucks, as well as the Interstate Highway System, fundamentally changed how people and goods move across the continent.

3. Learning Goal 4: Globalization and Development North America's economic base is now a service sector that is broad and varied. Its employment market has become oriented toward knowledge-intensive jobs that require more education and specialized professional training. Meanwhile, low-skill, mass-production industrial jobs are increasingly being moved abroad.

4. North America is a major promoter of globalization and free trade throughout the world. NAFTA is a major expression of this policy.

5. Flows of trade and investment between North America and Asia have increased dramatically in recent decades.

Sociocultural Issues

North American attitudes about urbanization, immigration, race, ethnicity, and religion are changing rapidly. Shifting gender roles are also redefining the North American family, and an aging population is raising new concerns about the future.

Urbanization and Sprawl

A dramatic change in urban spatial patterns has transformed the way most North Americans live. Since World War II, North America's urban populations have increased by about 150 percent, but the amount of land they occupy has increased almost 300 percent (Photo Essay 2.5). This is primarily because of suburbanization, a companion process to urbanization.

Today, close to 80 percent of North Americans live in **metropolitan areas**, cities of 50,000 or more plus their surrounding suburbs and towns. Most of these people live in car-dependent suburbs built since World War II that bear little resemblance to the central cities of the past.

In the nineteenth and early twentieth centuries, cities in Canada and the United States consisted of dense inner cores and less-dense urban peripheries that graded quickly into farmland. Starting in the early 1900s, central cities began losing population and investment while urban peripheries, the **suburbs**, began growing. Workers were drawn by the opportunity to raise their families in single-family homes with large lots in secure and pleasant surroundings. Some continued to work in the city, traveling to and from on streetcars. After World War II, suburban growth dramatically accelerated (Figure 2.14 on page 88) as cars became affordable to more workers.

As North American suburbs grew and spread out, nearby cities often coalesced into a single urban mass. The term **megalopolis** was originally coined to describe the 500-mile (800-kilometer) band of urbanization stretching from Boston through New York City, Philadelphia, and Baltimore, to the south of Washington, D.C. Other megalopolis formations in North America include the San Francisco Bay Area, Los Angeles and its environs, the region around Chicago, and the stretch of urban development from Eugene, Oregon, to Vancouver, British Columbia.

This pattern of urban sprawl requires residents to drive automobiles to complete most daily activities such as grocery shopping or commuting to work. Increasing air pollution and emission of greenhouse gases that contribute to climate change are two major side effects of the dependence on vehicles that comes with urban sprawl. Another important environmental consequence is habitat loss brought about by the invasion of farmland, forest, grassland, and desert by suburban development.

> **metropolitan areas** cities of 50,000 or more and their surrounding suburbs and towns
>
> **suburbs** populated areas along the peripheries of cities
>
> **megalopolis** an area formed when several cities expand so that their edges meet and coalesce

Farmland and Urban Sprawl Urban sprawl drives farmers from land that is located close to urban areas, because farmland on the urban fringe is very attractive to real estate developers. The land is cheap compared to urban land, and it is easy to build roads and houses on since farms are generally flat and already cleared. As farmland is turned into suburban housing, property taxes go up; soon all surrounding farmers can no longer afford to keep their land, so they sell it to housing developers. In North America each year, 2 million acres of agricultural and forestlands make way for urban sprawl.

Advocates of farmland preservation argue that beyond food and fiber, farms also provide economic diversity, soul-soothing scenery, and even habitat for some wildlife. For example, the town of Pittsford, New York (a suburb of Rochester), decided that farms were a positive influence on the community. Mark Greene's 400-acre, 200-year-old farm lay at the edge of town. As the population grew, the chances of the farm remaining in business for another generation looked dim. As land prices rose, so did property taxes, and the Greene family could not meet their tax payments. Residents in the new suburban homes, sprouting up on what had been neighboring farms, pushed local officials to halt normal farm practices, such as noisy nighttime harvesting or planting, spreading smelly manure, or importing bees to pollinate fruit trees. Pittsford, however, decided to stand by the farmers by issuing $410 million in bonds so that it could pay Greene and six other farmers for promises that they would not sell their 1200 acres to developers, but would instead continue to farm them.

Smart Growth and Livability The term *smart growth* has been coined for a range of policies aimed at stopping sprawl by making existing urban areas more *"livable."* *Livability* relates to factors that lead to a higher quality of life in urban settings, such as safety, good schools, affordable housing, quality health care, numerous and well-maintained parks offering recreational opportunities, and well-developed public transportation systems (see Reasons for Optimism B). Some indexes of livability for major cities worldwide are put out every year by *The Economist* magazine and the Mercer Quality of Living Survey.

The smart growth movement pushes for affordable and safe urban neighborhoods, where dwellings are much closer together. The vision is one of mass transit (buses, streetcars, and subways) replacing automobiles. In it, parks and other usable open spaces replace the large yards of suburbia, with more natural habitat preserved by the close-spaced dwellings. Spending on extensive

North America is highly urbanized, with 79 percent of the population living in cities. Some of the wealthiest cities in the world are located here, though only a few have achieved high levels of "livability." Many cities, especially those in the United States, are characterized by sprawling development patterns that require high levels of dependence on the automobile.

A Vancouver, Canada, is consistently ranked the most "livable" city in North America, and among the top four in the world. It is a leader in controlling sprawl. Over the past 10 years, the use of public transportation has risen by 50 percent, while the use of cars has fallen by 30 percent.

B New York City is the second wealthiest city in the world, the financial capital of North America, and its largest city. It is known for its extensive mass transit system (shown here), high population density (for North America), and world-class music and arts scene. In terms of livability, New York ranks toward the middle for the region.

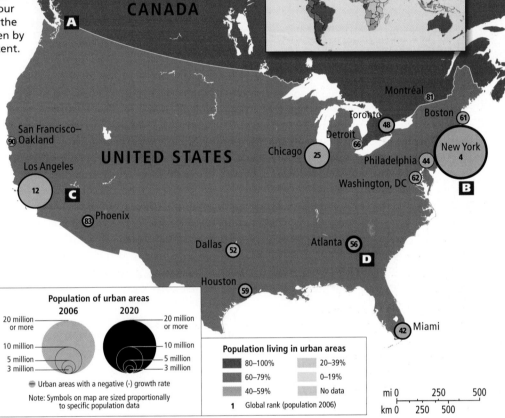

CANADA

Montréal 81
Toronto 48   Boston 61
Detroit 66
San Francisco– 90 Oakland
Chicago 25
UNITED STATES
Philadelphia 44
New York 4
Los Angeles
12
Washington, DC 62
Phoenix 83
Dallas 52   Atlanta 56
Houston 59
Miami 42

Population of urban areas

2006	2020
20 million or more	20 million or more
10 million	10 million
5 million	5 million
3 million	3 million

Urban areas with a negative (-) growth rate

Note: Symbols on map are sized proportionally to specific population data

Population living in urban areas

80–100%	20–39%
60–79%	0–19%
40–59%	No data

1 Global rank (population 2006)

mi 0   250   500
km 0   250   500

C Las Vegas, Nevada, was the fastest growing large city in North America between 2000 and 2007, with 31.8 percent growth. However, the recession that began in 2008 slowed growth for many months. In most rankings of livability Las Vegas comes in toward the lower end for North America.

D Morning traffic in Atlanta, Georgia, which added almost a million people between 2000 and 2007, more than any other city in North America. It is among the least dense and most sprawling cities in the region, with almost unversal dependence on cars for transport. In most rankings of livability, Atlanta comes in toward the lower middle for North America.

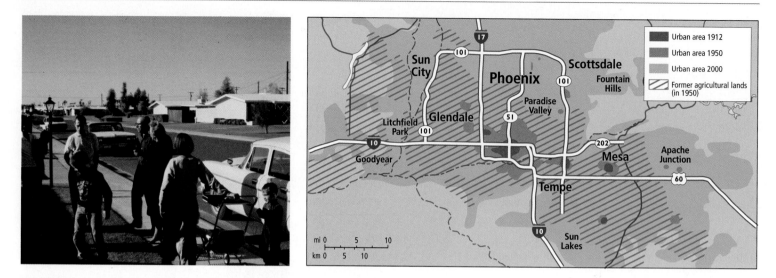

FIGURE 2.14 Urban sprawl in Phoenix, Arizona. Phoenix grew rapidly between 1950 and 2000. The photo was taken in 1964 in Sun City, then a new car-oriented suburb.

road infrastructure is shifted to spending on better schools, health care, and law enforcement. Many environmental benefits are envisioned as well. Air pollution and lower CO_2 emissions could result from more walking and greater use of mass transit, which is much more energy efficient than automobile-based transportation. More natural habitat could be preserved if dwellings were spaced closer together. While Smart Growth and the shift toward livability are just starting to take hold throughout North America, there are a few cities, such as Vancouver, British Columbia, and Portland, Oregon, where these principles are much further along in implementation (see Photo Essay 2.5).

Some older cities are enjoying a renaissance as people move back to them in search of the greater livability they sometimes offer. Places like New York City are arguably the furthest along in crucial aspects of Smart Growth and livability, given their high density and widespread use of mass transit. However, the high cost of living in these places can reduce their livability.

It may be a while before Smart Growth and livability make much of an impact in North America's fastest-growing cities, such as Atlanta and Las Vegas. Here, vast car-dependent suburbs have spread over enormous areas in recent years, and not surprisingly, neither city ranks high on livability indexes. Car dependence and rapid growth have brought Atlanta massive traffic jams and some of the worst air quality in the United States. Meanwhile, livability in the Las Vegas metro area suffers due to poor air and water quality as well as high crime rates.

The impact of sprawl on livability can also be seen in what is left behind in many North American cities. This is especially true in the United States, where many inner cities are dotted with large tracts of abandoned former industrial land, and neighborhoods debilitated by persistent poverty. Old industrial sites that once held factories or rail yards are called **brownfields**. Because they are often contaminated with chemicals and covered with obsolete structures, they can be very expensive to redevelop for other uses. Also left behind in the inner cities are the least-skilled and least-educated citizens, many of whom were drawn in generations ago by the promise of jobs that have long since moved out to the suburbs,

> **brownfields** old industrial sites whose degraded conditions pose obstacles to redevelopment
>
> **gentrification** the renovation of old urban districts by middle-class investment, a process that often displaces poorer residents

elsewhere in the region, or overseas. Often the majority population in these inner cities is a mixture of African-Americans, Asians, Hispanics, and other groups who identify themselves as nonwhite. Some are relatively affluent and are leading efforts to renew old city centers. Others are in great need of the very services—health care, schools, and social support (including churches, synagogues, and mosques)—that have moved to the suburbs.

The new emphasis on Smart Growth and livability has also reenforced the **gentrification** of old, urban residential districts. As affluent people invest substantial sums of money in renovating old houses and apartments, poor inner-city residents are often displaced in the process. The effect of gentrification on the displaced poor appears to be somewhat less harsh in Canada than in the United States, primarily because of Canada's stronger social safety net that better ensures housing and social services.

Some U.S. cities (for example, Knoxville, Tennessee; Portland, Oregon; and Charlotte, North Carolina) have initiated New Urbanism projects to re-house urban poor people in pleasant, newly built, walkable urban neighborhoods with conveniently located services.

Immigration and Diversity

Immigration has played a central role in populating both the United States and Canada. Most people in North America descend from European immigrants. However, new waves of migration from Middle and South America and parts of Asia promise to make this a region where people of non-European descent will be a majority (Figure 2.15). Most major North American cities are already characterized by ethnic diversity, and, in some, recent immigration has led to near majorities of foreign-born residents.

❙❙▶ 50. IMMIGRATION AND POPULATION REPORT

In the United States, the spatial pattern of immigration is also changing. For decades, immigrants settled mainly in coastal or border states such as New York, Florida, Texas, or California. However, since about 1990, immigrants are increasingly settling in interior states such as Illinois, Colorado, and Utah (Figure 2.15).

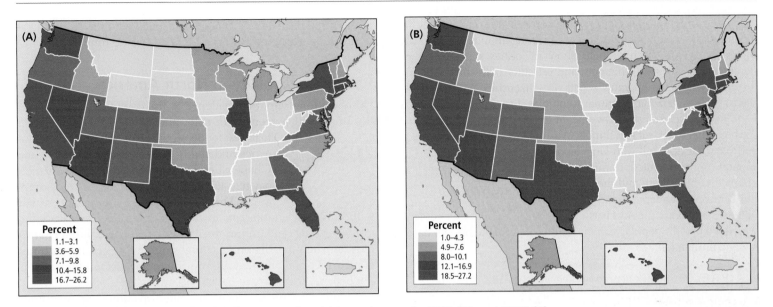

FIGURE 2.15 Percent of total foreign-born people within each state in 2000 (A) and 2007 (B).

Do New Immigrants Cost U.S. Taxpayers Too Much Money?
Many North Americans are concerned that immigrants burden schools, hospitals, and government services. And yet numerous studies have shown that, over the long run, immigrants contribute more to the U.S. economy than they cost. Legal immigrants have passed an exhaustive screening process that assures they will be self-supporting. As a result, most start to work and pay taxes within a week or two of their arrival in the country. Immigrants who draw on taxpayer-funded services such as welfare tend to be legal refugees fleeing a major crisis in their homeland, and are dependent only in the first few years after they arrive. More than one-third of immigrant families are firmly within the middle class, with incomes of $45,000 or higher. Even undocumented (illegal) immigrants play important roles as payers of payroll taxes, sales taxes, and indirect property taxes through rent. Because they fear deportation, illegal immigrants are also the least likely to take advantage of taxpayer-funded social services.

A 2004 study by the National Institutes of Health (NIH) reports that on average, immigrants are healthier and live longer than native U.S. residents. Hence, they represent less drain on the health-care and social service systems. The NIH attributes this difference to a stronger work ethic, a healthier lifestyle that includes more daily physical activity, and the more nutritious eating patterns of new residents compared to U.S. society at large. Unfortunately, these healthy practices tend to diminish the longer immigrants are in the country, and the more healthy status does not carry over to their children.

Do Immigrants Take Jobs Away from U.S. Citizens? The least-educated, least-skilled American workers are the most likely to end up competing with immigrants for jobs. In a local area, a large pool of immigrant labor can drive down wages in fields like roofing, landscaping, and general construction. Immigrants with little education now fill many of the very lowest paid service and agricultural jobs.

It is often argued that U.S. citizens have rejected these jobs because of their low pay, and hence immigrants are needed to fill them. Others argue that these jobs might pay more, and hence be more attractive to U.S. citizens, if there were not a large pool of immigrants ready to do the work for less pay. Research has failed to clarify the issue. Some studies show that immigrants have driven down wages by 7.4 percent for U.S. natives without a high school diploma. However, other studies show no drop in wages at all.

Professionals in the United States occasionally compete with highly trained immigrants for jobs, but such competition usually occurs in occupations where there is a scarcity of native-born people who are trained to fill these positions. The computer engineering industry, for example, regularly recruits abroad in such places as India, where there is a surplus of highly trained workers. In this case it is unclear whether these skilled immigrants are driving down wages. Until the recession caused jobs to be eliminated, it was quite clear that there were not enough sufficiently trained Americans to fill the available positions.

❙❙▶ 253. SKILLED FOREIGN WORKERS IN U.S. MAY HAVE TO LEAVE

Are Too Many Immigrants Being Admitted to the United States?
Many people are concerned that immigrants are coming in such large numbers that they will strain resources here. There is some truth to this point of view. Immigrants and their children accounted for 78 percent of the U.S. population growth in the 1990s. By the year 2050, the U.S. population is projected to exceed 440 million, with immigration accounting for the vast majority of the increase. But one of the problems with making these projections is that no one really knows how large undocumented (illegal) immigration really is. Most research indicates that it has reached unprecedented levels over the past 30 years, possibly having exceeded legal immigration since the mid-1990s. Estimates of the illegal immigrant population currently in the United States range from 7 to 20 million. This compares to around 20 million legal immigrants.

Undocumented immigrants tend to lack skills, and they are not screened for criminal background, as are all legal immigrants. However, research also shows that undocumented migrants are actually less likely to participate in criminal behavior than the general population, and that only tiny percentages of them have committed offenses.

❙❙▶ 47. AMERICAN CENSUS REPORT

❙❙▶ 51. IMMIGRATION LABOR SHORTAGE REPORT

There is less and less agreement or even clarity about what, if anything, should be done about illegal immigration to the United States. Numerous opinion polls show majorities supporting the deportation of illegal immigrants, and yet other polls show support for giving illegal immigrants a path to citizenship. A number of states, such as Arizona, Virginia, Kentucky, Tennessee, and South Carolina have passed strict laws meant to deter illegal immigration. And yet other states, such as Illinois, Washington, and New Mexico—all with large labor needs— support undocumented immigrants with programs that help them and their children. Even the stronger enforcement and the construction of fences and other physical barriers along the U.S.–Mexico border, which is a broadly popular response to illegal immigration, can have a tragic human toll.

ethnicity the quality of belonging to a particular culture group

Vignette Sheriff's deputy Michael Walsh works along the Arizona–Mexico border. Walsh describes a recent encounter with two bereaved young Mexican men holding the body of their relative: "Mostly you just find skeletons in the desert. This time there was a lot more emotion, family emotion. . . . Obviously they were close, they were crying. You have a name, you know that he had a brother, a cousin, a family in Mexico. . . ."

Matias Garcia, age 29, died after walking 32 miles through the desert. He was a Zapotec Indian who lived near Oaxaca (in southern Mexico) with his wife and three children, as well as his parents, younger brothers, and several cousins. Matias's cash crop of chili peppers was ruined by a spring frost just as they were ripening, leaving him in debt. He reluctantly decided to risk a trip to the United States to work in some vineyards where he had worked on and off since he was a teenager. From there he could send money home to his family to keep the house in repair and to send his children to school. But it takes money just to cross the border, and it took months for Matias, his younger brother, and a cousin to save the necessary amount.

Since the NAFTA agreement of 1994, miles of border fences have been built and patrols have increased, so the men decided to attempt to cross the more open but perilous Arizona desert on a route known as the "Devil's Highway." May is one of the hottest and driest months in this area. The men tried to avoid the worst of the heat by walking at night, but they failed to reach the highway on the Arizona side by dawn. They ran out of water, the sun became especially hot, and Matias began having seizures. His brother and cousin carried him, desperately looking for the highway, but he died shortly before they found it and could flag down someone to call for help. That's where Deputy Walsh found the two men grieving over Matias's body. [Find out more about Matias Garcia, why he migrated, how he died, and what his death has meant back in his village. Watch the Frontline/World video "Mexico: A Death in the Desert" on the web at http://www.pbs.org/frontlineworld/stories/mexico/.] ■

Research shows that the decision to migrate illegally into North America is largely a reluctant one undertaken because of a severe lack of economic opportunity at home. Ideally, NAFTA would have brought enough investment and jobs in Mexico to make illegal immigration to the United States unnecessary. And yet as we see in the case of NAFTA's impact on Mexican agriculture, incentives to immigrate to the United States have actually increased since the trade agreement came into effect. More and more, it seems that the problems associated with illegal immigration will be with North America for quite some time.

Race and Ethnicity in North America

Despite strong scientific evidence to the contrary, people across the world still perceive skin color and other visible anatomical features to be significant markers of intelligence and ability. As discussed previously (see Chapter 1 on pages 50–53), the science of biology tells us there is no validity to such assumptions. The same is true for **ethnicity**, which is the cultural counterpart to race, in that people may ascribe overwhelming (and unwarranted) significance to cultural characteristics, such as religion or family structure or gender customs. Hence, race and ethnicity are very important sociocultural factors not because they *have* to be, but because people *make* them so.

Numerous surveys show that large majorities of Americans of all backgrounds favor equal opportunities for minority groups. Nonetheless, in both the United States and Canada, many middle-class African-Americans, Native Americans, and Hispanics (and to a lesser extent, Asian-Americans) report experiencing both overt and covert discrimination that affects them economically as well as socially and psychologically. And indeed, even a cursory examination of statistics on access to health, education, and financial services shows that on average, Americans have quite uneven experiences based on their racial and ethnic characteristics.

Despite the increasing diversity of North America, in this region the term *race* has usually been used in relation to deeply embedded discrimination against African-Americans. Prejudice has clearly hampered the ability of African-Americans to reach social and economic equality with Americans of other ethnic backgrounds. In the United States, and to a somewhat lesser degree in Canada, despite the removal of legal barriers to equality, African-Americans as a group still experience lower life expectancies, higher infant mortality rates, lower levels of academic achievement, higher poverty rates, and greater unemployment than other groups.

In 2001, Hispanics overtook African-Americans as the largest minority group in the United States. Because of a higher birth rate and a high immigration rate, the Hispanic population increased by 58 percent in the 1990s. In 2010, Asian-Americans made up only 5 percent of the U.S. population, although their population had increased by 80 percent since 1990. Figure 2.16 shows the changes in the ethnic composition of the U.S. population in 1950, 2004, and projected for 2050.

In the United States, there are persistent discrepancies in income among Asian-Americans, Euro-Americans, Native Americans, Hispanics, and African-Americans (Figure 2.17). Over the past few decades, many non–Euro-Americans have joined the middle class, achieving success in the highest ranks of government and business. In particular, African-Americans have achieved advanced education in large numbers, and more than one-third now live in the suburbs. Yet overall, African-Americans, Hispanics, and Native Americans as groups remain the country's poorest people.

Is anything other than prejudice holding back some African-Americans, Hispanics, and Native Americans? Some argue yes. They point to the experience of first- and second-generation people of African descent, such as President Barack Obama (who is half

U.S. Population by Race and Ethnicity, 1950, 2006–2008, and 2050 (projected)

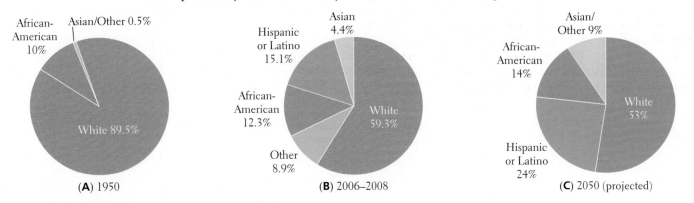

FIGURE 2.16 The changing U.S. ethnic composition.

African and half Euro-American) or African-Caribbean immigrants, such as Maya Angelou or Colin Powell, who have been notably successful even though their experience of discrimination and a family history of slavery is parallel to that in North America. A similar observation can be made about North Americans of Chinese and South Asian background. Although these groups often started out poor and suffered severe discrimination, they are now among the most prosperous in North America (Figure 2.17).

Observations such as these have led to the argument that some African-Americans, Hispanics, Native Americans, and persistently disadvantaged Anglo-Americans suffer from a *culture of poverty*. The argument is that poverty and low social status breed the perception among their victims that there is no hope for them and hence no point in trying to succeed. This perception is supported by the fact that, relative to wealthier North Americans of all races, the poorest, no matter what their background, have

fewer opportunities to get a decent education, a well-paying job, or a nice home to live in.

A major aspect of the culture of poverty is the single-parent family. In 2005, while only 23 percent of Euro-American children and 17 percent of Asian-American children lived in single parent families, 65 percent of African-Americans, 49 percent of Native Americans, and 36 percent of Hispanics did. In these situations, children usually stay with their mothers, and fathers often are not active in their support and upbringing. The enormous responsibilities of both child rearing and breadwinning are often left in the hands of undereducated young mothers who, in need themselves, are unable to help their children advance.

However, another explanation of persistent poverty among some of North America's minorities is that it is part of a larger problem of economic and spatial segregation based on class. In both the United States and Canada, the increasingly prosperous middle class, of whatever race or ethnicity, has moved to the suburbs. Hence, the very poor rarely have the chance to associate with models of success, and the successful no longer know anyone who is poor. Evidence for this class-based explanation is that when privileged Americans of any racial background share middle- and upper-class neighborhoods, workplaces, places of worship, and marriages, attention paid to skin color decreases markedly.

Religion

Because so many early immigrants to North America were Christian in their home countries, Christianity is currently the predominant religious affiliation claimed in North America. Seventy-six percent of Americans identified themselves as Christian in 2008. Nonetheless, virtually every medium-sized city has at least one synagogue, mosque, and Buddhist temple. In some localities, adherents of Judaism, Islam, or Buddhism are numerous enough to constitute a prominent cultural influence (Figure 2.18 on page 92).

There are many versions of Christianity in North America, and their geographic distributions are closely linked to the settlement patterns of the immigrants who brought them (see Figure 2.18). Roman Catholicism dominates in regions where Hispanic, French, Irish, and Italian people have settled—in southern Louisiana, the Southwest, and the far Northeast in the United States, and in Québec and other parts of Canada. Lutheranism is dominant where Scandinavian people have settled, primarily in Minnesota and the eastern Dakotas. Mormons dominate in Utah.

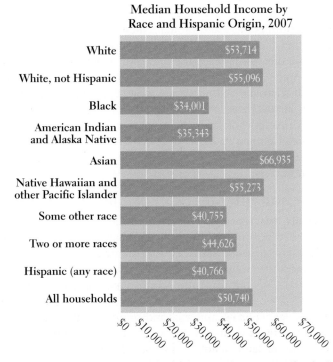

Median Household Income by Race and Hispanic Origin, 2007

White	$53,714
White, not Hispanic	$55,096
Black	$34,001
American Indian and Alaska Native	$35,343
Asian	$66,935
Native Hawaiian and other Pacific Islander	$55,273
Some other race	$40,755
Two or more races	$44,626
Hispanic (any race)	$40,766
All households	$50,740

FIGURE 2.17 Median household income by race and ethnicity, 2007. Amounts were reported in 2007 dollars.

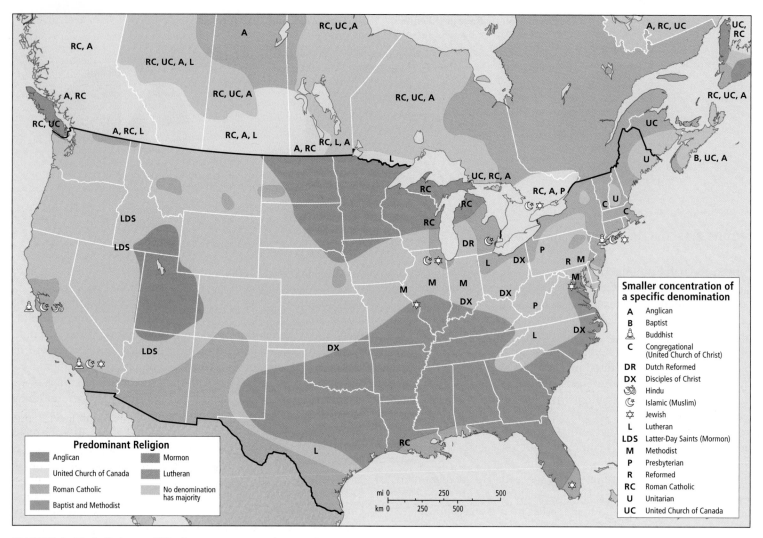

FIGURE 2.18 Religious affiliations across North America.

Baptists, particularly Southern Baptists and other evangelical Christians, are prominent in the religious landscapes of the South, which is known as the Bible Belt. Christianity—especially the Baptist version—is such an important part of community life in the South that newcomers to the region are frequently asked what church they attend.

According to data published by the U.S. Census Bureau, people in the United States who claim no religion ("nones") made up 15 percent of the population in 2008. The group includes atheists, agnostics, and the majority who simply claim no religion by choice.

The proper relationship of religion and politics has long been a controversial issue in the United States. This is true in large part because the framers of its Constitution, in an effort to ensure religious freedom, supported the idea that church and state should remain separate. In the past three decades, however, many conservative Christians have successfully pushed for a closer integration of religion and public life. Their political goals include banning abortion, promoting prayer in the public schools, teaching the biblical version of creation instead of evolution, and preventing gays and lesbians from marrying each other. The policies of conservative Christians have met with the most success in the southern United States, but their goals are shared by a minority scattered across the country.

> **nuclear family** a family consisting of a married father and mother and their children

New immigrants have brought their own faiths and belief systems, and they are contributing to the debate about religion and public life. Some leave the faith they came with and adopt another. Some 15 percent of Hispanic immigrants have left their traditional Catholic faith and are now evangelical Christians. National surveys have consistently indicated that a substantial majority of Americans favor the separation of church and state, and personal choice in belief and behavior—although the outcome of contentions in American religious and political life is not yet apparent.

Gender and the American Family

The family has repeatedly been identified as the institution most in need of support in today's fast-changing and ever more impersonal North America. A century ago, most North Americans lived in large extended families of several generations. Families pooled their incomes and shared chores. Aunts, uncles, cousins, siblings, and grandparents were almost as likely to provide daily care for a child as were the mother and father. The **nuclear family**, consisting of a married father and mother and their children, is a rather recent invention of the industrial age.

Beginning after World War I, and especially after World War II, many young people left their large kin groups on the farm and migrated to distant cities, where they established new nuclear families. Soon suburbia, with its many similar single-family homes, seemed to provide the perfect domestic space for the emerging nuclear family.

This small, compact family suited industry and business, too, because it had no firm ties to other relatives and hence was portable. Many North Americans born since 1950 have moved as many as ten times before reaching adulthood. The grandparents, aunts, and uncles who were left behind missed helping raise the younger generation, and they had no one to look after them in old age; therefore, nursing homes for the elderly proliferated.

In the 1970s, the whole system began to come apart. It was a hardship to move so often. Suburban sprawl meant onerous commutes to jobs for men and long, lonely days at home for women. Women began to want their own careers, and rising consumption patterns made their incomes increasingly useful to family economies. By the 1980s, 70 percent of the females born between 1947 and 1964 were in the workforce, compared with 30 percent of their mothers' generation.

Once employed, however, women could not easily move to a new location with an upwardly mobile husband. Nor could working women manage all of the family's housework and child care as well as a job. Some married men began to handle part of the household management and child care, but the demand for commercial child care grew sharply. With family no longer around to strengthen the marital bond, and with the new possibility of self-support available to women in unhappy marriages, divorce rates rose into the mid-1970s. Those who married after 1975, however, have a slowly declining rate of divorce. This may be related to the growing number of couples with college educations, as this group is less likely to divorce.

There is no longer a typical American household, only an increasing diversity of household forms (Figure 2.19). In 1960, the nuclear family—households consisting of married couples with children—comprised 74.3 percent of U.S. households. This number declined to 56 percent by 1990 and to 50 percent by 2008. Family households headed by a single person (female or male) rose from 10.7 percent in 1960 to 17 percent in 2008.

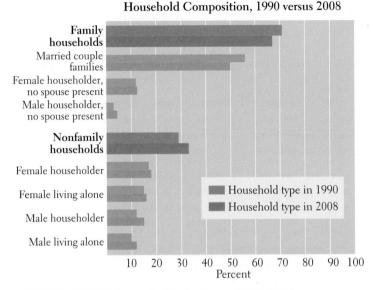

Household Composition, 1990 versus 2008

FIGURE 2.19 U.S. households by type, 1990–2008.

Nonfamily households (unrelated by blood or marriage) rose from 15 percent of the total in 1960 to 33 percent in 2008.

Some of these new family forms do not provide well for the welfare of children. In 2006, more than 29 percent of U.S. children lived in single-parent households. Although most single parents are committed to rearing their children well, the responsibilities can be overwhelming. Single-parent families tend to be hampered by economic hardship and lack of education (Figure 2.20A, B). The vast majority are headed by young women, whose incomes, on average, are more than one-third lower than those of single male heads of household (Figure 2.20A). Low income and low levels of education are closely linked, as the graph in Figure 2.20B demonstrates. A result of these patterns of single-parent households where education levels and incomes are low is that children in the United States are disproportionately poor. In 2007 in the United States, 18 percent of children lived in poverty, whereas only 11.4 percent of adults did. In Canada, 14.7 percent of children were poor. In Sweden, by comparison, 2.4 percent of children lived in poverty; in Ireland, 12.4 percent; and in Poland, 12.7 percent.

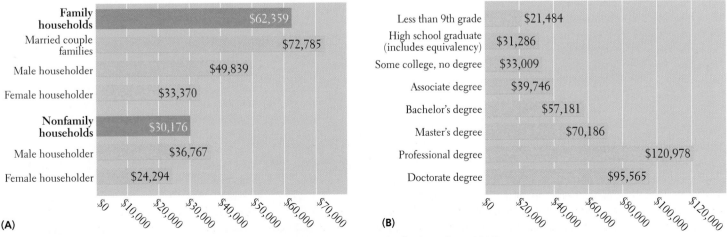

FIGURE 2.20 Median U.S. income by type of household and by level of education, 2007.

1. North America has some of the world's wealthiest cities, most of which are in the eastern United States. Nearly 80 percent of North Americans now live in metropolitan areas.

2. Since World War II, North America's urban populations have increased by about 150 percent, but the amount of land they occupy has increased almost 300 percent, a phenomenon known as urban sprawl.

3. A steady increase in migration from Middle and South America and parts of Asia promises to make North America a region where most people are of non-European descent.

4. Statistics on access to health, education, and financial services by race and ethnicity show that, on average, Americans have quite uneven experiences based on their racial and ethnic characteristics.

5. There is no longer a typical American household, only an increasing diversity of forms.

6. A large percentage of children in North America live in poverty, compared to other developed countries.

Population Patterns

The population map of North America (Figure 2.21) shows the uneven distribution of the more than 340 million people who

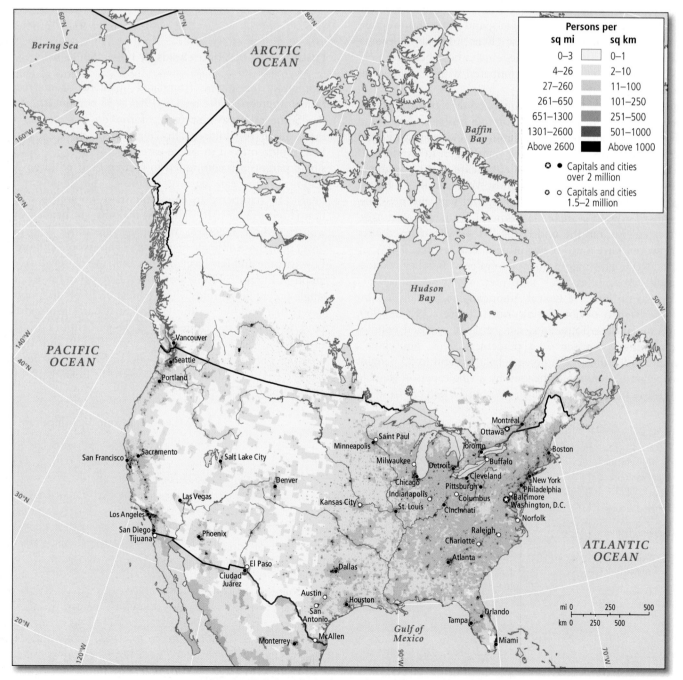

FIGURE 2.21 Population density in North America.

live on the continent. Canadians account for just under one-tenth (34 million) of North America's population. They live primarily in southeastern Canada, close to the border with the United States. The population of the United States is over 312 million, with the greatest concentration of people shifting away from the old economic core into other regions of the country that are now growing much faster. Just how this trend will play out, however, is uncertain. In 2004, the U.S. Census Bureau projected that between 2000 and 2030, the Northeast and Middle West would grow by under 10 percent while the South and West would grow at more than 40 percent. This general trend may persist, but the 2008 economic recession, water shortages, and the energy crisis have limited expansion, especially in the West, but also in the South.

In many farm towns and rural areas in the *Middle West* (the large central farming region of North America), populations are shrinking. As family farms are consolidated under corporate ownership, labor needs are decreasing and young people are choosing better-paying careers in cities. Middle Western cities are growing only modestly yet are becoming more ethnically diverse, with rising populations of Hispanics and Asians in places such as Indianapolis, St. Louis, and Chicago.

▌▌▶ 52. LATINO POLITICAL POWER

In the mountainous interior, settlement is generally light (see Figure 2.20). The principal reasons for this low population density are rugged topography, lack of rain, and in northern or high altitude zones a growing season that is too short to sustain agriculture. Some population clusters exist in irrigated agricultural areas, such as in the Utah Valley, and near rich mineral deposits and resort areas. The gambling economy and frenetic construction activity generated by real estate speculation account for several knots of dense population at the southern end of the region. Until the recession beginning in 2008, Las Vegas, Nevada, was the fastest-growing city in the United States. But by mid-2009, approximately 67,000 homes in Las Vegas were in foreclosure, tourist arrivals were sharply down, and hotel construction projects were halted.

Along the Pacific coast, a band of growing population centers stretches north from San Diego to Vancouver and includes Los Angeles, San Francisco, Portland, and Seattle. These are all port cities engaged in trade around the **Pacific Rim** (all the countries that border the Pacific Ocean). Over the past several decades, these North American cities have become centers of technological innovation.

The rate of natural increase in North America (0.6 percent per year) is low, less than half the rate of the rest of the Americas (1.4 percent). Still, North Americans are adding to their numbers fast enough through births and immigration that they could reach 440 million by 2050. Many of the important social issues now being debated in North America are linked to changing population patterns—issues such as legal and illegal immigration, which language should be used in public schools and in government offices, the geography of voting patterns, urbanization, cultural diversity, mobility, and the social effects of an aging population.

Mobility in North America

North Americans move more often than most other people. Every year, almost one-fifth of the U.S. population and two-fifths of Canada's population relocate. Some are changing jobs; others are

Learning Goal 5
Population and Gender: What changes have contributed to the aging of North America's population?

attending school, retiring to a warmer climate or a smaller city or town; others are merely moving across town or to the suburbs or to the countryside. Still other people are arriving from outside the region as immigrants.

Urbanization remains a powerful force behind this mobility. Dynamic economies and the search for lower production and living costs are drawing employers, employees, and retirees to urban areas. In Canada, they are going to cities in the southeast and on the West Coast. In the United States, people are moving to the South, Southwest, and Pacific Northwest. Cities in these areas, such as Atlanta and Washington D.C., have sprouted satellite or "edge" cities around their peripheries, often based on businesses dealing with technology and international trade. Suburbanization and urban sprawl are important features of this urbanization that have wider implications, such as the use of farmland for residential development and increased emissions of greenhouse gases as people commute longer distances. Urban sprawl is discussed on page 86.

Aging in North America

During the twentieth century, the number of older North Americans grew rapidly. One in 25 individuals was over the age of 65 in 1900; by 2008, the number was 1 in 8. By 2050, when most of the current readers of this book will be over 65, 1 in 5 North Americans will be elderly. This aging of the region's population relates to both longer life expectancy, which increases the number of older people, and a declining birth rate, which decreases the number of younger people. While life expectancy increased steadily throughout the twentieth century, birth rates declined significantly after the 1960s when more women chose to obtain more education and pursue careers, resulting in their having fewer children.

The number of people over the age of 65 is already high by global standards, and will increase dramatically over the next 20 years (see Thematic Concepts H). This will result from the combination of the marked jump in birth rate that took place after World War II, from 1947 to 1964, and the ensuing drop in birthrate that occurred afterward as women chose to have fewer children. The so-called *baby boomers* born in those years constitute the largest age group in North America, as can be seen in Figure 2.22 on page 96. As this group reaches age 65 and retires, the outflow of money from Social Security (the pool into which all workers pay to provide support for the elderly) and private pensions will be high, and medical costs are expected to leap upward. Moreover, because the baby boom was followed by much lower birth rates, fewer working-age people will be available to pay any medical costs that the baby boomers can't cover.

Aging populations in developed countries like Canada and the U.S. present us with an as-yet unresolved dilemma. On the one hand, it is widely agreed that population growth should be reduced to lessen the environmental impact of human life on earth, especially that of the societies that consume the most. On the other hand, slower population growth means that there will be fewer working-age people to keep the economy going and to provide the financial and physical help the increasing number of elderly people will require.

Pacific Rim a term referring to all the countries that border the Pacific Ocean

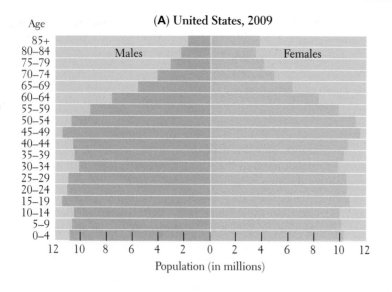

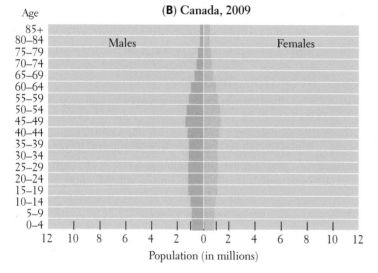

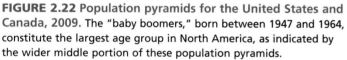

FIGURE 2.22 Population pyramids for the United States and Canada, 2009. The "baby boomers," born between 1947 and 1964, constitute the largest age group in North America, as indicated by the wider middle portion of these population pyramids.

THINGS TO REMEMBER

1. North Americans are highly mobile. Every year, almost one-fifth of the U.S. and two-fifths of the Canadian population move to a different location, for a variety of reasons.

2. North America's population is rapidly aging, and by 2050, one in five people will be over the age of 65. There will be fewer young people to work, pay taxes, and take care of the elderly.

3. **Learning Goal 5:** **Population and Gender** Birth rates in North America declined significantly after the 1960s as women chose to obtain more education and pursue careers, resulting in their having fewer children.

Reflections on North America

It is easy to rhapsodize about North America: the sheer size of the continent, its great wealth in natural and human resources, its

superior productive capacities, and its powerful political position in the world are attributes possessed by no other world region. North America enjoys this prosperity, privilege, and power as a result of fortunate circumstances, the hard work of its inhabitants, the diversity and creativity of its largely immigrant populations, astute planning and law writing on the part of early founders, and access to worldwide resources and labor at favorable prices. Perhaps the most important factor in North America's success has been its democratic institutions, in particular the Canadian and U.S. Constitutions, which allow for individual freedom and flexibility as times and circumstances change.

Yet life has not been good to everyone in North America, nor has the influence of North America on other parts of the world always been benign. Canada and the United States were created from lands forcibly taken from indigenous peoples. Many people of non-European background continue to suffer from racial prejudice and a "culture of poverty." People of all origins and their descendants had, and continue to have, a significant negative impact on the continent's environments. The standard of living now expected by all North Americans promises to increase the strain on the global environment.

There is no guarantee that North America will continue its leadership role into the future. In fact, in the post–9/11 world, challenges to that leadership occur in every corner of the globe. The recent U.S. tendency to use military force for its own ends has diminished respect for that country around the world. Meanwhile, North America's unwillingness to lead the world in responding to climate change is worsening many global environmental problems. There is also increasing resistance to globalization via models of economic development, such as free trade, that North America promotes but has not and does not observe itself.

And yet there is great reason for optimism about North America's future (Reasons for Optimism A, B, C). In recent decades, North America has shown its ability to adapt to economic changes and to come up with world-changing technological innovations in the process. As awareness of environmental problems is growing, so are potential solutions. Even the voracious appetite for resources of the North American lifestyle may be reduced in coming years as interest in "smart growth" increases and demand for food produced with sustainable methods grows. Indeed, perhaps no other region in the world has demonstrated its ability to adapt to change as has this one.

Learning Goals Review

1. **Climate Change and Urbanization:** How is North America's large per capita production of greenhouse gases related to its dominant pattern of urbanization?

How does urban sprawl influence the average North American's consumption of gasoline? What are some responses to these problems that you would be willing to take part in yourself?

2. **Water and Food:** How have North American food production systems contributed to water pollution?

How has "green revolution"–style agriculture introduced techniques that threaten the health of aquatic ecosystems?

3. **Democratization:** How has the ideal of democracy influenced the involvement of the United States in the affairs of foreign countries?

Reasons for Optimism in North America

Climate Change: While North America is far from leading the world in responding to climate change, it has technological capacity to do so. **A** *A wind farm outside of Los Angeles.* ▼

Urbanization: Renewed interest in livable cities and "smart growth" may reduce North America's demand for fossil fuels. **B** *A new high-population density, transit-oriented development in Portland, Oregon.* ▼

Food: Greater interest in sustainable food production may reduce impacts on the environment as more farms shift over to organic methods. **C** *A young gardener in Maryland.* ▼

Thinking Geographically: Why did the U.S. resist, until recently, efforts to reduce its consumption of fossil fuels?

Thinking Geographically: How would a shift towards mass transit help reduce demand for fossil fuels?

Thinking Geographically: What impacts on the environment may be avoided with a shift to organic agriculture?

Is democracy the driving force in these engagements, or do other factors have a greater influence?

4. Globalization and Development: How has globalization transformed economic development in North America, thereby changing the kinds of jobs available in this region?

How might this ongoing shift affect your education and future career?

5. Population and Gender: What changes have contributed to the aging of North America's population?

How have shifts in gender roles influenced the number of children women are having and, therefore, the age of the overall population? What are three ways in which you personally will be affected by the aging of North America's population?

Geographic Themes about North America

Look back at the Thematic Overview photos on page 57. Thinking geographically, answer the following questions about them:

(A) Climate Change: What is problematic about North America's production of greenhouse gases?

(B) Urbanization: How can you tell this suburb is on the urban fringe?

(C) Water: Can you see any evidence of efforts to contain pollution on this river?

(D) Food: What evidence of expensive investments in agriculture can be seen in this photo?

(E) Democracy: Why is health care related to democracy in North America?

(F) Globalization: Why might Kentucky Fried Chicken's presence in China be an aspect of globalization that the U.S. government would want to encourage?

(G) Development: What knowledge-intensive industry is evident in this photograph?

(H) Population: Is there any clue that this woman needs extra help in her daily life?

(I) Gender: How might this woman have paved the way for future generations of female workers to narrow the gender wage gap?

Key Terms

acid rain 66
agribusiness 80
aquifers 59
brownfields 88
clear-cutting 65
digital divide 83
economic core 72
ethnicity 90

gentrification 88
Hispanic 58
information technology (IT) 83
infrastructure 70
megalopolis 86
metropolitan areas 86
North American Free Trade
  Agreement (NAFTA) 82

nuclear family 92
Pacific Rim 95
service sector 82
social safety net 79
suburbs 86
trade deficit 83
urban sprawl 65

3 Middle and South America

Learning Goals

After you read this chapter, you will be able to answer the following questions:

1. **Climate Change and Deforestation:** How are the processes of global warming and deforestation linked?

2. **Water:** In this region of abundant water resources, what factors have come together to create a water crisis?

3. **Globalization, Development, and Democratization:** How has recent dissatisfaction with SAPs (structural adjustment policies) and globalization been expressed through the democratic process?

4. **Food and Urbanization:** How has the shift toward green revolution–style agriculture encouraged urbanization?

5. **Population:** What drove the population explosion that occurred in this region during the twentieth century?

6. **Gender:** What factors in the region's development are straining family ties and traditional gender roles?

FIGURE 3.1 Political map of Middle and South America.

Thematic Overview of Middle and South America

Climate Change: Deforestation in this region contributes to climate change by releasing large amounts of carbon dioxide into the atmosphere. **A** *Clearing along a road in Paraguay.* ▼

Water: Despite the region's abundant water resources, parts of the region are experiencing a water crisis due to unplanned urbanization, corruption, and inadequate water infrastructure. **B** *Swimming in the Amazon River.* ▼

Development: With the exception of a few small countries, the gap between rich and poor in this region is one of the largest in the world. **C** *High-rise apartments overlooking flooded slums in the city of João Pessoa, Brazil.* ▼

Globalization: Vast resources have been removed from the region over the past 500 years, most of these exported as cheap raw materials. **D** *Miners in Potosí, Bolivia, which has been mined for over 450 years.* ▼

Food: Food production is increasingly shifting away from small-scale to mechanized agriculture. **E** *Rice farming in southern Brazil.* ▼

Democratization: In the last 25 years, there have been repeated peaceful and democratic transfers of power in countries once dominated by dictators. **F** *Former Chilean president Eduardo Frei running for reelection.* ▼

Population: A population explosion in the twentieth century has given this region about 570 million people, more than 10 times the population of the region in 1492. **G** *A family in Guatemala.* ▼

Urbanization: Since the early 1970s, Middle and South America have led the world in migration from rural to urban communities. **H** *A public dance in Mexico City, the world's second-largest city.* ▼

Gender: Increasingly, women work outside the home, especially in export-oriented manufacturing. **I** *Women in Peru learning how to build computers.* ▼

Global Patterns, Local Lives The boat trip down the Aguarico River in Ecuador took me into a world of magnificent trees, river canoes, and houses built high up on stilts to avoid floods. I was there to visit the Secoya, a group of 350 indigenous people locked in negotiations with the U.S. oil company Occidental Petroleum over its plans to drill for oil on Secoya lands. Oil revenues supply 40 percent of the Ecuadorian government's budget and are essential to paying off its national debt. The government had threatened to use military force to compel the Secoya to allow drilling.

The Secoya wanted to protect themselves from pollution and cultural disruption. As Colon Piaguaje, chief of the Secoya, put it to me, "A slow death will occur. Water will be poorer. Trees will be cut. We will lose our culture and our language, alcoholism will increase, as will marriages to outsiders, and eventually we will disperse to other areas." Given all the impending changes, Chief Piaguaje asked Occidental to use the highest environmental standards in the industry. He also asked the company to establish a fund to pay for the educational and health needs of the Secoya people.

Like the Secoya, indigenous peoples around the world are facing environmental and cultural disruption arising from economic development efforts. Chief Piaguaje based his predictions for the future on what has happened in other parts of the Ecuadorian Amazon that have already experienced several decades of oil development.

The U.S. company Texaco was the first major oil developer to establish operations in Ecuador. From 1964 to 1992, its pipelines and waste ponds leaked almost 17 million gallons of oil into the Amazon Basin, enough to fill about 1900 fully loaded oil tanker trucks, or 35 Olympic-size swimming pools. Although Texaco sold its operations to the government and left Ecuador in 1992, its oil wastes continue to leak into the environment from hundreds of open pits (**Figure 3.2**).

In 1993, some 30,000 people sued Texaco in New York State, where the company (now owned by and called Chevron) is headquartered, for damages from the pollution. Those suing were both indigenous people and settlers who had established farms along Texaco's service roads. Several epidemiological studies concluded that oil contamination has contributed to higher rates of childhood leukemia, cancer, and spontaneous abortions among people who live near pollution created by Texaco.

Seven years after my visit to the Secoya, many of Chief Piaguaje's worries have been borne out. Occidental did establish a fund to help the Secoya deal with the disruption of oil development. However, part of this fund was distributed to individual households, with varying results. Some people invested their money in ecotourism and other commercial enterprises, and have prospered. Others simply spent their money and are now working for those who invested. Such employer–employee relationships are new to the Secoya, changing what was once an egalitarian culture.

Oil development has had many negative effects on the environment. Air and water pollution have increased rates of illness. The wildlife that the Secoya used to depend on, such as tapirs, have disappeared almost entirely due to overhunting by new settlers from the highlands who are working in the oil industry.

In 2002, the Ecuadorian suit against Chevron was dismissed by the U.S. Court of Appeals. It was refiled in Ecuador in 2003, and by 2010 it looked as if Chevron might lose the case and have to pay damages assessed now at between 40 and 90 billion dollars. *[Sources: Alex Pulsipher's field notes; Amazon Watch, 2006; Oxfam America, 2005; Juan Forero, "Rain Forest Residents, Texaco Face Off In Ecuador," National Public Radio, April 30, 2009, at http://www.npr.org/templates/story/story.php?storyId=103233560; Gonzalo Solano, "Damages in Chevron Ecuador Suit Jump Billions," San Francisco Chronicle, September 18, 2010, at http://www.sfgate.com/cgi-bin/article.cgi?f=/c/a/2010/09/17/BUKH1FFIOR.DTL.]* ■

FIGURE 3.2 Pollution from oil development in Ecuador. A worker samples one of the several hundred open waste pits that Texaco left behind in the Ecuadorian Amazon. Wildlife and livestock trying to drink from these pits are often poisoned or drowned. After heavy rains, the pits overflow, polluting nearby streams and wells.

The rich resources of Middle and South America have attracted outsiders since the first voyage of Christopher Columbus in 1492. Europe's encounter with this region marked a major expansion of the global economy. However, for several hundred years, Middle and South America occupied a disadvantaged position in global trade, supplying cheap raw materials that aided the Industrial Revolution in Europe. These industries did little to advance economic development in the region, as most profits went to foreign investors. In recent years, some countries of this region—Ecuador, Mexico, Bolivia, Brazil, and Venezuela—have taken control of their own resources, and are now supporting new and more profitable manufacturing and service industries. Meanwhile, trade blocs within the region are making countries better able to prosper from trade with each other.

Middle and South America differ from North America in several ways. Physically, the Middle and South America region is larger. Culturally, there are larger **indigenous** populations in Middle and South America, and more highly stratified social systems based on class, race, and gender. Economically, the gap between rich and poor is wider than in North America. Politically, the region's more than three dozen countries are more diverse, with governing ideologies that range from the socialism of Cuba to the capitalism of Chile. Yet there are many commonalities within the region, most arising from shared experience as colonies of Spain or Portugal.

THINGS TO REMEMBER

1. The region of Middle and South America is making a shift away from raw materials–based industries to more profitable manufacturing and service-based industries.

2. Politically, this region has a diverse array of governing ideologies, which range from socialist to capitalist.

I THE GEOGRAPHIC SETTING

Terms in This Chapter

In this book, **Middle America** refers to Mexico, Central America (the narrow ribbon of land that extends south of Mexico to South America), and the islands of the Caribbean (see Figure 3.1 on page 98). **South America** refers to the vast continent south of Central America. The term *Latin America* is not used in this book because it describes the region only in terms of the Roman (Latin-speaking) origins of the former colonial powers of Spain and Portugal. It ignores the region's large indigenous, African, Asian, and Northern European populations, as well as the many mixed cultures, often called *mestizo* cultures, that have emerged here. In this chapter, we use the term *indigenous groups* or *peoples* rather than *Native Americans* to refer to the native inhabitants of the region.

Physical Patterns

Middle and South America extend south from the midlatitudes of the Northern Hemisphere across the equator through the Southern Hemisphere, nearly to Antarctica (see Figure 3.3 on pages 102–103). This vast north-south expanse combines with variations in altitude to create the wide range of climates in the region. Tectonic forces have shaped the primary landforms of this huge territory to form an overall pattern of highlands to the west and lowlands to the east.

Landforms

There are a wide variety of landforms in Middle and South America, and this variety accounts for the many different climatic zones in the region. But for ease in learning, landforms are here divided into just two categories: highlands and lowlands.

Highlands A nearly continuous chain of mountains stretches along the western edge of the American continents for more

> **indigenous** native to a particular place or region
>
> **Middle America** in this book, a region that includes Mexico, Central America, and the islands of the Caribbean
>
> **South America** the continent south of Central America
>
> **subduction zone** a zone where one tectonic plate slides under another

than 10,000 miles (16,000 kilometers) from Alaska to Tierra del Fuego at the southern tip of South America. The middle part of this long mountain chain is known as the Sierra Madre in Mexico (see Figure 3.3A), by various names in Central America, and as the Andes in South America (see Figure 3.3B, C). It was formed by a lengthy **subduction zone,** which runs thousands of miles along the western coast of the continents (see Figure 1.21 on page 45). Here, two oceanic plates—the Cocos Plate and the Nazca Plate—plunge beneath three continental plates—the North American Plate, the Caribbean Plate, and the South American Plate.

In a process that continues today, the leading edge of the overriding plates crumples to create mountain chains. In addition, molten rock from beneath the earth's crust ascends to the surface through fissures in the overriding plate to form volcanoes. Such volcanoes are the backbone of the highlands that runs through Middle America and the Andes of South America (see Figure 3.3). Although these volcanic and earthquake-prone highlands have been a major barrier to transportation, communication, and settlement, people now live close to quiescent volcanoes and in earthquake-prone zones, which can pose deadly hazards (for example, the 2010 earthquake in Peru).

The chain of high and low mountainous islands in the eastern Caribbean is also volcanic in origin, created as the Atlantic Plate thrusts under the eastern edge of the Caribbean Plate. It is not unusual for volcanoes to erupt in this active tectonic zone. On the island of Montserrat, for example, people have been living with an active, and sometimes deadly, volcano for more than a decade (see Figure 3.3D). Eruptions have taken the form of violent blasts of superheated rock, ash, and gas (known as pyroclastic flows) that move down the volcano's slopes with great speed and force. The unusually strong earthquake in Haiti in January 2010 was also the result of plate tectonics.

USA

Tijuana
El Paso
Ciudad
Juárez

Hermosillo

Mulegé

Baja California

Sierra Madre Occidental

Gulf of California

MEXICO

Nuevo Laredo

Monterrey

Matamoros

Durango

Mazatlán

San Luis
Potosí

Tampico

Puerto Vallarta

Guadalajara

Sierra Madre Oriental

Bay of
Campeche

Mérida

Mexico
City

Veracruz

Puebla

Acapulco

Oaxaca

Gulf of
Tehuantepec

Yucatán
Peninsula

Rio Grande

San
Antonio

Houston

New
Orleans

Gulf of
Mexico

Jacksonville

Orlando

Tampa

Miami

Tropic of Cancer

PACIFIC

OCEAN

30°N

20°N

10°N

0° Equator

10°S

20°S

30°S

40°S

Havana

CUBA

Yucatán Channel

Cancún

BAHAMAS

ATLANTIC

OCEAN

Bermuda
(U.K.)

GREATER ANTILLES

DOMINICAN
REPUBLIC

Port-au-Prince

JAMAICA

Kingston

HAITI

Santo
Domingo

San Juan

Puerto Rico
(U.S.A.)

Montserrat
(U.K.)

Belize City
BELIZE
Belmopan

Guatemala
City

GUATEMALA

San Salvador
EL SALVADOR

HONDURAS
Tegucigalpa

NICARAGUA

Managua

Lake
Nicaragua

San José
COSTA RICA

Colón
Panamá

Panama
Canal

PANAMA

Gulf of
Panama

Caribbean Sea

LESSER ANTILLES

ST. LUCIA
ST. VINCENT

GRENADA

BARBADOS

TRINIDAD & TOBAGO
Port of Spain

Barranquilla

Maracaibo

Caracas

Valencia

Gulf of
Venezuela

Lake
Maracaibo

VENEZUELA

Llanos

Orinoco

Medellín

Bogotá

Cali

COLOMBIA

Quito

ECUADOR

Guayaquil

Gulf of
Guayaquil

Galápagos
Islands
(Ecuador)

Piura

Marañón

PERU

Callao
Lima

Cusco

Andes

Georgetown
Paramaribo

GUYANA

SURINAME

FRENCH
GUIANA
(France)

Guiana Highlands

Negro

Manaus

Amazon

E

B R A

Amazon Basin

Solimões

Ucayali

Madeira

Xingu

Mato
Grosso

BOLIVIA

La Paz

Sucre

Potosí

Cochabamba

Santa Cruz

Lake
Titicaca

Altiplano

B

Tropic of Capricorn

Antofagasta

Atacama
Desert

CHILE

Mtns.

PARAGUAY

Asunción

Paraná

Córdoba

Valparaíso

Santiago

Rosario

Porto Alegre

URUGUAY

Buenos Aires

La Plata

Montevideo

Río de la Plata

ARGENTINA

Pampas

Paraná

Monte
Verde

Puerto Montt

Patagonia

Comodoro Rivadavia

Falkland
Islands
(U.K.)

Stanley

Strait of
Magellan

Punta Arenas

Tierra
del Fuego

C

A Sierra Madre, Mexico

B Andes, Chile

C Tierra del Fuego, Argentina

H

D

I

Canary
Islands

Western
Sahara

MAURITANIA

CAPE
VERDE

SENEGAL

Praia

Dakar

D Volcano, Montserrat

G Rio de Janeiro, Brazil

H Yucatan Lowlands, Mexico (with Mayan temple)

Cayenne

ATLANTIC OCEAN

Belém

Z I L

Fortaleza

Araguaia

Tocantins

Recife

Land Elevations

meters	feet
4877	16,000
3353	11,000
2134	7000
914	3000
305	1000
152	500
0	0

São Francisco

Brasília

Brazilian Highlands

Salvador

mi 0   200   400   600
km 0   200   400   600   800   1000

1:37,000,000
Azimuthal Equidistant Projection

Belo
Horizonte

G

São Paulo

Rio de Janeiro

Curitiba

F Amazonian Basin, Brazil (from the air)

I Pampas, Argentina

E Amazonian Basin, Brazil (from space)

FIGURE 3.3 Regional map of Middle and South America.

30°W   20°W   10°W   0° Prime Meridian

Lowlands Vast lowlands extend over most of the land to the east of the western mountains. In Mexico, east of the Sierra Madre, a coastal plain borders the Gulf of Mexico (see Figure 3.3H). Farther south, in Central America, wide aprons of sloping land descend to the Caribbean coast. In South America, a huge wedge of lowlands, widest in the north, stretches from the Andes east to the Atlantic Ocean. These South American lowlands are interrupted in the northeast and the southeast by two modest highland zones: the Guiana Highlands and the Brazilian Highlands (see Figure 3.3G). Elsewhere in the lowlands, grasslands cover huge, flat expanses, including the llanos of Venezuela, Colombia, and Brazil and the pampas of Argentina (see Figure 3.3I).

The largest feature of the South American lowlands is the Amazon Basin, drained by the Amazon River and its tributaries (see Figure 3.3E, F). This basin lies within Brazil and the neighboring countries to its west. Here, the earth's largest remaining expanse of tropical rain forest gives the Amazon Basin global significance as a reservoir of **biodiversity**. Hundreds of thousands of plant and animal species live here.

The basin's water resources are also astounding. Twenty percent of the earth's flowing surface waters exist here, running in rivers so deep that ocean liners can steam 2300 miles (3700 kilometers) upriver from the Atlantic Ocean all the way to Iquitos, jokingly referred to as Peru's "Atlantic seaport." The vast Amazon River system starts as streams high in the Andes. These streams eventually unite as rivers that flow eastward toward the Atlantic. Once they reach the flat land of the Amazon Plain, their velocity slows abruptly, and fine soil particles, or **silt**, sink to the riverbed. When the rivers flood, silt and organic material transported by the floodwaters renew the soil of the surrounding areas, nourishing millions of acres of tropical forest. Not all of the Amazon Basin is rain forest, however. Variations in weather and soil types, as well as human activity, have created grasslands and seasonally dry deciduous tropical forests in some areas.

Climate

From the jungles of the Caribbean and the Amazon to the high, glacier-capped peaks of the Andes to the parched moonscape of the Atacama Desert, the climate variety of Middle and South America is enormous (Photo Essay 3.1). Climates are essentially the result of interactions between temperature and moisture. In this region, the wide range of temperatures reflects both the great distance the landmass spans on either side of the equator and the tremendous variations in altitude across the region's landmass (the highest point in the Americas is Aconcagua in Argentina, at 22,841 feet, or 6962 meters). Patterns of precipitation are affected both by the local shape of the land and by global patterns of wind and ocean currents that bring moisture in varying amounts.

Temperature-Altitude Zones Four main **temperature-altitude zones** are commonly recognized in the region (Figure 3.4 on page 106). As altitude increases, the temperature of the air decreases by about 1°F per 300 feet (1°C per 165 meters) of elevation. Thus temperatures are highest in the lowlands, which are known in Spanish as the *tierra caliente*, or "hot

land." The *tierra caliente* extends up to about 3000 feet (1000 meters), and in some parts of the region these lowlands cover wide expanses. Where moisture is adequate, tropical rain forests thrive, as does a wide range of tropical crops, such as bananas, sugarcane, cacao, and pineapples. Many coastal areas of the *tierra caliente*, such as northeastern Brazil, have become zones of plantation agriculture that support populations of considerable size.

Between 3000 and 6500 feet (1000 to 2000 meters) is the cooler *tierra templada* ("temperate land"). The year-round, spring-like climate of this zone drew large numbers of indigenous people in the distant past and, more recently, has drawn Europeans. Here, such crops as corn, beans, squash, various green vegetables, wheat, and coffee are grown.

Between 6500 and 12,000 feet (2000 to 3600 meters) is the *tierra fria* ("cool land"). Many crops such as wheat, fruit trees, potatoes, and cool-weather vegetables—cabbage and broccoli, for example—do very well at this altitude. Many animals—such as llamas, sheep, and guinea pigs—are raised for food and fiber. Several modern population centers are in this zone, including Mexico City, Mexico, and Quito, Ecuador.

Above 12,000 feet (3600 meters) is the *tierra helada* ("frozen land"). In the highest reaches of this zone, vegetation is almost absent, and mountaintops emerge from under snow and glaciers. A remarkable feature of such tropical mountain zones is that in a single day of strenuous hiking, one can encounter many of the climate types found on earth.

Precipitation The pattern of precipitation throughout the region is influenced by the interaction of global wind patterns with mountains and ocean currents (see Photo Essay 1.5, pages 48–49). The **trade winds** sweep off the Atlantic, bringing heavy seasonal rains to places roughly 23 degrees north and south of the equator (see the map in Photo Essay 3.1). Winds from the Pacific bring seasonal rain to the west coast of Central America, but mountains block that rain from reaching the Caribbean side, which receives heavy rainfall from the northeast trade winds.

The Andes are a major influence on precipitation in South America. They block the rains borne by the trade winds off the Atlantic into the Amazon Basin and further south, creating a rain shadow on the western side of the Andes in northern Chile and southwestern Peru (see the map in Photo Essay 3.1). Southern Chile is in the path of eastward-trending winds that sweep out of the Southern Ocean, bringing steady, cold rains that support forests similar to those of the Pacific Northwest of North America. The Andes block this flow of wet, cool air and divert it to the north. They thereby create another extensive rain shadow on the eastern side of the mountains along the southeastern coast of Argentina (Patagonia).

The pattern of precipitation is also influenced by the adjacent oceans and their currents. Along the west coasts of Peru and Chile, the cold surface waters of the Peru Current bring cold air that cannot carry much moisture. The combined effects of the Peru Current and the central Andes rain shadow have created what is possibly the world's driest desert, the Atacama of northern Chile (see Photo Essay 3.1B).

---

biodiversity the variety of life forms to be found in a given area

silt fine soil particles

temperature-altitude zones regions of the same latitude that vary in climate according to altitude

trade winds winds that blow from the northeast and the southeast toward the equator

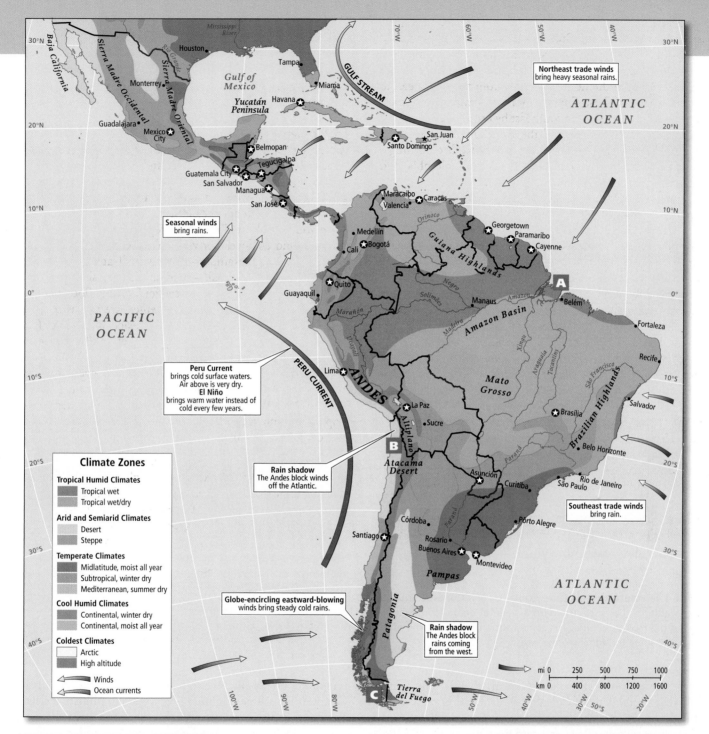

Baja California

Sierra Madre Occidental

Sierra Madre Oriental

Mississippi River

Houston

Rio Grande

Tampa

Gulf of Mexico

Miami

Yucatán Peninsula

Havana

GULF STREAM

Northeast trade winds bring heavy seasonal rains.

ATLANTIC OCEAN

Guadalajara

Monterrey

Mexico City

Belmopan

Tegucigalpa

Guatemala City
San Salvador

Managua

San José

San Juan

Santo Domingo

Maracaibo
Valencia

Caracas

Orinoco

Georgetown
Paramaribo
Cayenne

Seasonal winds bring rains.

Medellín

Cali

Bogotá

Guiana Highlands

Quito

Negro

Manaus

Amazon

Belém

A

PACIFIC OCEAN

Guayaquil

Marañón

Solimões

Fortaleza

Madeira

Amazon Basin

Xingu

Araguaia

Tocantins

Recife

ANDES

Lima

PERU CURRENT

Peru Current brings cold surface waters. Air above is very dry. **El Niño** brings warm water instead of cold every few years.

La Paz

Sucre

Altiplano

Mato Grosso

Brasília

São Francisco

Brazilian Highlands

Salvador

Belo Horizonte

B

Rain shadow The Andes block winds off the Atlantic.

Atacama Desert

Asunción

Curitiba

Paraná

São Paulo

Rio de Janeiro

Southeast trade winds bring rain.

Córdoba

Rosario
Buenos Aires

Santiago

Pampas

Pôrto Alegre

Montevideo

ATLANTIC OCEAN

Globe-encircling eastward-blowing winds bring steady cold rains.

Patagonia

Rain shadow The Andes block rains coming from the west.

C

Tierra del Fuego

Climate Zones

Tropical Humid Climates
- Tropical wet
- Tropical wet/dry

Arid and Semiarid Climates
- Desert
- Steppe

Temperate Climates
- Midlatitude, moist all year
- Subtropical, winter dry
- Mediterranean, summer dry

Cool Humid Climates
- Continental, winter dry
- Continental, moist all year

Coldest Climates
- Arctic
- High altitude

→ Winds
→ Ocean currents

| mi 0 | 250 | 500 | 750 | 1000 |
| km 0 | 400 | 800 | 1200 | 1600 |

A Tropical wet, Belém, Brazil

B Desert, Atacama, Chile

C Continental, moist all year, Tierra del Fuego

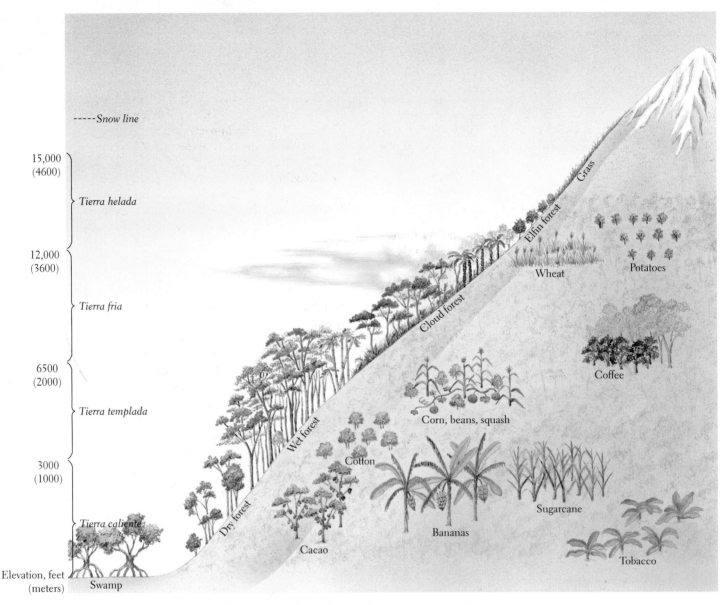

FIGURE 3.4 Temperature-altitude zones of Middle and South America. Temperatures tend to decrease as altitude increases. The natural vegetation on mountain slopes also changes, as shown along the slope of the mountain here. The same is true for crops, some of which are suited to lower, warmer elevations and some to higher, cooler ones.

El Niño One aspect of the Peru Current that is only partly understood is its tendency to change direction every few years (on an irregular cycle, possibly linked to sunspot activity). When this happens, warm water flows eastward from the western Pacific, bringing warm water and torrential rains, instead of cold water and dry weather, to parts of the west coast of South America. The phenomenon was named **El Niño**, or "the Christ Child," by Peruvian fishermen, who noticed that when it does occur, it reaches its peak around Christmastime.

El Niño also has global effects, bringing cold air and drought to normally warm and humid western Oceania and unpredictable weather patterns to Mexico and the southwestern United States. The El Niño

> **El Niño** periodic climate-altering changes, especially in the circulation of the Pacific Ocean, now understood to operate on a global scale

phenomenon in the western Pacific is discussed further in Chapter 11, where Figure 11.5 (page 414) illustrates its trans-Pacific effects.

Hurricanes In this region, many coastal areas are threatened by powerful storms that can create extensive damage and loss of life. These form annually, primarily in the Atlantic Ocean north of the equator and close to Africa. A tropical storm begins as a group of thunderstorms. A few hurricanes also form in the southeastern Pacific and can affect the western coasts of Middle America before turning west toward Hawaii. When enough warming wet air comes together, the individual storms organize themselves into a swirling spiral of wind that moves across the earth's surface. The highest wind speeds are found at the edge of the eye, or center, of the storm. Once wind speeds reach 75 miles (120.7 kilometers) per hour, such a storm is officially called a *hurricane*. Hurricanes usually last about 1 week; because they draw their energy from warm surface waters, they slow down and eventually dissipate as they move over cooler water or land. Human exposure to hurricanes is increasing as coastal populations increase. Some scientists also think that climate change is leading to an increase in the number and intensity of hurricanes.

THINGS TO REMEMBER

1. The rain forests of the Amazon Basin are planetary treasures of biodiversity that also play a key role in regulating the earth's climate.

2. There are four temperature-altitude zones in the region that influence where and how people live and the crops they can grow.

3. Significant environmental hazards include earthquakes, volcanic eruptions, and hurricanes.

Environmental Issues

Environments in Middle and South America have long inspired concern about the use and misuse of the earth's resources. Millenia before Europeans arrived in this region, human settlements had major environmental impacts. However, today's impacts are particularly severe because both population density and per capita consumption have increased so dramatically. Moreover, local environments now supply global demands.

Tropical Forests, Climate Change, and Globalization

Learning Goal 1
Climate Change and Deforestation: How are the processes of global warming and deforestation linked?

A period of rapid deforestation with global repercussions began in the 1970s in Brazil with the construction of the Trans-Amazon Highway. Migrant farmers followed the new road into the rain forest. They began clearing terrain to grow crops, prompting a few observers to warn of an impending crisis of deforestation. Initially, concern focused on the loss of plant and animal species, as the rainforests of the Amazon Basin are some of the most biodiverse on the planet. However, climate change now dominates concerns about deforestation in the Amazon and the rest of this region. As explained in Chapter 1 (see pages 24–25), forests release oxygen and absorb carbon dioxide, the greenhouse gas most responsible for global warming. Hence, the loss of large forests, such as those in the Amazon Basin, contributes to global warming by releasing carbon dioxide once locked in the woody bodies of trees. This carbon dioxide release happens as trees are burned to make way for crops and roads (see Thematic Overview A on page 99). Fewer trees also mean that less carbon dioxide can be taken out of the atmosphere. Together, all of the earth's tropical rainforests absorb around 18 percent of the CO_2 added to the atmosphere yearly by human activity. Because 50 percent of the earth's remaining tropical rainforests are in South America, keeping these forests intact is crucial to controlling climate change.

Middle and South American rain forests are being diminished by multiple human impacts (see Photo Essay 3.2 on page 108). Primary among them is the clearing of land to raise cattle and grow crops, such as soybeans (for animal feed) and sugarcane for ethanol. Brazil now ranks as the fourth-largest emitter of greenhouse gases in the world after the United States, China, and Indonesia. In the Amazon, Middle America, and elsewhere in the region, forests are cleared to create pastures for beef cattle, many of which are destined for the U.S. fast-food industry. If deforestation continues in Middle America at the present rates, the natural forest cover will be entirely gone in 20 years.

Also contributing to deforestation are the logging of hardwoods and the extraction of underlying minerals, including oil, gas, and precious stones. Investment capital comes from Asian multinational companies that have turned to the Amazon forests after having logged up to 50 percent of the tropical forests in Southeast Asia. Moreover, the construction of access roads to support these activities continues to accelerate deforestation by opening new forest areas to migrants. The governments of Peru, Ecuador, and Brazil encourage impoverished urban people to occupy cheap land along the newly built roads (see Photo Essay 3.2A), and encourage deforestation by supplying chainsaws to the settlers. However, the settlers have a difficult time learning to cultivate the poor soils of the Amazon. After a few years of farming, they often abandon the land, now eroded and depleted of nutrients, and move on to new plots. Ranchers may then buy the worn-out land from these failed small farmers to use as cattle pastures.

While Middle and South America are significantly contributing to global warming by adding CO_2 to the earth's atmosphere, many people in this very region are particularly vulnerable to the increasing threats of climate change. Photo Essay 3.3 on page 110 illustrates several such cases: shortages of clean water brought on by intensifying droughts and by glacial melting, vulnerability to rising sea levels and the effects of increasingly violent storms.

Major international efforts are under way to combat climate change by preserving the world's remaining forests, especially the crucial tropical rainforests found in this region. This strategy is deemed to be among the cheapest ways to reduce greenhouse gas emissions, with well-funded and carefully executed efforts potentially transforming forests throughout the region. However, implementing these complex programs, which will demand cooperation among multiple nations, will be an enormous challenge (see page 379 in Chapter 10).

Environmental Protection and Economic Development

In the past, governments in the region argued that economic development was so desperately needed that environmental regulations were an unaffordable luxury. Now, there are increasing attempts to embrace both economic development and

While there is a wide variety of human impacts on the environments of this region, here we focus on the processes of land cover and land use change that lead to the conversion of forests to grazing land or farmland. The forces guiding this process are complex, driven by poor people's need for livelihoods, governmental desire to assert control over lightly populated areas, and the demands for wood, meat, and food that arise in distant urban centers and the global market. Together these forces have led to a rapid loss of forest cover throughout much of this region. Brazil loses more forest cover each year than any country on the planet.

A A newly built road that connects to the Trans-Amazon Highway in Brazil.

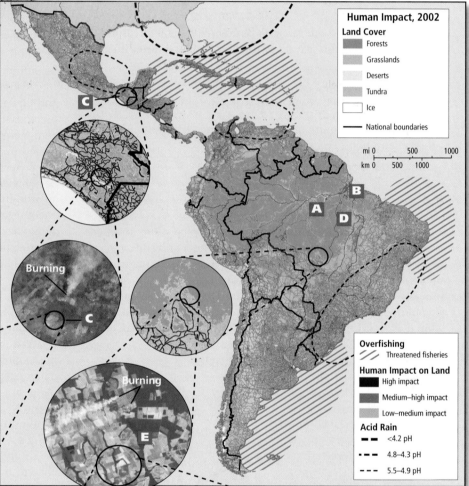

Human Impact, 2002

Land Cover
- Forests
- Grasslands
- Deserts
- Tundra
- Ice
- National boundaries

mi 0   500   1000
km 0 500 1000

Burning

Burning

C

C

E

Overfishing
- Threatened fisheries

Human Impact on Land
- High impact
- Medium–high impact
- Low–medium impact

Acid Rain
- <4.2 pH
- 4.8–4.3 pH
- 5.5–4.9 pH

B A farmer and his children play with a monkey outside their home in the Amazonian state of Pará, Brazil.

C A field in Chiapas, Mexico, recently cleared for agriculture by burning.

D Cattle on recently burned land in Pará, Brazil.

E Soy fields recently cleared of forest in Mato Grosso, Brazil.

Thinking Geographically

After you have read about Human Impact on the Biosphere in Middle and South America, you will be able to answer the following questions:

A What environmental impact is clearly visible in this photo?

B How did this farmer likely end up in the Amazon?

C In what way does burning of forests contribute to climate change?

D What land use likely preceded cattle grazing on the land shown in this photograph?

E Where are the soybeans grown on this land likely to end up?

environmental protection. One example of this new approach is ecotourism.

Ecotourism Many countries are now trying to earn money from the beauty of undegraded natural environments through **ecotourism**, which encourages visitors from wealthier areas to appreciate ecosystems and wildlife that do not exist where they live (see **Reasons for Optimism A** on page 134). Ecotourism is now the most rapidly growing segment of the global tourism and travel industry, which by some measures is the world's largest industry, with $3.5 trillion spent annually. Many Middle and South American nations also have spectacular national parks that can provide a basis for ecotourism.

Ecotourism has its downsides, however. Mismanaged, it can be similar to other kinds of tourism that damage the environment and return little to the surrounding community. While the profits of ecotourism can potentially be used to benefit local communities and environments, the profit margins may be small.

Vignette Puerto Misahualli, a small river boomtown in the Ecuadorian Amazon, is currently enjoying significant economic growth. Its prosperity is due to the many European, North American, and other foreign travelers who come for experiences that will bring them closer to the now-legendary rain forests of the Amazon.

The array of ecotourism offerings can be perplexing. One indigenous man offers to be a visitor's guide for as long as desired, traveling by boat and on foot, camping out in "untouched forest teeming with wildlife." His guarantee that they will eat monkeys and birds does not seem to promise the nonintrusive, sustainable experience the visitor might be seeking. At a well-known "eco-lodge," visitors are offered a plush room with a river view, a chlorinated swimming pool, and a fancy restaurant serving "international cuisine." All of this is on a private 740-acre nature reserve separated from the surrounding community by a wall topped with broken glass. It seems more like a fortified resort than an eco-lodge.

By contrast, the solar-powered Yachana Lodge has simple rooms and local cuisine. Its knowledgeable resident naturalist is a veteran of many campaigns to preserve Ecuador's wilderness. Profits from the lodge fund a local clinic and various programs that teach sustainable agricultural methods that protect the fragile Amazon soils while increasing farmers' earnings from surplus produce. The nonprofit group running the Yachana Lodge—the Foundation for Integrated Education and Development—earns just barely enough to sustain the clinic and agricultural programs. *[Source: Alex Pulsipher's field notes in Ecuador.]* ∎

> **ecotourism** nature-oriented vacations, often taken in endangered and remote landscapes, usually by travelers from affluent nations

The Water Crisis

> **Learning Goal 2**
> **Water:** In this region of abundant water resources, what factors have come together to create a water crisis?

Although this world region receives more rainfall than any other and has three of the world's six largest rivers (in volume), parts of it are experiencing a water crisis (see Thematic Overview B). Rapid urbanization combined with corruption, inadequate investment, and misguided policies have left many people without access to clean water or sanitation.

In the largest cities, the water infrastructure may be so inadequate that as much as 50 percent of freshwater is lost due to leaky pipes. Often as much as 80 percent of the population has no access to decent sanitation; toilets may be entirely lacking, leaving people to relieve themselves on the city streets. This poses a major health hazard and so pollutes local water resources that cities must bring in drinking water from distant areas (see Photo Essay 3.3A).

Behind these immediate problems lie systemic policy failures such as corruption and inadequate investment in infrastructure. For example, in the case of Cochabamba, Bolivia (discussed in Chapter 1 on page 37), efforts at improving the city's inadequate water supply system focused on selling the system to a group of foreign multinational corporations led by Bechtel of San Francisco, California. The hope was that Bechtel would make investments in infrastructure that Cochabamba's notoriously corrupt water utility would not. Unfortunately, Bechtel immediately increased water prices to levels that few urban residents could afford, while doing little to improve water supply or delivery systems. At one point Bechtel even charged urban residents for water taken from their own wells and rainwater harvested off their roofs! Popular protests forced Bechtel to abandon Cochabamba's water utility, which remains plagued by corruption and inadequate infrastructure.

THINGS TO REMEMBER

1. **Learning Goal 1:** **Climate Change and Deforestation** The rapid loss of forests in this region is a major driver of climate change. Trees absorb huge amounts of carbon dioxide in their bodies, which is released when forests are cleared through burning.

2. **Learning Goal 2:** **Water** Despite considerable water resources, parts of this region are experiencing a water crisis related to rapid urbanization, corruption, inadequate investment in infrastructure, and misguided policies. Many people have inadequate access to sanitation, and the dumping of waste in rivers is widespread.

Human Patterns over Time

The conquest of Middle and South America by Europeans set in motion a series of changes that helped create the ways of life found in this region today. The conquest wiped out much of indigenous

Adapting to the multiple stresses that climate change is bringing to this region is proving to be quite a challenge. Drought, hurricanes, and glacial melting are combining with growing populations and persistent poverty to create a complex landscape of vulnerability to climate change.

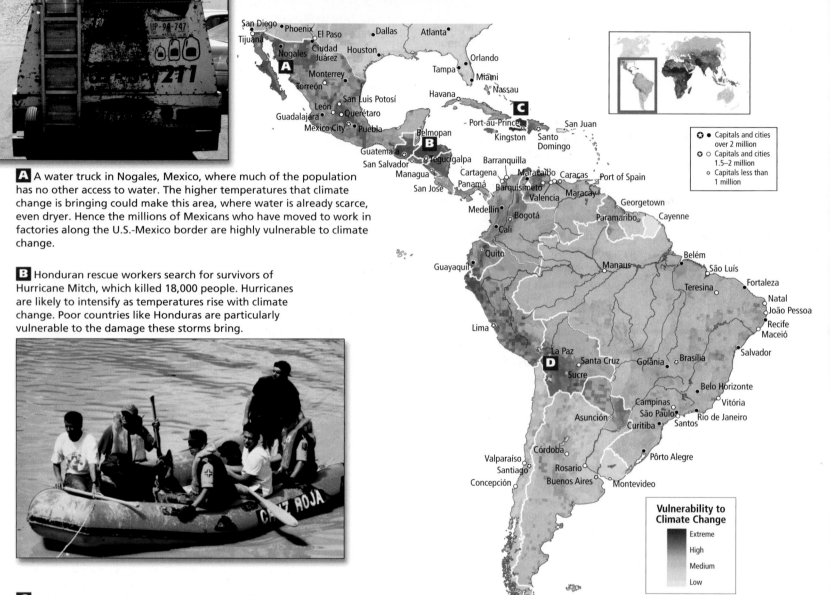

A A water truck in Nogales, Mexico, where much of the population has no other access to water. The higher temperatures that climate change is bringing could make this area, where water is already scarce, even dryer. Hence the millions of Mexicans who have moved to work in factories along the U.S.-Mexico border are highly vulnerable to climate change.

B Honduran rescue workers search for survivors of Hurricane Mitch, which killed 18,000 people. Hurricanes are likely to intensify as temperatures rise with climate change. Poor countries like Honduras are particularly vulnerable to the damage these storms bring.

C Haitians travel to the beach to receive supplies donated by foreign governments after flooding cut their village off from the rest of the country. So much of Haiti's forest cover has been removed that even mild tropical storms can cause catastrophic flooding. Much of Haiti's impoverished population is already dependent on foreign aid, a situation likely to get worse with climate change.

D La Paz and many smaller cities and towns in Bolivia's drought-prone highlands receive much of their drinking water from glaciers in the Andes. Higher temperatures are causing these glaciers to melt rapidly. Many have already disappeared and the rest could be gone in 15 years.

Vulnerability to Climate Change

- Extreme
- High
- Medium
- Low

- ● Capitals and cities over 2 million
- ○ Capitals and cities 1.5–2 million
- ○ Capitals less than 1 million

Thinking Geographically

After you have read about Vulnerability to Climate Change in Middle and South America, you will be able to answer the following questions:

A What does the existence of this water truck suggest about the municipal water distribution system in Nogales?

C In addition to Haiti's exposure to tropical storms and hurricanes, what else contributes to its vulnerability to climate change?

D How have other cities in Bolivia tried to improve their water distribution systems?

By 1492, there were 50 to 100 million indigenous people in Middle and South America. In some places, population densities were high enough to threaten sustainability. People altered the landscape in many ways. They modified drainage to irrigate crops, they terraced hillsides, and they built paved walkways across swamps and mountains. They constructed cities with sewer systems and freshwater aqueducts, and raised huge earthen and stone ceremonial structures that rivaled the pyramids of Egypt (see Timeline A on pages 114–115).

The indigenous people also practiced the system of **shifting cultivation** that is still common in wet, hot regions in Central America and the Amazon Basin. In this system, small plots are cleared in forestlands, the brush is dried and burned to release nutrients in the soil, and the clearings are planted with multiple crop species. Each plot is used for only 2 or 3 years and then abandoned for long enough to allow the forest to regrow. If there is sufficient land, this system is highly productive per unit of land and labor and is sustainable for long periods of time. However, if population pressure increases to the point that a plot must be used before it has fully regrown and its fertility has been restored, yields decrease drastically.

The **Aztecs** of the high central valley of Mexico had some technologies and social systems that rivaled or surpassed those of Asian and European civilizations of the time. Particularly well developed were urban water supplies, sewage systems, and elaborate marketing systems. Historians have concluded that, on the whole, by 1500 Aztecs probably lived more comfortably than their contemporaries in Europe.

In 1492, the largest state in the region was that of the **Incas**, stretching from southern Colombia to northern Chile and Argentina. The main population clusters were in the Andes

> **shifting cultivation** a productive system of agriculture in which small plots are cleared in forestlands, the dried brush is burned to release nutrients, and the clearings are planted with multiple species; each plot is used for only 2 or 3 years and then abandoned for many years of regrowth
>
> **Aztecs** indigenous people of high-central Mexico noted for their advanced civilization before the Spanish conquest
>
> **Incas** indigenous people who ruled the largest pre-Columbian state in the Americas, with a domain stretching from southern Colombia to northern Chile and Argentina

civilization and set up colonial regimes in its place. It introduced many new cultural influences and led to the disparities in the distribution of power and wealth that continue to this day.

The Peopling of Middle and South America

Recent evidence suggests that between 25,000 and 14,000 years ago, groups of hunters and gatherers from northeastern Asia spread throughout North America after crossing the Bering land bridge on foot, or moving along shorelines in small boats, or both. Some of these groups ventured south across the Central American land bridge, reaching the tip of South America by about 13,000 years ago.

highlands, where the cooler temperatures at these high altitudes eliminated the diseases of the tropical lowlands, while proximity to the equator guaranteed mild winters and long growing seasons. For several hundred years, the Inca Empire was one of the most efficiently managed empires in the history of the world. Highly organized systems of labor were used to construct paved road systems, elaborate terraces, irrigation systems, and great stone cities in the Andean highlands (see Timeline B). Incan agriculture was advanced, particularly in the development of crops, which included numerous varieties of potatoes and grains.

European Conquest

The European conquest of Middle and South America was one of the most significant events in human history (see Timeline C). It rapidly altered landscapes and cultures and ended the lives of millions of indigenous people through disease and slavery.

Columbus established the first Spanish colony in 1492 on the Caribbean island of Hispaniola (presently occupied by Haiti and the Dominican Republic). After learning of Columbus's exploits, other Europeans, mainly from Spain and Portugal on Europe's Iberian Peninsula, conquered the rest of Middle and South America.

The first part of the mainland to be invaded was Mexico, home to several advanced indigenous civilizations, most notably the Aztecs. The Spanish were unsuccessful in their first attempt to capture the Aztec capital of Tenochtitlán, but they succeeded a few months later after a smallpox epidemic decimated the native population. The Spanish demolished the grand Aztec capital in 1521 and built Mexico City on its ruins.

The conquest of the Incas in South America was achieved by a tiny band of Spaniards, again aided by a smallpox epidemic. Out of the ruins of the Inca Empire, the Spanish created the Viceroyalty of Peru, which originally encompassed all of South America except Portuguese Brazil. The newly constructed capital of Lima flourished, in large part as a transshipment point for enormous quantities of silver extracted from mines in the highlands of what is now Bolivia.

Diplomacy by the Roman Catholic Church prevented conflict between Spain and Portugal over the lands of the region. The Treaty of Tordesillas of 1494 divided Middle and South America at approximately 46° W longitude (see Figure 3.5 on page 112). Portugal took all lands to the east and eventually acquired much of what is today Brazil; Spain took all lands to the west.

The superior military technology of the Spanish and Portuguese speeded the conquest of Middle and South America. A larger factor, however, was the vulnerability of the indigenous people to diseases carried by the Europeans. In the 150 years following 1492, the total population of Middle and South America was reduced by more than 90 percent to just 5.6 million. To obtain a new supply of labor to replace the dying indigenous people, the Spanish initiated the first shipments of enslaved Africans to the region in the early 1500s.

By the 1530s, a mere 40 years after Columbus's arrival, all major population centers of Middle and South America had been conquered and were rapidly being transformed by Iberian colonial

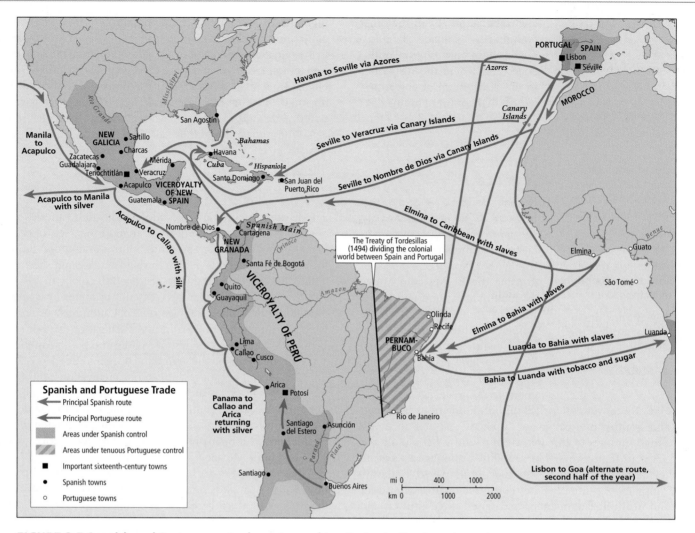

FIGURE 3.5 Spanish and Portuguese trade routes and territories in the Americas, circa 1600. The major trade routes from Spain to its colonies led to the two major centers of its empire, Mexico and Peru. The Spanish colonies could trade only with Spain, not directly with one another. By contrast, there were direct trade routes from Portuguese colonies in Brazil to Portuguese outposts in Africa. Many millions of Africans were enslaved and traded to Brazilian plantation and mine owners (as well as to Spanish, British, French, and Dutch colonies in Middle and South America).

policies. The colonies soon became part of extensive trade networks within the region and globally with Europe, Africa, and Asia.

A Global Exchange of Crops and Animals

From the earliest days of the conquest, plants and animals were exchanged between Middle and South America, Europe, Africa, and Asia via the trade routes illustrated in Figure 3.5. Many plants essential to agriculture in Middle and South America today—rice, sugarcane, bananas, citrus, melons, onions, apples, wheat, barley, and oats, for example—were all originally imports from Europe, Africa, or Asia. When disease decimated the native populations of the region, the colonists turned much of the abandoned land into pasture for herd animals imported from Europe, including sheep, goats, oxen, cattle, donkeys, horses, and mules.

Plants first domesticated by indigenous people of Middle and South America have changed diets everywhere and have become essential components of agricultural economies around

Creoles people mostly of European descent born in the Caribbean

mestizos people of mixed European, African, and indigenous descent

mercantilism the policy by which European rulers sought to increase the power and wealth of their realms by managing all aspects of production, transport, and commerce in their colonies

the globe. The potato, for example, had so improved the diet of the European poor by 1750 that it fueled a population explosion. Manioc (cassava) played a similar role in West Africa. Corn, peanuts, vanilla, and cacao (the source of chocolate) are globally important crops to this day, as are peppers, pineapples, and tomatoes.

The Legacy of Underdevelopment

In the early nineteenth century, wars of independence left Spain with only a few colonies in the Caribbean (see Timeline D). Figure 3.6 shows the European colonizing country and the dates of independence for the countries in the region. The supporters of the nineteenth-century revolutions were primarily **Creoles** (people of mostly European descent born in the Americas) and relatively wealthy **mestizos** (people of mixed European, African, and indigenous descent). The Creoles' access to the profits of the colonial system had been restricted by **mercantilism**, and the mestizos

FIGURE 3.6 The colonial heritage of Middle and South America. Most of Middle and South America was colonized by Spain and Portugal, but important and influential small colonies were held by Britain, France, and the Netherlands. Nearly all the colonies had achieved independence by the late twentieth century. Those for which no date appears on the map are still linked in some way to the colonizing country.

were excluded by racist colonial policies. Once these groups gained power, however, they became a new elite that controlled the state and monopolized economic opportunity. They did little to expand economic development or access to political power for the majority of the population (see Timeline E).

Today, the economies of Middle and South America are much more complex and technologically sophisticated than they once were. Nevertheless, disparities persist: Around 30 percent of the population is poor and lacks access to land, adequate food, shelter, water and sanitation, and basic education. Meanwhile, a small elite class enjoys levels of affluence equivalent to those of the very wealthy in the United States. These conditions are in part the lingering result of colonial economic policies that favored the export of raw materials and fostered undemocratic privileges for elites and outside investors who often spent their profits elsewhere rather than reinvesting them within the region. These policies will be further discussed in the Economic and Political Issues section below.

THINGS TO REMEMBER

1. Shifting cultivation of small plots is a traditional and potentially sustainable agricultural strategy still used by some throughout the region.

2. Spain and Portugal were the primary colonizing countries of Middle and South America, with smaller colonies held by Britain, France, Denmark, and the Netherlands. By the mid-nineteenth century, most of the larger colonies had gained independence.

II CURRENT GEOGRAPHIC ISSUES

Historically, wealth and political and economic power in the countries of Middle and South America have been concentrated in the hands of a few. Globalization, economic modernization, and the transformation from rural to urban societies have not changed that reality, although recent shifts toward democratization are beginning to give more political power to the majority. Culturally, this region retains diverse influences from Europe, Africa, and indigenous peoples. Meanwhile, shifting gender roles and new religious movements are powerful agents of change.

Economic and Political Issues

Although Middle and South America are not as poor on average as sub-Saharan Africa,

income disparity the gap in income between rich and poor

South Asia, or Southeast Asia, they face serious economic challenges. The unique colonial and postcolonial history of this region has resulted in wide economic inequality and income disparity, as well as persistent political disempowerment of poor people. Because these issues are so connected, overall discussions of economic and political issues are combined in this chapter.

Economic Inequality and Income Disparity

Many economic challenges in this region are related to **income disparity**. With the exception of a few small countries in Middle America, the gap between rich and poor is one of the largest on earth (see Thematic Overview C). According to the most recent UN figures, the richest 10 percent of the

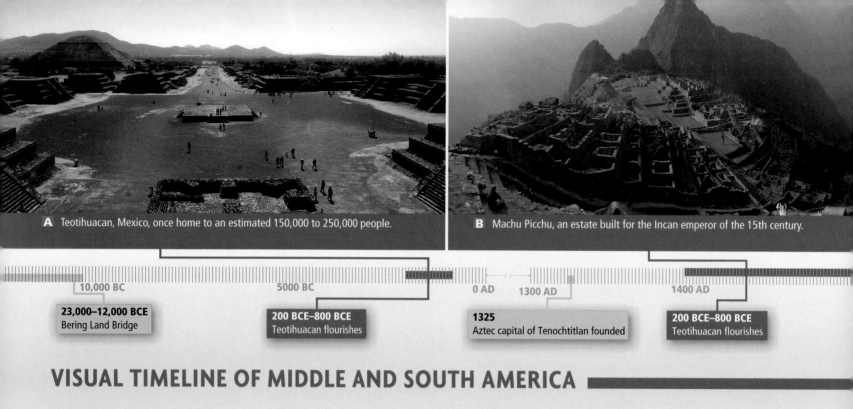

A Teotihuacan, Mexico, once home to an estimated 150,000 to 250,000 people.

B Machu Picchu, an estate built for the Incan emperor of the 15th century.

| 10,000 BC | 5000 BC | 0 AD | 1300 AD | 1400 AD |

23,000–12,000 BCE
Bering Land Bridge

200 BCE–800 BCE
Teotihuacan flourishes

1325
Aztec capital of Tenochtitlan founded

200 BCE–800 BCE
Teotihuacan flourishes

VISUAL TIMELINE OF MIDDLE AND SOUTH AMERICA

Thinking Geographically

After you have read about the human history of Middle and South America, you will be able to answer the following questions:

A Relative to their contemporaries in Europe in 1500, how well do historians think Aztecs lived?

B Why were the main population centers of the Incas in the Andea highlands?

population was between 19 and 94 times richer than the poorest 10 percent, depending on which country was being analyzed. In recent years, disparities have been shrinking in Brazil, Chile, Mexico, and Venezuela, primarily because of new government economic and social policies. However, elsewhere in the region, disparities have increased. While poverty for the region as a whole is around 33 percent, in the poorer countries (Haiti, Guatemala, Honduras, Nicaragua, Peru, Bolivia, and Paraguay), more than half the population now lives in poverty.

Phases of Economic Development

The current economic and political situation in Middle and South America derives from the region's history, which can be divided into three major phases: the early extractive phase, the import substitution industrialization (ISI) phase, and the current structural adjustment and marketization phase. All three phases have helped entrench wide income disparities despite a consensus that more egalitarian development is desirable.

Early Extractive Phase The **early extractive phase** began with the European conquest and lasted until the early twentieth century in parts of the region. Economic development was guided by a policy of mercantilism, in which European rulers sought to increase the power and wealth of their realms by extracting resources and controlling much of the economic activity in the colonies for the benefit

> **early extractive phase** a phase in Central and South American history, beginning with the Spanish conquest and lasting until the early twentieth century, characterized by a dependence on the export of raw materials
>
> **haciendas** a large agricultural estate in Middle or South America, more common in the past; usually not specialized by crop and not focused on market production

of the "mother" country. Even after independence was granted, many extractive enterprises were owned by foreign investors who had few incentives to build stable local economies.

A small flow of foreign investment and manufactured goods entered the region, but a vast flow of raw materials left for Europe and beyond. The money to fund the farms, plantations, mines, and transportation systems that enabled the extraction of resources for export came from abroad, first from Europeans and later from North Americans and other international sources. One example is the still highly lucrative Panama Canal. The French first attempted to build the canal in 1880. It was later completed and run by the United States, and finally turned over to Panamanian control in 1999. The profits from these ventures were usually banked abroad, depriving the region of investment that could have made it more economically independent. Industries were slow to develop in the region, so even essential items, such as farm tools and household utensils, had to be purchased from Europe and North America at relatively high prices. Many people simply did without.

A number of economic institutions arose in Middle and South America to supply food and raw materials to Europe and North America. Large rural estates called **haciendas** were granted to colonists as a reward for conquering territory and people for Spain. For generations these estates were then passed down through the families of those colonists. Over time, the owners, who

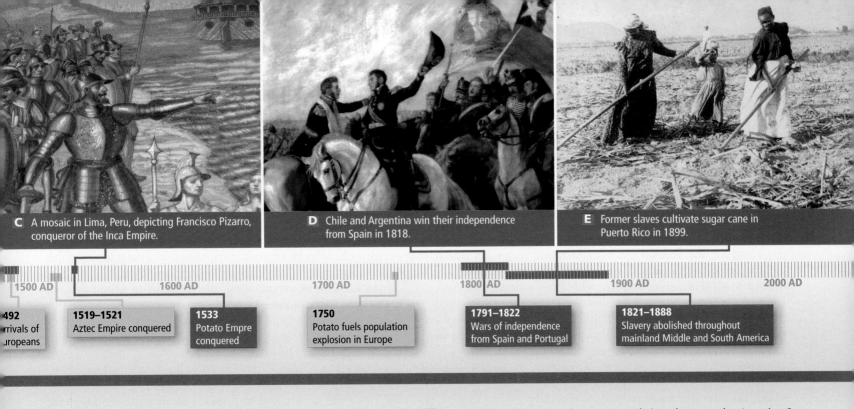

C A mosaic in Lima, Peru, depicting Francisco Pizarro, conqueror of the Inca Empire.

D Chile and Argentina win their independence from Spain in 1818.

E Former slaves cultivate sugar cane in Puerto Rico in 1899.

1500 AD — 1600 AD — 1700 AD — 1800 AD — 1900 AD — 2000 AD

492 rrivals of uropeans

1519–1521 Aztec Empire conquered

1533 Potato Empre conquered

1750 Potato fuels population explosion in Europe

1791–1822 Wars of independence from Spain and Portugal

1821–1888 Slavery abolished throughout mainland Middle and South America

C What assisted the Spanish conquest of the Incas?

D How did the Creoles depicted in this painting behave once they gained power?

E Why did the Spanish and Portuguese bring slaves to the Americas?

often lived in a distant city or in Europe, lost interest in the day-to-day operations of the haciendas. Hence, productivity was generally low and hacienda laborers remained extremely poor. Nevertheless, haciendas produced a diverse array of products (cattle, cotton, rum, sugar) for local consumption and export.

Plantations were large factory farms, meaning that in addition to growing crops such as sugar, coffee, cotton, or (more recently) bananas, some processing for shipment was done on site. Plantation owners made larger investments in equipment, and these farms were more efficient and profitable than haciendas. Because most profits were either reinvested in the plantations themselves or invested abroad, few connections with local economies or ancillary industries developed. And instead of employing local populations, they imported slave labor from Africa.

First developed by the European colonizers of the Caribbean and northeastern Brazil in the 1600s, plantations became more common throughout Middle and South America by the late nineteenth century. Unlike haciendas, which were often established in the continental interior in a variety of climates, plantations were for the most part situated in tropical coastal areas with year-round growing seasons. Their coastal and island locations gave them easier access to global markets via ocean transport.

As markets for meat, hides, and wool grew in Europe and North America, the livestock ranch emerged, specializing in raising cattle and sheep. Today, commercial ranches serving such markets as the

> **plantation** a large factory farm that grows and partially processes a single cash crop
>
> **import substitution industrialization (ISI)** policies that encouraged local production of machinery and other items that previously had been imported at great expense from abroad

fast-food industry are found in the drier grasslands and savannas of South America, Central America, northern Mexico, and even in the wet tropics on freshly cleared rain forest lands.

Mining was another early extractive industry. Important mines (at first, primarily gold and silver) were located on the island of Hispaniola in north-central Mexico, the Andes, the Brazilian Highlands, and in many other locations. Extremely inhumane labor practices were common in all of these mines. Today, oil and gas have been added to the mineral extraction industry, but rich mines throughout the region continue to produce gold, silver, copper, tin, precious gems, titanium, bauxite, and tungsten.

Profits from the region's mines, ranches, plantations, and haciendas continued to leave the region, even after the countries gained independence in the nineteenth century. A main reason for this was that wealthy foreign investors retained control of many extractive enterprises.

Import Substitution Industrialization Phase In the 1950s, there were waves of protests against the continuing domination of the economy and society by local elites and foreign investors. Many governments—Mexico and Argentina most prominent among them—proclaimed themselves socialist democracies. To boost their economic independence, these governments tried to keep money and resources within the region through policies of **import substitution industrialization (ISI)**. Subsidies, among other

115

measures, encouraged local production of machinery and other commonly imported items. To some extent these policies did help reduce consumer costs and retain profits at home, but generally they failed to lift the region out of poverty.

To encourage local people to buy manufactured goods from local suppliers, governments placed high tariffs on imported manufactured goods. The money and resources kept within each country were expected to provide the basis for further industrial development. This would create well-paying jobs, raise living standards for the majority of people, and ultimately replace the extractive industries as the backbone of the economy.

However, the state-owned manufacturing sectors on which the success of ISI depended were never able to produce high-quality goods that could compete with those produced in Asia, Europe, and North America. This was largely because they lacked the technological and managerial skills to run globally competitive factories. Moreover, local populations were not large or prosperous enough to support these industries as consumers. With both domestic consumer demand and exports weak, employment stagnated, tax revenues remained low, and social programs were not adequately funded.

Not all state-owned corporations were losing propositions, however. ISI still survives in some countries, and is being considered for reimplementation in a few others, such as Bolivia and Venezuela. Brazil—with its aircraft, armament, and auto industries—and Mexico—with its oil and gas industries—both had success with some ISI programs. Interest in state-supported manufacturing industries continues, although the general trend is now toward more market-oriented management and global competitiveness. Dependence on the export of raw materials persists.

Beginning in the early 1970s, a debt crisis brought an end to the ISI phase. Sharp increases in oil prices and decreases in global prices of raw materials ended a period of global prosperity that had begun in the early 1950s. Reluctant to let go of the prospect of rapid development, governments and private interests continued to pursue ambitious plans to modernize and industrialize their national economies. They did this even as prices for raw material exports—their main source of income—fell. With false optimism, they paid for these projects by borrowing millions of dollars from major international banks, most of which were in North America or Europe. When a global economic recession in 1980 put a halt to the development plans of governments in the region, many were unable to repay their loans, some of which are still considerable (**Figure 3.7**).

> **Learning Goal 3**
> **Globalization, Development, and Democratization:** How has recent dissatisfaction with SAPs and globalization been expressed through the democratic process?

Structural Adjustment and the Marketization Phase In the 1980s, alarmed over the mounting debt of their clients, foreign banks that had made loans to governments in the region took action.

The International Monetary Fund (IMF) developed and enforced policies that mandated profound changes in the organization of national economies. In order to ensure that sufficient money would be available to repay loans that governments took out from foreign banks to finance the now-discredited ISI phase, the IMF required **structural adjustment policies (SAPs)**. These SAPs were based on concepts of **privatization** (the selling of formerly government-owned industries and firms to private investors), **marketization** (the development of a *free market* economy in support of *free trade*), and *globalization* (see Chapter 1, pages 31–33). At the time, these concepts were considered the soundest ways for countries to achieve economic expansion and thus to repay the debts to banks in North America, Europe, and Asia.

In the case of privatization, the investors to whom government firms were sold were multinational corporations located in Asia, North America, and Europe. Hence, the SAP era resulted in industries across the region being returned to foreign ownership.

The other main SAP policy, marketization, meant that in order to obtain further loans, governments were required to remove tariffs on imported goods of all types, thus putting local industries at risk.

Crucially, SAPs also reversed the ISI-era trend of expanding government social programs and building infrastructure. To free up funds for debt repayment, governments were required to fire many civil servants and drastically reduce spending on public health, education, job training, day care, water systems, sanitation, and infrastructure building and maintenance. While severely cutting these badly needed government programs, SAPs encouraged the expansion of industries that were already earning profits by lowering taxes on their activities, thus further shrinking revenues to pay for government services. In the countries of Middle and South America, as in most developing countries, the most profitable industries remained those based on the extraction of raw materials for export (see Thematic Overview D).

▋▶ 61. PERUVIANS STRUGGLE TO GAIN HEALTH CARE ACCESS

Export Processing Zones A major component of SAPs was the expansion of manufacturing industries in **Export Processing Zones (EPZs)**, also known as *free trade zones*—specially created areas within a country where, in order to attract foreign-owned factories, taxes on imports and exports are not charged. The main benefit to the host country is the employment of local labor, which eases unemployment and brings money into the economy. Products are often assembled strictly for export to foreign markets.

EPZs exist in nearly all countries on the Middle and South American mainland and on some Caribbean islands (in many other world regions also). However, the largest of the EPZs is the conglomeration of assembly factories, called **maquiladoras**, which are located along the Mexican side of the U.S.–Mexico border (see Figure 3.7; see also **Figure 3.8** on page 117). Although these factories do provide employment, they might not be as beneficial to the Mexican economy as once thought, as the following vignette illustrates.

---

structural adjustment policies (SAPs) policies that require economic reorganization toward less government involvement in industry, agriculture, and social services; sometimes imposed by the World Bank and the International Monetary Fund as conditions for receiving loans

privatization the selling of formerly government-owned industries and firms to private companies or individuals

marketization the development of a free market economy in support of *free trade*

Export Processing Zones (EPZs) specially created legal spaces or industrial parks within a country where, to attract foreign-owned factories, duties and taxes are not charged

maquiladoras foreign-owned, tax-exempt factories, often located in Mexican towns just across the U.S. border from U.S. towns, that hire workers at low wages to assemble manufactured goods which are then exported for sale

FIGURE 3.7 Debt, export processing zones, and trade blocs in Middle and South America. The high rate of debt in the region stems from countries borrowing money to finance industrial and agricultural development projects that are intended to replace imports and reduce poverty by providing jobs. Debt is presented as total debt service (repayment of public and private loans) as a percentage of a country's GDP. Export processing zones have played a major role in efforts to reduce government debts. Countries in the region are using the creation of trade blocs such as NAFTA and Mercosur (see page 119) as another means to reduce debt by lowering tariffs and reducing dependence on higher-cost imports from outside the region.

Vignette Orbalin Hernandez has just returned to his self-built shelter in the town of Mexicali on the border between Mexico and the United States. Recently fired for taking off his safety goggles while loading TV screens onto trucks at Thompson Electronics, he had just gone to the personnel office to ask for his job back. Because there were no previous problems with him, he was rehired at his old salary of $300 per month ($1.88 per hour). Thompson is a French-owned electronics firm that took advantage of NAFTA (see page 119)

FIGURE 3.8 Living standards along the U.S.–Mexico border. An outhouse in Nogales, Mexico, that serves a family employed at the maquiladoras in the area. Many Mexicans have moved to the U.S.–Mexico border zone in search of a better life, only to find low wages and poor living conditions.

Thinking Geographically: In this photo, what kind of infrastructure is Nogales, Mexico, lacking?

when it moved to Mexicali from Scranton, Pennsylvania, in 2001. There, its 1100 workers had been paid an average of $20.00 per hour. Now 20 percent of the Scranton workers are unemployed, and many feel bitter toward the Mexicali workers.

Nonetheless, in 2006, Orbalin and his fellow Mexicali workers were being told that their wages were too high to allow their employers to compete with companies located in China, where in 2005, workers with the same skills as Orbalin earned just $0.35 an hour. Indeed, that year, 14 Mexicali plants closed and moved to Asia. Those firms remaining in Mexicali cut wages and reduced benefits.

Though Orbalin's salary at Thompson-Mexicali is barely enough for him to support his wife, Mariestelle, and four children, they are grateful for the job. They are originally from a farming community in the Mexican state of Tabasco, where, for people with no high school education, wages averaged $60 per month.

By 2009, it appeared that the global recession would affect Chinese–Mexican relations in new ways. Rising costs for fuel and storage, as well as slightly lower-quality goods coming from China, meant that it made less sense to move a Mexican factory to China. It is not yet clear whether this trend is short-lived or the wave of the future. *[Source: NPR reports by John*

> **nationalize** to seize private property and place under government ownership, with some compensation

Idste, August 14, 2001, August 25, 2003, and August 16, 2003, and by Gary Hadden, August 27, 2003; David Bacon, "Anti-China Campaign Hides Maquiladora Wage Cuts," ZNet, February 2, 2003, at http://www.zcommunications.org/anti-china-campaign-hides-maquiladora-wage-cuts-by-david-bacon; April 2006, http://www.maquilaportal.com/cgi-bin/public/index.pl; "How Rising Wages Are Changing the Game in China," Bloomberg Business Week, March 27, 2006, at http://www.businessweek.com/magazine/content/06_13/b3977049.htm; Pete Engardio, Bloomberg Businessweek, June 4, 2009, "So Much for the Cheap 'China Price,'" at http://www.businessweek.com/magazine/content/09_24/b4135054963557.htm.] ∎

Outcomes of SAPs and EPZs The SAP/EPZ era did not produce the sustained economic growth that was expected to relieve the debt crisis and create broader prosperity. SAPs increased some kinds of economic activity in Middle and South America, resulting in modest rates of total national GDP growth. But overall, the SAP era was a time of backsliding. During the last two decades of the ISI phase (1960 to 1980), per capita GDP grew by 82 percent for the region as a whole. When SAPs were implemented across the region (1980 to 2000), growth in per capita GDP slowed to just 12.6 percent; between 2000 and 2005, it slowed to only 1 percent.

A major reason that SAPs failed to stimulate economic growth was that they encouraged greater dependence on exports of raw materials just when prices for these items were falling in the global market. Ironically, prices fell partly because of SAPs! Since SAPs forced indebted countries around the globe to boost raw materials exports simultaneously, a global glut was created and prices were driven down. After 2008, prices fell also because demand for raw materials decreased during the global recession.

> **Learning Goal 3**
> **Globalization, Development, and Democratization:** How has recent dissatisfaction with SAPs and globalization been expressed through the democratic process?

Voter Backlash Against SAPs From the late 1990s, millions of voters within this region registered their opposition to SAPs. Since 1999, presidents who explicitly opposed SAPs were elected in eight countries in the region: Argentina, Bolivia, Brazil, Chile, Ecuador, Nicaragua, Uruguay, and Venezuela. Both Brazil and Argentina made considerable sacrifices to pay off their debts and liberate themselves from the restrictions of SAPs (see Figure 3.7). Venezuela, under President Hugo Chávez, has emerged as a leader of the SAP backlash. In 2005, the Chávez administration **nationalized** foreign oil companies operating in Venezuela. Since then it has also used its own considerable oil wealth to help other countries, such as Argentina, Bolivia, and Nicaragua, pay off their debts. Bolivia, led by Evo Morales (often an ally of Chávez), reversed the SAP policy of privatization by nationalizing that country's natural gas industry. Loans taken from the IMF dropped from $48 billion in 2003 to just $1 billion in 2008. During this same period, IMF policy reversed itself from promoting SAPs to promoting policies that attempted to combine marketization with strong investment in social programs. This new IMF position is holding, suggesting that voters in this region have had an important impact on a major global institution (see page 32 in Chapter 1).

II ▶ 63. NEW BOLIVIAN ENERGY POLICY CAUSES CONCERN

II ▶ 67. U.S.–VENEZUELA ENERGY TIES ENDURE DESPITE DETERIORATING POLITICAL RELATIONS

The Role of Remittances

Migration for the purpose of supporting family members back home with remittances is an extremely important economic activity throughout this region (see further discussion on page 129). Each year, remittances amount to roughly double the U.S. foreign aid to the region. Most such remittance migrations are within countries. While the amounts of cash sent home are small, these remittances are crucial for families that may not have much other income.

One report (by geographer Dennis Conway and anthropologist Jeffrey H. Cohen) suggests that couples that migrate to the United States from indigenous villages in Mexico typically work at menial jobs and live frugally in order to save a substantial nest egg. Then they may return home for several years to build a house (usually a family self-help project) and buy furnishings. When the money runs out, the couple, or just the husband or wife, may migrate again to save up another nest egg.

The Informal Economy

For centuries, low-profile businesspeople throughout Middle and South America have operated in the informal economy, meaning that they support their families through inventive entrepreneurship but are not officially recognized in the statistics and do not pay business, sales, or income taxes. Most are small-scale operators involved with street vending or recycling used items such as clothing, glass, or waste materials. In some instances, the informal economy can serve as an incubator for new businesses that may expand, eventually providing legitimate jobs for family and friends.

Throughout this region, nearly every citizen depends on the informal economy in some way as either a buyer or a seller. Critics argue that informal workers are just treading water, making too little to ever expand their businesses significantly. Moreover, the bribes they have to pay to avoid arrest or fines are much less beneficial to the economy as a whole than the taxes paid by legitimate businesses. Work in the informal economy is also risky because there is no protection of workers' health and safety and no sick leave, retirement, or disability benefits (also often lacking in the formal economy). Nevertheless, the informal economy can be a lifesaver during times of economic recession. For example, after a recession hit Peru in 2000, sixty-eight percent of urban workers were in the informal sector. They generated 42 percent of the country's total GDP, mostly through street-vending work.

Unfortunately, the informal economy goes hand in hand with corruption (further discussed on page 123) because entrepreneurs operating on the fringe of legality are vulnerable to unscrupulous agents who may demand protection money or extract high-priced favors for not turning vendors in to the authorities. On the other hand, those in the informal economy may actually have chosen the illegal route precisely because corrupt government officials demanded high bribes to dispense legal licenses to operate.

> **North American Free Trade Agreement (NAFTA)** a free trade agreement made in 1994 that added Mexico to the 1989 economic arrangement between the United States and Canada
>
> **Mercosur** a free trade zone created in 1991 that links the economies of Brazil, Argentina, Uruguay, and Paraguay to create a common market

Regional Trade and Trade Agreements

The growth in international trade and foreign investment in the region, encouraged in part by SAPs, has been joined by growth in regional free trade agreements. Such agreements reduce tariffs and other barriers to trade among a group of neighboring countries. The two largest free trade agreements within the region are NAFTA and Mercosur (see Figure 3.7).

The **North American Free Trade Agreement (NAFTA)**, a free trade bloc consisting of the United States, Mexico, and Canada, and containing more than 450 million people, was created in 1994. The main goal of NAFTA is to reduce barriers to trade, thereby creating expanded markets for the goods and services produced in the three countries. Since 1994, the value of the economies of these three countries has grown steadily, and by 2011 was worth at least $18 trillion. A subsequent effort by the United States to create a NAFTA-like trade bloc for all of the Americas (FTAA) has stalled in recent years, largely because of dissatisfaction with the effects of globalization.

Mercosur links the economies of Argentina, Brazil, Paraguay, and Uruguay to create a common market with 250 million potential consumers and an economy worth nearly $3 trillion per year (Bolivia, Chile, Colombia, Ecuador, and Peru are associate Mercosur members; these countries are not included in these figures). Mercosur is smaller than NAFTA (in terms of both area and GDP) and older, having been created in 1991.

II ▶ 62. MANY VENEZUELANS UNCERTAIN ABOUT CHÁVEZ'S TWENTY-FIRST CENTURY SOCIALISM

The overall record of regional free trade agreements so far is mixed. While they have increased trade, the benefits of that trade usually are not spread evenly among regions or among all sectors of society. In Mexico, for instance, the benefits of NAFTA have gone mainly to wealthy investors, and they have been concentrated in the northern states that border the United States. There, the agreement facilitated the growth of maquiladoras by easing cross-border finances and transit. But significantly, as many as one-third of small-scale farmers throughout Mexico have lost their jobs as a result of increased competition from U.S. corporate agriculture, which now, with NAFTA, has unrestricted access to Mexican markets. In particular, corn imports from the United States have driven down the price of corn in local markets, impoverishing small farmers or driving them out of business. These former farmers make up a significant number of the undocumented workers coming into the United States.

THINGS TO REMEMBER

1. Three phases of economic development—extractive, import substitution industrialization (ISI), and structural adjustment policy (SAP)—dominated the region into the twenty-first century.

2. **Learning Goal 3: Globalization, Development, and Democratization** In the 1990s, voters across the region rebelled and began electing governments opposed to SAPs.

3. Since the 1990s, free trade policies were adopted in some industries, and regional trading blocs were formed, with mixed results.

Food Production and Contested Space

The agricultural lands of Middle and South America (Figure 3.9) are examples of what geographers call **contested space**, where various groups are in conflict over the right to use a specific territory as each sees fit. The conflicts take place at various scales, as the following four case studies demonstrate, and all are in one way or another related to inequities that have been in place at least since colonial times. They also relate to a region-wide shift toward "green revolution" agriculture (see Chapter 1 on page 23; see also Thematic Overview E).

> **contested space** any area that several groups claim or want to use in different and often conflicting ways, such as the Amazon or Palestine

Local Scale: A Banana Worker in Costa Rica For centuries, while people labored for very low wages on haciendas, the owners would at least allow them a bit of land to grow a small garden or some cash crops. In recent years, however, SAPs encouraged a shift to large-scale, mechanized, export-oriented agriculture, also known as green revolution agriculture. The rationale was that these operations could earn larger profits that could help countries pay off debts faster. SAPs made it easier for foreign multinational corporations, such as Del Monte, to use their large financial resources to buy up many haciendas and other farms, converting them to plantations. However, this also forced many rural people off lands they once cultivated and onto plantations, where they now work as migrant laborers for low wages, as the following story illustrates.

Vignette Aguilar Busto Rosalino used to work on a Costa Rican hacienda. He had a plot on which to grow his own food and in return worked 3 days a week for the hacienda. Since a banana plantation took over the hacienda, he rises well before dawn and works 5 days a week from 5:00 A.M. to 6:00 P.M., stopping only for a half-hour lunch break. Because he doesn't have time to farm, he now has to buy most of his food.

Aguilar places plastic bags containing pesticide around bunches of young bananas. He prefers this work to his last assignment of spraying a more powerful pesticide, which left him and 10,000 other plantation workers sterile. He works very hard because he is paid according to how many bananas he treats. Usually he earns between $5.00 and $14.50 a day.

Right now Aguilar is working for a plantation that supplies bananas to the Del Monte corporation, but he thinks that in a few months he will be working for another plantation nearby. It is common practice for these banana operations to fire their workers every 3 months so that they can avoid paying the employee benefits that Costa Rican law mandates. Although Aguilar makes barely enough to live on, he has no plans to press for higher wages because he knows that he would be put on a "blacklist" of people that the plantations agree not to hire. [Source: Adapted from Andrew Wheat, "Toxic Bananas," Multinational Monitor 17 (9), September 1996, pp. 6–7, updated 2007, at http://multinationalmonitor.org/hyper/mm0996.04.html http://www.magney.org/photofiles/CostaRica-Bananas1.htm.] ■

Provincial Scale: The Zapatista Rebellion In the southern Mexican state of Chiapas, agricultural activists and indigenous leaders have mobilized armed opposition to the economic and political systems that have left them poor and powerless. The Mexican government redistributed some hacienda lands to poor farmers early in the twentieth century, but most fertile land in Chiapas is still held by a wealthy few who use green revolution agriculture to grow cash crops for export. The poor majority farm tiny plots on infertile hillsides. In 2000, about three-fourths of the rural population was malnourished, and one-third of the children did not attend school.

The Zapatista rebellion (named for the Mexican revolutionary hero Emiliano Zapata) began on the day the North American Free Trade Agreement took effect in 1994. The Zapatistas view NAFTA as a threat because it diverts the

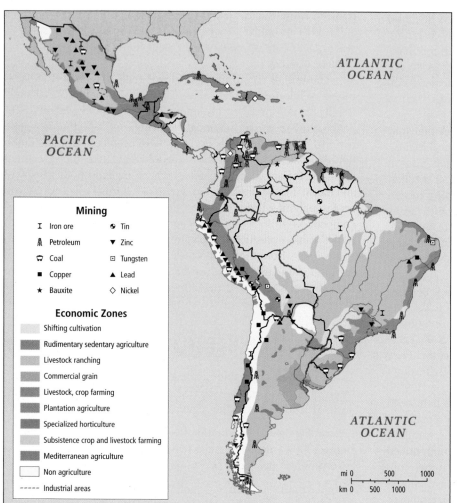

Mining

I	Iron ore	◔	Tin
A	Petroleum	▼	Zinc
⏚	Coal	▣	Tungsten
■	Copper	▲	Lead
★	Bauxite	◇	Nickel

Economic Zones

- Shifting cultivation
- Rudimentary sedentary agriculture
- Livestock ranching
- Commercial grain
- Livestock, crop farming
- Plantation agriculture
- Specialized horticulture
- Subsistence crop and livestock farming
- Mediterranean agriculture
- Non agriculture
- ----- Industrial areas

PACIFIC OCEAN

ATLANTIC OCEAN

ATLANTIC OCEAN

mi 0   500   1000
km 0   500   1000

FIGURE 3.9 Agricultural and mineral zones in Middle and South America.

support of the Mexican federal government from land reform to large-scale, export-oriented green revolution agriculture. The Mexican government used the army to suppress the rebellion.

In 2003, after 9 years of armed resistance, the Zapatista movement redirected part of its energies. It began a nonviolent political campaign to democratize local communities by setting up people's governing bodies parallel to local official governments. Before the national elections in 2006, the Zapatistas toured the country in an effort to turn the political climate against globalization and toward greater support for indigenous people and the poor. The election ended in a near tie, but in the end the Zapatista-favored candidate lost. The Zapatistas continue to be active in efforts to reform Mexican politics to include more grassroots participation.

National Scale: Brazil's Landless Movement The trends in agriculture described above for Middle America have sparked rural resistance in Brazil as well. Sixty-five percent of Brazil's arable and pasture land is owned by wealthy farmers who make up just 2 percent of the population. Since 1985, more than 2 million small-scale farmers have been forced to sell their land to larger farms that practice green revolution agriculture. Because the larger farms specialize in major export items, such as cattle and soybeans, they have been favored by governments, which want to increase exports. As a result, many poor farmers have been forced to migrate.

To help these farmers, organizations such as the Movement of Landless Rural Workers (MST) began taking over unused portions of some large farms. Since the mid-1980s, the MST has coordinated the occupation of more than 51 million acres of Brazilian land (an area about the size of Kansas). Some 250,000 families have gained land titles, while the elite owners have been paid off by the Brazilian government and have moved elsewhere. Movements with goals similar to those of MST now exist in Ecuador, Venezuela, Colombia, Peru, Paraguay, Mexico, and Bolivia.

> **Learning Goal 4**
> Food and Urbanization: How has the shift toward green revolution agriculture encouraged urbanization?

World Regional Scale: The Persistence of Green Revolution Agriculture Instances of workers contesting the use of agricultural land are exceptions to overall trends that favor the growth of large-scale green revolution agriculture. Because of the moneymaking potential and political power of large-scale agriculture, conflicts are only rarely resolved in favor of small farmers and agricultural workers. Moreover, in rapidly urbanizing countries, green revolution agriculture is seen as the only way to supply cheap food for the millions of city dwellers. Ironically, many of these new urbanites were once farmers capable of feeding themselves; they moved to the cities because they were unable to compete with the new agricultural systems.

Zones of modern mixed farming, including areas that produce meat, vegetables, and specialty foods for sale in urban centers, are located on large and small farms around most major urban centers. Many areas have had large farms for centuries. For example,

a wide belt of green revolution agriculture, similar to that found in the Midwestern United States, stretches through the Argentine pampas (Figure 3.9). And, as we saw in the earlier discussion of environmental issues (see page 107), soybeans, which are often produced by green revolution farms on once-forested lands in and around the Amazon Basin, are a major export for Brazil. Despite the recent political tensions, large-scale agriculture appears to have a solid place in the future of this region.

▐▐ ▶ 57. HAITI'S RISING COST OF FOOD WORRIES AID GROUPS

THINGS TO REMEMBER

1. Food production is undergoing a shift away from small-scale, often subsistence-oriented, production, toward large-scale, green revolution agriculture that is aimed at earning cash through exports.

2. Learning Goal 4: Food and Urbanization This shift toward large-scale green revolution agriculture has forced many small farmers, who cannot afford the investment in machinery or chemicals, off their land and into the cities.

3. Major political conflicts, some violent and some peaceful, have developed in certain areas in response to the shift toward large-scale agriculture.

4. Many governments now see large-scale agriculture as the only way to feed the region's large urban populations.

Political Reform and Democratization

After decades of elite and military rule, almost all countries in the region now have multiparty political systems and democratically elected governments. In the last 25 years, there have been repeated peaceful and democratic transfers of power in countries once dominated by rulers who seized power by force (see Thematic Overview F; and see Reasons for Optimism C on page 134). These **dictators** often claimed absolute authority, governing with little respect for the law or the rights of their citizens. Their authority was based on alliances between the military, wealthy rural landowners, wealthy urban entrepreneurs, foreign corporations, and even foreign governments such as the United States.

Although democracy has transformed the politics of the region, problems remain (see Photo Essay 3.4 on page 122). Elections are sometimes poorly or unfairly run, and their results are frequently contested. Elected governments are sometimes challenged by citizen protests or threatened with a **coup d'état**, in which the military takes control of the government by force. Such coups are usually a response to policies that are unpopular with large segments of the population, powerful elites, the military, or the United States. In the last decade, coups have threatened Honduras, Venezuela, Colombia, Ecuador, and Bolivia.

> **dictator** a ruler who claims absolute authority, governing with little respect for the law or the rights of citizens
>
> **coup d'état** a military- or civilian-led forceful takeover of a government

Hugo Chavez and Venezuela's Fragile Democracy In Venezuela, the landslide election of Hugo Chávez as president in 1998 appeared to advance the cause of democracy there. Elites had long dominated Venezuelan politics, and profits from the country's rich

Democratization is well underway in Middle and South America, although many barriers remain. Decades of elite and military rule combined with rampant corruption have led to popular revolutionary movements aimed at taking control of the government by force. The result has been war and brutal repression by both governments and militants that denies citizens the freedoms essential to democracy. The drug trade has also emerged as a major obstacle to democracy, with huge amounts of illicit cash being used to pay off elected officials, civil servants, police forces, and the military. Nevertheless, recent decades have seen a dramatic expansion of democracy and a decline in violence for the region as a whole.

A Federal agents in Hermosillo in Sonora, Mexico, arrest drug traffickers who killed another federal agent. Mexico's ongoing drug war has led to over 28,000 deaths over the past several years and corruption that has eroded local democratic institutions, especially near the U.S. border. Most of Mexico's drug production supplies the United States.

B Victims of Colombia's ongoing conflict march in protest in Bogotá. This complex conflict, the worst in the region, is driven by popular revolutionary movements and an ongoing drug war. Democratic institutions and the rule of law have been badly damaged, with many high-ranking officials linked to extreme abuses of human rights and civil liberties.

Democratization and Conflict
Democratization Index
- Full democracy
- Flawed democracy
- Hybrid regime
- Authoritarian regime
- No data

Armed Conflicts and Genocides with High Death Tolls Since 1945
- Ongoing conflict
- 1000–20,000 deaths
- 20,000– 50,000 deaths
- 50,000–100,000 deaths
- 100,000–200,000 deaths

C School children in León, Nicaragua, play in front of a mural commemorating students killed by government troops in 1959 during a protest against then-president Anastasio Somoza. Sixty thousand Nicaraguans died during decades of conflict between Somoza's military dictatorship and the pro-democracy political opposition.

D Cuban revolutionaries Che Guevara (right) and Camillo Cienfuegos (left) are immortalized in wax in the Museum of the Revolution in Havana. Since coming to power in 1959, Cuba's communist government has been criticized for jailing and sometimes executing leaders of the political opposition.

Thinking Geographically

After you have read about Democratization and Conflict in Middle and South America, you will be able to answer the following questions:

A What are two drugs that dominate the drug trade in this region?

B What side of Colombia's ongoing civil war does the drug trade finance?

D Cuba's alliance with the Soviet Union resulted in hostile relations with what nearby country?

oil deposits had not trickled down to the poor. One-third of the population lived on $2 (PPP) a day, and per capita GDP (PPP) was lower in 1997 than it had been in 1977. Chávez campaigned as a champion of the poor, calling for the government to provide jobs, community health care, and subsidized food for the large underclass. These policies, all of which were enacted soon after his election, eroded Chávez's original support from the small middle class and elites, who feared a turn toward general government control. When the U.S. government expressed alarm, Chávez became a major critic of U.S. and other foreign dominance in the region. Chávez sought to lead the region-wide backlash against SAPs and condemned the long history of U.S. intervention in the internal affairs of countries across the region.

▐▐▶ 66. SOCIAL PROGRAMS AT ROOT OF CHÁVEZ'S POPULARITY

Chávez was reelected in 2000, and then briefly deposed in a coup d'état in 2001. He was quickly reinstated, however, by a groundswell of popular support among the large underclass, who stood to benefit from his policies. His presidency was sustained in a referendum in 2004 and again in the landslide election of 2006. In 2009, voters approved a constitutional amendment removing presidential term limits, thereby allowing Chávez to run for president indefinitely. Some saw this as a triumph of the *populist movement* (see definition on page 132), others as a loss for democracy (see the map in Photo Essay 3.4).

▐▐▶ 68. WILL CHÁVEZ INHERIT CASTRO'S REVOLUTIONARY MANTLE?
▐▐▶ 69. VENEZUELANS REJECT CONSTITUTIONAL CHANGES

The Drug Trade, Conflict, and Democracy

The international illegal drug trade is a major factor contributing to corruption, violence, and subversion of democracy throughout the region. Cocaine and marijuana are the primary drugs of trade, with most drugs produced in or passing through northwestern South America, Central America, and Mexico. Figure 3.10 illustrates the geographic distribution of cocaine seizures of more than 10 kilograms in 2006. (Note: Such seizures may not reveal activity in areas where law enforcement is lax or entirely co-opted by drug traders.)

Most coca growers are small-scale farmers of indigenous or mestizo origin in remote locations who can make a better

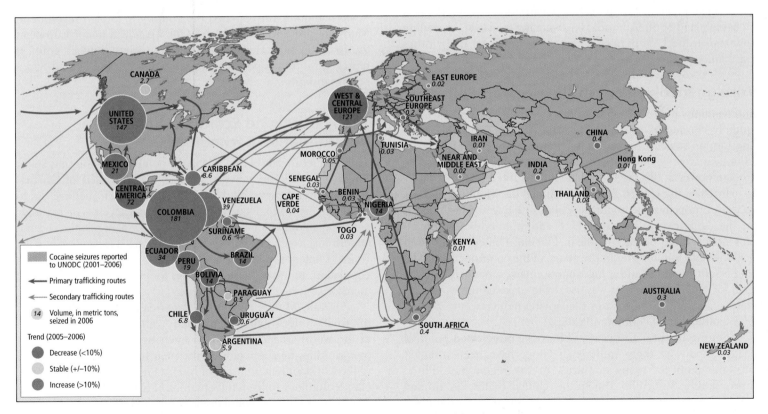

FIGURE 3.10 Linkages: Cocaine sources, trafficking routes, and seizures worldwide. Colombia, Peru, and Bolivia are the most important sources for the cultivation and production of cocaine in the world; Colombia produces 55 percent of the world's cocaine; Peru produces the second most, and Bolivia, the third. The big cocaine markets are in the Americas; use has declined by 36 percent in the United States since 1998 and increased in South America. Cocaine use in Western Europe and West Africa is on the rise, due in part to new trade routes through West Africa to Europe.

income for their families from these plants than from other cash crops. Production of addictive drugs is illegal in all of Middle and South America. However, public figures, from the local police on up to high officials, are paid to turn a blind eye to the industry. Those who oppose the cartels, whether farmers, journalists, or law enforcement officials, often end up kidnapped, tortured, and brutally murdered.

In Colombia, the illegal drug trade has financed all sides of an ongoing civil war that has threatened the country's democratic traditions and displaced more than 1.5 million people over the past several decades (see Photo Essay 3.4B). Meanwhile, competing drug-smuggling rings and paramilitaries bribe, intimidate, kidnap, and murder Colombian citizens and government officials who try to reduce or eradicate the drug trade. Mexico also now faces threats to its political stability (see Photo Essay 3.4A) from drug cartels that control a hugely profitable U.S.-oriented drug trade.

‖▶ 72. THOUSANDS KIDNAPPED IN COLOMBIA IN LAST DECADE

U.S. policy has emphasized stopping the production of illegal drugs in Middle and South America rather than curtailing the demand for drugs in the United States. As a result, the U.S. *war on drugs* has led to a major U.S. presence in the region that is focused on supplying intelligence, eradication chemicals, military equipment, and training to military forces in the region. One consequence of this is that U.S. military aid to Middle and South America is now about equal to U.S. aid for education and other social programs in the region. Meanwhile, drug production in Middle and South America is now greater than ever, exceeding demand in the U.S. market, where street prices for many drugs have fallen in recent years.

Foreign Involvement in the Region's Politics

Interventions in the region's politics by outside powers have frequently compromised democracy and human rights. Although the former Soviet Union, Britain, France, and other European countries have wielded much influence, by far the most active foreign power in this region has been the United States.

In 1823, the United States proclaimed the Monroe Doctrine to warn Europeans that no further colonization would be tolerated in the Americas. Subsequent U.S. administrations interpreted this policy more broadly to mean that the United States itself had the right to intervene in the affairs of the countries of Middle and South America, and it has done so many times. The official goal for such interventions was usually to make countries safe for democracy, but in most cases the driving motive was to protect U.S. political and economic interests.

During the past 150 years, U.S.-backed unelected political leaders, many of them military dictators, have been installed at some point in many countries in the region (see Photo Essay 3.4C). Since the 1960s, the United States has funded armed interventions in Cuba (1961), the Dominican Republic (1965), Nicaragua (1980s), Grenada (1983), and Panama (1989). Perhaps the most infamous intervention occurred in Chile in 1973. With U.S. aid, the elected socialist-oriented government of Salvador Allende was overthrown. Allende was killed, and a military dictator, General Augusto Pinochet, installed in his

place. Over the next 17 years, the Pinochet regime imprisoned and killed thousands of Chileans who protested the loss of democracy.

‖▶ 70. NORIEGA LAWYERS CONTINUE TO FIGHT AGAINST EXTRADITION
‖▶ 74. EX-CHILEAN DICTATOR AUGUSTO PINOCHET DIES

The Cold War and Post–Cold War Eras Since 1959, the Caribbean island of Cuba has been governed by a radically socialist government led first by Fidel Castro (see Photo Essay 3.4D) and then, starting in 2006, by Fidel's brother Raúl. The revolution that brought Fidel Castro to power transformed a plantation and tourist economy once known for its extreme income disparities into one of the most egalitarian in the region. However, because Castro adopted socialism and allied Cuba with the Soviet Union, relations with the United States became extremely hostile. The United States has funded many efforts to destabilize Cuba's government, and it actively discourages other countries from trading with Cuba. The United States maintains an ongoing embargo against Cuba, and so far international efforts to lift the embargo have failed.

‖▶ 73. EU SPLIT OVER DEVELOPMENT COMMISSIONER'S PROPOSAL TO NORMALIZE RELATIONS WITH CUBA

With help from the Soviet Union, Castro managed to dramatically improve the country's life expectancy, literacy, and infant mortality rates. Unfortunately, he also imprisoned or executed thousands of Cubans who disagreed with his policies, many of whom fled to southern Florida. Following the demise of the Soviet Union, Castro opened the country to foreign investment. Many countries responded, and Cuba is now a major European tourist destination. However, the Castro regime is intact, political repression in Cuba persists, and relations with the United States remain distant.

The Political and Economic Impacts of Information Technology

The Internet already plays a role in the politics of Middle and South America by helping activists, such as the Zapatistas of Chiapas and the MST (pages 120–121) of Brazil, get their message out to the rest of the world. However, while Internet access is increasingly common among the middle classes, very few poor people have the opportunity to get online. Nevertheless, access is expanding rapidly. In 2000, Internet users made up only 3.2 percent of the region's population; but that figure had risen to 34.8 percent by 2010, an increase of more than 1000 percent.

Overall, Middle and South America are more advanced in information technology than many other developing regions of the world—a fact that could open up many new economic opportunities for the region in the near future. Brazil ranks fifth in the world for total number of Internet users, and has two world-class technology hubs in the environs of São Paulo. Mexico, with 24.8 percent of its population connected (Figure 3.11), has recently launched a program to give its citizens access to training and higher education via the Internet. Similar efforts throughout the region are part of a long-term shift, especially in urban areas, toward more technologically sophisticated and better-paid service sector employment (see Reasons for Optimism B).

FIGURE 3.11 Internet use in Middle and South America, September 2009. The numbers on the map indicate the number of Internet users in each country; percents indicate the percentage of the country's population that uses the Internet. By September 2010, over 203 million people (34.8 percent of the population in Middle and South America) were using the Internet; the number of users more than doubled since 2005. Chile has the highest percentage of users (50.4) in South America; Antigua and Barbuda (75.9 percent) the highest in the Caribbean; and Costa Rica (34.3 percent) the highest in Middle America.

THINGS TO REMEMBER

1. Democracy has transformed the politics of the region, yet problems remain. Elected governments are at times challenged by citizen protests or with a coup d'état because policies are unpopular with large segments of the population, powerful elites, or the military.

2. The drug trade has emerged as a major obstacle to democracy, with huge amounts of illicit cash being used to pay off elected officials, civil servants, police forces, and the military.

3. Rapid growth of Internet use is part of a long-term shift toward more technologically sophisticated service sector employment.

Sociocultural Issues

Under colonialism, a series of social structures evolved that guided daily life—standard ways of organizing the family, community, and economy. They included rules for gender roles, race relations, and religious observance. These social structures, combined with economic systems, influenced population distribution and growth. Traditional social structures, still widely accepted in the region, are nonetheless changing in response to urbanization, economic development, migration, and globalization. The results are varied. In the best cases, change is leading to a new sense of initiative on the part of women, men, and the poor. In the worst cases, the result is loss of economic viability and community cohesion and the breakdown of family life, with some of the most extreme impacts felt by indigenous communities.

Population Patterns

Today, populations in Middle and South America continue to grow, but at a slower pace than in the past, due to lower birth rates. At the same time, a major migration from rural to urban areas is transforming traditional ways of life. A second, international migration trend is also growing as many people leave their home countries, temporarily or permanently, to seek opportunities in the United States and Europe. As of 2011, about 590 million people were living in Middle and South America, close to 10 times the population of the region in 1492.

Population Distribution The population density map of this region reveals a very unequal distribution of people. Comparing the population map (**Figure 3.12**) with the landforms map (see Figure 3.3), you will find areas of high population density in a

FIGURE 3.12 Population density in Middle and South America.

FIGURE 3.13 Trends in natural population increases, 1990–2010. The orange and blue columns show that rates of natural increase have declined steadily throughout the region. Nevertheless, in many countries, natural population increase rates remain high enough to outstrip efforts at improving standards of living. Note that the rates of natural increase given here are for ranges of time.

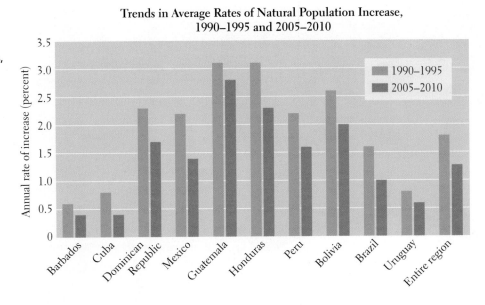

Trends in Average Rates of Natural Population Increase, 1990–1995 and 2005–2010

variety of environments. Some of the places with the highest densities, such as those around Mexico City and in Colombia and Ecuador, are in highland areas. But high concentrations are also found in lowland zones along the Pacific coast of Central America and especially along the Atlantic coast of South America. The cool uplands (*tierra templada*) were densely occupied even before the European conquest. Most coastal lowland concentrations, in the *tierra caliente*, are near seaports with vibrant economies and a cosmopolitan social life that attract people.

Learning Goal 5
Population: What drove the population explosion that occurred in this region during the twentieth century?

Population Growth Rates of natural population increase were high in Middle and South America in the twentieth century. They are now declining, as shown in Figure 3.13; however, because 30 percent of the region's population is under age 15, the population will continue to grow as this large group reaches the age of reproduction, even if couples choose to have only one or two children. Population projections for the year 2050 were recently revised downward from 778 million to 724 million—an increase overall of 18 percent. Though significantly lower than previous forecasts, for a region struggling to increase living standards, supplying 18 percent more people with food, water, homes, schools, and hospitals will be a challenge.

During the twentieth century, cultural and economic factors combined with improvements in health care to create a population explosion (see Thematic Overview G). Most of the population growth was due to longer life expectancy, not to higher birth rates. Improved living conditions (food, shelter, sanitation habits) and improved education and medical care all contributed to this change.

High birth rates were sustained for a while because the Roman Catholic Church discourages systematic family planning, and cultural mores encouraged men and women to reproduce prolifically. In agricultural areas, children were seen as sources of wealth because they could do useful farm and household work at a young age and eventually would care for their aging elders. Worry over infant death rates persisted, so some parents had four or more children to be sure of raising at least a few to adulthood.

By the 1930s, access to medical care was improving and incomes and education rates rose as more people moved to the cities. By 1975, death rates were one-third what they had been in 1900.

By the 1980s, migration out of the region and lower birth rates were curbing population growth. Now the region is undergoing a demographic transition (see Figure 1.11 on page 17). Between 1975 and 2011, the annual rate of natural increase for the entire region fell from about 1.9 percent to 1.3 percent—still a higher rate of growth than the world average of 1.2 percent. The low death rate continues to contribute to population growth in most of the region.

Cultural Diversity

The region of Middle and South America is culturally complex because many distinct indigenous groups were already present when the Europeans arrived, and many cultures were introduced during and after the colonial period. From 1500 to the early 1800s, some 10 million people from many different parts of Africa were brought to plantations on the islands and in the coastal zones of Middle and South America. After the emancipation of African slaves in the British-controlled Caribbean islands in the 1830s, more than half a million Asians were brought there from India, Pakistan, and China as indentured agricultural workers. Their cultural impact remains most visible in Trinidad and Tobago, Jamaica, and the Guianas. In some parts of Mexico, Central America, the Amazon Basin, and the Andean Highlands, indigenous people have remained numerous. To the unpracticed eye, they may appear little affected by colonization, but this is not the case.

FIGURE 3.14 Cultural diversity in Brazil. Japanese-Brazilian country singer Mauricio Miya sings in Portuguese, but his music sounds like it came from Nashville. Brazil is now home to 1.4 million people of Japanese ancestry, the largest such population outside of Japan. For some of Miya's music, go to http://www.mauriciomiya.com/.

and a high-status mate, may become recognized as upper class.

Nevertheless, the ability to erase the significance of skin color through one's actions is not quite the same as race having no significance at all. Overall, those who are poor, less educated, and of lower social standing tend to have darker skin than those who are educated and wealthy. And while there are poor people of European descent throughout the region, most light-skinned people are middle and upper class. Indeed, race and skin color have not disappeared as social factors in the region. In some countries—Cuba, for example, where overt racist comments are socially unacceptable—it is common for a speaker to use a gesture (tapping his or her forearm with two fingers) to indicate that the person referred to in the conversation is of African descent.

The Family and Gender Roles

The basic social institution in the region is the **extended family**, which may include cousins, aunts, uncles, grandparents, and more distant relatives. It is generally accepted that the individual should sacrifice many of his or her personal interests to those of the extended family and community and that individual well-being is best secured by doing so.

The arrangement of domestic spaces and patterns of socializing illustrate these strong family ties. Families of adult siblings, their mates and children, and their elderly parents frequently live together in domestic compounds of several houses surrounded by walls. Social groups in public spaces are most likely to be family members of several generations rather than unrelated groups of single young adults or married couples, as would be the case in Europe or the United States. A woman's best friends are likely to be her female relatives. A man's social or business circles will include male family members or long-standing family friends.

Gender roles in the region have been strongly influenced by the Roman Catholic Church. The Virgin Mary is held up as the model for women to follow through a set of values known as **marianismo**. *Marianismo* emphasizes chastity, motherhood, and service to the family. The ideal woman is the day-to-day manager of the house and of the family's well-being. She trains her sons to enter the wider world and her daughters to serve within the home. Over the course of her life, a woman's power increases as her skills and sacrifices for the good of all are recognized and enshrined in family lore.

Her husband, the official head of the family, is expected to work and to give most of his income to his family. Men have much more autonomy and freedom to shape their lives than women because they are expected to move about the larger community and establish relationships, both economic and personal. A man's social network is deemed just as essential to the family's prosperity and status in the community as his work.

Today in the Caribbean and along the east coast of Central America as well as the Atlantic coast of Brazil, the population majority is of a mixture of African and European ancestry. *Mestizos* are now the majority in Mexico, Central America, and much of South America. In some areas, such as Argentina, Chile, and southern Brazil, people of Central European descent are also numerous. The Japanese, though a tiny minority everywhere in the region, increasingly influence agriculture and industry, especially in Brazil (Figure 3.14), the Caribbean, and Peru. Alberto Fujimori, a Peruvian of Japanese descent, campaigned on his ethnic "outsider" status to become president of Peru during a time of political upheaval.

In some ways, diversity is increasing as the media and trade introduce new influences from abroad. At the same time, the processes of **acculturation** and **assimilation** are also accelerating, meaning that diversity is being erased to some extent as people adopt new ways. This is especially true in the biggest cities, where people of widely different backgrounds live in close proximity.

> **acculturation** adaptation of a minority culture to the host culture enough to function effectively and be self-supporting; cultural borrowing
>
> **assimilation** the loss of old ways of life and the adoption of the lifestyle of another culture
>
> **extended family** a family that consists of related individuals beyond the nuclear family of parents and children
>
> **marianismo** a set of values based on the life of the Virgin Mary, the mother of Jesus, that defines the proper social roles for women in Middle and South America

Race and the Social Significance of Skin Color

People from Middle and South America, especially those from Brazil, often proudly claim that race and color are of less consequence in this region than in North America. They are right in certain ways. Skin color is less associated with status than in the colonial past. A person of any skin color, by acquiring an education, a good job, a substantial income, the right accent,

In addition, there is an overt double sexual standard for males and females. While all expect strict fidelity from a wife to a husband in mind and body, a man is much freer to associate with the opposite sex. Males measure themselves by the model of **machismo**, which considers manliness to consist of honor, respectability, fatherhood, household leadership, attractiveness to women, and the ability to be a charming storyteller. Traditionally, the ability to acquire money was secondary to other symbols of maleness. Increasingly, however, a new market-oriented culture prizes visible affluence as a desirable male attribute.

> **machismo** a set of values that defines manliness in Middle and South America
>
> **primate city** a city, plus its suburbs, that is vastly larger than all others in a country and in which economic and political activity is centered
>
> **brain drain** the migration of educated and ambitious young adults to cities or foreign countries, depriving the communities from which the young people come of talented youth in whom they have invested years of nurturing and education

Learning Goal 6

Gender: What factors in the region's development are straining family ties and traditional gender roles?

Many factors are transforming these family and gender roles. With infant mortality declining steeply, couples are now having only two or three children instead of five or more. Moreover, because most people still marry young, parents are generally free of child-raising responsibilities by the time they are 40. Left with 30 or more years of active life to fill in other ways (life expectancies in the region average in the mid-70s), middle-aged mothers and grandmothers are increasingly working in urban factory or office jobs that put to use the organizational and problem-solving skills they perfected while managing a family. Employment outside the home allows women greater independence and helps them contribute to the needs of the extended family. Some women are even moving into high-level jobs, such as management positions, professorships, and positions that traditionally went to men—for example, the presidencies of Chile, Argentina, and Brazil.

Migration and Urbanization

Since the early 1970s, Middle and South America have seen very rapid urban growth as rural people migrate to cities and towns (see Thematic Overview H). Now more than 70 percent of the people in the region live in settlements of at least 2000. Increasingly, one city, known as a **primate city**, is vastly larger than all the others, accounting for a large percentage of the country's total population (see Photo Essay 3.5 on page 130). Examples in the region are Mexico City, Mexico (with 23 million people, a bit more than 21 percent of Mexico's total population); Managua, Nicaragua (33 percent of that country's population); Lima, Peru (29 percent); Santiago, Chile (34 percent); and Buenos Aires, Argentina (36 percent).

The concentration of people into just one or two large cities in a country leads to uneven spatial development and to government policies and social values that favor urban areas. Wealth and power are concentrated in one place, while distant rural areas, and even other towns and cities, have difficulty competing for talent, investment, industries, and government services. Many provincial cities languish as their most educated youth leave for the primate city.

Brain Drain It always takes some resourcefulness to move from one place to another. Those people who already have some years of education and strong ambition are the ones who migrate to cities. The loss of these resourceful young adults in which the community has invested years of nurturing and education is referred to as **brain drain**. Brain drain happens at several scales in Middle and South America: through rural-to-urban migration from villages to regional towns, and from towns and small cities to primate cities. There is also international brain drain when migrants move to North America and Europe; for example, one in four U.S. doctors are foreign born. Families often encourage their children to migrate so that they can benefit from the remittances, goods, and services that migrants send back to their home communities.

Unplanned Urban Neighborhoods A lack of planning to accommodate the massive rush to the cities has created urban landscapes that are very different from the common U.S. pattern. In the United States, a poor, older inner city is usually surrounded by affluent suburbs, with clear and planned spatial separation of residences by income as well as separation of residential, industrial, and commercial areas. In Middle and South America, by contrast, both affluent and working-class areas have become unwilling neighbors to unplanned slums filled with poor migrants. Too destitute even to rent housing, these "squatters" occupy parks and small patches of vacant urban land wherever they can be found, as depicted in the diagram in Figure 3.15.

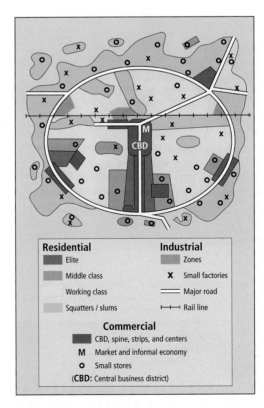

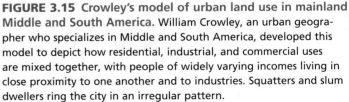

FIGURE 3.15 Crowley's model of urban land use in mainland Middle and South America. William Crowley, an urban geographer who specializes in Middle and South America, developed this model to depict how residential, industrial, and commercial uses are mixed together, with people of widely varying incomes living in close proximity to one another and to industries. Squatters and slum dwellers ring the city in an irregular pattern.

Urbanization is at the heart of much of the transformation in this region. Cities here have grown extremely quickly in the last several decades, and 77 percent of the region's population now lives in cities. New opportunities have opened up for migrants from rural areas, but new stresses have emerged as well. The low-skilled jobs people can get often don't pay well, and many are forced to live in unplanned neighborhoods on the outskirts of huge cities.

A Santiago is one of the region's primate cities, home to 34 percent of Chile's population. It dominates Chile's economy, generating 40 percent of the country's GDP.

B Below, children in Lima, Peru, search through a garbage can in search of food to eat. Some families disintegrate after the move to the city as parents are forced to work long hours at low-paying jobs. Even in those that remain intact, children are sometimes forced to beg for money and food instead of going to school.

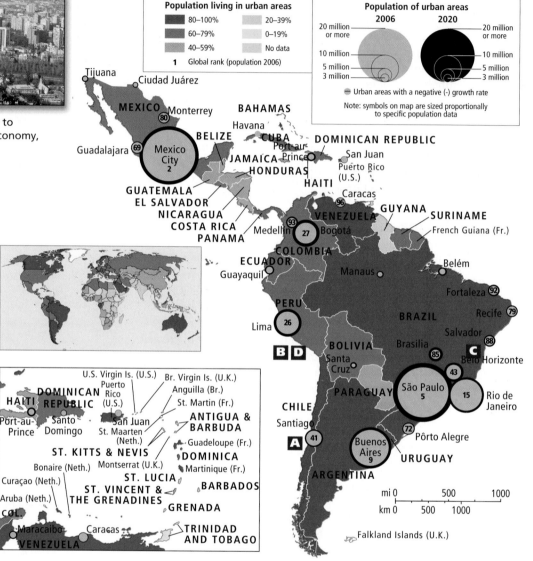

Population living in urban areas
- 80–100%
- 60–79%
- 40–59%
- 20–39%
- 0–19%
- No data

1 Global rank (population 2006)

Population of urban areas
2006   2020
20 million or more
10 million
5 million
3 million

Urban areas with a negative (-) growth rate

Note: symbols on map are sized proportionally to specific population data

C A woman works as a street vendor in Salvador, Brazil. Many new urban migrants work in similar informal sector jobs that generally pay less and are much less secure than formal sector jobs.

D An informal squatter settlement forming outside of Lima, Peru. Structures made out of woven grass mats are the first to go up, followed by wooden shacks and eventually cement or brick houses. Water and electricity may eventually be extended to such settlements, but only after years or decades of occupation.

Thinking Geographically

After you have read about Urbanization in Middle and South America, you will be able to answer the following questions:

A What about this picture suggests that there may be considerable wealth in Santiago?

B What changes in agriculture are pushing people out of rural areas?

C Why might work in the informal economy be considered risky?

D What is the defining characteristic of "informal" squatter settlements?

The best known of these unplanned communities are Brazil's **favelas** (Figure 3.16). In other countries they are known as *slums, shantytowns, colonias, barrios,* or *barriadas.* The settlements often spring up overnight, without city-supplied water or electricity and with housing made out of whatever is available. Once they are established, efforts to eject squatters usually fail. The impoverished are such a huge portion of the urban population that even those in positions of power will not challenge them directly. Hence, nearby wealthy neighborhoods simply barricade themselves with walls and security guards.

The squatters frequently are enterprising people who work hard to improve their communities. They often organize to press governments for social services. Some cities, such as Fortaleza in northeastern Brazil, even contribute building materials so that favela residents can build more permanent structures with basic indoor plumbing. Over time, as shacks and lean-tos are transformed through self-help initiatives into crude but livable suburbs, the economy of favelas can become quite vibrant, with much activity in the informal sector. Housing may be intermingled with shops, factories, warehouses, and other commercial enterprises. Favelas and their counterparts can become centers of pride and support for their residents, where community work, folk belief systems, crafts, and music (for example, Favela Funk) flourish. Many of the best steel bands of Port of Spain, Trinidad, have their homes in the city's shantytowns.

> **favelas** Brazilian urban slums and shantytowns built by the poor; called colonias, barrios, or barriadas in other countries

Vignette Favelas are everywhere in Fortaleza, Brazil. The city grew from 30,000 to 300,000 residents during the 1980s. By 2006, there were more than 3 million residents, most of whom had

FIGURE 3.16 A favela in Rio de Janeiro. (A) Favela Dona Marta, indicated by the white circle, clings to a once-vacant hillside that was considered too steep for apartment buildings or even for road access. Now home to 7000 people, it is still served only by footpaths and narrow alleyways. **(B)** Water is only available sporadically, so many residents store water in tanks on their roof. **(C)** Electrical connections consist of dangerous tangles of wires, such as those hanging over this man's head, illegally connected to a main power line. **(D)** Favelas are often full of children who spend much of the day under the supervision of a friend or relative while their parents work in distant parts of the city.

Thinking Geographically: (A) Why are favelas often located on steep hills? **(B)** Why do favela residents often store water on their roofs? **(C)** Why are dangerous tangles of power lines common in favelas?

fled drought and rural poverty in the interior. City parks of just a square block or two in middle-class residential areas were suddenly invaded by squatters. Within a year, 10,000 or more people were occupying crude, stacked concrete dwellings in a single park, completely changing the ambience of the surrounding upscale neighborhood. In the early days of the migration, lack of water and sanitation often forced the migrants to relieve themselves on the street.

One day, while strolling on the Fortaleza waterfront, Lydia Pulsipher chanced to meet a resident of a beachfront favela who invited her to join him on his porch. There, he explained how he and his wife had come to the city 5 years before, after being forced to leave the drought-plagued interior when a newly built irrigation reservoir flooded the rented land their families had cultivated for generations. With no way to make a living, they set out on foot for the city. In Fortaleza, they constructed the building they used for home and work from objects they collected along the beach. Eventually they were able to purchase roofing tiles, which gave the building an air of permanency. At the time of the visit, he maintained a small refreshment stand and his wife a beauty parlor that catered to women from the beach settlements. *[Source: Lydia Pulsipher's field notes on Brazil, updated with the help of John Mueller, Fortaleza, Brazil, 2006.]* ∎

Gender and Urbanization Interestingly, rural women are just as likely as rural men to migrate to the city. This is especially true when employment is available in foreign-owned factories that produce goods for export (see Thematic Overview I). Companies prefer women for such jobs because they are a low-cost and usually passive labor force. Factors that push people out of rural areas, such as the shift toward green revolution agriculture, often affect women intensely because women are rarely considered for jobs such as farm-equipment operators or mechanics. In urban areas, unskilled women migrants usually find work as street vendors (see Photo Essay 3.5C) or domestic servants.

Male urban migrants tend to depend on short-term, low-skilled day work in construction, maintenance, small-scale manufacturing, and petty commerce. Many work in the informal economy as street vendors, errand runners, car washers, and trash recyclers—and some turn to crime. The loss of family ties and village life is sorely felt by both men and women, and the chances for recreating the family life they once knew are extremely low in the urban context.

Families constructed in the cities may disintegrate because of extreme poverty and malnutrition, long commutes for both working parents, and poor quality of day care for children (see Figure 3.16D on page 131). At too young an age, children are left alone or sent into the streets to scavenge for food or to earn money for the family (see Photo Essay 3.5B; see also Figure 3.16D).

Religion in Contemporary Life

Judaism, Islam, Hinduism, and indigenous beliefs are found across the region, but the majority are at least nominal Christians, most of them Roman Catholic. While the Roman Catholic Church remains highly influential and relevant to the lives of believers (Figure 3.17), it has had to contend both

populist movements popularly based efforts, often seeking relief for the poor

FIGURE 3.17 Catholics in Mexico City praying for the end of swine flu. The devotees are wearing masks to reduce the risk of contagion while praying in Mexico City's metropolitan cathedral.

Thinking Geographically: What was the relationship between the Catholic Church and early colonial governments in Middle and South America?

with popular efforts to reform it and with increasing competition from other religious movements. From the beginning of the colonial era, the church was the major partner of the Spanish and Portuguese colonial governments. It received extensive lands and resources from colonial governments, and in return built massive cathedrals and churches throughout the region, sending thousands of missionary priests to convert indigenous people. The Roman Catholic Church encouraged the poor converts to accept their low status, obey authority, and postpone rewards for their hard work until heaven.

Poor people throughout the region embraced the faith, and they still make up the majority of the church members. Many are indigenous people who put their own spin on Catholicism, creating multiple folk versions of the Mass with folk music, greater participation of women in worship services, and interpretations of Scripture that vary greatly from European versions. A range of African-based belief systems (Candomblé, Umbanda, Santería, Obeah, and Voodoo) combined with Catholic beliefs is found in Brazil, northern South America, and in Middle America and the Caribbean—wherever the descendants of Africans have settled. These African-based religions have attracted adherents of European or indigenous backgrounds as well, especially in urban areas.

The power of the Roman Catholic Church began to erode in the nineteenth century in places such as Mexico. **Populist movements** seized and redistributed some church lands. They also canceled the high fees the clergy had been charging for simple rites of passage such as baptisms, weddings, and funerals. Over the years, the Catholic Church became less obviously connected to the elite and more attentive to the needs of poor and non-European people. By the mid-twentieth century, the church was abandoning many previous racist policies and ordaining indigenous and clergy of African descent. Women were also given a greater role in religious ceremonies.

In the 1970s, a radical Catholic movement known as **liberation theology** was begun by a small group of priests and activists. They sought to reform the church into an institution that could combat the extreme inequalities in wealth and power common in the region. The movement portrayed Jesus Christ as a social revolutionary who symbolically spoke out for the redistribution of wealth when he divided the loaves and fish among the multitude. The perpetuation of gross economic inequality and political repression was viewed as sinful, and social reform as liberation from evil.

At its height in the 1970s and early 1980s, liberation theology was the most articulate movement for region-wide social change. It had more than 3 million adherents in Brazil alone. But the Vatican objected to this popularized version of Catholicism, and today its influence is diminished. In countries such as Guatemala and El Salvador, governments targeted liberation theology participants, vilifying them as communist collaborators. Liberation theology has also had to compete with newly emerging evangelical Protestant movements.

II ▶ 79. POPE OPENS TOUR OF LATIN AMERICA IN BRAZIL

Evangelical Protestantism has diffused from North America into Middle and South America, and is now the region's fastest-growing religious movement. About 10 percent of the population, or at least 50 million people, are adherents. The movement is growing rapidly in Brazil and Chile, in the Caribbean, Mexico, and Middle America, especially among poor and middle classes in both rural and urban settings. It does not, however, share liberation theology's emphasis on combating the region's extreme inequalities in wealth and power. Some evangelical Protestants teach a "gospel of success," stressing that those who are true believers and give themselves to a new life of hard work and clean living will experience prosperity of the body (wealth) as well as of the soul.

The movement is *charismatic*, meaning that it focuses on personal salvation and empowerment of the individual through miraculous healing and psychological transformation. Evangelical Protestantism is not hierarchical in the same way as the Roman Catholic Church, and there is usually no central authority; rather, there are a host of small, independent congregations led by entrepreneurial individuals who may be either male or female.

> **liberation theology** a movement within the Roman Catholic Church that uses the teachings of Jesus to encourage the poor to organize to change their own lives and the rich to promote social and economic equity
>
> **evangelical Protestantism** a Christian movement that focuses on personal salvation and empowerment of the individual through miraculous healing and transformation; some practitioners preach the "gospel of success"— that a life dedicated to Christ will result in prosperity for the believer—to the poor

THINGS TO REMEMBER

1. **Learning Goal 5:** Population A population explosion occurred during the early twentieth century as improved health care lowered death rates.

2. Birth rates started to decline by the mid-1980s as more women began to delay childbearing in order to pursue work outside the home. At the same time, populations became more urbanized, which reduced the need for large families to operate farms.

3. Gender roles in the region have been strongly influenced by the Catholic Church through the ideals of *marianismo* and *machismo*.

4. **Learning Goal 6:** Gender Middle-aged women are increasingly working in factory or office jobs once dominated by men; they put to use the organizational and problem-solving skills perfected while supervising a family.

5. Urbanization is putting new pressures on the extended family and also opening up new opportunities for women.

6. Evangelical Protestantism came to the region from North America and is now the fastest-growing religious movement in the region.

Reflections on Middle and South America

European colonialism in Middle and South America launched the modern global economic system. It was in this region that large-scale extractive industries were inaugurated. Raw materials were shipped at low prices to distant locales, where they were turned into high-priced products, the profits of which went to Europe.

After the massive outflows of raw materials during the colonial era, the region tried a number of economic development strategies, with mixed results. In a few cases, import substitution industries (ISIs) worked to lower the rate of imports, but the industries failed to lift the region out of poverty. The huge debts that many governments accumulated while inefficiently pursuing these projects later led to the imposition of structural adjustment programs. SAPs also failed to produce economic growth as they reduced investment in human capital and relied heavily on raw materials exports just when prices for these exports were falling on global markets. Now deforestation, much of it driven by the need for export earnings, is contributing to both global climate change and the loss of the region's great biodiversity.

Massive rural-to-urban and international migrations resulted from the shift from small-scale agriculture for local consumption to large-scale green revolution agriculture oriented toward export production. Most cities were unprepared for the influx of migrants, and water crises developed as existing supply and delivery systems were overwhelmed. While the informal economy has been a lifesaver for many poor urbanites, its ability to lift people out of poverty is limited. Traditional gender roles have also changed with urbanization, giving new opportunities to some but also straining family ties.

However, there are many positive signs in the region (see Reasons for Optimism on page 134). Alternatives to deforestation hold the potential to reduce pressure on the region's forests

Reasons for Optimism in Middle and South America

Climate Change: Alternatives to deforestation are emerging that have the potential to preserve this region's biodiversity and reduce emissions of greenhouse gases. **A** *An ecotourism lodge in Brazil.* ▼

Development: Rapid expansion in Internet access is part of a shift toward better-paid and more technologically sophisticated service sector jobs. **B** *Techniques for increasing processor speed are explored at a computer conference in Mexico.* ▼

Democracy: Peaceful democratic elections are now the norm in this region where military dictatorships and coups d'état were once common. **C** *Women in Guatemala display their printed fingers, showing they have voted.* ▼

Thinking Geographically: What is the world's largest industry (by some measures)?

Thinking Geographically: Between 2000 and 2010, how many people gained access to the Internet in this region?

Thinking Geographically: Against which policies did a democratic backlash occur in many Middle and South American countries, beginning in the late 1990s?

and lessen the emission of greenhouse gases. Meanwhile, the rapid growth of Internet use throughout the region is part of an overall shift toward more technologically sophisticated and better-paid service sector employment. Finally, nearly all countries now hold regular peaceful democratic elections, and when leaders make moves reminiscent of old-style dictators, they must eventually answer to an ever-more-informed public.

Learning Goals Review

1. Climate Change and Deforestation: How are the processes of global warming and deforestation linked?

Why is deforestation having such an impact on the emissions of greenhouse gases? What factors are driving the cycle of deforestation in this region?

2. Water: In this region of abundant water resources, what factors have come together to create a water crisis?

How does the water crisis relate to patterns of urbanization?

3. Globalization, Development, and Democratization: How has recent dissatisfaction with SAPs (structural adjustment policies) and globalization been expressed through the democratic process?

Why has there been so much dissatisfaction with SAPs? Why has this region struggled for so long to develop its manufacturing and service-based industries?

4. Food and Urbanization: How has the shift toward green revolution agriculture encouraged urbanization?

How have new food production techniques changed the demand for agricultural labor? What kind of political responses have occurred? How have cities responded to the waves of migrants from rural areas?

5. Population: What drove the population explosion that occurred in this region during the twentieth century?

What accounted for the drop in death rates?

6. Gender: What factors in the region's development are straining family ties and traditional gender roles?

How are women's roles changing? How does migration to the city stress family relationships?

Geographic Themes about Middle and South America

Look back at the Thematic Overview photos on page 99. Thinking geographically, answer the following questions about them:

Thinking Geographically

(A) Climate Change: Where does a tree store the carbon dioxide it absorbs from the atmosphere?

(B) Water: Relative to other world regions, how much rainfall do Middle and South America receive?

(C) Development: In what countries has the gap between rich and poor been shrinking in recent years?

(D) Globalization: During the "early extractive phrase" of economic development, where did most of the raw materials taken from this region go?

(E) Food: Why does large-scale mechanized agriculture often push farmers off their land?

(F) Democratization: When was the last major U.S. intervention in Chilean politics?

(G) Population: What was the primary factor that drove the population explosion in the early twentieth century?

(H) Urbanization: What change in rural areas has increased migration to cities?

(I) Gender: Why do some export-oriented manufacturers prefer a largely female labor force?

Key Terms

acculturation 128

assimilation 128

Aztecs 111

biodiversity 104

brain drain 129

contested space 120

coup d'état 121

Creoles 112

dictator 121

early extractive phase 114

ecotourism 109

El Niño 106

evangelical Protestantism 133

Export Processing Zones (EPZs) 116

extended family 128

favelas 131

haciendas 114

import substitution industrialization (ISI) 115

Incas 111

income disparity 113

indigenous 101

liberation theology 133

machismo 129

maquiladoras 116

marianismo 128

marketization 116

mercantilism 112

Mercosur 119

mestizos 112

Middle America 101

nationalize 118

North American Free Trade Agreement (NAFTA) 119

plantation 115

populist movements 132

primate city 129

privatization 116

shifting cultivation 111

silt 104

South America 101

structural adjustment policies (SAPs) 116

subduction zone 101

temperature-altitude zones 104

trade winds 104

4 Europe

Learning Goals

After you read this chapter, you will be able to answer the following questions:

1. Water: What factors complicate efforts to reduce water pollution in the Mediterranean Sea?

2. Climate Change: In what ways is Europe responding to climate change?

3. Globalization and Development: How did Europe's colonization of much of the world over the past 400 years accelerate globalization and transform economic development in Europe?

4. Urbanization and Democratization: How has urbanization influenced the development of democracy in Europe?

5. Food: How is food production in Europe changing?

6. Population and Gender: How is Europe's aging population linked to changing gender roles?

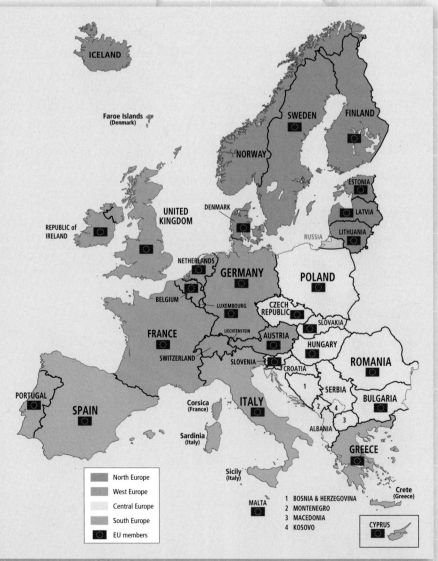

FIGURE 4.1 Political map of Europe.

North Europe	
West Europe	
Central Europe	
South Europe	
EU members	

1 BOSNIA & HERZEGOVINA
2 MONTENEGRO
3 MACEDONIA
4 KOSOVO

Thematic Overview of Europe

Climate Change: The EU is leading the world in responding to global warming. The EU's goal is to cut greenhouse gas emissions by 20 percent by 2020. **A** *Wind turbines off the coast of Denmark.* ▼

Water: The Mediterranean Sea is threatened by water pollution from the 460 million people now living in the countries that surround it. **B** *Naples, Italy, where beaches closed due to pollution.* ▼

Globalization: Most countries in the world have been ruled by a European power at some point in their history. **C** *A painting of Napoleon during his failed invasion of Egypt. The British went on to rule Egypt.* ▼

Democratization: While democratization has been ongoing for centuries in parts of Europe, some areas didn't democratize until after the fall of the Soviet Union. **D** *A pro-democracy demonstration in Serbia in 2007.* ▼

Development: Central Europe's economies have grown in recent years, thanks largely to investment from wealthier EU member states in western Europe. **E** *The construction of an EU-funded project in Poland.* ▼

Food: Agriculture is heavily subsidized throughout much of Europe. Agricultural subsidies account for almost half of the budget of the European Union. **F** *Vineyards in France, a major recipient of EU subsidies.* ▼

Urbanization: In western, northern, and southern Europe, around 80 percent of the population lives in urban areas. **G** *Karaoke in a crowded public park in Berlin, Germany, a city of more than 3 million people.* ▼

Population: Europe's population is aging, as families are choosing to have ever fewer children. **H** *Retirees rest while making a pilgrimage to a church in Spain.* ▼

Gender: Throughout the EU, women are paid on average 15 percent less than men for equal work. **I** *Female doctors perform laser surgery in a hospital in Dublin, Ireland.* ▼

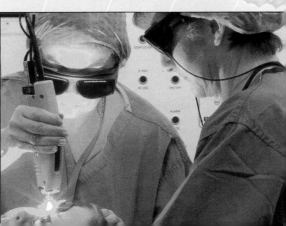

Global Patterns, Local Lives

Grigore Chivu (a pseudonym) wanders through the empty pig pens on his farm near Lugoj in western Romania. For generations his family has made a meager but rewarding living by raising and processing hogs and selling the meat. A few years ago, just before Romania joined the European Union (EU), he had more than 250 pigs. At Christmas, he and thousands of pig farmers across the country would slaughter a certain number of their pigs and preserve the meat, using age-old methods. Pig slaughtering was a time of high spirits, community cooperation, and celebration, as the farmers contemplated the coming feast and their profits. Flavorful sausages, which they smoked and hung high in the rafters of their kitchens for further drying, were slowly parceled out to select customers, providing a steady income well into summer.

When Romania entered the European Union in 2007, all farmers were required to conform to EU standards for processing meat. The old methods were no longer allowed, and the new standards were prohibitively expensive.

At first the farmers were uncertain just how to respond. Before they could organize a butchering cooperative that conformed to EU standards, and for which development funds were available, an American company stepped into the breach. Smithfield, a Fortune 500 meat company based in Virginia, was expanding operations into Central Europe, where it planned to produce pork and pork products and market them globally. Eager to enter the EU market, Smithfield enlisted the help of Romanian politicians and got permission and even subsidies to establish a conglomerate that included feed production, pig breeding, modernized sanitary barns for fattening thousands of hogs, and slaughterhouses.

As the old picturesque Romanian agricultural landscape is transformed into one of huge factory farms, the number of pig farmers has been reduced by more than 90 percent (**Figure 4.2**). Unable to compete with the lower prices Smithfield can charge, Grigore Chivu, like thousands of his fellow pig farmers, is thinking of migrating to Western Europe. Stronger economies there mean better-paying jobs, but because of his traditional farming background, Chivu will probably be eligible for only menial labor. [Source: Conversations with geographer Margareta Lelea, a Romanian specialist, Visiting Assistant Professor, Bucknell University; Doreen Carvajal and Stephen Castle, "A U.S. Hog Giant Transforms Eastern Europe," New York Times, May 5, 2009, at http://www.nytimes.com/2009/05/06/business/global/06smithfield.html?ref=europe.] ■

Grigore Chivu is faced with disruptive change as his country adjusts to new circumstances in the **European Union (EU)**, but he is also part of a global revolution in food production and marketing that is changing food patterns across the world. For example, Smithfield pork trimmings, produced and processed at low cost in Romania, are now marketed in West Africa, where the low prices are putting African pig farmers out of business.

The European Union is a supranational organization that unites most of the countries of West, South, North, and Central Europe (Figure 4.1 on page 136). In principle, throughout the European Union, people, goods, and money can move freely. Many people, like Grigore Chivu, decide

European Union (EU) a supranational organization that unites most of the countries of West, South, North, and Central Europe

FIGURE 4.2 Pig farms in Romania and the United States.
(A) Shepherds on a farm in Romania that raises small numbers of cows, sheep, and pigs (shown in **B**). **(C)** An industrial hog farm in Georgia, U.S., capable of raising thousands of pigs at a time.

Thinking Geographically: What about **(B)** suggests small-scale production of pigs?

to migrate only reluctantly, and often do so because the long-standing inequities that persist across Europe make life especially difficult for those from poorer regions. Adapting to the changes that have come in Europe is wrenching for many, especially for poor, rural people.

> **cultural homogenization** the tendency toward uniformity of ideas, values, technologies, and institutions among associated culture groups

There are now 27 countries in the European Union, the most recent joining in 2007, with more hoping to join in the next several years. The new members (with the exceptions of Malta and Cyprus) are formerly Communist countries in Central Europe, with lower standards of living and higher unemployment rates than countries in western Europe. As their own economies have faltered, hundreds of thousands of workers from Central Europe have taken advantage of the EU principle that (for the most part) citizens of any member state may move to any other EU state.

Some EU residents fear that the migration of workers is bringing very different people into close contact with each other, resulting in political tensions and increased costs for social and educational services. Others lament the **cultural homogenization** that is occurring. Across the European Union, as in Romania, distinctive local ways of life are disappearing as a result of new, well-intentioned EU regulations. Meanwhile, economic planners and employers across Europe argue that allowing workers to migrate is crucial to economic growth and competitiveness on a global scale.

THINGS TO REMEMBER

1. Throughout the European Union, smaller family-run farms are giving way to larger farms run by corporations. This change is strongest in those Central Europe states that have been recently admitted to the EU, and has resulted in many farmers and their families migrating to other EU countries in search of work.

I THE GEOGRAPHIC SETTING

Terms in This Chapter

This book divides Europe into *North*, *West*, *South*, and *Central Europe* (see Figure 4.2).

For convenience, we occasionally use the term Central Europe to refer to all the countries that were part of the experiment with communism in the Soviet sphere and Yugoslavia.

Physical Patterns

Europe is a region of peninsulas upon peninsulas (see **Figure 4.3** on pages 140–141). The entire European region is one giant peninsula extending off the Eurasian continent. Its very long coastline has many peninsular appendages, large and small. Norway and Sweden share one of the larger appendages. The Iberian Peninsula (shared by Portugal and Spain), Italy, and Greece are other large peninsulas. One result of these many fingers jutting into oceans and seas is that much of Europe feels the climate-moderating effect of the large bodies of water that surround it.

Landforms

Although European landforms are fairly complex, the basic pattern is mountains, uplands, and lowlands, all stretching roughly west to east in wide bands. As you can see in Figure 4.3, Europe's largest mountain chain stretches west to east through the middle of the continent, from southern France through Switzerland and Austria. It extends into the Czech Republic and Slovakia, and curves southeast into Romania. The *Alps* are the highest and most central part of this formation. This network of mountains is mainly the result of pressure from the collision of the northward-moving African Plate with the southeasterly moving Eurasian Plate (see Figure 1.21 on page 45). Europe lies on the westernmost extension of the Eurasian Plate.

South of the main Alps formation, mountains extend into the peninsulas of Iberia and Italy, and along the Adriatic Sea through Greece to the east. The northernmost mountainous formation is shared by Scotland, Norway, and Sweden. These northern mountains are old (about the age of the Appalachians in North America) and have been worn down by glaciers and millions of years of erosion.

Extending northward from the central mountain zone is a band of low-lying hills and plateaus curving from Dijon (France) through Frankfurt (Germany) to Krakow (Poland). These uplands (see Figure 4.3B) form a transitional zone between the high mountains and lowlands of the *North European Plain*, the most extensive landform in Europe (see Figure 4.3C). The plain begins along the Atlantic coast in western France and stretches in a wide band around the northern flank of the main European peninsula, reaching across the English Channel and the North Sea to take in southern England, southern Sweden, and most of Finland. The plain continues east through Poland, then broadens to the south and north to include all the land east to the Ural Mountains in Russia.

The coastal zones of the North European Plain are densely populated all the way east through Poland. Crossed by many rivers and holding considerable mineral deposits, this coastal lowland is an area of large industrial cities and densely occupied rural areas (see Figure 4.3D). Over the past thousand years, people have transformed the natural seaside marshes and vast river deltas into farmland, pastures, and urban areas by building dikes and draining the land with wind-powered pumps. This is especially true in the low-lying Netherlands, where concern over climate change and sea level rises is considerable.

The rivers of Europe link its interior to the surrounding seas. Several of these rivers are navigable well into the upland zone, and Europeans have built large industrial cities on their banks. The Rhine carries more traffic than any other European river, and the course it has cut through the Alps and uplands to the North Sea also serves as a route for railways and motorways (see Figure 4.3D). The area where the Rhine flows into the North Sea is considered

A Alps, Kühtai, Austria

B Uplands, Saarschleife, Germany

C North European Plain, Schleswig-Holstein, Germany

D Rhine River, Heldsburg, Switzerland

E Danube River, Budapest, Hungary

FIGURE 4.3 Regional map of Europe.

the economic core of Europe. Here Rotterdam, Europe's largest port, is located. The larger and much longer Danube River flows southeast from Germany, connecting the center of Europe with the Black Sea. As the European Union expands to the east, the economic and environmental roles of the Danube River basin, including the Black Sea, are getting increased attention (see Figure 4.3E).

Vegetation

Nearly all of Europe's original forests are gone, some for more than a thousand years. Today, forests with very large and old trees exist only in scattered areas, especially on the more rugged mountain slopes (see Figure 4.3 A, B) and in the northernmost parts of Norway, Sweden, and Finland. Forests are now regenerating where some small farms have been abandoned, and today cover about one-third of Europe. The dominant vegetation is crops and pasture grass.

Europe has three main climate types: temperate midlatitude, Mediterranean, and humid continental (Photo Essay 4.1). The **temperate midlatitude climate** dominates in western Europe, where the influence of the Atlantic Ocean is very strong. A broad warm-water ocean current called the **North Atlantic Drift** brings large amounts of warm water to the coasts of Europe. It is really just the easternmost end of the Gulf Stream, which carries water from the Gulf of Mexico north along the eastern coast of North America and across the North Atlantic to Europe (see Photo Essay 2.1 on page 63).

The air above the North Atlantic Drift is warm and wet. Eastward-blowing winds push it over North and West Europe and the North European Plain, bringing moderate temperatures and rain deep into the Eurasian continent. These factors create a climate that, although still fairly cool, is much warmer than elsewhere in the world at similar latitudes. To minimize the effects of heavy precipitation runoff, people in these areas have developed elaborate drainage systems for their houses and communities. Forests are both evergreen and deciduous. There is some concern that global warming could weaken the North Atlantic Drift, leading to a significantly cooler Europe.

Farther to the south, the **Mediterranean climate** prevails—warm, dry summers and mild, rainy winters. In the summer, warm, dry air from North Africa shifts north over the Mediterranean Sea as far north as the Alps, bringing high temperatures and clear skies. Crops grown in this climate, such as olives and grapes, citrus, apple and other fruits, and wheat, must be drought-resistant or irrigated. In the fall, this warm, dry air shifts to the south and is replaced by cooler temperatures and rainstorms sweeping in off the Atlantic. Overall, the climate here is mild, and houses along the Mediterranean coast are often open and airy to afford comfort in the hot, sunny summers.

In Central Europe, without the moderating influences of the Atlantic Ocean and the Mediterranean Sea, the climate is more extreme. In this region of **humid continental climate**,

temperate midlatitude climate as in south-central North America, China, and Europe, a climate that is moist all year with relatively mild winters and long, mild to hot summers

North Atlantic Drift the easternmost end of the Gulf Stream, a broad warm-water current that brings large amounts of warm water to Europe

Mediterranean climate a climate pattern of warm, dry summers and mild, rainy winters

humid continental climate a midlatitude climate pattern in which summers are fairly hot and moist, and winters become longer and colder the deeper into the interior of the continent one goes

Biosphere the global ecological system that integrates all life on this planet

summers are fairly hot, and the winters become longer and colder the farther north or deeper into the interior of the continent one goes. Here, houses tend to be well insulated, with small windows, low ceilings, and steep roofs that can shed snow. Crops must be adapted to much shorter growing seasons and include corn and other grains; a wide variety of vegetables, including root crops and cabbages adapted to cold; and fruit trees.

THINGS TO REMEMBER

1. Europe is a region of peninsulas upon peninsulas.

2. Europe has three main landforms—mountain chains, uplands, and the vast North European Plain.

3. Europe has three principal climates—the temperate midlatitude, the Mediterranean climate, and the humid continental climate.

Environmental Issues

Having dramatically transformed their environments over the past 10,000 years, Europeans are now increasingly taking action on environmental issues at the local and global scales. Nevertheless, Europe's air, seas, and rivers remain some of the most polluted in the world, and the EU still has a long way to go to meet its stated environmental goals of clean air and water; sustainable development in agriculture, industry, and energy use; and maintenance of biodiversity.

Europe's Impact on the Biosphere

There is a geographic pattern to the ways that human activity over time has transformed Europe's landscapes. Western Europe shows the effects of dense population and heavy industrialization, and Central Europe reveals the results of long decades of willful disregard for the environment. At present, this region continues to have a major impact on the **Biosphere** through the air and water pollution it generates (see Photo Essay 4.2 on page 144). Furthermore, Europe is itself especially vulnerable to a number of the potential effects of climate change, drier climates, and wider climate variability from year to year.

Europe's Energy Resources Europe's main energy sources have shifted over the years from coal to petroleum and natural gas, and in some countries to nuclear power. Increasingly, alternative energy sources are being pursued in response to rising energy costs and efforts to cut greenhouse gas emissions.

The 27 members of the European Union (often referred to as the EU-27) get a large portion of their fuel supplies from Russia—32 percent of their crude oil and 42 percent of their natural gas, as of 2010. Most of the gas now comes via pipelines through Belarus and Ukraine, but Turkey is negotiating to supply Russian gas to Europe via the Black Sea, and it already hosts a

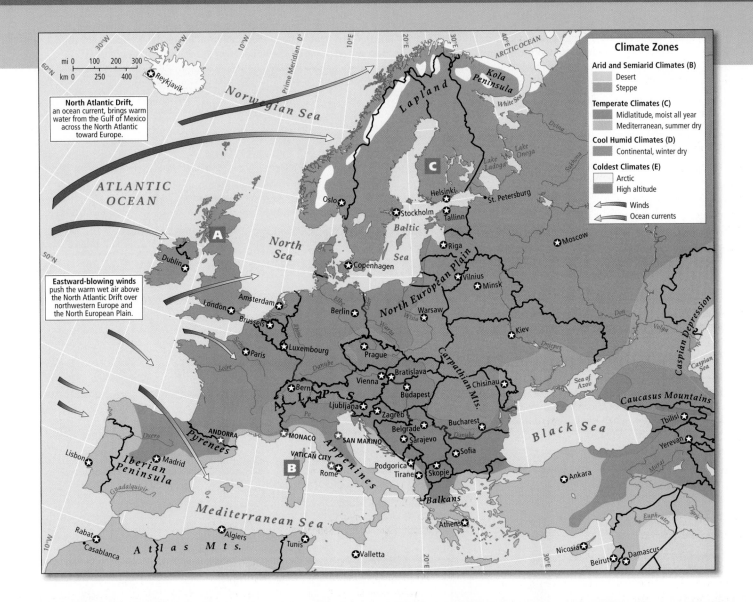

Climate Zones

Arid and Semiarid Climates (B)
- Desert
- Steppe

Temperate Climates (C)
- Midlatitude, moist all year
- Mediterranean, summer dry

Cool Humid Climates (D)
- Continental, winter dry

Coldest Climates (E)
- Arctic
- High altitude

→ Winds
→ Ocean currents

North Atlantic Drift, an ocean current, brings warm water from the Gulf of Mexico across the North Atlantic toward Europe.

Eastward-blowing winds push the warm wet air above the North Atlantic Drift over northwestern Europe and the North European Plain.

A Midlatitude, moist all year, Scotland

B Mediterranean, summer dry, Corsica

C Continental, winter dry, Finland

143

Most of Europe has been transformed by human activity. Western Europe has some of the most heavily impacted landscapes and ecosystems, but the formerly communist countries of Central Europe also have severe environmental problems. Meanwhile, agricultural intensification is creating new problems in South Europe. The map shows some of the impacts on Europe's land, sea, and air.

A The Garzweiler open-pit coal mine in Germany, one of the largest in the world, covers 25.3 square miles (66 square kilometers). Its operations have forced the abandonment of 12 villages and towns, and threaten local groundwater resources. The coal this mine produces is a highly polluting variety called lignite. Its high sulfur content is responsible for much of Europe's acid rain problem.

B A coal-fired power plant at an industrial facility in Poznan, Poland, in 2009. Central Europe has thousands of similar small but highly polluting facilities, many of them built during or before the era of Soviet domination, when environmental safeguards were rare.

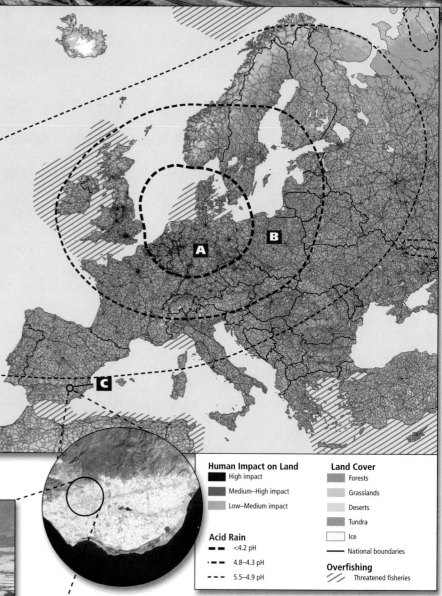

Human Impact on Land
- ⬛ High impact
- 🟥 Medium–High impact
- 🟧 Low–Medium impact

Acid Rain
- – – – <4.2 pH
- –·–· 4.8–4.3 pH
- – – – 5.5–4.9 pH

Land Cover
- Forests
- Grasslands
- Deserts
- Tundra
- Ice
- —— National boundaries

Overfishing
- ⧫ Threatened fisheries

C A sea of plastic greenhouses covers most of Campo de Dalía, a coastal plain in Spain's Almería province that has seen intense growth in year-round export-oriented vegetable production over the past several decades. Pesticides and fertilizers are heavily used, and health effects are beginning to be seen among the mainly Moroccan migrant workers and their families who work and live among the greenhouses. Meanwhile, water shortages are so severe that a desalination plant is being built nearby.

Thinking Geographically

After you have read about the Human Impact on the Biosphere in Europe, you will be able to answer the following questions:

A How might this large, open pit threaten groundwater resources?

B Is there any clue in the picture that large numbers of people may be exposed to this pollution?

C How might the large areas covered by plastic contribute to a water shortage?

pipeline that carries gas from the Caspian Sea to Europe. Russia is negotiating for yet another trans-Turkey pipeline to carry oil and gas to the Mediterranean. Europeans fear that Russia will use this dependency against them, withholding and releasing flows at will as it did in 2008 by controlling the gas flow through Ukraine and Belarus. Another 30 percent of the gas and oil the European Union consumes comes from the Middle East. Large oil and gas deposits in the North Sea, most controlled by Norway (not an EU member), have alleviated Europe's dependence on "foreign" sources of energy, but the production of oil from the North Sea has already peaked and is expected to run out by 2018.

❚❚▶ 82. WINTER ON THE WAY

The use of nuclear power to generate electricity has been more common in Europe than in North America. The EU-27 depend on nuclear power for 30 percent of their total needs. In France, 78 percent of the electricity is generated by nuclear power (compared with only 20 percent in the United States). However, support for nuclear power has declined, partly in response to the disastrous nuclear accidents in 1986 in Chernobyl, Ukraine, and in 2011 in Fukushima, Japan. The safe disposal of nuclear waste products is also a concern.

The European Union wishes to increase its use of renewable energy in order to reduce fuel imports and thereby increase energy security, stimulate the economy with new energy-related jobs, and combat climate change. The goal is to reduce CO_2 emissions by 20 percent, increase use of renewable energy by 20 percent, and power 60 percent of EU homes with renewably generated electricity, all by 2020. A wide array of alternative energy projects is now attracting massive investment. While solar power is growing, windpower is generally the favored technology (see **Thematic Overview A** on page 137). Denmark, Germany, and Spain are leading the way; Germany, for instance, has the most windfarms and the most windpower capacity, as well as the most installed solar panels.

Air Pollution At present, there is significant air pollution in much of Europe, but it is particularly heavy over the North European Plain. This is a region of heavy industry, dense transportation routes, and large and affluent populations. The intense fossil fuel use associated with such lifestyles results not only in the usual air pollution but also in *acid rain*, which can fall far from where it was generated.

High levels of air pollution are also found in the former communist states of Central and North Europe. Mines in Central Europe produce highly polluting soft coal (Photo Essay 4.2B) that is burned in out-of-date factories. This region produces high per capita emissions from burning oil and gas, and it also receives air pollution blown eastward from western Europe. In Upper Silesia, Poland's leading coal-producing area, acid rain has destroyed forests, contaminated soils and the crops grown on them, and raised water pollution to deadly levels. Residents have higher rates of birth defects, high rates of cancer, and lowered life expectancies. Industrial pollution was one of Poland's greatest obstacles to entry into the European Union, which it barely gained.

Central Europe's severe environmental problems developed in part because the Marxist theories and policies promoted by the Soviet Union portrayed nature as existing only to serve human needs. During the Soviet era, there was little opportunity for public protest against pollution, but the recent shift to democracy has enabled greater action in places like Hungary, where popular protest has resulted in reductions in air pollution.

The new market economies in the former Soviet bloc countries are improving energy efficiency and reducing emissions. Power plants, factories, and agriculture are polluting less, and the countries with the worst emissions records, such as Poland, have been making the most progress.

> **Learning Goal 1**
> **Water:** What factors complicate efforts to reduce water pollution in the Mediterranean Sea?

Freshwater and Seawater Pollution Sources of water pollution in Europe include insufficiently treated sewage, chemicals and silt in the runoff from agricultural plots and residential units, consumer packaging litter, petroleum residues, and industrial effluent. Most inland waters contain a variety of such pollutants. Any pollutants that enter Europe's wetlands, rivers, streams, and canals quickly reach Europe's surrounding coastal waters. Thus far, the Atlantic Ocean, the Arctic Ocean, and the North Sea are able to disperse most pollutants dumped into them because they are part of, or closely connected to, the circulating flow of the world ocean. In contrast, the Baltic, Mediterranean, and Black seas are nearly landlocked bodies of water that do not have the capacity to flush themselves out quickly; thus, all three are prone to accumulating pollution (see the red areas in **Figure 4.4** on page 146).

In the Mediterranean, the effect of the pollution that pours in from rivers, adjacent cities, hotel resorts, and farms is exacerbated by the fact that there is just one tiny opening to the world ocean (see Thematic Overview B; see also Figure 4.4). At the surface, seawater flows in from the Atlantic through the narrow Strait of Gibraltar and moves eastward. At the bottom of the sea, water exits through the same narrow opening, but only after it has been in the Mediterranean for 80 years! The natural ecology of the Mediterranean is attuned to this lengthy cycle, but the nearly 460 million people now living in the countries surrounding the sea have upset the balance. Their pollution stays in the Mediterranean for decades. As a result, fish catches have declined, beloved seaside resorts have become unsafe for swimmers, and agricultural workers have become sick.

Europe's Vulnerability to Climate Change Europe's wealth, technological sophistication, and well-developed emergency response systems make it more resilient to the consequences of climate change than most regions of the world. Nevertheless, some areas are much more vulnerable than others, due to their location, dwindling water resources, rising sea levels, or the

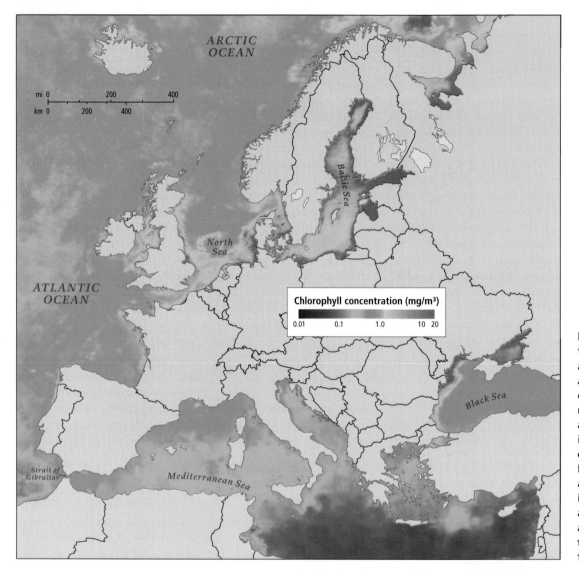

Chlorophyll concentration (mg/m³)
0.01   0.1   1.0   10 20

Thinking Geographically

After you have read about Vulnerability to Climate Change in Europe, you will be able to answer the following questions:

A What other clues can you see in this photo to indicate that the reservoir is at a very low level?

B Judging from the way these structures are arranged, how would you guess they work?

C How might a car decrease a family's sensitivity to flooding?

FIGURE 4.4 Pollution of the seas. Europe's exceptionally long and convoluted coast affords easy access to the world's oceans. However, pollution of the nearly landlocked Baltic, Black, and Mediterranean seas is causing increasing concern. Chlorophyll concentration, one measure of pollution levels, is significant in the Atlantic and Arctic oceans as well. Chlorophyll is used to measure the amount of algae in a body of water. Excessive algae growth is an indicator of pollution from fertilizers and sewage that fuel the growth of those organisms.

effects of poverty. Photo Essay 4.3 illustrates some of the consequences that have been predicted.

Learning Goal 2
Climate Change: In what ways is Europe responding to climate change?

European Leadership in Response to Global Climate Change Europe leads the world in response to global climate change, with EU governments having agreed to cut greenhouse gas emissions by 20 percent by 2020. Europe has been more willing than any other region to act on climate change, largely because it recognizes the economic sense in doing so. Recent research suggests that investments in energy conservation, alternative energy, and other measures would cost EU governments 1 percent of their GDP. By contrast, doing nothing about global warming would *shrink* GDP by 20 percent.

II ▶ 83. GERMAN INVESTMENTS IN CLEAN ENERGY PAY OFF

Europe's increasing concern about global warming may also be influenced by public alarm at recent abnormal weather. The summer of 2003 broke all high-temperature records for Europe.

Crops failed, freshwater levels sank, forests burned, and deaths soared. In France alone, 3000 people died. On the other hand, in 2002 and 2006, rainfall and snowfall in Central Europe reached record levels. In the spring of 2006, the rivers of Central Europe—the Elbe, the Danube, and the Morava—flooded for weeks. Then in the winter of 2006–2007, snowfall was unusually light.

Progress in Green Behavior By global standards, Europeans use large amounts of resources and contribute about one-quarter of the world's greenhouse gas emissions. However, one European resident averages only one-half the energy consumption of the average North American resident. Europeans live in smaller dwellings, which need less energy to heat, and air conditioning is rare. They drive smaller, more fuel-efficient cars and use public transportation widely. Because communities are denser, many people walk or bicycle to their appointments.

Green environmentally conscious

These energy-saving practices are related in part to the high population densities and social customs of the region, but also to widespread explicit support for ecological principles. **Green** political parties influence national policies in all European countries,

Europe's wealth, technological sophistication, and well-developed emergency response systems make it more resilient to climate change than most regions. Nevertheless, some areas are much more vulnerable than others due to poverty or dwindling water resources.

A A once-submerged bridge and an old church reemerge from a sinking reservoir in Catalonia, Spain. As temperatures rise, Spain's climate will become drier and scarce water resources will shrink further, threatening agriculture and drinking water, especially along the Mediterranean coast.

B This is part of the massive "delta works" that protect the Netherlands from rising sea levels during storms. A response to massive flooding in 1953, the delta works significantly reduce the Netherlands' sensitivity to sea level rise. Hence, the Netherlands is only slightly more vulnerable to climate change than the rest of Europe, even though 60 percent of its population lives below sea level.

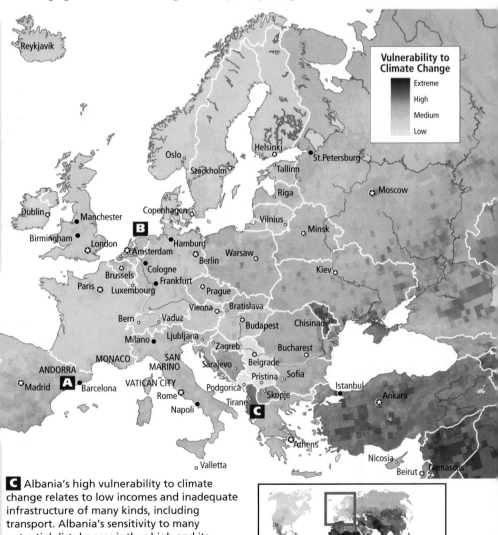

Vulnerability to Climate Change

Extreme
High
Medium
Low

C Albania's high vulnerability to climate change relates to low incomes and inadequate infrastructure of many kinds, including transport. Albania's sensitivity to many potential distubances is thus high and its resilience is low. Many problems stem from misguided government policies. For example, until 1992 it was illegal for Albanians to own private automobiles. Even today many poorer Albanians depend on horse carts as their main mode of transportation.

147

and Green policies are central to the agenda of the European Union. Results include strong regional advocacy for emission controls, community recycling programs now well entrenched across the European Union, and grassroots work on local environmental concerns.

- **II ▶ 84.** ICELAND: ENERGY TO SPARE
- **II ▶ 85.** BRITAIN REQUIRES ENERGY INSPECTIONS FOR HOME SALES
- **II ▶ 86.** MCD'S GARBAGE HEATS AND LIGHTS BRITISH CITY
- **II ▶ 87.** LONDON LEADS BY EXAMPLE TO CURB POLLUTION, CLIMATE CHANGE

Changes in Transportation Europeans have long favored fast rail networks for both passengers and cargo rather than private cars, trucks, and multilane highways. Now, rising fuel costs and CO2 emissions are encouraging the design of *multimodal transport* that links high-speed rail to road, air, and water transportation (see Photo Essay 4.5D on page 162). An EU transportation report notes that 1 kilogram of gasoline can move 50 tons of cargo a distance of 1 kilometer by truck, but the same amount of gasoline can move 90 tons of cargo 1 kilometer by rail and 127 tons by waterway. Another EU report shows that private cars used for passenger transportation produce 3 times more CO_2 emissions than rail-based public transportation.

The low cost of water transportation and Europe's reliance on global trade have been a boon for the development of ocean ports that cater to container ships: Helsinki, Riga, Hamburg, Copenhagen, Antwerp, Rotterdam, Plymouth, Southampton, La Havre, Marseille, Barcelona, and Koper are only a few of Europe's modern container ports. Soon, newly built ports in Morocco, Algeria, Tunisia, Malta, and Egypt, all catering at least in part to European markets, will join these ports—one-third of the world's container traffic now goes through the Mediterranean. The container ship industry, attuned to the CO_2 and climate change debates, now carefully calculates the optimal (often slower) speed for container ships in order to minimize fuel consumption and emissions while maximizing profits.

THINGS TO REMEMBER

1. **Learning Goal 1: Water** The ecology of the Mediterranean Sea is threatened by water pollution from the 460 million people now living in the countries that surround the sea.

2. **Learning Goal 2: Climate Change** Europe has emerged as the world's leader in combating global warming. The European Union has a goal of cutting greenhouse gas emissions by 20 percent by 2020.

3. European residents average only one-half the energy consumption of the average North American, and fuel costs and CO_2 emissions are increasingly being considered in the design of multimodal transport that links high-speed rail to road, air, and water transportation.

Human Patterns over Time

Over the last 500 years, Europe has profoundly influenced how the world trades, fights, thinks, and governs itself. Attempts to explain this influence are wide-ranging. One argument is that Europeans are somehow a superior breed of humans. Another is that Europe's many bays, peninsulas, and navigable rivers have promoted commerce to a greater extent there than elsewhere. In fact, much of Europe's success is based on technologies and ideas it borrowed from elsewhere. For example, the concept of the peace treaty, so vital to current European and global stability, was first documented not in Europe but in ancient Egypt. In an effort to explain how Europe gained the leading global role it continues to play, it is worth taking a look at the broad history of this area.

Sources of European Culture

Starting about 10,000 years ago, the practice of agriculture and animal husbandry gradually spread into Europe from the Tigris and Euphrates river valleys in Southwest Asia and from farther east in Central Asia and beyond. Mining, metalworking, and mathematics also came to Europe from these places and from North Africa. All of these innovations increased possibilities for trade and economic development in Europe.

The first European civilizations were ancient Greece (800 to 86 B.C.E.) and Rome (753 B.C.E. to 476 C.E.). Located in southern Europe, both Greece and Rome initially interacted more with the Mediterranean rim, Southwest Asia, and North Africa than with the rest of Europe, which then had only a small and relatively poor rural population. Later European traditions of science, art, and literature were heavily based on Greek ideas, which were themselves derived from yet earlier Egyptian and Southwest Asian sources.

The Romans, after first borrowing heavily from Greek culture, also left important legacies in Europe. Many Europeans today speak *Romance* languages, such as Spanish, Portuguese, Italian, French, and Romanian, all of which are largely derived from Latin, the language of the Roman Empire. Rome was the origin of European laws that determine how individuals own, buy, and sell land. These laws have been spread throughout the world by Europeans.

Roman practices used in colonizing new lands also shaped much of Europe. After a military conquest, the Romans secured control in rural areas by establishing large plantation-like farms. Politics and trade were centered on new Roman towns built on a grid pattern that facilitated both commercial activity and military repression of rebellions (see Timeline A on pages 152–153). These same systems for taking and holding territory were later used when Europeans colonized the Americas, Asia, and Africa.

The influence of Islamic civilization on Europe is often overlooked. After the fall of Rome, while Europe experienced a period known as the *Early Middle Ages* (roughly 450–1000 C.E.), pre-Muslim (Arab and Persian) and then Muslim scholars preserved learning from Rome and Greece. Muslim Arabs originally from North Africa ruled Spain from 711 to 1492 (see Timeline B). From the 1400s through the early 1900s, the Ottoman Empire (based in what is now Turkey) dominated much of southern Central Europe and Greece. The Arabs, Persians, and Turks all brought new technologies, food crops, architectural principles, and textiles to Europe from Arabia, Persia, Anatolia, China, India, and Africa. Arabs also brought Europe its numbering system, mathematics, and significant advances in medicine and engineering, building on ideas they picked up in South Asia.

Beginning 500 years ago, Europe began to draw on a host of cultural features from the various colonies that were established in the Americas, Asia, and Africa. For example, the food crops now popular in Europe came mostly from the Americas (potatoes, corn, peppers, tomatoes, beans, and squash) and Southwest and Central Asia (wheat, leafy greens, onions, and apples).

humanism a philosophy and value system that emphasizes the dignity and worth of the individual

mercantilism a strategy for increasing a country's power and wealth by acquiring colonies and managing all aspects of their production, transport, and trade for the colonizer's benefit

The Inequalities of Feudalism

As the Roman Empire declined, a social system known as *feudalism* evolved during the *medieval period* (450–1500 C.E.). This system originated from the need to defend rural areas against local bandits and raiders from Scandinavia and the Eurasian interior. The objective of feudalism was to have a sufficient number of heavily armed, professional fighting men, or knights, to defend a much larger group of *serfs*, who were legally bound to live on and cultivate plots of land for the knights. Over time, some of these knights became a wealthy class of warrior-aristocrats, called the *nobility*, who controlled certain territories. Some nobles gained power over other knights, amassing vast kingdoms.

The often lavish lifestyles and elaborate castles of the wealthier nobility were supported by the labors of the serfs. Most serfs lived in poverty outside castle walls and, much like slaves, were legally barred from leaving the lands they cultivated for their protectors.

The Role of Urbanization in the Transformation of Europe

While rural life followed established feudal patterns, new political and economic institutions were developing in Europe's towns and cities. Here, thick walls provided defense against raiders, and commerce and crafts supplied livelihoods. Thus, the people could be more independent from feudal knights and kings.

Located along trade routes, Europe's urban areas were exposed to new ideas, technologies, and institutions from Southwest Asia, India, and China. Some institutions, such as banks, insurance companies, and corporations, provided the foundations for Europe's modern economy. Over time, Europe's urban areas established a pace of social and technological change that left the feudal rural areas far behind.

Urban Europe flourished in part because of laws that granted basic rights to urban residents. With adequate knowledge of these laws, set forth in legal documents called *town charters*, people with few resources could protect their rights even if challenged by those who were more wealthy and powerful. Town charters provided a basis for European notions of *civil rights*, which have proved hugely influential throughout the world. With strong protections for their civil rights, some of Europe's townsfolk grew into a small middle class whose prosperity moderated the feudal system's extreme divisions of status and wealth (see Timeline C). A related outgrowth of urban Europe was a philosophy known as **humanism**, which emphasized the dignity and worth of the individual regardless of wealth or social status.

The liberating influences of European urban life transformed the practice of religion. Since late Roman times, the Catholic Church had dominated not just religion but also politics and daily life throughout much of Europe. In the 1500s, however, a movement known as the *Protestant Reformation* arose in the urban centers of the North European Plain. Reformers such as Martin Luther challenged Catholic practices—such as holding church services in Latin, which only a tiny educated minority understood—that stifled public participation in religious discussions. Protestants also promoted individual responsibility and more open public debate of social issues, which altered the perception of the relationship between the individual and society. These ideas spread faster with the invention of the European version of the printing press, which enabled widespread literacy.

European Colonialism: An Acceleration of Globalization

Learning Goal 3
Globalization and Development:
How did Europe's colonization of much of the world over the past 400 years accelerate globalization and transform economic development in Europe?

A direct outgrowth of the greater openness of urban Europe was the exploration and subsequent colonization of much of the world by Europeans. The increased commerce and cultural exchange began a period of accelerated *globalization* that persists today (see the discussion in Chapter 1 on pages 27–33).

In the fifteenth and sixteenth centuries, Portugal took advantage of advances in navigation, shipbuilding, and commerce to set up a trading empire in Asia and a colony in Brazil. Spain soon followed, founding a vast and profitable empire in the Americas and the Philippines. By the seventeenth century, however, England, the Netherlands, and France had seized the initiative from Spain and Portugal (see Thematic Overview D and Timeline D). All European powers implemented **mercantilism**, a strategy for increasing a country's power and wealth by acquiring colonies and managing all aspects of their production, transport, and trade for the colonizer's benefit. Mercantilism supported the Industrial Revolution in Europe (page 130) by supplying cheap resources from around the globe for Europe's new factories. The colonies also supplied markets for European manufactured goods (see Figure 4.5 on page 150).

II ▶ 88. ART EXHIBIT ENCOMPASSES THE WORLD OF PORTUGUESE EXPLORATIONS

By the nineteenth century, the English, French, and Dutch (the people who live in the Netherlands) overshadowed the Spanish and Portuguese colonial empires and extended their influence into Asia and Africa. By the twentieth century, European colonial systems had strongly influenced nearly every part of the world.

The overseas empires of England, the Netherlands, and eventually France were the beginnings of the modern global economy. The riches they provided shifted wealth, investment, and general economic development away from southern Europe and the Mediterranean and toward western Europe.

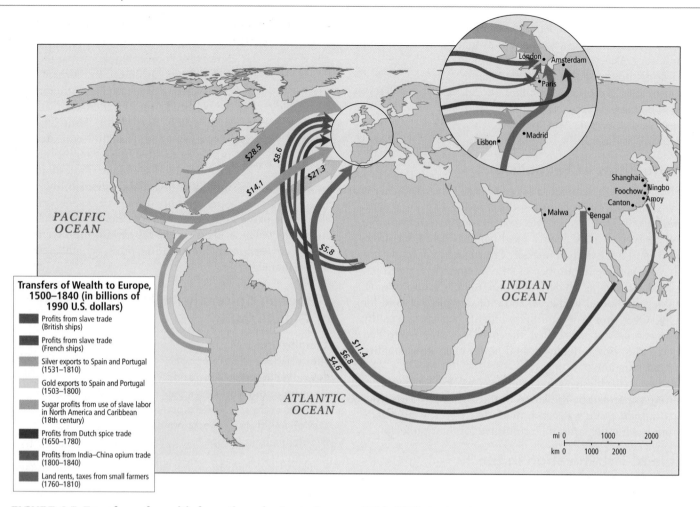

Transfers of Wealth to Europe, 1500–1840 (in billions of 1990 U.S. dollars)

- Profits from slave trade (British ships)
- Profits from slave trade (French ships)
- Silver exports to Spain and Portugal (1531–1810)
- Gold exports to Spain and Portugal (1503–1800)
- Sugar profits from use of slave labor in North America and Caribbean (18th century)
- Profits from Dutch spice trade (1650–1780)
- Profits from India–China opium trade (1800–1840)
- Land rents, taxes from small farmers (1760–1810)

FIGURE 4.5 Transfers of wealth from the colonies to Europe, 1500–1840. During the period of mercantilism, Europe received billions of dollars of income from its overseas colonies.

By the mid-1700s, wealthy merchants in London, Amsterdam, Paris, Berlin, and other western European cities were investing in new industries. Workers from rural areas poured into urban centers in England, the Netherlands, Belgium, France, and Germany to work in new manufacturing industries and mining. Wealth and raw materials flowed in from colonial ports in the Americas, Asia, and Africa. Some cities, such as Paris and London, were elaborately rebuilt in the 1800s to reflect their roles as centers of global empires.

By 1800, London and Paris, each of which had a million inhabitants, were Europe's largest cities, a status that eventually brought them to their present standing as *world cities* (cities of worldwide economic or cultural influence). London is a global center of finance, and Paris is a cultural center that has influence over global consumption patterns, from food to fashion to tourism.

Urban Revolutions in Industry and Democracy

The wealth derived from Europe's colonialism helped fund two of the most dramatic transformations in a region already characterized by rebirth and innovation: the industrial and democratic revolutions. Both first took place in urban Europe.

The Industrial Revolution Europe's Industrial Revolution—particularly Britain's ascendancy as the leading industrial power of the nineteenth century—was intimately connected with colonial

expansion. In the seventeenth century, Britain developed a small but growing trading empire in the Caribbean, North America, and South Asia, which provided it with access to a wide range of raw materials and to markets for British goods.

Sugar, produced by British colonies in the Caribbean, was an especially important trade crop (Figure 4.5). Sugar production was a complex process requiring major investments in equipment. Slaves were forcibly brought in from Africa to help produce the sugar. The skilled management and large-scale organization needed later provided a production model for the Industrial Revolution. The mass production of sugar also generated enormous wealth that helped fund industrialization.

By the late eighteenth century, Britain was introducing mechanization into its industries, first in textile weaving and then in the production of coal and steel. By the nineteenth century, Britain was the world's greatest economic power, with a huge and growing empire, expanding industrial capabilities, and the world's most powerful navy. Its industrial technologies spread throughout continental Europe (see Timeline F), North America, and elsewhere, and transformed millions of lives in the process.

Urbanization and Democratization Extremely low living standards in Europe's cities created tremendous pressures for change in the political order that ultimately led to democratization.

Learning Goal 4
Urbanization and Democratization: How has urbanization influenced the development of democracy in Europe?

Industrialization led to massive growth in urban areas in the eighteenth and nineteenth centuries. Most industrial jobs were dangerous and unhealthy, demanding long hours and offering little pay. Poor people were packed into tiny, airless spaces that they shared with many others. Children were often weakened and ill due to poor nutrition, industrial pollution, inadequate sanitation and health care, and child labor. Water was often contaminated, and most sewage ended up in the streets. Opportunities for advancement through education were restricted to the tiny few who could afford it.

The ideas and information gained by the few key people who did learn to read gave them the incentive to organize and protest for change. After lengthy struggles, democracy was expanded to Europe's huge and growing working class, and power, wealth, and opportunity were distributed more evenly throughout society. However, the road to democracy in Europe was rocky and violent, just as it is now in many parts of the world (see Timeline E).

In 1789, the French Revolution led to the first major inclusion of common people in the political process in Europe. Angered by the extreme disparities of wealth in French society, and inspired by news of the popular revolution in North America, the poor rebelled against the monarchy and the elite-dominated power structure that still controlled Europe. As a result, the general populace, especially in urban areas, became involved in governing through democratically elected representatives. The democratic expansion created by the French Revolution ultimately proved short-lived as elite-dominated governments soon regained control in France. Nevertheless, the French Revolution provided crucial inspiration to later urban democratic political movements in many parts of the world.

nationalism devotion to the interests or culture of a particular country, nation, or cultural group; the idea that a group of people living in a specific territory and sharing cultural traits should be united in a single country to which they are loyal and obedient

welfare state a government that accepts responsibility for the well-being of its people, guaranteeing basic necessities such as education, employment, and health care for all citizens

The Impact of Communism During the struggles that resulted ultimately in the expansion of democracy, popular discontent erupted periodically in the form of new revolutionary political movements that threatened the established civic order. The political philosopher and social revolutionary Karl Marx framed the mounting social unrest in Europe's cities as a struggle between socioeconomic classes. His treatise *The Communist Manifesto* (1848) helped social reformers across Europe articulate ideas on how wealth could be more equitably distributed. East of Europe in Russia, Marx's ideas inspired the creation of a revolutionary communist state in 1917, the Union of Soviet Socialist Republics, often referred to as the USSR or the Soviet Union. Eventually the Soviet Union extended its ideology and state power throughout Central Europe.

In West, North, and South Europe, communism gained some popularity, but this tended to lead to the formation of communist political parties that operated peacefully within the context of democratizing political systems. In countries like France, Italy, and Spain, communist political parties are still influential and often form governing coalitions with other parties.

Popular Democracy and Nationalism Throughout Europe, innumerable struggles between urban working class people and the authorities continued during the nineteenth and early twentieth centuries, yielding two main results: workers throughout Europe gained the right to form unions that could bargain with employers and the government for higher wages and better working conditions; and workers' struggles created pressure for democratization, with most national governments eventually deriving their authority from competitive elections in which all adult citizens could vote.

During the nineteenth and twentieth centuries, the development of democracy was also linked to the idea of **nationalism**, or allegiance to the state. The notion spread that all the people who lived in a certain area formed a nation, and that loyalty to that nation should supersede loyalties to family, clan, or individual monarchs. Eventually, the whole map of Europe was reconfigured, and the mosaic of kingdoms gave way to a collection of nation-states. All of these new nations were transformed into democracies by the political movements arising in Europe's industrial cities. Nationalism was a major component of World Wars I and II; so in the post-war era, and as the European Union was constructed, nationalism was deemphasized.

Democracy and the Welfare State Channeled through the democratic process, public pressure for improved living standards moved most European governments toward becoming **welfare states** by the mid-twentieth century. Such governments accept responsibility for the well-being of their people, guaranteeing basic necessities such as education, employment, and health care for all citizens. In time, government regulations on wages, hours, safety, and vacations established more harmonious relations between workers and employers. The gap between rich and poor declined somewhat, and overall civic peace increased. And although welfare states were funded primarily through taxes, industrial productivity and overall economic activity did not decline.

Modern Europe's welfare states have yielded generally adequate to high levels of well-being for all citizens. Just how much support the welfare state should provide is still a subject of debate, however, and one that is being resolved in different ways across Europe.

Two World Wars and Their Aftermath

Despite Europe's many advances in industry and politics, by the beginning of the twentieth century, the region still lacked a system of collective security that could prevent war among its rival nations. Between 1914 and 1945, two horribly destructive world wars left Europe in ruins, no longer the dominant region of the world. At least 20 million people died in World War I (1914–1918) and 70 million in World War II (1939–1945). During World War II, Germany's Nazi government killed 15 million civilians during its failed attempt to conquer the Soviet Union. Eleven million civilians died at the hands of the

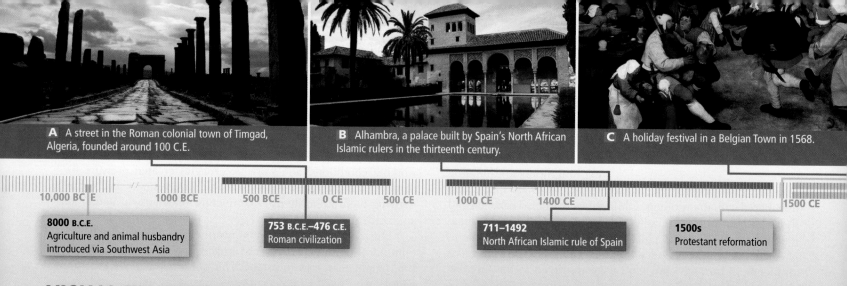

A A street in the Roman colonial town of Timgad, Algeria, founded around 100 C.E.

B Alhambra, a palace built by Spain's North African Islamic rulers in the thirteenth century.

C A holiday festival in a Belgian Town in 1568.

8000 B.C.E.
Agriculture and animal husbandry introduced via Southwest Asia

753 B.C.E.–476 C.E.
Roman civilization

711–1492
North African Islamic rule of Spain

1500s
Protestant reformation

10,000 BCE | 1000 BCE | 500 BCE | 0 CE | 500 CE | 1000 CE | 1400 CE | 1500 CE

VISUAL TIMELINE OF EUROPE

Thinking Geographically

After you have read about the human history of Europe, you will be able to answer the following questions:

A What about this photo suggests that Timgad might be laid out on a grid system?

B Can you see any clues that the builders of the Alhambra were technologically sophisticated?

Nazis during the **Holocaust**, a massive execution of 6 million Jews and 5 million gentiles (non-Jews), including ethnic Poles and other Slavs, **Roma** (Gypsies), disabled and mentally ill people, gays, lesbians, transgendered people, and political dissidents.

II ▶ 89. THE WORLD REMEMBERS VICTIMS OF HOLOCAUST

The defeat of Germany, the country seen as the instigator of both world wars, resulted in a number of enduring changes in Europe. After World War II ended in 1945, Germany was divided into two parts. West Germany became an independent democracy allied with the rest of western Europe—especially Britain and France—and the United States. The Soviet Union controlled East Germany and the rest of Central Europe (Latvia, Lithuania, Estonia, Poland, Czechoslovakia, Hungary, Romania, Bulgaria, Ukraine, Moldova, and Belarus). The line between East and West Germany was part of what was called the **iron curtain**, a long, fortified border zone that separated western Europe from Central Europe.

The Cold War The division of Europe created a period of conflict, tension, and competition between the United States and the Soviet Union known as the **Cold War**, which lasted from 1945 to 1991. During this time, once-dominant Europe, and indeed the entire world, became a stage on which the United States and the Soviet Union competed for dominance. The central issue was the competition between **capitalism**—characterized by privately owned businesses and industrial firms that adjusted

Holocaust during World War II, a massive execution by the Nazis of 6 million Jews and 5 million gentiles (non-Jews), including ethnic Poles and other Slavs, Roma (Gypsies), disabled and mentally ill people, gays, lesbians, transgendered people, and political dissidents.

Roma the now-preferred term in Europe for Gypsies

iron curtain a long, fortified border zone that separated western Europe from (then) eastern Europe during the Cold War

Cold War a period of conflict, tension, and competition between the United States and the Soviet Union that lasted from 1945 to 1991

capitalism an economic system characterized by privately owned businesses and industrial firms that adjust prices and output to match the demands of the market

communism an ideology and economic system, based largely on the writings of the German revolutionary Karl Marx, in which, on behalf of the people, the state owns all farms, industry, land, and buildings (a version of socialism)

central planning a communist economic model in which a central bureaucracy dictates prices and output with the stated aim of allocating goods equitably across society according to need

prices and output to match the demands of the market—and **communism** (actually a version of socialism), in which the state owned all farms, industry, land, and buildings. After 1945, in most of what we now call Central Europe, the Soviet Union forcibly implemented a communist economic model known as **central planning**: a central bureaucracy dictated prices and output with the stated aim of allocating goods and services equitably across society according to need. This system was successful in alleviating long-standing poverty and illiteracy among the working classes, but it was rife with waste, corruption, and bureaucratic bungling. It ultimately collapsed in the 1990s due to inefficiency, insolvency, high levels of environmental pollution, and public demands for more democratic participation.

In the rest of Europe, especially western Europe, a successful demonstration of capitalism was mounted, underwritten by financial support from the United States under the *Marshall Plan*. Basic facilities, such as roads, housing, and schools were rebuilt, and governments were aided in reestablishing free market economic systems. Economic reconstruction proceeded rapidly in the following decades.

II ▶ 91. MARSHAL PLAN'S 61ST ANNIVERSARY

Decolonization, Democratization, and Conflict in Modern Europe Europe's decline during the two world wars also led to the loss of its colonial empires. Many European colonies had participated in the wars, with some (India, the Dutch East Indies,

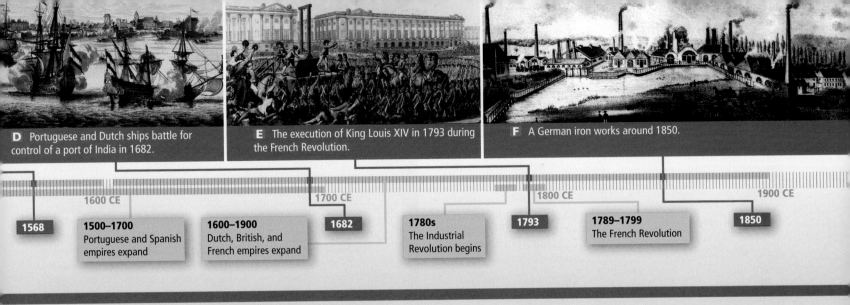

D Portuguese and Dutch ships battle for control of a port of India in 1682.

E The execution of King Louis XIV in 1793 during the French Revolution.

F A German iron works around 1850.

1600 CE		1700 CE		1800 CE	1900 CE

1568

1500–1700 Portuguese and Spanish empires expand

1600–1900 Dutch, British, and French empires expand

1682

1780s The Industrial Revolution begins

1793

1789–1799 The French Revolution

1850

C Is there anything to suggest relative prosperity in this painting?

D What does this image suggest about the value of trade with India to the Portuguese and the Dutch?

E What inspired the French Revolution and the execution of Louis XIV?

F Can anything about the spread of the Industrial Revolution through continental Europe be deduced from this image?

Burma, and Algeria) suffering extensive casualties, and almost all emerging economically devastated. After the war, Europe was less able to assert control over its colonies and demands for independence grew. By the 1960s, most former European colonies had gained independence, often after bloody wars fought against European powers and their local allies.

Democracy's long history in Europe (if not in Europe's colonies) entered a new era after World War II, when most of West, South, and North Europe reorganized more strongly around democratic principles and humanitarian ideals. This process was supported by the United States, though many undemocratic European regimes, such as that of Spanish dictator Francisco Franco, who ruled from 1936–1975, also received U.S. support. Other undemocratic European countries during the post-war period included Portugal, which remained a dictatorship and colonial power in Africa until 1974 (see Photo Essay 4.4D on page 154). Elsewhere in Europe, democratization was compromised by violence, as in the case of Northern Ireland, which until the late 1990s had sectarian violence between Catholics and Protestants so severe that free and fair democratic elections were not possible (see Photo Essay 4.4C). In Central Europe, democratization did not arrive until the Soviet Union dissolved in 1991, when its former satellites, one by one, declared independence (see Thematic Overview E). Yugoslavia, a large communist republic in southern Central Europe outside

the Soviet bloc, also dissolved beginning in the early 1990s. There democratization was hampered by a rising tide of ethnic xenophobia and violence (see Photo Essay 4.4 A, B).

THINGS TO REMEMBER

1. During the medieval period, political and economic transformations in Europe's towns and cities challenged the feudal system that dominated rural areas.

2. **Learning Goal 3: Globalization and Development** As European powers like Spain, the United Kingdom, and the Netherlands conquered vast overseas territories, they created trade relationships that laid the foundation for the modern global economy.

3. Most of the countries in the world have been ruled by a European colonial power at some point in their history.

4. **Learning Goal 4: Urbanization and Democratization** Extremely low living standards in Europe's cities created tremendous pressure for change in the political order; this ultimately led to democratization.

5. The road to democracy in Europe was rocky and at times violent.

II CURRENT GEOGRAPHIC ISSUES

Europe today is in a state of transition as a result of two major changes that occurred during the 1990s: the demise of the Soviet Union (discussed above and in Chapter 5) and the rise of the

European Union. These developments have already brought greater peace, prosperity, and world leadership to Europe. Nonetheless, problems and tensions remain.

Economic and Political Issues

At the end of World War II, European leaders felt that closer economic ties among the European nation-states would prevent the kind of hostilities that had led to two world wars. Over the next 50 years, a series of steps were taken that increasingly united Europe economically, socially, and to a certain extent politically. More recently, political unification has been tentatively attempted.

The European Union: A Rising Superpower

The original plan after the trauma of World War II was simply to work toward a level of economic and social integration that would make possible the free flow of goods and people across national borders. While this goal remains central, some Europeans believe that the European Union, which is already a global economic power, should become a global counterforce to the United States in political and military affairs.

Steps in Creating the European Union The first major step in achieving economic unity took place in 1958, when Belgium, Luxembourg, the Netherlands, France, Italy, and West Germany formed the *European Economic Community (EEC)*. The members of the EEC agreed to eliminate certain tariffs against one another and to promote mutual trade and cooperation. Denmark, Ireland, and the United Kingdom joined in 1973; Greece, Spain, and Portugal in the 1980s; and Austria, Finland, and Sweden in 1995, bringing the total number of member countries to 15. In 1992, the concept of the EEC was expanded to that of the European Union, which is concerned with more than just economic policy.

The European Union expanded into Central Europe after the demise and fall of the Soviet Union. As early as the 1980s, Soviet control over Central Europe began to falter in the face of a workers' rebellion in Poland, known as *Solidarity*. By 1990, East Germany was reuniting with West Germany. The dissolution of the Soviet Union in 1991 brought the collapse of many economic and political relationships in Central Europe. Many workers lost their jobs. In some of the poorest countries, such as Romania, Bulgaria, and Serbia, social turmoil and organized crime threatened stability. Membership in the EU became especially attractive to political leaders and citizens who thought it would spur economic development throughout Central Europe as a result of investment by wealthier EU member countries in West and North Europe (see Thematic Overview C).

II ▶ 90. POLES CELEBRATE THE 25TH ANNIVERSARY OF SOLIDARITY

Standards for EU membership, however, are rather specific. A country must achieve political stability and have a democratically elected government. Each country has to adjust its

---

Thinking Geographically

After you have read about Democratization and Conflict in Europe, you will be able to answer the following questions:

A What clues in the photo suggest that this place has suffered from war and not some other calamity?

C Why would the boy depicted in the mural likely be wearing a gas mask?

D What allowed Portugal to gain colonies like Angola?

Patterns of democratization and conflict in Europe and in former European colonies generally support the notion that countries with lower levels of democratization also suffer the most from violent conflict (see Photo Essay 1.3 on pages 34–35 for more). Since World War II, the most violent conflicts in Europe have occurred in Central Europe, where repressive governments frequently denied their populations basic democratic freedoms. Outside Europe, many long and brutal wars were fought to gain independence from European powers. In part these wars stemmed from the inequitable extraction of resources from the colonies by Europeans, but another factor was the denial of basic democratic freedoms to the local non-European population. The map shows how low levels of democratization persist in the countries that fought wars of independence even today, decades after the formal end of colonial rule. Other factors worked against democratization in these countries, but European imperialism helped set many on an anti-democratic trajectory, which has only recently started to change.

D Portuguese soldiers loading a bomb into a plane in Angola in the 1960s. Portugal, at the time an undemocratic authoritarian dictatorship, fought three major wars in Africa during the 1960s and 1970s. The rising human and financial costs of these wars eventually brought a revolution in Portugal itself in 1974, resulting in independence for the colonies and the eventual establishment of democracy in Portugal. None of Portugal's colonies has achieved a high level of democratization, due in part to authoritarian governing structures put in place by the Portuguese as well as the lengthy civil wars that followed decolonization.

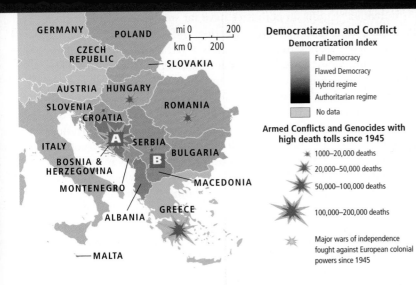

Democratization and Conflict
Democratization Index

- Full Democracy
- Flawed Democracy
- Hybrid regime
- Authoritarian regime
- No data

Armed Conflicts and Genocides with high death tolls since 1945

- 1000–20,000 deaths
- 20,000–50,000 deaths
- 50,000–100,000 deaths
- 100,000–200,000 deaths
- Major wars of independence fought against European colonial powers since 1945

A The aftermath of Bosnia's civil war in a suburb of Sarajevo in 1996. Attempts by Serbia to limit democratic decision making within the former Yugoslavia resulted in several wars during the 1990s. In Bosnia, a civil war that pitted different ethnicities against each other resulted in 175,000 deaths.

B Members of the Kosovo Liberation Army turn in their weapons to NATO forces at the end of the war in 1999. Like other parts of the former Yugoslavia, Kosovo fought for independence after Serbia's anti-democratic policies denied basic political freedoms.

C Two murals in Derry, Northern Ireland, marking an entrance to a "nationalist" neighborhood. The mural on the right depicts a boy wearing a gas mask and holding a homemade "petrol" bomb. Both murals were created during "the Troubles," a time of violent conflict lasting from 1969 to 1998. One aspect of the still-sensitive conflict is competing claims about Northern Ireland's democratic legitimacy. The "nationalists" (who are usually Catholic) claim that Northern Ireland was undemocratically separated from the rest of Ireland in 1920 by the government of the United Kingdom during Ireland's war of independence from the UK. The "loyalists" (usually Protestant) see Northern Ireland as an expression of the will of the voters in the state, most of whom are Protestant and want to remain part of the UK.

constitution to EU standards that guarantee the rule of law, human rights, and respect for minorities. Each must also have a functioning market economy that is open to investment by foreign-owned companies and that has well-controlled banks. Finally, farms and industries must comply with strict regulations governing the finest details of their products and the health of environments.

A few countries have chosen not to join the European Union: Switzerland, Norway, and Iceland. These three countries have long treasured their neutral role in world politics. Moreover, as wealthy countries, they have been concerned about losing control over their domestic economic affairs. However, during the 2008–2009 recession, Iceland, much reduced in wealth and highly indebted to the International Monetary Fund, began discussions to enter the European Union.

Several countries on the perimeter of Europe may eventually join over the next few decades. Of these, Turkey is the most likely candidate, but Turkey's strained relations with the island country of Cyprus (which was admitted to the European Union in 2004), its history of human rights violations against minorities (especially against its large Kurdish population), and some issues regarding separation of religion and the state are strikes against it. Ukraine, and perhaps even the Caucasian republics (Armenia, Azerbaijan, and Georgia), may at some time be invited to join. However, there is strong opposition to this within Europe, and Europe's huge and potentially powerful neighbor, Russia, opposes expansion of the European Union into what it regards as its sphere of influence.

▌▌▶ 93. EU TELLS TURKEY TO DEEPEN REFORMS

EU Governing Institutions Somewhat similar to the United States, the European Union has one executive branch and two legislative bodies. The *European Commission* acts like an executive branch of government, proposing new laws and implementing decisions. Each of the 27 member states gets one commissioner, who is appointed for a 5-year term, subject to the approval of the European Parliament; but after 2014 the size of the Commission will be reduced to 18, with the right to appoint rotating commissioners. Commissioners are expected to uphold common interests and not those of their own countries. The entire Commission must resign if censured by Parliament. The European Commission also includes about 25,000 civil servants who work in Brussels to administer the European Union on a day-to-day basis.

The *European Parliament* is directly elected by EU citizens, with each country electing a proportion of seats based on its population, much like the U.S. House of Representatives. The Parliament elects the president of the European Commission, who serves for 2½ years as a head of state and head of foreign policy. Laws must be passed in Parliament by 55 percent of the member states, which must contain 65 percent of the EU total population. In other words, a simple majority does not rule. The *Council of the European Union* is similar to the U.S. Senate in that it is the more powerful of the two legislative bodies. However, its members are not elected but consist of one minister of government from each EU country.

Economic Integration and a Common Currency Economic integration has solved a number of problems in Europe.

Individual European countries have far smaller populations than their competitors in North America and Asia. This means that they have smaller internal markets for their products, so their companies earn lower profits than those in large countries. The European Union solved this problem by joining European national economies into a common market. Companies in any EU country now have access to a much larger market and can potentially make larger profits through **economies of scale** (reductions in the unit costs of production that occur when goods or services are produced in large amounts, resulting in increased profits per unit). Before the European Union, when businesses sold their products to neighboring countries, their earnings were diminished by tariffs and other regulations, as well as through currency exchanges.

The EU economy now encompasses over 500 million people (out of a total of 540 million in the whole of Europe)—roughly 200 million more than live in the United States. Collectively, the EU countries are wealthy. In 2010, their joint economy was almost $15 trillion (PPP), about 5 percent larger than that of the United States, making the European Union the largest economy in the world. Members' trade with each other amounts to about twice the monetary value of trade with the outside world. And the combined EU total external trade (imports and exports with non-EU countries) was 19 percent of the world's total, equal to that of the United States (Figure 4.6). However, whereas the United States usually imports far more than it exports, resulting in a trade deficit, the European Union often has a trade balance—in which the values of imports and exports are roughly equal. The average 2010 GDP per capita (PPP) for the European Union ($32,600) was significantly less than that of the United States ($46,400). The disparity of wealth, though, was also lower by about one-third than in the United States, resulting in lower levels of poverty in the EU. Careful regulation of EU economies results in generally slower economic growth than in the United States. But these regulations have also made the EU more resilient to the global financial crises, such as that of 2008, largely due to stronger regulations of the banking industry and financial markets than in the United States.

The official currency of the European Union is the **euro (€)**. Sixteen EU countries now use the euro: Austria, Belgium, Cyprus, Finland, France, Germany, Greece, Ireland, Italy, Luxembourg, Malta, the Netherlands, Portugal, Slovakia, Slovenia, and Spain. Countries that use the euro have a greater voice in the creation of EU economic policies, and the use of a common currency greatly facilitates trade, travel, and migration within the European Union. All non-euro member states except Sweden and the United Kingdom have currencies whose value is determined by that of the euro. Depending on global financial conditions, either the euro or the U.S. dollar is the preferred currency of international trade and finance.

The Euro and Debt Crises

In recent years, there have been major challenges to the euro, primarily in the form of a debt crisis that has challenged the traditional mechanisms that governments use to maintain economic stability.

> **economies of scale** reductions in the unit cost of production that occur when goods or services are efficiently mass produced, resulting in increased profits per unit
>
> **euro** the official (but not required) currency of the European Union as of January 1999

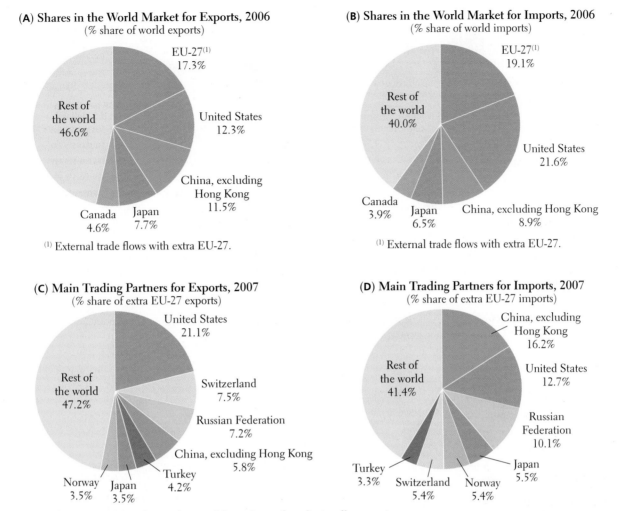

(A) Shares in the World Market for Exports, 2006
(% share of world exports)

EU-27[1]
17.3%

United States
12.3%

China, excluding
Hong Kong
11.5%

Japan
7.7%

Canada
4.6%

Rest of
the world
46.6%

[1] External trade flows with extra EU-27.

(B) Shares in the World Market for Imports, 2006
(% share of world imports)

EU-27[1]
19.1%

United States
21.6%

China, excluding Hong Kong
8.9%

Japan
6.5%

Canada
3.9%

Rest of
the world
40.0%

[1] External trade flows with extra EU-27.

(C) Main Trading Partners for Exports, 2007
(% share of extra EU-27 exports)

United States
21.1%

Switzerland
7.5%

Russian Federation
7.2%

China, excluding Hong Kong
5.8%

Turkey
4.2%

Japan
3.5%

Norway
3.5%

Rest of
the world
47.2%

(D) Main Trading Partners for Imports, 2007
(% share of extra EU-27 imports)

China, excluding
Hong Kong
16.2%

United States
12.7%

Russian
Federation
10.1%

Japan
5.5%

Norway
5.4%

Switzerland
5.4%

Turkey
3.3%

Rest of
the world
41.4%

FIGURE 4.6 The EU's shares in world trade and main trading partners.

The debt crisis first emerged in Greece, where government spending has long outpaced tax revenues and where a trade deficit grew to unsustainable levels due in part to the global economic recession that began in 2008. Normally, governments respond to high levels of debt by borrowing from other countries and then reducing the value of their currency, relative to that of the lenders, which makes these debts easier to pay. However, this wasn't an option for Greece because its main lender, Germany, uses the same currency as Greece, the euro. As Greece became less able to repay its loans, the economies of other countries in similarly indebted circumstances (such as Ireland, Italy, Portugal, and Spain) also began to falter as international investors started to pull out.

In response to these crises, the EU has developed mechanisms to limit government debt in all countries that use the euro. The potential for crisis remains, though, as many euro countries object to giving the EU more control over their spending.

The European Union, Globalization, and Development The European Union is pursuing a number of economic development strategies designed to ensure that it continues to be competitive with the United States, Japan, and the developing economies of Asia, Africa, and South America. A primary focus is on lowering the cost of producing goods in Europe. One

strategy is to shift labor-intensive industries from the wealthiest EU countries in western Europe, where wages are high, to the relatively poorer, lower-wage member states of Central Europe (see **Figure 4.7** on page 158). Generally this strategy has worked well, helping poorer European countries prosper while keeping the costs of doing business low enough to restrain European companies from moving outside Europe to developing countries where costs are lower still. However, while the economies of Central Europe have grown, in Europe's wealthiest countries, the resultant reduction of industrial capacity (*deindustrialization*) has led to higher unemployment. Despite these efforts, some EU firms have moved abroad to cut costs.

Like the United States, the European Union currently has a large share of world trade (see Figure 4.6) and therefore exerts a powerful influence on the global trading system. The European Union often negotiates privileged access to world markets for European firms and farmers. The European Union also employs protectionist measures that favor European producers by making goods from outside the European Union more expensive. These higher-priced goods create added expense for European consumers but help ensure jobs and control over supplies. Of course, non-European producers shut out of EU markets are unhappy, and increasingly they have united to protest the European

FIGURE 4.7 **The EU economy: GDP per capita (2007).** GDP per capita varies considerably among the 27 countries in the European Union.

Union's failure to open its economies to foreign competition. So far such protests have met with little success.

NATO and the Rise of the European Union as a Global Peacemaker A new role for the European Union as a global peacemaker and peacekeeper is developing through the **North Atlantic Treaty Organization (NATO)**, which is based in Europe. During the Cold War, European and North American countries cooperated militarily through NATO to counter the influence of the Soviet Union. NATO originally included the United States, Canada, the countries of western Europe, and Turkey; it now includes almost all the EU countries as well.

Since the breakup of the Soviet Union, other nations came to assist the United States in the difficult task of addressing global security issues only after a major failure to avert a bloody ethnic conflict during the breakup of Yugoslavia in 1991. NATO

> **North Atlantic Treaty Organization (NATO)** a military alliance between European and North American countries that was developed during the Cold War to counter the influence of the Soviet Union; since the breakup of the Soviet Union, NATO has expanded membership to include much of Eastern Europe and Turkey, and is now focused mainly on providing the international security and cooperation needed to expand the European Union

has since focused mainly on providing the international security and cooperation needed to expand the European Union. When the United States invaded Iraq in 1993, most EU members opposed the war. As worldwide opposition to the United States built, the global status of the European Union was elevated. With the United States preoccupied with Iraq, NATO assumed more of a role as a global peacemaker. It now provides a majority of the troops in Afghanistan, but enthusiasm for this war is waning in Europe.

In addition to the major role they now play in Afghanistan, NATO and the European Union are also helping patrol the world ocean. In the spring of 2009 during the Somali pirate crisis off the northeast coast of Africa, NATO reported that both French and Portuguese naval vessels foiled attempts by pirates to seize merchant ships. Also in May of 2009, France announced that in accordance with its role in NATO, it had established a base in Abu Dhabi.

▌▌▶ 94: NATO'S FUTURE ROLE DEBATED

▌▌▶ 95. NATO TO PROJECT DIFFERENT PHILOSOPHY

▌▌▶ 96. CONCERN OVER COMMON VALUES AT THE US–EU SUMMIT

▌▌▶ 97. NATO LEADERS, PUTIN MEET IN BUCHAREST

Food Production and the European Union

Most food is now produced on large mechanized farms that are efficient: they require less labor and are more productive per acre. One result is that only about 2.3 percent of Europeans are now engaged in full-time farming. A second result of the mechanization of agriculture is that farmland has declined as a land use in Europe since the mid-1990s, while forestlands have increased.

The Common Agricultural Program (CAP)

The drastic decline of labor and land in farming worries Europeans, who see it as endangering the goal of self-sufficiency in food. Toward this end, the European Union established its wide-ranging **Common Agricultural Program (CAP)**, meant to guarantee secure and safe food supplies at affordable prices (see Thematic Overview F). The CAP helps farmers by placing tariffs on imported agricultural goods and by giving **subsidies** (payments to farmers) to underwrite their costs of production. Subsidies are expensive—payments to farmers are the largest category in the EU budget, accounting for 48 percent of expenditures. While these policies do ensure a safe and sufficient food supply and provide a decent living standard for farmers, they also effectively raise food costs for millions of consumers. Moreover, because subsidies are based on the amount of land under cultivation, these payments favor large, often corporate-owned farms. It was this aspect of the CAP that the Smithfield company (with its European partners) took advantage of in setting up its giant pig farms in Romania (recall the opening vignette).

Protective agricultural policies like tariffs and subsidies—also found in the United States, Canada, and Japan—are unpopular in the developing world. Tariffs lock farmers in poorer countries out of major markets. Subsidies encourage overproduction in rich countries (in order to collect more payments). The result is a glut of farm products that are then sold cheaply on the world market, as is the case with Smithfield pork. This practice, called *dumping*, lowers global prices and thus hurts farmers in developing countries while it aids those in developed countries by keeping their supplies low and prices high.

> **Common Agricultural Program (CAP)** a European Union program, meant to guarantee secure and safe food supplies at affordable prices, that places tariffs on imported agricultural goods and gives subsidies to EU farmers
>
> **subsidies** monetary assistance granted by a government to an individual or group in support of an activity, such as farming, that is viewed as being in the public interest

> **Learning Goal 5**
> Food: How is food production in Europe changing?

Growth of Corporate Agriculture and Food Marketing As small family farms disappear in the European Union—just as they did several decades ago in the United States—smaller farms are being consolidated into larger, more profitable operations run by European and foreign corporations (see Photo Essay 4.2C). These farms tend to employ very few laborers and use more machinery and chemical inputs.

The move toward corporate agriculture is strongest in Central Europe. When communist governments gained power in the mid-twentieth century, they consolidated many small, privately owned farms into large collectives. After the breakup of the Soviet Union, these farms were rented to large corporations, which in turn further mechanized the farms and laid off all but a few laborers. Rural poverty rose and small towns shrank as farmworkers and young people left for the cities. With EU expansion, the CAP has provided further incentives for large-scale mechanized agriculture.

Green Food Production, a Case Study: Slovenia During the communist era, Slovenia was unlike most of the rest of Central Europe in that the farms were not collectivized. As a result, the average farm size is just 8.75 acres (3.5 hectares). Although Slovenia has plenty of rich farmland, as standards of living have risen, it has become a net importer of food. Nonetheless, Slovenia's new emphasis on private entrepreneurship, combined with a growing demand throughout Europe for organic foods, has encouraged some Slovene farmers to carve out a niche for themselves in local markets. The case of Vera Kuzmic is illustrative.

Vignette Vera Kuzmic (a pseudonym) lives 2 hours by car south of Ljubljana, Slovenia's capital. For generations, her family has farmed 12.5 acres (5 hectares) of fruit trees near the Croatian border. After Slovenia became independent in 1991, Vera and her husband lost their government jobs, due to economic restructuring. The Kuzmic family decided to try earning its living in vegetable market gardening because vegetable farming could be more responsive to market changes than fruit tree cultivation. By 2000, the adult children and Mr. Kuzmic were working on the land, and Vera was in charge of marketing their produce and that of neighbors whom she had also convinced to grow vegetables.

Vera secured market space in a suburban shopping center in Ljubljana, where she and one employee maintained a small vegetable and fruit stall (Figure 4.8). Her produce had to compete with much less expensive Italian-grown produce sold in the same shopping

FIGURE 4.8 Vera Kuzmic in her market stall in Ljubljana.

Thinking Geographically: How can you tell that Vera is involved in small-scale agriculture?

center—all of it produced on large corporate farms in northern Italy and trucked in daily. But Vera gained market share by bringing her customers special orders and by guaranteeing that only animal manure, no pesticides or herbicides, was used on the fields. For a while, her special customer services and her organically grown produce kept her in business. But when Slovenia joined the European Union in 2004, she had to do more to compete with produce growers and marketers across Europe.

Anticipating the challenges to come, the Kuzmics' daughter Lili completed a marketing degree at the University of Ljubljana. The family incorporated their business, and Lili is now its Ljubljana-based director, while Vera manages the farm. Lili's market research shows that it would be wisest to diversify. The Kuzmics continue to focus on Ljubljana's expanding professional population, who are willing to pay extra for fine organic vegetables and fruits. But now, in a recently built banquet facility on the farm, Vera also prepares special dinners for bus-excursion groups interested in traditional Slovene dishes made from homegrown organic crops. [*Source: Lydia Pulsipher's conversations with Vera Kuzmic and Dusan Kramberger, 1993 through 2009.*] ∎

Europe's Growing Service Economies

As industrial jobs have declined across the region, most Europeans (about 70 percent) have found jobs in the service economy. *Services* such as the provision of health care, education, finance, tourism, and information technology are now the engine of Europe's integrated economy, drawing hundreds of thousands of new employees to the main European cities. For example, financial services located in London and serving the entire world play a huge role in the British economy, and many multinational companies are headquartered in London.

A major component of Europe's service economy is *tourism*. Europe is the most popular tourist destination in the world, and one job in eight in the European Union is related to tourism. Tourism generates 13.5 percent of the EU's gross domestic product and 15 percent of its taxes, although this varies with global economic conditions. Europeans are themselves enthusiastic travelers, frequently visiting one another's countries as well as many distant locations throughout the world. This travel is made possible by the long paid vacations—usually 4 to 6 weeks—that Europeans are granted by employers. The most popular holiday destinations among EU members in 2006 included Austria, Ireland, and the traditional Mediterranean destinations of Cyprus, Malta, Spain, and Italy (2005).

Service occupations increasingly involve the use of technology. While Europe has lagged behind North America in the development and use of personal computers and the Internet, it leads the world in cell phone use. The information economy is especially advanced in West Europe, though in South Europe and Central Europe, where personal computer ownership is lowest, public computer facilities in cafés and libraries are common.

THINGS TO REMEMBER

1. The European Union's original plan was to reach a level of economic and social integration that would make possible the free flow of goods and people across national borders; for the most part, those goals have been reached among the current 27 members.

2. The European Union has one executive branch—the European Commission—and two legislative branches—the European Parliament, directly elected by EU citizens, and the Council of the European Union, whose members consist of one minister from each EU country.

3. The European Union joined the members' national economies into a common market. By 2008, the EU's economy was almost $15 trillion (PPP), about 5 percent larger than that of the United States, making the European Union the largest economy in the world.

4. A new role for the European Union as a global peacemaker and peacekeeper is developing through the North Atlantic Treaty Organization (NATO), which is based in Europe.

5. **Learning Goal 5: Food** Throughout the European Union, smaller family-run farms are giving way to larger corporate farms.

6. As industrial jobs have declined across the region, most Europeans (about 70 percent) have found jobs in the service economy.

Sociocultural Issues

The European Union was conceived primarily to promote economic cooperation and free trade, but its programs have social implications as well. As population patterns change cross Europe, attitudes toward immigration, gender roles, and social welfare programs are also evolving. Religion and language, once divisive issues in the region, are now fading as a focus of disputes. Immigration, however, continues to be a source of tension.

Population Patterns

Population patterns in Europe foretell processes that are emerging around the world. Europe's high population density and urbanization trends are long-standing. A newer phenomenon, the aging of the populace, is well advanced in Europe and affects all manner of social policies.

Population Distribution and Urbanization There are currently about 540 million Europeans. Of these, 500 million live within the European Union. The highest population densities stretch in a discontinuous band from the United Kingdom south and east through the Netherlands and central Germany into northern Switzerland (Figure 4.9). Northern Italy is another zone of high density, along with pockets in many countries along the Mediterranean coast. Overall, Europe is one of the more densely occupied regions on earth, as shown on the population density map in Figure 1.8 on page 14. Most of this population now lives in cities.

Today, Europe is a region of cities surrounded by well-developed rural hinterlands. These cities are the focus of the modern European economy, which, though long grounded in agriculture, trade, and manufacturing, is now primarily service based. In West, North, and South Europe, more than 75 percent of the population lives in urban areas. Even in Central Europe, the least urbanized part of the region, around 70 percent of people live in cities. As noted earlier, many European cities began as trading centers more than a thousand years ago and still bear the architectural marks of medieval life in their historic centers (Photo Essay 4.5A). These old cities are located either on navigable rivers

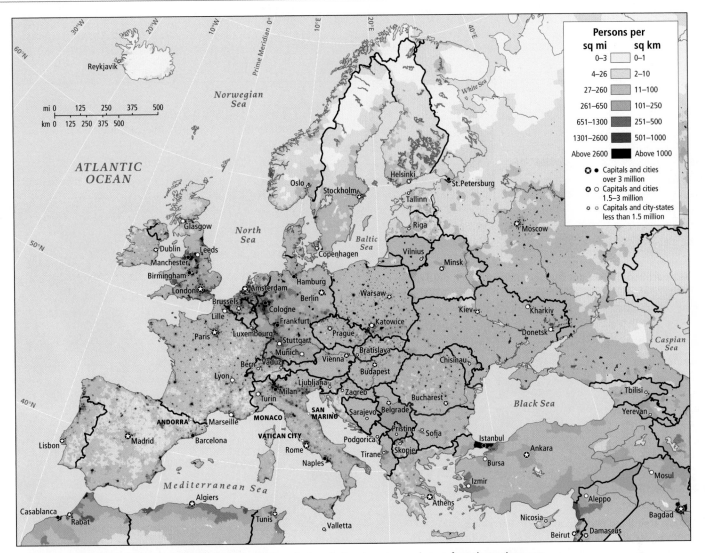

FIGURE 4.9 Population density in Europe. Europe's population is not growing, a fact that raises concerns about future economic conditions as the population ages. Many governments now encourage large families, with generous maternity and paternity leave (up to 10 months with full pay), free day care, and other incentives. However, few countries have seen much change in their population growth rate. A major reason is that more women now are working or want to work than ever before, so they are delaying childbearing and families remain small. As this trend shows no signs of decreasing, some officials are looking to immigration as a possible solution. While Europe has been attracting large numbers of immigrants for decades now, many Europeans are wary of hosting large populations of foreigners who might not share their values.

in the interior or along the coasts because water transportation figured prominently (as it still does) in Europe's trading patterns.

Since World War II, nearly all the cities in Europe have expanded around their perimeters in concentric circles of apartment blocks (see Photo Essay 4.5B on page 162). Usually, well-developed rail and bus lines link the blocks to one another and to the old central city. Land is scarce and expensive in Europe, so only a small percentage of Europeans live in single-family homes, although the number is growing. Even single-family homes tend to be attached or densely arranged on small lots. Except in public parks, which are common (see Thematic Overview G), one rarely sees the sweeping lawns that many North Americans are accustomed to. Publicly funded transportation is widely available, so many people live without cars in apartments near city centers (see Photo Essay 4.5 C, D). However, many others commute

by car daily from suburbs or ancestral villages to work in nearby cities. Although deteriorating housing and slums do exist, substantial public spending (on sanitation, water, utilities, education, housing, and public transportation) help most people maintain a generally high standard of urban living.

Europe's Aging Population Europe's population is aging as families are choosing to have fewer children and life expectancies are increasing. Between 1960 and 2010, the proportion of those 14 years and under declined from 27 percent to 16 percent, while those over 65 increased from 9 percent to 16 percent. By comparison, in the global population, the overall share of young persons was 27 percent in 2009, while older generations accounted for 8 percent. Life expectancies now range close to 80 years in North, West, and South Europe, and this is reflected in urban landscapes, where the elderly are more common than children (see Thematic Overview H).

Europe's cities are world famous for their architecture, economic dynamism, and cultural vitality. Many are quite ancient, but have expanded in recent decades with large apartment blocks connected to the old city centers by public transportation. Despite their global draw, many European cities will shrink in the next several decades, due to declining national populations (as in Italy and parts of Central Europe), as well as to deliberate efforts to shift urban growth to other urban centers, as in the case of London.

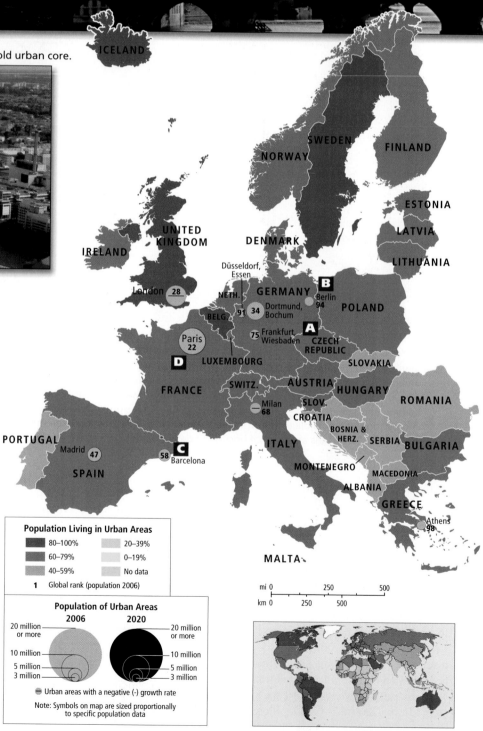

A The old urban core of Prague, Czech Republic, founded in the ninth century C.E. on the banks of the Vltava River.

B High-rise apartment blocks dot the periphery of Berlin's old urban core.

C La Rambla, a tree-lined pedestrian mall in Barcelona, Spain, is a popular strolling place among locals and tourists.

D A tram in Paris, France. The city's metropolitan transit system moves roughly 6 million people per day, and is the second-largest system in Europe, after London's.

Population Living in Urban Areas

80–100%	20–39%
60–79%	0–19%
40–59%	No data

1 Global rank (population 2006)

Population of Urban Areas

2006   2020

20 million or more
10 million
5 million
3 million

⊖ Urban areas with a negative (-) growth rate

Note: Symbols on map are sized proportionally to specific population data

mi 0   250   500
km 0   250   500

Thinking Geographically

After you have read about Urbanization in Europe, you will be able to answer the following questions:

A Why is Prague located along a river?

B How does the amount of greenhouse gas produced by people who live in European-style cities (where people often live in apartment blocks connected to the urban core via public transportation) compare to that of people who live in U.S.-style cities (where people live in spread-out suburbs and depend on private automobiles)?

C How might walking as a mode of transportation affect the environment in Barcelona?

D How does the amount of greenhouse gas produced by rail-based public transportation compare to that of private automobile-based transportation?

Learning Goal 6
Population and Gender: How is Europe's aging population linked to changing gender roles?

Overall, Europe is now close to a negative rate of natural increase (<0.0), the lowest in the world. Birth rates are lowest in Central Europe—the countries that were allied with the former Soviet Union or part of Yugoslavia. Increasingly, the one-child family is common throughout Europe, except among immigrants, who are the major source of population growth. However, once they have assimilated to European life, immigrants, too, choose to have small families.

The declining birth rate is illustrated in the population pyramids of European countries (see Figure 4.10 on page 164), which look more like lumpy towers than pyramids. The population pyramid of Sweden and of the whole European Union are examples (see Figure 4.10 B, C). The pyramids' narrowing base indicates that for the last 35 years, far fewer babies have been born than in the 1950s and 1960s, when there was a postwar baby boom across Europe. By 2000, twenty-five percent of Europeans were choosing to have no children at all.

The reasons for these trends are complex. For one thing, more and more women want professional careers. This alone could account for late marriages and lower birth rates; 25 percent of German women are choosing to remain unmarried well into their thirties. Historically, many governments have made few provisions for working mothers beyond paid maternity leave. In Germany, for example, there is insufficient day care available for children under 3, and school days run from 7:30 A.M. to noon or 1:00 P.M. Many German women choose not to become mothers because they would have to settle for part-time jobs in order to be home by 1:00 P.M. To encourage higher birth rates, there is a move within the European Union to give one parent (mother or father) a full year off with reduced pay after a child is born or adopted, provide better preschool care, lengthen school days, and serve lunch at school (see the discussion of gender on pages 164–165).

A stable population with a low birth rate has several consequences. Because fewer consumers are being born, economies may contract over time. Demand for new workers, especially highly skilled ones, may go unmet. Further, the number of younger people available to provide expensive and time-consuming health care for the elderly, either personally or through tax payments, will decline. Currently, for example, there are just two German workers for every retiree. Immigration provides one solution to the dwindling number of young people. However, Europeans are reluctant to absorb large numbers of immigrants, especially from distant parts of the world where cultural values are very different from those in Europe.

Immigration and Migration: Needs and Fears

Until the mid-1950s, the net flow of migrants was out of Europe, to the Americas, Australia, and elsewhere. By the 1990s, the net flow was into Europe. In the 1990s, most of the European Union (plus Iceland, Norway, and Switzerland) implemented the **Schengen Accord**, an agreement that allows free movement of people and goods across common borders. The accord has facilitated trade, employment, tourism, and most controversially, migration within the European Union. The Schengen Accord has also indirectly increased both the demand for immigrants from outside the European Union and their mobility once they are in the European Union.

▶ 92. AFTER 50 YEARS, EUROPE STILL COMING TOGETHER

Attitudes Toward Internal and International Migrants and Citizenship Like citizens of the United States, Europeans have ambivalent attitudes toward migrants. The internal flow of migration is mostly from Central Europe into North, West, and South Europe. These Central European migrants are mostly treated fairly, although prejudices against the supposed backwardness of Central Europe are still evident. Immigrants from outside Europe, so-called *international immigrants*, meet with varying acceptance.

International immigrants often come legally and illegally from Europe's former colonies and protectorates across the globe. Many Turks and North Africans come legally as **guest workers** who are expected to stay for only a few years, fulfilling Europe's need for temporary workers in certain sectors. Other immigrants are refugees from the world's trouble spots, such as Afghanistan, Iraq, Haiti, and Sudan. Many also come illegally from all of these areas.

While some Europeans see international immigrants as important contributors to their economies—providing needed skills and making up for the low birth rates—many oppose recent increases in immigration. Studies of public attitudes in Europe show that immigration is least tolerated in areas where incomes and education are low. Central and South Europe are the least tolerant of new immigrants, possibly because of fears that immigrants may drive down wages that are already relatively low. North and West Europe, with higher incomes and generally more stable economies, are the most tolerant. In response, the European Union is increasing its efforts to curb illegal immigration from outside Europe while at the same time encouraging EU citizens to be more tolerant of legal migrants.

Schengen Accord an agreement signed in the 1990s by the European Union and many of its neighbors that called for free movement across common borders

guest workers legal workers from outside a country who help fulfill the need for temporary workers but who are expected to return home when they are no longer needed

Rules for Assimilation: Muslims in Europe In Europe, culture plays as much of a role in defining differences between people as race and skin color. An immigrant from Asia or Africa may be accepted into the community if he or she has gone through a comprehensive

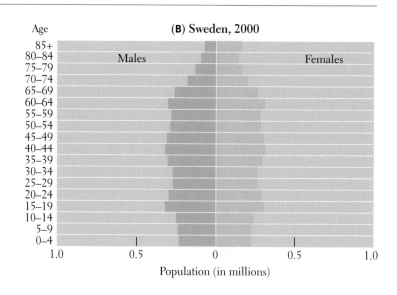

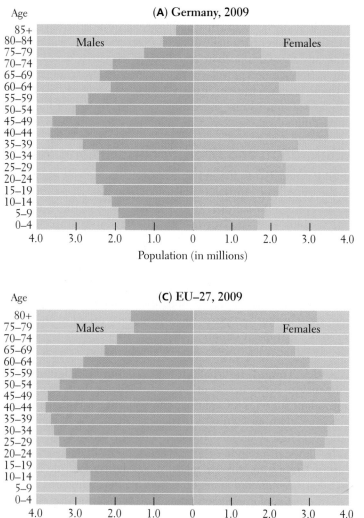

FIGURE 4.10 Population pyramids for Germany, Sweden, and the EU-27 in 2009–2010. These population pyramids have quite different shapes, but all exhibit a narrow base, indicating low birth rates. The EU-27 pyramid **(C)** shows the percent of the total population, not the numbers of people.

change of lifestyle. **Assimilation** in Europe usually means giving up the home culture and adopting the ways of the new country. If minority groups—such as the Roma (see page 152)—who have been in Europe for thousands of years—maintain their traditional ways, it is nearly impossible for them to blend into mainstream society.

> **assimilation** the loss of old ways of life and the adoption of the lifestyle of another country

II ▶ 101. THE ART OF INTEGRATION IN GERMANY

Europe's small but growing Muslim immigrant population (see Figure 4.11 on page 165) is presently the focus of assimilation issues in the European Union. Muslims come from a wide range of places and cultural traditions, including North Africa, Turkey, and South Asia. Some of these immigrants maintain traditional dress, gender roles, and religious values, while others have assimilated to European culture. The deepening alienation that has boiled over in recent years among some Muslim immigrants and their children—resulting in protests, riots, and occasional terrorism—relates primarily to the systematic exclusion of these less-assimilated Muslims from meaningful employment, from social services, and from higher education. The riots that broke out in Paris in the fall of 2005 were not about Islam, but rather about lack of access to higher education, jobs, and housing for less-assimilated Muslims. Subsequent investigations by the French media revealed that the protestors' complaints were indeed legitimate. However, many

Europeans, unaware of the context, viewed the Paris riots as linked to deadly terrorist bombings in Madrid in 2004 and London in 2005. Both of those violent events were carried out by young Muslims, some of them born in Europe, protesting British and Spanish involvement in the war in Iraq.

In some cases, conflicts have also arisen over European perceptions of cultural aspects of Islam, such as the *hijab* (one of several traditional coverings for women). For example, in France in 2004, wearing of the hijab by observant Muslim schoolgirls became the center of a national debate about civil liberties, religious freedom, and national identity. French authorities wanted to ban the hijab but were wary of charges of discrimination. Eventually the French declared all symbols of religious affiliation illegal in French schools (including crosses and yarmulkes, the Jewish head covering for men).

II ▶ 102. RELIGIOUS TOLERANCE FACING TEST IN BRITAIN

II ▶ 103. LONDON'S MEGA MOSQUE STIRS CONTROVERSY

Changing Gender Roles

Gender roles in Europe have changed significantly from the days when most women married young and worked in the home or on the family farm. Increasing numbers of European women are working outside the home, and the percentage of women

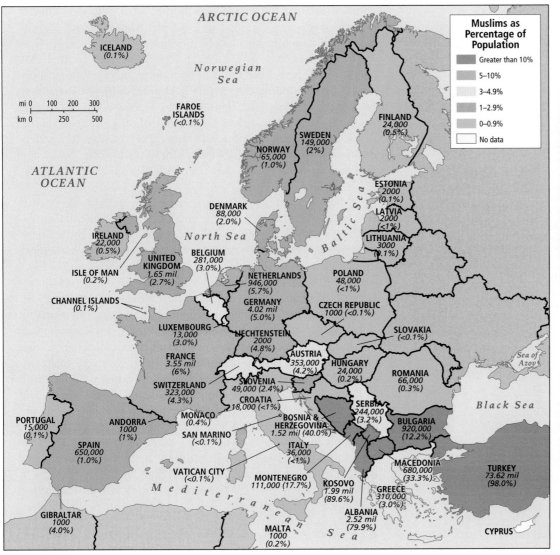

(A)

(B)

(C)

FIGURE 4.11 Muslims in Europe. (A) Muslims are a small minority in most European countries, but many intend to make the region their permanent home. Just how, or if, the assimilation of Muslims into largely secular Europe will proceed is a major topic of public debate. **(B)** Who is French? Youths shown here are waiting for a bus in a Paris suburb. There is a growing recognition that French society discriminates against immigrants and their children, especially when it comes to hiring practices. The unemployment rate is 9 percent for those of French ancestry, but 14 percent for those of foreign ancestry. NGOs like SOS Racisme have documented systematic discrimination in private sector hiring practices. The tensions periodically explode into violent episodes, such as the riots of 2005, which occurred throughout France, especially in the poorer suburbs that are home to most immigrants and children of immigrants. **(C)** A Muslim woman is interviewed during a demonstration in Brussels, Belgium, by a group calling itself Stop the Islamization of Europe. She is Belgian and has converted to Islam.

Thinking Geographically: (B) In France, how does the unemployment rate for those of French ancestry compare to those of foreign ancestry? **(C)** If this woman were a student in a French high school, would she be able to wear a headscarf?

in professional and technical fields is growing (see Figure 4.12 on page 166). Nevertheless, European public opinion among both women and men largely holds that women are less able than men to perform the types of work typically done by men, and that men are less skilled at domestic, caregiving, and nurturing duties. In most places, men have greater social status, hold more managerial positions, earn on average about 15 percent more pay for doing the same work, and have greater

double day the longer workday of women with jobs outside the home who also work as caretakers, housekeepers, and/or cooks for their families

autonomy in daily life than women—more freedom of movement, for example (see Thematic Overview I). These male advantages have a stronger hold in Central and South Europe today than they do in West and North Europe.

Although younger men are now assuming more domestic duties, women who work outside the home usually face what is called a **double day** in that they are expected to do most of the domestic work in the evening in

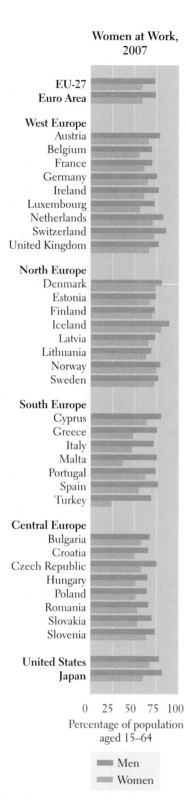

Women at Work, 2007

EU-27
Euro Area

West Europe
Austria
Belgium
France
Germany
Ireland
Luxembourg
Netherlands
Switzerland
United Kingdom

North Europe
Denmark
Estonia
Finland
Iceland
Latvia
Lithuania
Norway
Sweden

South Europe
Cyprus
Greece
Italy
Malta
Portugal
Spain
Turkey

Central Europe
Bulgaria
Croatia
Czech Republic
Hungary
Poland
Romania
Slovakia
Slovenia

United States
Japan

0   25   50   75   100

Percentage of population
aged 15–64

■ Men
■ Women

FIGURE 4.12 Women at work in Europe (and selected other countries for comparison), 2007. A majority of women in Europe work outside the home, and their numbers are increasing. A related trend is declining birth rates in these countries. The exceptions are in Greece, Italy, and Malta. Included for comparison is Turkey, where slightly less than 25 percent of women work outside the home.

addition to their job outside the home during the day. United Nations research shows that in most of Europe, women's workdays, including time spent in housework and child care, are 3 to 5 hours longer than men's. (Iceland and Sweden reported that women and men there share housework equally.) Women burdened by the double day generally operate with somewhat less efficiency in a paying job than do men. They also tend to choose employment that is closer to home and that offers more flexibility in the hours and skills required. These more flexible jobs (often erroneously classified as part-time) almost always offer lower pay and less opportunity for advancement, though not necessarily fewer working hours.

Many EU policies encourage gender equality. Managers in the EU bureaucracy are increasingly female, and well over half the university graduates in Europe are now women. However, the political influence and economic well-being of European women lag behind those of European men. In most European national parliaments, women make up less than a third of elected representatives. Only in North Europe do women come anywhere close to filling 50 percent of the seats in the legislature, and there several women have served as heads of government. In West Europe, the United Kingdom has had numerous women officials in high office, but elsewhere in the region this trend is only beginning. In 2009, Germany reelected Angela Merkel as its first woman chancellor (prime minister), and in France since the 2007 election of President Sarkozy (he defeated a female opponent), women have occupied a number of French cabinet-level positions. Still, in much of Europe, women generally serve only in the lower ranks of government bureaucracies.

Although change is clearly underway in the European Union, economic empowerment for women has been slow on many fronts. For example, in 2006, female unemployment was higher than male unemployment in all but a few countries (the United Kingdom, Germany, the Baltic Republics, Ireland, Norway, and Romania), where the differences were slight—and 32 percent of women's jobs were part time, as opposed to only 7 percent of men's jobs. Throughout the European Union, women are paid less than men for equal work, despite the fact that young women tend to be more highly educated than young men.

Norway leads Europe in redefining gender in society. It directs much of its most innovative work on gender equality toward advancing men's rights in traditional women's arenas. For example, men now have the right to at least a month of paternity leave with a newborn or adopted child. This policy recognizes a father's responsibility in child rearing and provides a chance for father and child to bond early in life. At the same time, equal rights for women are closer to being a reality in Norway than in any other country. In 1986, Gro Harlem Brundtland, a physician, became the country's first female prime minister, and appointed women to 44 percent of all cabinet posts. Norway became the first country in modern times to have such a high proportion of women in important government policy-making positions. By 2009, women made up 36 percent of Norway's parliament (Sweden had 47 percent; Finland had 42 percent). Norway also requires a minimum of 40 percent representation by each sex on all public boards and committees. Even though the rule has not yet been achieved in all cases, female representation averages over 35 percent for state and municipal agencies. In addition, employment ads are required to be gender neutral, and all advertising must be nondiscriminatory.

Social Welfare Systems and Their Outcomes

In nearly all European countries, tax-supported systems of **social welfare** or **social protection** (the EU term) provide all citizens with basic health care; free or low-cost higher education; affordable housing; old age, survivor, and disability benefits; and generous unemployment and pension benefits. Europeans generally pay much higher taxes than North Americans (the rate for EU countries is about 40 percent of GDP; for the United States, 27 percent; and for Canada, 30 percent), and in return they expect more in services. In some cases, European governments are able to deliver services more cheaply than the "free market" does elsewhere. For example, the European Union spends on average about $3000 per person for health care, while the United States spends more than $7000. However, the European Union has more doctors and acute-care hospital beds per citizen and better outcomes than the United States in terms of life expectancy and infant mortality.

> **social welfare** (in the European Union, **social protection**) in Europe, tax-supported systems that provide citizens with benefits such as health care, pensions, and child care

Europeans do not agree on the goals of these welfare systems, or on just how generous they should be. Some argue that Europe can no longer afford high taxes if it is to remain competitive in the global market. Others maintain that Europe's economic success and high standards of living are the direct result of the social contract to take care of basic human needs for all. The debate has been resolved differently in different parts of Europe, and the resulting regional differences have become a source of concern in the European Union. With open borders, unequal benefits can encourage those in need to flock to a country with a generous welfare system and overburden the taxpayers there.

European welfare systems can be classified into four basic categories (Figure 4.13). *Social democratic welfare systems*, common in Scandinavia, are the most generous systems. They attempt to achieve equality across gender and class lines by providing extensive health care, education, housing, and child and elder care benefits to all citizens from cradle to grave. Child care is widely available, in part to help women enter the labor market. But early childhood training, a key feature of this system,

is also meant to ensure that in adulthood every citizen will be able to contribute to the best of his or her highest capability, and that citizens will not develop criminal behavior or drug abuse. While finding comparable data is very difficult, surveys of crime victims in Scandinavia and the European Union show that generally the rate of crime in Scandinavia is lower than in other parts of Europe.

The goal of *conservative* and *modest welfare systems* is to provide a minimum standard of living for all citizens. These systems are common in the countries of West Europe. The state assists those in need but does not try to assist upward mobility. For example, college education is free or heavily subsidized for all, but strict entrance requirements in some disciplines can be hard for the poor to meet. State-supported health care and retirement pensions are available to all, but although there are movements to change these systems, they still reinforce the traditional "housewife contract" by assuming that women will stay home and take care of children, the sick, and the elderly. The "modest" system in the United Kingdom is considered slightly less generous than the "conservative" systems found elsewhere in West Europe. The two are combined here but not in Figure 4.13.

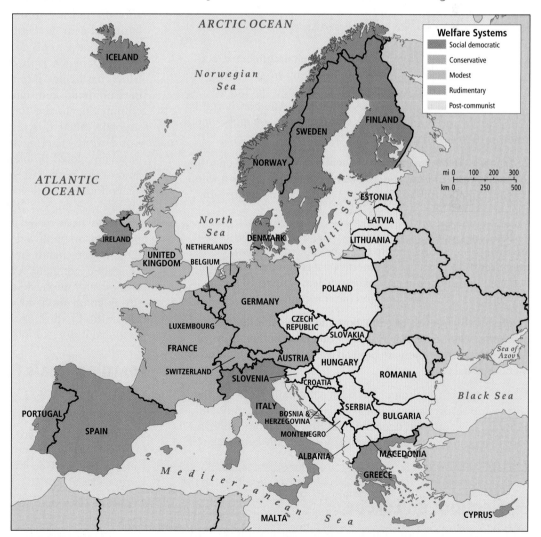

FIGURE 4.13 European social welfare/protection systems. The basic categories of social welfare systems shown here and described in the text should be taken as only an informed approximation of the existing patterns.

Rudimentary welfare systems do not accept the idea that citizens have inherent rights to government-sponsored support. They are found primarily in South Europe and in Ireland. Here, local governments provide some services or income for those in need, but the availability of such services varies widely, even within a country. The state assumes that when people are in need, their relatives and friends will provide the necessary support. The state also assumes that women work only part time, and thus are available to provide child care and other social services for free. Such ideas reinforce the custom of paying women lower wages than men.

Post-communist welfare systems prevail in the countries of Central Europe. During the communist era, these systems were comprehensive, resembling the cradle-to-grave social democratic system in Scandinavia, except that women were pressured to work outside the home. Benefits often extended to nearly free apartments, health care, state-supported pensions, and subsidized food and fuel. However, in the post-communist era, state funding has collapsed, forcing many to do without basic necessities. In many post-communist countries, welfare systems are being reviewed, but usually with an eye to reducing benefits and to extending work lives past age 65.

THINGS TO REMEMBER

1. Europe's cities are world famous for their architecture, economic dynamism, and cultural vitality. Many are quite ancient but have expanded in recent decades with large apartment blocks connected to the old city centers by public transport. In West, North, and South Europe, around 80 percent of the population lives in urban areas.

2. **Learning Goal 6: Population and Gender** Europeans are choosing to have fewer children, and as a result the population as a whole is aging. Small families are in part a result of women pursuing careers.

3. Because Europe's population is not growing and consumers are dying faster than they are being born, economies may contract over time. Many governments now encourage large families with generous maternity and paternity leave (up to 10 months with full pay), free day care, and other incentives.

4. Despite high levels of education, the political influence and economic well-being of European women lag behind those of European men; women are paid on average 15 percent less than men for equal work.

5. Anti-foreigner sentiment has been a hindrance to acquiring citizenship across Europe, but as the advantages of a multicultural, multiracial, and multigenerational Europe become more apparent, requirements for citizenship are gradually being relaxed. Future trends are uncertain.

Reflections on Europe

For better or for worse, Europe has transformed the world we live in, perhaps more than any other region on Earth. Europe's colonization of most of the planet initiated a wave of accelerated globalization that is still going on today. Places all over the globe were transformed as they produced raw materials for colonial Europe's growing industries. During the nineteenth and twentieth centuries, these industries produced fabulous wealth and dire poverty, especially in Europe's cities. The political transformations that resulted changed not only Europe, but also much of the planet, often influencing the shape of independence movements in Europe's colonies.

Though hundreds of years have passed since the period of colonization began, as a whole Europe today has the world's largest economy—but it faces many difficulties common to rich areas. How can economic growth be maintained without further damage to already stressed ecosystems without further degrading air and water resources? Will small family farms be able to prosper by using more sustainable production techniques, or will they be completely replaced by large corporate farms? Will Europe's economies decline as its aging populations strain the region's ability to maintain high productivity? Will gender roles and support for education have to change further to encourage optimal population growth? Will enough well-paying jobs be available to employ workers from industries that have relocated abroad? And as immigrants come, how can they be absorbed fairly and with the least disruption?

Europe's ability to solve these problems is bolstered by its long experience with democracy (Reasons for Optimism A), which has helped it to reach high levels of economic and social development, and has facilitated international cooperation, culminating in the formation of the European Union. These strengths lifted this region from the ashes of two devastating world wars, enabling it to provide inspiration and leadership on many global problems. While Europe still consumes huge amounts of resources, and faces massive problems of water and air pollution, it is leading the world in responding to climate change (Reasons for Optimism B). Moreover, several European countries have made unprecedented advances in helping the poor and unemployed attain a better life at the same time as these nations maintain very high overall levels of prosperity and economic competitiveness. Indeed, given Europe's long history of innovation, democracy, and cooperation, along with its strong connections to so many parts of the world (Reasons for Optimism C), this region is in a position to maintain and extend its global influence into the future.

Learning Goals Review

1. **Water:** What factors complicate efforts to reduce water pollution in the Mediterranean Sea?

How does Europe's physical geography make this problem harder to solve? What human factors are involved? What could the European Union do to reduce its contribution to pollution of Europe's seas?

2. **Climate Change:** In what ways is Europe responding to climate change?

What goals have governments set? On a more local level, what evidence is there that European ways of life contribute less to global warming than North American ways?

Reasons for Optimism in Europe

Democratization: Europe's long history of democracy has facilitated high levels of social and economic development and has helped the countries of the region work cooperatively with each other. **A** *A pro-democracy uprising in Germany in 1848.* ▼

Climate Change: Europe leads the world in responding to climate change, with EU governments having agreed to cut greenhouse gas emissions by 20 percent by 2020. **B** *A solar thermal power plant in Spain.* ▼

Globalization: Strong connections with many parts of the world put Europe in a position to maintain and extend its global influence. **C** *A Caribbean-style carnival in Rotterdam, the Netherlands.* ▼

Thinking Geographically: What does the presence of this raucous mob suggest about the transition to democracy in Europe?

Thinking Geographically: What does recent European research suggest would be the economic consequence of doing nothing about climate change?

Thinking Geographically: From whom did the inspiration for this carnival likely come?

---

3. Globalization and Development: How did Europe's colonization of much of the world over the past 400 years accelerate globalization and transform economic development in Europe?

How was Europe's Industrial Revolution related to its overseas colonies?

4. Urbanization and Democratization: How has urbanization influenced the development of democracy in Europe?

How was the Industrial Revolution central to the changes in urban Europe that inspired calls for greater democratization?

5. Food: How is food production in Europe changing?

How are these changes influenced by the EU's large subsidies for agriculture? Why are these subsidies criticized by developing countries?

6. Population and Gender: How is Europe's aging population linked to changing gender roles?

What are some of the potential consequences of negative population growth rates in Europe?

Geographic Themes about North America

Look back at the Thematic Overview photos on page 137. Thinking geographically, answer the following questions about them:

(A) Climate Change: How do wind turbines help reduce greenhouse gas emissions?

(B) Water: What aspect of Europe's geography in this picture could complicate efforts to reduce pollution in the Mediterranean?

(C) Development: How would a construction project contribute to economic development in Poland?

(D) Globalization: What clues can you see in this picture that suggest it depicts a non-European place?

(E) Urbanization: What clue is there that this photo was taken in a city?

(F) Population: How might an older population strain an economy?

(G) Gender: How do women in Europe who work affect the population growth of their region?

Key Terms

assimilation 164
Biosphere 142
capitalism 152
central planning 152
Cold War 152
Common Agricultural Program (CAP) 159
communism 152
cultural homogenization 139
double day 165
economies of scale 156
euro 156

European Union (EU) 138
Green 146
guest workers 163
Holocaust 152
humanism 149
humid continental climate 142
iron curtain 152
Mediterranean climate 142
mercantilism 149
nationalism 151
North Atlantic Drift 142

North Atlantic Treaty Organization (NATO) 158
Roma 152
Schengen Accord 163
social welfare (in the European Union, social protection) 167
subsidies 159
temperate midlatitude climate 142
welfare state 151

5 Russia and the Post-Soviet States

Learning Goals

After you read this chapter, you will be able to answer the following questions:

1. Environment, Development, and Urbanization: Why does this region have such severe environmental problems, especially in urban areas?

2. Climate Change, Food, and Water: Where in this region are food production systems and water resources most threatened by climate change? What parts of this region produce the most greenhouse gases?

3. Globalization and Development: How have economic reform and globalization changed patterns of development in this region since the fall of the Soviet Union?

4. Democratization: What are the obstacles to democratization in this region?

5. Population: Why are populations shrinking in some parts of this region?

6. Gender: How have the massive economic and political changes in this region affected men and women differently?

FIGURE 5.1 Political map of Russia and the post-Soviet states.

Thematic Overview of Russia and the Post-Soviet States

Climate Change: Although Russia and other countries have made moves to reduce their greenhouse gas emissions, this region is still a major contributor of greenhouse gases. **A** *Traffic in Moscow, where car ownership has boomed.* ▼

Water: Scarce water resources have been used intensively for many years in Central Asia, a region that is prone to drought and water scarcity. **B** *A river in Tajikistan flanked by irrigated agriculture.* ▼

Globalization: Fossil fuel exports have introduced a new wave of globalization. **C** *A statue of Ukrainian poet Taras Shevchenko overlooks billboards for Gazprom, Russia's largest energy company, and Coca Cola.* ▼

Development: Bureaucratic tangles and corruption can make starting a business outlandishly expensive in this region, forcing many to cut corners whenever possible. **D** *An abacus in a grocery store in Russia.* ▼

Democratization: While elections are now common, many forms of authoritarian control remain, including limits on media freedom. **E** *A demonstration against media censorship in Uzbekistan.* ▼

Food: Compared to Soviet times, food production has started to recover in recent years after falling by 20 to 30 percent in the 1990s. **F** *A small-scale farmer sells her produce in Tbilisi, Georgia.* ▼

Urbanization: Urban life is still shaped by the legacies of central planning from the communist era. **G** *Apartment blocks on the outskirts of Moscow.* ▼

Population: Much of this region is experiencing population decline as women have fewer children and as older people pass on. **H** *An elderly man in the village of Touroukhansk, Krasnoyarsk, Russia.* ▼

Gender: Soviet policy that encouraged all women to work for wages outside the home has transformed this region. **I** *Female workers in a garment factory in Moldova.* ▼

Global Patterns, Local Lives It's June 2009, and despite the gathering crowds, the Rustaveli Cinema in Tbilisi, Georgia, is closed. Located just in front of the parliament building, the cinema usually runs 20 showings a day, but since April, protesters demanding the resignation of Georgian president Mikhail Saakashvili have blocked the cinema entrance. Surrounding businesses have shuttered, and local economic losses due to the demonstrations are mounting as foreign investors, once pleased with opportunities in Georgia, pull out.

The irony of President Saakashvili's predicament is striking. In 2001, then age 34, he had resigned as Minister of Justice and accused then-president Eduard Shevardnadze of corruption and human rights abuses. By 2003, Saakashvili was leading a grassroots pro-democracy and anticorruption political campaign featuring mass nonviolent demonstrations. What became known as the "Rose Revolution" ultimately brought Saakashvili, an American-educated lawyer, to the presidency (**Figure 5.2**). But by 2007, thousands of protestors were taking to the streets *against* Saakashvili, charging that he was no better than the leaders he replaced. Seemingly proving their point, Saakashvili ordered Georgian police to shut down privately owned TV stations and to clear the streets by using batons, rubber bullets, and water cannons on protestors.

After the 2007 demonstrations, eager to salvage his image as a popular democratic leader, Saakashvili called for elections within 2 months (January 2008). He won by a large margin, but his popularity did not last. Faced with the everyday difficulties of managing a democracy, Saakashvili turned once again to authoritarian measures. By April 2009, mass demonstrations were again demanding his resignation, calling him a bloodthirsty tyrant and an embarrassment to Georgia. This time, sections of the military were taking

> **Union of Soviet Socialist Republics (USSR)**
> the multi-national union formed from the Russian empire in 1922 and dissolved in 1991
>
> **Soviet Union** see **Union of Soviet Socialist Republics**

part. In May 2009, a Georgian army tank battalion attempted, unsuccessfully, to organize a mutiny to remove Saakashvili from office. The 2009 demonstrations grew in size, and by the middle of June 2009, the success story that had been Georgia was disintegrating. [*Sources: Compiled with the aid of the following authors: Nana Sajaia, "Rustaveli Ave. Blues," June 16, 2009, Transitions Online, at http://www.tol.cz/look/TOL/article.tpl?IdLanguage=1&IdPublication=4&NrIssue=325&NrSection=1&NrArticle=20641&tpid=10; Sreeram Chaulia, "Democratisation, NGOs and 'Colour Revolutions,'" January 19, 2006, Open Democracy, at http://www.opendemocracy.net/globalization-institutions_government/colour_revolutions_3196.jsp; and Irakliy Khaburzaniya, "Neither White Knight nor Blackguard," June 2, 2009, Transitions Online, at http://www.tol.org/client/article/20620-neither-white-knight-nor-blackguard.html.*] ∎

While the protestors are clearly outraged by ongoing corruption, the fact that the protests are happening at all is a testimony to the pace of democratization in Georgia and other former Soviet states. After years of hard work by civil servants, NGOs, and citizens, free speech and the right to assemble in public are constitutionally guaranteed in Georgia. Thus the situation of leaders like Saakashvili is rather ironic. They came to power leading political movements aimed at ending corruption and promoting public participation. But because progress is often slow and only partial, they now find themselves the target of similar movements. Their predicament shows both how hard democratization is to achieve, and how badly so many people want to achieve it.

Georgia is a tiny country in a huge region that has changed its political and economic systems entirely in just a few short years. Barely two decades ago, the **Union of Soviet Socialist Republics (USSR)**, more commonly known as the **Soviet Union**, was the largest political unit

(A) **(B)**

FIGURE 5.2 Globalization and democratization in Georgia. (A) Student activists chant their demands during Georgia's "Rose Revolution" of 2003. Like other "color revolutions" that have occurred in this region, the Rose Revolution was funded by international NGOs, such as the Open Society Institute of world-renowned financier George Soros (which itself receives funding from the U.S. State Department), and followed models codified by U.S. political scientist Gene Sharp. **(B)** A picture of Georgian president Mikhail Saakashvili, who was brought to power by the Rose Revolution, graces a bus in Tbilisi during the 2007–2008 presidential elections.

Thinking Geographically: (A) Why is it ironic that Saakashvili was brought to power by the "Rose Revolution"?

on earth, stretching from Central Europe to the Pacific Ocean. It covered one-sixth of the earth's land surface. In 1991, the Soviet Union broke apart, ending a 70-year era of nearly complete government control of the economy, society, and politics. Over the course of a few years, attempts were made to replace economic control by government bureaucrats with capitalist systems similar to that of the United States, based on competition among private businesses. The conversion has proven difficult.

Politically, the Soviet Union has been replaced by a loose alliance of Russia and 11 post-Soviet states—Ukraine; Belarus; Moldova; the Caucasian states of Georgia, Armenia, Azerbaijan; and the Central Asian states of Kazakhstan, Kyrgyzstan, Tajikistan, Turkmenistan, and Uzbekistan (Figure 5.1 on page 170). Three former Soviet republics, the Baltic states of Lithuania, Latvia, and Estonia, are now part of the European Union. Russia, which was always the core of the Soviet Union, remains predominant in this alliance and influential in the world because of its size, population, military, and huge oil and gas reserves.

Geopolitically, this region is still sorting out its relationships. The Cold War between the Soviet Union and the United States and its allies is over. Many former Soviet allies in Central Europe have already joined the European Union, and some other countries in this region may eventually do the same. Trade with Europe, especially in oil and gas, is booming, if fraught with conflict. Meanwhile, the Central Asian states currently allied with Russia may align themselves with neighbors in Southwest Asia or South Asia. The far eastern parts of Russia—and Russia as a whole—are already finding common trading ground with East Asia and Oceania.

After 70 years of authoritarian rule, elections are becoming the norm throughout this region. Whether these are actually free and fair elections is debatable, as opposition candidates are increasingly marginalized by lack of access to both print and broadcast media. Economic and political instability have led many voters to support strong leaders who tolerate little criticism. An explosion of crime and corruption has many wondering if the new democracies in this region are strong enough to endure.

THINGS TO REMEMBER

1. The former Soviet Union has been replaced by a loose alliance of Russia and 11 post-Soviet states.

2. Elections have become common in the region, although whether they are free and fair is uncertain.

3. Democracies in the region are threatened by crime and corruption.

I THE GEOGRAPHIC SETTING

Terms in This Chapter

There is no entirely satisfactory new name for the former Soviet Union. In this chapter we use *Russia and the post-Soviet states*. Russia is still closely associated with all of these states economically, but they are independent countries, and their governments are legally separate from Russia's. Russia itself is formally known as the **Russian Federation** because it includes more than 30 (mostly) ethnic *internal republics*—such places as Chechnya, Ossetia, Tatarstan, and Buryatya—which all together constitute about one-tenth of its territory and one-sixth of its population. *Federation* should not be taken to mean that the internal republics share power equally with the Russian government.

Physical Patterns

The physical features of Russia and the post-Soviet states vary greatly over the huge territory they encompass. The region bears some resemblance to North America in size, topography, climate, and vegetation. Russia is the largest country in the world, nearly twice the size of the second-largest country, Canada.

Landforms

Because the region is so complex physically, a brief summary of its landforms is useful (see Figure 5.3 on pages 174–175). Moving west to east, there is first the eastern extension of the North European Plain, including the alluvial plain of the Volga river (see Figure 5.3 A, C), then the Ural Mountains (see Figure 5.3D), then the West Siberian Plain (see Figure 5.3E), followed by an upland zone called the Central Siberian Plateau (see Figure 5.3F), and finally, in the far east, a mountainous zone bordering the Pacific (see Figure 5.3G). To the south of these territories from west to east is an irregular border of mountains (the Caucasus; see Figure 5.3B); semiarid grasslands, or **steppes** (in western Central Asia, not pictured); and barren uplands and high mountains (in eastern Central Asia, not pictured). The adjacent regions of Southwest Asia, South Asia, and East Asia skirt this region to the south.

The eastern extension of the North European Plain rolls low and flat from the Carpathian Mountains in Ukraine and Romania 1200 miles (about 2000 kilometers) east to the Ural Mountains (see Figure 5.3 A, C, D). The part of Russia west of the Urals is often called *European Russia* because the Ural Mountains are traditionally considered part of the indistinct border between Europe and Asia. European Russia is the most densely settled part of the entire region and is its agricultural and industrial core. Its most important river is the Volga, which flows into the Caspian Sea. The Volga River is a major transportation route that connects many parts of the North European Plain to the Baltic and White seas in the north and to the Black and Caspian seas in the south (see Figure 5.3C).

The Ural Mountains extend in a fairly straight line south from the Arctic Ocean into Kazakhstan (see the map in Figure 5.3). The Urals, a low-lying range similar in elevation to the Appalachians, are not much of a barrier to humans and are only partially a barrier to nature (some European tree species do not extend east of the Urals). There are several easy passes across the mountains, and winds carry moisture all the way from the Atlantic and Baltic across the Urals and into Siberia. Much of the Urals' once-dense forest has been felled to build and fuel new industrial cities.

> **Russian Federation** Russia and its political subunits, which include 30 internal republics and more than 10 so-called autonomous regions
>
> **steppes** semiarid, grass-covered plains

A North European Plain, Belarus

E West Siberian Plain, Russia

B Caucusus Mountains, Georgia

C Volga River, Russia

ARCTIC OCEAN

UNITED KINGDOM
Manchester
Birmingham
London

North Sea

FRANCE
BELGIUM
Brussels
Amsterdam
NETHERLANDS
LUX.
Luxembourg
Hamburg
Copenhagen
Berlin
GERMANY
Prague
CZECH REP.
Vienna
Bratislava
SLOVAKIA
Budapest
HUNGARY
POLAND
Warsaw
Kaliningrad

SWEDEN
Stockholm
Gulf of Bothnia
Baltic Sea
FINLAND
Helsinki
Tallinn
ESTONIA
Gulf of Finland
Gulf of Riga
Riga
LATVIA
LITHUANIA
Vilnius
Brest
A Minsk
BELARUS

Svalbard

NORWAY

Murmansk
Kola Peninsula
Kirovsk

Barents Sea

Franz Josef Land
North Land

Novaya Zemlya

Kara Sea

Laptev Sea

Taymyr Peninsula

Dikson

ROMANIA
MOLDOVA
Chisinau
UKRAINE
Kiev
Odesa
Mykolaiv
Kherson
Dnipropetrovsk
Kharkiv
Donetsk
Sevastopol
Rostov-na-Donu
Sea of Azov

Lviv
Pripyat
Chernobyl
Homyel
Smolensk

St. Petersburg
Lake Ladoga
Lake Onega
Novgorod
Tver
Moscow
Bryansk
Kursk
Tula
Ivanovo
Yaroslavl
Moscow Canal
Lipetsk
Voronezh
Nizhniy Novgorod
Penza
Kazan
Saratov
Tolyatti
Samara
Volgograd

White Sea
Arkhangelsk
Northern Dvina Canal
Divina

North European Plain

Pechora
Vorkuta
Salekhard
D
Yamal Peninsula
Pechora Basin
Urengoy
E
West Siberian Plain

Norilsk
North Siberian Lowland
F
Central Siberian Plateau

Vilyuy Res.

Lena R.

Olenek

Black Sea
Caucasus Mts.
B Batumi
GEORGIA
Tbilisi
Groznyy
ARMENIA
Yerevan
AZER.
AZERBAIJAN
Baku
TURKEY
Lake Van
IRAQ
Lake Urmia
IRAN
Tehran
Esfahan
Mashhad
Zagros Mts.
Persian Gulf
Elburz Mts.

Astrakhan
Caspian Depression
Caspian Sea
Steppes
Aral Sea
KAZAKHSTAN
Aktogay
Astana
Qaraghandy
Leninsk

Ufa
Perm
Kama
Orenburg
Orsk
Nizhniy Tagil
Yekaterinburg
Tyumen
Chelyabinsk
Magnitogorsk
Ural Mountains

RUSSIAN FEDERATION
(RUSSIA)

Tunguska Basin

Omsk
Seversk
Tomsk
Novosibirsk
Novokuznetsk
Krasnoyarsk
Gladkaya
Bratsk
Bratsk Res.
Tayshet
Angara
E. Sayan Mts.
Irkutsk Basin
Lake Baikal
Angarsk
Irkutsk
Ulan-Ude
W. Sayan Mts.
Kyzyl
Altai Mts.

TURKMENISTAN
Ashkhabad
UZBEKISTAN
Samarkand
Tashkent
Dushanbe
TAJIKISTAN
Pamirs
Hindu Kush
AFGHANISTAN
PAKISTAN
Islamabad
Faisalabad
Lahore
INDIA
New Delhi

Lake Balkhash
Shymkent
Bishkek
KYRGYZSTAN
Almaty
Lake Yssyk
Tien Shan
Junggar Basin
Urumqi
Tarim Basin
Tarim
Gobi

MONGOLIA
Ulan Bator

C H
Lanzhou
Xian
Baotou
Chengdu

Land Elevations

meters	feet
4877	16,000
3353	11,000
2134	7000
914	3000
305	1000
152	500
0	0

mi 0 100 200 300 400 500
km 0 200 400 600 800

1:26,000,000
Azimuthal Equidistant Projection

ALASKA

Chukchi Sea

East Siberian Sea

New Siberian Islands

Wrangel Island

Bering Strait

St. Lawrence Island

Mys Shmidta

Anadyr

Bering Sea

Commander Islands

D Urals, Russia

Delta

Tiksi

Kolyma Range

Verkhoyansk Range

Verkhoyansk

Cherskiy Range

Kolyma

Kamchatka Peninsula

Shelikhov Gulf

Magadan

G

Petropavlovsk-Kamchatskiy

Okhotsk

Sea of Okhotsk

Yakutsk

Lena

Aldan

Olekminsk

Aldan

Stanovoy Range

Yablonovyy Range

Chita

Onon

Argun

Amur

Sakhalin

Nikolayevsk

Komsomolsk

Sovetskaya Gavan

Raychikhinsk

Khabarovsk

Tatar Strait

Sikhote Alin

Yuzhno-Sakhalinsk

Kuril Islands

La Perouse Strait

Hokkaido

Sapporo

Lake Khanka

Qiqihar

Harbin

Jilin

Changchun

Ussuriysk

Vostochny
Nakhodka

Vladivostok

Honshu

Tsugaru Strait

JAPAN

Sea of Japan

Shenyang

Fushun

Anshan

N. KOREA

P'yongyang

Tokyo

Nagoya

Northeast China Plain

Beijing

Tangshan

Dalian

Korea Bay

Seoul

Taegu

Pusan

S. KOREA

Taejon

Osaka-Kobe-Kyoto

Hiroshima

Kitakyushu

Fukuoka

Shikoku

Bo Hai Bay

Tianjin

Shijiazhuang

Jinan

Qingdao

Kyushu

Taiyuan

I N A

Zhengzhou

Nanjing

Shanghai

Wuhan

Hangzhou

Desert

F Central Siberian Plateau, Russia

G Pacific Mountain Zone, Russia

FIGURE 5.3 Regional map of Russia and the post-Soviet states.

The West Siberian Plain, east of the Urals, is the largest wetlands plain in the world (see Figure 5.3E). A vast, mostly marshy lowland about the size of the eastern United States, it is drained by the Ob River and its tributaries, which flow north into the Arctic Ocean. Long, bitter winters mean that in the northern half of this area a layer of permanently frozen soil (**permafrost**) lies just a few feet beneath the surface. In the far north, the permafrost comes to within a few inches of the surface. Because water does not percolate through this dense layer, swamps and wetlands form in the summer months, providing habitats for many migratory birds. In the far north lies the **tundra**, where because of the extreme cold, the shallow soils, and the permafrost, only mosses and lichens can grow. The West Siberian Plain has some of the world's largest reserves of oil and natural gas, although their extraction is made difficult by the harsh climate and the permafrost.

The Central Siberian Plateau and the Pacific Mountain Zone farther to the east together equal the size of the United States (see Figure 5.3 F, G). Permafrost prevails except along the Pacific coast. There, the ocean moderates temperatures, and additional heat is supplied by the many active volcanoes created as the Pacific Plate sinks under the Eurasian Plate. Places warmed by these forces are the Kamchatka Peninsula, Sakhalin Island, and Sikhote-Alin on Russia's southeastern Pacific coast. Only lightly populated, these places are havens for wildlife.

To the south of the West Siberian Plain is an irregular band of steppes and deserts stretching from the Caspian Sea to the mountains bordering China. To the south and east of these grasslands is a wide, curving band of mountains, including the Caucasus (see Figure 5.3B), Elburz, Hindu Kush, Pamir, Tien Shan, and Altai. Their rugged terrain has not deterred people from crossing these mountains. In this Caucasus region and adjacent areas, for literally tens of thousands of years, people have exchanged plants (apples, onions, citrus, rhubarb, wheat), animals (horses, sheep, goats, cattle), technologies (cultivation, animal breeding, portable shelter construction, rug and tapestry weaving), and religious belief systems (principally Islam and Buddhism, but also Christianity and Judaism).

Climate and Vegetation

The climates and associated vegetation in this large region are varied but not as much as in other regions because so much territory is taken up by expanses of midlatitude grasslands, and northern taiga forest and tundra that are cold much of the year. No inhabited place on earth has as harsh a climate as the northern part of the Eurasian landmass occupied by Russia and particularly by Siberia (see the Photo Essay 5.1 map; see also Figure 5.3C). The winters are long and cold, with only brief hours of daylight. Summers are short and cool to hot, with long days. Precipitation is moderate, coming primarily from the west. Farthest north, the natural vegetation is tundra grasslands. The major economic activities here are the extraction of oil, gas, and some minerals, as well as reindeer herding by the local indigenous population. Just south of the tundra lies a vast cold-adapted coniferous forest known as **taiga** that stretches from northern European Russia to the Pacific. The largest portion of taiga lies east of the Urals, and here

forestry—often unrestrained by ecological concerns—is a dominant economic activity. The short growing season and large areas of permafrost in this northern zone generally limit crop agriculture, except in the southern West Siberian Plain, where grain is grown.

Because massive mountain ranges in Central and East Asia to the south block access to warm wet air from the Indian Ocean, most rainfall in the entire region comes from storms that blow in from the Atlantic Ocean far to the west. By the time these initially rain-bearing air masses arrive, most of their moisture has been squeezed out over Europe. A fair amount of rain does reach Ukraine, Belarus, European Russia (Photo Essay 5.1A), and the Caucasian republics, and these regions are especially important food production (vegetables, fruits, and grain) areas. The natural vegetation in these western zones is open woodlands and steppes (grassland), though in ancient times, forests were common.

East of the Caucasus Mountains, the lands of Central Asia have semiarid to arid climates (Photo Essay 5.1B) influenced by their location in the middle of a very large continent. The summers are scorching and short, the winters intense. In the desert zones, daytime-to-nighttime temperatures can vary by 50 degrees or more. Northern Kazakhstan produces grain and grazing animals. The more southern areas (southern Kazakhstan, Uzbekistan, and Turkmenistan), support grasslands (steppes), which are also used for herding. In modern times, these grasslands have been used for irrigated commercial agriculture that is dependent upon glacially fed rivers, but most land is not useful for farming. The climates in the area where Kazakhstan, Uzbekistan, Tajikistan, and Kyrgyzstan meet are varied, supporting a number of small-scale agricultural activities, some of them commercial.

The construction of land transportation systems has long been held back by the region's climate. It is difficult to build during the long, harsh winters, which eventually give way to a period called the *rasputitsa* or the "quagmire season," when melting permafrost turns many roads and construction sites into impassable mud pits. Huge distances between populated places, and complex topography, especially in Siberia, add to the problem. As a result, few roads or railroads have ever been built outside of western Russia.

Environmental Issues

Soviet ideology held that nature was the servant of industrial and agricultural progress, and that humans should dominate nature on a grand scale. While this sentiment was common throughout much of the world at the time of the formation of the USSR, it does seem to have been taken further here than elsewhere. Joseph Stalin, General Secretary of the Communist Party of the USSR from 1922 to 1953, and a major architect of Soviet policy, is famous for having said, *"We cannot expect charity from Nature. We must tear it from her."* Hence, during the Soviet years huge dams, factories, and other industrial facilities were built without regard for their effect on the environment or public health. Now Russia and the post-Soviet states have some of the worst environmental problems in the world. By 2000, more than 35 million people in the region (15 percent of the population) were living in areas where the soil was poisoned and the air was dangerous to breathe (see Photo Essay 5.2 A, B, C on page 179).

---

permafrost permanently frozen soil just a few feet beneath the surface

tundra a treeless area, between the ice cap and the tree line of arctic regions, where the subsoil is permanently frozen

taiga subarctic forests

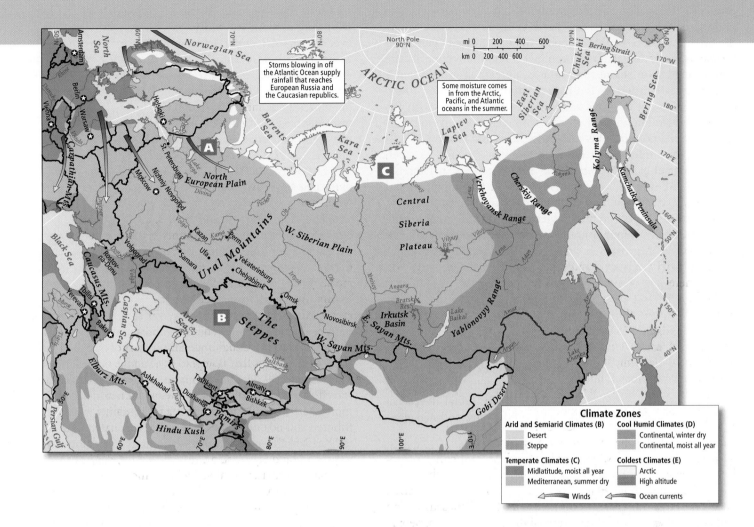

Storms blowing in off the Atlantic Ocean supply rainfall that reaches European Russia and the Caucasian republics.

Some moisture comes in from the Arctic, Pacific, and Atlantic oceans in the summer.

Climate Zones

Arid and Semiarid Climates (B)
- Desert
- Steppe

Temperate Climates (C)
- Midlatitude, moist all year
- Mediterranean, summer dry

Cool Humid Climates (D)
- Continental, winter dry
- Continental, moist all year

Coldest Climates (E)
- Arctic
- High altitude

⇐ Winds   ⇐ Ocean currents

A Continental, moist all year

B Steppe, Kazakhstan

C Arctic, Russia

Since the collapse of the Soviet Union, the region's governments, beset with myriad problems, have been reluctant to address environmental issues. As one Russian environmentalist put it, "When people become more involved with their stomachs, they forget about ecology." Pollution controls are complicated by a lack of funds and by an official unwillingness to correct even a few of the past environmental abuses. Photo Essay 5.2 shows just a few examples of human impact on the region's environment, most of which are related to ongoing reckless industrial pollution (as in Norilsk, in Photo Essay 5.2 A, B) or failure to properly shut down factories made obsolete when a more market-based economy was introduced (as in Donetsk, Ukraine, in Photo Essay 5.2C). Now, oil and gas extraction and their sale on the global market are a major source of income for Russia and several of the post-Soviet states, yet still there is little attention to environmental impacts (Photo Essay 5.2D).

Urban and Industrial Pollution

> **Learning Goal 1**
> **Environment, Development, and Urbanization: Why does this region have such a severe environmental problems, especially in urban areas?**

Urban and industrial pollution was ignored during Soviet times as cities expanded quickly—with workers flooding in from the countryside—to accommodate new industries, which generated lethal levels of pollutants. Today this region has some of the most polluted cities on the planet.

It is often difficult to link urban pollution directly to health problems because the sources of contamination are multiple and difficult to trace. Such **nonpoint sources of pollution** include untreated automobile exhaust, raw sewage, and agricultural chemicals that drain from fields into urban water supplies. Moscow, for example, is located at the center of a large industrial area that has a complex mixture of pollutants of diffuse origins. In all urban areas of the region, air pollution resulting from the burning of fossil fuels is skyrocketing as more people purchase cars and as the industrial and transport sectors of the economy continue to grow (see Thematic Overview A on page 171).

> **nonpoint sources of pollution** diffuse sources of environmental contamination, such as untreated automobile exhaust, raw sewage, or agricultural chemicals that drain from fields into water supplies

Some cities were built around industries that produce harmful by-products. The former chemical weapons manufacturing center of Dzerzhinsk is listed in the *Guinness Book of World Records* as the most chemically polluted city in the world. Men here have a life expectancy of 42 and women, 47. In the city of Norilsk, not a single living tree exists within 30 miles of the world's largest metal smelting complex (see Photo Essay 5.2B). Male life expectancy is only 50 years, despite the city's general prosperity and availability of free health care and sports clubs.

The Globalization of Nuclear Pollution

Russia and the post-Soviet states are also home to extensive nuclear pollution, and its effects have spread globally. The world's worst nuclear disaster occurred in Ukraine in 1986, when the Chernobyl nuclear power plant exploded. The explosion severely contaminated a vast area in northern Ukraine, southern Belarus, and Russia. It spread a cloud of radiation over much of Central Europe, Scandinavia, and eventually the entire planet. While precise figures are debated, 5000 people or more died, another 30,000 were disabled, and 100,000 people were evacuated from their homes.

▋▶ 278. SAFETY AT THE CENTER OF NUCLEAR POWER OPERATIONS WORLDWIDE

The Arctic Ocean and the Sea of Okhotsk in the northwestern Pacific (see the map in Figure 5.3 on pages 174–175; see Photo Essay 5.2D) are also polluted with nuclear waste dumped at sea. Although the Soviet government signed an international antidumping treaty, it later sank 14 nuclear reactors and dumped thousands of barrels of radioactive waste in these seas that flow into the Pacific and Atlantic oceans.

From 1949 to 1989, remote areas of eastern Kazakhstan served as a testing ground for Soviet nuclear devices (including more than 100 nuclear bombs). The residents were sparsely distributed and no one took the trouble to evacuate them. A museum in Kazakhstan now displays hundreds of deformed human fetuses and newborns. According to the Nuclear Threat Initiative (an international NGO), Kazakhstan has 237.2 metric tons of radioactive waste waiting for disposal. The cost of disposal and reclamation of the poisoned land will cost $1.1 billion, but only $1 million has been budgeted. One proposal is for Kazakhstan to earn the money by taking in the nuclear waste of other countries, such as France, which has large nuclear power industries. Estimates are that the country could earn from $30 to $40 billion by becoming a commercial repository for radioactive waste, and could use the profits to dispose of its own waste. But no reliably safe system has yet been found for storing nuclear waste until it is no longer radioactive, and strong opposition has been raised by environmental and human rights NGOs that point to the region's poor environmental track record.

The Globalization of Resource Extraction and Environmental Degradation

Russia and the post-Soviet states have considerable natural and mineral resources (see Figure 5.9 on page 189). Russia itself has the world's largest natural gas reserves, major oil deposits, and forests that stretch across the northern reaches of the continent. Russia also has major deposits of coal and industrial minerals such as iron ore and nickel. The Central Asian states share substantial deposits of oil and gas, which are centered on the Caspian Sea but extend east toward China. The value of all these resources

> ## Thinking Geographically
>
> *After you have read about the human impact on the Biosphere in Russia and the post-Soviet states, you will be able to answer the following questions:*
>
> **A** What is male life expectancy in Norilsk?
>
> **C** Why have many factories in this region closed since the 1990s?
>
> **D** In addition to oil pollution, from what other form of pollution has the arctic portion of this region suffered?

This region is home to some of the worst pollution on the planet. More than 70 years of industrial development with few environmental safeguards have wreaked havoc on ecosystems. Industries are attempting to become cleaner, but decades may pass before significant improvements are realized.

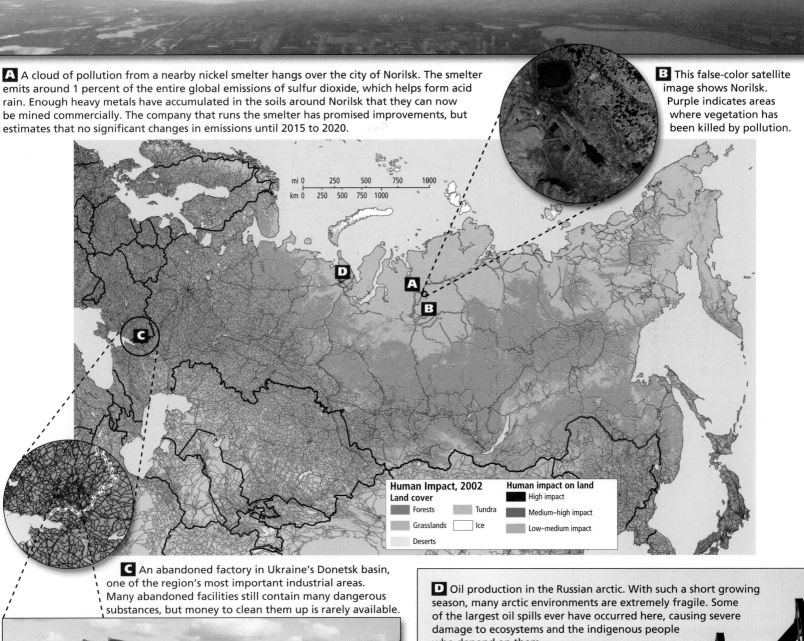

A A cloud of pollution from a nearby nickel smelter hangs over the city of Norilsk. The smelter emits around 1 percent of the entire global emissions of sulfur dioxide, which helps form acid rain. Enough heavy metals have accumulated in the soils around Norilsk that they can now be mined commercially. The company that runs the smelter has promised improvements, but estimates that no significant changes in emissions until 2015 to 2020.

B This false-color satellite image shows Norilsk. Purple indicates areas where vegetation has been killed by pollution.

Human Impact, 2002

Land cover

- Forests
- Grasslands
- Deserts
- Tundra
- Ice

Human impact on land

- High impact
- Medium–high impact
- Low–medium impact

C An abandoned factory in Ukraine's Donetsk basin, one of the region's most important industrial areas. Many abandoned facilities still contain many dangerous substances, but money to clean them up is rarely available.

D Oil production in the Russian arctic. With such a short growing season, many arctic environments are extremely fragile. Some of the largest oil spills ever have occurred here, causing severe damage to ecosystems and the indigenous people who depend on them.

is determined in the global marketplace, and their extraction affects the global environment.

Despite obvious environmental problems, cities like Norilsk (see Photo Essay 5.2A), which sits on huge mineral deposits, continue to attract foreign investment crucial to the new Russian economy. Thirty-five percent of the world's nickel supply, 10 percent of its copper, and 40 percent of its platinum are in the Norilsk area. By 2006, Norilsk Nickel, the privatized company that runs the smelter, was producing 2 percent of the Russian GDP and had attracted major investment from U.S. and European banks. The success of Norilsk Nickel led the company to globalize despite its shockingly poor environmental record. The company has now purchased mining operations in Australia, Botswana, Finland, the United States, and South Africa.

Water Issues

Water, once assumed to be inexhaustible, usable for endless purposes, yet in need of little maintenance, is now recognized by most countries in this region as a vital resource.

Rivers, Irrigation, and the Loss of the Aral Sea To the south of Russia in the Central Asian states, the glacially fed rivers of the Syr Darya and Amu Darya have long served an irrigation role, supporting commercial cotton agriculture (see the discussion of the effects of global climate change on glaciers in Chapter 1 on page 41; see also Thematic Overview B). For millions of years, the landlocked Aral Sea, once the fourth-largest lake in the world, was fed by the Syr Darya and Amu Darya rivers, which bring snowmelt and glacial melt from mountains in Kyrgyzstan. Water was first diverted from the two rivers in the 1960s when the Soviet leadership ordered its use to irrigate millions of acres of cotton in Kazakhstan and Uzbekistan. So much water was consumed by these projects that within a few years, the Aral Sea had shrunk measurably (Figure 5.4). By the early 1980s, no water at all from the two rivers was reaching the Aral Sea, and by early 2001, the sea had lost 75 percent of its volume and had shrunk into three smaller lakes. The huge fishing industry, on which the entire USSR population depended for some 20 percent of their protein, died out due to increasing water salinity.

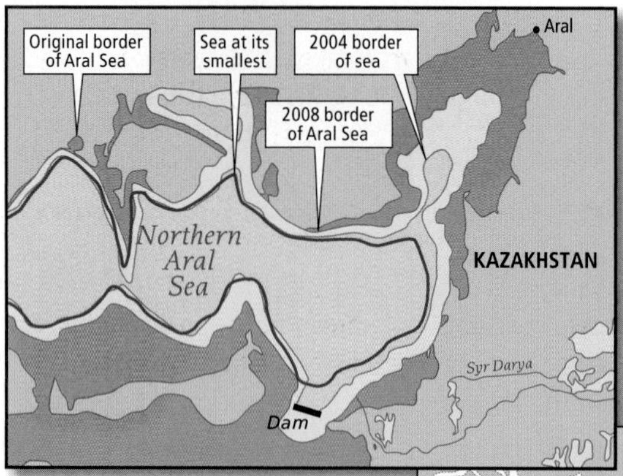

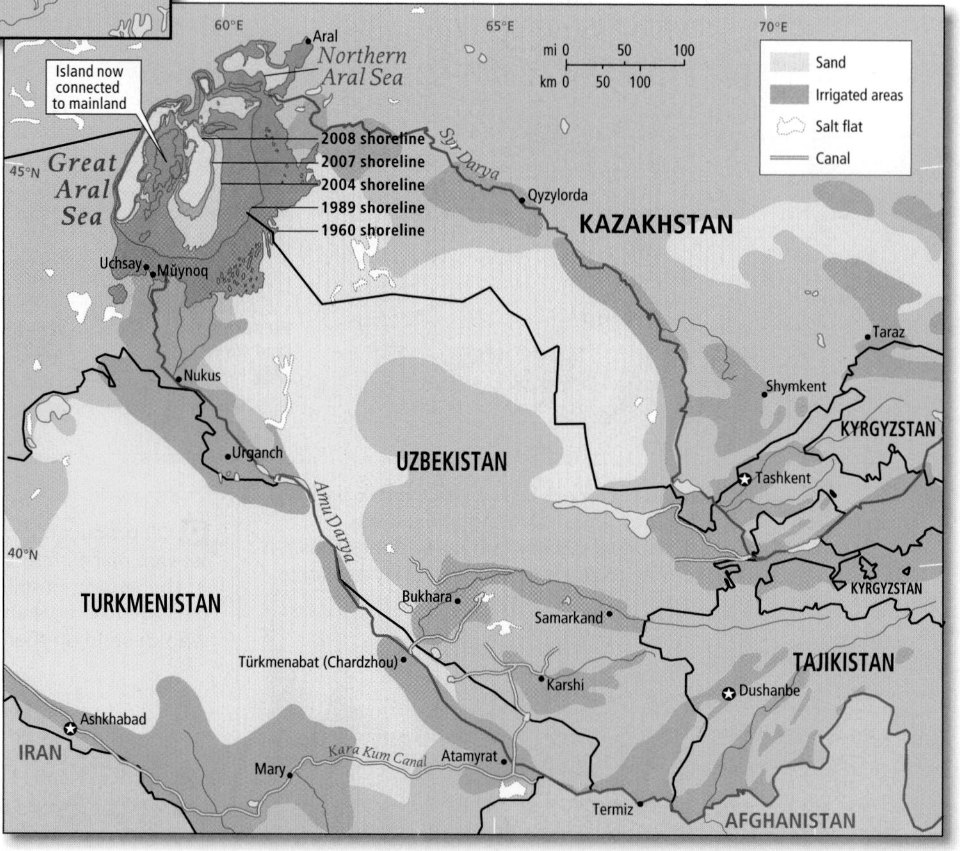

FIGURE 5.4 The decline and disappearance of the Aral Sea, 1990–2008.
(A) Once the fourth-largest lake in the world, the Aral Sea is disappearing as a result of large-scale irrigation projects in Central Asia. Since 2005, however, Kazakhstan officials have made efforts to improve the situation, and the northern Aral Sea (inset) is slowly increasing in size and depth.

Once-active port cities were marooned many kilometers from the water. The decline and disappearance of the Aral Sea has been described as the largest human-made ecological disaster on earth.

The shrinkage of the Aral Sea caused many human health problems. Winds that sweep across the newly exposed seabed pick up salt and chemical residues, creating poisonous dust storms. At the southern end of the Aral Sea in Uzbekistan, 69 of every 100 people report chronic illness, caused by air pollution and underdeveloped water and sanitation. In some villages, life expectancy is just 38 years (the national average is 66 years).

Efforts to increase water flows to the sea have been hampered by politics, because the Aral Sea is shared by the now-independent countries of Uzbekistan and Kazakhstan. Uzbekistan, in particular, wants to continue massive irrigation programs, which have made it the world's fifth-largest cotton grower. Cotton accounts for one-third of its total exports and employs 40 percent of its labor force. Nevertheless, some actions have been taken to restore the Aral Sea. Kazakhstan, which now relies on oil more than cotton for income, has built a dam to keep fresh water in the northern Aral Sea (Figure 5.4A). By 2006, the water had risen 10 feet, and the fish catch was improving. Kazakh fishers note that there is now more open public debate about what to do next regarding the Aral Sea.

‖▶ 107. DYING SEA MAKES COMEBACK

Climate Change

> **Learning Goal 2**
> **Climate Change, Food and Water:** Where in this region are food production systems and water resources most threatened by climate change? What parts of this region produce the most greenhouse gases?

Climate change is an issue in this region for many reasons. High levels of CO_2 and other greenhouse gases are produced here, but Russia's forests also play a major role in absorbing CO_2. Portions of the region are also particularly vulnerable to the negative effects of a changing climate. Much of this region has a medium level of vulnerability to the negative effects of climate change (see the map in Photo Essay 5.3 on page 182), but Central Asia, which is prone to drought and water scarcity (see Figure 5.4A), has high to extreme levels of vulnerability to climate change effects.

Although Russia has recently begun to reduce its CO_2 and other greenhouse gas emissions, it is still a major contributor of such emissions (see Figure 1.20 on page 40). Many wasteful practices date from the Soviet era, such as burning off or "flaring" natural gas that comes to the surface when oil wells are drilled. Still, in some ways, Russia has shown more willingness to limit its own emissions than other big polluters have: Russia has signed international treaties, such as the Kyoto Protocol and the Copenhagen Accord, to reduce greenhouse gas emissions.

Central Asian dependence on irrigated agriculture, using water from rivers that are fed by glaciers in the Pamirs and other mountains in Kyrgyzstan and Tajikistan (see Photo Essay 5.3 A, B, C) leaves the region vulnerable to global climate change.

Recent decades have seen more extreme temperature ranges and decreased rain- and snowfall, which have caused glaciers to shrink. If the glaciers no longer act as water storage systems, supplying melted ice flow to the rivers, the rivers may run dry during the summer, when irrigation is most needed for growing crops. Because most rain falls in the mountains in the winter and spring, Central Asia's agricultural systems would either have to adapt to a spring growing season (winters are too cold), or farmers would have to store water for use in the summer. Either proposition demands complex and expensive changes on a massive scale.

Better Times Ahead?

While environmental pollution remains serious throughout this region, there may be reasons for optimism in the post-Soviet era. New patterns of economic development could bring rising living standards, especially for urban residents who could pressure governments for higher environmental standards. Moreover, while freedom of expression and the press is still limited throughout the region, it is at least better than it was in Soviet times. This opens up the opportunity to expose the worst cases of pollution and environmental negligence through the media. For example, in Sochi, Russia, a Black Sea resort town long cherished for its natural amenities (Sochi was chosen to host the 2014 winter Olympic games), numerous environmental protests about the exposure of high levels of urban pollution have gained global attention in recent years (see Reasons for Optimism A on page 205). Similarly, much of the information presented above has come to light as a result of greater press freedom since the collapse of the Soviet Union.

THINGS TO REMEMBER

1. The largest country in the world, Russia, is nearly twice the size of the second-largest country, Canada. The region encompasses numerous landforms and associated climates.

2. Soviet ideology held that nature was the servant of industrial and agricultural progress, and that humans should dominate nature—an ideology that wreaked havoc on ecosystems and human settlements.

3. Industries are attempting to become cleaner, but decades may pass before significant improvements are realized.

4. **Learning Goal 1: Environment, Development, and Urbanization** Urban and industrial pollution was ignored during Soviet times as cities expanded quickly to accommodate new industries.

5. Global climate change is also a rising concern, with Russia a major contributor of greenhouse gases.

6. **Learning Goal 2: Climate Change, Food, and Water** Central Asia is especially vulnerable to climate change given its dependence on irrigated agriculture that uses water from rivers fed by melting glaciers.

Vulnerability to climate change is at medium levels throughout much of this region, but at high and extreme levels in Central Asia, which is highly exposed to drought and water scarcity. Both problems will worsen as temperatures rise and glaciers in the Pamirs and other mountains melt. The dependence of so many Central Asians on agriculture makes them highly sensitive to these problems, and widespread poverty combined with poorly developed disaster response systems reduces overall resilience to these and other disturbances.

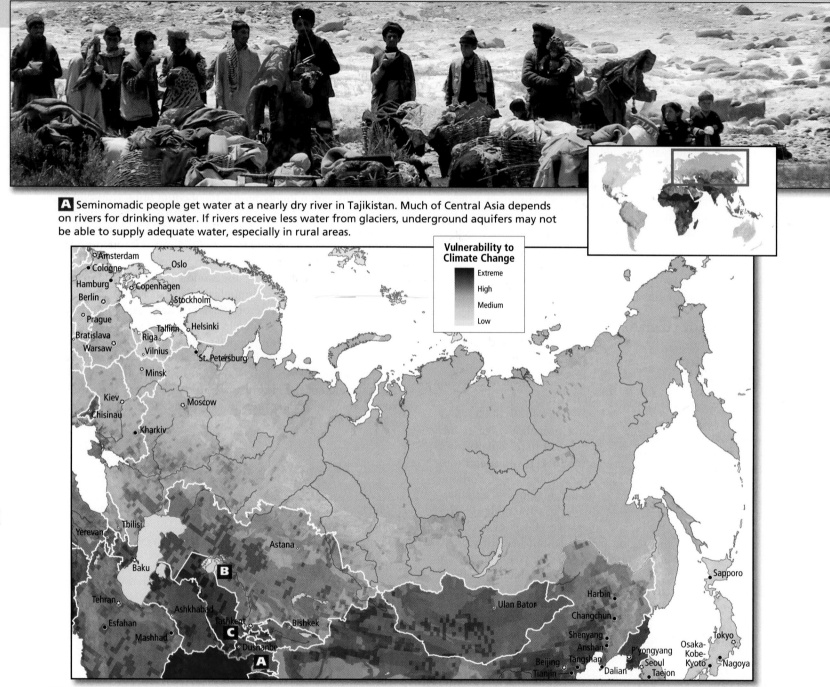

A Seminomadic people get water at a nearly dry river in Tajikistan. Much of Central Asia depends on rivers for drinking water. If rivers receive less water from glaciers, underground aquifers may not be able to supply adequate water, especially in rural areas.

Vulnerability to Climate Change

Extreme
High
Medium
Low

B A fishing boat stranded in what used to be part of the Aral Sea, which has shrunk dramatically due to overuse of the river waters that feed the sea irrigation of cotton and food crops. Even less water will reach the Aral Sea as glaciers continue melting.

C Workers in a cotton field in Uzbekistan. Forty-four percent of Uzbeks work in agriculture, and cotton is the country's leading export. If irrigation water becomes less available, the economy would be hurt and millions of Uzbeks, 33 percent of whom live below the poverty line, would be out of work.

Human Patterns over Time

The core of the entire region has long been European Russia, the large and densely populated homeland of the ethnic Russians. Expanding gradually from this center, the Russians conquered a large area inhabited by a variety of other ethnic groups. These conquered territories remained under Russian control as part of the Soviet Union (1917–1991), which attempted to create an integrated social and economic unit out of the disparate territories. The breakup of the Soviet Union has diminished Russian domination throughout this region, though Russia's influence is still considerable.

The Rise of the Russian Empire

For thousands of years, the militarily and politically dominant people in the region were **nomadic pastoralists** who lived on the meat and milk provided by their herds of sheep, horses, and other grazing animals, and used animal fiber to make yurts, rugs, and clothing. As possibly the first people to domesticate horses, their movements followed the changing seasons across the wide grasslands that stretch from the Black Sea to the Central Siberian Plateau. The nomads would often take advantage of their superior

horsemanship and hunting skills to plunder settled communities. To defend themselves, permanently settled peoples gathered in fortified towns.

Towns arose in two main areas: the dry lands of greater Central Asia and the forests of the Caucasus, Ukraine, and Russia. As early as 5000 years ago, Central Asia had settled communities that were supported by irrigated croplands (see **Timeline A** on page 184). These communities were enriched by trade along what became known as the Silk Road, that vast, ancient, interwoven ribbon of major and minor trading routes between the Pacific coast of China and the Mediterranean, with lesser connections to southern China, Russia, Southeast Asia, the Indian Ocean, the Black Sea, the Arabian Peninsula, sub-Saharan Africa, Turkey, and Europe (**Figure 5.5**; see also Figure 5.3 on page 174).

About 1500 years ago, the **Slavs**, a group of farmers including those known as the Rus (of possible Scandinavian origin), emerged in what is now Poland, Ukraine, and Belarus. They moved east, founding numerous settlements, including the towns of Kiev in about 480 and Moscow in 1100. By 600, Slavic trading towns were located along all the rivers west of the Ural Mountains. The Slavs prospered from a lucrative trade route over land and along the Volga River that connected Scandinavia (North Europe) and Central and Southwest Asia (via Constantinople, modern-day Istanbul) (see Figure 5.5). Powerful kingdoms developed in Ukraine and European Russia. In the mid-800s, Greek missionaries introduced both Christianity (now known as "Orthodox Christianity") and the Cyrillic alphabet, still used in most of the region's countries (see Timeline B).

> **nomadic pastoralists** people whose way of life and economy are centered on tending grazing animals who are moved seasonally to gain access to the best pasture
>
> **Slavs** a group of farmers who originated between the Dnieper and Vistula Rivers in modern-day Poland, Ukraine, and Belarus

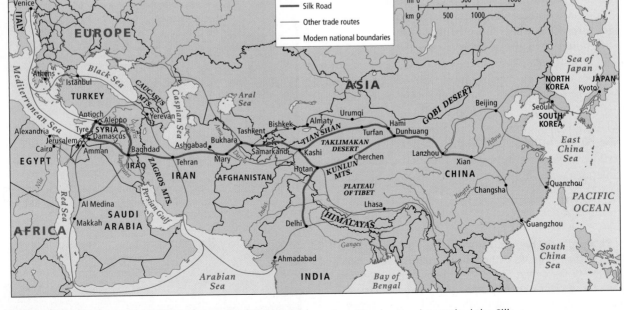

FIGURE 5.5 The ancient Silk Road and related trade routes. Merchants who worked the Silk Road rarely traversed the entire distance. Instead, they moved back and forth along part of the road, trading with merchants to the east or west.

A The fort at Bukhara, Uzbekistan, founded in 500 BCE along the Silk Road.

B An Orthodox Christian monastery in Russia, founded around 1500 CE.

C The summer palace of Czar Peter the built from 1714 to 1755.

4000 BCE   500 BCE   500 CE   1600 CE   1700 CE

3000 B.C.E.
Settled communities in Central Asia, supported by irrigated agriculture and trade on the Silk Road

500 B.C.E.
Fort at Bukhara founded

500 C.E.
Slaves emerge

1500 C.E.

1598–1689
Russian annexation of western Siberia

VISUAL TIMELINE OF RUSSIA AND THE POST-SOVIET STATES

Thinking Geographically

After you have read about the human history of Russia and the post-Soviet states, you will be able to answer the following questions:

Ⓐ In addition to trade along the Silk Road, what supported central Asian cities?

Ⓑ From where was Orthodox Christianity introduced into this region?

> **Mongols** a loose confederation of nomadic pastoral people centered in East and Central Asia, who by the thirteenth century had established by conquest an empire stretching from Europe to the Pacific

In the twelfth century, the Mongol armies of Genghis Khan conquered the forested lands of Ukraine and Russia. The **Mongols** were a loose confederation of nomadic pastoral people centered in East and Central Asia. Moscow's rulers became tax gatherers for the Mongols, dominating neighboring kingdoms and eventually growing powerful enough to challenge local Mongol rule. In 1552, the Slavic ruler Ivan IV ("Ivan the Terrible") conquered the Mongols, marking the beginning of the Russian empire. St. Basil's Cathedral, a major landmark in Moscow, commemorates the victory (Figure 5.6).

By 1600, Russians centered in Moscow had conquered many former Mongol territories, integrating them into their growing empire and extending it to the west as shown in Figure 5.7. The first major non-Russian area to be annexed was western Siberia (1598–1689). Russian expansion into Siberia (and even into North America) resembled the spread of European colonial powers throughout Asia and the Americas. Russian colonists forcibly took land and resources from the Siberian populations, and treated the people of those cultures as inferiors. Moreover, massive migrations of laborers from Russia to Siberia meant that indigenous Siberians were vastly outnumbered by the eighteenth

FIGURE 5.6 St. Basil's Cathedral, Moscow. Now the most recognized building in Russia, St. Basil's Cathedral was built between 1555 and 1561 to commemorate the defeat of the Mongols by Ivan the Terrible, the first Russian czar. A central chapel, capped by a pyramidal tower, stands amid eight smaller chapels, each with colorful "onion domes." Each chapel commemorates a saint on whose feast day Czar Ivan had won a battle.

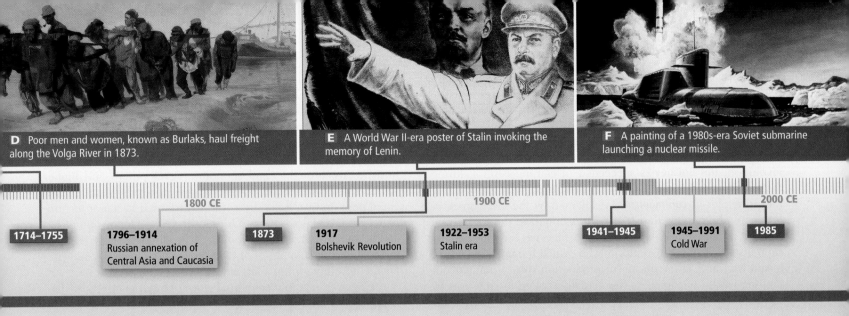

D Poor men and women, known as Burlaks, haul freight along the Volga River in 1873.

E A World War II-era poster of Stalin invoking the memory of Lenin.

F A painting of a 1980s-era Soviet submarine launching a nuclear missile.

| 1714–1755 | 1796–1914 Russian annexation of Central Asia and Caucasia | 1873 | 1917 Bolshevik Revolution | 1922–1953 Stalin era | 1941–1945 | 1945–1991 Cold War | 1985 |

1800 CE   1900 CE   2000 CE

D How does this painting convey the poverty of the Burlaks?

E How many Soviets were killed during World War II?

F How did the Cold War strain the USSR?

century. By the mid-nineteenth century, Russia had also expanded its control to the south in order to gain control of Central Asia's cotton crop, its major export.

The Russian empire was ruled by a powerful leader, the **czar**, who lived in splendor (along with a tiny aristocracy) while the vast majority of the people lived short brutal lives in poverty (see Timeline C and D). Many Russians were *serfs* who were legally bound to live on and farm land owned by an aristocrat. If the land was sold, the serfs were transferred with it. Serfdom was ended legally in the mid-nineteenth century. However, the brutal inequities of Russian society persisted into the twentieth

> **czar** title of the ruler of the Russian empire; derived from the word "caesar," the title of the Roman emperors

century, fueling opposition to the czar. By the early twentieth century, a number of violent uprisings were underway.

The Communist Revolution and Its Aftermath

Periodic rebellions against the czars and elite classes took place over the centuries. By the mid-nineteenth century, these rebellions were being led not only by serfs, but by members of the tiny educated and urban middle class. Resistance to repression grew in the pre–World War I era. Then, in 1917, at the height of Russian suffering during World War I, Czar Nicholas II was overthrown in a revolution. What followed was a civil war

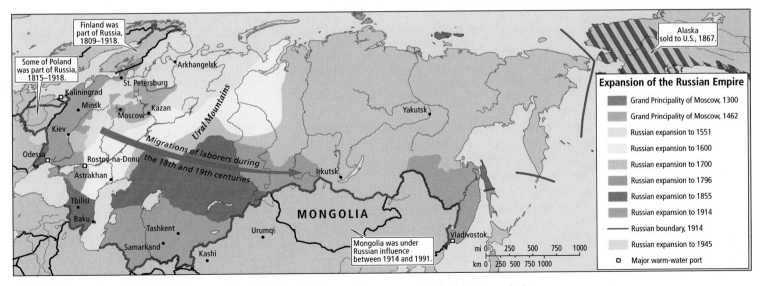

Finland was part of Russia, 1809–1918.

Some of Poland was part of Russia, 1815–1918.

Alaska sold to U.S., 1867.

Mongolia was under Russian influence between 1914 and 1991.

Expansion of the Russian Empire
- Grand Principality of Moscow, 1300
- Grand Principality of Moscow, 1462
- Russian expansion to 1551
- Russian expansion to 1600
- Russian expansion to 1700
- Russian expansion to 1796
- Russian expansion to 1855
- Russian expansion to 1914
- Russian boundary, 1914
- Russian expansion to 1945
- Major warm-water port

FIGURE 5.7 Russian imperial expansion, 1300–1945. A long series of powerful rulers expanded Russia's holdings across Eurasia to the west and east. Expansion was particularly vigorous after 1700 when Siberia, the largest area, was acquired.

between rival factions with different ideas about how the revolution should proceed. Eventually the revolution brought a complete restructuring of the Russian economy and society according to what at first promised to be a more egalitarian model—a model that was extended to adjacent countries.

Of the disparate groups that coalesced to launch the Russian revolution, the **Bolsheviks** succeeded in gaining control during the post-revolution civil war. The Bolsheviks were inspired by the principles of **communism** as explained by the German revolutionary philosopher Karl Marx. Marx criticized the societies of Europe as inherently flawed because they were dominated by **capitalists**—the wealthy minority who owned the factories, farms, businesses, and other means of production. Marx pointed out that the wealth of capitalists was actually created in large part by workers. The poverty of the workers was a result of their labor being undervalued by the capitalists. Under communism, workers were called upon to unite to overthrow the capitalists, take over the means of production (farms, factories, services), and establish a completely egalitarian society without government or currency. The philosophy held that people would work out of a commitment to the common good, sharing whatever they produced so that each had their basic needs met.

One Bolshevik leader, Vladimir Lenin, argued that the people of the former Russian empire needed a transition period in which to realize the ideals of communism. Accordingly, Lenin's Bolsheviks formed the **Communist Party**, which set up a powerful government based in Moscow. Lenin believed that the government should run the economy, taking land and resources from the wealthy and using them to benefit the poor majority.

The Stalin Era After Lenin suffered a series of strokes in 1922, Joseph Stalin stepped in to reorganize the society and economy in a much more authoritarian style than Lenin had envisioned. His 31-year rule of the Soviet Union brought a mixture of brutality and revolutionary change that set the course for the rest of the Soviet Union's history. He sought to increase general prosperity through rapid industrialization made possible by a **centrally planned**, or **socialist**, **economy**. The state would own all real estate and means of production, while government bureaucrats in Moscow would direct all economic activity. This system became known as the *command economy*. In this economy, central control was used to locate all factories, apartment blocks, and transport infrastructure, and to manage all production, distribution, and pricing of products. The idea was that under a socialist system the economy would grow more quickly, which would hasten the transition to the idealized communist state in which everyone shared equally. This notion was reflected in the new name chosen for the country: the Union of Soviet Socialist Republics ("Soviet" means, roughly, "council").

Stalin used the powers of the command economy with fervor and cruelty. The centerpiece of Stalin's vision was massive

Bolsheviks a faction of communists who came to power during the Russian Revolution

communism an ideology, based largely on the writings of the German revolutionary Karl Marx, that calls on workers to unite to overthrow capitalists and establish an egalitarian society where workers share what they produce

capitalists usually a wealthy minority that owns the majority of factories, farms, businesses, and other means of production

Communist Party the political organization that ruled the USSR from 1917 to 1991; other communist countries, such as China, Mongolia, North Korea, and Cuba, also have communist parties

centrally planned, or **socialist**, **economy** an economic system in which the state owns all real estate and means of production, while government bureaucrats direct all economic activity, including the locating of factories, residences, and transportation infrastructure

Cold War the contest that pitted the United States and western Europe, espousing free market capitalism and democracy, against the USSR and its allies, promoting a centrally planned economy and a socialist state

government investment in gigantic development projects, such as factories, dams, and chemical plants, some of which are still the largest of their kind in the world. To increase agricultural production, he forced farmers to join large government-run collectives.

Stalin's strategy of government control met many of his expectations. The labor of millions, guided by thousands of bureaucrats, brought rapid economic development and higher standards of living. Schools were provided for previously uneducated poor and rural children, and contributed to major technological and social advances. During the Great Depression of the 1930s, the Soviet Union's industrial productivity grew steadily even while the economies of other countries stagnated, allowing for the development of a large military.

However, Stalin's economic development model also had deep flaws. With production geared largely toward heavy industry (the manufacture of machines, transport equipment, military vehicles, and armaments), less attention was paid to the demand for consumer goods and services that could have further improved living standards.

The most destructive aspect of Stalin's rule, however, was his ruthless use of the secret police, starvation, and mass executions to silence anyone who dared to oppose him. Those who resisted were relocated to Siberia, where they were forced to work in prison labor camps. Here they mined for minerals and then built and worked in newly constructed industrial cities. Millions were executed or died from overwork under harsh conditions. Between 34 and 60 million people were killed as a result of Stalin's policies.

World War II and the Cold War

During World War II, the Soviet Union did more than any other country to defeat the armies of Nazi Germany (see Timeline E). A failed attempt to conquer the Soviet Union exhausted Hitler's war machine. In the process of defeating Germany on the Eastern Front, 23 million Soviets were killed, more than all the other European combatants combined. After the war, Stalin was determined to erect a buffer of allied communist states in Central Europe that would be the battleground of any future war with Europe. Because they were unwilling to risk another war, the leaders of the United States and the United Kingdom ceded control of much of Central Europe to Russia while they busied themselves with reconstructing Europe. However, when Stalin's intention to utterly dominate Central Europe became clear, the United States and its allies organized to stop the Soviets from extending their power even further into Europe and elsewhere. The result was the **Cold War**, a nearly 50-year-long global geopolitical rivalry that pitted the Soviet Union and its allies against the United States and its allies (Figure 5.8).

A wide variety of economic and political problems led to the eventual collapse of the Soviet Union. Fundamental flaws with Soviet central planning led to innumerable inefficiencies

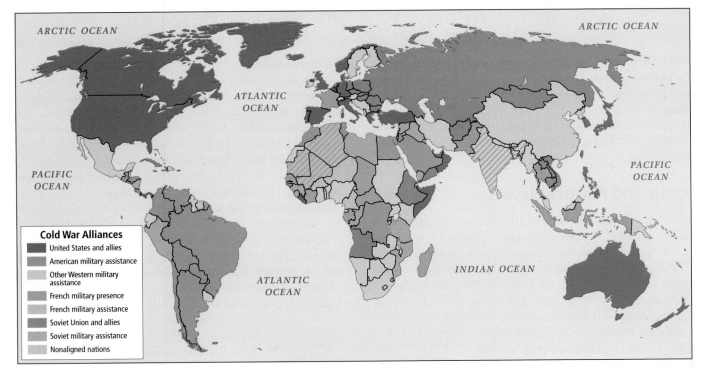

FIGURE 5.8 The Cold War in 1980. In the post–World War II contest between the Soviet Union and the United States, both sides enticed allies through economic and military aid. The group of countries militarily allied with the Soviet Union was known as the Warsaw Pact. Some countries remained unaligned.

and chronic mismanagement. Many problems also were related to the diversion of scarce resources to the military and away from much-needed economic and social development. Some of this related to the Cold War and the arms race with the United States (see Timeline F). Soviet efforts to promote communism in China, Mongolia, North Korea, Afghanistan, Cuba, Vietnam, Nicaragua, and various African nations were also expensive, as was extensive political, economic, and military coercion in Central Europe and Central Asia. As problems multiplied, so did resistance to Soviet control, especially in Central Europe in the late 1980s. Eventually, this resistance spread to the republics that made up the Soviet Union itself, and by 1991 the USSR had formally dissolved.

A war in Afghanistan also helped to stretch the USSR past the breaking point (see Photo Essay 5.4A on page 195). In the 1970s in Afghanistan, which lies just to the south of the Central Asian states, there was contention between a small group who favored a Western-style democratic government, those who favored communism, and Muslims who favored a theocratic state based on Islamic law. The Soviets feared that the strongly anti-communist Muslim movement would influence Central Asian Soviet republics to unite under Islam and rebel against the USSR. In response, the USSR backed an unpopular communist government in return for permission to build Soviet military bases in Afghanistan. This brought strong resistance from the Muslim fundamentalist Mujahedeen, and in 1979 Russia launched a war to support the Afghan government. A bloody, decade-long conflict ensued, concluding with the Soviets being defeated by highly motivated Afghan guerilla fighters who were financed, armed, and trained by the United States. Within

the USSR, the military, which had long been a centerpiece of national pride and political control, was in disgrace. Within 2 years of the Soviet withdrawal from Afghanistan, the USSR disintegrated, largely due to economic strains that had been made worse by the war. The Soviet–Afghanistan conflict and the aftermath are discussed further in Chapter 8.

THINGS TO REMEMBER

1. As early as 5000 years ago, Central Asia had settled communities that were supported by irrigated croplands and enriched by trade along what became known as the Silk Road.

2. The Bolsheviks, a group inspired by the principles of communism, were the dominant leaders of the Russian Revolution of 1917. Their goal, never achieved, was an egalitarian society where people would work out of a commitment to the common good, sharing whatever they produced.

3. Taking control in 1922 while Lenin was ill, Joseph Stalin brought to the country a mixture of brutality and revolutionary change that set the course for the rest of the Soviet Union's history. He established a centrally planned, or socialist, economy in which the state owned all property and means of production, while government bureaucrats in Moscow directed all economic activity. This system became known as the command economy.

4. In the aftermath of World War II, a nearly 50-year-long global geopolitical rivalry known as the Cold War pitted the Soviet Union and its allies against the United States and its allies.

II CURRENT GEOGRAPHIC ISSUES

The goals of the Soviet experiment begun in 1917 were unique in human history: to reform quickly and totally an entire human society for the benefit of the people. The Soviet Union's collapse brought an even more rapid shift from centrally planned economies to market economies. The transition remains incomplete and for many, life is proceeding amid economic, political, and social uncertainty.

Economic and Political Issues

When the Soviet Union disbanded in 1991 and the leaders of the post-Soviet states embraced various versions of a market economy, everyone had to learn how capitalism works. The first decade of this new era was characterized by a disorderly tumble into a market economy that gave the advantage to a few high-level Soviet bureaucrats who had run government industries in the old command economy. They often acquired previously publicly held assets at fire-sale prices, thereby becoming fabulously rich very quickly, and assumed the role of **oligarchs**—individuals who are so wealthy that they wield enormous, often clandestine, political control.

Oligarchs continue to exercise power in government and private enterprise. Recently, an attempt was made to refine their economic power by creating a master of business administration (MBA) program on a lavish new campus near Moscow. With a "who's who" of oligarchs serving as funders, board members, and even teachers, the Moscow School of Management, Skolkovo, opened in 2009. The curriculum addresses the fact that many of the oligarchs of the 1990s lacked the management skills necessary to lead firms that were profitable, responsible, and sustainable. Its emphasis is on entrepreneurial leadership honed in actual business situations.

Oil and Gas Development: Fueling Globalization

Learning Goal 3
Globalization and Development:
How have economic reform and globalization changed patterns of development in this region since the fall of the Soviet Union?

Natural gas and crude oil have emerged as the region's most lucrative exports, introducing a new wave of globalization and fossil fuel–dependent economic development. While control of Russia's oil and gas resources now lies largely in the hands of the Russian government, the struggle to control Central Asia's oil and gas resources has become global and could be a source of conflict for years to come (Figure 5.9).

By the year 2000 (after an initial decade of economic instability), rising revenues from oil and gas began to finance Russia's economic recovery. In response, Russia's central government

> **oligarchs** in Russia, those who acquired great wealth during the privatization of Russia's resources and who use that wealth to exercise power
>
> **Gazprom** in Russia, the state-owned energy company; it is the tenth-largest oil and gas entity in the world
>
> **Group of Eight (G8)** an organization of eight countries with large economies: France, the United States, Britain, Germany, Japan, Italy, Canada, and Russia

tightened control over the entire oil and gas sector. Today, Russia is the world's largest exporter of natural gas, and taxes on energy companies fund more than half of the federal budget (see Thematic Overview C and Figure 5.10). While this does guarantee that at least some of the profits from the new fossil fuel–based industries are broadly distributed, the remaining profits are concentrated in the hands of a few oligarchs.

Overall, Russia's relationships with Europe are stabilizing further as the lucrative oil and gas trade continues. Russia's majority state-owned **Gazprom**, one of the world's largest oil and gas companies, now dominates Russia's relationship with the EU. About 60 precent of Gazprom's profits are from sales to EU member states, which in turn receive about 25 percent of their natural gas from Gazprom. Russia occasionally tries to use its gas exports to manipulate the politics and economies of countries in Central Europe as well as in Ukraine and Belarus. These countries are all much more dependent on gas from Gazprom than is the rest of Europe. Russia has in the past curtailed access to Gazprom's gas in response to policies those countries enact that Russia opposes. Recently, Gazprom has tried to improve its image in the EU by sponsoring major European soccer teams (see Reasons for Optimism B).

Meanwhile, in Central Asia, stability has yet to emerge as a tug-of-war has evolved between Russia and foreign multinational energy corporations over the rights to develop, transport, and sell this area's oil and gas resources. The new Central Asian states do not have the capital to develop the resources themselves, yet individuals and special interests in the Central Asian countries are reaping enormous financial rewards from selling to outsiders the rights to exploit oil and gas. (Figure 5.10).

Contention hinges primarily on pipeline routes (see Figure 5.9). Russia, the United States, the European Union, Turkey, China, and India have all developed or proposed various routes, and by 2009 China was building a pipeline from the east side of the Caspian Sea to China's Xinjiang Uygur Autonomous region, a distance of 1900 miles (3000 km). By 2006, the United States had installed military bases to protect its interests in Georgia and Uzbekistan, both of which have pipelines to Europe. Russia has built pipelines that traverse its own territory and reach into Europe. Each of these parties has made numerous efforts to encourage Central Asian countries to use their pipelines instead of those of their competitors.

II ▶ 108. ENERGY REVENUES AND CORRUPTION INCREASE IN RUSSIA

Russia's Relations with the Global Economy Both the European Union and the United States want to ensure that Russia's oil and gas wealth does not finance a return to the hostile relations of the Cold War. For this reason, Russia was invited to join the **Group of Eight (G8)**, an organization of eight countries with large economies, as well as the World Trade Organization.

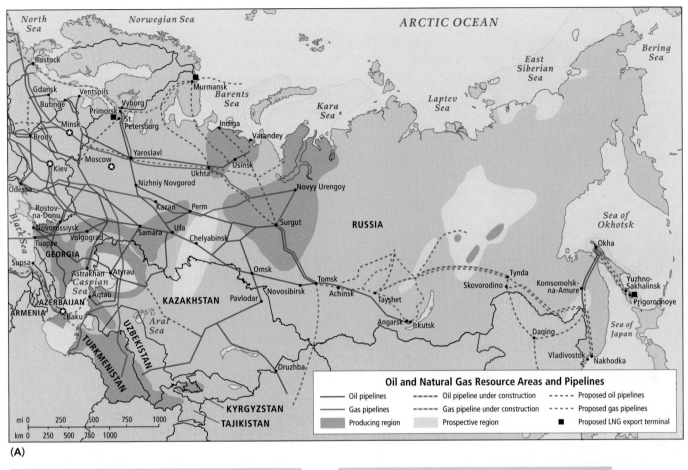

(A)

Oil, 2006	
Oil reserves	**80 billion barrels**
Oil reserves, as percentage of world	7 percent
Saudi Arabian reserves	264 billion barrels
U.S. reserves	30 billion barrels
Oil production	**10 million barrels per day**
Oil production, as percentage of world	12 percent
U.S. oil production	7 million barrels per day
Oil exports	**7 million barrels per day**
Oil exporter, rank	2
Oil exports, to U.S.	370,000 barrels per day

(B)

Natural Gas, 2006	
Gas reserves	**48 trillion cubic meters**
Gas reserves, as percentage of world	26 percent
Iranian reserves	28 trillion cubic meters
U.S. reserves	6 trillion cubic meters
Gas production	**612 billion cubic meters**
Gas production, as percentage of world	21 percent
U.S. gas production	524 billion cubic meters
Gas exports	**263 billion cubic meters**
Gas exporter, rank	1
Gas exports, to Europe	151 billion cubic meters

(C)

FIGURE 5.9 Oil and natural gas: Russia and the post-Soviet states' resources and pipelines. For additional information and maps, see http://www.eia.doe.gov/emeu/cabs/ and click on Newly Independent States on the map.

Looking to form an economic counterweight to the United States and the European Union, Russia has recently been meeting with Brazil, India, and China to form a trade consortium known as BRIC. All four countries have had spectacular economic growth over the last decade, though Russia is arguably the weakest and least diverse economy in the grouping due to its dependence on fossil fuel exports. In June of 2009, Russia hosted the first meeting of BRIC.

Economic Reforms in the Post-Soviet Era

To fully understand the transition from an economy run by bureaucrats and workers educated in the ideas of Marx, Lenin, and Stalin, to an economy dominated by oligarchs and entrepreneurs, we must first know more about the Soviet institutions that were previously in place.

The Former Command Economy Although the long-term goal of the command economy was to achieve a communist society, the shorter-term goal was to end the severe deprivation suffered under the czars. To a large extent, the Soviets met this goal. Their economy grew rapidly until the 1960s, and abject poverty was largely eradicated. Basic necessities such as housing, food, health care, education, and transportation were all provided for free or at very low cost—a remarkable accomplishment for any country.

FIGURE 5.10 Pipeline construction in Kazakhstan. Large oil and gas reserves are driving rapid economic growth in Kazakhstan. GDP per capita (PPP) nearly tripled between 2000 and 2008, and the government plans for similar growth in the future, based largely on pipeline deals signed with companies from China, Russia, the EU, and the United States.

Thinking Geographically: What does the amount of heavy equipment shown in the photo suggest about pipeline construction in Kazakhstan?

Nonetheless, the Soviet centrally planned economy was notably less efficient than market economies in allocating resources. Because of the lack of competition in the economy, producers had no incentive to use more efficient production methods or to come up with products that were more effective or attractive. Quality also suffered because hard work and innovation in the workplace were rarely rewarded. Promotions generally went to those with connections in the Communist Party. As a result, most consumer goods, such as cars or washing machines, were of poor quality and available only at high cost to a privileged few. Until the 1980s, punishment was severe for privately producing and selling (on the *black market*) food, liquor, or consumer products.

During Mikhail Gorbachev's tenure as president of the USSR (1985 to 1991), a private market economy began to be permitted. Under an overall policy of **perestroika** or "restructuring," private producers of food, especially meat, were allowed to legally sell their goods, but their prices were beyond the means of most. Gorbachev also introduced policies collectively known as **glasnost** or "openness," which encouraged greater transparency, openness, and publicity in the workings of all levels of Soviet government.

▐▐ ▶ 104. FORMER SOVIET UNION LAUNCHED SPACE AGE 50 YEARS AGO

Privatization and the Lifting of Price Controls The Soviet economy consisted almost entirely of industries owned and operated by the government. These have now been sold to private companies or individuals in a process called **privatization**. The hope is that they would be run more efficiently in a competitive free market setting. By 2000, approximately 70 percent of Russia's economy was in private hands, an earthshaking change from the 100 percent state-owned economy of 1991.

Another major reform has been the abandonment of government price controls that once kept goods affordable to all. Instead, prices are now determined by private business owners. The lifting of price controls led to skyrocketing prices in the 1990s for the many goods that were in high demand but also in short supply. While a tiny few grew rich, many people had to use their savings to pay for basic necessities such as food. Eventually, as opportunities opened and competition developed, the supply of goods increased and prices fell. But in the interim, many people suffered.

Unemployment and the Loss of Social Services Since becoming privatized, most formerly state-owned industries have cut many jobs in an attempt to compete against more-efficient foreign companies. Losing a job is especially devastating in a former Soviet country because one also loses the subsidized housing, health care, and other social services that were often provided along with the job. The new companies that have emerged rarely offer benefits to employees. There is little job security because most small private firms appear quickly and often fail. Discrimination is also a problem, given the absence of equal opportunity laws. Job ads often contain wording such as "only attractive, skilled people under 30 need apply."

Unemployment figures for the region vary widely but tend toward the high end, as has been common since the end of the Soviet era. By early 2010, official unemployment rates were just 1.9 percent in Belarus and 9.2 percent in Russia, but 27 percent in parts of Caucasia and Central Asia and in some recent years in Turkmenistan as high as 60 percent. Actual unemployment may be higher because many remaining state-owned firms cannot pay employees still listed as workers. The rate of **underemployment**, which measures the number of people who are working too few hours to make a decent living or who are highly trained yet working at menial jobs, is even higher in all of these countries.

The Difficult Legacies of Soviet Regional Development Schemes One of the reasons

perestroika literally, "restructuring"; the restructuring of the Soviet economic system in the late 1980s in an attempt to revitalize the economy

glasnost literally, "openness"; the policies instituted in the late 1980s under Mikhail Gorbachev that encouraged greater transparency, openness, and publicity in the workings of all levels of Soviet government

privatization the sale of industries that were formerly owned and operated by the government to private companies or individuals

underemployment the condition in which people are working too few hours to make a decent living or are highly trained but working at menial jobs

The heaviest concentrations of industrial sites are near natural resources.

Baikal-Amur Mainline (BAM)

The **Trans-Siberian Railroad** is the chief link between Moscow and the east.

Industrial Regions

▲ Ferrous ores and metals   ▬▬ Main trunk line, Trans-Siberian Railroad

■ Nonferrous metals   ─── Other railroads

▰ Industrial area

mi 0   250   500   750   1000

km 0   250   500   750   1000

FIGURE 5.11 Principal industrial areas and land transport routes of Russia and the post-Soviet states. The industrial, mining, and transport infrastructure is concentrated in European Russia and adjacent areas. The main trunk of the Trans-Siberian Railroad and its spurs link industrial and mining centers all the way to the Pacific, but the frequency of these centers decreases with distance from the borders of European Russia.

that so many state-owned firms have had to cut jobs is that they were components of largely failed regional development schemes. Numerous huge industrial areas were located in far-flung eastern portions of the Soviet Union, in large part to facilitate the political control of these areas. These projects were always held back by the vast distances and challenging physical geography that separated them from the main population centers in the west of the region. Because the region's rivers run mainly north-south, there was a need for east-west-running land transport systems such as railroads and highways. These proved exceedingly expensive to build and maintain, and many have fallen into disrepair in the post-Soviet era. Only one poorly paved road, and only one main rail line—the Trans-Siberian Railroad (**Figure 5.11**)—runs the full east-west length, connecting Moscow with Vladivostok, the main port city on the Russian Pacific coast.

By contrast, the road and rail network is fairly dense within western Russia, Belarus, and Ukraine, with Moscow as the main hub. This has had the effect of concentrating new economic development in these western areas. Hence, the industrial cities

of eastern Russia have lost many jobs and their populations are shrinking).

Small Businesses and Institutionalized Corruption Crippling bureaucratic tangles and corruption make starting a business outlandishly expensive in this region. A business license officially costs under $100, but bribes push actual costs to as much as $10,000. Even finding space is very difficult. The government still owns many buildings, but efforts to fairly allot space also end in bribery. Even those who make it through the labyrinth of setting up a business must then face protection racketeers. The result is that many businesses have to scramble just to stay afloat, cutting corners wherever possible (see Thematic Overview D). And yet, small businesses may indeed eventually be the economic salvation of this region, as they are in so many countries where they are the chief engine of job growth. But for this to happen, high-ranking officials will need to come under unrelenting pressure to combat corruption.

The Growing Informal Economy To some extent, the new informal economy in the region is an extension of the old one that flourished under communism. The black market of that time

FIGURE 5.12 The informal economy. Women inspect shirts for sale by a street vendor in Moscow before the winter holidays. Like almost all businesses in Moscow, street vendors pay "protection" money to gangsters. A much smaller number pay taxes.

Thinking Geographically: Why are informal economy enterprises generally unpopular with governments?

was based on currency exchange and the sale of hard-to-find luxuries. In the 1970s, for example, savvy Western tourists could enjoy a vacation on the Black Sea paid for by a pair or two of smuggled Levi's blue jeans and some Swiss chocolate bars. Today, many people who have lost stable jobs due to privatization now depend on the informal economy for their livelihood (Figure 5.12).

Workers in the informal economy tend to be very young unskilled adults, retirees on miniscule pensions, or those with only a low-level education. The majority of these workers operate out of their homes, selling cooked foods, vodka made in their bathtubs, clothing, or electronics they have smuggled into the country. Such a large percentage of the economy is now in the informal sector that in many countries of the region, people may actually be better off financially than official GDP per capita figures suggest.

Despite the fact that the informal economy helps people survive, it is not popular with governments. Unregistered enterprises do not pay business and sales taxes and usually are so underfinanced that they tend not to grow into job-creating formal sector businesses. In many cases, informal businesspeople must pay protection money to local gangsters (the so-called *mafia tax*) to keep from being reported to the authorities.

Vignette Natasha is an engineer in Moscow. She has managed to keep her job and the benefits it carries, but in order to better provide for her family, she sells secondhand clothes in a street bazaar on the weekends. "Everyone is learning the ropes of this capitalism business," she laughs. "But it can get to be a heavy load. I've never worked so hard before!" Asked about her customers, Natasha says, "Many are former officials and high-level bureaucrats who just can't afford the basics for their families any longer." Some are older retired people whose pensions are so low that they resort to begging on Moscow's elegant shopping streets close to the parked Rolls-Royces of Moscow's rich. These often highly educated and only recently poor people buy used sweaters from Natasha and eat in nearby soup kitchens. *[Source: This composite story is based on work by Alessandra Stanley, David Remnick, David Lempert, and Gregory Feifer.]* ∎

Food Production in the Post-Soviet Era

Across most of the region, agriculture is precarious at best, either hampered by a short growing season and boggy soils or requiring expensive inputs of labor, water, and fertilizer. Because of Russia's harsh climates and rugged landforms, only 10 percent of its vast expanse is suitable for agriculture. The Caucasus mountain zones are some of the only areas in the region where rainfall adequate for agriculture coincides with a relatively warm climate and long growing seasons. Together with Ukraine and European Russia, this area is the agricultural backbone of the region (Figure 5.13). The best soils are in an area stretching from Moscow south toward the Black and Caspian seas, and extending west to include much of Ukraine and Moldova. In Central Asia, irrigated agriculture is extensive, especially where long growing seasons support cotton, fruit, and vegetables.

Changing Agricultural Production During the 1990s, agriculture went into a general, if temporary, decline across the entire region. In most of the former Soviet Union, yields dropped by 20 to 30 percent compared to previous production levels. This was due mostly to the collapse of the subsidies and trade arrangements of the Soviet era. Many large and highly mechanized collective farms suddenly found themselves without access to equipment, fuel, or fertilizers. Since that time, many of the massive collective farms of the Soviet era have been privatized, with better management resulting in larger harvests. Thousands of collective farms have been broken up into small tracts and sold to the people who worked on them. These independent farmers now have a much greater stake in their productivity and are making better decisions about what to grow, how to grow it, and where to sell it.

In Central Asia, agriculture was reorganized with less emphasis on collective farming of export crops and more on smaller food-producing family farms. Grain and livestock farming (for meat, eggs, and wool) are increasing and production levels per acre and per worker are increasing. China, through the Asian Development Bank, has provided assistance in reducing the use of agricultural chemicals, which were extensively used during Soviet times.

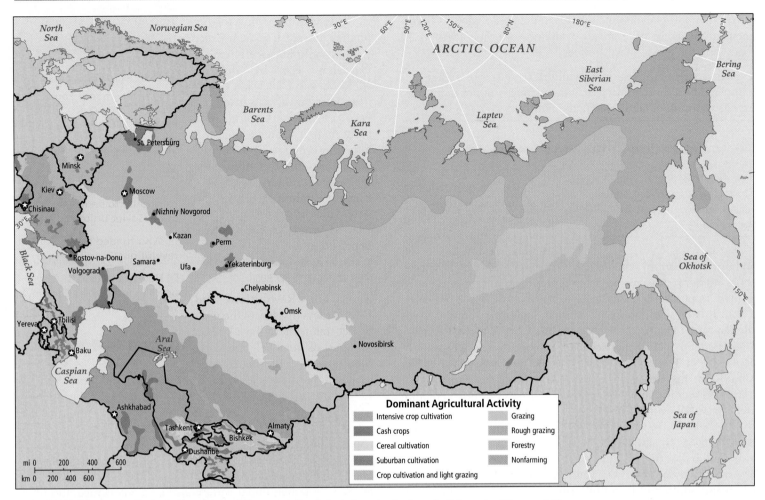

FIGURE 5.13 Agriculture in Russia and the post-Soviet states. Agriculture in this part of the world has always been a difficult proposition, partly because of the cold climate and short growing seasons, and partly because soil fertility or lack of rainfall are problems in all but a few places (Ukraine, Moldova, and Caucasia).

Georgian farmers, blessed with warm temperatures and abundant moisture from the Black and Caspian seas, can grow citrus fruits and even bananas; they do so primarily on family, not collective, farms (see Thematic Overview F; see also Figure 5.14). Before 1991, most of the Soviet Union's citrus and tea came from Georgia, as did most of its grapes and wine. Georgia still exports some food to Russia, but because of political tensions between the two countries, Georgia is increasing the amount it sells to Europe and other markets outside this region.

THINGS TO REMEMBER

1. **Learning Goal 3: Globalization and Development** Crude oil and natural gas have emerged as the region's most lucrative exports, introducing a new wave a globalization and fossil fuel–dependent economic development.

2. The Soviet centrally planned economy was less efficient than market economies in allocating resources. An enduring legacy of Soviet central planning is the location of huge industrial projects in the farthest reaches of Soviet territory.

FIGURE 5.14 Cabbage-planting season for Georgian family farmers.

Thinking Geographically: What about this photo suggests that this farm is relatively small?

3. The benefits of economic reforms in the post-Soviet era have been dampened by widespread corruption, lower incomes for many, and the loss of social services.

4. Across most of the region, agriculture is precarious, either hampered by a short growing season and boggy soils or requiring expensive inputs of labor, water, and fertilizer.

Democratization in the Post-Soviet Years

Learning Goal 4
Democratization: What are the obstacles to democratization in this region?

Democratization in this region has not proceeded as fast as has the introduction of a market economy. While several countries have held elections, many forms of authoritarian control remain. Elected representative bodies often act as rubber stamps for very strong presidents and exercise only limited influence on policy.

In Russia, Vladimir Putin, a former high official in the *KGB* (Russia's intelligence agency, during the Cold War, now known as the Federal Security Service or FSB), took over as acting president in 1999. It was hoped that he would rescue Russia after the tumultuous first decade of reform led by Boris Yeltsin, who, as historian Peter Rutland (2007) says, was given to "drunken pranks and erratic policy shifts."

In 2000, Putin was elected as president and served the two terms allowed by law, during which he exercised tight control over political and economic policy and consolidated political power in Moscow. He was very popular for bringing relative peace and prosperity and for restoring Russia's image of itself as a world power, despite disallowing meaningful democratic reforms at the local, state, and national levels, and extending government control over the press and the media (see page 197). Ostensibly, criticism of the government is now allowed, but behind the scenes, critics fear retribution from the *securocrats* or *siloviki* (former FSB functionaries loyal to Putin) who control many governing institutions.

The graph in Figure 5.15, developed by the political analyst Olga Kryshtanovskaya, gives insight into the role that the military and security personnel (siloviki) played as advisors to Putin. She assessed the percent of siloviki in government and elite groups during the Gorbachev, Yeltsin, and Putin administrations, and found that there has been a striking increase over time. Other analysts corroborate these findings and assert that this does not bode well for democracy because the siloviki see themselves as an elite, superior to ordinary citizens and therefore the rightful "bosses." By 2006, Kryshtanovskaya found that 77 percent of the people in the top 1016 government positions had backgrounds in security agencies.

In 2008, Putin tapped a former aide, Dmitri Medvedev, as the heir apparent to the presidency. Medvedev was then elected president with little opposition and Putin engineered his own appointment to the office of prime minister, which has no term limits. Many see these moves as an effort by Putin to hold on to power indefinitely.

▐▐ ▶ 117. PUTIN CONFIRMATION

▐▐ ▶ 123. NEW RUSSIAN LEADER

Elsewhere in the region, progress toward true democracy is similarly limited (see Thematic Overview E). In Belarus, Moldova, the Caucasian republics (Chechnya, discussed on page 196), and Central Asia, authoritarian leaders unaccustomed to criticism or to sharing power are still the norm (see the map in Photo Essay 5.4). Elections often result in the systematic intimidation of voters and arrests of the political opposition.

▐▐ ▶ 110. KAZAKHSTAN'S PRESIDENT TO MEET WITH GEORGE W. BUSH

The Color Revolutions and Democracy In a few of the post-Soviet states, there have been a number of relatively peaceful transitions to somewhat greater democracy—the so-called *color revolutions*, including the Rose Revolution in Georgia (2003–2004, discussed in the opening vignette), the Orange Revolution in Ukraine (January 2005; see Photo Essay 5.4B), and the Tulip Revolution in Kyrgyzstan (April 2005).

The democratization movements were named the "color revolutions" because they were led by coalitions of educated young adults who rallied under various symbolic colors meant to unite them despite differences. Funded largely by foreign NGOs and

Thinking Geographically

After you have read about democratization and conflict in Russia and the post-Soviet states, you will be able to answer the following questions:

A How did the USSR's war in Afghanistan ultimately contribute to democratization in the Soviet Union?

B Who funded the "color revolutions"?

C How does Russia's unwillingness to let Chechnya secede relate to fossil fuel resources?

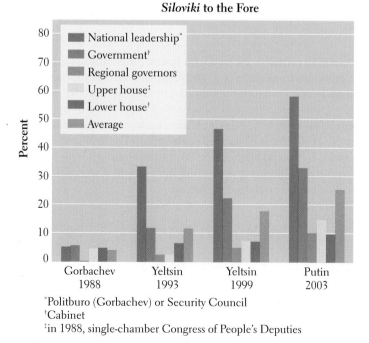

Siloviki **to the Fore**

Legend:
- National leadership*
- Government†
- Regional governors
- Upper house‡
- Lower house†
- Average

(X-axis: Gorbachev 1988, Yeltsin 1993, Yeltsin 1999, Putin 2003; Y-axis: Percent)

*Politburo (Gorbachev) or Security Council
†Cabinet
‡in 1988, single-chamber Congress of People's Deputies

FIGURE 5.15 Government officials with siloviki ties. Percent of people with military/security (siloviki) backgrounds in elite groups during the administrations of Gorbachev, Yeltsin, and Putin.

Since World War II, most conflicts in this region have been in Caucasia and Central Asia. Most occurred just before or shortly after the breakup of the Soviet Union. Since then, pro-democracy movements have emerged in several countries, where they hold the promise of political change without violent conflict. However, their long-term effectiveness remains to be seen.

A A fighter plane in Afghanistan lies among other pieces of wreckage dating from the Soviet Union's 10-year war in that country. The Soviets intervened in 1979 to prop up an authoritarian and unpopular Afghan government that had initiated radical reforms and murdered much of the political opposition. The war resulted in at least 1.5 million deaths and a humiliating defeat for the USSR. Fourteen thousand Soviet soldiers died, billions of dollars were spent, and practically nothing was gained. Public outrage over the Afghan war played a major role in the breakup of the Soviet Union. Since the fall of the USSR, democratization has proceeded at a much more rapid pace throughout the region.

Democratization and Conflict

Armed Conflicts and Genocides with High Death Tolls Since 1945

- ❗ Ongoing conflict
- ✳ 1000–5000 deaths
- ✳ 5000–50,000 deaths
- ✳ 50,000–300,000 deaths
- ✳ 300,000–1,000,000 deaths

Democratization Index

- Full democracy
- Flawed democracy
- Hybrid regime
- Authoritarian regime
- No data

mi 0   500   1000
km 0   500   1000

B Protesters pushing for a transition from an authoritarian to a more democratic government in Ukraine decorated police barricades with balloons and flowers during the Orange Revolution of 2004–2005. The movement followed a model pioneered in Serbia and Georgia, where student groups, funded by NGOs and governments in Europe and the United States, spearheaded grassroots political campaigns that featured massive nonviolent demonstrations. A similar "color revolution" occurred in Kyrgyzstan, and one is developing in Belarus.

C A bombed-out apartment building in Grozny, Chechnya, where separatist rebels have fought two unsuccessful wars and have conducted ongoing terror campaigns against Russia since the fall of the Soviet Union. Political instability in Caucasia is one of the greatest forces working against democratization in the region.

governments, including the United States, that wanted to hasten democratization, it appeared for a time that these movements would result in significant reforms. However, while elections have taken place, and greater freedom of expression is increasingly tolerated, many forms of authoritarian control and practices that disrupt democratic procedures remain.

In Ukraine, which is now the most democratized part of this region, the Orange Revolution occurred during the presidential elections of 2004 (see Photo Essay 5.4B). Before the election, candidate Victor Yuschenko, who favored closer ties with the European Union and Ukrainian membership in NATO, was poisoned with dioxin. He recovered but was horribly disfigured. Viktor Yanukovych (who favored closer ties with Russia) won the election, but the results were so obviously rigged by the government that widespread protests went on for months. The protests, combined with international pressure, resulted in a new vote in which Yuschenko won.

However, Yuschenko's term was characterized by years of conflict between two deeply divided, equally powerful factions. Yuschenko's faction favored closer ties with the European Union, while the opposition, buttressed by a large Russian-speaking minority, favored close ties with Russia. In the 2009 election, Yuschenko finished far back in the ranks of candidates for president. Yanukovych, the pro-Russian original winner in 2004, came in first. While this dealt a blow to the Orange Revolution, many of its leaders remained prominent in Ukrainian politics. One such leader, Yulia Tymoshenko, is among the region's few prominent female politicians (see Reasons for Optimism C).

Cultural Diversity and Democracy Russia's long history of expansion into neighboring lands has left it with an exceptionally complex internal political geography. As the Russian czars and the Soviets pushed the borders of Russia eastward toward the Pacific Ocean over the past 500 years, they conquered a number of small non-Russian areas. These 30 republics and 10 autonomous regions now constitute 25 percent of the Russian Federation's land area (see Figure 5.7). Many have significant ethnic minority populations that are descendant from indigenous people or trace their origins to long-ago migrations from Germany, Turkey, or Persia.

Both the czars and the USSR had a policy of **Russification**, whereby large numbers of ethnic Russian migrants were settled in non-Russian ethnic areas and given the best jobs and most powerful positions in regional governments. The goal was to force potentially rebellious regional minorities to conform to the state's goals. However, even during the Soviet period, but especially after, minorities organized to resist Russification and to enhance local ethnic identities.

> **Russification** the assimilation of all minorities to Russian (Slavic) ways

Shortly after the breakup of the Soviet Union, several internal republics demanded greater autonomy, and two of them, Tatarstan and Chechnya, declared outright independence. Tatarstan has since been placated with greater economic and political autonomy. However, Chechnya's stronger resistance to Moscow's authority has led to the worst bloodshed of the post-Soviet era and raised many doubts about Russia's commitment to human rights.

Chechnya, located on the fertile northern flanks of the Caucasus Mountains, is home to 800,000 people (see the inset map in **Figure 5.16** on page 197). The Chechens converted to the Sunni branch of Islam in the 1700s, partially in response to Russian oppression. Since then, Islam has served as an important symbol of the Chechen identity and an emblem of resistance against the Orthodox Christian Russians, who annexed Chechnya in the nineteenth century.

In 1942, when the Germans invaded Russia during World War II, a group of Chechens rebels simultaneously waged a guerilla war against the Soviets. Near the end of the war, Stalin acted out of revenge by deporting the majority of the Chechen population (as many as 500,000 people) to Kazakhstan and the Russian Far East (Siberia). Here they were held in concentration camps, with many dying of starvation. The Chechens were finally allowed to return to their villages in 1957, but a heavy propaganda campaign portrayed them as traitors to Russia.

In 1991, as the Soviet Union was dissolving, Chechnya declared itself an independent state. Russia saw this as a dangerous precedent that could spark similar demands by other cultural enclaves throughout its territory. Russia also wished to retain the agricultural and oil resources of the Caucasus, and was planning to build pipelines across Chechnya to move oil and gas to Europe from Central Asia. Russia has responded to acts of terrorism by Chechen guerrillas with bombing raids and other military operations that have killed tens of thousands and created 250,000 refugees (see Photo Essay 5.4C). Presently, most Chechen guerillas have given up, most Russian combat troops have been pulled out of Chechnya, and Russia has begun making substantial investments in rebuilding the capital city of Grozny. While this shift is a welcome change, some Chechen rebels have continued their struggle by carrying out brutal terrorist attacks, many of which now take place in Moscow and other areas outside Chechnya. Hence, the conflict in Chechnya continues to raise doubts about Russia's ability to peacefully address internal political dissent.

II▶ 116. CHECHNYA HANGS ON TO UNEASY PEACE

Just south of Chechnya, conflicts between Russia and Georgia over the ethnic republics of South Ossetia and Abkhazia have worsened in the post-Soviet era. Both republics became part of Georgia at the behest of Joseph Stalin (a native Georgian). He then moved Georgians and Russians into Ossetia and Abkhazia so that the native people became a minority in their own place. More recently, Russia has strategically supported the ethnic Ossetian and Abkhazian populations' agitation for secession from Georgia, even granting Russian citizenship to between 60 to 70 percent of the non-Georgian population. Russia may have done this in retaliation for Georgia's increasingly close relations with the United States and the European Union, as evidenced by its candidacy for NATO membership. A pipeline that links the oil fields of Azerbaijan with the Black Sea via Georgia is another source of contention. Russia would like to enhance its control over the oil resources of the Caspian Sea region by routing pipelines through Russian-controlled territory.

In August 2008, Georgia's military, attempting to gain control over rebelling parts of South Ossetia, engaged with the Russian army on Russian territory under circumstances that are in contention. South Ossetia had become a major smuggling center between Russia and Georgia, including for the smuggling of nuclear materials (discussed on page 197). Russia

FIGURE 5.16 Ethnic character of Russia and percentage of Russians in the post-Soviet states. Russia, with all of its internal republics and autonomous regions, is formally called the Russian Federation. The ethnic character of many of the more than 30 internal republics was changed by the policy of central planning, so that Russians now form significant minorities in most republics. As of 2002, the ethnic makeup of Russia was 79.8 percent Russian, 3.8 percent Tatar, 2.0 percent Ukrainian, 1.2 percent Bashkir, 1.1 percent Chuvash, and 12.1 percent other.

responded to the Georgian attack with its much larger military, driving Georgia's forces out of South Ossetia and following them back into Georgia. At the same time, fighters in Abkhazia drove Georgia's military out of that province, and Russian planes bombed a town near a Georgian pipeline. At present, it is not clear if Abkhazia and South Ossetia will break away and become independent countries, become provinces within the Russian Federation, or be satisfied by offers of greater autonomy within Georgia's federal structure.

The Media and Political Reform   In the Soviet era, all communication media were under government control. There was no free press, and public criticism of the government was a punishable offense. However, toward the end of the Soviet era, many journalists risked retribution for criticizing public officials and policies. Between 1991 and the early 2000s, the communications industry was a center of privatization, and several media tycoons emerged to challenge the Communist Party. Privately owned newspapers and television stations regularly criticized the policies of various leaders of Russia and the other states. It appeared that a free press was developing.

Vladimir Putin's rise to power was a turning point, after which the most critical newspapers and TV stations in Russia were shut down. Since then, critical analysis of the government

has become rare throughout Russia. Journalists openly critical of government policies have been treated to various forms of censorship, exile, or violence. From 2000 to 2008, more than 80 journalists were killed.

Corruption and Organized Crime

Any potential benefits of political and economic reforms in the post-Soviet era have been dampened by widespread corruption and the growth of organized crime. Many oligarchs became closely connected to the so-called *Russian Mafia*, a highly organized criminal network dominated by former KGB (the Soviet intelligence agency and counterpart to the CIA) and military personnel who control the thugs on the streets. The Russian Mafia extended its influence into nearly every corner of the post-Soviet economy, especially in illegal activities and the arms trade.

In the post–9/11 world, concern spread that military corruption in this region could put nuclear weapons from the former Soviet arsenal in the hands of terrorists. After huge military funding cuts, weapons, uniforms, and even military rations were routinely sold on the black market. In 2007, Russian smugglers were caught trying to sell nuclear materials on the black market. Recent developments are more encouraging. In an effort to fight corruption and the temptation to sell nuclear materials in the black market, the

Russian government has given impoverished military personnel long-delayed pay raises or termination compensation. Moreover, all countries in the region are now cooperating with the International Atomic Energy Agency in controlling nuclear material.

THINGS TO REMEMBER

1. Learning Goal 4: Democratization While several countries have held elections, many forms of authoritarian control remain. Elected representative bodies often act as rubber stamps for very strong presidents and exercise only limited influence on policy.

2. Since World War II, most conflicts in this region occurred in Caucasia and Central Asia, mainly just before or shortly after the breakup of the Soviet Union. In the years after the breakup, pro-democracy movements have emerged in several countries, where they hold the promise of political change without violent conflict. However, their long-term effectiveness remains unclear.

3. There continue to be limitations on the free press and related media across the region.

4. The potential benefits of political and economic reforms in the post-Soviet era have been dampened by official corruption and the growth of organized crime.

Sociocultural Issues

When the winds of change began to blow through the Soviet Union in the 1980s, few anticipated the rapidity and depth of the transformations or the social instability that resulted. On the one hand, new freedoms have encouraged self-expression, individual initiative, and cultural and religious revival; on the other hand, the post-Soviet era has brought very hard times to many as their jobs and the social safety net were obliterated.

> **Caucasia** the mountainous region between the Black Sea and the Caspian Sea

Population Patterns

This region shares some of the same population characteristics as the United States and Europe—low birth rates and an aging population—both usually features of wealthy and developed societies (see Thematic Overview H). But in Russia and the post-Soviet states, low birth rates and the rate of aging in the population are more extreme and are coupled with high mortality rates for middle-aged males. Several circumstances contribute to this situation.

Population Distribution and Urbanization The region of Russia and the post-Soviet states is one of the largest on earth but is the least densely populated, with only about 278 million people. A broad area of moderately dense population forms an irregular triangle that stretches from Ukraine on the Black Sea north to St. Petersburg on the Baltic Sea and east to Novosibirsk, the largest city in Siberia. In this triangle, settlement is highly urbanized, but the cities are widely dispersed. The capital city of Moscow, with 11 million people, is a primate city in Russia, as are all other capital cities of the nations in this region (**Figure 5.17**).

Beyond Novosibirsk on the West Siberian Plain, settlement follows an irregular and sparse pattern of industrial and mining development in widely spaced cities, stretching east across Siberia. These economic activities and the cities they support are linked primarily to the route of the Trans-Siberian Railroad (see Figure 5.11). Although Siberia is a desolate, lonely place, nearly 90 percent of its people are concentrated in a few relatively large urban areas. The costs of maintaining these settlements in this remote and harsh environment are considerable.

A secondary spur of dense settlement extends south from Russia into **Caucasia**, the mountainous region between the Black Sea and the Caspian Sea, where there are several primate cities of well over 1 million people each. In Central Asia, another patch of relatively dense settlement is centered on the cities of Tashkent and Almaty (both over 1 million in population) and the new capital of Kazakhstan, Astana, which has 750,000 people (**Photo Essay 5.5C** on page 200). Along major Central Asian rivers during the Soviet period, the development of irrigated cotton farming and mineral extraction resulted in patches of high rural density, fueled partly by ethnic Russian immigration.

Urban life is still shaped by the legacies of central planning during the communist era. Giant apartment blocks, built for industrial workers according to the designs of central planners, dominate most cities, especially on the fringes of pre-Soviet urban cores (see Thematic Overview G). The housing shortages of the Soviet era have not abated since 1991, and cramped drab apartments badly in need of repair, with shared kitchens and bathrooms, are common. At the community scale, inadequate sewage, garbage, and industrial waste management pose serious long-term health and environmental threats. Meanwhile, urban parks have fallen into decline in the post-Soviet era as state funding for these once-beautiful urban amenities has diminished (see Photo Essay 5.5B).

Housing shortages are particularly severe in Moscow, which has grown rapidly since 1991 because it has been the focus of new investment in the region. Many other cities face housing shortages even though they have begun to shrink in size due to population decline. While the loss of population may eventually help some cities resolve their housing shortage, the larger and often still-growing cities are demanding, and in some cases receiving, greater investment from national governments to increase their housing stock.

> **Learning Goal 5**
> **Population:** Why are populations shrinking in some parts of this region?

Shrinking Populations: High Death Rates, Low Birth Rates This region is experiencing a unique variant of the demographic transition (see Chapter 1, page 16). In all but the Caucasian and Central Asian states, populations are shrinking faster than in any other world region. During the Soviet era, the increase in women's opportunities to become educated and work outside the home curtailed population growth. Severe housing shortages were an additional disincentive to having children, and many families chose to have only one or two children. Free health care and adequate retirement pensions also helped lower

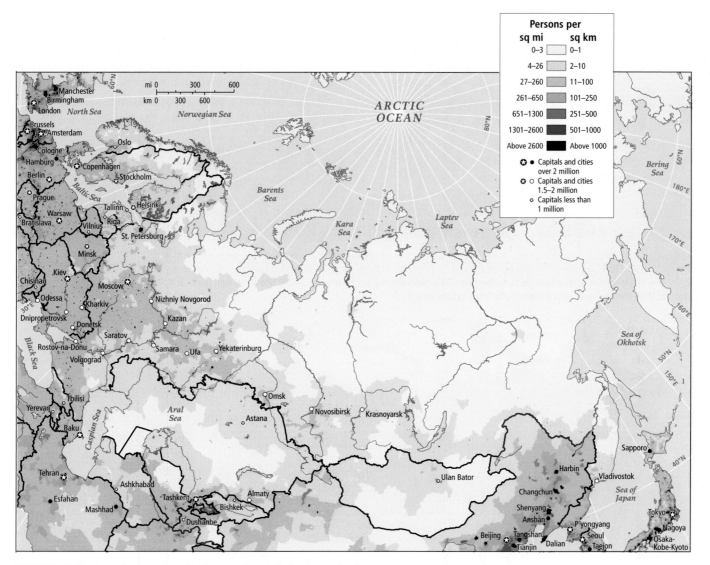

FIGURE 5.17 Population density in Russia and the post-Soviet states. Population trends in this region are highly uneven. While Central Asia and some countries in Caucasia are growing, populations in the rest of the region are shrinking. Some of this may be related to cultural differences or to varying levels of dependence on social welfare institutions that collapsed with the end of the Soviet Union.

incentives for large families. During the last years of the Soviet Union, the population was shrinking, but this trend accelerated during the economic crisis brought on by the breakup of the USSR. Since then, Russia's population has shrunk more than 5 percent, to 142 million. The United Nations predicts that Russia will drop to just 116 million people by 2050. Populations are also shrinking in Belarus, Moldova, and Ukraine. Even in Central Asia and the Caucasian states, where higher birth rates should be resulting in overall growth, countries are still losing population due to emigration.

In the parts of this region that have low birth rates, surveys suggest that people are now choosing to have fewer children primarily out of concern over gloomy economic prospects for the near future and also out of the desire to make money and have fun. Russia is attempting to reverse population loss by paying couples to raise more children and by attracting back Russians and their dependents who live abroad. In 2007 and 2008, Russia spent $300 million to send emissaries to the far corners of the earth (Brazil, Egypt, Germany, and all the post-Soviet states) to lure "returnees." Only 10,300 were recruited.

In addition to negative birth and migration rates, much of the reason for the population decline is the declining life expectancy. In Russia, for example, between 1990 and 2008, male life expectancy dropped from 63.9 to 60 years, the shortest in any industrialized country. Female life expectancy dropped less, from 74.4 to 73 years. Major causes of declining life expectancies in the region are the loss of health care, which was usually tied to employment, and the physical and mental distress caused by lost jobs and social disruption. The higher male death rate is explained in large part by alcohol abuse and related suicides (women are less prone to both but tend to smoke in excess). In Russia, approximately 500,000 deaths per year are alcohol related, which translates into an overall rate of alcohol-related deaths that is 14 times that of the United States.

An uneven pattern of urbanization is developing in this region. In several countries, the largest cities are growing rapidly, as they are centers of new development and globalization. Meanwhile, in Russia and elsewhere, many cities are shrinking as the overall population declines.

A Moscow is already the region's primate city, as its population of almost 11 million people is larger than the next three largest cities combined. Moscow produces 20 percent of Russia's GDP, and this number is growing, fueled by the city's status as the headquarters of Russia's globalizing economy. By 2020, Moscow will gain 900,000 inhabitants, five times what any other city in the region will gain.

Population Living in Urban Areas

- 80–100%
- 60–79%
- 40–59%
- 20–39%
- 0–19%
- No data

1 Global rank (population 2006)

Population of Urban Areas

2006 — 2020

20 million or more
10 million
5 million
3 million

⊖ Urban areas with a negative (-) growth rate

Note: Symbols on map are sized proportionally to specific population data

B A public park in Kharkov, Ukraine, suffers from neglect, a symptom of the city's decline. Kharkov has lost 10 percent of its population since 1989, and will lose another 10 percent by 2020. This is related both to Ukraine's overall population decline and to the closing of many of Kharkov's defense-related industries after the collapse of the USSR.

C The presidential palace and other public buildings in Astana, Kazakhstan, which doubled its population in 10 years after being designated the new capital. Astana's growth has been funded by Kazakhstan's new oil and gas wealth.

Population pyramids for several countries (Figure 5.18) show the overall population trends in the region and reflect geographic differences in patterns of family structure and fertility. The pyramids for Belarus and Russia resemble those of European countries (for example, Germany, as shown in Figure 4.10A on page 164). They are significantly narrower at the bottom, indicating that birth rates in the last several decades have declined sharply. The narrower point at the top for males in all four pyramids shows their much shorter average life span.

The base of the pyramid for Kazakhstan in Central Asia is also narrowing as a drop in birth rates accompanies urbanization. However, in all countries but Kazakhstan, there has been a slight rebound in birth rates in the youngest age group.

Gender: Challenges and Opportunities in the Post-Soviet Era

Soviet policies that encouraged all women to work for wages outside the home have transformed this region (see Thematic Overview I). By the 1970s, ninety percent of able-bodied women were working full time, giving the Soviet Union the highest rate of female paid employment in the world. However, the traditional attitude that women are the keepers of the home persisted. The result was the *double day* for women. Unlike men, most women worked in a factory or office or on a farm for 8 or more hours and then returned home to cook, care for children, shop daily for food, and do the housework (without the aid of household

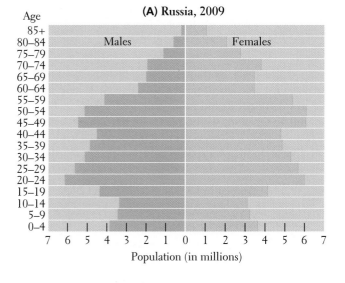
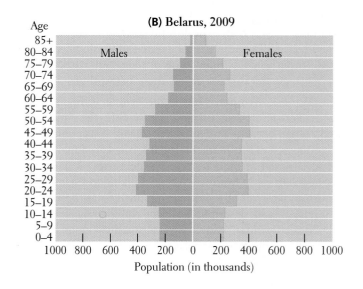
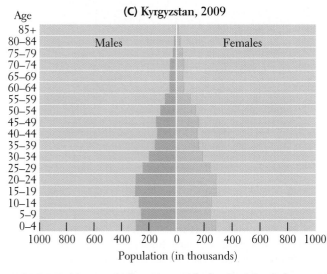
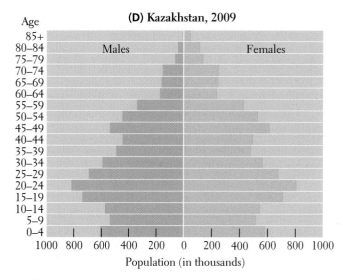

FIGURE 5.18 Population pyramids for Russia, Belarus, Kyrgyzstan, and Kazakhstan. Note that the pyramid for Russia is at a different scale (millions) than the other three pyramids (thousands) because of Russia's larger total population. This difference does not significantly affect the pyramid's shape.

appliances). Because of shortages (the result of central-planning miscalculations), they also had to stand in long lines to procure necessities for their families.

> **Learning Goal 6**
> **Gender:** How have the massive economic and political changes in this region affected men and women differently?

When the first market reforms in the 1980s reduced the number of jobs available to all citizens, President Gorbachev encouraged women to go home and leave the increasingly scarce jobs to men. Many women lost their jobs involuntarily, and by the late 1990s, seventy percent of the registered unemployed were women, despite the fact that due to illness, death, or divorce, many if not most were the sole support of their families. Consequently, many had to find new jobs.

On average, the female labor force in Russia is now better educated than the male labor force. The same pattern is emerging in Belarus, Ukraine, Moldova, and parts of Muslim Central Asia. In Russia, the best-educated women commonly hold professional jobs, but they are unlikely to hold senior supervisory positions; they also are paid less than their male counterparts. As recently as 2005, the wages of women professional workers averaged 36 percent less than those of men. Still, this region ranks higher in gender income equity than many others.

The Trade in Women During the economic boom stimulated by marketization and oil and gas wealth, the "marketing" of women had become one of the less savory entrepreneurial activities. One part of this market is the Internet-based mail-order bride services aimed at men in Western countries. A woman in her late teens or early twenties, usually seeking to escape economic hardship, pays about $20 to be included in an agency's catalog of pictures and descriptions. (One Internet agency advertises 30,000 such women.) She is then assessed via email or Facebook by the prospective groom, who then travels—usually to Russia or Ukraine—to choose from the women he has selected from the catalog.

In recent years, there has been a large increase in sex work. Precise numbers are hard to come by. However, in 2000, the *Economist* estimated that 300,000 women were smuggled each year into the European Union, where the sex trade then generated $9 billion annually (Figure 5.19). The business of supplying sex workers is dominated by members of the Russian Mafia, who have been known to kidnap schoolgirls or deceive women seeking jobs as domestic servants or waitresses in Europe, then forcing them to work as prostitutes.

The Political Status of Women The most effective way for women to address institutionalized discrimination is to gain access to positions where they can affect wide-ranging policies—usually as elected officials. Although women were granted equal rights in the Soviet constitution, they never held much power. In 1990, women accounted for 30 percent of Communist Party membership, but just 6 percent of the governing Central Committee. The very few in party leadership often held these positions at the behest of male relatives.

Ironically, since the fall of the Soviet Union, the political empowerment of women has advanced the least where democracy

FIGURE 5.19 The trade in women: Ukrainian sex workers in Amsterdam, Netherlands. Over 100,000 Ukrainian women work in the sex industry outside of Ukraine, many of them in West Europe. Most were lured abroad with the promise of a good job in an office only to find themselves sold into prostitution.

has developed the most, perhaps because of long-standing cultural biases against women in positions of power. Where governments are less democratic and more authoritarian, as in Belarus, Moldova, and Kyrgyzstan, women actually hold more legislative positions (Figure 5.20). However, in those countries, leaders who wanted to appear more progressive in the eyes of international

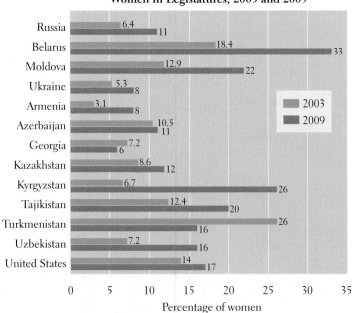

FIGURE 5.20 Women legislators, 2003 and 2009. This graph shows the percentage of legislators in Russia and the post-Soviet states (with the United States for comparison) who are women. Belarus and Kyrgyzstan stand out as having the most women lawmakers, but both are authoritarian societies in which true democratic participation is rare.

donors may have promoted them undemocratically, choosing women who were least likely to work for change. Ironically, support among women for women's political movements is not widespread, as many fear being seen as anti-male or against traditional feminine roles.

Religious Revival in the Post-Soviet Era

The official Soviet ideology was atheism, and religious practice and beliefs were seen as obstacles to revolutionary change. The Orthodox Church was tolerated, but few went to church, in part because the open practice of religion could be harmful to one's career. Now, religion is a major component of the general cultural revival across the former Soviet Union. Throughout the Russian Federation, especially among indigenous ethnic minorities, people are turning back to ancient religious traditions. For example, the Buryats, from east of Lake Baikal in Siberia, who are related to the Mongols, are relearning the prayers and healing ceremonies of the Buryat version of Tibetan Buddhism, which they adopted in the eighteenth century. The shamans who lead them have organized into a guild to give official legitimacy to their spiritual work. They now pay taxes on their clergy income.

In Russia, Ukraine, Moldova, Belarus, Georgia, and Armenia, most people have some ancestral connection to Orthodox Christianity. Those with Jewish heritage form an ancient minority, mostly in the western parts of Russia and the Caucasus, where they trace their heritage back to 600 B.C.E. Religious observance by both groups increased markedly in the 1990s, and many sanctuaries that had been destroyed or used for nonreligious purposes by the Soviets were rebuilt and restored.

A countertrend to the robust revival of Orthodox Christianity is the spread of evangelical Christian sects from the United States (Southern Baptists, Adventists, and Pentecostals). Evangelical Christianity first came to Russia in the eighteenth century, but after 1991, American missionary activity increased markedly. The notion often promoted by this movement—that with faith comes economic success—may be particularly comforting both to those struggling with hardship and to those adjusting to new prosperity. Figure 5.21 is a sculptor's humorous attempt to show the jarring cultural change that has transpired over the last few decades.

Vignette Valerii, age 35, once a government research scientist, now makes a comfortable living importing and exporting goods in the informal economy of Moscow. Although he has to bribe officials and pay protection money to mobsters, his income has made his family much wealthier than their longtime friends. Valerii's wife Nina is the only woman among them who does not work outside the home. In search of values that will guide them in these new circumstances, both have recently been baptized in an evangelical Christian sect. They say they chose this particular religious group because it promotes modesty, honesty, and commitment to hard work. [Source: Adapted from Timo Piirainen, Towards a New Social Order in Russia: Transforming Structures and Everyday Life (Aldershot, UK, and Brookfield, Vermont: Dartmouth Press, 1997), pp. 171–179; updated in 2007.] ∎

FIGURE 5.21 Globalization and cultural change. A sculptor's depiction of some of the cultural transformations occurring in this region since the fall of the Soviet Union. Vladimir Lenin, symbolizing the Soviet Era, holds hand with Mickey Mouse, representing cultural globalization, and Jesus, who personifies the resurgence of Christianity throughout the region since 1991.

In Central Asia, Islam was long repressed by the Soviets, who feared Islamic fundamentalist movements would cause rebellion against the dominance of Russia. Today, Islam is openly practiced and increasingly important politically across Central Asia, Azerbaijan, and some of the Russian Federation's internal ethnic republics, such as Chechnya and Tatarstan. Especially in the Central Asian states, however, the return to religious practices is often a subject of contention. Some local leaders still view traditional Muslim religious practices as obstacles to social and economic reform.

Both devout Muslims and more secular reformists are wary of religious extremists from Iran and Saudi Arabia who may pose a security risk. Between 1992 and 1997, Tajikistan fought a civil war against Islamic insurgents with links to the Taliban in Afghanistan—and the conflict in Chechnya involves combatants with similar links to Saudi militants. In 2000, Uzbekistan and Kyrgyzstan joined forces to eliminate an extremist Islamic movement. But many religious leaders and some human rights groups say that the fervor to eliminate radical insurgents has resulted in the persecution of ordinary devout Muslims, especially men, and that this persecution is recruiting anti-Western enthusiasts. Human Rights Watch, an organization that monitors human rights abuses worldwide, reports that at least 4000 Muslim men demanding economic reforms have been arrested and detained in Uzbekistan alone. Those protesting were routinely labeled Muslim extremists and dealt with violently; perhaps as many as 700 have been killed.

THINGS TO REMEMBER

1. **Learning Goal 5: Population** Population trends in this region are highly uneven. Russia and the countries bordering Europe have the most rapidly declining populations on earth. Meanwhile, the Central Asian and Caucasian countries are growing, though they are still losing people to emigration.

2. An uneven pattern of urbanization is developing in this region. In several countries, the largest cities are growing rapidly, as they are centers of new economic development and globalization. Meanwhile, in Russia and elsewhere, many cities are shrinking as the overall population declines.

4. **Learning Goal 6: Gender** The Soviet Union gave strong incentives for women to work and become involved in politics, but in the post-Soviet era, gaps have been increasing between men and women in employment rates and political representation. New threats to women have also emerged, such as international prostitution rings, now active in some countries.

5. Official Soviet ideology was atheism, and religious practice and beliefs were seen as obstacles to revolutionary change. In the post-Soviet era, religion is a major components of the general cultural revival.

Reflections on Russia and the Post-Soviet States

This is a region of dramatic changes. During the Soviet era, the countries here amazed the world with their rapid industrialization and societal transformations. Since the collapse of the Soviet system, they have attempted an even more rapid transformation to a free market economy. Through it all, Russia has retained its leadership position, though its power and influence do not compare with that of the Soviet Union. Some former Soviet allies in Central Europe are now members of the increasingly powerful European Union, and some of the region's most important agricultural producers (Ukraine and Georgia) are beginning to market their products outside the region. The Central Asian states are likewise drawn outward, toward their neighbors in Southwest, South, and East Asia.

While oil wealth is transforming relationships with the rest of the world, the benefits within the region have been uneven. Many are still reeling from the loss of jobs and social services since the collapse of the Soviet system. Women in particular have suffered from the high unemployment rate. Meanwhile, corruption and crime have exploded. Partially in response to all of these problems and the uncertain futures they create, birth rates are falling and populations are shrinking in much of the region.

In the face of these challenges, it is not surprising that central governments have remained strong and democracy weak. While some countries are eager to learn from the world's wealthy democracies, there is also widespread resistance to change and to excessive influence from abroad. Many see stronger, more authoritarian government control as necessary to maintain stability, even if it restricts freedom of speech and weakens democratic processes.

All parts of the region will be hampered for years to come by an aging and inefficient infrastructure and by severe environmental degradation. As modernization and privatization proceed, the rapid development of the region's rich resource base will probably lead to more pollution. Greenhouse gas emissions are likely to increase, as are water and air pollution.

However, there are also many reasons for optimism about the future of this region (see Reasons for Optimism A, B, C). Rising living standards may create pressure for better environmental protection, especially in highly polluted but increasingly wealthy urban areas. Meanwhile, once-hostile relationships between the USSR and Europe have transitioned into more amicable relationships dominated by the lucrative oil and gas trade between Russia and the European Union. And if societies continue to open, there will be exhilarating opportunities for entrepreneurism, self-expression, and political empowerment. This trend has already been seen in events like the Orange Revolution, which was a major step in the democratization of Ukraine that brought new leaders to power and prominence. Perhaps this region's experience with rapid and profound transformation will serve it well in the coming years.

Learning Goals Review

1. Environment, Development, and Urbanization: Why does this region have such severe environmental problems, especially in urban areas?

How did Soviet attitudes toward the environment contribute these problems? Are any solutions emerging?

2. Climate Change, Food and Water: Where in this region are food production systems and water resources most threatened by climate change?

Where are irrigated agriculture systems particularly vulnerable to rising temperatures? How can they adapt?

What parts of this region produce the most greenhouse gases?

What signs are there that reductions can be made?

3. Globalization and Development: How have economic reform and globalization changed patterns of development in this region since the fall of the Soviet Union?

How widely have the benefits from Russia's new fossil fuel–based economic development been shared? How does the situation in Russia compare with that in Central Asia? Why has access to health care changed for so many in this region since the fall of the USSR?

4. Democratization: What are the obstacles to democratization in this region?

How has economic and political instability influenced the kind of leaders that have emerged? How have the color revolutions influenced democratization?

5. Population: Why are populations shrinking in some parts of this region?

What Soviet-era developments helped curtail population growth? What is being done in the post-Soviet era to encourage population growth?

Reasons for Optimism in Russia and the Post-Soviet States

Development and Pollution: Higher living standards may increase pressure for environmental protection. **A** *Vacationers in Sochi, Russia, site of the 2010 Olympic Winter Games and a growing venue for environmental protests about urban pollution.* ▼

Globalization: Once-hostile relations with Europe are now stabilizing due to the lucrative oil and gas trade. **B** *German soccer star Jefferson Farfan. In an attempt to improve Russia's image in the EU, Gazprom took over sponsorship of Farfan's team.* ▼

Democratization: While authoritarianism remains strong, democratization is taking hold in places like Ukraine. **C** *Yulia Tymoshenko, a leader of the Orange Revolution and a former prime minister of Ukraine.* ▼

Thinking Geographically: Why does Sochi make a particularly effective venue for environmental protests?

Thinking Geographically: Why would Gazprom be concerned about improving its and Russia's image within the EU?

Thinking Geographically: In what sense was the Orange Revolution a pro-democracy movement?

---

6. Gender: How have the massive economic and political changes in this region affected men and women differently?

How has the gap between men and women changed in terms of economic and political empowerment during the post-Soviet era? What new threats to women have emerged?

Geographic Themes about Russia and the Post-Soviet States

Look back at the Thematic Overview photos on page 171. Thinking geographically, answer the following questions about them:

(A) Climate Change: What has Russia done to reduce its greenhouse gas emissions?

(B) Water: What inland sea has significantly shrunk as a result of overuse of river water for irrigation in Central Asia?

(C) Globalization: Gazprom earns most of its profits from selling to _____?

(D) Development: Why are small businesses so important in many post-Soviet countries?

(E) Democratization: What do the national leaders in Central Asia have in common in terms of how they govern?

(F) Food: Why is Caucasia such an important agricultural area in this region?

(G) Urbanization: Why is Moscow's housing shortage particularly severe?

(H) Population: What are the major causes of declining life expectancy in this region?

(I) Gender: In the 1970s, what percentage of able-bodied females were working outside the home?

Key Terms

Bolsheviks 186

capitalists 186

Caucasia 198

centrally planned, or socialist, economy 186

Cold War 186

communism 186

Communist Party 186

czar 185

Gazprom 188

glasnost 190

Group of Eight (G8) 188

Mongols 184

nomadic pastoralists 183

nonpoint sources of pollution 178

oligarchs 188

perestroika 190

permafrost 176

privatization 190

Russian Federation 173

Russification 196

Slavs 183

Soviet Union 172

steppes 173

taiga 176

tundra 176

underemployment 190

Union of Soviet Socialist Republics (USSR) 172

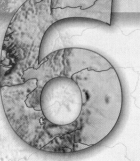

North Africa and Southwest Asia

Learning Goals

After you read this chapter, you will be able to answer the following questions:

1. Water, Food, and Climate Change: How will people get enough water to grow food in the future, especially if climate change makes this dry region even dryer?

2. Gender and Population: How does the low status of women contribute to this region's high population growth?

3. Urbanization and Globalization: How has globalization shaped different patterns of urbanization throughout the region?

4. Development and Globalization: How have the huge fossil fuel reserves of some countries transformed economic development and driven globalization in this region?

5. Democratization: Are there any signs that this region may be democratizing?

FIGURE 6.1 Political map of North Africa and Southwest Asia. The occupied Palestinian Territories are not shown here because of the scale of the map.

Thematic Overview of North Africa and Southwest Asia

Water: By 2005, few countries in the region will have enough water to support human development. Currently, most water resources support irrigated agriculture. **A** *Date palms irrigated by Iraq's Tigris River.* ▼

Food: Most countries are highly dependent on imported food. Population growth has outpaced increases in agricultural productivity in this dry region. **B** *A bakery in Israel, where 90 percent of grains are imported.* ▼

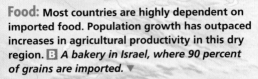

Climate Change: Rising temperatures could accelerate desertification, the conversion of nondesert lands into deserts. **C** *In Morocco, plants are used to stabilize dunes that could spread as temperatures rise.* ▼

Population: This region has the second-highest population growth rate in the world, after sub-Saharan Africa. **D** *Palestinian children play in Gaza, which has one of the highest population growth rates in the world.* ▼

Urbanization: Migration from rural to urban areas is increasing, with 58 percent of the population now living in urban areas. **E** *A street vendor sells fruit in Cairo. Many recent migrants work as street vendors.* ▼

Globalization: The sale of fossil fuels and the investment of resulting profits have linked many economies to global flows of money, resources, and people. **F** *Construction workers from South Asia in Dubayy.* ▼

Development: Dependence on fossil fuel development has left economies unstable, due to fluctuating prices for oil and gas on global markets and the fact that most profits go to a privileged few. **G** *A labor camp in front of oil facilities in Oman.* ▼

Democratization: Many countries are ruled by kings or authoritarian leaders. However, a region-wide shift toward democratization is underway. **H** *Pro-democracy demonstrations in Iran in 2009.* ▼

Gender: Women are generally less educated than men and tend not to work outside the home. Hence, childbearing remains crucial to a woman's status. **I** *A mother and her child in Iraq.* ▼

Global Patterns, Local Lives

Amos Oz, a well-known Israeli author (**Figure 6.2**), has written a novel about the founding of the state of Israel. It has no heroes, but rather tells an honest tale about Jewish settlers and the Palestinians they displaced. The book, *A Tale of Love and Darkness*, has gained wide acclaim for depicting both sides of the Israeli–Palestinian story with compassion and insight. Oz is frequently interviewed on the Arab network Al Jazeera, the most popular broadcaster in North Africa and Southwest Asia.

Elias Khoury is a prominent Palestinian lawyer. His son George, a student at Hebrew University, was jogging on the West Bank and was shot and killed by fellow Palestinians who mistook him for a Jew. In his grief, Elias Khoury searched for a fitting memorial to his son, who had been noted for his open, multicultural views and for his friendships with both Jews and Arabs. Elias Khoury had been touched by Amos Oz's book and proposed to pay for translating it into Arabic and distributing it so that sensitive, open-minded Arabs across the region could read it.

Amos Oz was touched by this proposal and agreed. Oz strongly argues for partition and a two-state solution to the Israeli–Palestinian dispute—a space for the Israeli state and a separate space for a Palestinian state.

Al Jazeera's coverage of Oz and the Khoury affair is itself highly controversial. Based in Doha, Qatar (with offices in Washington, D.C., Kuala Lumpur, and London), and supported both by advertising and by an income from the Qatari emir, Sheikh Hamad bin Khalifa, Al Jazeera tries to remain neutral on the Israeli–Palestinian conflict. Many broadcasts reflect the sentiments of its mainly Arab viewers, who tend to side with the Palestinians. When Al Jazeera tries to cover the views of Israelis, even those with views as sensitive to Palestinian issues as Amos Oz, it risks the sharp criticism of Arabs who accuse it of having a "pro-Israeli" bias. *[Sources: Adapted from broadcasts available at Al*

> **Islam** a monotheistic religion that emerged in the seventh century C.E. when, according to tradition, the archangel Gabriel revealed the tenets of the religion to the Prophet Muhammad
>
> **Islamism** a grassroots religious revival in Islam that seeks political power to curb what are seen as dangerous secular influences; also seeks to replace secular governments and civil laws with governments and laws guided by Islamic principles
>
> **fossil fuel** a source of energy formed from the remains of dead plants and animals

Jazeera.com, February 19, 2009: http://english.aljazeera.net/programmes/ rizkhan/2009/02/200921865142701932.html; and CNN.com: Christiane Amanpour, "The Role of Literature in the Path to Peace," March 11, 2010, at http://transcripts.cnn.com/TRANSCRIPTS/1003/11/ampr.01.html.] ∎

Freedom of the press, which is essential for democracy, is a fragile concept in any part of the world, and in North Africa and Southwest Asia it is only beginning to develop. This is in part because the 21 countries in the region (see Figure 6.1 on page 206) did not become politically independent until the twentieth century. Most countries in the region now have parliaments and elections, but the parliaments have few powers and elections are controlled to ensure that a ruler or a particular party is reseated.

North Africa and Southwest Asia is a region of striking continuities but equally striking contrasts. Although the vast majority of people practice **Islam**, a monotheistic religion that emerged between 601 and 632 C.E., it is a faith with multiple aspects. Only a minority of Muslims are drawn to ultra-fundamentalist Islam. Most Muslims are moderate in their thinking, accepting the validity of other beliefs, especially regarding Christianity, as Mary and Jesus both play important roles in Islam.

The term **Islamism** refers to grassroots religious revivals that seek political power to curb what they see as dangerous secular influences that are spreading because of globalization. Some such movements characterize Western influence as corrupt and destructively self-serving. But Islamist movements also vary greatly, are unevenly distributed across the region, and have waxed and waned in influence, often in sync with economic recessions and booms.

Access to oil wealth is also highly uneven. **Fossil fuel** reserves, consisting of oil and natural gas formed from the fossilized remains of dead plants and animals, are found mainly around the Persian Gulf. These fossil fuels are extracted and exported throughout the world for tremendous profits, but these profits are not equitably distributed, especially within oil-rich Gulf states (see the discussion of wealth disparities below and in Photo Essay 6.4 A, B on page 230).

There are many other sources for the variety in this region. Most countries in this area share an arid climate, but the degree of aridity varies widely. While Arabic culture and language is widespread, the second and third most populous countries in the region, Turkey and Iran, as well as many minority populations such as the Kurds, Berbers, and Jews, are not Arab. In some countries, nearly everyone lives in a city, while in others life is still primarily rural. Even in a forward-looking country like Turkey, women may be highly educated, outspoken, and active in commerce, public life, and government, or they may lead secluded domestic lives with few educational opportunities. Under all of these cultural umbrellas, local ethnicity can add another variable, and migration, within and from without the region, brings yet another component of diversity.

FIGURE 6.2 Amos Oz and his wife Nily.

THINGS TO REMEMBER

1. While most countries have some elected bodies of government, democratization has been slow.

2. Al Jazeera, the most popular media source in the region, covers sensitive political issues, advocates free speech and political reform, and exposes political corruption—practices for which

many governments in the region have banned it. Many of its broadcasts favor the Arab perspective, but it aspires to remain neutral. When Al Jazeera tries to cover Israeli views, it risks criticism from Arabs who accuse it of having a pro-Israeli bias.

3. The region's oil and gas deposits are concentrated around the Persian Gulf, though some North African countries also have deposits.

I THE GEOGRAPHIC SETTING

Terms in This Chapter

The common term *Middle East* is not used in this chapter because it describes the region from a European perspective. To someone in Japan, the region lies to the far west, and to a Russian, it lies to the south. The *Arab world* is also not used because not all people in the region are of Arab ethnicity. In this book, the term used for this region is *North Africa* and *Southwest Asia*.

We use the term **occupied Palestinian Territories (OPT)** to refer to Gaza and the West Bank, since that is how the United Nations currently refers to those areas where Israel still exerts control despite treaty agreements; the word "occupied" is not capitalized to show the supposed temporary quality of the occupation. The U.S. Department of State uses the term *Palestinian Territories*.

> **occupied Palestinian Territories (OPT)** Palestinian lands occupied by Israel in 1967

Physical Patterns

Landforms and climates are particularly closely related in this region. The climate is dry and hot in the vast stretches of relatively low flat land; it is somewhat moister where mountains are able to capture rainfall. The lack of vegetation perpetuates aridity. Without plants to absorb and hold moisture, the rare but occasionally copious rainfall simply runs off, evaporates, or sinks rapidly into underground aquifers.

Climate

No other region in the world is as dry as North Africa and Southwest Asia (see Photo Essay 6.1A). A belt of dry air that circles the planet between roughly 20° and 30° N creates desert climates in the Sahara of North Africa, the Arabian Peninsula, Iraq, and Iran.

The Sahara's size and location under this high-pressure belt of dry air make it a particularly hot desert region. In some places, temperatures can reach 130°F (54°C) in the shade at midday. With little water or moisture-holding vegetation to retain heat, nighttime temperatures can drop quickly to below freezing. Nevertheless, in even the driest zones, humans survive at scattered oases, where they maintain groves of drought-resistant plants such as date palms. Desert inhabitants often wear light-colored, loose, flowing robes that reflect the sunlight, retain body moisture during the day, and provide warmth at night.

In the uplands and at the desert margins, enough rain falls to nurture grass, some trees, and limited agriculture. Such is the case in western Morocco; northern Algeria and Tunisia; the highlands of Yemen; Turkey; and the northern parts of Iraq and Iran. The rest of the region, generally too dry for cultivation, has long been the prime herding lands for nomads, such as the Kurds of Southwest Asia, the Berbers in North Africa, and the Bedouin of the steppes and deserts on the Arabian Peninsula. Recently, most nomads have been required to settle down and some of their lands are now irrigated for commercial agriculture, but the general aridity of the region means that sources of irrigation water are scarce.

Landforms and Vegetation

The rolling landscapes of rocky and gravelly deserts and steppes cover most of North Africa and Southwest Asia (see **Figure 6.3** on page 211; see also Photo Essay 6.1). In a few places, mountains capture moisture, allowing plants, animals, and humans to flourish. In northwestern Africa, the Atlas Mountains stretch from Morocco on the Atlantic coast to Tunisia on the Mediterranean coast. They block and lift damp winds from the Atlantic Ocean, creating rainfall of more than 50 inches (127 centimeters) per year on windward slopes. In some Atlas Mountain locations, there is enough snowfall to support a skiing industry.

Africa and Southwest Asia are separated by a rift formed between two tectonic plates—the African Plate and the Arabian Plate—that are moving away from each other (see Figure 1.21 on page 45). The rift, which began to form about 12 million years ago, is now filled by the Red Sea. The Arabian Peninsula lies to the east of this rift. There, mountains bordering the rift in the southwestern corner rise to 12,000 feet (3658 meters).

Behind these mountains to the east lies the great desert region of the Rub'al Khali. Like the Sahara, it has virtually no vegetation. The sand dunes of the Rub'al Khali, which are constantly moved by strong winds, are among the world's largest, some reaching more than 2000 feet (610 meters).

The landforms of Southwest Asia are more complex than those of North Africa. The Arabian Plate is colliding with the Eurasian Plate and pushing up the mountains and plateaus of Turkey and Iran (see Figure 1.21). Turkey's mountains lift damp air passing over Europe and the Mediterranean from the Atlantic, resulting in considerable rainfall. Only a little rain makes it over the mountains to the interior of Iran, which is very dry. The tectonic movements that create mountains also create earthquakes, a common hazard in Southwest Asia.

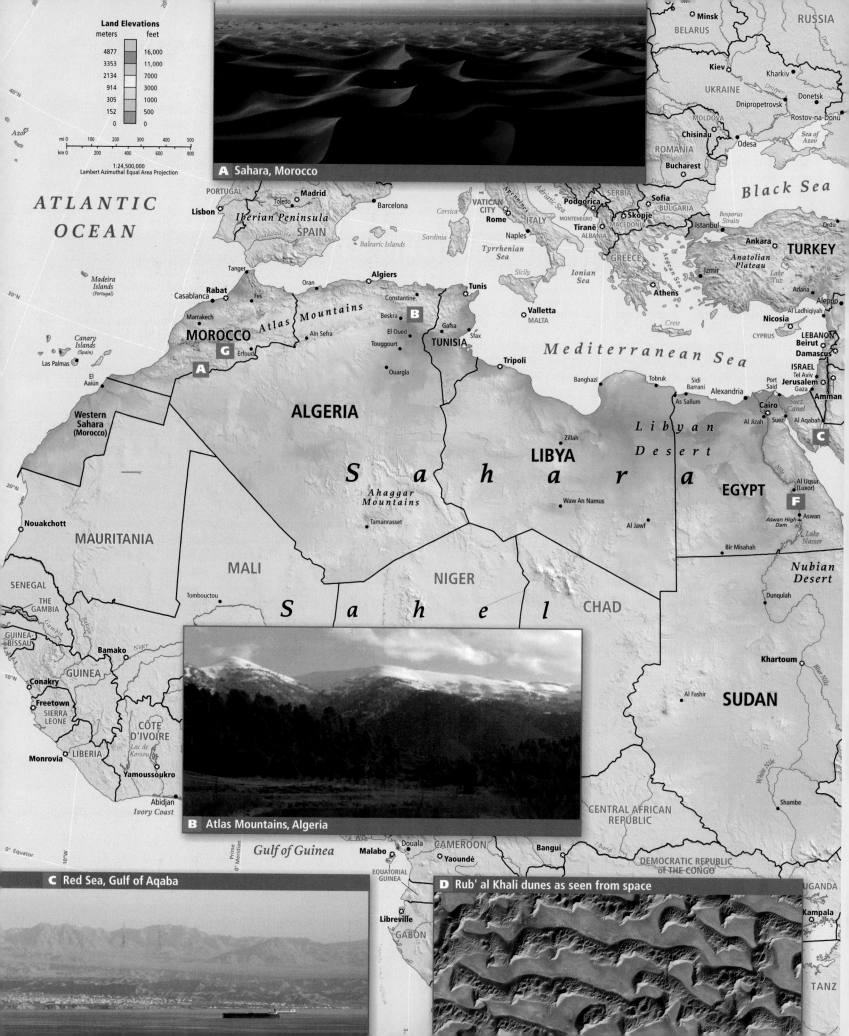

Land Elevations

meters	feet
4877	16,000
3353	11,000
2134	7000
914	3000
305	1000
152	500
0	0

mi 0   100   200   300   400   500
km 0   200   400   600   800

1:24,500,000
Lambert Azimuthal Equal Area Projection

A Sahara, Morocco

B Atlas Mountains, Algeria

C Red Sea, Gulf of Aqaba

D Rub' al Khali dunes as seen from space

ATLANTIC OCEAN

Black Sea

Mediterranean Sea

Azores

Madeira Islands (Portugal)

Canary Islands (Spain)
Las Palmas

PORTUGAL
Lisbon
Toledo
Madrid
Barcelona
Iberian Peninsula
SPAIN
Guadalquivir

Corsica
VATICAN CITY
Rome
ITALY
Naples
Sardinia
Balearic Islands

Tyrrhenian Sea
Sicily
MALTA
Valletta

Apennines
Adriatic Sea
MONTENEGRO
Podgorica
SERBIA
Skopje
MACEDONIA
ALBANIA
Tiranë
GREECE
Athens
Ionian Sea
Crete
Aegean Sea

SOFIA
BULGARIA
ROMANIA
Bucharest
Danube

MOLDOVA
Chisinau
Odesa
Sea of Azov

BELARUS
Minsk
UKRAINE
Kiev
Dnieper
Kharkiv
Dnipropetrovsk
Donetsk
Rostov-na-Donu

RUSSIA
Don

Bosporus Straits
Istanbul
TURKEY
Ankara
Anatolian Plateau
Izmir
Lake Tuz
Adana
Al Ladhiqiyah
Aleppo
Nicosia
CYPRUS
LEBANON
Beirut
Damascus
ISRAEL
Tel Aviv
Jerusalem
Gaza
Amman
Ordu

Tanger
Rabat
Casablanca
Fes
Marrakech
MOROCCO
Erfoud
Oran
Algiers
Atlas Mountains
Ain Sefra
Constantine
Beskra
El Oued
Touggourt
TUNISIA
Tunis
Gafsa
Sfax
Tripoli
Banghazi
Tobruk
Sidi Barrani
As Sallum
Alexandria
Port Said
Cairo
Al Jizah
Suez
Al Aqabah
Suez Canal

Western Sahara (Morocco)
El Aaiún

ALGERIA

Sahara

Ahaggar Mountains
Tamanrasset

LIBYA
Zillah
Waw An Namus
Al Jawf

Libyan Desert

EGYPT
Al Uqsur (Luxor)
Aswan High Dam
Aswan
Lake Nasser
Bir Misahah

Nile

Nubian Desert
Dunqulah

Nouakchott
MAURITANIA

MALI

Sahel

NIGER

CHAD

SUDAN
Khartoum
Al Fashir

Tombouctou

SENEGAL
THE GAMBIA
GUINEA-BISSAU
Bamako
GUINEA
Conakry
Freetown
SIERRA LEONE
LIBERIA
Monrovia
CÔTE D'IVOIRE
Lac de Kossou
Yamoussoukro
Abidjan
Ivory Coast

Gulf of Guinea

Niger
Gambia
Bakoy

Prime Meridian
0° Meridian
0° Equator
10°W
10°N
20°N
30°N
40°N

Blue Nile
White Nile

CENTRAL AFRICAN REPUBLIC

Malabo
Douala
CAMEROON
Yaoundé
Bangui
EQUATORIAL GUINEA
Libreville
GABON
DEMOCRATIC REPUBLIC OF THE CONGO
Ubangi
UGANDA
Kampala
TANZ

Shambe

Red Sea
Gulf of Aqaba

E Mt. Ararat, Turkey

F Nile River, Egypt

G A Wadi, Morocco

FIGURE 6.3 Regional map of North Africa and Southwest Asia.

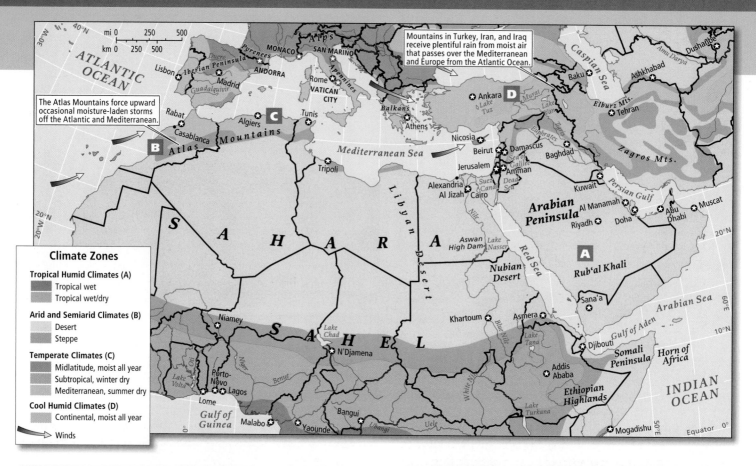

Mountains in Turkey, Iran, and Iraq receive plentiful rain from moist air that passes over the Mediterranean and Europe from the Atlantic Ocean.

The Atlas Mountains force upward occasional moisture-laden storms off the Atlantic and Mediterranean.

Climate Zones

Tropical Humid Climates (A)
- Tropical wet
- Tropical wet/dry

Arid and Semiarid Climates (B)
- Desert
- Steppe

Temperate Climates (C)
- Midlatitude, moist all year
- Subtropical, winter dry
- Mediterranean, summer dry

Cool Humid Climates (D)
- Continental, moist all year
- Winds

A Desert, Saudi Arabia

B Steppe, Morocco

C Mediterranean, summer dry, Algeria

D Continental, moist all year, Turkey

There are only three major river systems in the entire region, and all have attracted human settlement for thousands of years. The Nile flows north from the moist central East African highlands. It crosses arid Sudan and desert Egypt and forms a large delta on the Mediterranean. The Euphrates and Tigris rivers both begin with the rain that falls in the mountains of Turkey; the rivers flow southeast to the Persian Gulf. A fourth and much smaller river, the Jordan, starts as snowmelt in the uplands of southern Lebanon and flows through the Sea of Galilee to the Dead Sea. Most other streams are dry riverbeds, or wadis, most of the year, carrying water only after generally light rains that fall between November and April.

North Africa and Southwest Asia were home to some of the very earliest agricultural societies. Today, rain-fed agriculture is practiced primarily in the highlands and along parts of the Mediterranean coast, where there is enough precipitation to grow citrus fruits, grapes, olives, and many vegetables, though supplemental irrigation is often needed. Irrigated agriculture is more widespread but concentrated in the valleys of the major rivers, where seasonal flooding fills irrigation channels, and where aquifers are tapped by wells to water cotton, wheat, barley, vegetables, and fruit trees.

Environmental Issues

Environmental concerns are only beginning to be an overt focus in this region. This is in part because for thousands of years people here have confronted the challenges of a naturally arid environment.

> **Qur'an (or Koran)** the holy book of Islam, believed by Muslims to contain the words Allah revealed to Muhammad through the archangel Gabriel

An Ancient Heritage of Water Conservation

The **Qur'an** (or **Koran**), the holy book of Islam, guides believers to avoid spoiling or degrading human and natural environments and to share resources, especially water, with all forms of life. In actual practice, the residents of this region conserve water better than most people in the world. Daily bathing is a religious requirement, and it often takes place in public baths where water use is minimized. For millennia, mountain snowmelt has been captured and moved to dry fields and villages via constructed underground water conduits called *qanats*. Likewise, traditional architectural designs are used to create buildings that stay cool by maximizing shade and airflow.

Despite their long history of water-conserving technologies and practices, however, this region's 477 million residents now have such limited water resources that even clever combinations of ancient and modern measures are no longer sufficient to ensure an adequate supply. Rapidly growing populations and unwise modern usages of water virtually guarantee that water shortages will be more extreme in the future (see Photo Essay 6.2 B, C, D on page 215). According to the map in Figure 6.4, Turkey has the most plentiful water resources, followed by Iran, Iraq, and Syria. However, all other countries in the region suffer from freshwater scarcity, meaning that they have less water than the minimum the United Nations considers necessary to support basic human development—1000 cubic meters per person per year (see Figure 6.4).

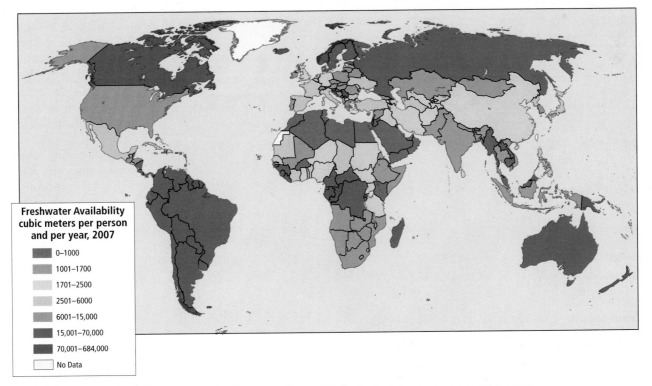

FIGURE 6.4 Global freshwater availability and stress, 2007. Freshwater stress and scarcity occur when so much water is withdrawn from rivers, lakes, and aquifers that not enough water remains to meet human and ecosystem requirements for sustainability.

Water and Food Production

Learning Goal 1
Water, Food, and Climate Change: How will people get enough water to grow food in the future, especially if climate change makes this dry region even drier?

The greatest use of water in North Africa and Southwest Asia is for irrigated agriculture, even though agriculture does not contribute significantly to national economies (see Thematic Overview A on page 207). In Tunisia, for example, agriculture accounts for just 10 percent of GDP but 86 percent of all the water used. Only 13 percent of water is used in homes, and only 1 percent by industry. Nonetheless, when irrigated agriculture is measured in terms of its value to local diets, jobs, family budgets, and rural economies, it emerges as essential in this water-stressed region; hence, most countries subsidize irrigation for food and fiber crops in some way.

Imported Food and Virtual Water This region's substantial human population and its arid climate mean that almost all people consume imported food (see Thematic Overview B). The water used to produce this imported food must be added to the virtual water consumption of the citizens of this region. *Virtual water*, now a widely accepted term in water scarcity discussions, is the volume of water used to produce all that a person consumes in a year (see Chapter 1 on page 37). For example, 1 kilogram (2.2 pounds) of beef requires 15,500 liters (4094 gallons) of water

to produce; 1 kilogram of goat meat requires 4000 liters (1,056 gallons); and 1 kilogram of corn requires 900 liters (238 gallons). Beef and corn are common imports in the region, while goat meat is produced locally and has a somewhat less significant water component than beef.

Until the twentieth century, agriculture was confined to a few coastal and upland zones where rain could support agriculture, and to river valleys (such as the Nile, Tigris, and Euphrates valleys) where farms could be irrigated with simple gravity-flow technology. However, to accommodate population growth and development, Libya, Egypt, Saudi Arabia, Tunisia, Turkey, Israel, and Iraq all now have ambitious mechanized irrigation schemes that have expanded agriculture deep into formerly uncultivated desert environments (Photo Essay 6.2 B, C, D; see also Figure 6.5).

Over time, irrigation projects damage soil fertility through **salinization**. When irrigation is used in hot, dry environments, the water evaporates, leaving behind a salty residue of minerals or other contaminants. Over time, so much residue accumulates that the plants are unable to grow or even survive. This human-made process is one of the largest environmental water issues that the world faces today in arid and semiarid regions.

Israel has developed relatively efficient techniques of drip irrigation that dramatically reduce the amount of water used, thereby limiting salinization and freeing up water for other uses. Until very recently, however, poorer states have been unable to afford this somewhat complex technology, which requires an extensive network of hoses and pipes to deliver water to each plant. Other countries are wary of depending on a technology developed by Israel, a country they deeply distrust.

Strategies for Increasing Access to Water Some strategies have been developed for increasing supplies of fresh water, but each presents a set of difficulties. All

> **salinization** a process that occurs when large quantities of water are used to irrigate areas where evaporation rates are high, leaving behind dissolved salts and other minerals

Thinking Geographically

After you have read about the human impact on the Biosphere in North Africa and Southwest Asia, you will be able to answer the following questions:

B What other dam in the region is described in the text as having created problems after it was constructed?

C How might the need for this type of irrigation be related to the Aswan dams?

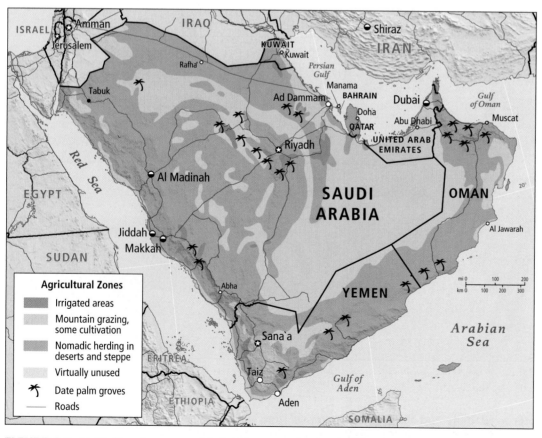

Agricultural Zones
- Irrigated areas
- Mountain grazing, some cultivation
- Nomadic herding in deserts and steppe
- Virtually unused
- Date palm groves
- Roads

FIGURE 6.5 Agricultural zones and irrigation areas in the Arabian Peninsula.

War and water scarcity have resulted in major impacts on the Biosphere in this region. The growth of populations promises to increase these stresses on ecosystems.

A An attack on an oil pipeline near Kirkuk, Iraq, resulted in a large spill that quickly caught fire. The largest oil spill in history occurred during the first Gulf War (1991) when the Iraqi government deliberately spilled 300 million gallons of oil (30 times the amount spilled by the *Exxon Valdez*) into the Persian Gulf in order to thwart a land invasion by the United States. Lebanon suffered a smaller, though similarly devastating, spill in 2006 as a result of Israeli bombing raids.

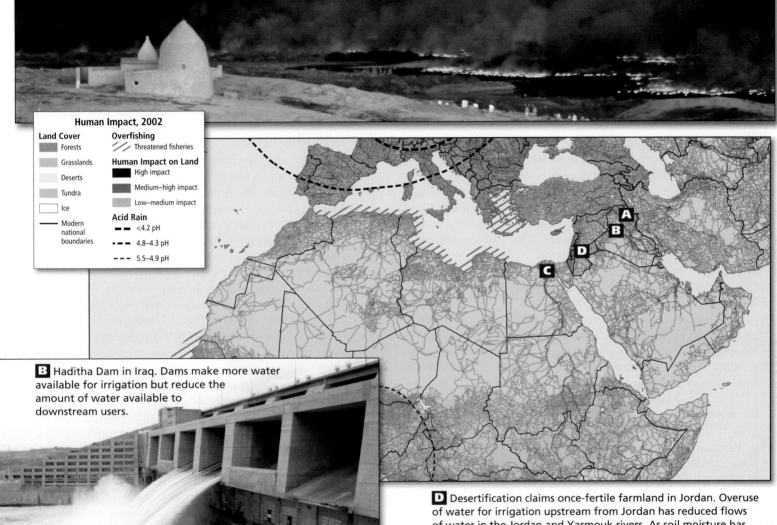

Human Impact, 2002

Land Cover
- Forests
- Grasslands
- Deserts
- Tundra
- Ice
- — Modern national boundaries

Overfishing
- /// Threatened fisheries

Human Impact on Land
- High impact
- Medium–high impact
- Low–medium impact

Acid Rain
- – – <4.2 pH
- –·– 4.8–4.3 pH
- – – – 5.5–4.9 pH

B Haditha Dam in Iraq. Dams make more water available for irrigation but reduce the amount of water available to downstream users.

D Desertification claims once-fertile farmland in Jordan. Overuse of water for irrigation upstream from Jordan has reduced flows of water in the Jordan and Yarmouk rivers. As soil moisture has decreased, many plants have died off.

C In Egypt, a gas-powered engine pumps water from a ditch onto a field. This type of irrigation has increased soil salinity and reduced soil fertility.

of them are expensive, some have enormous potential to cause increased wasting of water, and most do not include guarantees of equal access to the increased supply.

Seawater desalination The fossil fuel–rich countries of the Persian Gulf have invested heavily in **seawater desalination** technologies that remove the salt from seawater, making it suitable for drinking or irrigating. Seventy percent of Saudi Arabia's drinking water is supplied by desalination plants, and some of its wheat fields are irrigated with desalinated water. However, the process of desalination uses huge amounts of energy, potentially contributing to global warming. Furthermore, if all the costs of producing food with desalinated water were counted—the energy to desalinate, the irrigation equipment, and the fact that irrigated soil inevitably loses productivity due to soil salinization—the wheat so produced would be exorbitantly expensive. Currently the water used to grow food is heavily subsidized.

Groundwater pumping Many countries pump groundwater from underground aquifers to the surface for irrigation or drinking water. Libya has invested some of its earnings from fossil fuels into one of the world's largest groundwater pumping projects, known as the *Great Man-Made River*. This project draws on ancient fossil water, deposited at least 14,000 years ago, to supply almost 2 billion gallons of water per day to Libya's coastal cities, and to 600 square miles of agricultural fields. With this irrigated agriculture, Libya hopes to grow enough food to end its current food imports, and even supply food to the EU. However, the aquifer that the project depends on is not being replenished, making this use of groundwater unsustainable. By 1982, hydrologists were reporting fissures in the land surface above the aquifer of widths up to 16 inches (100 cm) that are associated with land sinking, or *subsidence*, as the water is withdrawn. Another problem is seawater intruding as fresh water is withdrawn, so that aquifers and wells in coastal zones are rendered brackish and useless.

Dams and reservoirs Dams and reservoirs have been built on the regions' major river systems to increase water supplies, but these projects have created new problems (see Photo Essay 6.2B). In Egypt, for example, the natural cycles of the Nile River have been altered by the construction of the Aswan dams. Downstream of these two dams, water flows have been reduced and floods no longer deposit fertility-enhancing silt on the land. As a result, expensive fertilizers are now necessary. Moreover, with less water and silt coming downstream, parts of the Nile delta are sinking into the sea. Upstream, the artificial reservoir created by the dams has flooded villages, fields, wildlife habitats, and historic sites, and created still-water pools that harbor parasites. Dams can also cause hardship across borders. The Aswan Dam sits on Egypt's border with Sudan, long an area of contention.

Water-acquisition strategies strain cross-border relations The need to share the water of rivers that flow across or close to national boundaries often impedes national development efforts. Turkey's Southeastern Anatolia Project, which involves the construction of several large dams on the Euphrates River for hydropower and irrigation purposes, has reduced the flow of water to the downstream

countries of Syria and Iraq. In negotiations over who should get Euphrates water, for example, Turkey argues that it should be allowed to keep more water behind its dams because the river starts in Turkey and most of its water originates there as mountain rainfall (see Figure 6.6 on page 218). Meanwhile, Iraq points out that the Euphrates travels the longest distance in Iraq.

▐▐ ▶ 285. THE JORDAN RIVER IS DYING

Vulnerability to Climate Change

North Africa and Southwest Asia are especially vulnerable to both the effects of climate change (especially global warming) and the world's attempts to reduce greenhouse gas emissions, which will ultimately reduce revenues from the sale of oil and gas. As the climate warms, a sea level rise of a few feet could severely impact the Mediterranean coast, especially the Nile delta, one of the poorest and most densely populated areas in the world (Photo Essay 6.3A). Elsewhere, shifting rainfall patterns resulting from changes in the regional climate could drastically reduce water availability where it is already scarce and where people are impoverished (see Photo Essay 6.3 B, C), causing more nondesert lands to become transformed into deserts. Conversely, under some climate-change scenarios, periods of unusually intense rainfall could increase, causing flooding.

Independent of any environmental changes, global efforts to reduce fossil fuel consumption could devastate oil- and gas-based economies and transform the region's geopolitics, leaving it with much less income and much less power. Despite the enormous wealth from fossil fuel sales that has flowed into some of these countries over the past 40 years, few countries have undertaken significant economic diversification to prepare for reduced global consumption of fossil fuels.

Thinking Geographically

After you have read about the vulnerability to climate change in North Africa and Southwest Asia, you will be able to answer the following questions:

A What about Alexandria's location makes it particularly vulnerable to sea level rise?

B What is complicating emergency response and planning efforts in Sudan that might increase its resilience to climate change?

C What does Yemen receive, from outside of its borders, that makes it more vulnerable to the impacts of climate change?

> **seawater desalination** the removal of salt from seawater—usually accomplished through the use of expensive and energy-intensive technologies—to make the water suitable for drinking or irrigating
>
> **desertification** a set of ecological changes that converts nondesert lands into deserts

Climate Change and Desertification

Climate change could accelerate **desertification**, the conversion of nondesert lands into deserts (see Photo Essay 6.2D; see also Thematic Overview C). As temperatures increase, soil moisture decreases because of evaporation. As a result, plant cover is reduced, which causes less protection for arid soil. Bare patches of soil can become badly eroded by wind, and eventually sand dunes can blow onto formerly vegetated land.

Water scarcity, sea level rise, poverty, and political instability are some of the factors that make parts of this region highly vulnerable to climate change.

A Alexandria, Egypt, located on the Mediterranean coast of the Nile delta region, is a low-lying city that highly exposed to rising sea levels. Efforts to improve the sea wall around the city, visible above, may be outpaced by rising sea levels. The entire Nile delta is struggling to adapt to salt water that is moving up rivers and permeating soils, making agriculture extremely difficult, and polluting freshwater resources, that cities depend on. Half of Egypt's population lives in the Nile delta, and 80 percent of the country's imports and exports run through Alexandria.

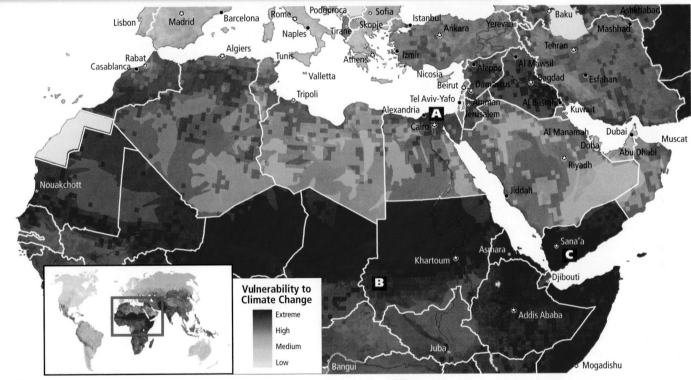

Vulnerability to Climate Change

Extreme
High
Medium
Low

B Refugees line up for food and water in Darfur, Sudan. This water-scarce area is very susceptible to drought. Meanwhile, widespread poverty leaves the population quite sensitive to any disruption in food or water supplies. Political instability complicates emergency response and longer-term planning efforts that might lead to greater resilience.

C Residents on the outskirts of San'a, Yemen, ride out a sandstorm by the side of the road. The same factors that make Darfur, Sudan, very vulnerable to climate change also affect Yemen. However, Yemen also receives large numbers of refugees from Somalia and other parts of Africa, which further stretches resources and frustrates planning efforts.

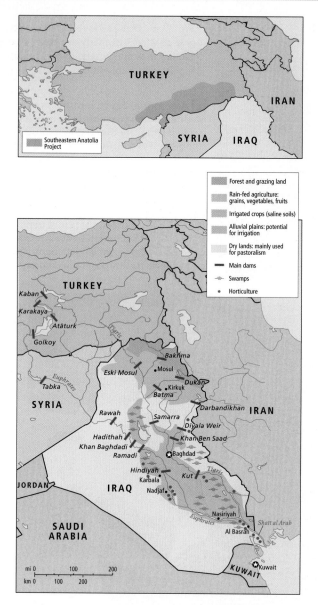

FIGURE 6.6 Dams on the Tigris and Euphrates drainage basins. Turkey's projects to manage the Tigris and Euphrates river basins through dam construction have international implications. Water that is retained in Turkey will not reach its neighbors. The main map shows Turkey's dams on the headwaters of the two rivers, as well as dams built in Syria and Iraq, all of which will have environmental effects, especially on the lower reaches of both rivers in Iraq. The smaller map shows the full extent of Turkey's Southeastern Anatolia Project.

In the grasslands (steppes) that often border deserts, a wide array of land use changes can contribute to the general drying. For example, as groundwater levels fall due to pumping for cities and irrigated agriculture, plant roots can no longer reach sources of moisture. In other cases, international development agencies have encouraged nomadic herders to take up settled cattle ranching of the type practiced in western North

> **Fertile Crescent** an arc of lush, fertile land formed by the uplands of the Tigris and Euphrates river systems and the Zagros Mountains, where nomadic peoples began the earliest known agricultural communities

America. The thinking is that nomadism is dysfunctional in the modern world where records must be kept, taxes collected, and children must be sent to school. All of these activities are complicated by the mobility of nomads. However, ranching on fragile grasslands has led to overgrazing and excessive water use. Irrigation projects using groundwater were established to support ranching, but these projects also depleted water resources, resulting in desertification. Meanwhile encouraging nomads to settle may increase their vulnerability to climate change, as mobility has long helped them adapt to climate variability.

THINGS TO REMEMBER

1. Landforms and climates are particularly closely related in this region, and it is the driest region in the world. Rolling deserts and steppes cover most of the region. In a few places, mountains capture moisture, allowing plants, animals, and humans to flourish—but sustainability is becoming more difficult.

2. **Learning Goal 1:** Water, Food, and Climate Change As the population of such a dry region increases, obtaining enough cultivable land and water for agriculture will become harder. Many countries are dependent on imported food, which causes food insecurity when global prices rise. Climate change could reduce food output both locally and globally. New technologies may offer solutions, but are expensive.

Human Patterns over Time

Important developments in agriculture, societal organization, and urbanization took place long ago in this part of the world. Three of the world's great religions were born here: Judaism, Christianity, and Islam.

Agriculture and the Development of Civilization

Between 10,000 and 8000 years ago, formerly nomadic peoples founded some of the earliest known agricultural communities in the world. These communities were located in an arc formed by the uplands of the Tigris and Euphrates river systems (in modern Turkey and Iraq) and the Zagros Mountains of modern Iran (see Timeline A on page 222). This zone is often called the **Fertile Crescent** (Figure 6.7) because of its plentiful fresh water, open forests and grasslands, abundant wild grains, and fish, goats, sheep, wild cattle, and other large animals.

The skills of these early people in domesticating plants and animals allowed them to build ever more elaborate settlements. The settlements eventually grew into societies based on widespread irrigated agriculture along the base of the mountains and in river basins, especially along the Tigris and Euphrates. Nomadic herders living in adjacent grasslands traded animal products for the grain and other goods produced in the settled areas.

Over the next several thousand years, agriculture spread to the Nile Valley, west across North Africa, east to the mountains of Persia (modern Iran), and ultimately influenced other cultivation systems worldwide. Eventually, the agricultural settlements took on urban qualities: dense populations, specialized occupations, concentrations of wealth, and centralized government and bureaucracies. For example, the agricultural villages of Sumer (in modern southern Iraq), which existed 5000 years ago, gradually turned into city-states that extended their influence over the surrounding territory. The Sumerians developed wheeled vehicles, oar-driven ships, and irrigation technology.

At times, nomadic tribes who had adopted the horse as a means of conquest banded together and, with devastating cavalry raids, swept over settlements. They then set themselves up as a ruling class, but soon they adopted the settled ways and cultures of the peoples they conquered and thus themselves would become vulnerable to attack.

Agriculture and Gender Roles

Increasing research evidence suggests that the dawning of agriculture may have marked the transition to markedly distinct roles for men and women. Archaeologist Ian Hodder reports that at the 9000-year-old site of Çatalhöyük, near Konya in south-central Turkey (see Figure 6.7), where the economy was primarily hunting and gathering, there is little evidence of gender differences. Families were small and men and women performed similar chores in daily life. Both had comparable status and power, and both played key roles in social and religious life.

Scholars think that after the development of agriculture, as wealth and property became more important in human society, a concern with family lines of descent and inheritance emerged. This led in turn to the idea that women's bodies needed to be controlled so that a woman could not become pregnant by a man other than her mate and thus confuse lines of inheritance.

FIGURE 6.7 The Fertile Crescent, one of the earliest known agricultural sites. About 10,000 years ago, people in the Fertile Crescent began domesticating cereal grains, legumes, and animals, especially sheep and goats. The uses of domesticated animals spread into Europe and Africa as agricultural peoples traded their surpluses for other goods or moved into other regions. Three major empires developed successively in the eastern part of the Fertile Crescent: the Sumerian, the Babylonian, and the Assyrian. **(A)** Sheep graze near the ruins of Apamea, founded about 300 B.C.E. by one of Alexander the Great's generals in Syria.

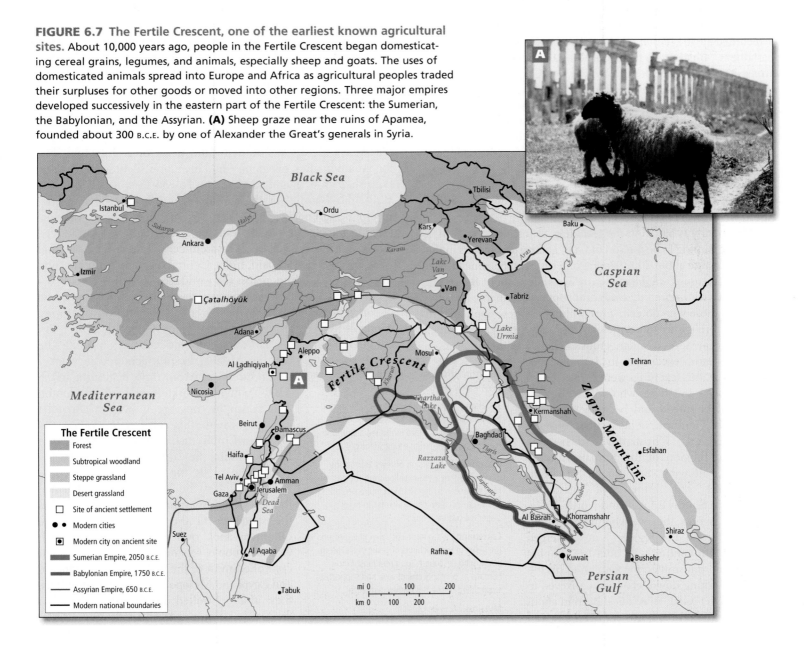

The Coming of Monotheism: Judaism, Christianity, and Islam

The very early religions of this region were based on a belief in many gods who controlled natural phenomena; such was the case through the Greek era and into the Roman period. Several thousand years ago, **monotheistic** belief systems—those based on one god—began to emerge. The three major monotheistic world religions—Judaism, Christianity, and Islam—all have connections to the eastern Mediterranean where the city of Jerusalem is sacred to all three. Muslims also revere Makkah (Mecca) and Al Madinah (Medina) in Saudi Arabia. All three religions have a connection to a sacred text: the Old Testament of the Bible for Jews; the Old and New Testaments for Christians; and for Muslims both the Bible and the Qu'ran are sacred.

Judaism was founded approximately 4000 years ago. According to tradition, the patriarch Abraham led his followers from Mesopotamia (modern Iraq) to the shores of the eastern Mediterranean (modern Israel and the occupied Palestinian Territories) where he founded Judaism. Jewish religious history is recorded in the Torah (the first five books of the Bible's Old Testament). Judaism is characterized by the belief in one God, a strong ethical code summarized in the Ten Commandments, and an enduring ethnic identity reinforced by dietary and religious laws.

After the Jews rebelled against the Roman Empire, which culminated in their expulsion in 73 C.E. from the eastern Mediterranean, some were enslaved by the Romans and most migrated to other lands in a movement known as the **diaspora** (the dispersion of an originally localized people). Many Jews dispersed across North Africa and Europe, and others went to various parts of Asia. After 1500, Jews were among the earliest European settlers in all parts of the Americas.

Christianity is based on the teachings of Jesus of Nazareth, a Jew who, claiming to be the son of God, gathered followers in the area of Palestine about 2000 years ago. Jesus, who became known as Christ (meaning *anointed one* or *Messiah*), taught that there is one God, who primarily loves and supports humans, but who will judge those who do evil. This philosophy grew popular, and both Jewish religious authorities and Roman imperial authorities of the time saw Jesus as a dangerous challenge to their power.

After Jesus' execution in Jerusalem in about 32 C.E., his teachings were written down (the Gospels) by those who followed him, and his ideas spread and became known as Christianity. Centuries of persecution ensued, but by 400 C.E., Christianity had become the official religion of the Roman Empire. However, following the spread of Islam after 622 C.E., only remnants of Christianity remained in Southwest Asia and North Africa.

Islam is now the overwhelmingly dominant religion in the region. Islam emerged in the seventh century C.E., after the Prophet Muhammad transmitted the Qur'an to his followers by writing down what was conveyed to him by Allah ("God" in Arabic). Born in about 570 C.E., Muhammad was a merchant and caravan manager in the small trading town of Makkah (Mecca) on the Arabian Peninsula near the Red Sea. Followers of Islam, called **Muslims**, believe that Muhammad was the final and most important in a long series of revered prophets, which includes Abraham, Moses, and Jesus.

Unlike many versions of Christianity, Islam has virtually no central administration and only informal religious hierarchy (this is somewhat less true of the Shi'ite version of Islam; see the discussion on page 223). The world's 1 billion Muslims may communicate directly with God (Allah). A clerical intermediary is not necessary, though there are numerous clerical leaders who help their followers interpret the Qur'an. An important result of the lack of a central authority is that the interpretation of Islam varies widely within and among countries and from individual to individual.

The Spread of Islam

Among the first converts to Islam were the Bedouin—nomads of the Arabian Peninsula. By the time of Muhammad's death in 632 C.E., they were already spreading the faith and creating a vast Islamic sphere of influence. Over the next century, Muslim armies built an Arab–Islamic empire over most of Southwest Asia, North Africa, and the Iberian Peninsula of Europe (Figure 6.8).

While most of Europe was stagnating during the medieval period (450–1500), the Arab–Islamic empire nurtured learning and economic development. Muslim scholars traveled throughout Asia and Africa, advancing the fields of architecture, history, mathematics, geography, and medicine. Centers of learning flourished from Baghdad (Iraq) to Toledo (Spain). During the early Arab–Islamic era, the development of banks, trusts, checks, receipts, and bookkeeping fostered vibrant economies and wideranging trade. The traders founded settlements and introduced new forms of living spaces. The architectural legacy of Arabs and Muslims lives on in Spain, India, Central Asia, the Americas, and in countless buildings across the world (see Timeline B).

By the end of the tenth century, the Arab–Islamic empire had begun to break apart. From the eleventh to the fifteenth centuries, Mongols from eastern Central Asia (converted to Islam by 1330) conquered parts of the Arab-controlled territory, forming the Muslim Mughal Empire centered in what is now north India. Meanwhile, beginning in the 1200s, nomadic Turkic herders from Central Asia began to converge in western Anatolia (Turkey) where they eventually forged the Ottoman Empire, which became the greatest Islamic empire the world has ever known.

By the 1300s, the Ottomans had become Muslims and by the 1400s they had defeated the Christian Byzantine Empire, which was the successor to the Roman Empire. The Ottomans took over the Byzantine capital, Constantinople, renamed it Istanbul, and soon controlled most of the eastern Mediterranean, and Egypt and Mesopotamia. By the late 1400s, they also controlled much of southeastern and central Europe. At about the same time,

monotheistic pertaining to the belief that there is only one god

Judaism a monotheistic religion characterized by the belief in one God, Yahweh, a strong ethical code summarized in the Ten Commandments, and an enduring ethnic identity

diaspora the dispersion of Jews around the globe after they were expelled from the eastern Mediterranean by the Roman Empire beginning in 73 C.E.; the term can now refer to other dispersed culture groups

Christianity a monotheistic religion based on the belief in the teachings of Jesus of Nazareth, a Jew, who described God's relationship to humans as primarily one of love and support, as exemplified by the Ten Commandments

Muslims followers of Islam

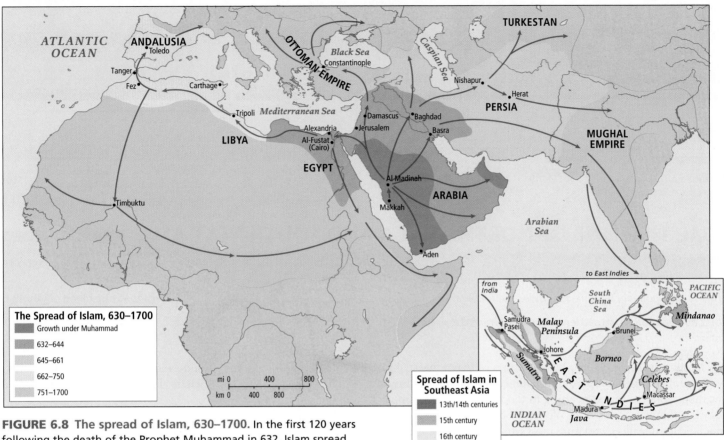

FIGURE 6.8 The spread of Islam, 630–1700. In the first 120 years following the death of the Prophet Muhammad in 632, Islam spread primarily by conquest. Over the next several centuries, Islam was carried to distant lands by both traders and armies.

the Arab Muslims lost their control of the Iberian Peninsula to Christian kingdoms. Today, Islam still dominates in a huge area that stretches from Morocco to western China and includes northern India, and Malaysia and Indonesia in Southeast Asia.

Once a location was completely conquered, the Ottoman Empire, like the Arab–Islamic empire before it, encouraged religious tolerance toward the conquered peoples so long as they adhered to a religion with a sacred text. Jews, Christians, Buddhists, and Zoroastrians were allowed to practice their religions, although there were attractive economic and social advantages to converting to Islam. Multicultural urban life in Ottoman cities facilitated vast trading networks spanning the known world, and Istanbul became a cosmopolitan capital with elaborate buildings and lavish public parks that outshined anything in Europe until the nineteenth century (see Timeline C).

Western Domination and State Formation

The Ottoman Empire ultimately withered in the face of a Europe made powerful by colonialism and the Industrial Revolution. Throughout the nineteenth century, North Africa provided raw materials for Europe in a trading relationship dominated by European merchants. In 1830, France became the first European country to exercise direct control over a North African territory (Algeria; see Timeline D). France took control of Tunisia in 1881 and Morocco in 1912; Britain gained control of Egypt in 1882

and Sudan in 1898; and Italy took control of Libya in 1912 (see Figure 6.9 on page 224).

World War I (1914–1918) brought the fall of the Ottoman Empire, which in a strategic error had linked itself with Germany. At the end of the war, the victorious Allied powers dismantled the Ottoman Empire and all of the former Ottoman territories; only Turkey was recognized as an independent country. The rest of the formerly Ottoman-controlled territory was allotted to France and Britain as protectorates (see Figure 6.9B). On the Arabian Peninsula, Bedouin tribes were consolidated under Sheikh Ibn Saud in 1932, and Saudi Arabia began to emerge as an independent country.

The aftermath of World War II further affected the political development of North Africa and Southwest Asia. Most famously, in the aftermath of the Holocaust in Europe, the Jewish state of Israel was created in the eastern Mediterranean on land inhabited by Arab farmers and nomadic herders as well as some Jews (see the discussion on pages 239–243, and see Timeline E).

By the 1950s, European and U.S. energy companies played a key role in deciding who ruled Iran and Saudi Arabia—where vast oil deposits were to become especially lucrative. European textile companies played a similar role in Egypt, a major cotton exporter. The governments of these countries showed their loyalty to the foreign companies with low taxes on oil and cotton exports and easy access to land. While a tiny ruling elite in these countries

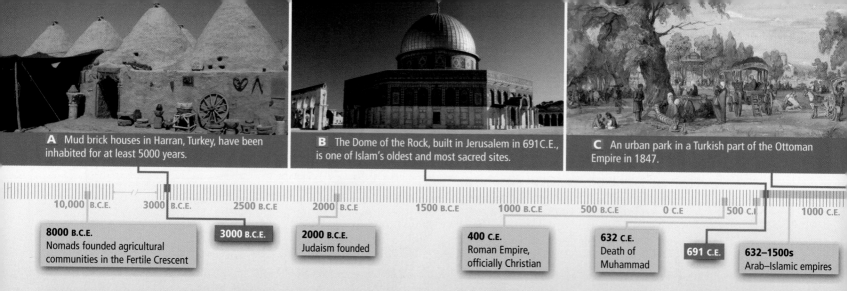

A Mud brick houses in Harran, Turkey, have been inhabited for at least 5000 years.

B The Dome of the Rock, built in Jerusalem in 691C.E., is one of Islam's oldest and most sacred sites.

C An urban park in a Turkish part of the Ottoman Empire in 1847.

| 10,000 B.C.E. | 3000 B.C.E. | 2500 B.C.E | 2000 B.C.E | 1500 B.C.E | 1000 B.C.E | 500 B.C.E | 0 C.E | 500 C.I | 1000 C.E |

8000 B.C.E.
Nomads founded agricultural communities in the Fertile Crescent

3000 B.C.E.

2000 B.C.E.
Judaism founded

400 C.E.
Roman Empire, officially Christian

632 C.E.
Death of Muhammad

691 C.E.

632–1500s
Arab–Islamic empires

VISUAL TIMELINE OF NORTH AFRICA AND SOUTHWEST ASIA

Thinking Geographically

After you have read about the human history of North Africa and Southwest Asia, you will be able to answer the following questions:

A What allowed for urban settlements like Harran to develop?

B What is a major difference between Islam and many versions of Christianity?

C What Turkish city had lavish parks that outshined anything in Europe until the nineteenth century?

grew fabulously wealthy, the oil-tax revenues were not invested in creating opportunities for poor or middle-class people. Over time, ever more political power accrued to the foreign energy companies. The United States and Western Europe supported those autocratic local leaders who were most sympathetic to the interests of their energy companies and their Cold War strategic interests, as opposed to those of the Soviet Union (see Timeline F). As a result of these concerns, both the Europeans and the Americans supported undemocratic governments and stood in the way of reforms that would have resulted in a more educated populace that would be able to participate in democracy.

THINGS TO REMEMBER

1. About 10,000 years ago in the Fertile Crescent, formerly nomadic peoples founded some of the world's earliest known agricultural communities. The domestication of plants and animals allowed them to build ever more elaborate settlements that eventually grew into societies based on widespread irrigated agriculture.

2. The three major monotheistic world religions—Judaism, Christianity, and Islam—all have their origins in this region in the eastern Mediterranean. Islam is by far the largest in numbers of adherents in the region, and it is the principal faith in all of the region's countries except Israel.

3. Beginning in the nineteenth century and continuing through the end of World War II, European colonial powers ruled or controlled most countries in the region.

4. Following World War II, the state of Israel was created, and in countries with oil deposits, the United States and Western Europe supported those autocratic local leaders most likely to maintain a friendly attitude toward U.S. and European business and strategic interests.

II CURRENT GEOGRAPHIC ISSUES

For decades, social and political change in this region lagged behind economic development. Why did the wealth generated by oil not result in a spreading of opportunities, such as broad public education and an opening up of public discourse? Most analysts say that social change, such as the expansion of educational opportunities for women, is actually underway and is just beginning to gain momentum. Recently, political reform has accelerated rapidly, and will likely continue along with further social change and economic development.

Sociocultural Issues

This section explores the basics of Islam and examines the broad social changes occurring in the region with regard to family values, gender roles and gendered spaces, demographic change, urbanization and migration, and patterns of human well-being.

Religion in Daily Life

Ninety-three percent of the people in the region are followers of Islam; for them, the Five Pillars of Islamic Practice embody the

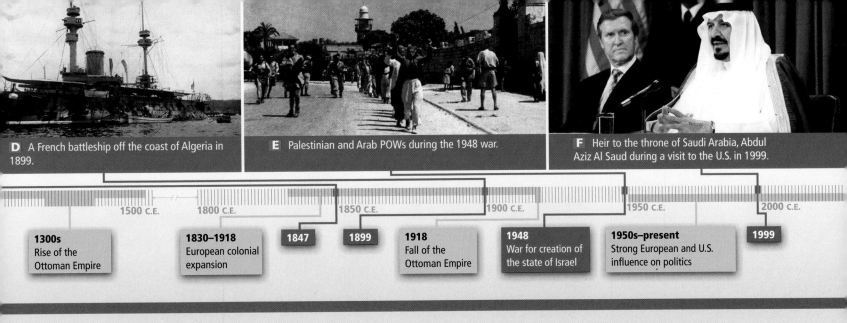

D A French battleship off the coast of Algeria in 1899.

E Palestinian and Arab POWs during the 1948 war.

F Heir to the throne of Saudi Arabia, Abdul Aziz Al Saud during a visit to the U.S. in 1999.

1500 C.E.	1800 C.E.	1850 C.E.	1900 C.E.	1950 C.E.	2000 C.E.	

1300s Rise of the Ottoman Empire

1830–1918 European colonial expansion

1847

1899

1918 Fall of the Ottoman Empire

1948 War for creation of the state of Israel

1950s–present Strong European and U.S. influence on politics

1999

D When did France first take control of territory in what is now Algeria?

E The state of Israel was formally created in the aftermath of what event in Europe?

F Why did European and U.S. companies become involved in deciding who ruled Iran and Saudi Arabia?

central teachings of Islam. Not all Muslims are fully observant, but the Pillars have an impact on daily life for all.

The Pillars of Muslim Practice

1. A testimony of belief in Allah as the only God and in Muhammad as his messenger (prophet).

2. Daily prayer at five designated times (daybreak, noon, mid-afternoon, sunset, and evening). Although prayer is an individual activity, Muslims are encouraged to pray in groups and in mosques. The call to prayer, broadcast five times a day in all parts of the region, is a constant reminder to all people to reflect on their beliefs.

3. Obligatory fasting (no food, drink, or smoking) during the daylight hours of the month of Ramadan, followed by a light celebratory family meal after sundown.

4. Obligatory almsgiving (*zakat*) in the form of a "tax" of at least 2.5 percent. The alms are given to Muslims in need. *Zakat* is based on the recognition of the injustice of economic inequity. Although it is usually an individual act, the practice of government-enforced *zakat* is returning in certain Islamic republics.

5. Pilgrimage (**hajj**) at least once in a lifetime to the Islamic holy places, especially Makkah (Mecca), during the twelfth month of the Islamic calendar.

Saudi Arabia occupies a prestigious position in Islam, as it is the site of two of Islam's three holy shrines: Makkah, the birthplace of the Prophet Muhammad and of Islam; and Al Madinah (Medina), the site of the Prophet's mosque and his burial place. (The third holy shrine is in Jerusalem.) The fifth pillar of Islam has placed Makkah and Al Madinah at the

hajj the pilgrimage to the city of Makkah (Mecca) that all Muslims are encouraged to undertake at least once in a lifetime

shari'a literally, "the correct path"; Islamic religious law that guides daily life according to the interpretations of the Qur'an

Sunni the larger of two major groups of Muslims, with different interpretations of shari'a

Shi'ite (or Shi'a) the smaller of two major groups of Muslims, with different interpretations of shari'a; Shi'ites are found primarily in Iran and southern Iraq

heart of Muslim religious geography. Each year, a large private sector service industry, owned and managed by members of the huge Saud family, organizes and oversees the 5- to 7-day hajj for more than 2.5 million foreign visitors.

Islamic Religious Law and Variable Interpretations Beyond the Five Pillars, Islamic religious law, called **shari'a**, "the correct path," guides daily life according to the principles of the Qur'an. There are many interpretations of the Qur'an and a wide variety of versions of the observant Muslim life. Some Muslims believe that no other legal code is necessary in an Islamic society, as shari'a provides guidance in all matters of life, including worship, finance, politics, marriage, diet, hygiene, war, and crime. Insofar as the interpretation of shari'a is concerned, the Muslim community is split into two major groups: **Sunni** Muslims, who today account for 85 percent of the world community of Islam, and **Shi'ite** (or **Shi'a**) Muslims, who live primarily in Iran but also in southern Iraq and southern Lebanon. Shi'ites recognize an authoritative priestly class whom they call *mullahs*.

The Sunni–Shi'ite split dates from shortly after the death of Muhammad, when divisions arose over who should succeed the Prophet and have the right to interpret the Qur'an for all Muslims. This division continues today. The original disagreements have been exacerbated by countless local disputes over land, resources, and philosophies. In Iraq, for example, conflict between Sunnis and Shi'ites has been intensified by the rivalry over political power and fossil fuel resources that followed the U.S. invasion in 2003.

Family Values and Gender

Perhaps because Islam has so little religious hierarchy, the family is the most important

223

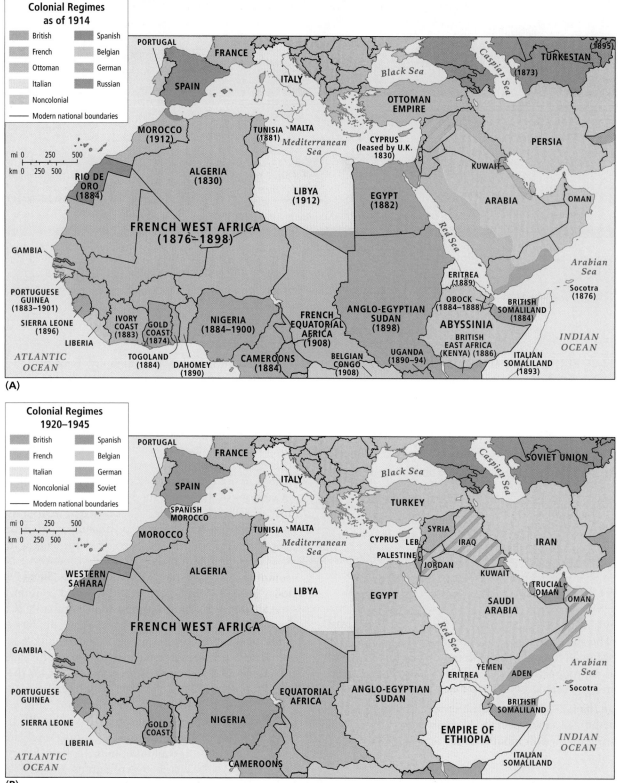

FIGURE 6.9 Colonial regimes in North Africa and Southwest Asia. (A) European powers began influencing the affairs of the region in the nineteenth century and expanded their control by 1914 at the beginning of World War I. The dates on the map indicate when the Europeans took control of each country. **(B)** Between 1920 and 1945, what was left of the Ottoman Empire in the eastern Mediterranean became protectorates administered by the British and French. The striped areas reflect the British colonial practice of allowing local rulers to govern while controlling many of their policies and actions.

institution in this region. Although the role of the family is changing, a person's family is still such a large component of personal identity that the idea of individuality is almost a foreign concept. Each person is first and foremost part of a family, and the defining parameter of one's role in the family is gender identity (see Thematic Overview I).

The head of the family is nearly always a man; even when a woman is widowed or divorced, she is under the tacit supervision of a male, perhaps her father or her son. An educated unmarried woman with a career outside the home is likely to live in the home of her parents or a brother and defer to them in decision making. Traditionally, men are considered more capable of making decisions and thus, it is thought, they should be in charge. These ideas, often labeled **patriarchal**, are now changing.

> **patriarchal** relating to a social organization in which the father is supreme in the clan or family
>
> **female seclusion** the requirement that women stay out of public view
>
> **Gulf states** Saudi Arabia, Kuwait, Bahrain, Oman, Qatar, and the United Arab Emirates

Gender Roles and Gendered Spaces

Carefully specified gender roles are common in many cultures, and there is often a spatial component to these roles. In the region of North Africa and Southwest Asia, in both rural and urban settings, the ideal is for men and boys to go forth into *public spaces*—the town square, shops, the market. Women are expected to inhabit primarily *private spaces*. But there may be many exceptions to those ideals.

To facilitate this ideal, traditional family compounds included a courtyard that was usually a private, female space within the home; the only men who could enter were relatives. For the urban upper classes, female space was an upstairs set of rooms with latticework or shutters at the windows, which increased the interior ventilation and from which it was possible to look out at street life without being seen. Today, the majority of people in the region live in urban apartments, yet even here there is a demarcation of public and private space. One or two formally furnished reception areas are reserved for non-family visitors, and rooms deeper into the dwelling are for family-only activities. When guests are present, women in the family are absent or present only briefly. Today, many women as well as men go out into public spaces, but how women enter these spaces remain an issue. Customs vary not only from country to country, but also from rural to urban settings and by social class.

The requirement that women stay out of public view (also known as **female seclusion**) is most strictly enforced in the more conservative Islamic countries of the **Gulf states** (Saudi Arabia, Kuwait, Bahrain, Oman, Qatar, and the United Arab Emirates). Here women are generally expected to stay out of public spaces except when on important business, accompanied by a male relative. In the more secular Islamic countries—Morocco, Tunisia, Libya, Egypt, Turkey, Lebanon, and Iraq—women regularly engage in activities that place them in public spaces. Some wear conservative religious clothing; others dress in western styles. Increasingly, female doctors, lawyers, teachers, and businesspeople are found in even the most conservative societies. Figure 6.10 compares the various levels of restrictions on women across the region.

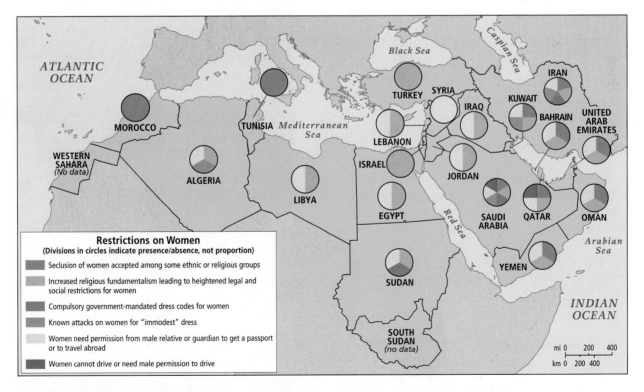

FIGURE 6.10 Variations in restrictions on women. The restrictions placed on women vary from country to country. Women's rights are perhaps most strongly protected in Turkey and Israel, where equality for women is constitutionally guaranteed, though religious fundamentalists are working to repeal these guarantees in both countries.

Affluent urban women may observe seclusion either very little (especially if they are highly educated), or even more strictly than do rural women. Although rural women are often more traditional in their outlook, they have many tasks that they must perform outside the home: agricultural work, carrying water, gathering firewood, and buying or selling food from markets. Meanwhile upper-class women can afford servants to perform daily tasks in public spaces.

Many women in this region use clothing as a way to create private space (Figure 6.11). This is done with the many varieties of the **veil**, which may be a garment that totally covers the person's body and face, or just a scarf that covers the person's hair. In

> **veil** the custom of covering the body with a loose dress and/or of covering the head—and in some places the face—with a scarf

some cultures, even prepubescent girls wear the veil; in others they go unveiled until their transition into adulthood is observed. The veil allows a devout Muslim woman to preserve a measure of seclusion when she enters a public space, thus increasing the space she may occupy with her honor preserved. A young woman may choose to wear a headscarf with jeans and a T-shirt in order to signal to the public that she is both a modern woman and an observant Muslim.

There is considerable debate about the origin and validity of female seclusion and veiling as specifically Muslim customs. Scholars of Islam say that these ideas predate Islam by thousands of years and do not derive from the teachings of the Prophet Muhammad. In fact, Muhammad may have been reacting against such customs when he advocated equal treatment of males and females. Muhammad's first wife, Khadija, did not practice seclusion, and worked as an independent businesswoman whose counsel Muhammad often sought.

▐▐▶ 146. EDUCATION, ECONOMIC EMPOWERMENT ARE KEYS TO BETTER LIFE FOR MUSLIM WOMEN

The Rights of Women in Islam

In the Gulf states, women cannot travel independently or even drive a car or shop without male supervision. Yet even here changes have recently been made. In the decade of the 2000s, women in the Gulf states became noticeably more active in public life, education, and business. Some Saudi women even staged mini-demonstrations by posting on YouTube videos of themselves driving. In Qatar and the United Arab Emirates (UAE), a few female activists, such as Sheikha Moza, a wife of the emir of Qatar, along with other female lawyers and journalists, have challenged persistent patriarchal attitudes. Saudi Arabia remains the most restrictive country, but even there it is now possible for a woman to register a business without first proving that she has hired a male manager.

While political equality with men may seem far off for the women of this region, economic equality may be even further away. With the sole exception of Israel, women in this region are paid, as a group, only about 40 percent of what men earn for comparable work. As a region, only South Asia has a similarly large gap in what is paid to men and women.

FIGURE 6.11 Variations on the veil. There is an almost infinite variety of interpretations of the veil. **(A)** An Iraqi woman wears a scarf through which her hair can be seen. **(B)** A Turkish woman wears a headscarf that covers her hair completely. **(C)** Schoolchildren in Iran wear a uniform that covers their hair completely and a suit that covers most of their body. **(D)** An Egyptian woman covers all but her eyes and her hands.

Thinking Geographically Why do these women wear a veil, no matter what style of veil they choose?

Vignette What is a well-to-do, highly educated young woman in Saudi Arabia to do if she wants to avoid an arranged marriage and find a compatible mate for herself in a society that allows virtually no contact between the sexes during adolescence and young adulthood? In a daring novel, *The Girls of Riyadh*, Rajaa al-Sanea, herself such a girl, chronicles the daily lives of four pre-med and pre-dentistry college women, all of whom are empowered by the Internet to make "virtual" contact with men. A series of romantic, and in some cases sexual, encounters with men ensues. Al-Sanea explores the complexities, including divorce and social ostracism that arise for women (but generally not men) who cross the boundaries set by modern Saudi society. Banned in Saudi Arabia, the book has become a bestseller throughout the region. *[Sources: Claudia Rot Pierpont, "Found in Translation: The Contemporary Arabic Novel," New Yorker, January 18, 2010, pp. 74–80; and the Complete Review, at http://www.complete-review.com/reviews/arab/alsanea.htm.]* ■

A practice that is a source of contention within this region and abroad is **polygyny**—the taking by a man of more than one wife at a time. Although the Qur'an allows a man up to four wives, it generally does not encourage it and imposes financial limits on the practice by requiring that each wife be given separate living quarters. While legal in most of the region, it is relatively rare, with less than 4 percent of males in North Africa practicing it.

polygyny the taking by a man of more than one wife at a time

THINGS TO REMEMBER

1. For Muslims (93 percent of the population in this region), the Five Pillars of Islamic Practice embody the central teachings of Islam. Some Muslims are fully observant, some are not; but the Pillars have an impact on daily life for all. The call to prayer, broadcast five times a day in all parts of the region, is a reminder to all people to reflect on their beliefs.

2. Given the wide diversity of thought, beliefs, and practices of Islam, the family is probably the most important societal institution for Muslims in the region.

3. Most Muslim families here are patriarchal; and the role of women, the spaces they occupy, and in some cases, the clothes they wear, are carefully defined and sometimes rigidly enforced.

Changing Population Patterns

Although the region as a whole is nearly twice as large as the United States, most of the population is concentrated in the few areas that are useful for agriculture. Vast tracts of desert are virtually uninhabited, while the region's 477 million people are packed into coastal zones, river valleys, and mountainous areas that capture orographic rainfall (compare Figure 6.12 on page 228 with Photo Essay 6.1). Population densities in these areas can be quite high. For example, some of Egypt's urban neighborhoods have over 260,000 people per square mile (100,000 people per square kilometer), a density 4 times higher than that of New York City, the densest city in the United States.

Although fertility rates have dropped significantly since the 1960s, at 3.1 children per woman in 2009, they are still higher than the world average of 2.6 (see Thematic Overview D). Only sub-Saharan Africa, at 5.4 children per woman, is growing faster. At present growth rates, the population of the region will reach 540 million by 2025. This growth is severely straining supplies of fresh water and food, and worsening shortages of housing, jobs, medical care, and education.

Population Growth and Gender Status

Learning Goal 2
Gender and Population: How does the low status of women contribute to this region's high population growth?

This region's high population growth rate can be explained in part by the persistent low status of women. As noted in many world regions, population growth rates are higher in societies where women are not accorded basic human rights, are less educated, and work primarily inside the home. In places where women have opportunities to work or study outside the home, they usually choose to have fewer children (see Figure 6.12). Figure 6.13 on page 228 shows that as of 2005, considerably less than 50 percent of women across the region (except in Israel and Kuwait) worked outside the home at jobs other than farming. Moreover, on average only 74.8 percent of adult females can read, whereas 88 percent of adult males can.

For uneducated women who work only at home or in family agricultural plots, children remain the most important source of personal status, family involvement, and power. This may partially explain why in 2009 only about 35 percent of women in this region were using modern methods of contraception, and only 54 percent were using any method of contraception at all. Both of these numbers are well below the world average of 55 and 62 percent, respectively. Other factors in the low use of contraception include male dominance over reproductive decisions and the unavailability or high cost of effective birth control products.

The deeply entrenched cultural preference for sons in this region is both a cause and a result of women's lower social and economic standing. It also contributes to population growth, as families sometimes continue having children until they have a desired number of sons. Moreover, some young females may not survive because of malnutrition and associated illnesses or because female fetuses are sometimes aborted. The result is that males slightly outnumber females in several age cohorts of the population pyramids, even those above age 5 in Figure 6.14 on page 229 (see also the discussion in Chapter 1 on pages 15–16). In Qatar and the UAE, gender imbalance is extreme (see Figure 6.14B). The unusually large numbers of males in Qatar over the age of 15 is the result of the presence of a large number of male guest workers.

Urbanization, Globalization, and Migration

Urbanization is transforming this region, with two highly globalized patterns of urbanization having emerged in recent decades.

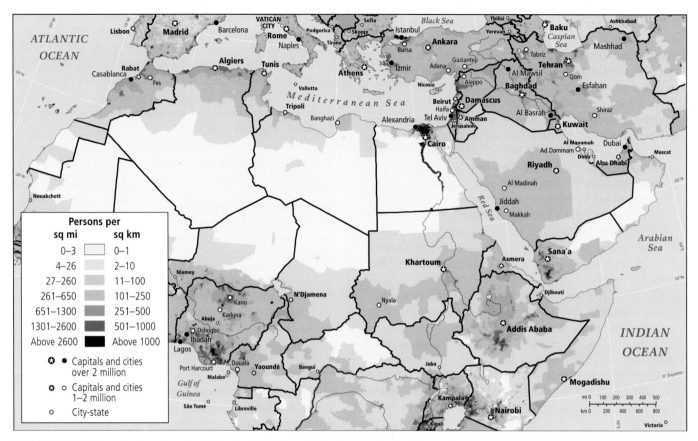

FIGURE 6.12 Population density in North Africa and Southwest Asia. This region has the second-highest population growth rate in the world and, at the present fertility rate, the population could reach 540 million by 2025. Population growth is putting a severe strain on water, food, and housing supplies. The high growth and fertility rates are due in part to the persistently low status of women in nearly all countries in the region: women are generally less educated than men and tend not to work outside the home. Population trends in this region are highly uneven.

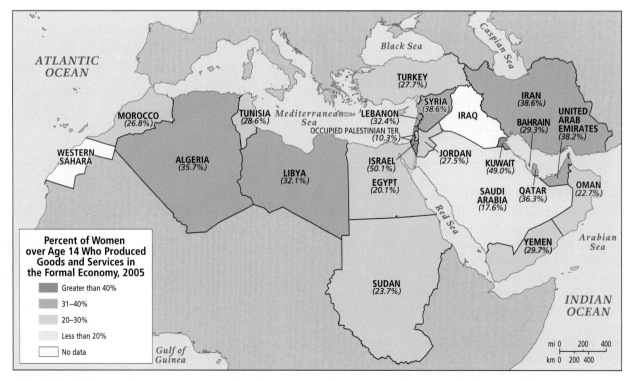

FIGURE 6.13 Percentage of women who are wage-earning workers in the region's countries.

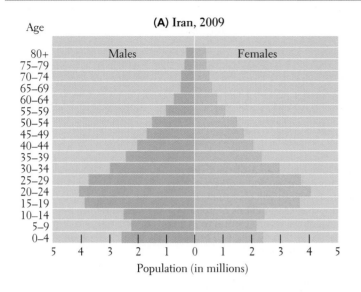

(A) Iran, 2009

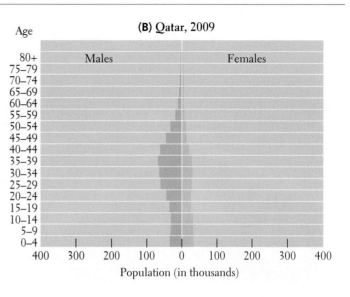

(B) Qatar, 2009

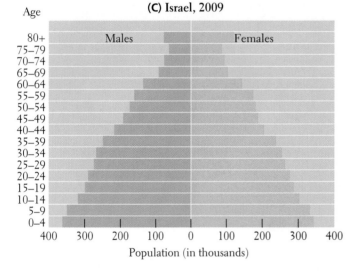

(C) Israel, 2009

FIGURE 6.14 Population pyramids for Iran, Israel, and Qatar. The population pyramid for Iran is at a different scale (millions) than those for Israel and Qatar (thousands). The imbalance of Qatar's pyramid in the 25–54 age groups is caused by the presence of numerous male guest workers. Note, too, that all three pyramids show missing females in the younger age groups. (This is most easily observed by drawing lines from the ends of the male and female age bars to the scale at the bottom of the pyramid and comparing the numbers.)

Learning Goal 3

Urbanization and Globalization: How has globalization shaped different patterns of urbanization throughout the region?

In the oil-rich countries, spectacular new luxury-oriented urban development is tied to global flows of money, goods, and people. Elsewhere, economic reforms aimed at improving global competitiveness have brought massive migration from rural areas, creating crowding and slums.

Until recently, most people lived in small settlements. Since the late 1970s, significant migration from rural villages to urban areas occurred in response to economic forces driven by oil wealth and globalization, and by agricultural modernization that displaced small farmers. By 2008, more than 70 percent of the region's people lived in urban areas (the definition of urban varies by country; see Photo Essay 6.4). By 2009, there were more than 434 cities with populations of at least 100,000, and 37 cities with more than 1 million people. The largest city in the region, and one of the largest in the world is Cairo, with 16.3 million residents (see Thematic Overview E).

The fossil fuel–rich Gulf states are now highly urbanized, and their modern cities draw money and workers from all over the world. In the Persian Gulf states, between 70 and 100 percent of the population now live in urban areas, which are extravagant in design. For example, Dubayy (one of the United Arab Emirates), has built elaborate (and as yet largely unoccupied) new condominiums for the rich on Palm Jumeirah, a fanciful palm tree–shaped island and peninsula off of its Persian Gulf coastline (see Photo Essay 6.4A). The wealthier emirates subsidize the living standards of those who do not have oil wealth, and for the most part poverty is apparent only among foreign contract laborers (see Thematic Overview G).

Outside the Gulf states, urban growth has been massive and less well-financed, driven by rapid rural-to-urban migration (see Photo Essay 6.4B). For example, in 1950, Cairo had around 2.4 million residents, while today it is home to over 16.3 million. Cairo has had to provide for millions of new residents who live in huge makeshift slums. Cairo's middle class occupies the medieval interiors of the old city, where streets are narrow pedestrian pathways (see Photo Essay 6.4C).

In Egypt, as in most of the region outside the Gulf states, rural-to-urban migration is pushed by a shift towards green revolution–style agriculture, which usually reduces the amount of labor needed to produce food. This shift has been a part of economic reforms

Globalization has brought two distinct patterns of urbanization to this region. In fossil fuel–rich countries, populations are more urbanized and cities have undertaken lavish building booms, drawing in laborers and highly skilled workers from across the globe. In countries without fossil fuel wealth, cities are receiving massive flows of poor rural migrants. This is partially a result of economic reforms aimed at making rural and urban economies more globally competitive.

A Palm Jumeirah is an artificial palm-shaped island in Dubayy, UAE. Built by more than 40,000 workers, mostly low-wage migrants from South Asia, Palm Jumeirah will eventually house 65,000 people, mostly in villas and condos that cost millions of dollars each. Palm Jumeirah and several similar developments are central to Dubayy's efforts to build a globalized tourism-based economy that will prosper long after the region's fossil fuel resources are exhausted.

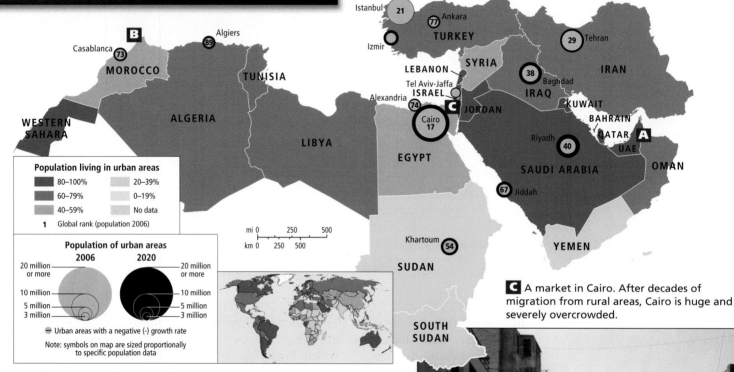

Population living in urban areas

■ 80–100%	▨ 20–39%
■ 60–79%	▨ 0–19%
■ 40–59%	▨ No data

1 Global rank (population 2006)

Population of urban areas

2006	2020
20 million or more	20 million or more
10 million	10 million
5 million	5 million
3 million	3 million

⊖ Urban areas with a negative (-) growth rate

Note: symbols on map are sized proportionally to specific population data

mi 0 — 250 — 500
km 0 — 250 — 500

C A market in Cairo. After decades of migration from rural areas, Cairo is huge and severely overcrowded.

B Tensions run high at the fishing docks in Casablanca, Morocco. The city's population is growing quickly, due largely to migration from Morocco's drought-stricken interior, but a lack of jobs is resulting in a higher crime rate.

designed to reduce the price of Egyptian exports on the world market. In the case of agricultural commodities (for example, cotton), the use of machinery, fertilizers, and pesticides can produce more with less labor, allowing the commodity to be sold for less.

To some extent, rural people are also pulled into Cairo and other cities outside the Gulf states by the prospect of jobs. However, stable well-paying jobs are often very hard to find and many migrants end up working in the informal economy (see page 119 in Chapter 3) as street vendors.

Immigrants come from all over the world to be temporary guest workers in the Gulf states, working as laborers on construction sites and as low-wage workers throughout the service economy (see Thematic Overview F). In some countries, such as the UAE, they make up 85 percent of the labor force. Most employers prefer Muslim guest workers, and over the last two decades, several hundred thousand Muslim workers have arrived from Palestinian refugee camps in Lebanon and Syria, as well as Egypt, Pakistan, and India. Some female domestic and clerical workers come from Muslim countries in South and Southeast Asia, including the Philippines. Overwhelmingly, these immigrant workers are temporary residents with no job security and no rights to earn citizenship. They remit most of their income to their families at home, and often live in stark conditions alongside the opulent lifestyles of those enriched by oil and gas.

A final category of migrants is the refugees: This region has the largest number of refugees in the world. Usually they are escaping human conflict, but environmental disasters such as earthquakes or long-term drought also displace many people. When Israel was created in 1949, many Palestinians were placed in refugee camps in Lebanon, Syria, Jordan, the West Bank, and the Gaza Strip. Palestinians still constitute the world's oldest and largest refugee population, numbering at least 5 million. Elsewhere, Iran is sheltering more than a million Afghans and Iraqis because of continuing violence and instability in their home countries. Across the region, even more people are refugees within their own countries: 1.7 million Iraqis are internal refugees, and in Sudan, between 5 and 6 million Sudanese are living in refugee camps.

Refugee camps often become semipermanent communities of stateless people in which whole generations are born, mature, and die. Although residents of these camps can show enormous ingenuity in creating a community and an informal economy, the cost in social disorder is high. Tension and crowding create health problems. Because birth control is generally unavailable, refugee women have an average of 5.78 children each. Disillusionment is widespread. Years of hopelessness, extreme hardship, and lack of employment take their toll on youth and adults alike, leading some to adopt extremist attitudes against those they see as responsible for their suffering. Moreover, even though international organizations provide basic services for refugees, the refugees constitute a huge drain on the resources of their host countries. In Jordan, for example, native Jordanians are a minority in their own country because Palestinian refugees and their children account for well over half the total population of the country, and their presence has changed life for all Jordanians. Since the beginning of the Iraq war in 2003, over 750,000 Iraqis have also fled to Jordan.

THINGS TO REMEMBER

1. **Learning Goal 2:** Gender and Population This region has the second-highest population growth rate in the world, after sub-Saharan Africa. Part of the reason for this is that women are generally less educated than men and tend not to work outside the home. Hence, childbearing remains crucial to a woman's status, a situation that encourages large families.

2. **Learning Goal 3:** Urbanization and Globalization Two globalized patterns of urbanization have emerged in the region. In the oil-rich countries, spectacular luxury-oriented urban development is tied to flows of money, goods, and people. Elsewhere, economic reforms aimed at improving global competitiveness have brought massive migration of the rural poor into cities, creating crowding and slums.

Economic and Political Issues

There are major economic and political barriers to peace and prosperity within North Africa and Southwest Asia today. As the human well-being maps indicate, wealth from fossil fuel exports remains in the hands of a few elites. Most people are low-wage urban workers or relatively poor farmers or herders. The economic base is unstable because the main resources—fossil fuels and agricultural commodities—are subject to wide price fluctuations on world markets. Meanwhile, in many poorer countries, large national debts are forcing governments to streamline production and cut jobs and social services.

Political and economic cooperation in the region has been thwarted by a complex tangle of hostilities between neighboring countries. Many of these hostilities are the legacy of outside interference by Europe and the United States in regional politics, including colonial intrusions in earlier times, and more recently, activities by global corporations, including oil, gas, industrial, and agricultural enterprises. The Israeli–Palestinian conflict, which has profoundly affected politics throughout the region, is the result of the massive migrations of Jews from Europe and the former Soviet Union in the aftermath of the Holocaust and World War II. The Iran–Iraq war of 1980–1988 and the Gulf War of 1990–1991 were instigated in part by pressures from outside the region. The long siege of violence in Iraq, though begun by a homegrown dictator, did not come to a head until the U.S. invasion and occupation of Iraq, which began in the spring of 2003.

Globalization, Development, and Fossil Fuel Exports

Learning Goal 4
Development and Globalization: How have the huge fossil fuel reserves of some countries transformed economic development and driven globalization in the region?

The vast fossil fuel resources of a few countries have transformed economic development and driven globalization in this region. In these countries, economies have become intrinsically linked to global flows of money, resources, and people. Politics have also become globalized, with Europe and the United States strongly influencing many governments.

This region is the major fossil fuel supplier to the world, and it has two-thirds of the world's known reserves of oil and natural gas. Most oil and gas reserves are located around the Persian Gulf (Figure 6.15) in the countries of Saudi Arabia, Kuwait, Iran, Iraq, Oman, Qatar, and the United Arab Emirates (UAE). The North African countries of Algeria, Libya, and Sudan also have oil and gas reserves. It is important to note, however, that several countries in this region have minimal oil and gas resources or are net importers of fossil fuels: Morocco, Tunisia, Syria, Lebanon, Jordan, Israel, and Turkey. Meanwhile, Egypt and Sudan produce enough to supply most of their own needs, but export little.

European and North American companies were the first to exploit the region's fossil fuel reserves early in the twentieth century. These companies paid governments a small royalty for the right to extract oil (natural gas was not widely exploited in this region until the 1960s, though gas extraction has grown rapidly since then). Oil was processed at onsite refineries owned by the foreign companies and sold at very low prices, primarily to Europe and the United States and eventually to other countries, such as Japan.

Global oil and gas prices have risen and fallen dramatically since 1973, when governments in the Gulf states raised the price of oil and gas for several reasons. On the one hand, the Gulf states were responding to U.S. support for Israel in the 1973 Yom Kippur War between Israel and neighboring Arab countries (discussed further below). For a few months, there was a total halt of oil shipments from the Gulf states to the United States and to a few other countries that supported Israel, followed by much higher prices for oil once shipments were resumed. However, the price increase was also a response to a long-term trend of European countries and the United States raising the prices of their exports to the Gulf states. The oil price rise was made possible by **OPEC (the Organization of Petroleum Exporting Countries)**, which was founded in 1960 as a **cartel**—a group of producers strong enough to control production and set prices for its products. OPEC now includes all of the states indicated in Figure 6.16 on page 234. OPEC members cooperate to periodically restrict or increase oil production, thereby significantly influencing the price of oil on world markets. A move to create an OPEC-like cartel for natural gas is currently underway.

OPEC countries, many of which were exceedingly poor just 40 years ago, have become vastly wealthier since 1973, though this wealth has not been widely shared within these countries or within the region. Oil income in Saudi Arabia shot up from U.S.$2.7 billion in 1971 to U.S.$110 billion in 1981. The Gulf states, though, were slow to invest their fossil fuel earnings at home in basic human resources, and they were slow to explore other economic strategies in case oil and gas ran out. Like their poorer neighbors, they have remained highly dependent on the industrialized world for much of their technology, manufactured goods, skilled labor, and expertise. However, most of the Gulf states have invested heavily in roads, airports, new cities, irrigated agriculture, and petrochemical industries. A good example of the scale of projects in the Gulf states is Palm Jumeirah (Palm Island), a huge artificial island in Dubayy, UAE (see Photo Essay 6.4A on page 230), intended to be the centerpiece of Dubayy's planned tourism economy—insurance against the day when the regional oil economy fails.

The recession that started in 2008 badly damaged the economies of the Gulf states and the many places that send people there to work. Most of the massive building projects underway across the Gulf states screeched to a halt, including Palm Island and similar projects in the UAE. The recession dealt a blow to Dubayy's tourism economy, as it showed how an expensive luxury like tourism is highly vulnerable to economic downturns. The effect of the recession on Dubayy and the Gulf states as a whole has had global effects. For example, with drastically fewer jobs available, countries that send workers to the Gulf states, such as Indonesia, Pakistan, the Philippines, and India, have seen a steep decline in remittances sent home.

Much less attention has been given to spreading the benefits of oil wealth broadly throughout society via support for education, social services, public housing, and health care. Libya, and Kuwait are somewhat exceptional, having developed ambitious plans for addressing all social service needs, with especially heavy investment in higher education. Iran is another possible exception, though reliable data are hard to obtain given the government's poor relations with most of the world. Meanwhile, the non–fossil fuel producing (non-OPEC) countries do not share directly in the wealth of the Persian Gulf states, and for the most part, the oil-rich countries have not helped their poorer neighbors develop.

Many factors outside of OPEC's control strongly influence world oil prices. Recent rapid industrialization in China and India has increased their demand for oil, creating long-term pressures that will raise the price of oil. Employing a theory known as "peak oil," some experts argue that in an era of diminished oil reserves—which these experts say has already begun—prices will be driven higher. However, efforts to combat climate change with renewable energy sources could dramatically reduce demand for oil. Current global oil flows are summarized in Figure 6.16.

> **OPEC (Organization of Petroleum Exporting Countries)** a cartel of oil-producing countries—including Algeria, Angola, Iran, Iraq, Kuwait, Libya, Nigeria, Qatar, Saudi Arabia, the United Arab Emirates, Ecuador, Venezuela, and Nigeria—that was established to regulate the production, and hence the price, of oil and natural gas
>
> **cartel** a group of producers strong enough to control production and set prices for its products

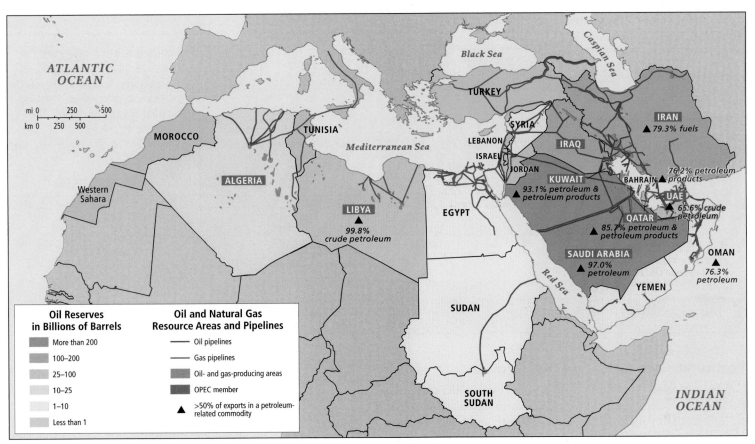

(A)

Oil Reserves in Billions of Barrels
- More than 200
- 100–200
- 25–100
- 10–25
- 1–10
- Less than 1

Oil and Natural Gas Resource Areas and Pipelines
- Oil pipelines
- Gas pipelines
- Oil- and gas-producing areas
- OPEC member
- ▲ >50% of exports in a petroleum-related commodity

Map labels:
ATLANTIC OCEAN, Black Sea, Caspian Sea, TURKEY, SYRIA, IRAN ▲ 79.3% fuels, 76.2% petroleum products, BAHRAIN ▲, UAE ▲ 65.6% crude petroleum, QATAR ▲ 85.7% petroleum & petroleum products, LEBANON, ISRAEL, IRAQ, JORDAN, KUWAIT ▲ 93.1% petroleum & petroleum products, MOROCCO, TUNISIA, Mediterranean Sea, ALGERIA, Western Sahara, LIBYA ▲ 99.8% crude petroleum, EGYPT, SAUDI ARABIA ▲ 97.0% petroleum, OMAN ▲ 76.3% petroleum, YEMEN, Red Sea, SUDAN, SOUTH SUDAN, INDIAN OCEAN

mi 0 250 500
km 0 250 500

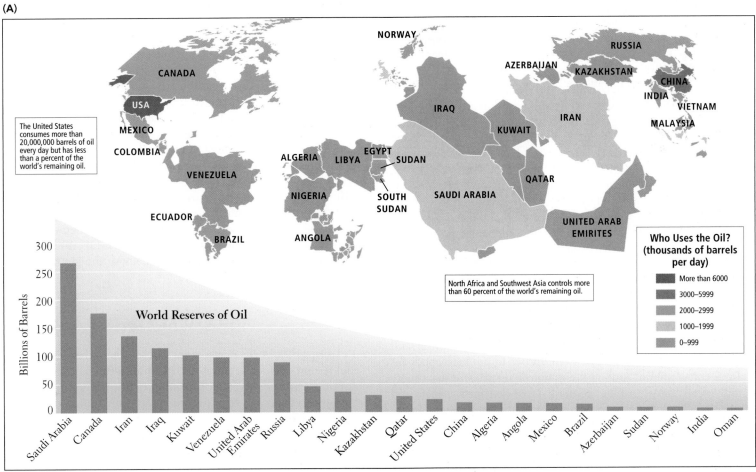

(B)

The United States consumes more than 20,000,000 barrels of oil every day but has less than a percent of the world's remaining oil.

North Africa and Southwest Asia controls more than 60 percent of the world's remaining oil.

Who Uses the Oil? (thousands of barrels per day)
- More than 6000
- 3000–5999
- 2000–2999
- 1000–1999
- 0–999

Map labels: NORWAY, RUSSIA, AZERBAIJAN, KAZAKHSTAN, CHINA, INDIA, VIETNAM, MALAYSIA, CANADA, USA, MEXICO, COLOMBIA, IRAQ, IRAN, KUWAIT, QATAR, ALGERIA, LIBYA, EGYPT, SUDAN, SOUTH SUDAN, SAUDI ARABIA, UNITED ARAB EMIRATES, VENEZUELA, NIGERIA, ECUADOR, BRAZIL, ANGOLA

World Reserves of Oil

Billions of Barrels (y-axis: 0, 50, 100, 150, 200, 250, 300)

x-axis: Saudi Arabia, Canada, Iran, Iraq, Kuwait, Venezuela, United Arab Emirates, Russia, Libya, Nigeria, Kazakhstan, Qatar, United States, China, Algeria, Angola, Mexico, Brazil, Azerbaijan, Sudan, Norway, India, Oman

FIGURE 6.15 Economic issues: Oil and gas resources in North Africa and Southwest Asia.
The size of countries corresponds to the amount of oil reserves.

233

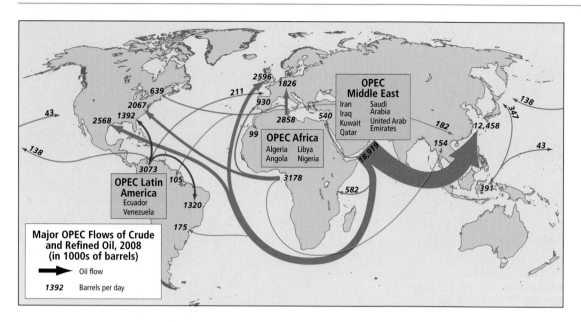

FIGURE 6.16 Major OPEC oil flows in 2008. The map shows the average number of barrels (in thousands) of crude and refined oil distributed per day by OPEC countries to the world.

Economic Diversification and Growth

Greater **economic diversification**—expansion into a wider range of types of economic development—could have a significant impact on the region. It could bring economic growth and broader prosperity that would limit the damage caused by a drop in the price of oil, gas, or other commodities on the world market. Except in a few countries, however, actual diversification has been meager, despite—in the case of most Gulf states—wildly excessive spending.

> **economic diversification** the expansion of an economy to include a wider array of activities

By far the most diverse economy in the region is that of Israel, which has a large knowledge-based service economy and a particularly solid manufacturing base. Israel's goods and services and the products of its modern agricultural sector are exported worldwide. Turkey is the next most diversified, in large part because—like Israel—it has never had oil income to fall back on. Egypt, Morocco, Libya, and Tunisia are also starting to move into many new economic activities. Some fossil fuel–rich countries have tried to diversify into other industries. In the UAE, for example, only 25 percent of GDP is based directly on fossil fuels; trade, tourism, manufacturing, and financial services are now dominant. Even so, most Gulf states are still highly dependent on fossil fuel exports and the spin-off industries they generate.

Diversification has also been limited by economic development policies that favored *import substitution* (see pages 115–116 in Chapter 3). Beginning in the 1950s, many governments, such as those of Turkey, Egypt, Iraq, Israel, Syria, Jordan, Tunisia, and Libya, established state-owned enterprises to produce goods for local consumption. Among the major products were machinery and metal items, textiles, cement, processed food, paper, and printing. These enterprises were protected from foreign competition by tariffs and other trade barriers. With only small local markets to cater to, profitability was low and the goods were relatively expensive. Without competitors, the products tended to be shoddy and unsuitable for sale in the global marketplace. Meanwhile, the extension of government control into so many parts of the economy nurtured corruption and bribery and

squelched entrepreneurialism. Almost all of the state-owned import substitution enterprises have subsequently either gone bankrupt or been sold to private investors.

Economic diversification and export growth were also limited by a lack of financing, private and public, from within the region. For example, members of the Saudi royal family generally prefer more profitable private investments in Europe, North America, and Southeast Asia. Much private and public investment has recently gone into lavish development in the UAE, particularly in Dubayy. Only recently have governments recognized that they need to plan for a time when oil and gas run out.

Finally, both international and domestic military conflicts and the ensuing political tensions have stymied economic diversification because they have resulted in some of the highest levels (proportionate to GDP) of military spending in the world. Military spending diverts funds from other types of development, such as health care and education (Figure 6.17). The top four spenders—Oman, Israel, Saudi Arabia, and Yemen—lead the world in military spending as a percentage of GDP. And all countries in the region, except Tunisia, are above the global average of 2 percent.

THINGS TO REMEMBER

1. **Learning Goal 4: Development and Globalization** The vast fossil fuel resources of a few countries have transformed economic development and driven globalization in this region. In these countries, economies have become powerfully linked to global flows of money, resources, and people. Politics have also become globalized, with Europe and the United States strongly influencing many governments.

2. Profits to the oil- and gas-producing countries were low until OPEC initiated price increases in 1973. Since then, wealth has accrued mainly to a privileged few.

Military, Health and Education Expenditures as Percent of GDP

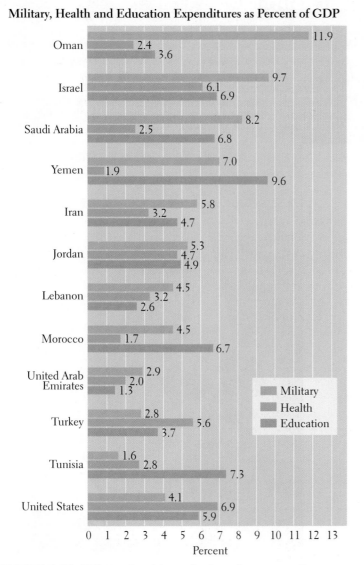

FIGURE 6.17 Military, health, and education expenditures as a percentage of GDP. These charts show only those countries in the region that have data for each variable. The United States is included for comparative purposes.

3. Until very recently, countries in the region, with the exception of Turkey and Israel, had not diversified their economies. In the Gulf states, diversification has been limited by oil wealth being invested abroad rather than at home. Elsewhere, import substitution policies failed to lift countries out of poverty.

Democratization and Islamism in a Globalizing World

> **Learning Goal 5**
>
> **Democratization:** Are there signs that this region may be democratizing?

From late 2010 through 2011, a wave of political protests swept through the region, toppling the regimes of dictators who had ruled for decades. The demonstrations were precipitated by high unemployment, rising food prices, poor living conditions, government corruption, and the absence of freedom of speech and other political freedoms. Beginning in Tunisia and spreading to every country in the region with the exception of Israel, Qatar, and the UAE, the protests resulted in new, more democratically oriented governments in Tunisia, Egypt, and Jordan, and diverse pro-democracy reforms in many other states.

While the immediate causes of the 2011 demonstrations may be economic, activists who have long promoted democratization as a path to peace in this region have been energized by the protests and continue their work to end authoritarian rule. Democratization is taking many forms, and there is no "best" system of government around which countries are converging; however, there are some broad trends regarding political freedoms in this region. In the least democratized countries, repression of the political opposition has brought violence. In the countries that have embraced democratization, opposition movements have been brought into the political process peacefully and productively. The case of Islamism (see page 208) helps to illustrate these points.

Since the attacks of September 11, 2001, Islamism has been viewed by outsiders as a great threat to the region's political stability and economic development. And yet Islamist movements are simply grassroots religious revivals that are intended to counteract the secular influences that have become widespread, thanks to globalization. They have tended to become violent mainly in response to repression by authoritarian governments. When given the opportunity, Islamists have been able to work productively within democratic processes.

Globalization is a particular problem for many conservative Muslims, who object to what they view as immoral secular influences emanating from Europe and North America. Some object to the liberalization of women's roles, especially to women being educated and active outside the home (see Figure 6.13). Many also lament what they see as the global spread of open sexuality, consumerism, and hedonism, transmitted in part by TV, movies, and popular music. While many see the advantages of bringing people into the computer age, most Islamists abhor Internet pornography and often oppose free access to non-Islamic political ideas via the Internet and print media. Islamist movements and the social services they often provide have a strong popular base in the slums of the largest cities, where the economic situation of millions has been worsened by globalization and, more recently, by the global recession.

‖ ▶ 139. WHAT MOTIVATES A TERRORIST?

‖ ▶ 140. NEW POLL OF ISLAMIC WORLD SAYS MOST MUSLIMS REJECT TERRORISM

‖ ▶ 144. JIHADIST IDEOLOGY AND THE WAR ON TERROR

Historically, Islamism is rooted in the mixture of governmental and religious authority common in this region, as well as in non-democratic authoritarian responses to popular political movements. With a few exceptions, authoritarian governments dominate this region (see Photo Essay 6.5 on pages 236–237). Some, such as such as Saudi Arabia, Yemen, the UAE, Oman, and Iran, are **theocratic states** in which Islam is the official religion and political leaders are considered to be divinely guided by both Allah and the teachings of the Qur'an. Many other countries are **secular states**, where the

> **theocratic states** countries that require all government leaders to subscribe to a state religion and all citizens to follow rules decreed by that religion
>
> **secular states** countries that have no state religion and in which religion has no direct influence on affairs of state or civil law

government is officially neutral on matters of religion, though in practice Islamic ideas influence many government policies. In either case, the authoritarian leanings of governments in the region mean that free speech and the right to hold public meetings have often been so severely limited that the only public spaces in which people have been allowed to gather are mosques, and the only public discussions free of censorship by the government have been religious discourses. In this context, much political dissent has been channeled into movements headed by religious leaders who focus on defending Islam against the challenges posed by globalization.

The militancy often associated with Islamism is not unique to it but is rather a shared characteristic of many popular political movements in this region that have challenged the authority of governments only to be violently repressed (see Photo Essay 6.5). In both secular and theocratic governments, minorities have been denied the right to engage in their own cultural practices and speak their own language. Journalists and private citizens have been harassed, jailed, or even killed for criticizing governments or exposing corruption. While the 2011 protests that swept through this region were in part a response to these conditions, the extent to which the new governments and political reforms will protect political freedoms remains to be seen.

Turkey has been the most successful at using the democratic process to constructively engage Islamist political parties. In fact, Turkey's government has been controlled by Islamists on and off for decades. Nevertheless, there have been no significant departures from the country's safeguards for freedom of religion or its steady improvement on women's rights, education, and freedom of the press. Nor has the country's long-term goal of becoming a modern, prosperous member of the European Union been irreparably compromised by theocratic politics, though it has certainly been jeopardized. Similarly, in Jordan and Algeria, the election of moderate Islamists has at times diminished discontent and decreased violence. However, in other cases (primarily Iran and Gaza), elected Islamists have incited their followers to violence, worsening relations with the Muslim and non-Muslim world.

II ▶ 153. TURKEY VOTES FOR STABILITY

The Role of the Press and Media

In some countries—Turkey, Israel, and Morocco, for example—the press is reasonably free and opposition newspapers are

Thinking Geographically

After you have read about democratization and conflict in North Africa and Southwest Asia, you will be able to answer the following questions:

A What was the role of the United States in the Iran–Iraq war of the 1980s?

B How did OPEC respond to U.S. support for Israel in the 1973 war?

C How many Sudanese are currently living in refugee camps?

D In addition to Turkey, what other countries have large Kurdish populations?

E When did violence peak during the U.S. occupation of Iraq?

This region is one of the least democratized in the world. It is no coincidence that it also suffers more than most from violent conflict. Wars and terrorism here have reduced the potential for democratization by giving authoritarian regimes an excuse to hold onto power and forcibly repress legitimate political opposition groups. Meanwhile, the absence of democratic freedoms has prolonged conflicts that might have been resolved peacefully through free elections.

A A monument in Baghdad to the Iran–Iraq War (1980–1988), in which over 1 million people died. One factor that motivated Iraq's then-dictator, Saddam Hussein, to invade Iran was that Iraq's majority Shi'ite population had been denied democratic freedoms that might have brought their leaders to power. Saddam feared that Iran's Shi'ite-dominated government would back Shi'ite rebellion in Iraq that would remove him from power.

E An Iraqi medic and a U.S. soldier move a victim of a terrorist attack to an ambulance. While some progress is being made, a complex and violent insurgency now severely limits the expansion of democracy.

B A painting in Cairo commemorating the 1973 war between Israel and Egypt. Ongoing tensions between Israel and its Arab neighbors have worked against democratization throughout the region. Many Arab governments use the conflict to justify repression of legitimate political parties. While historically far more democratic, Israel recently banned anyone who has visited "enemy countries" (Lebanon, Syria, Iraq, Iran, Saudi Arabia, Sudan, Libya, and Yemen) from running for national office for 7 years.

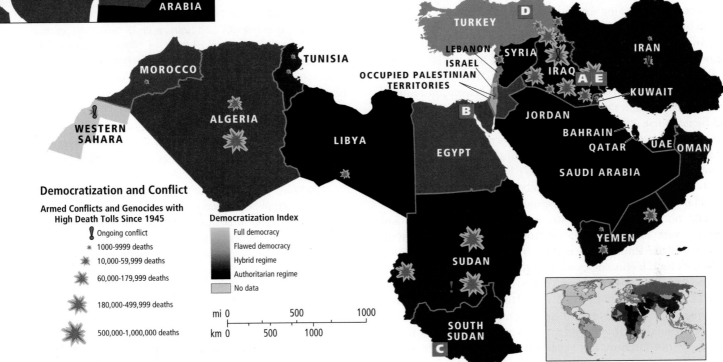

Democratization and Conflict

Armed Conflicts and Genocides with High Death Tolls Since 1945

❗ Ongoing conflict
✳ 1000-9999 deaths
✴ 10,000-59,999 deaths
✷ 60,000-179,999 deaths
✸ 180,000-499,999 deaths
✹ 500,000-1,000,000 deaths

Democratization Index
- Full democracy
- Flawed democracy
- Hybrid regime
- Authoritarian regime
- No data

mi 0   500   1000
km 0   500   1000

D A Kurdish rebel in southeastern Turkey. Kurds form a majority of the population here and in neighboring parts of Iraq and Iran. Kurds have long been denied democratic freedoms and in some cases even the right to speak their language. Though Kurdish militias still exist, democratization has recently reduced public support for violence, especially in northern Iraq.

C A rebel army mobilizes in southern Sudan. Cultural tensions between the north and the south have led to two bloody civil wars and ongoing violence. The absence of basic democratic freedoms in Sudan has frustrated many efforts at reconciliation.

aggressive in their criticism. But in much of the region, journalism can be a risky career, often leading to prison. Egypt has a checkered history where the press is concerned. For years, Hisham Kassem was editor of the *Cairo Times*, an independent English-language weekly. When he became too critical of the government, he lost his license to publish in Egypt. For a time he took great pains to write and print his paper outside the country and smuggle it into Egypt, always risking arrest. Business leaders in Cairo began to provide backing for Kassem, enabling him to publish *Al-Masry Al-Youm*, specializing in domestic issues of corruption, election fraud, and the need for an independent judiciary. Kassem and many independent journalists like him participated actively in the 2011 protests in Egypt that brought an end to the 30-year regime of Hosni Mubarak.

In Saudi Arabia, a dozen newspapers are on the newsstands every morning, but all are owned or controlled by the royal Saud family, and all journalists are constrained by the fact that they may not print anything critical of Islam or of the Saud family, which numbers in the tens of thousands. When accidents happen or some malfeasance by a public official is revealed, the story is blandly reported, with little effort to explore the causes of events or their effects or to hold responsible officials accountable.

The television broadcasting network Al Jazeera, introduced in the opening vignette, is credited with changing the climate for public discourse across the region and even with changing public opinion outside of the region. However, because it operates at the pleasure of the fairly liberal emir (Muslim ruler) of Qatar, it has apparently agreed not to cover sensitive issues in Qatar.

▌▌▶ 147. AL JAZEERA LAUNCHES GLOBAL BROADCAST OPERATION

The role of the Internet is emerging, as illustrated by the vignette on this page about women bloggers in Yemen, and together with other types of electronic communication (such as Twitter) the Internet is creating the climate for democratic reforms. These technologies also promise to open up the spaces for dissent, as was first demonstrated in 2009 when protests erupted in Iran after hotly contested elections appeared to have been rigged by the government (see Thematic Overview H). The fact that so many Iranians had video-capable cell phones meant that the world instantly saw via the Internet the brutality of the Iranian police and army.

Democratization and Women

Most countries in this region now allow women to vote, and two countries—Israel and Turkey—have elected female heads of state (prime ministers). Nevertheless, women who want to actively participate in politics still face many barriers. Women's political status is the lowest in the Gulf states, but circumstances vary from country to country. Oman, which in 1994 became the first Gulf country to grant some women the vote, gave all women the right to vote in 2003. In Kuwait in 2009, four highly educated women were elected to parliament; they quickly energized the pace of law making, often publicly criticizing their male colleagues, who were frequently absent for crucial votes. In Saudi Arabia, by contrast, women are denied the right to vote and only one woman serves as a public official. Moreover, across the region there is a tendency—even among women—to not support women candidates. Such

was the case in Bahrain's elections of 2002, when women were a majority of the electorate but elected no female candidates. Across the region in 2009, women made up less than 10 percent of national legislatures, the lowest of any world region and half the world average.

An important impetus for change is the fact that more women are becoming educated and employed outside the home. In a number of countries, women outnumber men in universities; most notably, in Saudi Arabia women make up 70 percent of university students (but only 5 percent of the workforce). Saudi women activists have characterized their country as practicing a sort of apartheid with its own women, circumscribing or forbidding all manner of public activities. Yemen is another country that sees its female citizens as creatures to be set apart, yet some Yemeni women and men are leading change, as the following vignette illustrates.

Vignette Yemen's most outspoken feminist, Raufa Hassan al-Sharki, is in her 40s and holds a Ph.D. in social communications from the University of Paris. In 1996, she founded the Empirical Research and Women's Studies (ERWS) Center at San'a University in Yemen. Her overarching goal was to help women learn how to vote independently. Yemen's Islamist party, Islah, supports her efforts.

Despite having the right to vote, few Yemeni women do so. Most live in rural areas, where only 4 in 10 women can read and few girls go to school. Most work at home, herding cattle, grinding wheat, and carrying water. The average adult woman bears six children.

For years, al-Sharki was careful to respect patriarchal customs when encouraging rural women to vote. Nevertheless in 1999, after she organized a successful international conference on women at ERWS, the government reorganized the center with an all-male staff and board of directors and removed al-Sharki as its director. Now she works as an independent consultant explaining gender issues of the Arabian Peninsula to audiences in American universities. ■

The Iraq War (2003–2010)

The U.S.-led invasion and occupation of Iraq has transformed the politics of this region. Anti-U.S. sentiment has exploded, together with criticism of leaders and governments, such as Israel's, who support the goals of the United States. Meanwhile, U.S. support for democracy in Iraq has led to many unexpected consequences.

The origins of the Iraq war are complex. In 1963, the United States backed a coup that installed a pro-U.S. government. For decades after that, the United States had an amicable relationship with the Iraqi government, driven in large part by Iraq's considerable oil reserves, the fourth-largest in the world after those of Saudi Arabia, Canada, and Iran. The United States publicly supported Iraq in its 1980–1988 war with Iran (see Photo Essay 6.5A), but also secretly supplied Iran with weapons when it appeared that Iraq might become more troublesome if it won the war. Relations with Iraq took a dramatic turn for the worse in 1990 when Saddam Hussein, Iraq's dictator, invaded Kuwait. The United States forced Iraq's military out of Kuwait in the Gulf

War of 1990–1991, and afterward placed Iraq under crippling economic sanctions.

After the terrorist attacks of September 11, 2001, the U.S. administration of George W. Bush first launched a war on Afghanistan, but by 2003, had shifted its focus to Iraq and its president, Saddam Hussein (see page 74 in Chapter 2). The Bush administration claimed that Iraq had or was creating an arsenal of weapons of mass destruction, and used this pretext to declare war on March 20, 2003, with the goals of confiscating Iraq's weapons of mass destruction, removing Saddam from power, and turning Iraq into a democracy. The United Kingdom and a few other countries joined the United States as allies, but most of the world objected to the war.

After the initial invasion met little resistance, President Bush declared the war won on May 1, 2003. However, terrorist bombs and insurgent attacks continued, with violence peaking between 2006 and 2007, after which it has gradually decreased. By February 2010, over 4380 U.S. troops had been killed and 31,500 wounded (see Photo Essay 6.5E). Furthermore, over 150,000 veterans of the conflict had been diagnosed with some form of serious mental disorder, including posttraumatic stress (PTSD). The death toll for Iraqis, including civilians, has been much higher, estimated as at least 103,500 and possibly as high as 1,340,000, when counting Iraqi deaths due to poor health and public safety conditions created by the war.

▌▶ 136. IRAQ WAR ENTERS SIXTH YEAR

▌▶ 152. FIVE YEARS AFTER 'MISSION ACCOMPLISHED' IN IRAQ, WAR CONTINUES

The war has had many unintended consequences. While Saddam has been removed from power and executed, weapons of mass destruction have never been found, nor have any links between the 9/11 attacks and Saddam's government been uncovered. The Bush administration's main stated justification for continuing the war was democratization, but the failure to understand and account for the long-simmering tensions between Iraq's major religious and ethnic groups crippled the democratization process. Sunnis in the central northwest had dominated the country under Saddam, although they constituted only 32 percent of the population. Shi'ites in the south, with 60 percent of the population, have dominated politics since the fall of Saddam. This has given greater influence to Iran, whose population is mostly Shi'ite. Meanwhile, Kurds in the northeast (see Figure 6.5D), once brutally suppressed under Saddam Hussein, have joined with Kurdish populations in Turkey and Syria and resist cooperating with the Iraqi national government in Baghdad.

▌▶ 134. KURDISH NATIONALISTS IN IRAQ, TURKEY SEEK LAND OF THEIR OWN

Regarding the still-incomplete move toward democracy in Iraq, 2010 elections took place with much less violence than expected, but the results were disputed and it took months for an Iraqi government to form. Recent studies of Iraqi public opinion indicate a number of important trends that will shape a more democratic Iraq. A majority of Iraqis want a strong central government that can protect them from violent insurgents and that can maintain control of the country's large fossil fuel reserves. Polls also show that the vast majority of Iraqis want all fighting to stop and all U.S. and allied military forces to leave. When elected in 2008, U.S. president Barack Obama promised to

remove all U.S. combat troops by January 2012. Many had left by summer 2010, though a large U.S. military presence will remain in Iraq for the foreseeable future.

THINGS TO REMEMBER

1. Learning Goal 5: Democratization: While the immediate causes of the 2011 demonstrations may be economic, activists who have long promoted democratization as a path to peace in this region have been energized by the protests and continue their work to end authoritarian rule.

2. Most countries now allow women to vote, and two countries have elected female heads of state. Even though women who seek active participation in politics still face barriers, patterns are changing as more women become educated and employed outside the home.

3. Demand for greater political freedoms is increasing, and public spaces for debate, other than mosques, are emerging. The press is reasonably free in some countries and severely curtailed in others. Al Jazeera is credited with changing the climate for public discourse across the region and with changing public opinion outside the region.

4. The situation in Iraq remains fluid as the United States continues withdrawing its troops and turns over power and policing operations to U.S.-trained Iraqi forces.

The State of Israel and the "Question of Palestine"

The Israeli–Palestinian conflict has lasted more than 60 years and has included several major wars and innumerable skirmishes. It is a persistent obstacle to political and economic cooperation within the region and it complicates the relations between many countries in this region and the rest of the world.

Of the two entities (Palestine has not yet been recognized as a state), Israel has by far the more modern and diversified economy; indeed, Israel leads the region in this regard. Since the 1950s, Israel's development has been facilitated by the immigration of relatively well-educated middle-class Jews from the United States, Europe, Russia, and South America. Israel's excellent technical and educational infrastructure, its diverse and prospering economy, and the large aid contributions it receives from the United States and others, public and private, have made it one of the region's wealthiest, most technologically advanced and militarily powerful countries.

▌▶ 154. 60 YEARS AFTER ISRAEL'S FOUNDING, PALESTINIANS ARE STILL REFUGEES

The Palestinian people, by contrast, are severely impoverished and undereducated after years of conflict, inadequate government, and meager living (see Table 6.1 on page 240), often in refugee camps. Through a series of events over the past 60 years, Palestinians have lost most of the lands on which they used to live. They now live in two main areas—the West Bank and the Gaza Strip, both of which are highly dependent on Israel's economy. Israel often takes military action in these

TABLE 6.1	Circumstances and state of human well-being among Palestinians and Israelis			
	Population (in millions), 2009	Infant mortality (per 1000 live births) 2009	Percentage of unemployed, 2005	Percentage of population in poverty, 2007–2009
Palestinians	3.9	25	26.7	39
Israelis	7.6	3.6	9	23.6*

*Israel's poverty line is U.S.$7.30 per person per day (2005).
Source: *United Nations Human Development Report 2005* (New York: United Nations Development Programme)

two zones in retaliation for Palestinian suicide bombings and rocket fire launched primarily from Gaza. The West Bank Palestinian territories continue to be encircled by security walls built by Israel to defend against violence and curtail Palestinian access.

❙❙▶ 131. ISSUES FROM 1967 ARAB–ISRAELI WAR REMAIN UNRESOLVED

The Creation of the State of Israel The idea of Israel as a Jewish homeland actually began in Europe in the late nineteenth century. In response to centuries of discrimination and persecution in Europe and Russia, a small group of European Jews, known as **Zionists**, began to purchase land in an area of the Ottoman Empire known at the time as Palestine. Most sellers were wealthy non-Palestinian Arabs and Ottoman Turks living outside of Palestine. These absentee landowners had long leased their lands to Palestinian tenant farmers and herders or granted them the right to use the land. Such rights were negated by the sales to Zionists, and historians still debate whether or not the Zionists or the former landlords adequately compensated the displaced indigenous inhabitants. In the aftermath of World War I (see page 221), the United Kingdom in 1922 received a Mandate from the League of Nations to administer the territory of Palestine (Figure 6.18A).

Earlier, in 1917, the British government had adopted the Balfour Declaration, which favored the establishment of a national home for Jews in the Palestine territory, and which explicitly stated that "nothing shall be done which may prejudice the civil and religious rights of existing non-Jewish communities in Palestine, or the rights and political status enjoyed by Jews in any other country." It is at this time that the word *Palestine*, with roots far back in history, began to be used officially.

On this newly purchased land, the Zionists established communal settlements called *kibbutzim*, into which a small flow of Jews came from Europe and Russia. While Jewish and Palestinian populations intermingled in the early years, tensions emerged as Zionist land purchases displaced more and more Palestinians, in seeming breach of the Balfour Declaration.

Efforts to secure a Jewish homeland in Palestine continued. By 1946, following the death of 6 million Jews in Nazi death camps during World War II, strong sentiment had grown throughout the world in favor of a Jewish homeland. Although a few other homeland sites were suggested, the only place seriously considered was the space in the eastern Mediterranean that Jews had shared in ancient times with Palestinians and other Arab groups and that was mentioned in a Biblical promise to Moses.

Zionists those who have worked, and continue to work, to create a Jewish homeland (Zion) in Palestine

intifada a prolonged Palestinian uprising against Israel

Hundreds of thousands of Jews migrated to Palestine, and against British policy, many took up arms to support their goal of a Jewish state.

In November, 1947, after intense debate at the second session of the United Nations General Assembly, the UN adopted the "Plan of Partition with an Economic Union" that ended the earlier Mandate and called for the creation of both Arab and Jewish states and for an international regime for the city of Jerusalem (see Figure 6.18B).

Sixty Years of Conflict The Palestinians and neighboring Arab countries fiercely objected to the establishment of a Jewish state and feared that they would continue to lose land and other resources upon which they were dependent. Then as now, the conflict between Jews and Palestinian Arabs was less about religion than control of land, settlements, and access to water.

On May 14, 1948, the British ended the Mandate and withdrew their forces. On that same day, the Jewish Agency for Palestine declared the state of Israel on land designated to them by the Plan of Partition. Warfare between the Jews and Palestinians began immediately. Neighboring Arab countries—Lebanon, Syria, Iraq, Egypt, and Jordan—supporting the Palestinian Arabs, invaded Israel the next day. Fighting continued for several months, during which Israel prevailed militarily. An armistice was reached in 1949, and as a result, the Palestinians' land shrank still further, with the remnants incorporated into Jordan and Egypt (Figure 6.18C).

In the repeated conflicts over the next decades—such as the Six-Day War in 1967 and the Yom Kippur War in 1973—Israel again defeated its larger Arab neighbors, expanding into territories formerly controlled by Egypt (Sinai), Syria (Golan Heights), and Jordan (West Bank of the Jordan River; Figure 6.19A). Since 1948, hundreds of thousands of Palestinians have fled the war zones, with many removed to refugee camps in nearby countries. Some Palestinians stayed inside Israel and became Israeli citizens, but they have not been treated by the state as equal to Jewish Israelis.

❙❙▶ 128. HEZBOLLAH—SERVING MUSLIMS WITH GOD AND GUNS

❙❙▶ 133. LEBANESE OIL SPILL—COLLATERAL DAMAGE OF THE BOMBINGS

In 1987, the Palestinians mounted the first of two prolonged uprisings, known as the **intifada**, characterized by escalating violence. The first ran until 1993, when the Oslo Peace Accords were concluded (Figure 6.19B). The Oslo Accords included Israeli withdrawal from parts of the Gaza Strip and the West Bank, and the creation of the Palestinian Authority, which

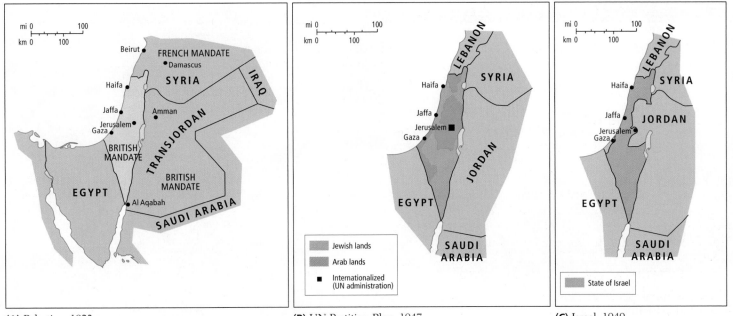

(A) Palestine, 1923

(B) UN Partition Plan, 1947

(C) Israel, 1949

FIGURE 6.18 Israel and Palestine, 1923–1949. (A) Palestine, 1923. Following World War I, Britain controlled what was called Palestine and is now Israel and Jordan (Transjordan was the precursor to Jordan). **(B)** The UN Partition Plan, 1947. After World War II, the United Nations developed a plan for separate Jewish and Palestinian (Arab) states. **(C)** Israel, 1949. The Jewish settlers did not agree to the partition plan; instead, they fought and won a war, creating the state of Israel.

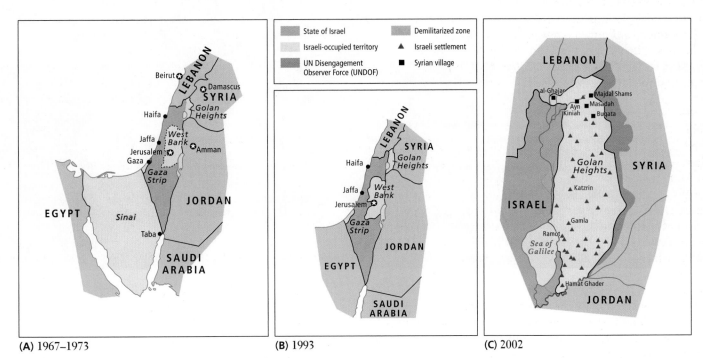

(A) 1967–1973

(B) 1993

(C) 2002

FIGURE 6.19 Israel and the Palestinian Territory after 1949. When the state of Israel was created in 1949, its Arab neighbors were opposed to a Jewish state. **(A)** In 1967, Israel soundly defeated combined Arab forces and took control over Sinai, the Gaza Strip, the Golan Heights, and the West Bank. **(B)** In subsequent peace accords, Sinai was returned to Egypt, but Israel maintained control over the Golan Heights and the West Bank, claiming that they were essential to Israeli security. **(C)** Although the Palestinians were granted some autonomy in the Gaza Strip and the West Bank, during the 1990s the Israelis, contrary to verbal agreements in the Oslo Accords, began building Jewish settlements in the West Bank and Golan Heights.

FIGURE 6.20 Conflict and cooperation between Israelis and Palestinians. Both conflict and cooperation between Israelis and Palestinians has increased in recent years. **(A)** The West Bank barrier and checkpoint (in the lower right) near Ramallah, West Bank. The barrier has hindered the flow of people and goods between Israel and the West Bank, severely damaging the latter's economy. **(B)** Palestinian residents of Gaza struggle to cross a border checkpoint with Egypt. Gaza is periodically plagued by shortages of food, water, and other necessities as Israel seals its borders in retaliation for rocket attacks into Israel from Gaza. **(C)** Palestinians harvest tomatoes from a farm in the West Bank. Food-processing industries have a large potential for growth throughout the West Bank, where much food is imported. **(D)** An Israeli food-processing company advertises its product at a holiday festival in northern Israel. Israeli food-processing companies have access to the expertise, equipment, and investment capital that could help food-processing industries in the West Bank and Gaza grow. While many partnerships are developing between Israeli and Palestinian companies, these suffer when violence forces border closures.

would enable Palestinians to govern themselves. The second intifada began in 2000 and continues into the present, primarily fueled by the issue of Israeli settlements expanding, in breach of the Oslo Accords, into Palestinian territories in Gaza, the West Bank, and the Golan Heights (see Figure 6.19C). Between 2000 and 2008, Palestinian suicide bombers targeted Israeli civilians, killing about 1000, maiming thousands, and psychologically impacting all Israelis. In response, the Israeli military has used deadly force to quell demonstrations and to punish the families and communities of the suicide bombers. According to B'Tselem, an Israeli group that educates the Israeli public about human rights violations in the occupied Palestinian Territories, more than 5000 Palestinians and 64 foreigners died in the same period.

Territorial Disputes When Israel occupied Palestinian lands in 1967, the UN Security Council passed a resolution requiring Israel to return those lands, known as the *occupied Palestinian Territories (OPT)*, in exchange for peaceful relations between Israel and neighboring Arab states. This *land-for-peace formula*, which sets the stage for an independent Palestinian state, has been only partially fulfilled.

Despite the land-for-peace agreement, between 1967 and 2005 Israel secured ever more control over the land and water

West Bank barrier a 25-foot-high concrete wall in some places and a fence in others that now surrounds much of the West Bank and encompasses many of the Jewish settlements there

resources of the occupied territories, placing hundreds of Jewish settlements in the West Bank and the Gaza Strip. Israel took a major step toward peace in 2005 when it removed all Jewish settlements from the Gaza Strip. However, in the eyes of Palestinians, this progress was negated by the blockade of Gaza's economy, the construction of the **West Bank barrier**, which began in 2003 (Figure 6.20A), and by continuing policies that constrain Palestinians' access to jobs and freedom of movement (Figure 6.20B).

The West Bank barrier (see Figure 6.20A) encircles many of the remaining Jewish settlements on the West Bank and separates approximately 30,000 Palestinian farmers from their fields. It also blocks roads that once were busy with small businesses, and effectively annexes 6 to 8 percent of the West Bank to Israel (see Figure 6.19C). It also severely limits Palestinian access to much of the city of Jerusalem, most of which is now on the Israeli side of the barrier. The barrier, declared illegal by the World Court and the United Nations, and opposed by the United States, is nonetheless very popular among Israelis because it has reduced the number of Palestinian suicide bombings.

▐▐ ▶ 132. WEST BANK BARRIER, NEW DIVIDE IN PALESTINIAN–ISRAELI CONFLICT

Can Economic Interdependence Be a Force for Peace? Economic relations with Israel are essential to Palestinian survival because Israel has long been the largest trading partner of the West Bank and Gaza Strip. Israel provides Palestinians with a currency (the Israeli shekel), electricity, and most imports. Israelis employ tens of thousands of Palestinian workers in fields, factories, and homes within Israel (Figure 6.20C, D). Though rarely covered in the media, economic cooperation is a fact of life for Israelis and Palestinians, and a crucial component for any "two-state solution" to the current conflict.

Currently, there are several cooperative industrial parks established jointly by the Palestinian Authority and the Israeli government in the West Bank and Gaza Strip. These developments have been dubbed "peace parks" because of their goal of using economic development to overcome conflict. Some critics regard this relationship as exploitative because Israel tightly controls Palestinian access to the world market. It may also take some time for the "peace parks" to earn the trust of Palestinians, given Israel's history of discouraging industrial development in the occupied territories. Nevertheless, the peace park idea has caught on and is now being pursued by neighboring Israeli and Palestinian cities such as Gilboa, Israel, and Jenin, West Bank.

In an ideal world, economic cooperation between Israel and Palestine would be only the beginning of what could become a transformation across the region—a "peace dividend." The Gulf states, Turkey, Egypt, Iran, and all the other countries in the region are developing their human capital through social services and education. Business-level cooperation with Israel, which has long had the most developed economy in the region, could facilitate rapid improvements everywhere.

THINGS TO REMEMBER

1. While the modern conflict between Israel and the Palestinians is decades old, the situation remains dynamic and complex and is a persistent obstacle to political and economic cooperation in the region.

2. Economic cooperation is a fact of life for Israelis and Palestinians, and a crucial component for any "two-state solution" to the current conflict.

3. The entire region would benefit from the peace dividend that would result from economic cooperation.

Reflections on North Africa and Southwest Asia

North Africa and Southwest Asia are frequently referred to as hopelessly mired in a morass of unsolvable problems. Many point to a gloomy future: violence, religious fundamentalism, and internal and external terrorism leading eventually to global disaster. If not that, then autocratic governments; economic inequities; water shortages; extreme gender discrimination; expanding populations; and for many countries, overdependence on energy resources that the world may soon declare obsolete. Yet many of the changes currently underway could provide a route through this daunting maze of issues.

Much evidence suggests that shifting toward democracy and away from repressive authoritarian rule can reduce violence throughout the region. Pressure for democracy is growing in the wealthier countries, where an increasingly educated, informed, and globally aware citizenry is demanding greater freedom of speech, a larger voice in politics, and more equality for women (Reasons for Optimism A, C). Al Jazeera has introduced a new era of open debate and hard-nosed analysis. Across the region, urbanization and education for women promises to curb population growth as it has in much of the rest of the world. But if urbanization is to also be a force for rising standards of living, then economic growth, increased employment, and upgrades to educational systems are essential (see Reasons for Optimism B).

Democracy and greater economic opportunities could reduce sectarian and ethnic conflict in places like Iraq, Israel, and Palestine. The most hopeful signs are the bottom-up peace initiatives and grassroots entrepreneurialism and constructive activism found across the region.

Climate change could complicate many of these positive shifts, but it could also reinforce them. For example, improvements in education and economic growth would be even more necessary to provide growing populations with food and water in a changing climate. The need to use new water-saving agricultural technologies, such as drip irrigation, would mean that farmers would need to become more educated and able to invest in sophisticated and expensive technologies. Similarly, climate change could also increase pressures for democratization as citizens faced with a wider range of environmental uncertainties could demand a level of responsiveness and accountability on the part of governments that only democracies can deliver.

Reasons for Optimism in North Africa and Southwest Asia

Democratization: Shifting toward democracy and away from repressive authoritarian rule may reduce violence throughout the region. **A** *Volunteers working for a candidate in Turkey's 2007 elections.* ▼

Development: Greater economic opportunities could calm sectarian and ethnic conflict in many places. **B** *A desk factory in Iraq that employs 300 people and produces 1500 desks per day.* ▼

Gender: Greater educational opportunity for women promises to curb population growth as it has in much of the rest of the world. **C** *Nurses training in the West Bank.* ▼

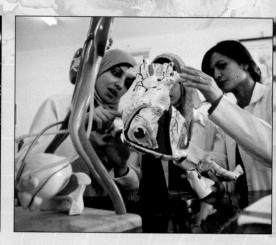

Thinking Geographically: What type of political party has controlled Turkey's government on and off for decades?

Thinking Geographically: What is a major benefit of economic diversification?

Thinking Geographically: How do greater educational opportunities for women help to decrease population growth?

Learning Goals Review

1. Water, Food, and Climate Change: How will people get enough water to grow food in the future, especially if climate change makes this dry region even dryer?

What factors could make it difficult to obtain enough water for agriculture? What technologies could provide potential solutions? How crucial is imported food for this region?

2. Gender and Population: How does the low status of women contribute to this region's high population growth?

Relative to other regions, how high is this region's population growth rate? How do women's education levels influence their decisions about how many children to have?

3. Urbanization and Globalization: How has globalization shaped different patterns of urbanization throughout the region?

Why is urbanization so different in the Gulf states than in the rest of the region? How have economic reforms aimed at improving global competitiveness, especially in export-oriented agriculture, influenced urbanization outside the Gulf states?

4. Development and Globalization: How have the huge fossil fuel reserves of some countries transformed economic development and driven globalization in this region?

What flows of money, resources, and people characterize globalization in the Gulf states? How important have outside political influences become?

5. Democratization: Are there any signs that this region may be democratizing?

How have recent events changed the political landscape of this region?

What has repression of the political opposition often resulted in? What has been the effect of integrating political opposition movements into democratic process?

Geographic Themes About North Africa and Southwest Asia

Look back at the Thematic Overview photos on page 207. Thinking geographically, answer the following questions about them:

(A) Water: What about this photo suggests that the date trees have been planted?

(C) Climate Change: Why would stabilizing sand dunes help control desertification?

(E) Urbanization: In the states without oil wealth in this region, what factor tends to push people out of rural areas and into cities?

(F) Globalization: What percentage of workers in the United Arab Emirates, where Dubayy is located, are temporary guest workers?

(G) Development: What clue do you see in the photo that suggests the workers living in this camp are not being paid very well?

(H) Democratization: What inspired the demonstrations in Iran in 2009?

Key Terms

cartel 232
Christianity 220
desertification 216
diaspora 220
economic diversification 234
female seclusion 225
Fertile Crescent 218
fossil fuel 208
Gulf states 225
hajj 223
intifada 240

Islam 208
Islamism 208
Judaism 220
monotheistic 220
Muslims 220
occupied Palestinian Territories (OPT) 209
OPEC (Organization of Petroleum Exporting Countries) 232
patriarchal 225
polygyny 227

Qur'an (or Koran) 213
salinization 214
seawater desalination 216
secular states 235
shari'a 223
Shi'ite (or Shi'a) 223
Sunni 223
theocratic states 235
veil 226
West Bank barrier 243
Zionists 240

7 Sub-Saharan Africa

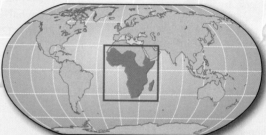

Learning Goals

After you read this chapter, you will be able to answer the following questions:

1. Climate Change, Food, and Water: Why are many African food production systems particularly vulnerable to climate change?

2. Globalization and Development: How has economic development been shaped by globalization?

3. Democratization: What forces are working for and against democracy in Africa?

4. Democratization and Gender: How are issues of gender influencing trends toward democratization?

5. Urbanization and Population: How is urbanization influencing population growth in Africa?

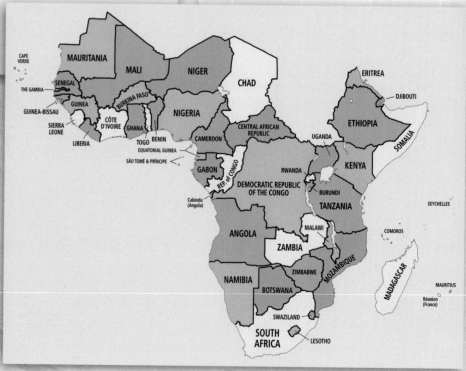

FIGURE 7.1 Political map of sub-Saharan Africa.

Thematic Overview of Sub-Saharan Africa

Climate Change: Because of widespread poverty in this region, people are less able to adapt to climate change. **A** *A nomadic woman in Niger helps her baby drink.* ▼

Food: Most people in this region practice subsistence agriculture, usually a small-farm practice which provides food for only the farmer's family. **B** *Plowing a field in Chad.* ▼

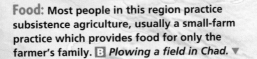

Water: Massive ecological changes created by large-scale irrigation projects have made many Africans appreciate irrigation on a smaller scale. **C** *Watering lettuce in the Central African Republic.* ▼

Development: Africa is the poorest region in the world, in part because it depends on exporting cheap raw materials to wealthier regions. **D** *A diamond mine in Sierra Leone.* ▼

Globalization: In recent years, sub-Saharan Africa has become a new frontier for resource acquisition and investment for Asia's large and growing economies, especially those of China and India. **E** *Chinese boats in Mauritius.* ▼

Democratization: Democratic elections have at times lessened tensions in Africa by enabling former combatants to effect change nonviolently. **F** *A wounded veteran participates in elections in a disputed area claimed by Kenya, Ethiopia, and Sudan.* ▼

Gender: Women across Africa are assuming positions of power after numerous crises have undermined faith in traditional male leadership. **G** *Asha-Rose Migiro, former minister of foreign affairs of Tanzania.* ▼

Population: African populations are growing faster than in any other region, having more than tripled in 50 years. **H** *Schoolchildren in South Africa.* ▼

Urbanization: Rapid and unplanned urbanization throughout Africa has resulted in many slums where living conditions are very poor. **I** *A flooded slum in Freetown, Sierra Leone.* ▼

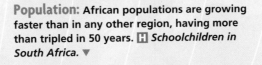

Global Patterns, Local Lives

Liberian environmental activist Silas Siakor is an affable and unassuming fellow. But his casual style conceals a fierce dedication to his homeland and a remarkable sleuthing ability. At great personal risk, Siakor uncovered evidence that 17 international logging companies were bribing Liberia's then-president, Charles Taylor, with cash and guns. Taylor looked the other way while the companies illegally logged Liberia's forests, which are home to many endangered species, such as forest elephants and chimpanzees. In return, the companies paid cash and weapons that Taylor used to equip his own personal armies. Made up largely of kidnapped and tragically abused children, Taylor's armies fought those of other Liberian warlords in a 14-year civil war that took the lives of 150,000 civilians (**Figure 7.2**). The war also spilled over into neighboring Sierra Leone, where another 75,000 people died. Meanwhile, the logging companies—based in Europe, China, and the Southeast Asia—reaped huge fortunes from the tropical forests.

▐▐▶ 175. 'EZRA,' TRAGIC TALE OF CHILD SOLDIERS IN AFRICA

Silas Siakor pulled together publicly available information that had been previously ignored by the international community and information provided by strategically placed informants in ports, villages, and lumber companies to prepare a clear, well-documented report substantiating the massive logging fraud. In response to Siakor's report, the UN Security Council voted to impose sanctions to stop the timber trade (**Figure 7.3**) and prosecute some of the people involved. Charles Taylor fled to Nigeria, but in 2006 he was turned over to face a war crimes tribunal in The Hague, Netherlands. His trial began in 2007 and is expected to end in 2011.

Democratic elections followed in Liberia, and in 2006 Ellen Johnson-Sirleaf took office, Africa's first elected woman president. In a move that was bold—given the poverty and political instability of her country—she cancelled all contracts with timber companies

FIGURE 7.2 Former child soldiers in Liberia. Former child soldiers hand in their weapons in 2004 at the end of the most recent civil war in Liberia.

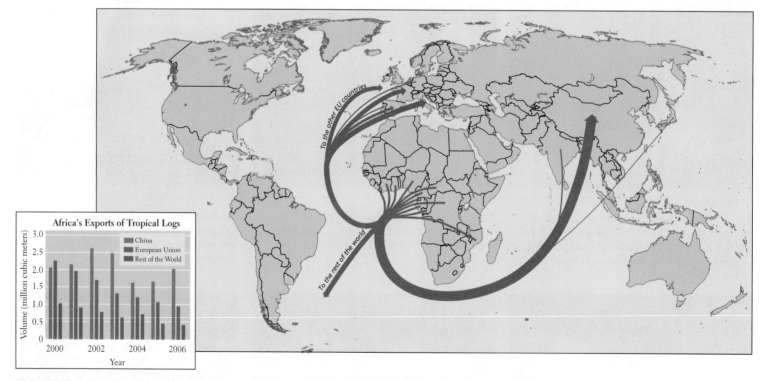

FIGURE 7.3 Destinations of Africa's exported tropical logs, 2000–2006. Most endpoint consumers of Africa's tropical woods are in North America and Europe. Though China is the largest single recipient of exported logs, much of that wood ends up in furniture sold by firms like Ikea in the United States and Europe. As of 2008, Ikea published standards that their wood suppliers must meet.

pending a revision of Liberian forestry law. *[Sources: Adapted from "Silas Kpanan'Ayoung Siakor: A Voice for the Forest and Its People," Goldman Environmental Prize, at http://www.goldmanprize.org/node/442; Scott Simon, "Reflections of a Liberian Environmental Activist," Weekend Edition Saturday, National Public Radio, April 29, 2006, at http://www. npr.org/templates/story/story.php?storyId=5370987; and author's personal communication with Siakor, September 2006.]* ∎

The story of Silas Siakor illustrates how Liberia, like much of this region, is still in the process of developing strong democratic institutions that can direct the sustainable development of the country's resources to the benefit of its citizens. Liberia also exemplifies how sub-Saharan Africa has failed to benefit substantially from economic globalization. Liberia has rich timber and mineral (diamond) resources, but most of the profits from these industries have gone to Liberian elites or to foreign companies. And yet, Silas Siakor's success illustrates how Africans can find solutions to these challenges. It shows as well that African efforts toward achieving peace and prosperity can be both aided and thwarted by the rest of the world.

Sub-Saharan Africa (Figure 7.4 on pages 250–251) is home to about 809 million people, a population that is growing at the rate of 2.5 percent per year. At this rate, the population will double in 28 years. The region contains several of the fastest-growing economies in the world, as well as some of the world's richest deposits of oil, gold, platinum, copper, and other strategic minerals. During the era of European colonialism (1850s–1950s), this massive wealth flowed out of Africa, providing little benefit to its people. Even after African countries achieved political independence in the 1950s, 1960s, and 1970s, wealth continued to flow out of Africa. It is not surprising, then, that much of Africa is impoverished and often at war with itself. But amid the turmoil, many hopeful signs suggest that the conflicts can be resolved and the well-being of most people enhanced.

THINGS TO REMEMBER

1. Europe's colonization of sub-Saharan Africa set up patterns of economic development that still leave many countries in a weak position in the global economy.

2. A shift toward democracy is occurring throughout Africa, but powerful forces continue to work against this trend in many countries.

I THE GEOGRAPHIC SETTING

Terms in This Chapter

We refer to this region as *sub-Saharan Africa* to recognize that North Africa is not included. Only occasionally do we refer to the whole continent and then we refer to it simply as Africa. Sub-Saharan Africa is defined by the countries shown in Figure 7.1 on page 246. Notice that Sudan is not included. Sudan's location on the Nile River and the strong Arab influence in the capital of Khartoum historically brought that country into closer association with North Africa and Southwest Asia, and it is discussed in Chapter 6.

The naming of African countries can often be confusing. For example, there are two neighboring countries called Congo—the Democratic Republic of the Congo and the Republic of Congo. Because these designations are both lengthy and easily confused, we abbreviate them in this text. The Democratic Republic of the Congo (formerly Zaire) carries the name of its capital in parentheses: Congo (Kinshasa). The Republic of Congo is called Congo (Brazzaville). Check the regional map (see Figure 7.4) to see the locations of these countries and capitals.

Physical Patterns

The African continent is big—the second largest after Asia and about 2 million square miles bigger than North America. But Africa's great size is not matched by its surface complexity. More than one-fourth of the continent is covered by the Sahara, many thousands of square miles of what, to the unpracticed eye, appears to be a homogeneous desert landscape (see Chapter 6). Africa has no major mountain ranges, but it does have several high volcanic peaks, including Mount Kilimanjaro (19,324 feet [5890 meters] high; see Figure 7.4B) and Mount Kenya (17,057 feet [5199 meters] high). At their peaks, both have permanent snow and ice, though these features have shrunk dramatically due to global climate change and associated local environmental changes.

Landforms

The surface of the continent of Africa can be envisioned as a raised platform, or plateau, bordered by fairly narrow and uniform coastal lowlands. The platform slopes downward to the north; it has an upland region with several high peaks in the southeast, and lower areas in the northwest. The steep escarpments (long cliffs) between plateau and coast have obstructed transportation and hindered connections to the outside world. Africa's long, uniform coastlines also provide few natural harbors.

Geologists usually place Africa at the center of the ancient supercontinent Pangaea (see Figure 1.21 on page 45). As landmasses broke off from Africa and moved away—North America to the northwest, South America to the west, and India to the northeast—Africa readjusted its position only slightly. Hence, it did not pile up long linear mountain ranges, as did the other continents when their plates collided with one another (see page 44).

Today, Africa continues to break apart along its eastern flank. There the Arabian Plate has already split away and drifted to the northeast, leaving the Red Sea, which separates Africa and Asia. Another split, known as the Great Rift Valley, extends south from the Red Sea more than 2000 miles (3200 kilometers) (see Figure 7.4C). In the future, Africa is expected to split again along these rifts.

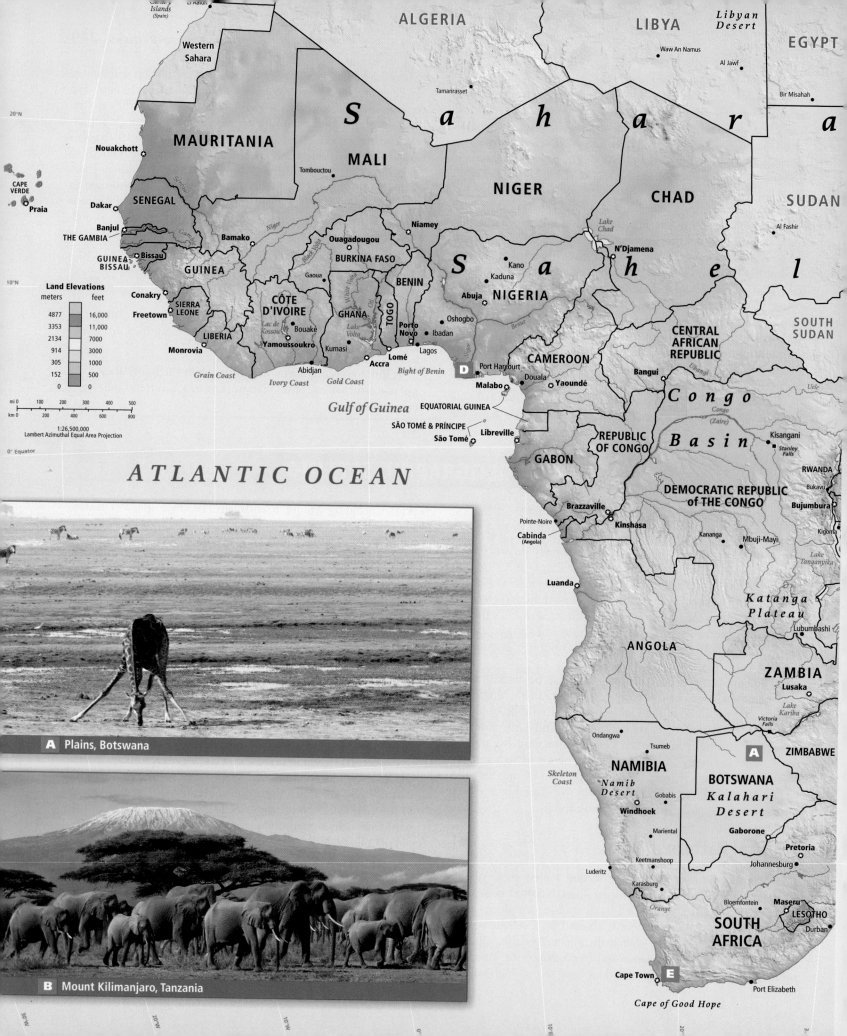

ALGERIA

LIBYA

Libyan Desert

EGYPT

Western Sahara

Waw An Namus

Al Jawf

Bir Misahah

20°N

S   a   h   a   r   a

MAURITANIA

Nouakchott

MALI

Tombouctou

NIGER

CHAD

SUDAN

CAPE VERDE

Praia

Dakar

SENEGAL

Banjul

THE GAMBIA

GUINEA BISSAU

Bissau

10°N

Bamako

Niamey

Ouagadougou

BURKINA FASO

Lake Chad

N'Djamena

Al Fashir

SOUTH SUDAN

Land Elevations

meters	feet
4877	16,000
3353	11,000
2134	7000
914	3000
305	1000
152	500
0	0

Conakry

GUINEA

Freetown

SIERRA LEONE

CÔTE D'IVOIRE

Gaoua

S   a   h   e   l

Kano

Kaduna

Abuja

NIGERIA

CENTRAL AFRICAN REPUBLIC

LIBERIA

Monrovia

Yamoussoukro

GHANA

Bouaké

Kumasi

TOGO

BENIN

Porto Novo

Oshogbo

Ibadan

Lagos

Lac de Kossou

Lake Volta

Benue

mi 0   100   200   300   400   500

km 0   200   400   600   800

1:26,500,000
Lambert Azimuthal Equal Area Projection

Abidjan

Grain Coast

Ivory Coast

Gold Coast

Accra

Lomé

Bight of Benin

D

Port Harcourt

CAMEROON

Douala

Yaoundé

Bangui

Uele

Congo

Kisangani

Stanley Falls

0° Equator

Gulf of Guinea

EQUATORIAL GUINEA

Malabo

SÃO TOMÉ & PRÍNCIPE

São Tomé

Libreville

GABON

REPUBLIC OF CONGO

Congo (Zaire)

Basin

RWANDA

Bukavu

Bujumbura

ATLANTIC OCEAN

Brazzaville

Kinshasa

DEMOCRATIC REPUBLIC of THE CONGO

Kigoma

Pointe-Noire

Cabinda (Angola)

Kananga

Mbuji-Mayi

Lake Tanganyika

Luanda

Katanga Plateau

Lubumbashi

ANGOLA

ZAMBIA

Lusaka

Lake Kariba

Victoria Falls

Ondangwa

Tsumeb

A

ZIMBABWE

NAMIBIA

Skeleton Coast

Namib Desert

Gobabis

BOTSWANA

Kalahari Desert

Windhoek

Mariental

Gaborone

Pretoria

Johannesburg

Lüderitz

Keetmanshoop

Karasburg

Bloemfontein

Maseru

LESOTHO

Orange

SOUTH AFRICA

Durban

Cape Town

E

Port Elizabeth

Cape of Good Hope

A Plains, Botswana

B Mount Kilimanjaro, Tanzania

C Great Rift Valley, Tanzania

D Niger River Delta, Nigeria

E Escarpment at Cape Town, South Africa

FIGURE 7.4 Regional map of sub-Saharan Africa.

Climate and Vegetation

Most of sub-Saharan Africa has a tropical climate (Photo Essay 7.1). Average temperatures generally stay above 64°F (18°C) year-round everywhere except at the more temperate southern tip of the continent and in cooler upland zones (hills, mountains, high plateaus). Seasonal climates in Africa differ more by the amount of rainfall than by temperature.

Most rainfall comes to Africa by way of the **intertropical convergence zone (ITCZ)**, a band of atmospheric currents that circle the globe roughly around the equator (see the inset map in Photo Essay 7.1). At the ITCZ, warm winds converge from both the north and the south and push against each other. This causes the air to rise, cool, and release moisture in the form of rain. The rainfall produced by the ITCZ is most abundant in Africa near the equator (see Photo Essay 7.1D). Here, dense tropical rainforests flourish in places such as the Congo Basin (Photo Essay 7.1A).

The ITCZ shifts north and south seasonally, generally following the area of earth's surface that has the highest average temperature at any given time. Thus, during the height of summer in the Southern Hemisphere in January, the ITCZ might bring rain far enough south to water the dry grasslands, or steppes, of Botswana. During the height of summer in the Northern Hemisphere in August, the ITCZ brings rain as far north as the southern fringes of the Sahara—an area called the **Sahel**, where steppe and savanna grasses grow. Poleward of both of these extremes, the belt of air that has dumped its moisture while rising and cooling is now drier. At roughly 30° N latitude (the Sahara) and 30° S latitude (the Namib and Kalahari deserts) the drier air descends, forming a subtropical high-pressure zone that shuts out lighter, warmer, moister air. As a result of this system (which is in no way precise), deserts tend to be found in Africa (and on other continents) in these zones about 30 degrees north and south of the Equator.

The tropical wet climates that support equatorial rain forests are bordered on the north, east, and south by seasonally wet/dry subtropical woodlands (Photo Essay 7.1B). These give way to moist tropical savannas or steppes, where tall grasses and trees intermingle in a semiarid environment. These tropical wet, wet/dry, and steppe climates have provided suitable land for agriculture for thousands of years. Farther to the north and south lie the true desert zones of the Sahara and the Namib and Kalahari (Photo Essay 7.1C). This banded pattern of African ecosystems is modified in many areas by elevation and wind patterns.

Without mountain ranges to block them, wind patterns can have a strong effect on climate in Africa. Winds blowing north along the east coast keep ITCZ-related rainfall away from the **Horn of Africa**, the triangular peninsula that juts out from northeastern Africa below the Red Sea. As a consequence, the Horn of Africa is one of the driest parts of the continent (see the map in Photo Essay 7.1). Along the west coast of the Namib Desert, moist air from the Atlantic is blocked from moving over the desert

by cold air above the northward-flowing water of the Benguela Current. Like the Peru Current off South America, the Benguela is an oceanic current that is chilled by its passage past Antarctica. Rich in nutrients, it supports a major fishery along the west coast of Africa.

Environmental Issues

While Africans have generally contributed very little to the build-up of greenhouse gases in the atmosphere, deforestation in Africa is amplifying global climate change. Because of Africa's poverty, its people are much less able to adapt to climate change than those in other regions. Yet in the face of these problems, many Africans are developing strategies to cope with uncertainties related to the region's changing climate.

▐▌▶ 158: DISAPPEARING GLACIERS ON MT. KILIMANJARO RAISE ENVIRONMENTAL CONCERNS

Deforestation and Climate Change

Deforestation is sub-Saharan Africa's main contribution to CO_2 emissions and potential climate change. Trees absorb CO_2 as they photosynthesize, thus removing carbon from the air and storing it as biomass, a process known as **carbon sequestration**. When trees die, as they decompose or are burned, they release the stored CO_2 into the atmosphere. Hence, deforestation is a major contributor to carbon buildup in the atmosphere and ultimately to climate change.

African countries lead the world in the *rate* of deforestation, the percentage of total forest area lost. Of the eight countries that had the world's highest rates of deforestation between 1990 and 2005, six are African (Burundi, Togo, Nigeria, Benin, Uganda, and Ghana). This constitutes a major human environmental impact across the region (see Photo Essay 7.2 on page 255). In terms of emissions from deforestation, the leaders were Brazil and Indonesia, but Nigeria and Congo (Kinshasa), ranked third and fourth.

▐▌▶ 168: PLAN TO CLEAR-CUT UGANDAN FOREST RESERVE FOR GROWING SUGAR CANE SPARKS CONTROVERSY

Most of Africa's deforestation is driven by the growing demand for farmland and fuelwood, although logging by international timber companies is also increasing. Africans use wood (or charcoal made from wood) to supply nearly all their domestic energy (see Photo Essay 7.2 A, B). Wood remains the cheapest fuel available, in part because of African traditions that consider forests to be a free resource held in common. Even in Nigeria, a major oil producer, most people use fuelwood because they cannot afford petroleum products.

For a decade or more, many in sub-Saharan Africa have recognized the need to use fuelwood more sustainably. Solar ovens and fuel-efficient wood stoves have been widely promoted. While many users find them less convenient than old-fashioned cooking fires, the growing scarcity of fuelwood is driving greater acceptance of these technologies.

---

intertropical convergence zone (ITCZ) a band of atmospheric currents that circle the globe roughly around the equator; warm winds from both north and south converge at the ITCZ, pushing air upward and causing copious rainfall

Sahel a band of arid grassland, where steppe and savanna grasses grow, that runs east-west along the southern edge of the Sahara

Horn of Africa the triangular peninsula that juts out from northeastern Africa below the Red Sea and wraps around the Arabian Peninsula

carbon sequestration the removal and storage of carbon taken from the atmosphere

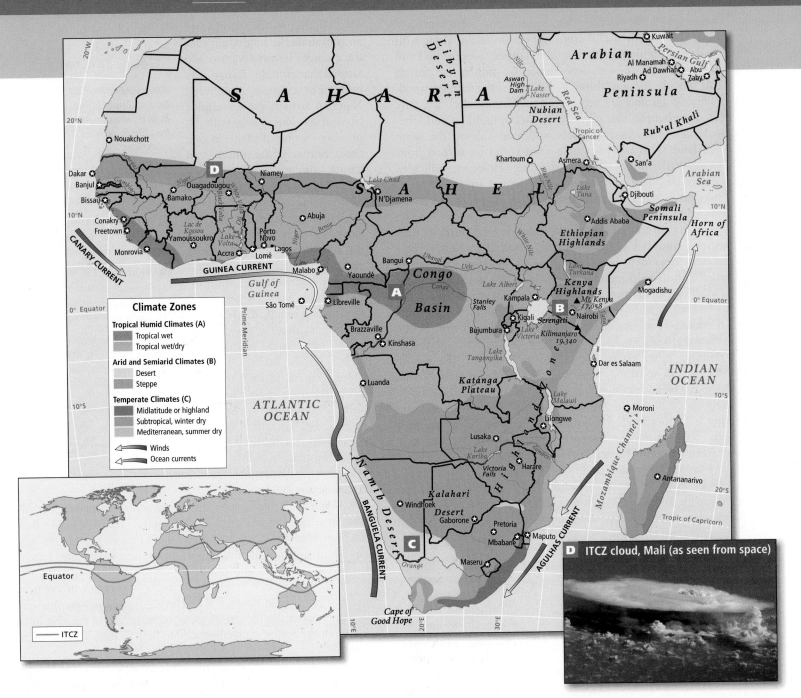

Climate Zones

Tropical Humid Climates (A)
- Tropical wet
- Tropical wet/dry

Arid and Semiarid Climates (B)
- Desert
- Steppe

Temperate Climates (C)
- Midlatitude or highland
- Subtropical, winter dry
- Mediterranean, summer dry

→ Winds
→ Ocean currents

Equator

ITCZ

D ITCZ cloud, Mali (as seen from space)

A Tropical wet, Gabon

B Subtropical, winter dry, Kenya

C Desert in Namibia

Some governments are also encouraging **agroforestry**—growing economically useful crops of trees on farms, along with the usual plants and crops. Using these farmed trees for fuel and construction helps reduce dependence on nonfarmed forests and can provide income through the sale of the farmed wood. By practicing agroforestry on the fringes of the Sahel, a family in Mali, for example, can produce fuelwood, building and fencing materials, medicinal products, and food—all on the same piece of land. This would double what could be earned from single-crop agriculture or animal herding. However, critics of agroforestry note that sometimes the trees used are *invasive species*, nonnative plants that threaten many African ecosystems by outcompeting indigenous species.

A strong promoter of agroforestry is the Green Belt Movement, founded by Dr. Wangari Maathai of Kenya. The Green Belt Movement helps rural women plant trees for use as fuelwood and to help reduce soil erosion. The movement grew from Maathai's belief that a healthy environment is essential for democracy to flourish. Thirty years and 30 million trees after she began, Maathai was awarded the 2004 Nobel Peace Prize for her contributions to sustainable development, democracy, and peace.

Multiple other efforts are underway to reduce the extent of deforestation resulting from logging in sub-Saharan Africa. One approach is to require logging companies to use methods that reduce clear-cutting and damage to the surrounding forest. Another is to minimize the building of logging roads, because after the logging companies have left, poor farmers often use these roads to permanently occupy deforested areas. Under pressure from consumers in the EU, some European logging companies now use more sustainable practices, such as fewer logging roads and less clear-cutting. However, the Asian logging companies, which are expanding rapidly in sub-Saharan Africa, are generally not adopting such practices.

Food Production, Water Resources, and Vulnerability to Climate Change

> **Learning Goal 1**
> **Climate Change, Food, and Water:**
> Why are many African food production systems particularly vulnerable to climate change?

Africa's food production systems have both advantages and disadvantages in helping the region adapt to climate change (see Photo Essay 7.3 on page 256).

Agricultural Systems Most sub-Saharan Africans practice **subsistence agriculture**, which provides food for only the farmer's family and is usually done on small farms of about 2 to 10 acres (1 to 4 hectares). Most African farmers also practice **mixed agriculture**, raising a diverse array of crops and a few animals as livestock (see Photo Essay 7.3C). Many also fish, hunt, herd, and gather some of their food from forest or grassland areas (see Thematic Overview A and B on page 247).

These widely practiced food acquisition techniques have advantages and disadvantages in a world of climate change. Mixed

> **agroforestry** growing economically useful crops of trees on farms, along with the usual plants and crops, to reduce dependence on trees from nonfarmed forests and to provide income to the farmer
>
> **subsistence agriculture** farming that provides food for only the farmer's family and is usually done on small farms
>
> **mixed agriculture** the raising of a variety of crops and animals on a single farm, often to take advantage of several environmental riches
>
> **commercial agriculture** farming in which crops are grown deliberately for cash rather than solely as food for the family

agriculture provides a diverse array of strategies for coping with the changes in temperature and rainfall that climate change may bring. For example, traditional Nigerian farmers grow complex tropical gardens, often with 50 or more species of plants at one time. Some of the plants can handle drought, while others can withstand intense rain or heat.

However, the subsistence nature of most African farming can also leave families without much cash. If harvests are too low to provide sellable surpluses, and hunting and gathering fail to provide enough food to feed the family, there is insufficient money to buy food. While such situations can lead to famine, it is important to note that the most serious famines in Africa have occurred not because of low harvests but rather because of political instability that disrupts economies and food growing and distribution systems.

▐▐ ▶ 162: FELLOWSHIP PROGRAM AIMS TO OPEN DOORS FOR AFRICAN WOMEN IN AGRICULTURAL RESEARCH

Much of Africa is now shifting over to **commercial agriculture**, in which crops are grown deliberately for cash rather than solely as food for the farm family. This type of production also has advantages and disadvantages with respect to climate change. If harvests are good and prices for crops are adequate, farmers can earn enough cash to get them through a year or two of poor harvests (see Photo Essay 7.3D). Having some cash income can also allow poor families to invest in other means of earning a living, such as opening a grinding mill to help process other farmers' harvests.

However, some of the most common commercial crops, such as peanuts, cacao beans (used for making chocolate), rice, and tea and coffee, are often less well adapted to environments outside their native range. They are more likely to produce smaller yields if temperatures increase or water becomes scarce. Moreover, to maximize profits, these crops are often grown in large fields of only a single plant species. This can leave crops vulnerable to pests and if crops fail, farmers have no other garden foods to rely on.

The potential for commercial agriculture to help farmers adapt to the uncertainties of global climate change is also limited by the instability of prices for commercial crops. Prices can rise or fall dramatically from year to year because of overproduction or crop failures

Thinking Geographically

After you have read about human impact on the Biosphere in sub-Saharan Africa, you will be able to answer the following questions:

A Other than by reading the labels, how can you tell that vegetation has been burned?

B What about this photo suggests that the trees being logged are old?

C What about the fishing methods depicted in the photo suggest that catches are fairly small in size?

D What, if any, clues in this photo suggest that this irrigation project is large in scale?

Deforestation, desertification, and increasing water use have had major impacts on sub-Saharan Africa's ecosystems.

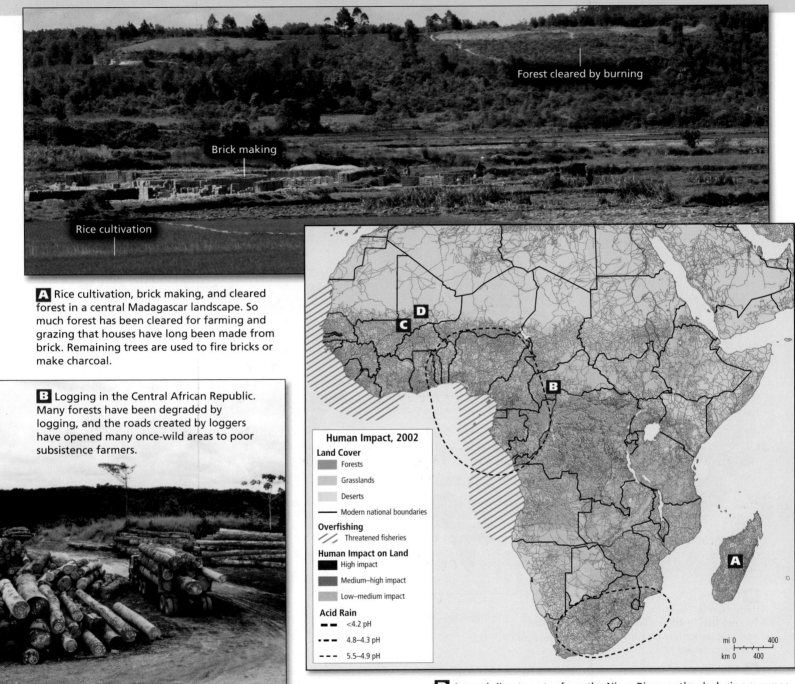

A Rice cultivation, brick making, and cleared forest in a central Madagascar landscape. So much forest has been cleared for farming and grazing that houses have long been made from brick. Remaining trees are used to fire bricks or make charcoal.

B Logging in the Central African Republic. Many forests have been degraded by logging, and the roads created by loggers have opened many once-wild areas to poor subsistence farmers.

Human Impact, 2002

Land Cover
- Forests
- Grasslands
- Deserts
- ⎯⎯ Modern national boundaries

Overfishing
- ⫽⫽ Threatened fisheries

Human Impact on Land
- High impact
- Medium–high impact
- Low–medium impact

Acid Rain
- – – <4.2 pH
- – - - 4.8–4.3 pH
- - - - 5.5–4.9 pH

mi 0   400
km 0   400

D A canal diverts water from the Niger River wetlands during summer flooding in order to supply farmers with irrigation water. The project increases food security for farmers elsewhere in Mali, while threatening the livelihoods of fishers, farmers, and pastoralists in the Niger wetlands.

C A fisher casts his net in the Niger River wetlands, where carefully synchronized resource use patterns have developed over millennia.

Much of this region is highly exposed to drought, flooding, and other climatic disturbances. Poverty and having little access to cash increase sensitivity to climate change, while political instability makes it harder for governments to boost resilience with effective disaster management.

A Children bring water back to their home in Somalia. Drought, poverty, and political instability come together in Somalia to create extremely high vulnerability.

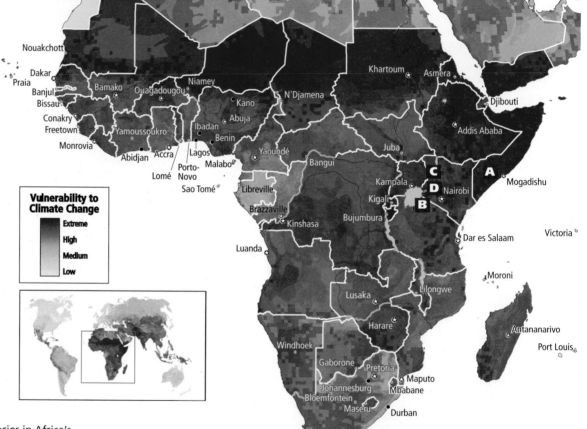

Vulnerability to Climate Change

Extreme
High
Medium
Low

B The annual wildebeest migration in Tanzania and Kenya. Twenty five to 40 percent of the species in Africa's national parks may become endangered because of climate change.

C A landscape of subsistence and mixed agriculture in Kenya. These diversified systems lend farmers some resilience to climatic disturbances, but they provide no cash to help farmers buy food they might need.

D Tea harvested for export in Kenya. Commercial crops like tea can increase the farmer's resilience to climate change by providing cash to buy any needed food. But tea, like many commercial crops, requires expensive and polluting fertilizers, and prices for tea are unstable.

Thinking Geographically

After you have read about vulnerability to climate change in sub-Saharan Africa, you will be able to answer the following questions:

A How is poverty evident in this photo?

B What caused the loss of many wildebeest in 2007?

C How is agricultural diversity illustrated in this photo?

D What kind of climate change in Africa most often causes a reduced yield for a crop like tea?

both in Africa and abroad. Another downside of commercial production is that it is usually based on the permanent conversion of forests to fields. This can put a greater strain on soil and water resources. Like tropical soils everywhere (see page 107 in Chapter 3), African soils rapidly lose their fertility when cultivated. This is especially true in the warmer, wetter areas, where the nutrient-carrying organic matter in the soil decays rapidly. To maintain soil quality in such a climate, subsistence farmers have long used **shifting cultivation**, in which patches of forest are cleared and cultivated for just 3 years or so and then left to regrow. After a few decades, the soil naturally replenishes its organic matter and nutrients and is ready to cultivate again. However, when fallow periods are reduced, as is happening in many rural areas with increasingly high population densities, the soil can become degraded.

When fields are cleared permanently, as many commercial agriculture systems demand, soil fertility declines and crops become more likely to fail even under ideal climatic conditions. Chemical fertilizers can compensate for this loss, but are often unaffordable for many farmers—and after repeated use, they may lose their effectiveness. Moreover, rains almost always wash much of the fertilizer into nearby waterways, thus polluting them. This may ultimately hurt farmers and fishers by reducing the quantity of available fish, which are important sources of protein for rural people.

These aspects of commercial agricultural systems make them poorly adapted even to ideal climatic conditions in Africa, let alone the hotter and more drought-prone environments that climate change may bring. Foreign agricultural "experts" who often push commercial systems can be woefully ignorant of local conditions and the value of local cultivation techniques. For example, in Nigeria, it is women who grow most of the food for family consumption and who are guardians of the knowledge that makes Nigeria's complex farming systems possible. However, women were rarely included or even consulted by the British experts who promoted commercial agricultural development projects. At best, women were employed as field laborers. Consequently, diverse subsistence agricultural systems based on numerous plant species and deep agricultural knowledge were replaced by less-stable commercial systems based on a single plant species.

Agricultural scientists now recognize past mistakes and are trying to incorporate the knowledge of African farmers, female and male, into more diverse commercial agriculture systems. For example, scientists at Nigeria's International Institute for Tropical Agriculture are developing cultivation systems that, like traditional systems, use many species of plants that help each other cope with varying climatic conditions. Most of these systems are designed for both subsistence and commercial agriculture, and thus can give families both a stable food supply and cash to help them ride out crop failures and pay school fees for their children.

Irrigation and Water Management Alternatives Because climate change is likely to result in changing rainfall patterns, many are looking to irrigation to provide more stability to agricultural systems. However, the massive ecological changes created by large-scale irrigation projects have made some Africans appreciate irrigation on a smaller scale (see Thematic Overview C).

Africa has a number of major rivers—the Nile, the Congo, the Zambezi, and the Limpopo—and climate change will affect all of them. Here we focus on the Niger, one of Africa's most important rivers, which flows through very different ecological zones. The Niger rises in the tropical wet Guinea Highlands and carries summer floodwaters northeast into the normally arid lowlands of Mali and Niger (see Photo Essay 7.2 C, D). There the waters spread out into lakes and streams that nourish wetlands. For a few months of the year (June through September), the wetlands ensure the livelihoods of millions of fishers, farmers, and pastoralists. These people share the territory in carefully synchronized patterns of land use that have survived for millennia. Wetlands along the Niger produce 8 times more plant matter per acre than the average wheat field, provide seasonal pasture for millions of domesticated animals, and are an important habitat for wildlife.

The governments of Mali and Niger now want to dam the Niger River and channel its water into irrigated agriculture projects that will help feed the more than 26 million people in the two countries. Rising food prices in global markets, mounting population pressure, and the threat of changing rainfall patterns due to climate change are driving the two governments to undertake the massive project. However, the dams will forever change the seasonal rise and fall of the river. The irrigation systems may also pose a threat to human health. Systems that rely on surface storage lose a great deal of water through evaporation and leakage, and the standing pools of water they create often breed mosquitoes that spread tropical diseases such as malaria and harbor the snails that host schistosomiasis.

Many smaller-scale alternatives are available. In some parts of Senegal, for example, farmers are using hand- or foot-powered pumps to bring water from rivers or ponds directly to the individual plants that need it. This is in some ways a more modern version of traditional African irrigation practices whereby water is delivered directly to the roots of the plants by human water brigades. Smaller-scale projects provide the same protection against drought that larger systems offer, but are much cheaper and simpler to operate for small farmers. They also avoid the large-scale ecological disruption of larger projects. Already in successful use for 15 years, these pumps will help farmers adjust to the drier conditions that may come with climate change.

> **shifting cultivation** a productive system of agriculture in which small plots are cleared in forestlands, the dried brush is burned to release nutrients, and the clearings are planted with multiple species; each plot is used for only 2 or 3 years and then abandoned for many years of regrowth

Herding and Desertification Herding, or **pastoralism**, is practiced by millions of Africans, primarily in savannas, on desert margins, or in the mixture of grass and shrubs called *open bush*. Herders use the milk, meat, and hides of their animals, and they typically circulate seasonally through wide areas, taking their animals to available pasturelands. They trade with settled farmers for grain, vegetables, and other necessities.

Many traditional herding areas in Africa are now experiencing **desertification**, the process by which arid conditions spread to areas that were previously moist, often as a result of the loss of native vegetation. Traditional herding may be partially to blame, but economic development schemes that encourage cattle raising have also created problems. Cattle need more water and forage than traditional herding animals and hence can place greater stress on native grasslands than goats or camels. Agricultural intensification in the Sahel also contributes to desertification, as scarce water resources are diverted to irrigation, leaving soils exposed to wind and water erosion.

> **pastoralism** a way of life based on herding; practiced primarily on savannas, on desert margins, or in the mixture of grass and shrubs called open bush
>
> **desertification** the process by which arid conditions spread to areas that were previously moist

The effects of desertification are most dramatic in the Sahel (Arabic for "shore" of the desert). This band of arid grassland, 200 to 400 miles (320 to 640 kilometers) wide, runs east-west along the southern edge of the Sahara (see the map in Photo Essay 7.1 on page 253). Over the last century, desertification has shifted the Sahel to the south. For example, the *World Geographic Atlas* in 1953 showed Lake Chad situated in a forest well south of the southern edge of the Sahel. By 1998, the Sahara itself was encroaching on Lake Chad (see Photo Essay 7.1) and the lake had shrunk to a tenth of the area it occupied in 1953.

Wildlife and Climate Change

Africa's world-renowned wildlife faces multiple threats from both human and natural forces, all of which could become more severe due to global climate change. The Intergovernmental Panel on Climate Change, a body of scientists tasked by the United Nations with assessing scientific information relevant to understanding climate change, estimates that 25 to 40 percent of the species in Africa's national parks may become endangered as a result of climate change.

Wildlife managers need to develop new management techniques to help animals survive. For example, one of the greatest wildlife spectacles on the planet took a tragic turn in 2007. The annual 1800-mile-long natural migration of more than a million wildebeest, zebras, and gazelles in Kenya's Masai Mara game reserve requires animals to traverse the Mara River (see Photo Essay 7.3B). In the best of times, this is a difficult migration that usually results in a thousand or so animals drowning. In the past, the park's managers have taken a hands-off approach to the migration, considering the losses normal. However, in 2007 extremely heavy rains, which some scientists think are related to global climate change, swelled the Mara River to record levels. When the animals tried to cross, 15,000 drowned. Park managers are now considering taking a more active role in helping the migrating animals cope with unusual climatic conditions that may worsen with climate change. This may involve stopping animals from attempting a river crossing where many have already drowned and directing them to a safer crossing.

II ▶ 169: PROTECTING NATURE IN GUINEA COLLIDES WITH HUMAN NEEDS

The dependence of farmers on hunting wild game (*bushmeat*) for part of their food and income is already a major threat to wildlife in much of Africa. For example, farmers who need food or extra income are killing endangered species such as gorillas and chimpanzees in record numbers in the Congo Basin. If crop harvests are diminished by global climate change, many farmers will become even more dependent on bushmeat. The threat to wild populations of various species has led to calls to expand protected areas for wildlife and establish new ones. However, Africa's existing parks are already struggling to deal with poaching (illegal hunting) within the parks by members of surrounding communities. Poaching is also fueled by demand for exotic animal parts (tusks, hooves, penises) as medicines and aphrodisiacs, especially in Asia. Some park managers have reduced poaching by promoting alternative livelihoods in the poachers' home communities.

Ecotourism already makes money for Africa's national parks, which constitute one-third of the world's preserved national park land. Some parks are now using profits from ecotourism to sponsor economic development in nearby communities that once depended economically in part on poaching. For example, in 1985 Kasanka National Park in Zambia was plagued by poaching that threatened its wildlife populations. Park managers decided to generate employment for the villagers through tourism-related activities. They built tourist lodges and wildlife-viewing infrastructure and started cottage industries to make products to sell to the tourists. Funding is also provided to local schools and clinics, and students are included in research projects within the park. Local farming has expanded into alternative livelihoods, such as beekeeping and agroforestry. Today, poaching in Kasanka is very low, its wildlife populations are booming, tourism is growing, and local communities have an ongoing stake in the park's success.

THINGS TO REMEMBER

1. The surface of the continent of Africa can be envisioned as a raised platform, or plateau, bordered by fairly narrow and uniform coastal lowlands. Most rainfall comes to Africa by way of the intertropical convergence zone (ITCZ), a band of atmospheric currents that circle the globe roughly around the equator.

2. Deforestation is Africa's main contribution to CO_2 emissions and potential climate change, and most of Africa's deforestation is driven by the growing demand for farmland and fuelwood, although logging by international timber companies is also increasing.

3. Long-standing physical challenges, such as deforestation, desertification, and increasing water scarcity, have had a major impact on sub-Saharan Africa's ecosystems. These challenges are

likely to increase with climate change. The effects of desertification are most dramatic in the region called the Sahel.

4. **Learning Goal 1: Climate Change, Food, and Water** Africa's food production systems, still largely subsistence based with some mixed agriculture, are threatened by climate change. Commercial food production for Africa's cities is needed, but commercial agriculture has many ecological drawbacks in Africa.

Human Patterns over Time

Africa's rich past has often been misunderstood and dismissed by people from outside the region. European slave traders and colonizers called Africa the "Dark Continent" and assumed it was a place where little of significance in human history had occurred. The substantial and elegantly planned cities of Benin in western Africa, Djenné in the Niger River basin, and Loango in the Congo basin, which European explorers encountered in the 1500s, never became part of Europe's image of Africa. Even today, most people outside the continent are unaware of Africa's internal history or its contributions to world civilization.

The Peopling of Africa and Beyond

Africa is the original home of humans (see Timeline A on page 260). It was in eastern Africa (in what is today the Ethiopian, Kenyan, and Ugandan highlands) that the first human species evolved more than 2 million years ago, although they differed anatomically from humans today. These early, tool-making humans (*Homo erectus*) ventured out of Africa, reaching north of the Caspian Sea and beyond as early as 1.8 million years ago. Anatomically, modern humans (*Homo sapiens*) evolved from earlier hominoids in eastern Africa about 200,000 years ago, and by about 90,000 years ago, they had reached the eastern Mediterranean. Like the migrations of earlier humans, modern human migrations radiated out of Africa, spreading into Europe and across mainland and island Eurasia.

Early Agriculture, Industry, and Trade in Africa

In Africa, people began to cultivate plants as early as 7000 years ago in the Sahel and the highlands of present-day Sudan and Ethiopia. Agriculture was brought south to equatorial Africa 2500 years ago and to southern Africa about 1500 years ago. Trade routes spanned the African continent, extending north to Egypt and Rome, and east to India and China. Gold, elephant tusks, and timber from tropical Africa were exchanged for salt, textiles, beads, and a wide variety of other goods.

About 3400 years ago, people in northeastern Africa learned how to smelt iron and make steel. By 700 C.E., when Europe was still recovering from the collapse of the Roman Empire, a remarkable civilization with advanced agriculture, iron production, and gold-mining technology had developed in the highlands of southeastern Africa in what is now Zimbabwe. This empire, now known as the Great Zimbabwe Empire (see Timeline B), traded with merchants from Arabia, India, Southeast Asia, and China, exchanging the products of its mines and foundries for silk, fine porcelain, and exotic jewelry. The Great Zimbabwe Empire collapsed around 1500 for reasons not yet understood.

Complex and varied social and economic systems existed in many parts of Africa well before the modern era. Several influential centers made up of dozens of linked communities developed in the forest and the savanna of the western Sahel. There, powerful kingdoms and empires rose and fell, such as that of Ghana (700–1000 C.E.), centered in what is now Mauritania and southwestern Mali, and the Mali Empire (1250–1600 C.E.), centered a bit to the east on the Niger River in the famous Muslim trading and religious center of Tombouctou (Timbuktu). Some rulers periodically sent large entourages on pilgrimages through Tombouctou to Makkah (Mecca), where their opulence was a source of wonder (see Timeline C).

Africans also traded slaves. Long-standing customs of enslaving people captured during war fueled this trade. The treatment of slaves within Africa was sometimes brutal and sometimes reasonably humane. Long before the beginning of Islam, a slave trade developed with Arab and Asian lands to the east (see Figure 7.5 on page 260). After the spread of Islam began around 700 C.E., the slave trade continued, and close to 9 million African slaves were exported to parts of Southwest, South, and Southeast Asia by Muslim traders. When slaves were traded to non-Africans, indigenous checks on brutality were lost. For example, to ensure sterility and to help promote passivity, Muslim traders often preferred to buy castrated male slaves.

Europeans and the African Slave Trade

The course of African history shifted dramatically in the mid-1400s, when Portuguese sailing ships began to appear off Africa's west coast. The names given to stretches of this coast by the Portuguese and other early European maritime powers reflected their interest in Africa's resources: the Gold Coast, the Ivory Coast, the Pepper Coast, and the Slave Coast.

By the 1530s, the Portuguese had organized a slave trade with the Americas. The trading of slaves by the Portuguese, and then by the British, Dutch, and French, was more widespread and brutal than any trade of African slaves that preceded it. African slaves became part of the elaborate production systems supplying the raw materials and money that fueled Europe's Industrial Revolution.

To acquire slaves, the Europeans established forts on Africa's west coast and paid nearby African kingdoms with weapons, trade goods, and money to make slave raids into the interior. Some slaves were taken from enemy kingdoms in battle. Many more were kidnapped from their homes and villages in the forests and savannas. Most slaves traded to Europeans were male because they brought the highest prices and the raiding kingdoms preferred to keep captured women for their reproductive capacities. Between 1600 and 1865, about 12 million captives were packed aboard cramped and filthy ships and sent to the Americas. One-quarter or more of them died at sea. Of those who arrived in the Americas, about 90 percent went to plantations in South America and the Caribbean. Between 6 and 10 percent were sent to North America (see Figure 7.5).

The European slave trade severely drained parts of Africa of human resources and set in motion a host of damaging social responses within Africa that are not completely understood

A Early humans (not *Homo sapiens*) in Southern Africa.

B The ruins of Great Zimbabwe.

C A trading party approaching Tombouctou.

2,000,000 B.C.E.	5000 B.C.E.	2500 B.C.E.		1000 B.C.E.	500 C.E.	1000 C.E.		1500 C.E.

2 million years ago
First human species evolve in eastern Africa

5000 B.C.E.
Early agriculture in Sahel

1400 B.C.E.
Iron and steelmaking in northeastern Africa

700–1500 C.E.

1250–1600
Mali Empire

VISUAL TIMELINE OF SUB-SAHARAN AFRICA

Thinking Geographically

After you have read about the human history of sub-Saharan Africa, you will be able to answer the following questions:

A What clues are there that these are not modern humans?

B What about this photo suggests a sophisticated village?

C When did Europe first establish a slave trade with the Americas?

even today (Timeline D). The trade enriched those African kingdoms that could successfully conquer their neighbors, and impoverished or enslaved more peaceful or less powerful kingdoms. It also encouraged the slave-trading kingdoms to be dependent on European trade goods and technologies, especially guns.

Slavery persists in modern Africa and is a growing problem that some argue exceeds the trans-Atlantic slave trade of the past. Today, slavery is most common in the Sahel region, where several countries have made the practice officially illegal only in the past few years. People may become enslaved during war; or are sold by their parents or relatives to pay off debts; or are forced into slavery

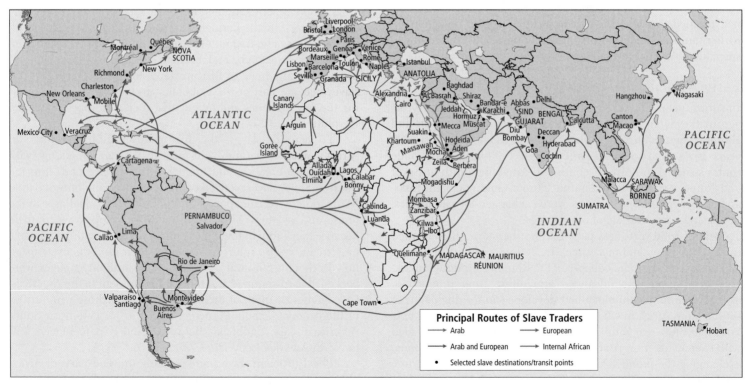

FIGURE 7.5 The African slave trade.

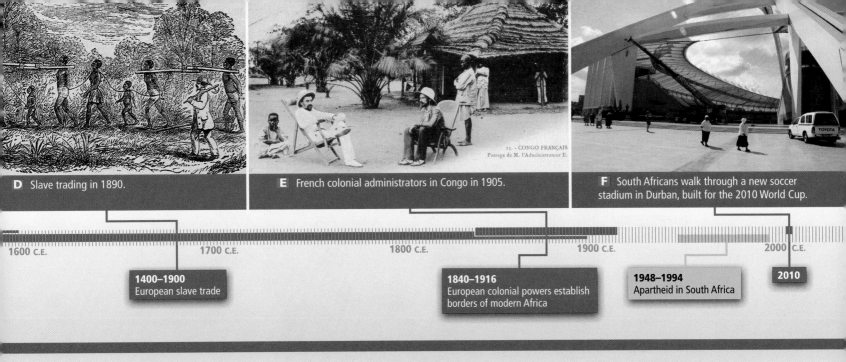

D Slave trading in 1890.

E French colonial administrators in Congo in 1905.

F South Africans walk through a new soccer stadium in Durban, built for the 2010 World Cup.

1600 C.E. 1700 C.E. 1800 C.E. 1900 C.E. 2000 C.E.

1400–1900
European slave trade

1840–1916
European colonial powers establish borders of modern Africa

1948–1994
Apartheid in South Africa

2010

D Are there clues of European involvement in the slave trade in this image?

F How does this stadium convey the wealth of South Africa?

E What about this photo suggests that the Europeans are in a dominant position?

when trying to migrate or find a job in a city (Figure 7.6) or even in Europe. Slaves often work as domestic servants or as prostitutes, and they are increasingly used in commercial agriculture, mining, and war. It is hard to know exactly how many Africans are currently enslaved, but estimates range from several million to more than 10 million.

The Scramble to Colonize Africa

The European slave trade wound down by about the mid-nineteenth century, as Europeans found it more profitable to use African labor within Africa to extract raw materials for Europe's growing industries.

European colonial powers competed avidly for territory and resources, and by World War I, only two areas in Africa were still independent (see Figure 7.7 on page 262). Liberia, on the west coast, was populated by former slaves from the United States. Ethiopia (then called Abyssinia), in East Africa, managed to defeat early Italian attempts to colonize it. Otto von Bismarck, the German chancellor who convened the 1884 Berlin Conference at which the competing European powers formalized the partitioning

FIGURE 7.6 Modern slavery in Niger. Here, slaves collect water at a well in western Niger. Though slavery was "abolished" in Niger in 1960, and made a criminal offense in 2003, it remains deeply embedded in society. At least 43,000 people are enslaved in Niger, most of them born into the status. Slaves have no rights, little access to education, and will spend most of their lives herding cattle, tending fields, or working in their "master's" house.

Thinking Geographically: What does this photo suggest about the lives of these child slaves?

261

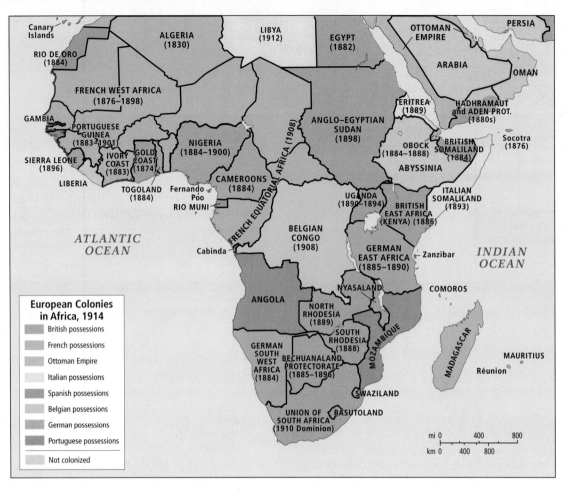

FIGURE 7.7 The European colonies in Africa in 1914. The dates on the map indicate the beginning of officially recognized control by the European colonizing powers. Countries without dates were informally occupied by colonial powers for a few centuries.

of Africa, declared, "My map of Africa lies in Europe." With some notable exceptions, the boundaries of most African countries today derive from the colonial boundaries set up between 1840 and 1916 by European administrators and diplomats (see Timeline E). These territorial divisions lie at the root of many of Africa's current problems.

In some cases, the boundaries were purposely drawn to divide tribal groups and thus weaken them. In other cases, groups who used a wide range of environments over the course of a year were forced into smaller, less diverse lands or forced to settle down entirely, thus losing their livelihoods. As access to resources shrank, hostilities between competing groups developed. Colonial officials often encouraged these hostilities, purposely removing strong leaders with leaders who could be manipulated. Food production, which was the mainstay of most African economies until the colonial period, was discouraged in favor of activities that would support European industries: cash-crop production (cotton, rubber, palm oil) and mineral and wood extraction. Eventually, many formerly prosperous Africans were hungry and poor.

One of the main objectives of European colonial administrations in Africa, in addition to extracting as many raw materials

> **apartheid** a system of laws mandating racial segregation. South Africa was an apartheid state from 1948 until 1994.

as possible, was to create markets in Africa for European manufactured goods. The case of South Africa provides insights into European expropriation of African lands and the subjugation of African peoples. In this case, these aims ultimately led to the infamous system of racial segregation known as **apartheid**.

Case Study: The Colonization of South Africa

The Cape of Good Hope is a rocky peninsula sheltering a harbor on the southwestern coast of South Africa. Portuguese navigators seeking a sea route to Asia first rounded the Cape in 1488. The Portuguese remained in nominal control of the Cape until the 1650s, when the Dutch took possession with the intention of establishing settlements. Dutch farmers, called Boers, expanded into the interior, bringing with them herding and farming techniques that used large tracts of land and depended on the labor of enslaved Africans. The British were also interested in the wealth of South Africa, and in 1795 they seized control of areas around the Cape of Good Hope. When slavery was outlawed throughout the British Empire in 1834, large numbers of slave-owning Boers migrated to the northeast. There, in what became known as the Orange Free State and the Transvaal, the Boers often came into intense and violent conflict with African populations.

In the 1860s, extremely rich deposits of diamonds and gold were unearthed in these areas, securing the Boers' economic future. Africans were forced to work in the diamond and gold mines under extreme hardship and unsafe conditions and minimum wages. They lived in unsanitary compounds that travelers of the time compared to large cages.

Britain, eager to claim the wealth of the mines, invaded the Orange Free State and the Transvaal in 1899, waging the Boer War. This brutal war gave the British control of the mines briefly, until resistance by Boer nationalists forced the British to grant independence to South Africa in 1910. This independence, however, applied to only a small minority of whites: the Boers (who have since been known as Afrikaners) and some British who chose to remain. Not until 1994 would full political rights (voting, freedom of expression, freedom of assembly, freedom to live where one chose) be extended to black South Africans—who made up more than 80 percent of the population—and to Asians and mixed race or "coloured" people (2 percent and 8 percent of the population respectively).

In 1948, the long-standing segregation of South African society was reinforced by apartheid, a system of laws that required everyone except whites to carry identification papers at all times and to live in racially segregated areas. Eighty percent of the land was reserved for the use of white South Africans, who at that time made up just 10 percent of the population. Black, Asian, and "coloured" people were assigned to ethnically based "homelands." The homelands were considered by the South African government to be independent enclaves within the borders of, but not legally part of, South Africa. Nevertheless, the South African government exerted strong influence in them. Democracy theoretically existed throughout South Africa, but nonwhites were allowed to vote only in the homelands.

The African National Congress (ANC) was the first and most important organization fighting to end racial discrimination in South Africa. Formed in 1912 to work nonviolently for civil rights for all South Africans, the ANC grew into a movement with millions of supporters, most of them black, but some of them white. Its members endured decades of brutal repression by the white minority. One of the most famous members to be imprisoned was Nelson Mandela, a prominent ANC leader, who was jailed for 27 years and was finally released in 1990.

Violence increased throughout South Africa until the late 1980s, when it threatened to engulf the country in civil war. The difficulties of maintaining order, combined with international pressure, forced the white-dominated South African government to initiate reforms that would end apartheid. A key reform was the dismantling of the homelands. Finally, in 1994, the first national elections in which black South Africans could participate took place. Nelson Mandela, the long-jailed ANC leader, was elected the country's president. He was awarded the Nobel Peace Prize in 1993. Today in South Africa, the difficult process of dismantling systems of racial discrimination still continues.

The Aftermath of Independence

The era of formal European colonialism in Africa was relatively short. In most places, it lasted for about 80 years, from roughly the 1880s to the 1960s. In 1957, Ghana became the first sub-Saharan African colonial state to achieve its independence. The last sub-Saharan African country to gain independence was South Sudan in 2011, although South Sudan won its independence not from a European power but from its neighbor Sudan after more than 40 years of civil war.

Africa entered the twenty-first century with a complex mixture of enduring legacies from the past and looming challenges for the future. Although it has been liberated from colonial domination, most old colonial borders remain intact (compare the colonial borders in Figure 7.7 with the modern country borders in Figure 7.4). Often these borders exacerbate conflicts between incompatible groups by joining them into one resource-poor political entity; other borders divide potentially powerful ethnic groups, thus diminishing their influence, or cut off nomadic people from resources used on a seasonal basis.

With few exceptions, governments continue to mimic the colonial bureaucratic structures and policies that distance them from their citizens. Corruption and abuse of power by bureaucrats, politicians, and wealthy elites stifle individual initiative, **civil society**, and entrepreneurialism, creating instead frustration and suspicion. Too often, coups d'état have been used to change governments. Democracy, where it exists, is often weakly connected only to elections and not to true participation in policy making at the local and national levels.

Other relics of the colonial era are the dependence of African economies on relatively expensive imported food and manufactured goods, and the production of agricultural and mineral raw materials

> **civil society** the totality of voluntary civic and social organizations and institutions that form the basis of a functioning society, as opposed to the structures of a state that are backed by force (regardless of that state's political system) and commercial institutions of the market

for which profit margins are low and prices on the global market highly unstable. Hence, many sub-Saharan African countries remain economically entwined in trade relationships that rarely work to their advantage.

The wealthiest country in the region remains South Africa, where the economy has highly profitable manufacturing and service sectors. However, South Africa also suffers from widespread poverty. This was demonstrated during the summer of 2010 when South Africa hosted the 2010 World Cup for soccer, the world's largest sporting event and a first for this region that up to that point had never hosted an Olympics or an event of similar scale. The preparations involved demolishing the homes of tens of thousands of poor urban residents who were relocated to make way for the new stadiums (see Timeline F).

THINGS TO REMEMBER

1. Anatomically, modern humans (*Homo sapiens*) evolved from earlier hominoids in eastern Africa about 200,000 years ago, and by about 90,000 years ago, they had reached the eastern Mediterranean.

2. In Africa, people began to cultivate plants as early as 7000 years ago in the Sahel and the highlands of present-day Sudan and Ethiopia.

3. The European slave trade severely drained the African interior of human resources and set in motion a host of damaging social responses within Africa with dimensions that are difficult to quantify even today. Slavery persists in modern Africa and is a growing problem that some argue exceeds the trans-Atlantic slave trade of the past.

4. The main objectives of European colonial administrations in Africa were to extract as many raw materials as possible and to create markets in Africa for European manufactured goods and food exports.

5. The borders and administrative units of the African colonies were designed so that strong groups would be divided, groups hostile to one another would be under the same jurisdiction, and leaders would be weakened.

6. Now-independent governments continue to mimic colonial policies that distance them from their citizens. Abuse of power stifles individual initiative, civil society, and entrepreneurialism. Democracy, where it exists, is often weak.

II CURRENT GEOGRAPHIC ISSUES

Most countries of sub-Saharan Africa have been independent of colonial rule for about 50 years. While many countries are still struggling with the lingering effects of the colonial era, others have moved forward.

Economic and Political Issues

Sub-Saharan Africa emerged from the exploitation and dependence of the colonial era just in time to get sucked into the vortex of globalization, where small or weak countries are often left with little control over their fates.

Commodity-Based Economic Development and Globalization in Africa

> **Learning Goal 2**
> **Globalization and Development:**
> How has economic development been shaped by globalization?

Sub-Saharan Africa has had a centuries-long role in the global economy as a producer of human labor and raw materials, but the profits from turning these human resources and raw materials into higher-value manufactured and processed products have gone to wealthier countries. When traded, raw materials are called **commodities**. The profits of commodity production are usually too low to lift poor countries out of poverty (see Thematic Overview D). Because many countries are often producing the same commodities, competition between them on the world market often drives down profits.

> **commodities** raw materials that are traded, often to other countries, for processing or manufacturing into more valuable goods

Commodity prices are also subject to wide fluctuations. This is because commodities are of more or less uniform quality and are traded as such on a global basis. For example, on the major commodities exchanges, a pound of copper may sell for U.S.$3 to $5, regardless of whether it is mined in South Africa, Chile, or Papua New Guinea. Because of the global scale of commodity trading, a change anywhere in the world that influences the supply or demand for a particular commodity can cause immediate instability in the pricing of that commodity. For example, an earthquake in Chile could damage a major copper mine, sending up the price of copper on world markets. In response, governments that make money from sales of raw copper may overestimate their revenues and commit to expensive infrastructure projects. But other forces could decrease the demand for copper—for example, innovations in the construction industry could reduce the use of copper in building wiring and plumbing—resulting in sudden drops in the price of copper on world markets. As a result, governments that depend on revenues from copper may need to cut essential services, such as electricity, road maintenance, or health care.

Succeeding Eras of Globalization

At least three eras of globalization have transformed Africa over the past several centuries. This section examines the economic and political impacts of these three eras and the current factors moving some African countries away from raw materials exports and toward new types of economic development and regional integration.

The Era of Colonialism and Early Independence Most early European colonial administrations in Africa (Britain, France, Germany, Belgium) evolved directly out of private resource-extracting corporations, such as the German East Africa Company. The welfare of Africans and their future development was a relatively low priority for most colonial administrators. Education and health care for Africans were generally neglected by colonial governments guided by *mercantilism* (see page 115 in Chapter 3). The Africans who worked in European-owned mines or on plantations were poorly paid, as were those who grew cash crops on their own lands. Moreover, the colonial governments of Africa strongly discouraged any other economic activities that might compete with Europe. Hence, it was impossible for African economies to make a transition to the more profitable manufacturing-based industries that were transforming Europe and North America.

Not all European colonial powers had the same policies. The British in Nigeria, for example, tended to use education and Christianizing as tools to encourage Africans to comply with British extractive economic policies. On the other hand, the

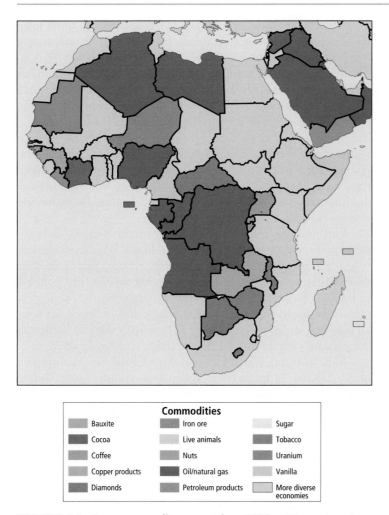

Commodities

Bauxite	Iron ore	Sugar
Cocoa	Live animals	Tobacco
Coffee	Nuts	Uranium
Copper products	Oil/natural gas	Vanilla
Diamonds	Petroleum products	More diverse economies

FIGURE 7.8 One-commodity countries, 2000. African countries depended on only one commodity for more than 50 percent of their export earnings in 2000. By 2010, most countries in Africa had diversified their economies sufficiently to no longer depend on only one main commodity.

Belgians in the Congo River basin held their colonial subjects at arm's length and used extreme brutality, rather than education, to control them.

When African nations started gaining independence in the 1950s and 1960s, most economies remained focused on producing commodities for export (Figure 7.8). Profits generally were not fairly distributed or wisely invested. Like their colonial predecessors, some leaders and bureaucrats considered it their right to enrich themselves with government tax revenues derived from commodity exports. Government jobs and contracts were handed out to cronies or allies as a form of political patronage. Often, enormous sums of borrowed money went to fund government-run development projects that were plagued by corruption and poor planning. Meanwhile, farmers and miners remained so poorly paid that their spending couldn't drive economic development to the extent that it does in wealthier areas. As a result, widespread

> **dual economy** an economy in which the population is divided by economic disparities into two groups, one prosperous and the other near or below the poverty level

poverty and the lack of a market for local products and services still characterize even sub-Saharan Africa's largest, most prosperous economies.

South Africa's economy is an exception to commodity dependence but its success contains a warning. It is the only sub-Saharan African country with a strong manufacturing base. Early on, profits from its commodity exports (mainly minerals) were reinvested in related manufacturing of mining and railway equipment. The country also has a well-developed service sector, with particular strengths in finance and communications that developed in part to support the mining and manufacturing industries. With only 6 percent of sub-Saharan Africa's population, South Africa today produces 30 percent of the region's economic output.

However, South Africa is a classic example of a **dual economy**. For centuries, it has had a well-off minority white population, whose skills and external connections fostered economic development within South Africa. After independence from Britain in 1910, the Dutch (Afrikaners or Boers) and British who remained in the country continued to dominate the economy. Even though the labor of black South Africans was essential to the country's prosperity, most black South Africans remained poor. Under the apartheid system, 84 percent of black South Africans lived at a bare subsistence level. This pattern began to change at the end of apartheid, as black South Africans benefited from government-backed loans and other programs designed to encourage entrepreneurship. In 2009, however, 50 percent of black South Africans still lived below the poverty line (compared with just 7 percent of white South Africans), and only 22 percent of black South Africans had finished high school (compared with 70 percent of white South Africans). The average income for a white South African is still more than 5 times that of a black South African. These inequalities contribute to high rates of crime and violence that have encouraged many skilled South Africans of all races to emigrate to Europe and North America.

The Era of Structural Adjustment By the 1980s, most African countries remained poor and dependent on their volatile and relatively low-value commodity exports (see Figure 7.9 on page 266). Generally, attempts at investing in manufacturing industries had failed due to a number of factors, which are discussed below.

Unfortunately, sub-Saharan African countries had taken out massive loans for these projects and were struggling to repay them. A breaking point came in the early 1980s when an economic crisis swept through sub-Saharan Africa and much of the rest of the developing world, leaving most countries unable to repay their debts. In response, the IMF and the World Bank designed *structural adjustment programs* (SAPs; see pages 116–118 in Chapter 3) to help countries repay their loans. SAPs did have some useful results. They tightened bookkeeping procedures and thereby curtailed corruption and waste in bureaucracies. They closed some corrupt state-owned monopolies in industries and services, and opened some sectors of the economy to medium- and small-scale business entrepreneurs, and they made tax collection more efficient. But

FIGURE 7.9 Commodity dependence. Roads are in disrepair immediately outside the bauxite refinery in Fria, Guinea. Despite the country's rich bauxite resources (aluminum is made from bauxite), most Guineans lack access to safe water, adequate sanitation, or electricity.

Thinking Geographically: What evidence of disrepair can you see in this photo?

overall, SAPs and their successors had many unintended consequences (discussed below), and surprisingly, they failed at their primary objective—reducing debt (Figure 7.10).

Investments in manufacturing failed in the 1980s partly because of corruption and civil war, but tariffs in developed countries also played a crucial role. The story of a shoe factory in Tanzania provides a case in point. In the 1980s, the World Bank gave a loan to the government of Tanzania to develop an export-oriented shoe factory. Tanzania's large supply of animal hides was to be used to manufacture high-fashion shoes for the European market, using expensive imported machinery. However, due to EU tariffs that protected shoemakers in Europe, the Tanzanian factory never managed to export shoes to Europe. Ideally, the factory would have stayed in business by producing shoes for the large African shoe market. However, its machinery could produce only fancy dress shoes, not the practical working shoes most Africans needed. Years went by and, as the unused factory deteriorated and produced no income, Tanzania struggled to repay its loan to the

currency devaluation the lowering of a currency's value relative to the U.S. dollar, the Japanese yen, the European euro, or another currency of global trade

informal economy all aspects of the economy that take place outside official channels

World Bank. Such experiences, repeated hundreds of times over, left African countries impoverished, highly indebted, and unable to repay their loans by the 1980s.

Structural adjustment programs had further negative effects across sub-Saharan Africa. To facilitate loan repayment, SAPs required governments to reduce their involvement in the economy by selling off government-owned enterprises, often at bargain-basement prices. Also, government payrolls, along with many social service, education, health, and agricultural programs, were slashed so that tax revenues could be devoted to loan repayment. If countries refused to implement SAP requirements, the international banks cut off any future lending for economic development.

Furthermore, prospective Western investors in sub-Saharan Africa were discouraged by problems that SAPs either ignored or made worse. Loss of public funds for schools perpetuated an underskilled workforce. As unemployment rose, so did political instability. Deteriorating infrastructure reduced the quality of remaining social services, transportation, and financial services, all of which scared away investors. Thus, SAPs made it harder for the poor majority to make a decent living and stay healthy.

SAPs also reduced the availability of food for consumption in sub-Saharan Africa because agricultural resources were shifted toward the production of cash crops for export. Between 1961 and 2005, per capita food production in sub-Saharan Africa actually decreased by 14 percent, making it the only region on earth where people are eating less well now than in the past (see Figure 1.13 on page 26).

The litany of hardships imposed by structural adjustment policies is long. In order to sell more export crops, sub-Saharan African countries were encouraged to devalue their currency relative to currencies of other countries selling similar commodities. But **currency devaluation** also makes all imports more expensive. Thus farmers growing food for the sub-Saharan African market spend more on seeds, fertilizers, pesticides, and farm equipment (not yet produced within the region), and so food costs have risen. The shift to export crops also left farmers with less time and space to grow food for local markets. Increasingly, sub-Saharan Africans must pay for expensive imported food.

SAPs actually worked against the creation of export-oriented manufacturing industries. Consider sub-Saharan Africa's ancient textile industries: with some investment, these textiles could have gained a global market because of the artistic distinctiveness of the cloth. Instead, SAPs forced sub-Saharan African countries to remove their own tariffs on textiles from other countries, resulting in a flood of imported cloth from China and, to a lesser degree, from India. Local weavers lost their local markets and were impeded from developing global markets. The overall result was factory closings and job losses in sub-Saharan African textile industries.

Ultimately, Africa's **informal economies** provided relief from the hardships created by SAPs. Informal economies in Africa are ancient and wide-ranging, providing employment and useful services and products. People may grow and sell garden produce, prepare food, make craft items and utensils, or tend to the sick.

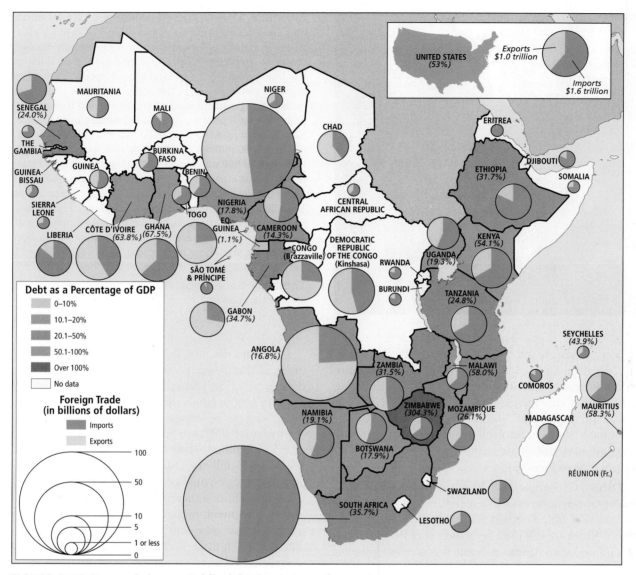

FIGURE 7.10 Economic issues: Public debt, imports, and exports. In most sub-Saharan African countries, imports exceed exports—even in the large diverse economy of South Africa. As these countries borrow money for development, their public debt increases. Even in Côte d'Ivoire, where exports are greater than imports, public debt is over 50 percent of the GDP. Oil producers Angola, Guinea, and Nigeria appear to have the most well-balanced trade, coupled with low debt.

Others, however, earn a living doing less socially redeeming things, like distilling liquor (see Figure 7.11 on page 268) or smuggling scarce or illegal items, such as drugs, weapons, endangered animals, bushmeat, or ivory. Because most activities take place "under the radar," informal jobs may involve criminal acts, wildly unsafe procedures, and hazardous substances.

In most African cities, informal economy once supplied one-third to one-half of all employment; now it often provides more than two-thirds. This creates problems for governments because the informal economy typically goes untaxed, so less money is available to pay for government services or to repay debts. Moreover, the profits of many informal businesses have declined over time as more people compete to sell goods and services to people with less disposable income. And although women typically dominate informal economies, when large numbers of men lose their jobs in factories or the civil service,

they crowd into the streets and bazaars as vendors, displacing the women and young people.

Many of the displaced women and girls have turned to, or been forced into, sex work—a growing informal sector—putting themselves at high risk of social rejection, emotional problems, and HIV-AIDS. Displacement also leads to the disintegration of many families, the children of which are often forced to fend for themselves on the streets. Cities such as Nairobi, Kenya, which had very few children living on the streets before SAPs, now have thousands.

In response to the now widely recognized failures of SAPs, and the overemphasis on the power of markets to guide development, the IMF and the World Bank in 2000 replaced SAPs with *Poverty Reduction Strategy Papers*, or *PRSPs*. While these policies are similar to SAPs in that they push market-based solutions intended to reduce the role of government in the economy, they

FIGURE 7.11 An informal economy of illegal liquor. Lucy Mugure is a single mother of five living in Kenya's Mathare slum where she brews an illegal and sometimes deadly liquor known as *chang'aa* in makeshift barrels. "I sell a glass at 10 shillings [U.S.$0.15]—less than half the price of legal beer. Although most of my customers drink on credit, paying at the end of the week or the month, some refuse to pay. I spend most of my time trying to avoid arrest by the police who will pour my liquor away before taking me to court where I am fined or jailed. During good months, I make a profit of between 500 and 1000 shillings [U.S.$7–$14], which is still not enough because I pay rent of 800 shillings [U.S.$12]. Sometimes I only make 50 shillings [U.S.$0.70] in a day, yet the children need to eat and we need to buy fuel. I usually end up sending the children to an eating place to spend 10 shillings [U.S.$0.15] each on a small plate of rice or a doughnut and beans to survive. The uniforms for my three school-going children are also expensive, and I have to pay examination fees. Sometimes the children are sent home from school, and I cannot afford to take them back to school; most of the time, at least one is out of school."

Thinking Geographically: What about this photo suggests an illegal operation?

differ in several ways. They focus on reducing poverty, rather than on just "development" per se, and are generally more democratic in the way reforms are implemented. They also include the possibility that a country may have all or most of its debt "forgiven" (paid off by the IMF, the World Bank, or the African Development Bank) if the country follows the PRSP rules. Thirty sub-Saharan countries had qualified for and been approved to receive debt relief as of July 2010.

The Era of Diverse Globalization The current wave of globalization is resulting in new sources of investment as well as new pressures on the prices of Africa's export commodities. While Europe and the United States are still the largest sources of investment in Africa, Asia's influence on African economies is increasing significantly (**Figure 7.12**).

Perhaps the most dramatic sign of change for African economies is that Africans working and living abroad, primarily in Europe and North America, are sending home more in *remittances* (money sent to family members) than Africa is receiving from all other sources of foreign investment. According to World Bank estimates for 2007, sub-Saharan countries received 4.5 percent (or U.S.$15.2 billion) of all the global remittances, which amounted to U.S.$337 billion. This money was used to build houses, start small businesses, fund education for children, and help the needy. Remittances are a more stable source of investment than foreign direct investment, in that they tend to come regularly from committed donors who will continue their support for years. They are also much more likely to reach poorer communities. Yet this increasing reliance on international migrants presents its own

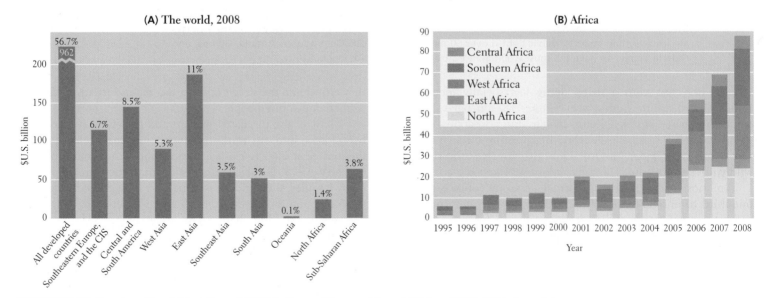

FIGURE 7.12 Foreign direct investment (FDI) inflows: The world and Africa, 2008. FDI inflows into developed countries declined in 2008, while those to developing countries continued to increase and reached the second-highest percentage ever at 43.3 percent of the global total.

form of vulnerability. Remittances tend to come in amounts too small to start anything but small projects, and while remittances are generally still more stable than foreign investments, they can abruptly cease in an economic recession, should the remitters lose their jobs.

In recent years, sub-Saharan Africa has become a new frontier for Asia's large and growing economies, especially those of China and India. Through their demand for Africa's export commodities, through direct investment in sub-Saharan Africa, and through the sale of their manufactured goods in sub-Saharan Africa, China and India now exert a more powerful influence on sub-Saharan Africa than ever before. Of the two, China's influence is by far the greater (see Thematic Overview E; see also Figure 7.13).

Together, China and India consume about 15 percent of Africa's exports, but their share is growing twice as fast as is that of any of Africa's other trading partners. However, the increasing Chinese and Indian demand for resources is a double-edged sword. While it has brought higher and more stable prices for sub-Saharan Africa's export commodities in recent years, the rising demand for food in China, India, and other parts of Asia has also helped drive up global food prices. This has been particularly painful for the many poor African countries in which expanding urban populations, located where there is no access to subsistence plots, are highly dependent on food imports.

Will Asian investments in sub-Saharan Africa ultimately prove beneficial or hurtful to sub-Saharan African economies? Much like Europe during the colonial era, China has invested mainly in commodity exports, especially minerals, timber, and oil, and in the infrastructure—roads, airports—to remove these raw materials from sub-Saharan Africa. The improved infrastructure could ultimately facilitate intra-African trade, but the construction phase has not provided jobs for Africans because the agreements the governments have made require that all work be done by Chinese companies using Chinese labor. Most controversial has been the willingness of Chinese companies to deal with brutal and corrupt local leaders, such as Liberia's Charles Taylor (see the opening vignette on page 248) or Zimbabwe's Robert Mugabe, who barter away their country's nonrenewable resources at bargain prices and use the profits to enrich themselves and to wage war against their own citizens.

New trade relationships with both China and India could have both positive and negative influences for Africa over the long term. China's involvement in road building, for example, could increase regional economic integration. China's investment in agriculture could result in more efficient production for export (and thus higher earnings) and greater food supplies for African internal markets, which would increase overall food security. In recent years, there has been an increase in the crucial manufacturing industries that Africa has needed for so long to lift itself out of poverty (see Reasons for Optimism B on page 284). While it is always possible that cheaper Chinese imports could undercut these industries, Chinese financing of infrastructure could also help the industries move forward.

▌▶ 160: SOMALILAND EXPATRIATES RETURN HOME TO HELP NATIVE LAND DEVELOP
▌▶ 161: AFRICAN UNION APPEALS TO DIASPORA TO AID HOMELANDS

Regional and Local Economic Development

Seeking alternatives to past development strategies, many African governments are focusing on regional economic integration similar to that of the European Union. Local agencies and public and private donors are pursuing grassroots development designed to foster very basic innovation.

Regional Integration Less than 20 percent of the total trade of sub-Saharan Africa is conducted between African countries. This is true partly because so many countries produce the same raw materials for export. And everywhere except South Africa, industrial capacity is so low that the raw materials cannot be absorbed within the continent, so African countries compete with each other and all other global producers to sell to their main buyers, which currently are in Europe and Asia.

Much of this internal trade takes place within the regional trading blocs formed over the last several decades. By combining the markets of several countries (as do NAFTA and the EU), regional trade blocs can create a market size sufficient to foster industrialization and entrepreneurialism. Africa's many different regional trade blocs share several goals: reducing tariffs between members, forming common

FIGURE 7.13 China in Angola. The man on the left is one of the estimated 20,000 Chinese workers in Angola. China has given loans and aid to Angola in excess of U.S.$4 billion since 2004, and in return, China has been guaranteed a large portion of Angola's future oil production. In addition, 70 percent of Angola's development projects have been given to Chinese companies, most of which import workers from China.

Thinking Geographically: What aspects of China's investments in Africa are illustrated by this photo?

currencies, reestablishing peace in war-torn areas, upgrading transportation and communication infrastructure, and building regional industrial capacity. Full-scale continent-wide economic union along the lines of the European Union is a long-term goal.

Local Development An increasingly common strategy for improving living standards in this region is **grassroots economic development**. Projects using this strategy provide sustainable livelihoods in rural and urban areas, often using simple technology that requires minimal or no investment in imported materials. One approach is **self-reliant development**, which consists of small-scale self-help projects that use local skills to create products or services for local consumption. Crucially, local control is maintained so that participants retain a sense of ownership and commitment in difficult economic times. One district in Kenya has more than 500 such self-reliant groups. Most members are women who terrace land, build water tanks, and plant trees. They also build schools and form credit societies.

> **grassroots economic development** economic development projects designed to help individuals and their families achieve sustainable livelihoods
>
> **self-reliant development** small-scale development schemes in rural areas that focus on developing local skills, creating local jobs, producing products or services for local consumption, and maintaining local control so that participants retain a sense of ownership

The issue of improving rural transportation illustrates how a focus on local African needs can generate unique solutions. When non-Africans learn that transportation facilities in Africa are in need of development, they usually imagine building and repairing roads for cars and trucks. But a recent study that analyzed village transportation on a local level found that 87 percent of the goods moved are carried via narrow footpaths on the heads of women! Women "head up" (their term) firewood from the forests, crops from the fields, and water from wells (Figure 7.14).

FIGURE 7.14 Rural water transport. A girl uses a footpath to bring home water in the Democratic Republic of Congo (Kinshasa). Some grassroots development efforts aimed at improving transportation in Africa are focusing on improving footpaths because so much material is transported along them.

Thinking Geographically: What is this footpath paved with?

An average adult woman spends about 1.5 hours each day moving the equivalent of 44 pounds (20 kilograms) more than 1.25 miles (2 kilometers).

Unfortunately, the often-dilapidated footpaths trod by Africa's load-bearing women have been virtually ignored by African governments and international development agencies, which tend to focus solely on roads for motorized vehicles (which are also badly needed). Grassroots-oriented nongovernmental organizations (NGOs) are now making much less expensive but equally necessary improvements to Africa's footpaths. Some women have been provided with bicycles, donkeys, and even motorcycles that can travel on the footpaths. This saves time and energy for women who can direct more of their efforts to education and generating income.

Africa's energy needs, which are currently unmet even at the most basic level of home electricity, can also be addressed by local solutions, as the following vignette illustrates.

Vignette In Malawi, 14-year-old William Kamkwamba was forced to drop out of school when a famine struck his country in 2001 and his family could no longer afford the $80 school fee. Depressed at the prospect of having no future, he went to a local library when he could. There he found a book in English called *Using Energy* that described an electricity-generating windmill. With an old bicycle frame, PVC pipes, and scraps of wood, he built a windmill that generated enough power to light his home, run a radio, and charge neighborhood cell phones.

Now known as "the boy who harnessed the wind," in 2009 William appeared on Jon Stewart's *Daily Show* in the United States to explain how he plans to start his own windmill company and other ventures that will bring power to remote places across Africa. He has returned to school, this time in the first pan-African prep school in South Africa. William's web page is http://williamkamkwamba.typepad.com/about.html. ∎

THINGS TO REMEMBER

1. **Learning Goal 2: Globalization and Development** Sub-Saharan Africa has had a centuries-long role in the global economy as a producer of human labor and raw materials, but the profits from turning these human resources and raw materials into higher-value manufactured and processed products go to wealthier countries.

2. Prospective investors in sub-Saharan Africa have been discouraged by problems that structural adjustment programs either ignored or worsened; yet in recent years, Africa has become a new frontier for Asia's large and growing economies, especially those of China and India.

3. Many African governments are focusing on regional economic integration along the lines of the European Union. Grassroots economic development is also being pursued.

4. Africa's informal economies have provided some relief from the hardships created by SAPs. These economies in Africa are ancient and wide-ranging, providing employment and useful services and products. The people in these economies do not pay taxes and therefore do not provide government revenues.

Democratization and Conflict

Learning Goal 3
Democratization: What forces are working for and against democracy in Africa?

After years of rule by corrupt elites and the military, signs of a shift toward democracy and free elections are now visible across Africa. Yet progress is often blocked by conflict, and even when democratic reforms are enacted and free elections established, violence often accompanies these elections.

Ethnic Rivalry Africa's democratic and economic progress has been held back by frequent civil wars that are in many ways the legacy of colonial era policies of **divide and rule**. Divisions and conflicts between ethnic or religious groups were deliberately intensified by European colonial powers. To make it hard for Africans to unite and overthrow their foreign rulers, the borders and administrative units of the African colonies were designed so that different and sometimes hostile groups would be put together under the same jurisdiction (**Figure 7.15**). After independence,

this made rule by Africans more difficult. African officials inevitably belonged to one local ethnic group or another and hence could not be seen as impartial in their attempts to resolve conflicts. Moreover, older traditions for ensuring ethical behavior, for resolving ethnic conflict, and for punishing greed on the part of leaders had been erased during the colonial era. The result has been years of carnage as ethnic and other hostilities, in the absence of democratic systems of conflict resolution, have developed into civil wars.

divide and rule the deliberate intensification of divisions and conflicts by potential rulers; in the case of sub-Saharan Africa, by European colonial powers

Case Study: Conflict in Nigeria

Long-standing troubles in Nigeria illustrate the roots of conflict. Nigeria was and remains a creation of British divide-and-rule imperialism. Many disparate groups—speaking 395 indigenous languages—have been joined into one unusually diverse country.

The British reinforced a north-south dichotomy that mirrored the physical north (dry)–south (wet) patterns. Among the Hausa and Fulani ethnic groups in the north, the British ruled via local Muslim leaders who did not encourage public education. In the south, the animist/Christian Yoruba–Igbo ethnic groups were ruled more directly by the British, with the help

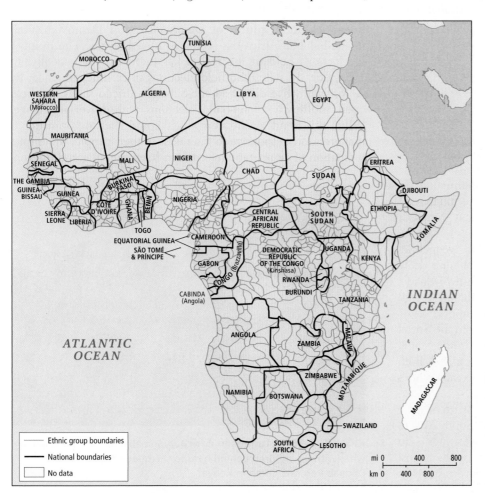

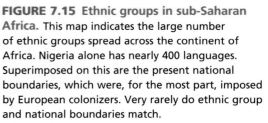

FIGURE 7.15 Ethnic groups in sub-Saharan Africa. This map indicates the large number of ethnic groups spread across the continent of Africa. Nigeria alone has nearly 400 languages. Superimposed on this are the present national boundaries, which were, for the most part, imposed by European colonizers. Very rarely do ethnic group and national boundaries match.

of Christian missionary schools that were open to the public. At independence, the south had more than 10 times as many primary and secondary school students as the north. The south was more prosperous, and southerners also held most government civil service positions. Yet the northern Hausa dominated the top political posts, in part because of their long collaboration with British colonial administrators. Over the years, bitter and often violent disputes have erupted between the southern Yoruba–Igbo and northern Hausa–Fulani regarding the distribution of economic development funds, jobs, and oil revenues, as well as over severe environmental damage from oil extraction and access to increasingly scarce clean water.

▐▶ 178: NEW NIGERIAN PRESIDENT INHERITS TURBULENT NIGER DELTA

The politics of oil have complicated the troubles in Nigeria. Nigeria is a major oil producer and exporter, and much of Nigeria's oil is located on lands occupied by the Ogoni people, which lie in the south along the edges of the Niger River delta. However, virtually none of the profits from oil production, and very little of the oil itself, goes to the states in which this land lies, much less to the Ogoni's homeland, Ogoniland. While receiving few benefits from oil extraction, Ogoniland has suffered gravely from the resulting pollution (see the photo in Figure 1.6 on page 8). Oil pipelines crisscross Ogoniland, and spills and blowouts are frequent; between 1985 and 2009 there were hundreds of spills, many larger than that of the Exxon Valdez disaster in Alaska. Natural gas, a by-product of oil drilling, is burned off, even though it could be used to generate electricity—something many Ogoni lack. Royal Dutch Shell, a multinational oil company that is very active in the Niger River delta, acknowledges that while it has historically netted $200 million in profit yearly from Nigeria, in 40 years it has paid only a total of $2 million to the Ogoni community.

Geographic strategies have often been used to reduce tensions in Nigeria. One approach has been to create more political states (Nigeria now has 30) and thereby reallocate power to smaller local units with fewer ethnic and religious divisions. Recently, large, wealthy states have been subdivided, reputedly to spread oil profits more evenly, but effectively the subdivisions have muted the voice and power of many local groups.

The Role of Cold War Geopolitics Cold War geopolitics between the United States and the former Soviet Union deepened and prolonged African conflicts that grew out of divide-and-rule policies. After independence, some sub-Saharan African governments turned to socialist models of economic development, often receiving economic and military aid from the Soviet Union. Other governments became allies of the United States, receiving equally generous aid (see Figure 5.8 on page 187). Both the United States and the USSR tried to undermine each other's African allies by arming and financing rebel groups.

In the 1970s and 1980s, southern sub-Saharan Africa became a major area of tension. The United States aided South Africa's apartheid government in military interventions against Soviet-allied governments in

genocide the deliberate destruction of an ethnic, racial, or political group

Namibia, Angola, and Mozambique. Another area of Cold War tension was the Horn of Africa, where Ethiopia and Somalia fought intermittently throughout the 1960s, 1970s, and 1980s. At different times, the Soviets and Americans funded one side or the other.

Conflict and the Problem of Refugees Conflicts create refugees, and refugees are commonplace across sub-Saharan Africa. With only 11 percent of the world's population, this region contains about 19 percent of the world's refugees (Figure 7.16), and if people displaced within their home countries are also counted, this region has about 28 percent of the world's refugee population. Women and children constitute three-fourths of Africa's refugees because many adult men who would be refugees are either combatants, jailed, or dead.

Throughout the last decade of the twentieth century, refugees from Somalia, Ethiopia, Uganda, Liberia, Sierra Leone, Congo (Kinshasa), Congo (Brazzaville), Rwanda, and Mozambique poured back and forth across borders and were displaced within their own countries. Often they were trying to escape **genocide**, efforts to murder an entire ethnic group.

As difficult as life is for these refugees, the burden on the areas that host them is also severe. Even with help from international agencies, the host areas find their own development plans complicated by the arrival of so many distressed people, who must be fed, sheltered, and given health care. Large portions of economic aid to Africa have been diverted to deal with the emergency needs of refugees.

Successes and Failures of Democratization Democratization in sub-Saharan Africa has produced mixed results. The number of elections held in the region has increased dramatically. In 1970, only 11 states had held elections since independence. By 2006, twenty-five out of 44 sub-Saharan African states had held open, multiparty, secret-ballot elections, with universal suffrage. While this trend toward democracy could lead to governments that are more responsive to the needs of their people, the implementation of democratic reforms has been irregular and uneven.

FIGURE 7.16 Refugees in Africa. A Somali man carries his ailing wife (with his children behind him) to the queue for admission to Dadaab Refugee Camp in Kenya.

At times, flawed elections have brought about massive violence. In Kenya, which had had several peaceful election cycles, an election in 2008 was so corrupt that deadly riots broke out (see Photo Essay 7.4A on page 274). More than 1000 people died, and 600,000 were displaced by mobs of enraged voters. By 2010, a new Kenyan constitution gave some hope that democracy would return. In 2008 in Nigeria, local elections sparked similar violence, and in the Congo (Kinshasa) in 2006, the first elections held there in 46 years resulted in violence that left more than 1 million Congolese refugees within their own country (see Photo Essay 7.4C).

Zimbabwe may represent the worst case of the failure of democratization. In the 1960s and 1970s, Robert Mugabe became a hero to many for his successful guerilla campaign in what was then called Rhodesia (now Zimbabwe). The white minority government was allied with apartheid South Africa but was not formally recognized by any other country because of its extreme racist policies. Mugabe was elected president in 1980, following relatively free, multiparty elections. Over the years, however, his authoritarian policies impoverished and alienated more and more Zimbabweans. In the 1990s, he implemented a highly controversial land-redistribution program that resulted in his supporters gaining control of the country's best farmland, much of which had been in the hands of white Zimbabweans. This move decimated agricultural production and contributed to a chronic food shortage and a massive economic crisis that left 80 percent of Zimbabweans unemployed (see Photo Essay 7.4D). The resulting political violence has created 3 to 4 million refugees, most of whom have fled to neighboring South Africa and Botswana. Mugabe held on to his office through the rigged elections of 2002 and 2008, but was then forced to share power in 2009 with Morgan Tsvangirai, who had won a plurality in 2008. Recently, modest economic growth has returned to Zimbabwe, following the implementation of some of Tsvangarai's policies.

▐▐ ▶ 159: ZIMBABWE'S ROBERT MUGABE—A PROFILE

In some places, however (for example, Rwanda and South Africa), elections have helped end civil wars, as the possibility of becoming respected elected leaders has induced former combatants to lay down their arms (see Thematic Overview F). In Sierra Leone and Liberia, public outrage against corrupt ruling elites has resulted in elections that brought a fortuitous change of leadership.

> **Learning Goal 4**
> **Democratization and Gender:** How are issues of gender influencing trends toward democratization?

Gender and Democratization One major characteristic of the democratization process has been the increase in the number of women across Africa who are assuming positions of power (see Thematic Overview G and Photo Essay 7.4B). This chapter opened with a description of the democratic and environmental reforms that President Ellen Johnson-Sirleaf of Liberia has been making. Rwanda, devastated by genocide and mass rapes in the 1990s, is now the first country in the world where women constitute a majority of the national legislature. And Rwandan women are also leaders at the local level, where they make up 40 percent of the mayors. In Mozambique, 34.8 percent of the parliament is female, and in South Africa, 32.8 percent. All together, there are 11 African countries where the percentage of women in national legislatures is above the world average of 18.2 percent (the U.S. figure is 17 percent).

▐▐ ▶ 170: WOMEN HAVE STRONG VOICE IN RWANDAN PARLIAMENT

In many cases, these statistics reflect government policies as much as the willingness of Africans to be led by women. All of the countries cited above have enacted quotas that guarantee a certain percentage of legislative seats for women, and fewer women are elected in the countries that do not have quotas. The quotas are often a response to national crises, especially civil wars, in which women fought alongside men in battle or suffered disproportionately from the chaos and destruction of war. Quotas are usually a reflection of a larger post-conflict commitment toward the empowerment of women, sometimes written into a country's constitution, that encourages female participation in politics and civil society. It is important to note, however, that many female leaders in Africa, such as Ellen Johnson-Sirleaf and at least half of Rwanda's female parliamentarians, were elected without quotas.

THINGS TO REMEMBER

1. **Learning Goal 3: Democratization** After years of rule by corrupt elites and the military, signs of a shift toward democracy and free elections are now visible across Africa. Yet progress has been blocked by conflict. Even when democratic reforms are enacted and free elections established, violence often accompanies these elections.

2. Africa's democratic and economic progress has been held back by frequent civil wars that are in many ways the legacy of colonial-era divide-and-rule policies.

3. Democratization is helping reduce the potential for civil conflict, and the most democratized countries have had noticeably fewer violent conflicts since 1945. However, in many countries, elections are still plagued by violence.

4. The trend toward democracy in sub-Saharan Africa could lead to governments that are more responsive to the needs of their people, but democratic reforms are not necessarily durable.

5. **Learning Goal 4: Democratization and Gender** One major characteristic of the democratization process has been the increase in the number of women across Africa who are coming into positions of power.

Sociocultural Issues

To the casual observer, it may appear that a majority of sub-Saharan Africans live traditional lives in rural villages. It is true that at just 35 percent urban, Africa is the world's most rural region. A closer look, however, reveals that migration and urbanization are occurring so rapidly that the impact on African life is monumental.

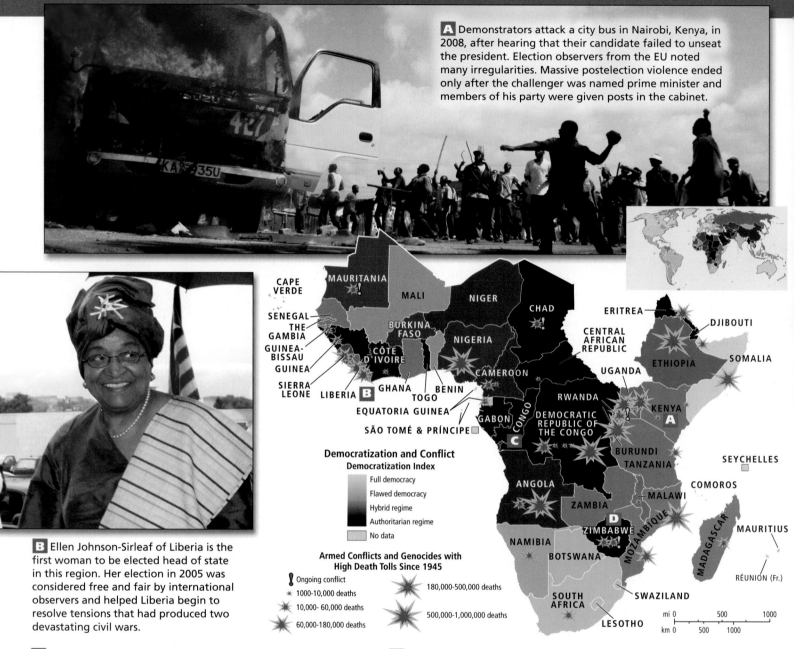

Democratization is helping reduce the potential for civil war throughout this region, and the most democratized countries have had noticeably fewer violent conflicts since 1945. However, in many countries, elections are still plagued by violence.

A Demonstrators attack a city bus in Nairobi, Kenya, in 2008, after hearing that their candidate failed to unseat the president. Election observers from the EU noted many irregularities. Massive postelection violence ended only after the challenger was named prime minister and members of his party were given posts in the cabinet.

B Ellen Johnson-Sirleaf of Liberia is the first woman to be elected head of state in this region. Her election in 2005 was considered free and fair by international observers and helped Liberia begin to resolve tensions that had produced two devastating civil wars.

Democratization and Conflict

Democratization Index

- Full democracy
- Flawed democracy
- Hybrid regime
- Authoritarian regime
- No data

Armed Conflicts and Genocides with High Death Tolls Since 1945

- ❗ Ongoing conflict
- ✴ 1000–10,000 deaths
- ✴ 10,000–60,000 deaths
- ✴ 60,000–180,000 deaths
- ✴ 180,000–500,000 deaths
- ✴ 500,000–1,000,000 deaths

mi 0 · · · 500 · · · 1000
km 0 · · · 500 · · · 1000

Map labels: CAPE VERDE, MAURITANIA, MALI, NIGER, CHAD, ERITREA, DJIBOUTI, SENEGAL, BURKINA FASO, CENTRAL AFRICAN REPUBLIC, THE GAMBIA, NIGERIA, ETHIOPIA, SOMALIA, GUINEA-BISSAU, CÔTE D'IVOIRE, GUINEA, CAMEROON, UGANDA, SIERRA LEONE, GHANA, BENIN, RWANDA, KENYA, LIBERIA, TOGO, EQUATORIA GUINEA, GABON, CONGO, DEMOCRATIC REPUBLIC OF THE CONGO, BURUNDI, SÃO TOMÉ & PRÍNCIPE, TANZANIA, SEYCHELLES, ANGOLA, MALAWI, COMOROS, ZAMBIA, ZIMBABWE, MOZAMBIQUE, MADAGASCAR, MAURITIUS, NAMIBIA, BOTSWANA, RÉUNION (Fr.), SOUTH AFRICA, SWAZILAND, LESOTHO

C Supporters of Joseph Kabila in Kinshasa, Congo, during the 2006 elections. Widespread irregularities combined with deep mistrust of the government to create postelection violence that left over 1 million people homeless.

D Shelves go empty in Harare, Zimbabwe, during an economic crisis created by years of authoritarian, antidemocratic rule. Fraudulent elections in 2008 discouraged foreign aid donors from helping, as there was no legitimate or trusted government to work with. Widespread hunger and disease resulted.

Thinking Geographically

After you have read about democratization and conflict in sub-Saharan Africa, you will be able to answer the following questions:

A How many people were displaced by the post-election violence in Kenya in 2008?

B Did Ellen Johnson-Sirleaf win her election with the aid of quotas that reserve certain elected positions for women?

C How many people were displaced by violence during the elections of 2006 in Congo?

D What has recently brought modest growth in Zimbabwe?

Population Patterns

While sub-Saharan Africa's population growth rates are the highest in the world, the rates have slowed recently as a result of urbanization. Even in rural areas they are declining. On the other hand, in some of the most developed countries in the area, life expectancy figures have deteriorated due to HIV-AIDS. Also, despite successful efforts to lower infant mortality rates, they still remain the highest on earth. How can all of this apparently conflicting information be explained?

Population Growth, Density, and the Demographic Transition

Despite generally declining birth rates, sub-Saharan African populations are growing faster than in any other region in the world. In fewer than 50 years, sub-Saharan Africa's population has more than tripled, growing from around 200 million in 1960 to 828 million in 2009. By 2050, the population of this region is projected to be just under 1.7 billion. Hence, many places that are relatively uncrowded now may change dramatically over the next few decades (Figure 7.17). How can this be, if population growth rates are lower now than in the past and birth rates are shrinking? The short answer is that while families are now much smaller than in the past, people are still choosing to have more children than would be necessary to maintain population numbers, and more people now survive long enough to reproduce than did in the past (see Thematic Overview H).

⏸▶ 167: AFRICA'S EXPECTED POPULATION BULGE THREATENS FUTURE SUSTAINABILITY

Birth rates are as high as they are because many Africans view children as both an economic advantage and a spiritual link between the past and the future. Childlessness is considered a tragedy, as children ensure a family's genetic and spiritual survival. Large families are also viewed as having economic value, as children and young adults still perform important work on family farms and in family-scale industries. Moreover, in this region of generally poor health care and the resulting incidence of high infant mortality,

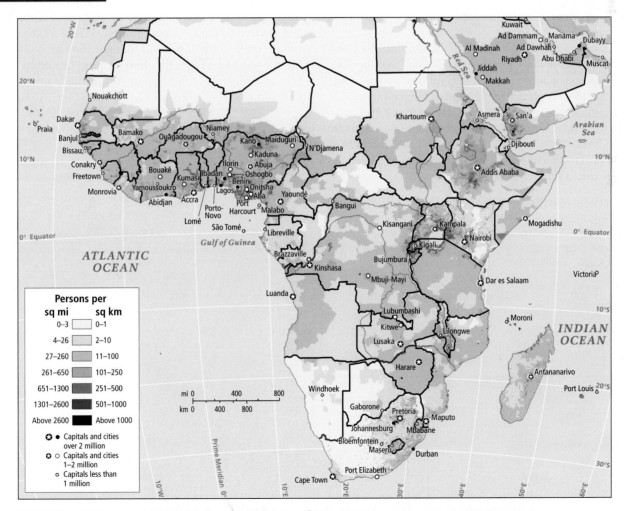

FIGURE 7.17 Population density in Sub-Saharan Africa. Of all the world regions, populations are growing fastest in sub-Saharan Africa. The demographic transition is taking hold in a few more prosperous countries where better health care and more economic and educational opportunities for women are allowing women to pursue careers and put off childbearing, resulting in smaller families.

parents have extra children in the hope of raising a few to maturity. In all but a few countries, the demographic transition—the sharp decline in births and deaths that usually accompanies economic development (see Figure 1.11 on page 17)—has barely begun.

Five countries in the region have gone through the demographic transition. In South Africa, Botswana, Seychelles, Réunion, and Mauritius (the last three being small island countries off Africa's east coast), circumstances have changed sufficiently to make smaller families desirable. In all five countries, per capita incomes are 5 to 10 times the sub-Saharan average of U.S.$2000. Advances in health care have cut the infant mortality rate to about half the regional average of 88 infant deaths per 1000 live births; because of this, parents can have only a few children and expect most to live to adulthood. The circumstances of women have also improved, as reflected in female literacy rates of around 80 to 90 percent, compared to the regional average of 54.4 percent. Research also shows that opportunities for women to work outside the home at decent-paying jobs are greater in these countries than in the rest of the region. Thus, many women are choosing to use contraception because they have life options beyond motherhood. Indeed, the percent of married women using contraception in these five countries is double or even triple the rate for sub-Saharan Africa as a whole, which is only 22 percent (about one-third of the world average).

The population pyramids (Figure 7.18) demonstrate the contrast between countries that are growing rapidly (such as Nigeria, Africa's most populous country, with 152.6 million people) and countries that are already going through the demographic transition, such as South Africa, which has a population of just 50.7 million. Nigeria's pyramid is very wide at the bottom because over half the population is under the age of 20. In just 15 years, this entire cohort will be of reproductive age. Only 12 percent of Nigerian women use any sort of birth control.

In contrast to Nigeria, South Africa's pyramid has contracted at the bottom because its birth rate has dropped from 35 per 1000 to 23 per 1000 over the last 20 years. This decrease, primarily the consequence of economic and educational improvements and social changes that have come about since the end of apartheid in the early 1990s, is likely to persist as the advantages of smaller families, especially to women, become clear. The birth rate decrease is all the more remarkable because contraception is used by only 60 percent of South African women. Part of the birth rate decrease in South Africa is probably a consequence of the spread of HIV among young adults there (see Figure 7.19B on page 277). HIV-AIDS is now a major cause of slowing population growth rates and declining life expectancies in Africa. Nigeria and South Africa represent the two extremes of growth rates in Africa.

Urbanization and Population Growth   Sub-Saharan Africa is undergoing a massive shift in population from tiny rural settlements of a few houses to substandard residential facilities in urban areas. This has major implications for population growth. On average, rural sub-Saharan African women give birth to about 6.6 children, while urban African women give birth to 4.7. Urban life strengthens all the factors that influence this demographic transition—increased economic development, better health care, and more educational opportunities. While sub-Saharan Africa's urban fertility rate of 4.7 children per woman is still almost double the world average, the demographic transition has only just begun, and urban sub-Saharan Africa's birth rates will continue to decline.

> **Learning Goal 5**
> **Urbanization and Population:**
> How is urbanization influencing population growth in Africa?

Part of the reason that urban fertility rates remain as high as they are is that compared to cities in other regions, sub-Saharan Africa's cities

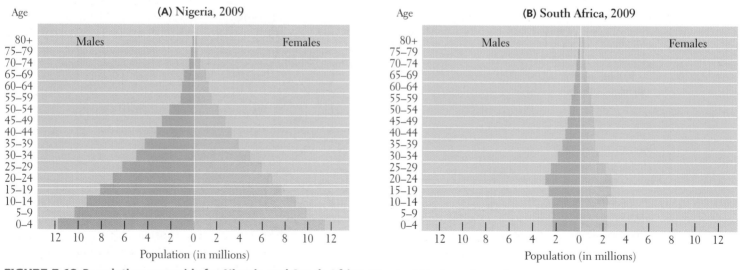

FIGURE 7.18 Population pyramids for Nigeria and South Africa. Nigeria, Africa's most populous country with a total population of 152.6 million, has a population growth rate of 2.6 percent, and has nearly twice the population of Ethiopia, Africa's second-most populous country. South Africa, with a population of 50.7 million, has a growth rate of 0.9 percent and has the fifth largest number of people on the continent.

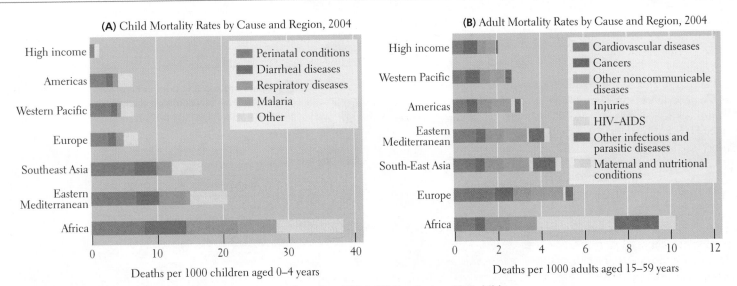

(A) Child Mortality Rates by Cause and Region, 2004

Legend:
- Perinatal conditions
- Diarrheal diseases
- Respiratory diseases
- Malaria
- Other

Regions (top to bottom): High income, Americas, Western Pacific, Europe, Southeast Asia, Eastern Mediterranean, Africa

Deaths per 1000 children aged 0–4 years

(B) Adult Mortality Rates by Cause and Region, 2004

Legend:
- Cardiovascular diseases
- Cancers
- Other noncommunicable diseases
- Injuries
- HIV–AIDS
- Other infectious and parasitic diseases
- Maternal and nutritional conditions

Regions (top to bottom): High income, Western Pacific, Americas, Eastern Mediterranean, South-East Asia, Europe, Africa

Deaths per 1000 adults aged 15–59 years

FIGURE 7.19 Mortality rates by major cause and region 2004. (A) Deaths per 1000 children aged 0–4. **(B)** Deaths per 1000 adults aged 15–59.

have delivered fewer improvements in living standards, especially with regard to access to clean water and sanitation (see Thematic Overview I and **Photo Essay 7.5 A, C** on page 278). With poverty and disease still widespread in sub-Saharan Africa's cities, and for a variety of other reasons—lack of access to birth control, low levels of education for women—urban families continue to have relatively large families.

Migration, not the birth rate, is the major factor behind sub-Saharan cities' growth rate of 5 percent per year, the highest rate in the world. In the 1960s, only 15 percent of sub-Saharan Africans lived in cities; now about 35 percent do (see Photo Essay 7.5). By 2030, the urban population of this region will have doubled to about 530 million and will account for about half of all Africans. In 1960, just one sub-Saharan African city—Johannesburg, South Africa—had more than one million people; in 2009, 52 do. The largest sub-Saharan African city is Lagos, Nigeria, where various estimates put the population at between 11 and 13 million; by 2020, Lagos is projected to have 20 million people (see the map in Photo Essay 7.5). Much of this growth is taking place in primate cities (see Chapter 3, page 129). For example, Kampala, Uganda, with 1.8 million people, is almost 10 times the size of Uganda's next largest city, Gulu.

Because governments and private investors have paid little attention to the need for affordable housing, most migrants have to construct their own dwellings using found materials (see Photo Essay 7.5). The vast unplanned one-story slums that result surround the older urban centers and house 72 percent of Africa's urban population. Transportation in these huge and shapeless settlements is a jumble of government buses and private vehicles. People often have to travel long hours through extremely congested traffic to reach distant jobs, getting most of their sleep while sitting on a crowded bus.

Public health is also a major concern, as many water distribution systems are contaminated with harmful bacteria from untreated sewage (see Photo Essay 7.5C). Only the largest sub-Saharan African cities have sewage treatment plants, and few of these extend to the slums that surround them. The result is frequent outbreaks of waterborne diseases such as cholera, dysentery, and typhoid.

Migration The migration of sub-Saharan people looking for work in North Africa, Europe, Turkey, or the United States is now a familiar phenomenon, but the vast majority of the African migrants who are seeking work or escaping violence move within the continent. In 2005, the United Nations estimated that there were 17 million such internal migrants. Most migrants go to West Africa and Southern Africa because jobs are more numerous. The five sub-Saharan African countries with the largest number of immigrants in 2005 were Côte d'Ivoire (2.4 million), Ghana (1.7 million), South Africa (1.1 million), Nigeria (1.0 million), and Tanzania (0.8 million). Together they have 40 percent of the migrants in Africa. Gabon, Gambia, and Côte d'Ivoire have the largest number of immigrants relative to their own populations (Gabon, 17.7 percent; Gambia, 15.3 percent; and Côte d'Ivoire, 13.1 percent). Even though the migrants typically make few demands, the impact on receiving countries that are unable to provide adequately for their own citizens is substantial. Like migrants who leave the continent, internal migrants live frugally and send much of their earnings home to families.

Population and Public Health Infectious diseases, including HIV-AIDS, are by far the largest killers in sub-Saharan Africa, responsible for about 50 percent of all deaths (Figure 7.19). Some diseases are linked to particular ecological zones. For example, people living between the 15th parallels north and south of the equator are most likely to be exposed to sleeping sickness (trypanosomiasis), which is spread among people and cattle by the bites of tsetse flies. The disease attacks the central nervous system and, if untreated, results in death. Several hundred thousand Africans suffer from sleeping sickness, and most of them are not treated

Urban populations are exploding due to migration from rural areas and relatively high birth rates in cities. By 2030, half of the people in the region will be living in cities, most in slums plagued by violence and inadequte access to water, sanitation, and education. Largely ignored by most governments, slum dwellers survive by helping themselves.

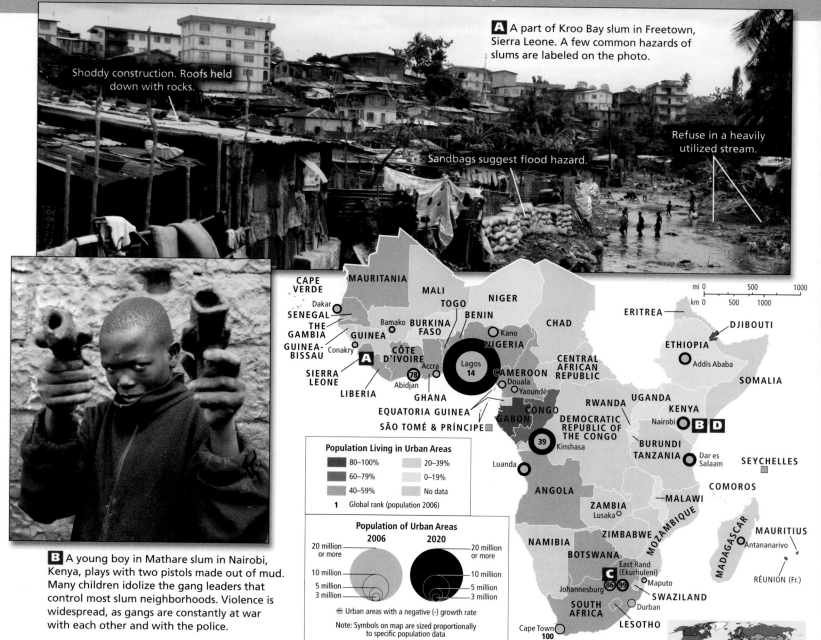

A A part of Kroo Bay slum in Freetown, Sierra Leone. A few common hazards of slums are labeled on the photo.

Shoddy construction. Roofs held down with rocks.

Sandbags suggest flood hazard.

Refuse in a heavily utilized stream.

Population Living in Urban Areas

- 80–100%
- 60–79%
- 40–59%
- 20–39%
- 0–19%
- No data

1 Global rank (population 2006)

Population of Urban Areas

2006 | 2020

20 million or more · 10 million · 5 million · 3 million

⊖ Urban areas with a negative (-) growth rate

Note: Symbols on map are sized proportionally to specific population data

B A young boy in Mathare slum in Nairobi, Kenya, plays with two pistols made out of mud. Many children idolize the gang leaders that control most slum neighborhoods. Violence is widespread, as gangs are constantly at war with each other and with the police.

C Children fetching water in Soweto, South Africa. Slum houses rarely have indoor plumbing. Water is obtained from a public spigot, and latrines provide the only sanitation.

D An organic garden in Kibera, one of many "self-help" projects found throughout Nairobi's slums.

because they cannot afford the expensive drug therapy.

Africa's most common chronic tropical diseases, schistosomiasis and malaria, are linked to standing fresh water. Thus, their incidence has increased with the construction of dams and irrigation projects. Schistosomiasis is a debilitating, though rarely fatal, disease that affects about 170 million sub-Saharan Africans. It develops when a parasite carried by a particular freshwater snail enters the skin of a person standing in water. Malaria, spread by the anopheles mosquito (which lays its eggs in standing water), is more deadly. The disease kills at least 1 million sub-Saharan

Africans annually, most of them children under the age of 5. Malaria also infects millions of adult Africans who are left feverish, lethargic, and unable to work efficiently because of the disease.

Until recently, relatively little funding was devoted to controlling the most common chronic tropical diseases. Now, however, major international donors are funding research in Africa and elsewhere. More than 60 research groups in Africa are working on a vaccine that will prevent malaria in most people. The distribution of simple, low-cost, and effective mosquito nets is also reducing the transmission rates of malaria.

HIV-AIDS in Sub-Saharan Africa A leading cause of death in Africa, and the leading cause of death for women of reproductive age, is acquired immunodeficiency syndrome (AIDS), caused by the human immunodeficiency virus (HIV) (Figure 7.20). As of 2010, more than 16 million sub-Saharan Africans had died of the disease, and as many as 80 million more AIDS-related deaths are expected by 2025. In 2010, sub-Saharan Africa had two-thirds of the estimated worldwide total of 33 million people living with HIV. In Botswana, one of the richest countries in southern Africa,

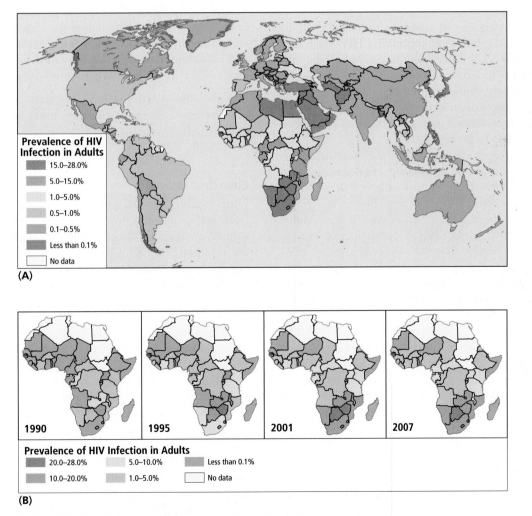

FIGURE 7.20 Global prevalence of HIV-AIDS, 2007.

nearly 25 percent of the adult population is infected. Southern Africa alone accounted for 38 percent of global HIV-AIDS deaths in 2007, and there the epidemic has significantly constrained economic development.

Worldwide, women account for half of all people living with HIV, but in Africa, HIV-AIDS affects women more than men: 59 percent of the region's HIV-infected adults are women. The reasons for this pattern are related to the social status of women in sub-Saharan Africa and elsewhere. Women are often infected by their mates, who may visit sex workers when traveling for work or business. Women have little power to insist that their partners use condoms.

The rapid urbanization of Africa has hastened the spread of HIV-AIDS, which is much more prevalent in urban areas. Many poor new urban migrants, removed from their village support systems, become involved in the sex industry. For some poor women, occasional sex work is part of what they do to survive economically. In some cities, virtually all sex workers are infected. Meanwhile, many urban men, especially those with families back in rural villages, visit prostitutes on a regular basis. These men often bring HIV-AIDS back to their rural homes. As transportation between cities and the countryside has improved, bus and truck drivers have also become major carriers of the disease to rural villages.

A number of myths and social taboos surround HIV-AIDS. This has exacerbated the problem of controlling HIV-AIDS and means that education is a key component to combating the epidemic. Many men think that only sex with a mature woman can result in infections, so very young girls are increasingly sought as sex partners. (This is sometimes referred to as "the virgin 'cure'.") Elsewhere, infection is considered such a disgrace that even those who are severely ill refuse to get tested.

Massive education programs are credited with stabilizing the HIV-AIDS epidemic by lowering rates of infection among those who can read and understand explanations of how HIV is spread. For example, Senegal started HIV-AIDS education in the 1980s, and levels of infection there have so far remained low. Major education campaigns in Uganda lowered the incidence of new HIV infections from 15 percent to just 4.1 percent between 1990 and 2004. By contrast, in areas such as South Africa, where a few top politicians have put forth untenable theories about the causes of HIV infection or have denied that HIV-AIDS was a problem, infection rates have soared.

Across the continent, the consequences of the HIV-AIDS epidemic are enormous. Millions of parents, teachers, skilled farmers, craftspeople, and trained professionals have been lost. More than 15 million children have been orphaned, many without any family left to care for them or to pass on vital knowledge and skills. The disease has severely strained the health-care systems of most countries. Demand for treatment is exploding, drugs are prohibitively expensive, and many health-care workers themselves are infected. Because so many young people are dying of AIDS, decades of progress in improving the life expectancy of Africans have been erased. For example, in 1990 adult life expectancy in South Africa was 63, but in 2010 it was 55.

THINGS TO REMEMBER

1. The overall low population density figures in sub-Saharan Africa—34.8 people per square kilometer, compared to the global average of 49 people per square kilometer—are misleading. Densities are extremely high in some places and very low in less habitable areas.

2. Of all the world regions, populations are growing fastest in sub-Saharan Africa. The demographic transition is taking hold in a few more prosperous countries where better health care and child survival, along with more economic and educational opportunities for women, are encouraging women to pursue careers and put off childbearing, resulting in smaller family sizes.

3. **Learning Goal 5: Urbanization and Population** Sub-Saharan Africa is undergoing a massive shift in population from tiny rural settlements to urban areas, resulting in slower population growth. Urban life strengthens all the factors that influence the demographic transition—increased economic development, better health care, and more educational opportunities.

4. Urban populations are exploding, mainly due to migration from rural areas. By 2030, half of the region will be living in cities, most in slums with inadequate access to water, sanitation, and education. Largely ignored by most governments, slum dwellers survive by providing for themselves.

5. Sub-Saharan Africa suffers more than any other world region from infectious diseases, with the world's worst epidemics of malaria, schistosomiasis, and HIV-AIDS.

Gender Issues

Long-standing African traditions dictate a fairly strict division of labor and responsibilities between men and women. In general, women are responsible for domestic activities, including raising the children, attending to the sick and elderly, and maintaining the house. Women collect water and firewood and prepare nearly all the food. Men are usually responsible for preparing land for cultivation. In the fields intended to produce food for family use, women sow, weed, and tend the crops as well as process them for storage. In the fields where cash crops are grown, men perform most of the work and retain control of the money earned.

When husbands in search of cash income migrate to work in the mines or in urban jobs, women take over nearly all of the agricultural work. They usually work with simple hand tools in the fields, and in the home, they often labor with a child or grandchild strapped on their back. When there are small agricultural surpluses or handcrafted items to trade, it is women who transport and sell them in the market. Throughout Africa, married couples often keep separate accounts and manage their earnings as individuals. When a wife sells her husband's produce at the market, she usually gives the proceeds to him.

A practice known as **female genital mutilation (FGM)** (formerly called female circumcision) has been documented in 27 countries throughout the central portion of the African continent (**Figure 7.21**). It also has been documented in Yemen, India, Indonesia, Iraq, Israel, Malaysia, the United Arab Emirates, and in some cases in the United States, with anecdotal reports in several other countries. The practice predates Islam and Christianity, and today occurs among all social classes and in Christian, Muslim, and animist religious traditions. In the procedure, which is usually performed without anesthesia, parts of the labia and the entire clitoris are removed from a young girl. In the most extreme cases (called infibulation), the vulva is stitched nearly shut. This mutilation far exceeds that of male circumcision, eliminating any possibility of sexual stimulation for the female and making urination and menstruation difficult. Intercourse is painful and childbirth is particularly devastating because the flesh scarred by the mutilation is inelastic. A 2006 medical study conducted with the help of 28,000 women in six African countries showed that women who had undergone FGM were 50 percent more likely to die during childbirth, and their babies were at similarly high risk. The practice also leaves women exceptionally susceptible to infection, especially HIV infection.

> **female genital mutilation (FGM)** removing the labia and the clitoris and sometimes the stitching the vulva nearly shut
>
> **animism** a belief system in which natural features carry spiritual meaning

▐▶ 171: FEMALE GENITAL MUTILATION STILL COMMON IN SOMALILAND

The practice is probably intended to ensure that a female is a virgin at marriage and that she thereafter has a low interest in intercourse other than for procreation. While in decline today, FGM is still widespread among some groups. Among the Kikuyu of Kenya, 40 years ago nearly all females would have had FGM, but today only about 40 percent do. Kikuyu women's rights leaders have had some success in curbing FGM by replacing it with right-of-passage ceremonies that joyously mark a girl's transition to puberty.

Many African and world leaders have concluded that the practice is an extreme human rights abuse, and it is now against the law in 16 countries. However, because it is so deeply ingrained in some value systems, the most successful eradication campaigns are those that emphasize the threat FGM poses to a woman's health and make it socially acceptable to not undergo this ritual. Well over 100 million girls and women now alive in Africa have suffered FGM, with about 2 million more subjected to FGM annually.

The World Health Organization in 2008 took a strong stand against FGM, saying:

Female genital mutilation has been recognized as discrimination based on sex because it is rooted in gender inequalities and power imbalances between men and women and inhibits women's full and equal enjoyment of their human rights. It is a form of violence against girls and women, with physical and psychological consequences.

WORLD HEALTH ORGANIZATION, *Eliminating Female Genital Mutilation—An Interagency Statement* (GENEVA: WORLD HEALTH ORGANIZATION PRESS, 2008), P. 10.

Religion

Africa's rich and complex religious traditions derive from three main sources: indigenous African belief systems (many of them animist), Islam, and Christianity.

Indigenous Belief Systems Traditional African religions, found in every part of the continent, are probably the most ancient on earth, since this is where human beings first evolved. **Figure 7.22A** on page 282 highlights the countries in which traditional beliefs remain particularly strong. Traditional beliefs and rituals often seek to bring departed ancestors into contact with living people, who in turn are the connecting links in a timeless spiritual community that stretches into the future. The future is reached only if present family members procreate and perpetuate the family heritage through storytelling.

Most traditional African beliefs can be considered **animist** in that spirits, including those of the deceased, are thought to exist everywhere—in trees, streams, hills, and art, for example. In return for respect (expressed through ritual), these spirits offer protection from sickness, accidents, and the ill will of others. African religions tend to be fluid and adaptable to changing circumstances. For example, in West Africa, Osun, the god of water,

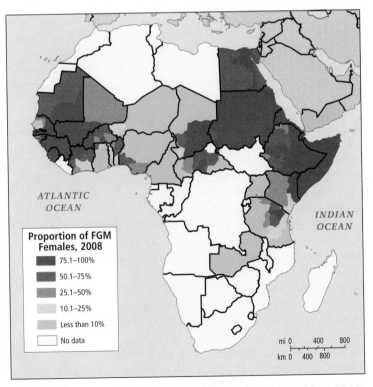

FIGURE 7.21 Female genital mutilation (FGM) in Africa, 2008. This practice, mapped here using 2008 data, occurs all across the center of the African continent despite government policies against it. FGM has been declared illegal in many countries, but enforcement is lax in most.

Proportion of FGM Females, 2008: 75.1–100%, 50.1–75%, 25.1–50%, 10.1–25%, Less than 10%, No data

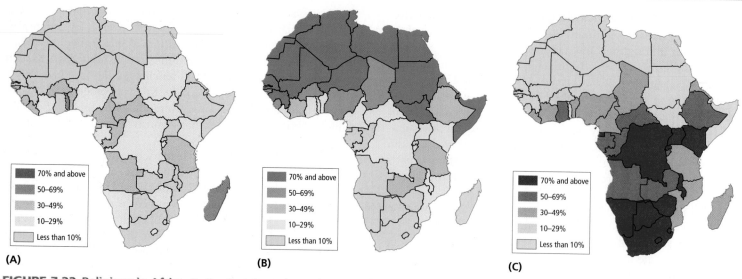

FIGURE 7.22 Religions in Africa. Notice that the various religions in Africa overlap in distribution; in many countries, one is dominant but others are present. **(A)** Distribution of traditional religions. **(B)** Distribution of Islam. **(C)** Distribution of Christianity.

traditionally credited with healing powers, is now also invoked for those suffering economic woes.

Religious beliefs in Africa, as elsewhere, continually evolve as new influences are encountered. If Africans convert to Islam or Christianity, they commonly retain aspects of their indigenous religious heritage. The three maps in Figure 7.22 show a spatial overlap of belief systems, but they do not convey the philosophical blending of two or more faiths, which is widespread. In the Americas, the African diaspora has influenced the creation of new belief systems developed from the fusion of Roman Catholicism and African beliefs. Voodoo in Haiti and Santería in Cuba are examples of this fusion.

Islam and Christianity Islam began to extend south of the Sahara in the first centuries after Muhammad's death in 632 C.E., and today, about one-third of sub-Saharan Africans are Muslim. As Figure 7.22B shows, Islam is now the predominant religion throughout the Sahel and parts of East Africa, where Muslim traders from North Africa and Southwest Asia brought the religion. Powerful Islamic empires have arisen here since the ninth century. The latest of these empires challenged European domination of the region in the late nineteenth century.

Today about half of Africans are Christian. Christianity first came to the region via Egypt and Ethiopia shortly after Jesus's time, well before it spread throughout Europe (Figure 7.22C). However, Christianity did not come to the rest of Africa until nineteenth-century missionaries from Europe and North America became active along the west coast. Many Christian missionaries provided the education and health services that colonial administrators had neglected. In the 1980s, old-line established churches began to gain adherents in Africa. The Anglican Church (Church of England) grew so rapidly that by 2000, Anglicans in Kenya, Uganda, and Nigeria outnumbered those in the United Kingdom. The Anglican Church in Africa attracts the educated urban middle class.

Evangelical Christianity and the Gospel of Success In contrast to the Anglican Church, modern evangelical versions of Christianity appeal to the less-educated, more recent urban migrants—the most rapidly growing African populations. The independent, evangelical sects interpret Christianity as being combined with African beliefs in such things as the importance of sacrificial gifts to spiritual leaders and ancestors and the power of miracles. The result is one of the world's fastest-growing Christian movements.

Vignette Preachers at the Miracle Center in Kinshasa describe the new Gospel of Success forcefully: "The Bible says that God will materially aid those who give to Him. . . . We are not only a church, we are an enterprise. In our traditional culture you have to make a sacrifice to powerful forces if you want to get results. It is the same here."

Generous gifts to churches are promoted as a way to bring divine intervention to alleviate miseries, whether physical or spiritual. Practitioners donate food, television sets, clothing, and money. One woman gave 3 months' salary in the hope that God would find her a new husband.

Like all religious belief systems, the gospel of success is best understood within its cultural context. Many of the believers, new to the city, feel isolated and are looking for a supportive community to replace the one they left behind. People view their material contributions to the church as similar to the labor and goods they previously donated to maintain standing in their home village. In return for these dues and volunteer services, members receive social acceptance and community assistance in times of need. ■

1. Gender relationships in sub-Saharan Africa are complex and variable, but in this region as in others, women are generally subordinate. Nonetheless, change is underway.

2. Female genital mutilation has been recognized as discrimination based on sex because it is rooted in gender inequalities and power imbalances between men and women.

3. Traditional African religions are among the most ancient on earth and are found in every part of the continent. Today, however, about one-third of sub-Saharan Africans are Muslim and about half are Christian, with many in the evangelical movement, a subset that promotes the Gospel of Success.

Reflections on Sub-Saharan Africa

Late one evening in a restaurant in Central Europe, after a lengthy conversation that touched on some of the world's perplexing problems, a friend leaned across the elegant white tablecloth and asked, "But don't you think that Africa is, after all, better off for having been colonized by Europeans?"

How does one politely reply to such a question? Of all regions on earth, sub-Saharan Africa is the poorest region of the world, yet the reasons for sub-Saharan Africa's poverty are not immediately apparent to the casual observer and do not necessarily lie in Africa. Sub-Saharan Africa is blessed with many kinds of resources—agricultural, mineral, and forest—but most of the market value is added to these resources only after they leave Africa. Sub-Saharan Africa is not densely occupied, and birth rates are dropping, but population growth, especially in urban areas (due to migration), is hindering efforts to improve standards of living.

The deep reasons for sub-Saharan Africa's poverty and social and political instability emerge only through an exploration of its history over the last several centuries. Africans were physically taken from this region in large numbers to supply labor for European colonialist enterprises in the Americas. Later, Africans were deprived of control of their own societies and resources, which were reoriented to benefit Europe. Even today, with colonialism officially dead for more than three decades, many sub-Saharan Africans are struggling with borders developed by Europeans and economies that are still focused on cheap and unstable raw material exports to Europe and elsewhere. In view of all this, it is hard to believe that anyone would think Africa is better off for having been colonized.

Sub-Saharan Africa is changing, however. Authoritarian political structures, many of which date to the colonial era, are giving way to more democratic systems of government. Patterns of economic development also show signs of change, as recent years have seen an increase in the crucial manufacturing industries that Africa has needed for so long to lift itself out of poverty. Moreover, the contributions of women to society, and the importance of gender issues in slowing population growth are increasingly being recognized (see Reasons for Optimism).

Sometimes good-willed people suggest that the rest of the world should just "leave Africa to the Africans," but that view is unrealistic. Three more likely strategies are currently emerging as a consensus among sub-Saharan Africa specialists. First, SAPs should be completely abandoned and remaining debts to foreign governments and international lending agencies cancelled. Tax revenues could then once again support schools, health care, and social services. Second, the developed world should lower tariffs against sub-Saharan African manufactured products to foster the development of African industries. In so doing, the wealthier countries would be obeying the advice they themselves give to African countries. Third, future aid to sub-Saharan Africa should be designed to take advantage of indigenous skills and knowledge, including African expertise in economic development and planning.

Looking forward, this last point could be of great importance in helping this region adapt to climate change. Greater respect for African knowledge about environments and how humans use them may prove essential in reducing Africa's high vulnerability to climate change. Intimate knowledge of African environments and food production systems will be central to efforts to adapt to growing climatic uncertainty. Much of this knowledge lies with the rural Africans who have arguably gained the least from Africa's "development" so far. Perhaps efforts to adapt to global climate change will bring greater benefits to this group.

Learning Goals Review

1. **Climate Change, Food, and Water:** Why are many African food production systems particularly vulnerable to climate change?

Why are African farmers and herders particularly sensitive to changes in temperature and water availability? How do poverty and having little access to cash influence the vulnerability of many Africans to climate change?

2. **Globalization and Development:** How has economic development been shaped by globalization?

How did Europe's colonization of this region result in patterns of economic development that still leave many countries in a weak position in the global economy? Are there any exceptions to the general pattern of dependence on exports of cheap raw materials?

3. **Democratization:** What forces are working for and against democracy in Africa?

To what extent can the weakness of democracy in Africa be traced to institutions put in place during the colonial era? What role have elections played in both diffusing and inciting violence?

Reasons for Optimism in Sub-Saharan Africa

Democratization: Authoritarian political structures, many of which date to the colonial era, are giving way to more democratic systems of government. **A** *Nigeria during recent elections.* ▼

Development: Recent years have seen growth in the manufacturing industries that could help Africa lift itself out of poverty. **B** *The manufacturing of mosquito nets in Congo (Kinshasa).* ▼

Gender: The contributions of women to society, and the importance of gender issues in slowing population growth, are increasingly being recognized. **C** *A girls' school in Zanzibar, Tanzania.* ▼

Thinking Geographically: What shortcoming of democratization in Nigeria does this photo suggest?

Thinking Geographically: How might manufacturing help lift Africa out of poverty?

Thinking Geographically: How might more girls' schools contribute to slower population growth in Tanzania?

---

4. Democratization and Gender: How are issues of gender influencing trends toward democratization?

Why have national crises led, in some cases, to greater political empowerment of women?

5. Urbanization and Population: How is urbanization influencing population growth in Africa?

What aspects of urban African life are slowing the demographic transition? How has rapid and unplanned urbanization encouraged larger families than are typically found in urban areas in other regions?

Geographic Themes about Sub-Saharan Africa

Look back at the Thematic Overview photos on page 247. Thinking geographically, answer the following questions about them:

(B) Food: What in this photo suggests that small-scale subsistence farming rather than larger-scale commercial agriculture is being practiced?

(C) Water: What about this photo suggests that small-scale irrigation is being used?

(D) Development: How does the photo illustrate the poverty of the workers in Sierra Leone's diamond mines?

(E) Globalization: Is there any clue on these ships that they are Chinese?

(F) Democratization: What is the veteran's wound?

(G) Gender: How do the numbers of elected women officials in African governments compare with the numbers in countries in the rest of the world?

(H) Population: What is a major contributor to the decreasing life expectancy figures in this region?

(I) Urbanization: Other than flooding, what about this slum suggests poor living conditions?

Key Terms

agroforestry 254

animism 281

apartheid 262

carbon sequestration 252

civil society 263

commercial agriculture 254

commodities 264

currency devaluation 266

desertification 258

divide and rule 271

dual economy 265

female genital mutilation (FGM) 281

genocide 272

grassroots economic development 270

Horn of Africa 252

informal economy 266

intertropical convergence zone (ITCZ) 252

mixed agriculture 254

pastoralism 258

Sahel 252

self-reliant development 270

shifting cultivation 257

subsistence agriculture 254

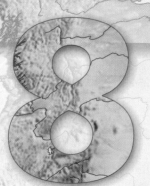

8 South Asia

Learning Goals

After you read this chapter, you will be able to answer the following questions:

1. Climate Change and Water: How is South Asia vulnerable to water-related problems associated with climate change?

2. Gender and Population: How have issues related to gender transformed South Asian populations in recent decades?

3. Food and Urbanization: In what ways have recent changes in South Asian food production systems contributed to urbanization?

4. Globalization and Development: How has globalization benefited some Indian workers in recent years?

5. Democratization: What role has the democratic process played in South Asia's many violent conflicts?

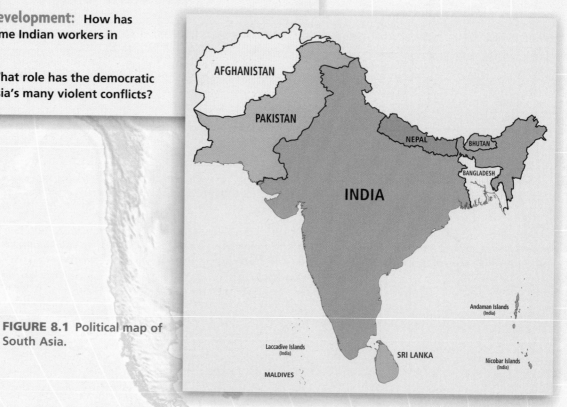

FIGURE 8.1 Political map of South Asia.

Thematic Overview of South Asia

Climate Change: In South Asia, climate change puts more lives at risk than in any other region in the world. **A** *Heavy rains associated with climate change brought devastating floods to Pakistan in 2010.* ▼

Food: The shift towards green revolution agriculture, which uses expensive new seeds, fertilizers, pesticides, and equipment, has benefited some farmers but not others. **B** *Pesticide application in Bangladesh.* ▼

Water: South Asia has more than 20 percent of the world's population but only 4 percent of its fresh water. **C** *Delivering water in Mumbai, India.* ▼

Globalization: The British controlled most of South Asia from the 1830s through 1947. Globalization was accelerated as South Asia was brought into the British Empire. **D** *Indian troops protecting British interests in China in 1902.* ▼

Gender: More males than females are surviving to adulthood because of customs that enable sons rather than daughters to contribute more to family incomes. **E** *A Bangladeshi woman living in seclusion, without a formal income.* ▼

Population: South Asia is the most densely populated region in the world, with more people than in China squeezed onto just over half as much land. **F** *A family compound in Afghanistan.* ▼

Urbanization: Changes in food production are forcing people into cities. The urban population could rise from around 460 million today to as much as 712 million by 2025. **G** *A slum in Dhaka, Bangladesh.* ▼

Development: To take advantage of India's large, college-educated, low-cost workforces, foreign companies in North America and Europe are outsourcing jobs to many Indian cities. **H** *A technology workshop for educators and workers in New Delhi, India.* ▼

Democratization: In India, completely opposed political ideologies, such as Communism and free market capitalism, often compete peacefully in democratic elections. **I** *A Communist party rally in the Indian state of Tripura.* ▼

Global Patterns, Local Lives

Global Patterns, Local Lives On April 16, 2006, Narendra Modi, the governor of the state of Gujarat in India, draped in garlands by well-wishers, was embarking on a hunger strike. He was protesting a decision by India's national government in New Delhi to limit the height of the Sardar Sarovar Dam on the Narmada River in the neighboring state of Madhya Pradesh.

Damming the river in neighboring Madhya Pradesh has become the centerpiece of Gujarat's efforts to deal with the periodic droughts that affect as many as 50 million of its citizens. Governor Modi saw limiting the height of the dam in the neighboring state as threatening his own state's many irrigation programs (**Figure 8.2B** and map).

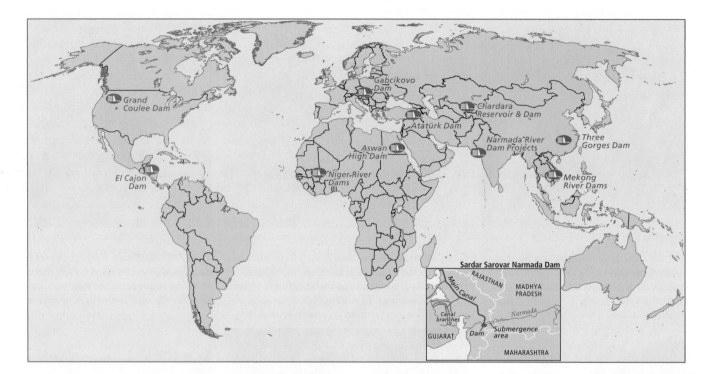

(A)

(B)

(C)

FIGURE 8.2 The Sardar Sarovar Dam on the Narmada and other major dams around the world. The Sardar Sarovar Dam on the Narmada River is only one of hundreds of thousands of dam projects throughout the world that together have displaced between 40 and 80 million people. The map shows some of the major dams around the world. **(A)** A village in Rajasthan that eventually may receive irrigation water from the Sardar Sarovar Dam. **(B)** The Sardar Sarovar Dam as it appeared in 2006, before the 5-meter height increase discussed in the opening vignette. **(C)** A resettlement camp in southern India, typical of those built in India for people whose homes have been flooded by dam reservoirs.

Thinking Geographically: **(A)** What does the vegetation around this village suggest? **(C)** What about the arrangement and construction of these dwellings suggest a large-scale resettlement project?

Far away in New Delhi, Medha Patkar, the leader of the "Save the Narmada River" movement, was in day 18 of her hunger strike in protest against the same dam. As water rose in the dam's reservoir, 320,000 farmers and fishers were being forced to relocate. Although Indian law required that these people be given land or cash to compensate them for what they had lost, so far only a fraction had received any compensation. Some of the farmers demonstrated their objection by forfeiting their right to compensation and refusing to move, even as the rising waters of the reservoir consumed their homes. Forcibly removed by Indian police, many have since relocated to crowded urban slums.

Environmental problems have been at the heart of the Sardar Sarovar controversy since the project began in 1961. The Narmada River was once a placid, slow-moving river—and one of India's most sacred. Now the project has disturbed the river's natural cycles, causing massive die-offs of aquatic life and high unemployment levels among fishers.

The benefits of the dam have also been called into question. A major justification for the dam was that it would provide irrigation waters to drought-prone areas in the adjacent states of Gujarat and Rajasthan (see Figure 8.2A and map). However, 80 percent of the areas in Gujarat most vulnerable to drought would not receive water from the project. Concerned that the economic benefits would be small and easily negated by the environmental costs, the World Bank withdrew its funding of the dam in the mid-1990s. Ecologists say that far less costly water-management strategies, such as rainwater harvesting, groundwater recharge (water that naturally replenishes the aquifer), and watershed management, would be better options for the drought-stricken farmers of Gujarat.

Less than a day after Gujarat's governor began his hunger strike, the Indian Supreme Court ruled that the Sardar Sarovar Dam could be raised higher, so, satisfied that he had won, Governor Modi ended the hunger strike. The following day, Medha Patkar ended her fast as well, because in the same decision the Supreme Court ruled that all people displaced by the dam must be adequately relocated (see Figure 8.2C). Furthermore, the decision confirmed that human impact studies are required for dam projects. Up to then, no such study had been done for the Sardar Sarovar Dam. By July 2009, the dam project was 5 years behind schedule and had stalled again in court. Both sides were using the Internet to promote their positions. *[Adapted from "Modi Goes on Fast over Narmada Dam," India eNews, April 16, 2006, http://www.indiaenews. com/politics/20060416/4531.htm; Rahul Kumar, "Medha Patkar Ends Fast After Court Order on Rehabilitation," One World South Asia; "Narmada's Revenge," Frontline 22(9), 2005, at http://www.hinduonnet.com/fline/ fl2209/stories/20050506002913300.htm; "Water Harvesting, Addressing the Problem of Drinking Water," at http://www.narmada.org/ALTERNATIVES/ water.harvesting.html; Friends of the River Narmada, August/September 2009, at http://www.narmada.org/; "Dam Delay," Indian Express.com, July 15, 2009, at http://www.indianexpress.com/news/dam-delay-sardar-sarovar-project-5-years-of/489337/.]* ■

The recent history of water management in the Narmada River valley highlights some key issues now facing South Asia and other regions that have developing economies. Across the world, large, poor populations are depending on increasingly overtaxed environments, and this dependency is only increasing with climate change. Improving their standard of living nearly always requires more water and energy. Efforts to meet these urgent needs often make neither economic nor environmental sense, but are driven to completion by political and social pressures. In the case of the Sardar Sarovar Dam, the wealthier, more numerous, and more politically influential farmers of Gujarat have tipped the scales in favor of a project that may be creating more problems than it is solving.

The countries that make up the South Asia region are Afghanistan and Pakistan in the northwest; the Himalayan states of Nepal and Bhutan; Bangladesh in the northeast; India (including the Indian territories of the Laccadives, Andaman, and Nicobar Islands); and the island countries of Sri Lanka and the Maldives (Figure 8.1 on page 286 and Figure 8.3 on pages 290–291).

THINGS TO REMEMBER

1. The Sardar Sarovar Dam and similar dams throughout the world are created primarily for irrigation, but they often create more problems than they solve.

2. Improving the standard of living of poor populations in many parts of the world nearly always requires more water and energy.

I THE GEOGRAPHIC SETTING

Terms in This Chapter

Because its clear physical boundaries set it apart from the rest of the Asian continent, the term **subcontinent** is often used to refer to the entire Indian peninsula, which includes Nepal, Bhutan, India, Pakistan, and Bangladesh.

South Asians have recently adopted new place names to replace the names given them during British colonial rule. The city of Bombay, for example, is now officially *Mumbai*, Madras

> **subcontinent** a term often used to refer to the entire Indian peninsula, including Nepal, Bhutan, India, Pakistan, and Bangladesh

is *Chennai*, Calcutta is *Kolkata*, Benares is *Varanasi*, and the Ganges River is the *Ganga River*.

Physical Patterns

Many of the landforms, and even the climates, of South Asia are the result of huge tectonic forces. These forces have positioned the Indian subcontinent along the southern edge of the Eurasian

C Indus Valley, Northern Pakistan

D Western Ghats, Kerala, India

E Deccan Plateau, Maharashtra, India

F Indo-Gangetic Plain, North India

FIGURE 8.3 Regional map of South Asia.

A

of Tibet

ayas

Mt. Everest
elev. 29,028

Lhasa
Brahmaputra

SIKKIM
Gangtok

ARUNACHAL PRADESH

Thimphu
BHUTAN
Paro Phuntsholing

Brahmaputra

Gauhati ASSAM

NAGALAND
Kohima

BIHAR

Khasi Hills

MEGHALAYA

Imphal
MANIPUR

BANGLADESH

Dhaka

Aizwal

TRIPURA

MIZORAM

Dhanbad
Asansol

Chandpur

Mandalay

WEST BENGAL

Khulna

Kolkata
(Calcutta)

Chittagong

BURMA
(Myanmar)

*Ganga-Brahmaputra
Delta*

Arakan Mts.

Irrawaddy

Mouths of the Ganga

Bhubaneswar

THAILAND

Bay of Bengal

Rangoon

*Gulf of
Martaban*

*Preparis
North Channel*

Land Elevations

meters	feet
4877	16,000
3353	11,000
2134	7000
914	3000
305	1000
152	500
0	0

mi 0   50   100   150   200   250

km 0   100   200   300   400

1:13,400,000
Lambert Azimuthal Equal Area Projection

*Andaman
Islands
(India)*

Port Blair

*Andaman
Sea*

Ten Degree Channel

*Nicobar
Islands
(India)*

OCEAN

Great Channel

*Strait of
Malacca*

Sumatra

continent, where the warm Indian Ocean surrounds it and the massive mountains of the Himalayas shield it from cold airflows from the north (see Figure 8.3).

Landforms

The Indian subcontinent and the territory surrounding it dramatically illustrate what can happen when two tectonic plates collide. Millions of years ago, the Indian-Australian Plate, which carries India, broke free from the eastern edge of the African continent and drifted to the northeast (see Figure 1.21 on page 45). As it began to collide with the Eurasian Plate about 60 million years ago, India became a giant peninsula jutting into the Indian Ocean. As the relentless pushing from the south continued, both the leading (northern) edge of South Asia and the southern edge of Eurasia crumpled and buckled. The result is the world's highest mountains—the Himalayas, which rise more than 29,000 feet (8800 meters)—as well as other very high mountain ranges to the east and west that curve away from the central impact zone (see Figure 8.3B and map). The continuous compression also lifted up the Plateau of Tibet, which rose up behind the Himalayas to an elevation of more than 15,000 feet (4500 meters) in some places. The compression and mountain-building process continues into the present.

South and southwest of the Himalayas are the Indus and Ganga river basins, also called the Indo-Gangetic Plain (see Figure 8.3 C, F). Still farther south is the Deccan Plateau, an area of modest uplands (1000–2000 feet (300–600 meters) in elevation) interspersed with river valleys (see Figure 8.3E). This upland region is bounded on the east and west by two moderately high mountain ranges, the Eastern and Western Ghats (see Figure 8.3D). These mountains descend to long but narrow coastlines interrupted by extensive river deltas and floodplains. The river valleys and coastal zones are densely occupied; the uplands only slightly less so.

Because of its high degree of tectonic activity and deep crustal fractures, South Asia is prone to devastating earthquakes, such as the magnitude 7.7 quake that shook the state of Gujarat in western India in 2001, the 7.6 quake that hit the India–Pakistan border region in 2005, and the 6.4 quake in Quetta Province, Pakistan, in 2008 that left 120,000 people homeless. Coastal areas are also vulnerable to tidal waves or *tsunamis* that are caused by undersea earthquakes. A massive tsunami originating off Sumatra in Southeast Asia wrecked much of coastal Sri Lanka and southern India (see Figure 8.3A) in 2004, killing tens of thousands of people there.

II ▶ 179. TOWN OF HAMBANTOTA, SRI LANKA, REGAINS SOME NORMALCY AFTER TSUNAMI A YEAR AGO

Climate and Vegetation

The end of the dry [winter] season [April and May] is cruel in South Asia. It marks the beginning of a brief lull that is soon overtaken by the annual monsoon rains. In the lowlands of eastern India and Bangladesh, temperatures in the shade are routinely

> **monsoon** a wind pattern in which, in summer months, warm, wet air coming from the ocean brings copious rainfall, and in winter months, cool, dry air moves from the continental interior toward the ocean

above a hundred degrees; the heat causes dirt roads to become so parched that they are soon covered in several inches of loose dirt and sand. Tornadoes wreak havoc, killing hundreds and flattening entire villages. . . .

This is also a time of hunger, as with each passing day thousands of rural families consume the last of their household stock of grain from the previous harvest and join the millions of others who must buy their food. Each new entrant into the market nudges the price of grain up a little more, pushing millions from two meals a day to one.

(ALEX COUNTS, *GIVE US CREDIT* (NEW YORK: TIMES BOOKS, 1996), P. 69

From mid-June to the end of October [summer] is the time of the river. Not only are the rivers full to bursting, but the rains pour down so relentlessly and the clouds are so close to village roofs that all the earth smells damp and mildewed, and green and yellow moss creeps up every wall and tree. . . .Cattle and goats become aquatic, chickens are placed in baskets on roofs, and boats are loaded with valuables and tied to houses. . . . As the floods rise, villages become tiny islands . . . self-sustaining outpost[s] cut off from civilization . . . for most of three months of the year.

(JAMES NOVAK, *BANGLADESH: REFLECTIONS ON THE WATER* (BLOOMINGTON: INDIANA UNIVERSITY PRESS, 1993), PP. 24–25

These two passages highlight the contrasts between South Asia's dominant winter and summer wind patterns, known as **monsoons** (Figure 8.4). In winter, cool, dry air flows from the Eurasian continent to the ocean. In summer, warm, moisture-laden air flows from the Indian Ocean over the Indian subcontinent, bringing with it heavy rains. The abundance of this rainfall is amplified by the *intertropical convergence zone (ITCZ)*. Air masses moving south from the Northern Hemisphere converge near the equator with those moving north from the Southern Hemisphere. As the air rises and cools, copious precipitation is produced. As described in Chapter 7 (see page 252), the ITCZ shifts north and south seasonally. The intense rains of South Asia's summer monsoon are likely caused by the ITCZ being sucked onto the land by a vacuum created when huge volumes of air over the Eurasian landmass heat up and rise into the upper troposphere.

The monsoons are a major influence on South Asia's climate. In early June, the warm, moist ITCZ air of the summer monsoon first reaches the mountainous Western Ghats. The rising air mass cools as it moves over the mountains, releasing rain that nurtures patches of dense tropical rain forests and tropical crops on the upper slopes of the Western and Eastern Ghats and in the central uplands. Once on the other side of India, the monsoon gathers additional moisture and power in its northward sweep up the Bay of Bengal, sometimes turning into tropical cyclones.

As the monsoon system reaches the hot plains of the Indian provinces of West Bengal and Bangladesh in late June, columns of warm, rising air create massive, thunderous cumulonimbus clouds that drench the parched countryside. Monsoon rains

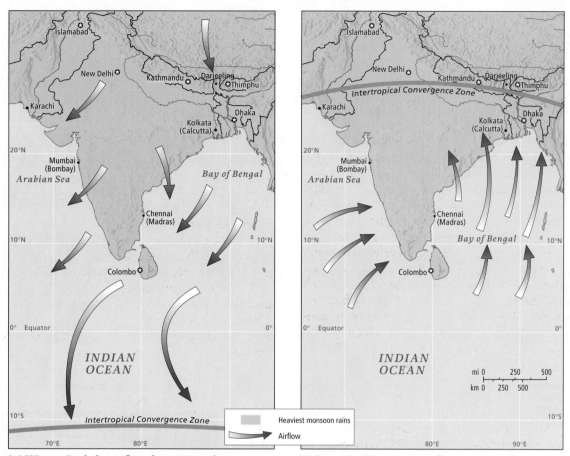

(A) **Winter.** Cool, dry air flows from Asian subcontinent.   (B) **Summer.** Warm, moist air flows to Asian subcontinent.

FIGURE 8.4 Winter and summer monsoons in South Asia. (A) In the winter, cool, dry air blows from the Eurasian continent south across India toward the ITCZ, which in winter lies far to the south. **(B)** In the summer, the ITCZ moves north across India, picking up huge amounts of moisture from the ocean, which are then deposited over India and Bangladesh.

run in a variable band parallel to the Himalayas that reaches across northern India and Pakistan, finally petering out over the Kabul Valley in eastern Afghanistan by July. Rainfall is especially intense in the east, north of the Bay of Bengal, where the town of Darjeeling holds the world record for annual rainfall—about 35 feet, even though no rain falls for half the year. These patterns of rainfall are reflected in the varying climate zones (see **Photo Essay 8.1** on page 294) and agricultural zones (see **Figure 8.16** on page 316) of South Asia. Although sufficient rain falls in central India to support forests, most land has long been cleared of forest (see the discussion on pages 296–298) and planted in crops. Patches of forest are now so fragmented that they no longer provide suitable habitat for India's wildlife.

Periodically, the monsoon seasonal pattern is interrupted and serious drought ensues. This happened in July and August of 2009, when the worst drought in 40 years struck much of South Asia (see the discussion below). Then in September, heavy rains came to central south India, causing crop-damaging floods that killed hundreds of people. Scientists are increasingly concluding that the extreme droughts and floods of recent years are not an anomaly but instead are part of the general global climate change.

By November each year, the cooling Eurasian landmass sends cooler, drier air over South Asia. This heavier air from the north pushes the warm, wet air back south to the Indian Ocean. Very little rain falls in most of the region during this winter monsoon. However, as the ITCZ retreats southward across the Bay of Bengal, it picks up moisture, which is then released as early winter rains over parts of southeastern India and Sri Lanka.

The monsoon rains deposit large amounts of moisture over the Himalayas, much of it in the form of snow and ice that add to the existing mass of huge glaciers (see page 41 in Chapter 1). Meltwater from these glaciers feeds the headwaters of the three river systems that figure prominently in the region: the Indus, the Ganga, and the Brahmaputra. All three rivers begin within 100 miles (160 kilometers) of one another in the Himalayan highlands near the Tibet–Nepal–India borders (see Figure 8.3; see also the map in Photo Essay 8.1). Forest vegetation is more common in the Himalayan highlands and in the foothills, but in many places human pressure has resulted in widespread deforestation and an increasingly patchy forest.

These rivers, and many of the tributaries that feed them, are actively wearing down the surface of the Himalayas. They carry enormous loads of sediment, especially during the rainy season. When the rivers reach the lowlands, their velocity slows and much of the sediment settles out as silt. It is then repeatedly picked up and deposited by successive floods. As illustrated in the diagram of the

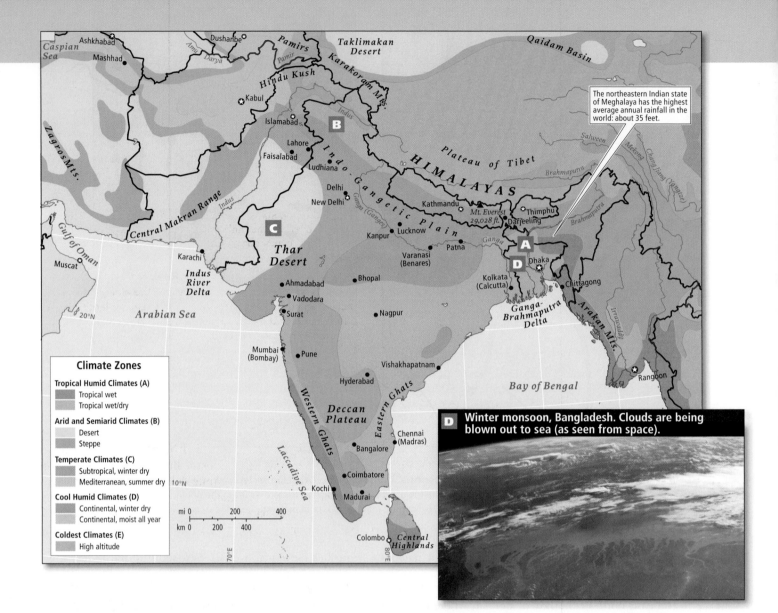

The northeastern Indian state of Meghalaya has the highest average annual rainfall in the world: about 35 feet.

Climate Zones

Tropical Humid Climates (A)
- Tropical wet
- Tropical wet/dry

Arid and Semiarid Climates (B)
- Desert
- Steppe

Temperate Climates (C)
- Subtropical, winter dry
- Mediterranean, summer dry

Cool Humid Climates (D)
- Continental, winter dry
- Continental, moist all year

Coldest Climates (E)
- High altitude

D Winter monsoon, Bangladesh. Clouds are being blown out to sea (as seen from space).

A Tropical wet, Meghalaya, India

B High altitude, Kashmir, India

C Desert, Jaisalmer, India

(A) Pre-monsoon Stage. The river flows in multiple channels across the flat plain.

(B) Peak Flood Stage. During peak flood stage, the great volume of water overflows the banks and spreads across fields, towns, and roads. It carves new channels, leaving some places cut off from the mainland.

FIGURE 8.5 The Brahmaputra River in Bangladesh at various seasonal stages. People who live along the river have learned to adapt their farms to a changing landscape, and along much of the river farmers are able to produce rice and vegetables nearly year-round.

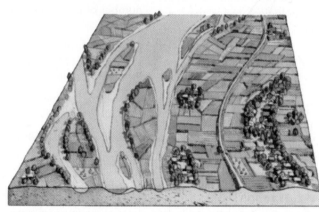

(C) Post-monsoon Stage. The river returns to its banks, but some of the new channels persist, changing the lay of the land. As the river recedes, it leaves behind silt and algae that nourish the soil. New ponds and lakes form and fill with fish.

Brahmaputra River in Figure 8.5, the movement of silt constantly rearranges the floodplain landscape, and this complicates human settlement and agricultural efforts. However, the seasonal replenishment of silt nourishes much of the agricultural production in the densely occupied plains of Bangladesh. The same is true on the Ganga and Indus plains.

THINGS TO REMEMBER

1. Because of its high degree of tectonic activity and deep crustal fractures, South Asia is prone to devastating earthquakes.

2. The monsoons are a major influence on South Asia's climate. In June, the warm, moist air of the summer monsoon reaches India, where coastal mountains and rising hot air force the moist monsoon air up, cooling it and causing rain. Monsoon rains run in a variable band parallel to the Himalayas that reaches across northern India and Pakistan, finally petering out over the Kabul Valley in eastern Afghanistan by July.

3. Precipitation from monsoons is especially intense in the eastern foothills of the Himalayas.

Environmental Issues

South Asia has been occupied by people for millennia, but as recently as 1700 (just prior to British colonization), population density and human environmental impacts were far lower than they are today. By the beginning of the twenty-first century, population density and human impacts had grown exponentially. As a result, today South Asia has a range of serious environmental problems.

South Asia's Vulnerability to Climate Change

Learning Goal 1
Climate Change and Water: How is South Asia vulnerable to water-related problems associated with climate change?

In South Asia, climate change—with the attendant sea level rise, water shortage, and crop failure—puts more lives at risk than in any other region in the world.

Sea Level Rise Bangladesh, where tens of millions of poor farmers and fishers live near sea level, has more people vulnerable to *sea level rise* (see page 43 in Chapter 1) than does any other country on earth (see Photo Essay 8.2A on page 297). With 162 million people already squeezed into a country the size of Iowa, as many as 17 million people might have to find new homes if sea levels rise by 5 feet. The greatest economic impacts of sea level rise, however, could come from the submergence of parts of South Asia's wealthiest city, Mumbai, India, and the many other large, coastal urban centers (see Photo Essay 8.2C).

Also threatened by continuing sea level rise are the Maldives islands in the Indian Ocean, 80 percent of which lie 1 meter or less above the sea. Beach erosion is so severe that homes built only a few years ago in this richest of South Asia's countries are falling into the sea.

Water Shortage South Asia's three largest rivers are fed by glaciers high in the Himalayas that are now melting rapidly because of climate change. Smaller glacial-melt streams serve Afghanistan (see Photo Essay 8.2B). As many as 703 million people, almost half of the region's population, depend on these glacially fed rivers for irrigation and drinking water. The immediate effect of glacial melting may be flooding. However, as glaciers shrink and provide less and less water to rivers each year during the dry season, the long-term effect will be severe water shortages. Unless dramatic action against climate change is successful, most Himalayan glaciers could eventually disappear. As a result, South Asia's largest rivers may run nearly dry during the winter monsoon when cool, dry air is flowing off the Eurasian continent (see Figure 8.4 on page 293).

Shifting rainfall patterns and rising temperatures are already resulting in drier conditions in much of South Asia, especially Afghanistan, Pakistan, and northwest India. However, shifting or abnormally heavy rainfall can also result in flooding, as was the case in Pakistan in 2010 (see Thematic Overview A on page 287). High Himalayan communities in Ladakh, on the India–China border, are attempting to retain autumn glacial melt in stone catchments, where the melt refreezes over the winter and is available for spring irrigation.

Responses to Global Climate Change South Asia is pioneering some innovative responses to the multiple threats posed by global climate change. India, by far the largest country in the region, has for some time had a well-developed system for storing grain. There is also a government-run emergency employment program that gives temporary jobs to those who lose their means of livelihood because of drought. These programs have a short history and their efficacy is not yet proven.

Public and private entities are taking measures to reduce contributions of CO_2 to the atmosphere. For example, India, which is the seventh-largest source of greenhouse gases in the world, is also home to the world's largest producer of plug-in electric cars (see Reasons for Optimism C on page 324). The Reva company sells its cars in India and the UK for about U.S.$8000. Even factoring in emissions from the plants that generate the electricity used to charge the cars, electric cars contribute substantially less to climate change than do gasoline- or diesel-powered cars. Several South Asian countries are investing private and public funds in solar energy, but the use of more affordable wind power has increased more rapidly. In 2010, wind supplied only about 3 percent of India's electricity, yet this was enough to make India the world's fifth-largest user of wind power.

The melting of Himalayan glaciers will likely require both water conservation and increased water storage in order for supplies to last through the dry winter monsoon. Fortunately, South Asia has a lot of experience with such technologies. The Indus Valley in Pakistan was home to innovative underground water management even in ancient times. More recently, India and Pakistan have both pioneered methods of increasing the rate at which water deposited during the summer monsoon percolates through the soil and into underground aquifers (Photo Essay 8.2D). More water is thus available for irrigation during the dry season. Because agriculture uses the most water of any human activity, drip irrigation technology, now thought to be the most efficient way to irrigate, would help conserve water. All South Asian countries have experience with drip irrigation, although the relatively high cost of modern efficient equipment has hampered widespread implementation of drip systems.

Deforestation

Deforestation has been occurring in South Asia since the first agricultural civilizations developed more than 5000 years ago. Ecological historians have shown that, as the forests vanished, the western regions of the subcontinent (from India to Afghanistan) became increasingly drier. The pace of deforestation has increased dramatically over the past 200 years. By the mid-nineteenth century, perhaps a million trees a year were felled for use in building railroads alone.

Causes of Deforestation In the twenty-first century, South Asia's forests are still shrinking due to commercial logging and expanding village populations that use wood for building and for fuel. Many of South Asia's remaining forests are in mountainous or hilly areas, where forest clearing dramatically increases erosion during the rainy season. One result is massive landslides that can destroy villages and close roads. With fewer trees and less soil to retain the water, rivers and streams become clogged with runoff, mud, and debris. The effects can reach so far downstream that increased flooding in the plains of Bangladesh is linked to deforestation in the Himalayas.

Environmental Activism and Resistance Unlike China and many other nations facing similar problems, the countries of South Asia have a healthy and vibrant culture of environmental activism that has alerted the public to the consequences of deforestation. In 1973, for example, in the Himalayan district of Uttarakhand, India, a sporting-goods manufacturer planned to cut down a grove of ash trees so that the factory, in the distant city of Allahabad, could use the wood to make tennis racquets. The trees were sacred to nearby villagers, however, and when their protests were ignored, a group of local women took dramatic action. When the loggers came, they found the women hugging the trees and refusing to let go until the manufacturer decided to find another grove.

The women's action grew into the *Chipko* (literally, "hugging"), or *Social Forestry Movement*. The movement has spread to other forest areas, slowing deforestation and increasing ecological awareness. Proponents of the movement argue that the management of forest resources should be turned over to local communities. They say that people living at the edges of forests

Thinking Geographically

After you have read about vulnerability to climate change in South Asia, you will be able to answer the following questions:

A Can you see anything in this photo that suggests the farmer is poor?

B Is there any indication in the photo that suggests that Afghanistan has a large number of poor, unemployed people?

C What about this photo suggests that Mumbai has drainage problems?

D How might these berms help to capture rainwater?

Climate change is putting more lives at risk in South Asia than in any other region. However, many responses to climate change are emerging, which could also increase resilience to climate-related hazards.

A A farmer in Bangladesh plows his field. Bangladesh is highly exposed to sea level rise, increased storm intensity, and flooding related to glacial melting. Bangladesh's large poor population makes it very sensitive to these climate hazards.

B In Afghanistan, a crew improves irrigation infrastructure that may help reduce the sensitivity of nearby areas to water shortage. Widespread poverty and political instability contribute to Afghanistan's extremely high vulnerability to climate hazards.

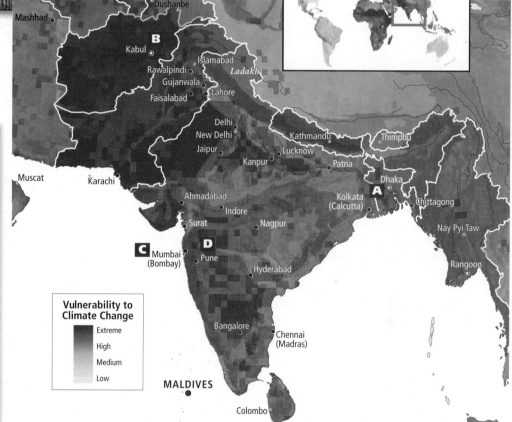

Vulnerability to Climate Change

Extreme

High

Medium

Low

C Monsoon rains during a high tide in Mumbai. Low-lying Mumbai is at high risk for sea level rise, and while its wealth and well-developed emergency response systems increase its resilience to climate hazards, large slum areas remain quite sensitive.

D Fields in Maharashtra, India, are equipped with earthen berms that help capture rainwater during the summer monsoon, allowing water to filter down into the ground and replenish local aquifers. More water can then be pumped to the surface for irrigation during the dry season, reducing sensitivity to water shortage.

possess complex local knowledge of those ecosystems that have been gained over generations—knowledge about which plants are useful as building materials, for food, medicines, and fuel. They are more likely to manage the forests carefully because they want their descendants to benefit from forests for generations to come.

▌▌▶ 180. TRIBAL PEOPLE IN INDIA WANT TO PROTECT INDIGENOUS WAY OF LIFE

Case Study: Nature Preserves in the Nilgiri Hills

The Mudumalai Wildlife Sanctuary and neighboring national parks in the Nilgiri Hills (part of the Western Ghats) harbor some of the last remaining forests in southern India. Here, in an area of about 600 square miles, live a few of India's last wild tigers and a dozen or more other rare species, such as sloth bears and barking deer.

▌▌▶ 181. INDIA'S NATIONAL SYMBOL BECOMING MORE DIFFICULT TO SPOT

Even much smaller forest reserves play important roles in conservation. At 287 acres, Longwood Shola is a tiny remnant of the ancient tropical forests that once covered the Nilgiris.

Phillip Mulley, a naturalist, Christian minister, and leader of the Badaga ethnic group, points out that the indigenous peoples of the Nilgiris must now compete for space with a growing tourist industry (1.7 million visitors in 2005). In addition, huge tea plantations were cut out of forestlands by the state government to provide employment for Tamil refugees from the conflict in Sri Lanka (Photo Essay 8.3D). So while the forestry department and citizen naturalists are trying to preserve forestlands, the social welfare department, faced with a huge refugee population, is cutting them down. [*Sources: Lydia and Alex Pulsipher's field notes, Nilgiri Hills, June 2000; Government of Tamil Nadu, Tamil Nadu Human Development Report (Delhi: Social Sciences Press), 2003, at http://data.undp.org.in/shdr/tn/TN%20HDR%20final.pdf.*]

Water

One of the most controversial environmental issues in South Asia today is the use of water. South Asia has more than 20 percent of the world's population, but only 4 percent of its fresh water. Access to water is often difficult, and disputes over water are increasingly common (see Thematic Overview C).

Conflicts Over the Use of Ganga River Water In recent years during the dry season, India has diverted 60 percent of the Ganga's flow to Kolkata to flush out channels where silt is accumulating and hampering river traffic (see the Ganga-Brahmaputra delta in Figure 8.3; see also the map in Photo Essay 8.1). However, the diversions deprive Bangladesh of normal freshwater flow. Less water in the Ganga-Brahmaputra delta allows salt water from the Bay of Bengal to penetrate inland, ruining agricultural fields. The diversion has also caused major alterations in Bangladesh's coastline, damaging its small-scale fishing industry. Thus, to serve the needs of Kolkata's 15 million people, the livelihoods of 40 million rural Bangladeshis have been put at risk, triggering protests in Bangladesh.

When two or more countries share a water basin, conflicts can become geopolitical in scale. In the late 1990s, India signed a treaty with Bangladesh promising a fairer distribution of water, but as of 2010, Bangladesh was still receiving a considerably reduced flow.

Similar water-use conflicts occur between states within India and Pakistan and between the wealthier and poorer sectors of the populations. For example, diverting water from the Ganga in the state of Haryana deprives farmers downstream in Uttar Pradesh of the means to irrigate their crops. And in Delhi, just 17 five-star hotels use about 210,000 gallons (800,000 liters) of water daily, which would be enough to serve the needs of 1.3 million slum dwellers.

▌▌▶ 182. COCA-COLA BLAMED FOR INDIA'S WATER PROBLEMS

Water Purity of the Ganga River Water purity has become an issue in historic religious pilgrimage towns such as Varanasi, where each year millions of Hindus come to die, be cremated, and have their ashes scattered over the Ganga River. The number of such final pilgrimages has increased with population growth and affluence; hence, wood for cremation fires has become scarce. As a result, incompletely cremated bodies are being dumped into the river, where they pollute water used for drinking, cooking, and ceremonial bathing (Photo Essay 8.3B). In an attempt to deal with this problem, the government recently installed an electric crematorium on the riverbank. It is attracting considerable business, as a cremation in this facility costs 30 times less than a traditional funeral pyre.

Of greater concern now is the large amount of industrial waste and sewage dumped into the river. Most sewage enters the river in raw form because city sewage systems (most built by the British early in the twentieth century) have long ago exceeded capacity. In Varanasi, pumps have been installed to move the sewage up to a new and expensive processing plant, but the plant is so overwhelmed by the volume of water during the rainy season that it can process only a small fraction of the city's sewage.

Veer Bhadra Mishra, a Brahmin priest (see Figure 8.10 on page 307) and professor of hydraulic engineering at Banaras Hindu University in Varanasi, is on a mission to clean up the Ganga. In 2008, after decades of lobbying and executing numerous demonstration projects, he finally received approval from India's central government to build a series of processing ponds that will use India's heat and monsoon rains to divert Varanasi's sewage

Thinking Geographically

After you have read about human impact on the Biosphere in South Asia, you will be able to answer the following questions:

A Is there any indication that the liquid coming out of the pipe is more than just water?

B What is the greater source of pollution for the Ganga River, incompletely cremated dead bodies or industrial pollution?

C Air pollution in New Delhi is so severe that it is equivalent to smoking how many cigarettes a day?

D What role has conflict in this region played in the expansion of tea plantations in the Nilgiri Hills?

South Asia's huge population and growing industries have had major impacts on water and air quality, as well as on the extent and health of remaining ecosystems, such as forests.

A One of 230 leather tanning facilities in Kasur, Pakistan, pumps effluent into a small river, where toxins are now 40 times the limit recommended by the World Health Organization. Local residents suffer dysentery, respiratory disorders, and skin diseases as a result. Still more tanneries are opening up in Kasur as companies try to avoid stricter environmental regulations on leather tanning in the European Union and China.

B Hindus flock to the highly polluted Ganga River in Varanasi, India, for ritual purification.

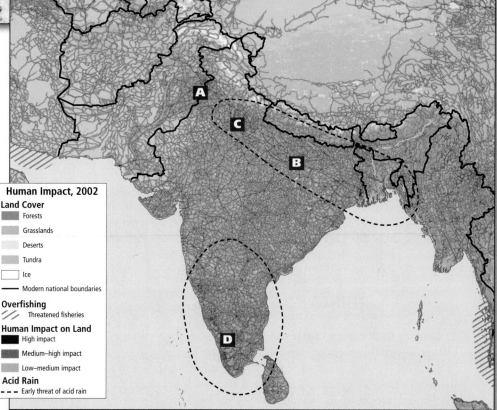

Human Impact, 2002

Land Cover
- Forests
- Grasslands
- Deserts
- Tundra
- Ice
- —— Modern national boundaries

Overfishing
- ///// Threatened fisheries

Human Impact on Land
- High impact
- Medium–high impact
- Low–medium impact

Acid Rain
- – – – Early threat of acid rain

C New Delhi, India, where air pollution now rivals that of Beijing, China, for the title of "world's worst." Sixty-seven percent of pollution comes from vehicles. Despite a shift to cleaner-burning compressed natural gas (CNG) and the banning of older vehicles, air quality continues to worsen as more people drive cars.

D A tea plantation in the Nilgiri Hills in Kerala, India. The expansion of tea plantations has resulted in habitat loss for many endangered species, as well as degraded water resources.

away from the river and clean it at half the cost of other methods. Mishra also preaches a contemporary religious message to the thousands who visit his temple on a bank of the sacred Ganga. The belief that the Ganga is a goddess who purifies all she touches leads many Hindus to think that it is impossible to damage this magnificent river. Mishra reminds them that because the Ganga is their symbolic mother, it would be a travesty to smear her with sewage and industrial waste.

Industrial Air Pollution

In many parts of South Asia, the air as well as the water is endangered by industrial activity. Emissions from vehicles and coal-burning power plants are so bad that breathing Delhi's air is equivalent to smoking 20 cigarettes a day (see Photo Essay 8.3C). The acid rain caused by industries up and down the Yamuna and Ganga rivers is destroying good farmland and such great monuments as the Taj Mahal.

M.C. Mehta, a Delhi-based lawyer, became an environmental activist partly in response to the condition of the Taj Mahal. For more than 20 years, he has successfully promoted environmental legislation that has removed hundreds of the most polluting factories from India's river valleys. His efforts are also a response to a disastrous event that took place in central India in 1984. At that time, an explosion at a pesticide plant in Bhopal produced a gas cloud that killed at least 15,000 people and severely damaged the lungs of 50,000 more. The explosion was largely the result of negligence on the part of the U.S.-based Union Carbide Corporation (which owned the plant) and the local Indian employees who ran the plant. To help address the tragedy, the Indian government launched an ambitious campaign to clean up poorly regulated factories.

> **Indus Valley civilization** the first substantial settled agricultural communities, which appeared about 4500 years ago along the Indus River in modern-day Pakistan and northwest India
>
> **Harappa culture** see Indus Valley civilization
>
> **Mughals** a dynasty of Central Asian origin that ruled India from the sixteenth to the nineteenth century

THINGS TO REMEMBER

1. **Learning Goal: Climate Change and Water** In South Asia, climate change—with the attendant sea level rise, water shortage, and crop failure—puts more lives at risk than in any other region in the world. Low-lying, densely populated areas such as coastal Bangladesh are particularly vulnerable to sea level rise. However, even more people will be impacted by the melting of glaciers in the Himalayas that feed the region's major rivers during the dry season.

2. The *Chipko* ("tree hugging") *Social Forestry Movement* Uttarakhand, India, has spread to other forest areas, slowing deforestation and increasing ecological awareness.

3. India has water-allocation problems that impact farmers, towns, and cities, especially along the Ganga; water management in India also affects millions of people in Bangladesh.

4. Pakistan, India, and Bangladesh have major air and water pollution problems related to industrial waste, lack of sewage treatment facilities, and end-of-life religious pilgrimages.

Human Patterns over Time

A variety of groups have migrated into South Asia, many of them as invaders who conquered peoples already there. Despite much blending over the millennia, the continued coexistence and interaction of many of these groups make South Asia both a richly diverse and an extremely contentious place.

The Indus Valley Civilization

There are indications of early humans in South Asia as far back as 200,000 years ago, but the first real evidence of modern humans in the region is about 38,000 years old. The first large agricultural communities, known as the **Indus Valley civilization** (or **Harappa culture**), appeared about 4500 years ago along the Indus River in modern-day Pakistan and northwest India. The architecture and urban design of this civilization were quite advanced for the time. Homes featured piped water and sewage disposal. Towns were well planned, with wide, tree-lined boulevards laid out in a grid. Evidence of a trade network that extended to Mesopotamia and eastern Africa has also been found.

Vestiges of the Indus Valley civilization's agricultural system survive to this day in parts of the valley, including techniques for storing monsoon rainfall to be used for irrigation in dry times (see Timeline A on page 304). Remnants of language, and possibly superficial biological traits such as skin color, survive today among the Dravidian peoples of southern India, who originally migrated from the Indus region.

The reasons for the decline of the Indus Valley civilization after about 800 years (3700 years ago) are debated. Some scholars believe that complex geologic (seismic) and ecological changes (drier conditions) brought about a gradual demise. Others argue that foreign invaders brought a swift collapse.

A Series of Invasions

The first recorded invaders to join the indigenous people of South Asia came from Central and Southwest Asia into the rich Indus Valley and Punjab about 3500 years ago. Many scholars believe that these people, referred to as Indo-European (a linguistic term; see page 305), in conjunction with the Harappa and other indigenous cultures, instituted some of the early elements of classical Hinduism, the major religion of India today. One of those elements was the still-influential caste system (discussed on pages 305–306), which divides society into hereditary hierarchical categories.

Other invaders included the Persians, the armies of the Greek general Alexander the Great, and numerous Turkic and Mongolian peoples. Defensive structures against these invaders can still be found across northwest South Asia. Jews came to the Malabar Coast of Southwest India more than 2500 years ago, Christians shortly after the time of Jesus. Arab traders came by land and sea to India long before the emergence of Islam; and then starting about 1000 years ago, Arab traders and religious mystics introduced Islam to what are now Afghanistan, Pakistan, and northwest India, and by sea to the coasts of southwestern India and Sri Lanka. In 1526, the invasion by the **Mughals**, a group of Turkic

Persian people from Central Asia, intensified the spread of Islam. The Mughals reached the height of their power and influence in the seventeenth century, controlling the north-central plains of South Asia. The last great Mughal ruler (Aurangzeb) died in 1707, but the legacy of the Mughals has remained as the power and range of the dynasty declined. One aspect of the legacy is the 520 million Muslims now living in South Asia. The Mughals also left a unique heritage of architecture, art, and literature that includes the Taj Mahal, miniature painting, and a rich tradition of lyric poetry (see Timeline B). The Mughals contributed to the evolution of the Hindi language, which became the language of trade of the northern subcontinent and which is still used by more than 400 million people.

As Mughal rule declined, a number of regional states and kingdoms rose and competed with one another. The absence of one strong power created an opening for yet another invasion. By the late 1700s, several European trading companies were competing to gain a foothold in the region. Of these, Britain's East India Company was the most successful. By 1857, the East India Company, acting as an extension of the British government, put down a rebellion against European intrusion and became the dominant power in the region.

Globalization and the Legacies of British Colonial Rule

The British controlled most of South Asia from the 1830s through 1947 (Figure 8.6). By making the region part of the British Empire, the British accelerated the process of globalization in South Asia (see Thematic Overview D). In the process, South Asia was transformed politically, socially, and economically. Even areas not directly ruled by the British felt the influence of their empire. Afghanistan repelled British attempts at military conquest, but the British continued to intervene there, trying to make Afghanistan a "buffer state" between British India and Russia's expanding empire. Nepal remained only nominally independent during the colonial period, and Bhutan became a protectorate of the British Indian government.

The Deindustrialization of South Asia As in their other colonies, the British used South Asia's resources primarily for their own benefit, often with disastrous results for South Asians. One example was the fate of the textile industry in Bengal (modern-day Bangladesh and the Indian state of West Bengal).

Bengali weavers, long known for their high-quality muslin cotton cloth, initially benefited from the greater access British

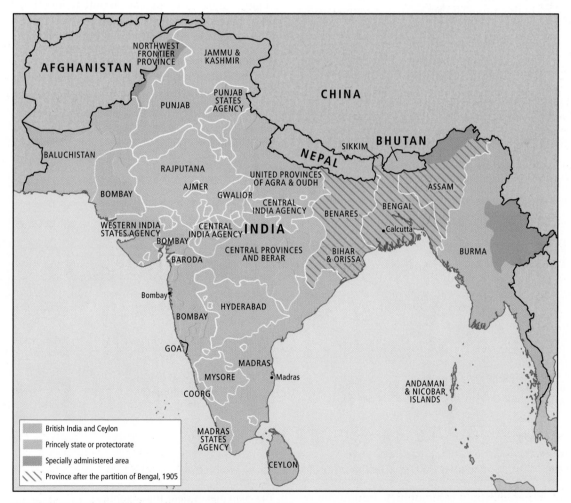

FIGURE 8.6 The British Indian Empire, 1860–1920. After winning control of much of South Asia, Britain controlled lands from Baluchistan to Burma, including Ceylon and the islands between India and Burma.

traders gave them to overseas markets in Asia, the Americas, and Europe. By 1750, South Asia had an advanced manufacturing economy that produced 12 to 14 times more cotton cloth than Britain alone and more than all of Europe combined. However, during the eighteenth and nineteenth centuries, Britain's own highly mechanized textile industry, based on cotton grown in India, various other colonies, and the American South, developed cheaper cloth that then replaced Bengali muslin. This shift occurred first in the British colonies in the Americas, then in Europe, and eventually throughout South Asia. The British textile industry was further aided by the British East India Company, which began severely punishing Bengalis who continued to run their own looms. As a result, the traditional South Asian textile industry held on in only a few areas (see Timeline C).

Many people who were pushed out of their traditional livelihood in textile manufacturing were compelled to find work as landless laborers. But rural South Asia already had an abundance of agricultural labor, so many migrated to emerging urban centers. In the 1830s, a drought worsened an already difficult situation, and more than 10 million people starved to death. Throughout the nineteenth century, similar events forced millions of South Asian workers to join a stream of indentured laborers and migrate to other British colonies in the Americas, Africa, Asia, and the Pacific (Figure 8.7).

▌▶ 183. CARIBBEAN BEAT PULSES WITH INDIAN ACCENTS

Economic development in South Asia was tightly controlled to benefit Britain, which encouraged the colonies to produce tropical agricultural raw materials, such as cotton, tea, sugar, and indigo (a widely used blue dye) in order to supply Britain and its other colonies. Industrial development, which might have competed with Britain's own industries, was discouraged.

Nonetheless, British rule did bring some beneficial changes. Trade with the rest of Britain's empire brought prosperity to a few areas, especially in the large British-built cities on the coast, such as Bombay (now Mumbai), Calcutta (Kolkata), and Madras (Chennai). The British built a railroad system that boosted trade within South Asia and greatly eased the burden of personal transportation. In addition, English became a common language for South Asians of widely differing backgrounds, assisting both trade and cross-cultural understanding.

Democratic and Administrative Legacies Contemporary South Asian governments retain institutions put in place by the British to administer their vast empire. These governments inherited many of the shortcomings of their colonial forebears, such as highly bureaucratic procedures, resistance to change, and a tendency to remain aloof from the people they govern. Nonetheless, the governments have proved functional over time, but not without occasional major civil disturbances in virtually every country. Democratic government was not instituted on a large scale until the final days of the Empire, but since independence in 1947 (see below), it has given people an outlet for voicing their concerns and has enabled many peaceful transitions of elected governments. Still, the struggle to retain and build democratic institutions continues.

Independence and Partition The tremendous changes brought by the British inspired many resistance movements among South Asians. Some of these were militant movements intent on pushing

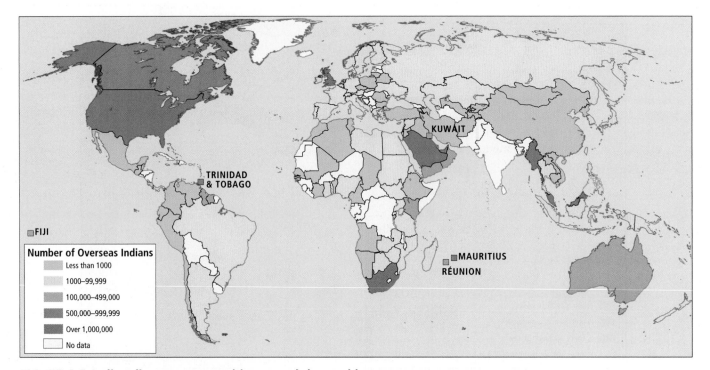

FIGURE 8.7 Indian diaspora communities around the world. This map makes clear that people of South Asian origin now live virtually everywhere on earth. They engage in many occupations, but are found particularly in the professions, technology development, commerce, and agriculture. Large-scale migrations out of South Asia began in the nineteenth century and continue on today.

the British out by force, such as the unsuccessful rebellion of South Asian soldiers employed by the British East India Company in 1857. Other movements such as the Indian National Congress (founded in 1885) focused on political agitation for greater democracy as a route to South Asian political independence. Although both types of action were brutally repressed, the democracy movements achieved worldwide attention and, after decades of struggle, success.

In the early twentieth century, Mohandas Gandhi, a young lawyer from Gujarat, emerged as a central political leader of South Asia's independence movement. His tactics of **civil disobedience** focused on the nonviolent violation of laws imposed by the British that discriminated against South Asians. Gathering a large group of peaceful protesters, he would notify the government that the group was about to break a discriminatory law. If the authorities ignored the act, the demonstrators would have made their point and the law would be rendered moot. If the government instead used force against the peaceful demonstrators, it would lose the respect of the masses. Throughout the 1930s and 1940s, this technique was used to slowly but surely undermine British authority across South Asia. The most famous example was the *Salt March* of 1930, when Gandhi led tens of thousands of people on a march to the sea where they made salt by evaporating sea water, thus breaking the law that made it illegal for South Asians to produce salt (see Timeline D). The purpose of the law was to facilitate British rule by controlling a vital human nutritional necessity. Breaking the law on such a massive scale created an international media frenzy that catapulted Gandhi to global notoriety; it moved millions of South Asians to support independence from Britain. The hunger strikes described in the vignette that opens this chapter about the Sardar Sarovar Dam on the Narmada are another strategy that reflect Gandhi's legacy of nonviolent political protest.

In 1947, Britain granted independence to British India, which was then divided into two independent countries: predominantly Hindu India and Muslim West and East Pakistan (Figure 8.8) (Afghanistan, Bhutan, and Nepal were never officially British colonies; Ceylon [now Sri Lanka] became independent in 1948). This division—called **Partition**—which Gandhi greatly lamented, was perhaps the most enduring and damaging outcome of colonial rule.

The idea of two nations was first suggested by Muslim political leaders concerned about the fate of a minority Muslim population in a united India with a Hindu majority. Though Partition was highly controversial, it became part of the independence agreement between the British and the Indian National Congress (India's principal nationalist party). Northwestern and northeastern India, where the population was predominantly Muslim, became a single country consisting of two parts, known as West and East Pakistan, separated by northern India. Although both India and Pakistan maintained secular constitutions, with no official religious affiliation, the general understanding was that Pakistan would have a Muslim majority and India a Hindu majority. Fearing that they would be persecuted if they did not move, more than 7 million Hindus and Sikhs migrated to India from their ancestral homes

> **civil disobedience** the breaking of discriminatory laws by peaceful means
>
> **Partition** the breakup following Indian independence that resulted in the establishment of Hindu India and Muslim Pakistan

in what had become Pakistan. A similar number of Muslims left their homes in India for Pakistan. In the process, families and communities were divided, looting and rape were widespread, and between 1 and 3.4 million people were killed in numerous local outbreaks of violence. In 1971, a bloody civil war led to the division of Pakistan into Bangladesh (formerly East Pakistan) and Pakistan (formerly West Pakistan).

‖▶ 184. 60TH ANNIVERSARY OF INDIA–PAKISTAN PARTITION ON AUGUST 15TH

Partition was the tragic culmination of the divide-and-rule tactics the British used throughout their empire (see pages 261–263 in Chapter 7). This approach heightened tensions between South Asian Muslims and Hindus, thus creating a role for the British as seemingly indispensable and benevolent mediators. Instead of relieving tensions, the partition of India and Pakistan laid the groundwork for the repeated wars and skirmishes, strained relations, and ongoing arms race between India and Pakistan.

The Postindependence Period In the more than 60 years since the departure of the British, South Asians have experienced both progress and setbacks. Democracy has expanded steadily, albeit somewhat slowly. India is now the world's most populous democracy. It is gradually dismantling age-old traditions that hold back poor, low-caste Hindus, women, and other disadvantaged groups. Pakistan has had repeated elections and some of its governments have been modestly effective, but militaristic authoritarian and corrupt regimes have been more

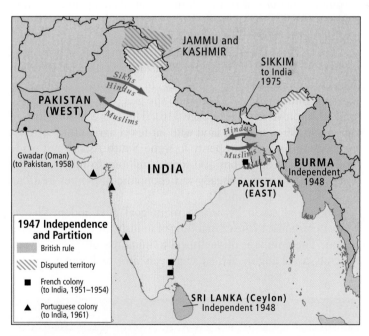

FIGURE 8.8 Independence and Partition. India became independent of Britain in 1947, and by 1948, the old territory of British India was partitioned into the independent states of India and East and West Pakistan. The Jammu and Kashmir region was contested space, and remains so today. Sikkim went to India, and both Burma and Sri Lanka became independent. Following additional civil strife, East Pakistan became the independent country of Bangladesh in 1971.

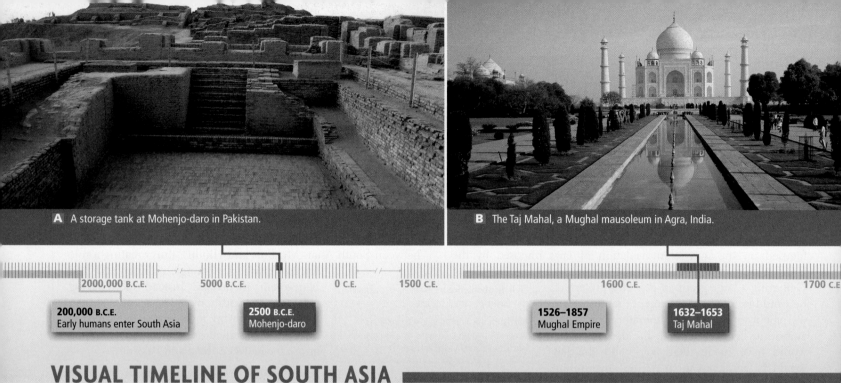

A A storage tank at Mohenjo-daro in Pakistan.

B The Taj Mahal, a Mughal mausoleum in Agra, India.

2000,000 B.C.E.	5000 B.C.E.	0 C.E.	1500 C.E.	1600 C.E.			1700 C.E.

200,000 B.C.E.
Early humans enter South Asia

2500 B.C.E.
Mohenjo-daro

1526–1857
Mughal Empire

1632–1653
Taj Mahal

VISUAL TIMELINE OF SOUTH ASIA

Thinking Geographically

After you have read about the human history of South Asia, you will be able to answer the following questions:

A This structure was designed to store water from what source?

C In 1750, South Asia produced how much more cotton than Britain and the rest of Europe combined?

the norm, with Pakistan's and India's long border feud (discussed on page 320) sapping resources and national spirit. Bangladesh, despite the damage it sustained in its war of independence from Pakistan in 1971, has actually had a more stable, responsive, and less militaristic government than Pakistan.

Under British rule, agricultural modernization lagged, and during World War II the Bengal Famine took an estimated 4 million lives. After independence, progress in agricultural production was slow until the late 1960s, when the green revolution brought marked improvements (see page 316). The move to modernized farming on large tracts of land with far fewer agricultural workers than before brought prosperity to some South Asians and made food exports possible, but also forced millions to migrate to the cities. This intensified already vast economic disparities in urban South Asia (see Timeline E).

Industrialization became a main goal after independence, partly in response to the dismantling of industry during the colonial period. The emphasis on technical training has produced several generations of highly skilled engineers and technocrats, whose

talents are in demand around the world. Now, in most countries in the region, urban-based industrial and service economies constitute a far larger share of GDP than agriculture, with the information technology (IT) sector growing especially rapidly in India.

THINGS TO REMEMBER

1. Following some 300 years of Mughal rule, the British controlled most of South Asia from the 1830s through 1947, profoundly influencing the region politically, socially, and economically.

2. In the early twentieth century, Mohandas Gandhi emerged as a central political leader of India's independence movement, using nonviolent civil disobedience that eventually led to independence.

3. Instead of relieving tensions, the partition of India and Pakistan laid the groundwork for the repeated wars and skirmishes, strained relations, and ongoing arms race between India and Pakistan.

II CURRENT GEOGRAPHIC ISSUES

The size and diversity of South Asia can be overwhelming. This section provides a glimpse of the whole region, first by looking at the texture of daily life in both village and city, and then by examining the cultural characteristics that touch all lives yet vary greatly in practice across the region.

Sociocultural Issues

Within the life of one South Asian village or urban neighborhood, there can be considerable cultural variety. Differences based on caste, economic class, ethnic background, religion, and even

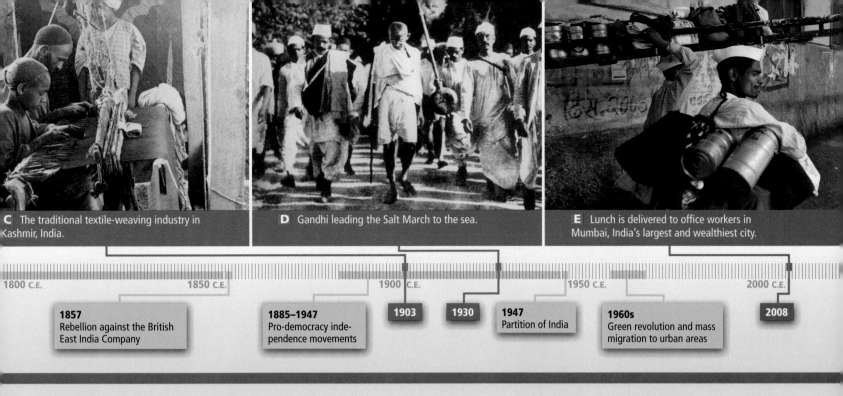

C The traditional textile-weaving industry in Kashmir, India.

D Gandhi leading the Salt March to the sea.

E Lunch is delivered to office workers in Mumbai, India's largest and wealthiest city.

1800 C.E.   1850 C.E.   1900 C.E.   1950 C.E.   2000 C.E.

1857
Rebellion against the British East India Company

1885–1947
Pro-democracy independence movements

1903

1930

1947
Partition of India

1960s
Green revolution and mass migration to urban areas

2008

D What was the immediate result of the Salt March?

E What does this photo suggest about Mumbai's economy?

language are usually accommodated peacefully by long-standing customs that guide cross-cultural interaction. That everyone in South Asia is, in one way or another, a minority is to some extent an equalizing factor—in part because this fact of ubiquitous minority status has become part of the national dialogue about difference.

Language and Ethnicity

There are many distinct ethnic groups in South Asia, each with its own language or dialect. In India alone, 18 languages are officially recognized, but hundreds of distinct languages actually exist. This complexity results partly from the region's history of multiple invasions from outside. However, some groups were isolated for long periods of time, which also contributed to this complexity. In **Figure 8.9** on page 306, the Dravidian language-culture group, represented by numbers 14–19, is an ancient group that predates the Indo-European invasions by a thousand years or more. Today, Dravidian languages are found mostly in southern India, but a small remnant of the extensive Dravidian past can still be found in the Indus Valley in south-central Pakistan.

By the time of British colonization, Hindi—an amalgam of Persian-based and Sanskrit-based Indo-European languages—was the dominant language throughout northern India and what is now Pakistan. Today, variants of Hindi serve as national languages for both India and Pakistan (there, called Urdu), though it is the first, or native, language of only a minority. English is a common second language throughout the region. For years, it was the language of the colonial bureaucracy, and it remains a language used at work by professionals. Between 10 and 15 percent of South Asians speak, read, and write English.

Religion, Caste, and Conflict

The main religious traditions of South Asia are Hinduism, Buddhism, Sikhism, Jainism, Islam, and Christianity (see **Figure 8.11** on page 308). (For a discussion of Islam, Christianity, and Judaism, see pages 220–221 in Chapter 6.)

Hinduism Hinduism is a major world religion practiced by approximately 900 million people, 800 million of whom live in India. It is a complex belief system, with roots in both ancient literary texts (known as the Great Tradition) and in highly localized folk traditions (known as the Little Tradition).

> **Hinduism** a major world religion practiced by approximately 900 million people, 800 million of whom live in India; a complex belief system, with roots both in localized folk traditions (known as the Little Tradition) as well as in a broader system based on literary texts (known as the Great Tradition)

The Great Tradition is based on 4000-year-old scriptures called the *Vedas*. Its major tenet is that all gods are merely illusory manifestations of the ultimate divinity, which is formless and infinite. Some devout Hindus worship no gods at all, and instead engage in meditation, yoga, and other spiritual practices designed to bring people to a state described as "infinite consciousness." The average person, however, is thought to need the help of personified divinities in the form of gods and goddesses (the Little Tradition). While some deities (such as Vishnu, Shiva, Ganesh, or Krishna) are recognized by all Hindus, many are found only in one region, one village, or even one family.

Some beliefs are held in common by nearly all Hindus. One is the belief in reincarnation, the idea that any living thing that

305

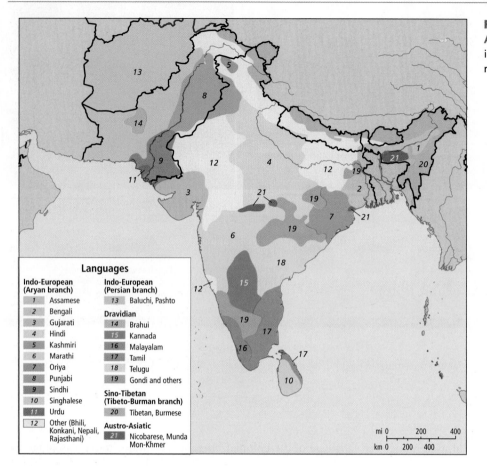

FIGURE 8.9 Major language groups of South Asia. The modern pattern of language distribution in South Asia is a testimony to the fact that this region has long been a cultural crossroads.

Languages

Indo-European (Aryan branch)
1. Assamese
2. Bengali
3. Gujarati
4. Hindi
5. Kashmiri
6. Marathi
7. Oriya
8. Punjabi
9. Sindhi
10. Singhalese
11. Urdu
12. Other (Bhili, Konkani, Nepali, Rajasthani)

Indo-European (Persian branch)
13. Baluchi, Pashto

Dravidian
14. Brahui
15. Kannada
16. Malayalam
17. Tamil
18. Telugu
19. Gondi and others

Sino-Tibetan (Tibeto-Burman branch)
20. Tibetan, Burmese

Austro-Asiatic
21. Nicobarese, Munda Mon-Khmer

mi 0   200   400
km 0   200   400

desires the illusory pleasures (and pains) of life will be reborn after it dies. A reverence for cows, which are seen as only slightly less spiritually advanced than humans, also binds all Hindus together. This attitude, along with the Hindu prohibition on eating beef, may stem from the fact that cattle have been tremendously valuable in rural economies as the primary source of transport, field labor, dairy products, fertilizer, and fuel (animal dung is often burned).

Caste Hinduism includes the **caste system**, a complex and ancient system of dividing society into hereditary hierarchical categories (Figure 8.10). One is born into a given subcaste, or community (called a ***jati***) that traditionally defined much of one's life experience—where one would live, where and what one could eat and drink, with whom one would associate, one's marriage partner, and often one's livelihood. The classical caste system has four main divisions or tiers, called ***varna***, within which are many hundreds of *jatis* and sub-*jatis*, which vary from place to place.

Brahmins, members of the priestly caste, are the most privileged in ritual status. Thus, they must conform to those behaviors that are considered most ritually pure (for example, strict vegetarianism and abstention from alcohol). As is the case with many castes, Brahmins are found in many occupations outside their place as priests in the *varna* system (as barbers and hairdressers, for example). In descending

> **caste system** an ancient Hindu system for dividing society into hereditary hierarchical classes
>
> ***jati*** in Hindu India, the subcaste into which a person is born, which largely defines the individual's experience for a lifetime
>
> ***varna*** the four hierarchically ordered divisions of society in Hindu India underlying the caste system: *Brahmins* (priests), *Kshatriyas* (warriors/kings), *Vaishyas* (merchants/landowners), and *Sudras* (laborers/artisans)

rank are *Kshatriyas*, who are warriors and rulers; *Vaishyas*, who are landowning (small-plot) farmers and merchants; and *Sudras*, who are low-status laborers and artisans. A fifth group, the *Dalits*— "the oppressed," or untouchables—is actually considered to be so lowly as to have no caste. Dalits perform those tasks that caste Hindus consider the most despicable and ritually polluting: killing animals, tanning hides, cleaning, and disposing of refuse. A sixth group, also outside the caste system, is the *Adivasis*, who are thought to be descendants of the region's ancient aboriginal inhabitants.

Although *jatis* are associated with specific subcategories of occupations, in modern economies, this aspect of caste is more symbolic than real. Members of a particular *jati* do, however, follow the same social and cultural customs, dress in a similar manner, speak the same dialect, and tend to live in particular neighborhoods or villages. This spatial separation arises from the higher-caste communities' fears of ritual pollution through physical contact or sharing of water or food with lower castes. When one stays in the familiar space of one's own *jati*, one is enclosed in a comfortable circle of families and friends that becomes a mutual aid society in times of trouble. The social and spatial cohesion of *jatis* helps explain the persistence and respect paid to a system that, to outsiders, seems to put a burden of shame and poverty on the lower ranks.

It is important to note that caste and class are not the same thing. Class refers to economic status, and there are class

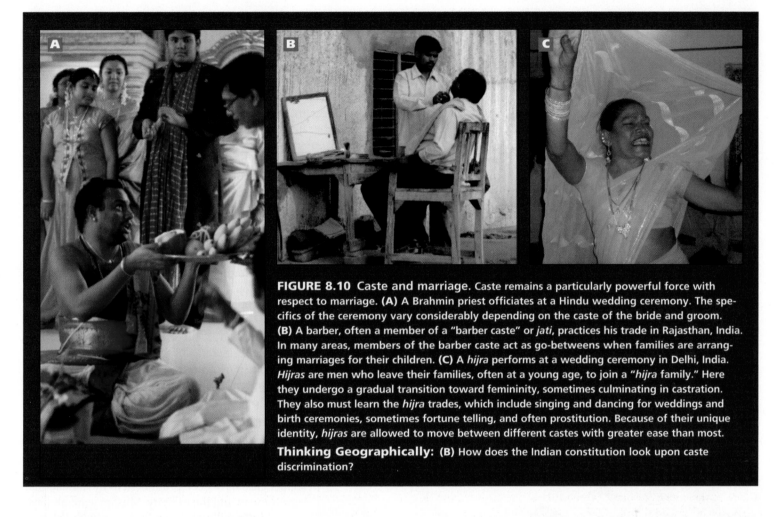

FIGURE 8.10 Caste and marriage. Caste remains a particularly powerful force with respect to marriage. **(A)** A Brahmin priest officiates at a Hindu wedding ceremony. The specifics of the ceremony vary considerably depending on the caste of the bride and groom. **(B)** A barber, often a member of a "barber caste" or *jati*, practices his trade in Rajasthan, India. In many areas, members of the barber caste act as go-betweens when families are arranging marriages for their children. **(C)** A *hijra* performs at a wedding ceremony in Delhi, India. *Hijras* are men who leave their families, often at a young age, to join a "*hijra* family." Here they undergo a gradual transition toward femininity, sometimes culminating in castration. They also must learn the *hijra* trades, which include singing and dancing for weddings and birth ceremonies, sometimes fortune telling, and often prostitution. Because of their unique identity, *hijras* are allowed to move between different castes with greater ease than most.

Thinking Geographically: (B) How does the Indian constitution look upon caste discrimination?

differences within caste groups because of differences in wealth. Historically, upper-caste groups (Brahmins and Kshatriyas) owned or controlled most of the land, and lower-caste groups (Sudras) were the laborers, so caste and class tended to coincide. There were many exceptions, however. Today, as a result of legally mandated expanding educational and economic opportunities, caste and class status are less connected. Some Vaishyas and Sudras have become large landowners and extraordinarily wealthy businesspeople, while some Brahmin families struggle to achieve a middle-class standard of living. By and large, however, Dalits remain very poor.

Caste, Politics, and Culture In the twentieth century, Mohandas Gandhi began an organized effort to eliminate discrimination against "untouchables." As a result, India's constitution bans caste discrimination. However, in recognition that caste is still hugely influential in society, upon independence from Britain, India began an affirmative action program. The program reserves a portion of government jobs, places in higher education, and parliamentary seats for Dalits and Adivasis. Together, the two groups now constitute approximately 23 percent of the Indian population and are guaranteed 22.5 percent of government jobs. In 1990, this program was extended to include other socially and educationally "backward castes" (the term used in India), such as disadvantaged *jatis* of the Sudras caste. An additional 27 percent of government jobs were reserved for these groups. However, reserving nearly half of government jobs in this way has resulted in considerable controversy. In 2006, medical students successfully protested against quotas in elite higher education institutions for lower-caste applicants. The Indian Supreme Court ruled in favor of the students.

At the local level, most political parties design their vote-getting strategies to appeal to subcaste loyalties. They often secure the votes of entire *jati* communities with such political favors as new roads, schools, or development projects. These arrangements fly in the face of the official ideologies of the major political parties and of Indian government policies, which actively work to eliminate discrimination on the basis of caste. Nonetheless, currently, the role of caste in politics seems to be increasing, as several new political parties that explicitly support the interests of low castes have emerged.

Among educated people in urban areas, the campaign to eradicate discrimination on the basis of caste may appear to have succeeded, but the reality is more complex. Throughout the country there are now at least some Dalits in powerful positions. Members of high and low castes ride city buses side by side, eat together in restaurants, and attend the same schools and universities. For some urban Indians—especially educated professionals who meet in the workplace—caste is deemphasized as the crucial factor in finding a marriage partner. However, less than 5 percent of marriages cross even *jati* lines, let alone the broader gulf of *varna*. Nearly everyone notices the tiny social clues that reveal an individual's caste, and in rural areas, where the majority of Indians still reside, the divisions of caste remain prevalent.

Geographic Patterns of Religious Beliefs The geographic pattern of religion in South Asia is complex and overlapping. As Figure 8.11 shows, there is a core Hindu area in central India, with other faiths more common on the fringes of the region.

The approximately 520 million Muslims in South Asia form the majority in Afghanistan, Pakistan, Bangladesh, and the Maldives; and Muslims are a large and important minority in India, where at 140 million, they form about 12 percent of the population. They live mostly in the northwestern and central Ganga River plain.

Buddhism began about 2600 years ago as an effort to reform and reinterpret Hinduism. Its origins are in northern India, where it flourished early in its history before spreading eastward to East and Southeast Asia. Only 1 percent of South Asia's population—about 10 million people—are Buddhists. They are a majority in Bhutan and Sri Lanka.

Jainism, like Buddhism, originated as a reformist movement within Hinduism more than 2000 years ago. Jains (about 6 million people, or 0.6 percent of the region's population) are found mainly in cities and in western India. They are known for their educational achievements, promotion of nonviolence, and strict vegetarianism.

> **Buddhism** a religion of Asia that originated in India in the sixth century B.C.E. as a reinterpretation of Hinduism; it emphasizes modest living and peaceful self-reflection leading to enlightenment
>
> **Jainism** originally a reformist movement within Hinduism, Jainism is a faith tradition that is more than 2000 years old; found mainly in western India and in large urban centers throughout the region, Jains are known for their educational achievements, nonviolence, and strict vegetarianism
>
> **Sikhism** a religion of South Asia that combines beliefs of Islam and Hinduism

Sikhism was founded in the fifteenth century as a challenge to both Hindu and Islamic systems. Sikhs believe in one god, hold high ethical standards, and practice meditation. Philosophically, Sikhism accepts the Hindu idea of reincarnation but rejects the idea of caste. (In everyday life, however, caste plays a role in Sikh identity.) The 21 million Sikhs in the region live mainly in Punjab, in northwestern India. Their influence in India is greater than their numbers because throughout India many Sikhs hold positions in the government, military and security forces, and the police.

The first Christians in the region are thought to have arrived in the far southern Indian state of Kerala with St. Thomas, the Apostle of Jesus, in the first century C.E. Today, Christians and Jews are influential but tiny minorities along the west coast of India. A few Christians live on the Deccan Plateau and in northeastern India.

Animism, the most ancient religious tradition, is practiced throughout South Asia, especially in central and northeastern India, where there are indigenous people whose occupation of the area is so ancient that they are considered aboriginal inhabitants. (For a discussion of animism, see pages 281–282 in Chapter 7.)

The Hindu–Muslim Relationship Indian independence leaders like Mohandas Gandhi and Jawaharlal Nehru (first prime minister of India, 1947–1964) emphasized the common cause that once united Muslim and Hindu Indians: throwing off British colonial rule. Since independence, members of the Muslim upper class have been prominent in Indian national government and the military. Muslim generals have served India willingly, even in its wars with Pakistan after Partition. Hindus and Muslims often interact amicably, and they occasionally marry each other.

But there is a darker side to the Hindu–Muslim relationship. At the community level in South Asia, relations between the region's two largest religious groups are often quite tense. In some Indian villages, Hindus may regard Muslims as members of low castes. Religious rules about food are often the source of discord because dietary habits are a primary means of distinguishing caste. Hindus forbid killing cows, so consumption of beef and the processing of cowhides into leather is permitted for only the lowest castes. Muslims, on the other hand, run slaughterhouses and tanneries (though discreetly), eat beef, and use cowhide to make shoes and other items. To some Hindus, therefore, some Muslims appear to have offensive customs. Also

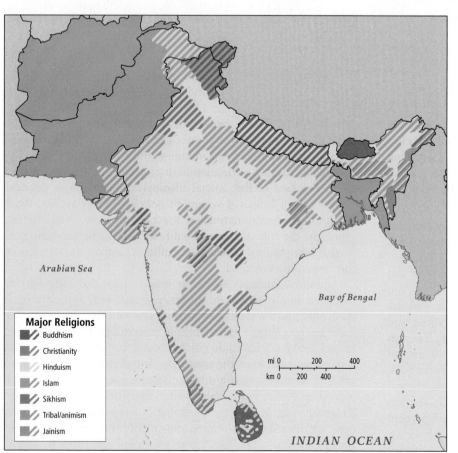

Arabian Sea

Bay of Bengal

Major Religions
- Buddhism
- Christianity
- Hinduism
- Islam
- Sikhism
- Tribal/animism
- Jainism

mi 0   200   400
km 0   200   400

INDIAN OCEAN

FIGURE 8.11 Major religions in South Asia. Notice the overlapping patterns in many parts of the region.

fueling this perception is the occasional conversion to Islam (or sometimes Christianity) of entire low-caste Hindu villages seeking to escape the hardships of being members of a disadvantaged social category.

The Hindu–Muslim relationship is no less complex in Bangladesh. After Partition in 1947, some wealthy Hindu landowners remained. In some Bangladeshi villages today, while Muslims may be a majority, Hindus are often somewhat wealthier. Although the two groups may coexist amicably for many years, they view each other as different, and conflict resulting from religious or economic disputes—euphemistically called **communal conflict**—can erupt over seemingly trivial events, as described in the following vignette.

> **communal conflict** a euphemism for religiously based violence in South Asia
>
> **purdah** the practice of concealing women from the eyes of nonfamily men

Vignette The sociologist Beth Roy, who studies communal conflict in South Asia, recounts an incident that she refers to as "some trouble with cows" in the village of Panipur (a pseudonym), Bangladesh. The incident started when a Muslim farmer either carelessly or provocatively allowed one of his cows to graze in the lentil field of a Hindu. The Hindu complained, and when the Muslim reacted complacently, the Hindu seized the offending cow. By nightfall, Hindus had allied themselves with the lentil farmer and Muslims with the owner of the cow. More Muslims and Hindus converged from the surrounding area, and soon there were thousands of potential combatants lined up facing each other. Fights broke out. The police were called. In the end, a few people died when the police fired into the crowd of rioters. [Source: Beth Roy, Some Trouble with Cows—Making Sense of Social Conflict (Berkeley: University of California Press, 1994), pp. 18–19.] ■

THINGS TO REMEMBER

1. There are many distinct ethnic groups in South Asia, each with its own language or dialect. Today, variants of Hindi are the principal languages of India and Pakistan, while Bengali is the official language of Bangladesh. English is the common second language throughout South Asia.

2. Caste and class are not the same thing: "class" refers to economic status, and "caste" to hereditary hierarchical social categories. There are class differences within caste groups because of differences in wealth. Even though India's constitution bans caste discrimination, caste is still hugely influential in Indian society.

3. Of the more than 900 million Hindus living in South Asia, most (800 million) live in India; of the 425 million Muslims living in the region, 140 million live in India. Relations between these religious groups, the region's two largest, are often quite tense, and the rise of religious nationalism has led to violent confrontation. Other religious traditions also play prominent roles in the life of South Asia.

4. There is a geographic pattern to religion in South Asia, but it is not absolute. People of different religions often live in close proximity, and both communal conflict and communal peace are common.

Geographic and Social Patterns in the Status of Women

The overall status of women in South Asia is notably lower than the status of men. Even so, young women today have greater educational and employment opportunities than did women of a generation ago, and a number of women hold and have held very high positions of power. However, women's literacy rates, social status, earning power, and welfare are generally lower than those of men, especially in the belt that stretches from the northwest in Afghanistan across Pakistan, western India, Nepal and Bhutan, and the Ganga Plain into Bangladesh. Women fare better in eastern and central India and considerably better in southern India and in Sri Lanka. In these latter regions, where literacy rates are higher (see **Figure 8.12** on page 310), different marriage, inheritance, and religious practices give women greater access to education and resources.

II▶ 187. SUFI ROCK SINGER FALU BLENDS OLD WITH NEW

Urban women in South Asia generally have greater individual freedom than rural women have, with many now pursuing professional careers and some becoming involved in politics. Generally speaking, rural women are less free than urban women. In rural India, middle- and upper-caste Hindu women are often more restricted in their movements than are lower-caste women because they have a status to maintain. Meanwhile, lower-caste women who go into public spaces may have to contend with sexual harassment and exploitation from upper-caste men.

The socioeconomic status of Muslim women in South Asia is notably lower than that of their Hindu and Christian counterparts. In India, some of this relates to the generally lower incomes and standard of living for Muslims, which also usually means lower educational levels. Muslim women also work outside the home less than non-Muslim women in India. Low rates of education and workforce participation for women also prevail in Muslim-dominated countries such as Pakistan and Afghanistan, although Bangladesh has a notably better record.

II▶ 191. REMEMBERING PAKISTAN'S FORMER PRIME MINISTER BENAZIR BHUTTO

Purdah The practice of concealing women from the eyes of nonfamily men, especially during their reproductive years, is known as **purdah**. It is observed in various ways across the region. The practice is strongest in Afghanistan and across the Indo-Gangetic Plain, where it takes the form of seclusion of women and of veiling or head covering in both Muslim and Hindu communities. Purdah is weaker in central and southern India, but even there, separation between unrelated men and women is maintained in public spaces. In general, low-caste Hindus do not observe this custom, but that is changing. In recent decades, as some low-status households have increased their incomes, they have adopted purdah as a sign of their surplus wealth.

Purdah practices have influenced the architecture of South Asia. Homes are often in walled compounds that seclude kitchens and laundries as women's spaces. In grander homes, windows to the street are usually covered with lattice screens known as *jalee* that allow in air and light but shield women from the view of outsiders.

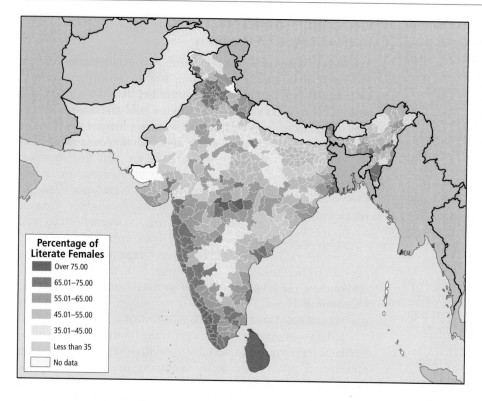

FIGURE 8.12 Female literacy in India, 2005, and in adjacent countries. Female literacy lags behind male literacy in nearly all parts of South Asia. Female literacy is crucial to improving the lives of children: women who can read often seize opportunities to earn some income and nearly always use this income to help their children. The white lines indicate district boundaries in India. Female literacy in India as a whole is 54.5 percent.

Percentage of Literate Females
- Over 75.00
- 65.01–75.00
- 55.01–65.00
- 45.01–55.00
- 35.01–45.00
- Less than 35
- No data

Marriage, Motherhood, and Widowhood Throughout South Asia, most marriages are arranged by the parents of the prospective bride and groom. Especially in wealthier, better-educated families, the wishes of the bride and groom are considered, but in some cases they are not. This is especially true of child marriage, when a young girl (often as young as 12 and in some places even younger) is married off to a much older man.

Usually a bride goes to live in her husband's family compound, where she becomes a source of domestic labor for her mother-in-law. Most brides work at domestic tasks for many years until they have produced enough children to have their own crew of small helpers, at which point they gain some prestige and a measure of autonomy.

Motherhood in South Asia determines much about a woman's life. A woman's power and mobility increase when she has grown children and becomes a mother-in-law herself. On the other hand, in some communities, the death of a husband, regardless of cause, is a disgrace to a woman and can completely deprive her of all support and even of her home, children, and reputation. Widows may be ritually scorned and blamed for their husband's death. Widows of higher caste rarely remarry, and in some areas, they become bound to their in-laws as household labor or may be asked to leave the family home. Most simply become low-status members of extended families and help with household duties of all sorts.

Dowry and Violence Against Females A **dowry** is a sum of money paid by the bride's family to the groom's family at the time of marriage. Dowries originated as an exchange of wealth between Muslim landowners or high-caste families that practiced purdah. With her ability to work reduced by purdah, a woman was considered a liability for

> **dowry** a price paid by the family of the bride to the groom (the opposite of *bride price*); formerly a custom practiced only by the rich

the family that took her in. Changing dowry customs appear to be a cause of the growing incidence of various kinds of domestic violence against females in Pakistan, India, and Bangladesh.

Until the last several decades, it was only wealthy families that gave the groom a dowry—in this case, a substantial sum that symbolized the family wealth, meant to give a daughter a measure of security in her new family.

Increasing affluence and education have reinforced the custom of dowry and made it much more common for all families. As more men became educated, their families felt that their diplomas increased their worth as husbands and gave them the power to demand larger and larger dowries. Soon, the practice spread through lower-caste families wanting to upgrade their status. Now the dowries they must pay to get their daughters married can cripple poor families. A village proverb captures this dilemma: "When you raise a daughter, you are watering another man's plant." Some poor families view the birth of a daughter as such a calamity that they are led to the desperate act of female infanticide—killing second and third daughters soon after birth.

Gender and Democratization As countries in South Asia have moved toward greater democracy, the status of women in the region has risen. India, Pakistan, Bangladesh, and Sri Lanka have all had female heads of state (prime ministers). However, it is important to note that all of these women were either wives or daughters of previous heads of state. Women have been notably less successful in elections; and at the parliamentary level, Indian and Sri Lankan women remain very poorly represented.

As of 2009, only 9 percent of the Indian parliament was female. A confederation of Muslim and Hindu women's groups has lobbied for legislation that would reserve one-third of the seats in the lower house of Parliament and in state assemblies for women for a 15-year trial period. Such quotas are already in place in Pakistan (21 percent) and Nepal (33 percent). Bangladesh, where women currently make up only 6 percent of the parliament, allocated one-third of the seats for women in March 2009, to take effect at the next election.

Women and the Taliban in Afghanistan Women in Afghanistan have sometimes suffered brutal repression since a conservative Islamist movement, the **Taliban**, gained control of the government there in the mid-1990s. Prior to that time, rights for women in Afghanistan were slowly but steadily improving, and upper-class women enjoyed many freedoms. The Taliban support strict and radical interpretations of Islamic law, forcing females, even urban professional women, to live in seclusion. In regions where the Taliban retain control, girls and women are not allowed to work outside the home or attend school. In virtually all parts of the country, despite the decline of Taliban control, women must wear a heavy, completely concealing garment, called a *burqa* (or burka), whenever they are out of the house. (Men also must follow a dress code, though a less restrictive one.) Although the Taliban were driven from official power in November of 2001, they maintain control of the southern provinces and mountainous zones near Pakistan, where cultural and religious conservatism continues to adversely affect Afghan women. Even efforts to provide women and girls with a basic education, such as that described in the vignette below about Radio Sahar, run into hostility, oftentimes violence.

> **Taliban** an archconservative Islamist movement that gained control of the government of Afghanistan in the mid-1990s

▌▶ 189. REPORT: DOMESTIC VIOLENCE WIDESPREAD IN AFGHANISTAN

▌▶ 190. "FRONTRUNNER" DOCUMENTARY TELLS STORY OF AFGHAN POLITICIAN WHO INSPIRES WOMEN

Vignette From behind her microphone at Radio Sahar ("Dawn"), Nurbegum Sa'idi speaks to a female audience on a wide range of topics. Located in the city of Herat, Radio Sahar is one in a network of independent women's community radio stations that has sprung up in Afghanistan since early 2003. Radio Sahar provides 13 hours of daily programming consisting of educational items that address cultural, social, and humanitarian matters as well as music and entertainment. For example, one recent broadcast, aimed at informing women of their legal rights, followed the life of a young woman who was physically abused by her husband and his entire family. The woman took the brave step of asking for a divorce. As a result, she was forced into hiding, where she was counseled on what steps she might take next. Another program explored the various concepts of what democracy is and how it might work in Afghanistan. A reported 600,000 Afghan women and youth listen to Radio Sahar while they do their chores.

The broadcasts allow women to connect with one another in this conservative, male-dominated society. As Sa'idi attests, "It's great when you feel you can bring about change. The feedback we have been getting from listeners tells us that Sahar is providing new hope for the women in Herat." *[Source: Internews Afghanistan, March 2008, at http://www.internews.org/bulletin/afghanistan/Afghan_200803. html.]* ▐

THINGS TO REMEMBER

1. South Asia has had a number of women in very high positions of power, and young women today have greater educational and employment opportunities than those of a generation ago. However, the overall status of women in the region is notably lower than the status of men, and women's influence on policy remains low even when the number of women in national parliaments rises above a tiny minority.

2. The socioeconomic status of Muslim women is significantly lower than that of their Hindu and Christian counterparts; this is especially so in those parts of Afghanistan and Pakistan where tribal customs include deeply ingrained patterns of violence against women and Taliban influence remains high.

Population Patterns

South Asia is the most densely populated region in the world (see Figure 8.13 on page 312). The region already has more people (1.59 billion) than China (1.33 billion), which has almost twice the land area of South Asia. U.S. Census Bureau projections indicate that by 2050, India alone, with 1.7 billion people, will overtake China's 1.4 billion (see Thematic Overview F). By 2050, China's population (if the one-child policy persists) will be shrinking at the rate of about 20 million people per decade, while India's will still be growing. And at current rates of growth, Pakistan and Bangladesh will have nearly doubled their populations. Nevertheless, the rate of population growth is slowing as families choose to have fewer children for a number of reasons, such as improved educational opportunities for women, better health care, and urbanization.

Population densities in this region are greatest in the cities that lie just south of the Hindu Kush and Himalayas, and the very greatest density is in the Ganga-Brahamaputra delta (see Figure 8.13), much of which is rural and yet extremely densely occupied.

Slowing Population Growth: Health Care, Urbanization, and Gender Efforts to slow population growth have been underway for over 50 years. Improvements in health care have resulted in fewer babies dying in infancy. Couples can then have fewer children, assured that many of the children will survive to adulthood and be able to care for their elderly parents. A major source of improved access to health care is the ongoing shift in population from rural to urban areas (see Photo Essay 8.4 on page 314). Urbanization also reduces the economic incentive for large families. In rural areas, children (staring at an early age) contribute their labor to farming, increasing the family income. Urban incomes are less likely to be boosted by the labor of children, who are more likely to go to school and thus represent a cost to the family. Improvements in the status of women are also slowing population. Research throughout South Asia and elsewhere has shown that women who have the opportunity to go to school and or get a paying job tend to delay childbearing and have fewer children. As a result of all of these factors, growth rates continue to decline in nearly every country.

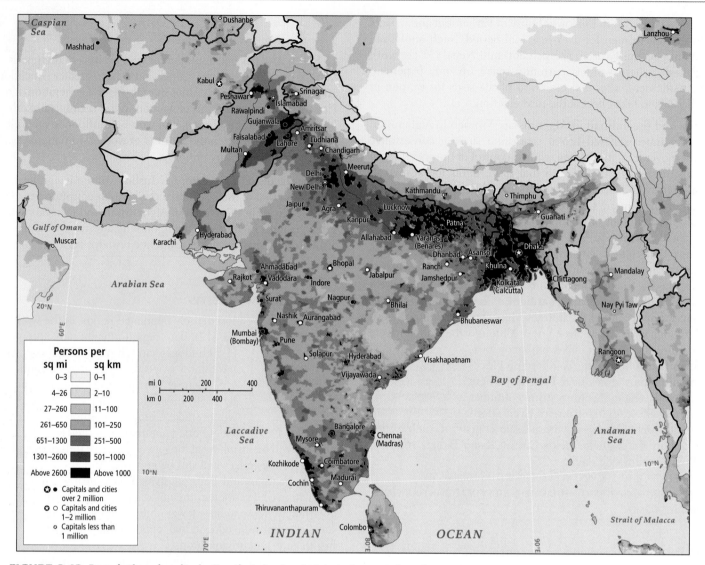

FIGURE 8.13 Population density in South Asia. South Asia is the most densely populated world region. While its population growth is slowing as most countries begin the demographic transition, all countries will keep adding large numbers of people for at least another 40 years.

A comparison of the population pyramids as well as some other statistics for Sri Lanka and Pakistan helps illustrate these points. Sri Lanka (**Figure 8.14B**) is far along but not completely through the demographic transition (see Figure 1.11 on page 17). Sri Lanka has a much higher GDP per capita (U.S.$4480) than Pakistan (U.S.$2700). Health care is generally better in Sri Lanka, as indicated by its much lower infant mortality rate (in 2009, Sri Lanka had 15 deaths per 1000 live births versus Pakistan's 67 per 1000) (**Figure 8.15**). Sri Lanka has also made greater efforts to educate and economically empower its women. Three key indicators of women's education and empowerment that are associated with reduced fertility rates are much higher in Sri Lanka than in Pakistan: female literacy (90 percent versus 30 percent), the percentage of young women who attend high school (89 percent versus 19 percent), and the percentage of women who work outside the home (36 percent versus 16 percent).

II ▶ 185. CHILD LABOR PERSISTS IN INDIA DESPITE NEW LAWS

> **Learning Goal 2**
> **Gender and Population:** How have issues related to gender transformed South Asian populations in recent decades?

Gender Imbalance A significant gender imbalance exists in South Asian populations because of cultural customs that make sons more likely than daughters to contribute to a family's wealth (see Thematic Overview E). A popular toast to a new bride is "May you be the mother of a hundred sons." Many middle-class couples who wish to have sons hire high-tech laboratories that specialize in identifying the sex of an unborn fetus. The intention is to abort a female fetus. This practice is now illegal, but it persists. Poorer South Asians may neglect the health of girl children, with some even committing female infanticide (the dowry implications are discussed above on page 310).

II ▶ 186. GIRLS PAY PRICE FOR INDIA'S PREFERENCE FOR BOYS

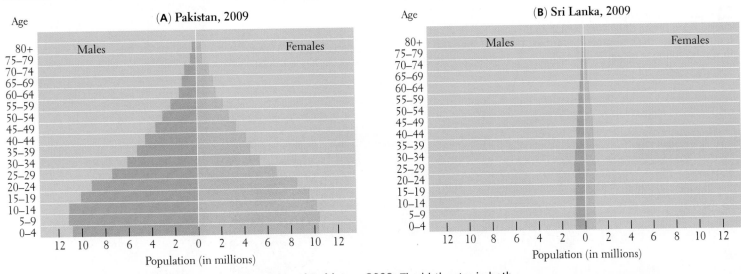

FIGURE 8.14 Population pyramids for Sri Lanka and Pakistan, 2009. The birth rates in both Pakistan and Sri Lanka have declined since 1960. However, the great size and youth of Pakistan's population ensure that it will continue to grow for years, while Sri Lanka's population is more or less stable. Note that the scales on the two pyramids are the same; Pakistan's population is about 9 times larger than Sri Lanka's.

Such practices have resulted in men significantly outnumbering women throughout this region. Elsewhere on earth, adult women normally outnumber adult men, because of their longer natural life spans; but the 2001 Indian census showed only 933 females for every 1000 males. After the 2001 census, India took a stronger stand against selecting for male children. A follow-up sex-ratio survey in five Indian states from 2004 to 2006 showed some improvement in all five states, with Tamil Nadu coming closest to a normal sex ratio, probably due to an especially aggressive campaign for women's health and against sex-selective abortions.

Gender imbalance of the magnitude India is facing could create serious problems, such as a surge in crime and drug abuse, due to the presence of so many young men with no prospect of having a family. The efforts of the state of Kerala to overcome gender imbalance through greater attention to women's development bear watching. Kerala's government has made women's development a priority, funding education for women well beyond basic levels. The results are reflected in its female literacy rate, which is the highest in India at 87.8 percent (the average for India as a whole is 54.5 percent), and in the number of women who work outside the home. In Kerala, daughters, far from being viewed as an economic liability to the family, are seen as an asset. Women can travel the public streets alone or in groups and commonly work in public places, many in high positions in government, education, health care, and other professions.

❚❚▶ 188. GLOBAL POPULATION BOOM PUTS 'MEGA' PRESSURE ON CITIES IN DEVELOPING WORLD

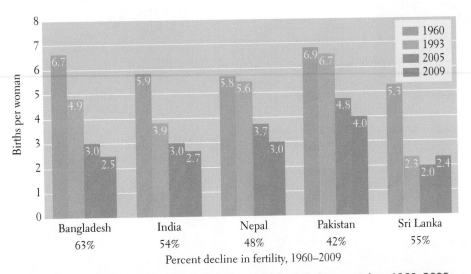

FIGURE 8.15 Total fertility rates in selected South Asia countries, 1960–2009. Most South Asian countries have had a substantial decline in fertility since 1960. Many factors have contributed to this decline. Observe, however, that Sri Lanka's fertility rate has increased slightly since 2005, possibly in response to civil disorder or the death toll of the tsunami in 2004. Note also that the decline in fertility shown here for Pakistan is also evident in Figure 8.14, with lower numbers for children ages 0–4 years old.

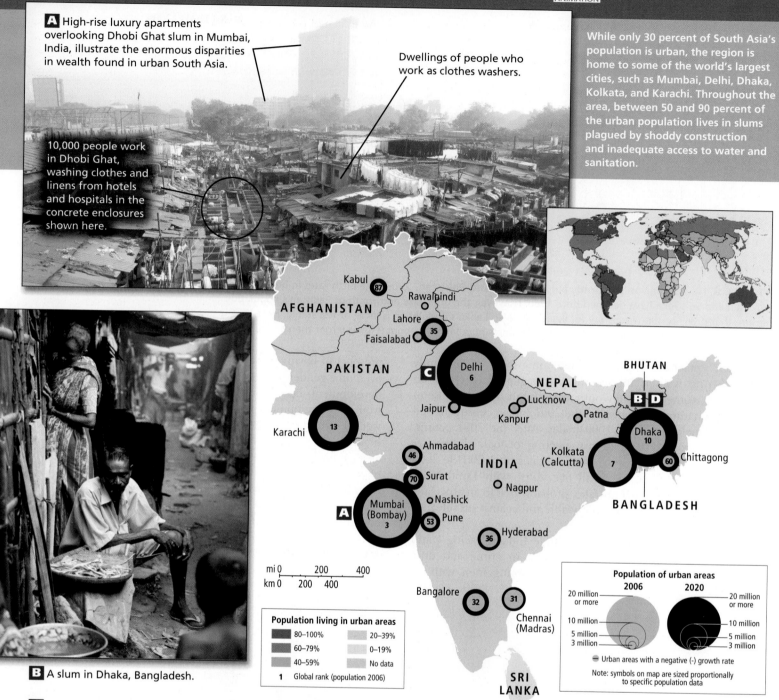

A High-rise luxury apartments overlooking Dhobi Ghat slum in Mumbai, India, illustrate the enormous disparities in wealth found in urban South Asia.

Dwellings of people who work as clothes washers.

10,000 people work in Dhobi Ghat, washing clothes and linens from hotels and hospitals in the concrete enclosures shown here.

While only 30 percent of South Asia's population is urban, the region is home to some of the world's largest cities, such as Mumbai, Delhi, Dhaka, Kolkata, and Karachi. Throughout the area, between 50 and 90 percent of the urban population lives in slums plagued by shoddy construction and inadequate access to water and sanitation.

Kabul 87
Rawalpindi
AFGHANISTAN
Lahore
Faisalabad 35
PAKISTAN
C
Delhi 6
NEPAL
BHUTAN
B D
Lucknow
Jaipur
Kanpur
Patna
Karachi 13
Ahmadabad 46
Surat 70
INDIA
Nagpur
Dhaka 10
Kolkata (Calcutta) 7
Chittagong 60
A
Mumbai (Bombay) 3
Nashick
53 Pune
Hyderabad 36
BANGLADESH
mi 0   200   400
km 0   200   400
Bangalore
32   31
Chennai (Madras)
SRI LANKA

Population living in urban areas
■ 80–100%	■ 20–39%
■ 60–79%	■ 0–19%
■ 40–59%	■ No data
1 Global rank (population 2006)	

Population of urban areas
2006   **2020**
20 million or more
10 million
5 million
3 million
20 million or more
10 million
5 million
3 million
Urban areas with a negative (-) growth rate
Note: symbols on map are sized proportionally to specific population data

B A slum in Dhaka, Bangladesh.

C A computer workshop at the University of Delhi. Many middle-class South Asians come to cities, and stay there, to take advantage of educational opportunities.

D Workers unload sand from a barge in Dhaka, Bangladesh. Most poor rural South Asians who come to cities work in low-paying, low-skill jobs.

Thinking Geographically

After you have read about urbanization in sub-Saharan Africa, you will be able to answer the following questions:

A What has forced most poor urban South Asians into cities?

C Many middle class South Asians come to cities in search of what?

D What about this photo suggests that this is a low-paying, low-skill job?

Urbanization Although only 30 percent of the region's population is urban, South Asia has several of the world's largest metropolitan areas. By 2010, Mumbai had 21.9 million people; Kolkata, 15.6 million; Delhi, 20.9 million; Dhaka, 14.3 million; and Karachi, 13.2 million. All of these cities have grown quickly, and South Asia's current urban population of around 460 million could expand to as much as 712 million by 2025.

Many middle-class South Asians move to cities for education, training, or business opportunities (Photo Essay 8.4C). Because people come to cities for reasons such as these, large cities have higher literacy rates than rural areas. Mumbai and Delhi both have literacy rates above 80 percent, approximately 25 percent above the average for the country; Dhaka, in Bangladesh, and Karachi, in Pakistan, both have 63 percent literacy, over 15 percent above the rate for each country as a whole.

South Asian cities increasingly have split identities. On the one hand, sleek modern skyscrapers bear the logos of powerful global corporations; and on the other hand, urban slums are chaotic, crowded, and violent, with overstressed infrastructures and dilapidated low-rise housing. These settlements are filled with millions of poor, once-rural people who have been pushed into urban migration by agricultural modernization (see Thematic Overview G). Other urban migrants are refugees who have left drought-stricken or flooded areas or have been displaced by development projects, such as the building of the dam on the Narmada discussed in this chapter's opening vignette.

Mumbai is South Asia's wealthiest city and a showcase of South Asian economic disparity. It is the largest deepwater harbor on India's west coast. Its metropolitan area, with over 21 million people, hosts India's largest stock exchange and the nation's central bank. Mumbai pays about a third of the taxes collected in the entire country and brings in nearly 40 percent of India's trade revenue. Its annual per capita GDP is 3 times that of India's next wealthiest city, the capital Delhi. Mumbai's wealth also extends into the realm of culture through the city's flourishing creative arts industries, including the internationally known "Bollywood" film industry.

Mumbai's wealth is most evident when one looks up at the elegant high-rise condominiums built for the city's rapidly growing middle class. But at street level, the urban landscape of large parts of Mumbai is dominated by the large numbers of people living on the sidewalks, in narrow spaces between buildings, or in large, rambling slums. The largest of these is Dharavi, which houses up to a million people on less than 1 square mile and rivals Orangi Township in Karachi, Pakistan, as the world's largest slum.

Implausibly, Dharavi is known for its inventive entrepreneurs. It is said to contain 15,000 one-room factories turning out thousands of products that make it into the global market, many of them from India's recycled plastic and metal. One young man, for example, collects and sells aluminum cans that once held ghee, a popular cooking oil in India. He says he makes about 15,000 rupees a month (U.S.$480), nearly twice the salary of the average teacher in India and much more than he made as a truck driver.

Indeed, while the poverty in slums like Dharavi is widespread and tragic, not all neighborhoods that may look like slums are actually impoverished. Many South Asian cities are composed of thousands of tightly compacted, reconstituted villages where standards of living are relatively high and daily life is intimate and familiar, not anonymous as in Western cities. Such a place is Koli.

Vignette Koli is an ancient fishing village that predates the city of Mumbai and is now squeezed between Mumbai's elegant coastal high-rises and the Bay of Mumbai, where some villagers still fish every day. Koli is a labyrinth of low-slung, tightly-packed homes ringed by fishing boats. The screeching of taxis and buses is soon lost in quiet calm as one ducks into a narrow covered passageway that winds through the village and branches in multiple directions. At first Koli appears impoverished, but inside, small but well-appointed homes, some with marble floors, TVs, and computers, open onto the dimly lit but pleasant alleys. The visitor soon learns that this is no warren of destitute shanties, no slum, but rather a community of educated bureaucrats, tradespeople, and artisans who constitute South Asia's rising urban middle class. ∎

THINGS TO REMEMBER

1. Population growth is definitely slowing across the region, but the momentum provided by a young population means growth will continue for the foreseeable future.

2. Learning Goal 2: Gender and Population A significant, and often extreme, gender imbalance exists in South Asia, originating in cultural customs that make sons more likely to contribute to a family's wealth than daughters. Too often this imbalance results from outright discrimination and oppression that affects women in every stage of their lives.

3. South Asia's current urban population of around 460 million could expand to as much as 712 million by 2025.

4. Mumbai is South Asia's wealthiest city and is a showcase of South Asian economic disparity.

Economic Issues

South Asia is a region of startling economic incongruities. India is a good example: It is home to hundreds of millions of desperately poor people, but it is also a global leader in the computer software industry. It is celebrated for being the largest democracy in the world and yet its poor are often left out of the political process, either bypassed or hurt by economic reforms.

Economic Trends

Agriculture still employs more than 50 percent of the workers in South Asia, but the contribution of agriculture to most national economies is much lower, averaging below 20 percent of the GDP for the region. The industrial sector employs far fewer people, yet produces between one-quarter and one-third of GDP in all countries except Nepal, Bhutan, and Afghanistan. The service sector has expanded across the region more rapidly than either agriculture or industry and now produces more than half of the GDP in every country except Afghanistan and Bhutan. Rapid development and self-sufficiency have been the dream for all parts of the region since the end of the colonial era (for India, Pakistan, and Bangladesh, this came via the end of British control in 1947 and in 1948 for Sri Lanka). Successful measures to encourage economic modernization include an emphasis on IT and other high-tech industries as well as the automotive industry in India, textile and leather manufacturing in Pakistan, and innovative microfinancing strategies pioneered in Bangladesh. Less successful,

but still promising, is the development of modernized agribusiness, which really involves all three sectors of agriculture, industry, and services. Especially in India, but also in Pakistan and Bangladesh, agribusiness has drastically increased the amount of food available.

Food Production and the Green Revolution

Agricultural production per unit of land has increased dramatically over the past 50 years, especially in parts of India; nonetheless, agriculture remains the least efficient economic sector, meaning that it has the lowest return on investments of land, labor, and cash. Figure 8.16 shows the distribution of agricultural zones in South Asia.

Until the 1960s, agriculture across South Asia was based largely on traditional small-scale systems that managed to feed families in good years, but often left them hungry or even starving in years of drought or flooding. Moreover, these systems did not produce sufficient surpluses for the region's growing cities. Even now, cities must import food from outside South Asia. Much of

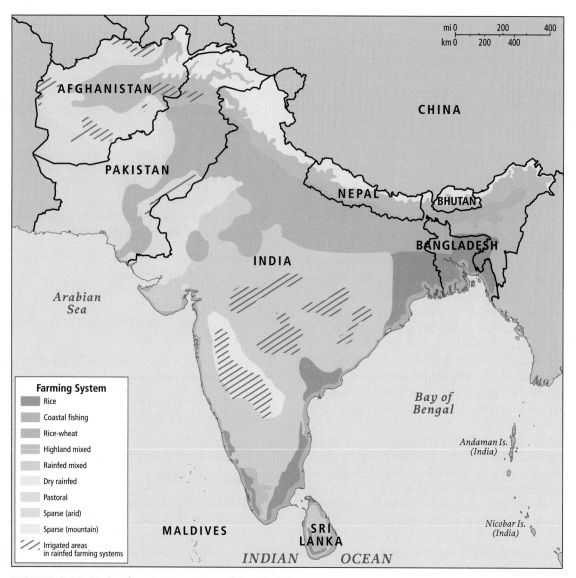

Farming System
- Rice
- Coastal fishing
- Rice-wheat
- Highland mixed
- Rainfed mixed
- Dry rainfed
- Pastoral
- Sparse (arid)
- Sparse (mountain)
- Irrigated areas in rainfed farming systems

FIGURE 8.16 Major farming systems of South Asia.

South Asia's agricultural land is still cultivated by hand, and for decades, agricultural development was neglected in favor of industrial development, especially in India. Nonetheless, by the 1970s, important gains in agricultural production had begun.

Beginning in the late 1960s, the green revolution (see page 23 in Chapter 1) boosted grain harvests dramatically through the use of new agricultural tools and techniques. Such innovations included seeds bred for high yield and for resistance to disease and wind damage; fertilizers; mechanized equipment; irrigation; pesticides; herbicides; and double-cropping (two crops produced consecutively per year). To a lesser extent, an increase in the amount of land under cultivation also contributed to a greater yield. Where the new techniques were used, yield per unit of farmland improved by more than 30 percent between 1947 and 1979, and both India and Pakistan became food exporters.

> **Learning Goal 3**
> **Food and Urbanization:** In what ways have recent changes in South Asian food production systems contributed to urbanization?

The benefits of the green revolution have been uneven, however. Some Indian states, such as Punjab and Haryana, which have extensive irrigation networks, have increased prosperity tremendously. Others have lagged behind. Many poor farmers who were unable to afford the special seeds, fertilizers, pesticides, and new equipment could no longer compete and had to give up farming. Most migrated to the cities, where their skills matched only the lowest-paying jobs (see Thematic Overview B).

South Asia needs alternatives to standard green revolution strategies because it will have difficulty maintaining current levels of food production over the long term. The green revolution's chemical fertilizers, pesticides, and high levels of irrigation all contribute to waterway pollution, increased erosion, and the loss of soil fertility over time through the buildup of salt in soils. Soil salinization (see page 214 in Chapter 6) is already reducing yields in many areas, such as the Pakistani Punjab, the country's most productive—but highly irrigated—agricultural zone.

> **agroecology** the practice of traditional, non-chemical methods of crop fertilization and the use of natural predators to control pests
>
> **microcredit** a program based on peer support that makes very small loans available to very low-income entrepreneurs

II▶ 192. TECHNOLOGY KEY TO PRODUCING MORE FOOD

A potential remedy for some of the failings of green revolution agriculture is **agroecology**. Traditional methods are being revived, such as fertilizing crops with animal manure, intercropping (planting several species together) with legumes to add nitrogen and organic matter, and using natural predators to control pests. Unlike green revolution techniques, the methods of agroecology are not disadvantageous to poor farmers because the necessary resources are readily available in most rural areas.

Malnutrition remains a problem despite the green revolution. Increased food supplies have reduced, but not eliminated, hunger and malnutrition in the region, because food goes to those with money to spend and, as noted previously, agricultural modernization can push the unskilled off the land and into precarious underemployment. Between 1970 and 2001, the amount of food produced per capita in South Asia increased 18 percent, and the proportion of undernourished people dropped

from 33 percent of the population to 22 percent. Nonetheless, 22 percent of nearly 1.6 billion is 352 million people—roughly half of the world's total undernourished population. Because of corruption and social discrimination, government programs that provide food to the poor have generally failed to reach those most in need. For example, in India, despite massive resources devoted to improve child nutrition, about half of the children show signs of malnutrition.

Microcredit: A South Asian Innovation for the Poor

Over the past four decades, a highly effective strategy for helping lift people out of extreme poverty has been pioneered in South Asia. **Microcredit** makes very small loans (generally under U.S.$100) available to poor would-be business owners. Throughout South Asia, as in most of the world, banks are generally not interested in administering the small loans that poor people, especially poor women, need. Instead, the poor must rely on small-scale moneylenders, who often charge interest rates as high as 30 percent or more *per month*. In the late 1970s, Muhammad Yunus, an economics professor in Bangladesh, started the Grameen Bank, or "Village Bank," which makes small loans, mostly to people in rural villages who wish to start businesses.

II▶ 195. NOBEL PEACE PRIZE GOES TO BANGLADESH'S "BANKER TO THE POOR"

The microcredit loans often pay for the start-up costs of small enterprises, such as cell phone–based services, chicken raising, small-scale egg production, or the construction of pit toilets. Potential borrowers (more than 90 percent are women) are organized into small groups that are collectively responsible for repaying any loans to group members. If one member fails to repay a loan, then everyone in the group will be denied loans until the loan is repaid. This system, reinforced with weekly meetings, creates incentives to repay. The repayment rate on the loans is extremely high, averaging around 98 percent, much higher than most banks achieve.

So far, the Grameen Bank has been an enormous success in Bangladesh, where it has loaned over U.S.$8 billion to more than 8 million borrowers. Similar microcredit projects have been established in India and Pakistan and throughout Africa, Middle and South America, North America, and Europe. In 2006, Dr. Yunus was awarded the Nobel Peace Prize for his work in microcredit. In 2009, he received the Presidential Medal of Freedom from President Obama.

Vignette In a small hamlet in Bangladesh, not too far from the Indian border, is the house of Mosamad Shonabhan, a 32-year-old married woman whose life has been changed by her 11-year participation in the Grameen Bank. Everyone agreed that she had been the smartest of her family's children. However, because her father earned only 50 cents a day as a farm laborer, she could not go to school, and was instead married at the age of 14 to a young barber. For a year, she lived in her father-in-law's house, but then financial problems forced her to move back into her father's home. There she faced increasingly dire circumstances as her father's health deteriorated.

After a few years, a local political leader suggested that Mosamad join the Grameen Bank's lending program. Eventually, she took out a loan for $40.00 that would allow her to set up a small rice-husking operation in her father's backyard.

Eleven years and 11 loans later, Mosamad earns about $1.50 every day—3 times what her father once made—and is a pillar of the local community. Her main source of income is a small shop inside her father's old house. She also leases an acre of land, which produces enough rice to feed her family and the numerous guests and friends who now come by to see her. [Source: Adapted from Alex Pulsipher's field notes, 2000.] ◼

Industry over Agriculture: A Vision of Self-Sufficiency

After independence from Britain in 1947, the new leaders in India, Pakistan, Bangladesh, and Sri Lanka tended to favor industrial development over agriculture. Influenced by socialist ideas, they believed that government involvement in industrialization was necessary to ensure the levels of job creation that would cure poverty. Another of their goals was to reduce the need to import manufactured goods from the industrialized world (see pages 115–116 in Chapter 3 for a discussion of import substitution industrialization). To accomplish this goal, governments took over the industries they believed to be the linchpins of a strong economy: steel, coal, transportation, communications, and a wide range of manufacturing and processing industries.

South Asian industrial policies in the decades after independence generally failed to meet their goals. The emphasis on industrial self-sufficiency was ill suited to countries that had such large agricultural populations. In India, for example, governments invested huge amounts of money in a relatively small industrial sector—even today industry employs only 14 percent of the population, compared with the 52 percent employed in agriculture. Since such a small portion of the population directly benefited from this investment, industrialization failed to significantly increase South Asia's overall prosperity.

Another problem was that the measures intended to boost employment often contributed to inefficiency. One policy encouraged industries to employ as many people as possible, even if they were not needed. So, for example, until recently, it took more than 30 Indian workers to produce the same amount of steel as 1 Japanese worker. Consequently, for years, Indian steel was not competitive in the world market. In addition, as in the former Soviet Union, decisions about which products should be produced were made by ill-informed government bureaucrats and were not driven by consumer demand. Until the 1980s, items that would improve daily life for the poor majority, such as cheap cooking pots or simple tools, were produced only in small quantities and were of inferior quality. At the same time, there was a relative abundance of large kitchen appliances and cars that only a tiny minority could afford.

Economic Reform: Globalization and Competitiveness

> **Learning Goal 4**
> **Globalization and Development:** How has globalization benefited some Indian workers in recent years?

Globalization has transformed economic development in this region, providing jobs for skilled workers in export-oriented industries such as software, manufacturing, call centers, and other types of "offshore outsourcing" enterprises. However, many low-skill workers, especially in rural areas, have not benefited from this type of development.

During the 1990s, much of South Asia began to undergo a wave of economic reforms. In many world regions, structural adjustment programs (SAPs; see pages 115–118 in Chapter 3) were mandated by the International Monetary Fund (IMF) and the World Bank; but in contrast, in India, due to an earlier financial crisis from the 1980s, the Indian government itself initiated economic reforms. Although privatization of India's public sector industries and banks has proceeded slowly, it has been arguably more successful than SAPs and similar reforms that are run by the IMF and the World Bank in other countries.

India also used the economic reforms to free its companies from a maze of regulations, which enabled both foreign and Indian companies to invest heavily in manufacturing and other industries (Figure 8.17). Drawn by India's large and cheap workforce, its excellent educational infrastructure (for the middle and upper classes), and especially by large, pent-up domestic demand for manufactured goods of all sorts, many companies are now setting up manufacturing headquarters in major Indian cities. For example, in addition to Indian auto companies (such as Tata Motors), nearly every major global automobile company is currently establishing significant manufacturing facilities somewhere in India—producing everything from economical first cars for Indian families to luxury brands such as Mercedes-Benz for Indian elites. So many global manufacturers have flocked to India in recent years that the country is challenging China as an exporter of manufactured goods. As the current recession subsides, many of India's urban poor will see their incomes rise as manufacturing jobs increase, and then they, too, will increase their consumption of Indian-made products. Since 2006, GDP per capita (PPP) has risen in all Indian states (Figure 8.17) and the average (PPP) income in India in 2009 was U.S.$2960, nearly ten times the income in 1990.

India's current manufacturing boom is benefiting from a previous boom in offshore outsourcing that started in the 1990s. **Offshore outsourcing** occurs when a company contracts to have some of its business functions performed in a country other than the one where its products or services are actually developed, manufactured, and sold. To take advantage of India's large, college-educated, and relatively low-cost workforce, companies in North America and Europe are outsourcing an increasing number of jobs to cities such as Bangalore, Mumbai, and Ahmadabad. Many types of jobs are being outsourced, including IT, engineering, telephone support,

> **offshore outsourcing** the contracting of certain business functions or production functions to providers where labor and other costs are lower

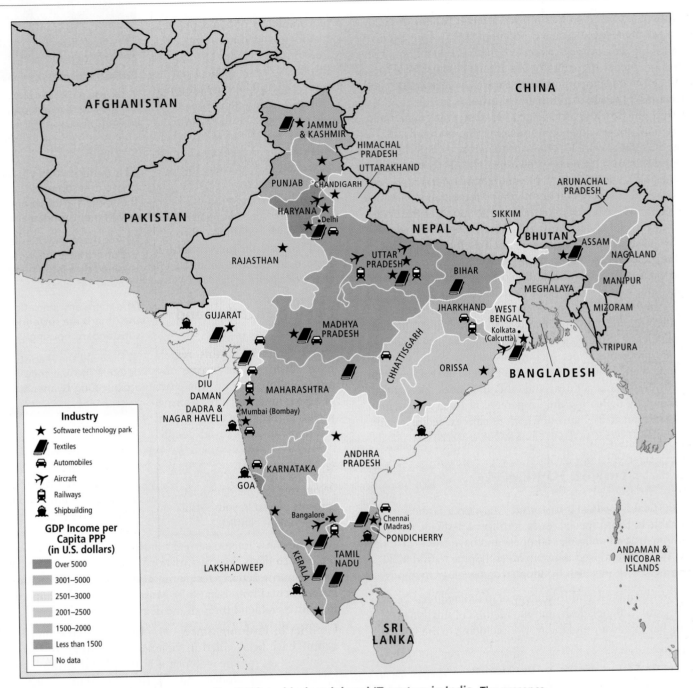

FIGURE 8.17 GDP income per capita (PPP) and industrial and IT centers in India. The presence of industry is often associated with higher incomes, and because of this, planners may try to bring industry to low-income places. In some areas, poverty may be so great that even fairly intensive industry is able to raise average incomes above base levels only slowly. (The per capita income (PPP) information is primarily from 2007–2009 data, but for five states the data is from 2005–2007.)

pharmaceutical research, and "back office" work. Many of the workers in these jobs are women.

Major global finance firms are increasingly seeking out the highly skilled workers of Mumbai's "Wall Street"—Dalal Street—to provide finance and accounting services. Stiff global competition makes cost cutting imperative for finance firms, and India's relatively low salaries provide a solution. While a junior analyst from an Ivy League school costs $150,000 a year in the United States, a graduate of a top Indian business school costs only $35,000 a year in India. Yet, for that Indian employee, this salary buys a much higher standard of living than the U.S. employee would have. In fact, well-educated Indian migrants in the United States are now returning home to take these seemingly low salaries because they can still live well and join this exciting development phase in their home country. This trend of return-migration could eventually happen across the South Asian region.

India's outsourcing growth reflects a broad pattern of steady expansion in the service sector throughout South Asia. Services now contribute well over 50 percent of the GDP in all countries of the region except Afghanistan and Bhutan. Bangladesh has a service sector that is proportionately nearly as large as India's, and it is increasingly the site of international investment.

Within the service sector, facilities that engage in trade, transportation, storage, and communication (including all facets of IT) show the most growth. Finance, insurance, real estate, business services, and tourism have also grown quickly. All these activities are connected in some way with international commerce and benefit from India's success at developing information technology (see Thematic Overview H). Yet because so much of this sector is linked to the global economy, all who are connected to it are vulnerable to downturns in demand such as those experienced during the global recession beginning in 2007.

The impact on wealth distribution of India's self-implemented economic reforms is the subject of much debate. The highly skilled and educated urban middle and upper classes have gained the most by far. India's middle class now stands at 50 million and is likely to grow dramatically in the near future. However, the new economic policies are producing wider income disparities. While many of India's more than 300 million urban poor people are gaining from recent growth in manufacturing and other industries, such growth is only indirectly benefiting the 72 percent of India's population still living in rural areas, many of whom remain exceedingly poor.

❚❚▶ 194. BHUTAN STRIVES TO DEVELOP "GROSS NATIONAL HAPPINESS"

THINGS TO REMEMBER

1. **Learning Goal 3: Food and Urbanization** Many poor farmers in India, unable to afford special seeds, fertilizers, pesticides, and new equipment, can no longer compete and have had to give up farming; some have moved to urban areas, hoping to find jobs. Other farmers have successfully adopted agroecology methods.

2. The Grameen Bank and its strategy of microcredit has been very successful in Bangladesh, and the method of small-loan financing has spread around the world, including to the United States.

3. **Learning Goal 4: Globalization and Development** To take advantage of India's large, college-educated, low-cost workforces, companies in North America and Europe are outsourcing an increasing number of jobs to Indian cities like Bangalore, Mumbai, and Ahmadabad. India's current manufacturing boom is benefiting from the offshore outsourcing that started in the 1990s. However, compared to urban areas, rural India is not benefiting nearly as much from these changes.

Political Issues

Since independence in 1947, many South Asian countries have used democracy to peacefully resolve conflicts, smooth potentially bloody transfers of power, and nurture vibrant public debate over the issues of the day. However, in many countries, the supporters of completely opposed political ideologies—communism and free market capitalism, for instance—have often tried to resolve their differences through violence. A potential path out of these conflicts is being charted in parts of India, where these same groups often compete peacefully in democratic elections (see Thematic Overview I). Nevertheless, every South Asian country has missed opportunities to resolve conflicts through democratic means, resorting instead to the use of force on many occasions.

Democratization and Conflict

Many conflicts in this region have been made worse by an unwillingness on the part of governments and warring parties to recognize the results of elections, or even to let people

> **Goal 5**
> **Democratization and Conflict:** What role has the democratic process played in South Asia's many violent conflicts?

vote. Meanwhile, some conflicts have been defused, at least in the short run, by holding elections and letting former combatants run for office.

The most intense armed conflicts in South Asia today are **regional conflicts**, in which nations dispute territorial boundaries or a minority actively resists the authority of a national or state government (Photo Essay 8.5). Most represent failed opportunities to resolve problems through the democratic processes.

Conflict in Kashmir Since 1947, between 60,000 and 100,000 people have been killed in violence in Kashmir. At the root of the violence is a struggle for territory between India and Pakistan, neither of which are willing to let the people of Kashmir resolve the dispute democratically (see the map in Photo Essay 8.5).

> **regional conflict** especially in South Asia, a conflict created by the resistance of a regional ethnic or religious minority to the authority of a national or state government

Kashmir has long been a Muslim-dominated area, and in 1947, some Kashmiris believed that it should be turned over to Pakistan. Although the maharaja (king) of Kashmir at the time, a Hindu, wanted Kashmir to remain independent, the most popular Kashmiri political leader and significant portions of the populace favored joining India. They preferred India's stated ideals of secular, nonreligiously based government to Pakistan's less robust safeguards for secularism. When Pakistan-sponsored raiders invaded western Kashmir in 1947, the maharaja quickly agreed to join India. A brief war between Pakistan and India resulted in a cease-fire line that became a tenuous boundary.

Thinking Geographically

After you have read about democratization and conflict in South Asia, you will be able to answer the following questions:

A Is there anything in this photo that suggests that the soldiers are part of a government-funded army?

B Are there any clues in the photo that this woman is joined by a large number of other refugees?

C Is there any indication that the Biharis are suffering from substandard housing?

D How did the Maoists describe the war they waged against King Gyanendra?

Most armed conflicts in South Asia have been sparked by breakdowns in the democratic process and by authoritarian behavior by governments. Supporters of political opposition groups have faced disenfranchisement, imprisonment, and sometimes execution. However, in some cases, democratization has paved the way for peaceful reconciliation between former combatants.

A Afghan troops, trained and supplied by the U.S., fire on Taliban positions near Kabul. The U.S.-led war in Afghanistan is largely a response to the September 11th attacks of 2001, although another goal of the war is to support democratization in Afghanistan.

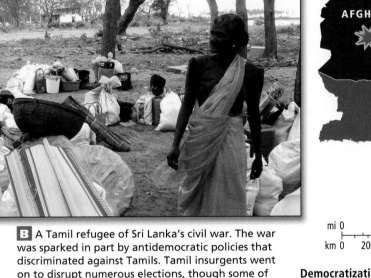

B A Tamil refugee of Sri Lanka's civil war. The war was sparked in part by antidemocratic policies that discriminated against Tamils. Tamil insurgents went on to disrupt numerous elections, though some of their supporters have now become peaceful elected politicians.

C In Dhaka, Bihari refugees of Bangladesh's 1971 war of independence from Pakistan look down from a building they occupy. The war was sparked by the refusal of West Pakistan (now Pakistan) to recognize elections that gave East Pakistan (now Bangladesh) control of the government. The Bihari community (some 600,000 people) sided with West Pakistan during the war and have since been denied citizenship in Bangladesh.

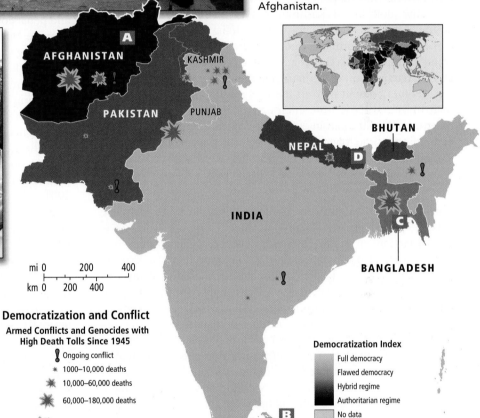

mi 0   200   400

km 0   200   400

Democratization and Conflict

Armed Conflicts and Genocides with High Death Tolls Since 1945

- Ongoing conflict
- * 1000–10,000 deaths
- 10,000–60,000 deaths
- 60,000–180,000 deaths
- 180,000–500,000 deaths
- 500,000–1,000,000 deaths

Democratization Index
- Full democracy
- Flawed democracy
- Hybrid regime
- Authoritarian regime
- No data

D A Nepali rupee depicting the former monarch, King Gyanendra, whose removal from power paved the way for democratization and the end of Nepal's civil war. Since then, warring parties, such as the Maoists, have become peaceful political parties.

321

Pakistan attempted another invasion of Kashmir in 1965, but was defeated. The two countries are technically still waiting for a UN decision about the final location of the border. In the meantime, Pakistan effectively controls the thinly populated mountain areas north and west of the densely populated valley known as the Vale of Kashmir. India holds nearly all the rest of the territory, where it maintains more than 500,000 troops. The Ladakh region of Kashmir (see the map in Figure 8.3 on pages 290–291) is the object of a more limited border dispute between India and China.

After years of military occupation, most Kashmiris now support independence from both India and Pakistan. However, neither country is willing to hold a vote on the matter. Much of the conflict in Kashmir has centered on the right of people to run their own affairs through local democratic processes. Both India and Pakistan's national leaders have often appointed their own favorites in an attempt to maintain tight control over their respective portions of Kashmir. Anti-Indian Kashmiri guerrilla groups equipped with weapons and training from Pakistan have carried out many bombings and assassinations of these appointees. Blunt counterattacks launched by the Indian government have killed large numbers of civilians and alienated many Kashmiris.

Another complication in the Kashmir dispute is the fact that both India and Pakistan—which came close to war against each other in 1999 and again in 2002—have nuclear weapons. Because of the nationalistic fervor of the protagonists, many see the conflict in Kashmir as more likely to result in the use of nuclear weapons than any other conflict in the world. Analysts agree that any use at all of nuclear weapons would have severe repercussions for all on earth.

War and Reconstruction in Afghanistan In the 1970s, political debate in Afghanistan became polarized. On one side were several factions of urban elites, who favored modernization and varying types of democratic reforms. Opposing them were rural conservative religious leaders, whose positions as landholders and ethnic leaders were threatened by the proposed reforms. Divisions intensified as successive governments, all of which came to power through military coups, became more and more authoritarian. Political opponents were imprisoned, tortured, and killed by the thousands, resulting in a growing insurgency outside the major cities. Some of the more authoritarian urban elites, inclined toward violence, allied themselves with the Soviet Union, which supported some of their goals and promised aid.

In 1979, fearing that a civil war in Afghanistan would destabilize neighboring Soviet republics in Central Asia, the Soviet Union invaded Afghanistan. Rural conservative leaders (often erroneously called "warlords") and their followers formed an anti-Soviet resistance group, the *mujahedeen*. As resistance to Soviet domination increased, the mujahedeen became ever more strongly influenced by militant Islamist thought and by Persian Gulf Arab activists who provided funding and arms. The United States, via its regional ally of Pakistan, supported the mujahedeen with billions of dollars for equipment and weapons. The United States ignored the increasingly militant Islamism of the mujahedeen, considering the mujahedeen an anti-Soviet movement and therefore a valuable Cold War ally. Moderate, educated Afghans who favored democratic reforms fled the country during this turbulent time, hoping to go back eventually when peace returned.

The mujahedeen proved to be tenacious fighters, and in 1989, the Soviets, after heavy losses—14,000 Soviet soldiers killed and

billions of dollars wasted—gave up and left Afghanistan. Anarchy prevailed for a time in Afghanistan as mujahedeen factions fought one another, adding to the 1.5 million civilians and combatants killed in the war with the Soviets.

In the early 1990s, the radical religious-political-military movement called the *Taliban* emerged from among the mujahedeen. For the most part, the Taliban were (and remain) illiterate young men from remote villages, led by students from the *talibs* (Islamist schools of philosophy and law). The Taliban wanted to control corruption and crime and minimize Western ways introduced in earlier decades by the urban elites and reinforced or made more extreme by the Russian occupation. Many of the most disliked Western ways related to the role, status, and dress of women. The behavior of Russian women accompanying the Russian army was seen as especially licentious by rural Afghans. The Taliban also wanted to strictly enforce shari'a, the Islamic social and penal code (see page 224 in Chapter 6). Efforts by the Taliban to purge Afghan society of non-Muslim influences included greatly restricting women (see page 311), promoting only fundamentalist Islamic education, and publicly banning the production of opium, to which many Afghan men had become addicted, while privately promoting its sale to raise funds for their side. By 2001, the Taliban controlled 95 percent of the country, including the capital, Kabul.

❚❚▶ 198. TALIBAN INSURGENCY FUELED BY POPPY CULTIVATION

The events of September 11, 2001, focused the United States and its allies on removing the Taliban, who were giving shelter to Osama bin Laden and his international Al Qaeda network. By late 2001, the Taliban were overpowered by an alliance of Afghans, supported heavily by the United States and the United Kingdom and, eventually, NATO (see Photo Essay 8.5A and map). The United Nations stepped in to help establish an interim coalition government. A national assembly was convened to designate a new national government and appoint a head of state in 2002, as well as to ratify a new constitution in 2004.

In 2003, the United States launched the war in Iraq that diverted national attention, troops, and financial resources away from Afghanistan. Conditions in Afghanistan deteriorated to desperate levels during a time when humane strategies might have won the support of the Afghan people. Almost immediately the Taliban were back again, effectively thwarting the ability of Afghanistan's new government to ensure security and to meet the needs of people outside Kabul. Based in rural areas in both Pakistan and Afghanistan, the Taliban are now aided by widespread distrust of the government in Kabul, which is seen as corrupt. Most Afghans see the international military forces and private security personnel stationed in Afghanistan since 2001 as hostile intruders responsible for a growing number of civilian casualties. Meanwhile, the vast majority of the people in the country remain desperately poor and uneducated despite the greatly increased attention from international donors since 2001. It appears that most people favor democratic government based on Muslim principles, but an election held in 2009 was so flawed that confidence in the Afghan government (especially President Karzai) was undermined at home and abroad.

In May of 2011, Bin Laden was killed in a raid by U.S. forces in the town of Abbottabad, Pakistan. This raid, along with another one a month later that killed the next highest Al Qaeda commander, was carried out without the knowledge or consent of Pakistani authorities, suggesting a weakening of the U.S.-Pakistani alliance in the War on Terror. With the threat from Al Qaeda seemingly diminished, calls within the United States for a faster withdrawal from Afghanistan increased (see Chapter 2 on page 76).

Sri Lanka's Civil War In Sri Lanka, violence between the majority Singhalese and minority Tamil communities has left 68,000 people dead, produced more than a million refugees, and severely impeded Sri Lanka's economic development. Though both groups coexisted peacefully for centuries, beginning in the 1970s, antidemocratic government policies that favored the Singhalese over Tamils resulted in violence.

The Singhalese have dominated Sri Lanka since their migration from Northern India several thousand years ago. Today they make up about 74 percent of Sri Lanka's population of 20.5 million. Most Singhalese are Buddhist. Tamils, a Hindu ethnic group from South India, make up about 18 percent of the total population of Sri Lanka. About half of these Tamils have been in Sri Lanka since the thirteenth century, when a Tamil Hindu kingdom was established in the northeastern part of the island. The other half were brought over by the British in the nineteenth century to work on tea, coffee, and rubber plantations. Some Tamils have done well, especially in urban areas, where they dominate the commercial sectors of the economy. However, many others have remained poor laborers isolated on rural plantations.

Upon its independence in 1948, Sri Lanka had a thriving economy, led by a vibrant agricultural sector and a government that made significant investments in health care and education. It was poised to become one of Asia's most developed economies. But Singhalese nationalism was already alienating many Tamils. Singhalese was made the only official language, and Tamil plantation workers were denied the right to vote. Efforts were also made to deport hundreds of thousands of Tamils to India. Protests against these moves were brutally repressed by the government. Conditions in rural areas took a drastic turn for the worse in the 1960s when global prices for the country's chief agricultural exports declined. In response, the government shifted investment away from agricultural development and toward urban manufacturing and textile industries, which were dominated by Singhalese.

In 1983, the Tamil minority, lacking the political power and influence to demand attention to their grievances, chose guerilla warfare against the Singhalese. A variety of Tamil groups united behind a guerilla army known as the Tamil Tigers.

For over 30 years the entire island was subjected to repeated terrorist bombings and kidnappings (see Photo Essay 8.5B), though the main battleground was in the north where the Tamils were hoping to establish an independent state. Peace agreements were attempted several times, but in the end it was an overwhelming military victory by the government, combined with an effective crackdown on international funding for the Tamils, that forced the Tamil surrender in May of 2009.

Despite long years of violence that severely curtailed the tourist industry, economic growth in other sectors has been surprisingly robust in Sri Lanka. Driven by strong growth in food processing, textiles, and garment making, Sri Lanka is today one of the wealthiest nations in South Asia on a per capita GDP basis, and provides best for the human well-being of its citizens.

Nepal's Rebels In Nepal, political turmoil has followed a civil war and associated protest movement that brought an end to generations of monarchy. An elected legislature and multiparty democracy were introduced into Nepal in 1990, but until 2008, Nepal was governed by a royal family that paid only superficial respect to these changes.

In 1996, Maoist revolutionaries inspired by the ideals of the late Chinese leader Mao Zedong (but with no apparent support from China) waged a "people's war" against the Nepalese monarchy. After a decade of civil war, during which 13,000 Nepalese died, the Maoists had both military control of much of the countryside and strong political support from most Nepalese, who objected to the dictatorial rule of the latest monarch, King Gyanendra. Massive protests forced Gyanendra to step down in 2006, and soon thereafter the Maoists declared a ceasefire with the government (see Photo Essay 8.5D). In 2008, the Maoists won sweeping electoral victories that gave them a majority in parliament and made their former rebel leader prime minister. Then, in May of 2009, when his many conditions for reforming Nepalese society remained unmet, the Maoist prime minister resigned and took his party into opposition against a new prime minister and his weak 22-party coalition. Recently, the Maoist opposition agreed to participate in writing a new constitution. Observers concur that if the various factions can see that democratic processes will give them a voice, a return to war will be unlikely.

II ▶ 200. FUTURE OF NEPAL'S KING GYANENDRA IN QUESTION

Religious Nationalism

Increasingly, people frustrated by government inefficiency, corruption, caste politics, and the failure of governments to deliver on their promises of broad-based prosperity are joining religious nationalist movements. **Religious nationalism** is the association of a particular religion with a political unit, be it a neighborhood, a city, or an entire country. Political control over a given territory is often the ultimate goal of such movements.

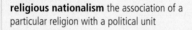

religious nationalism the association of a particular religion with a political unit

Although both India and Pakistan were formally created as secular states, religious nationalism has long been a reality, shaping relations between people and their governments in those countries. Rather than embracing the idea of multiculturalism, India is increasingly thought of as a Hindu state, while Pakistan calls itself an Islamic Republic and Bangladesh a People's Republic. In each country, many people in the dominant religious group strongly associate their religion with their national identity.

Hindu nationalism in India (sometimes called Hindutva) is supported predominantly by urban men from middle- and upper-caste groups. Its proponents fear the erosion of their castes' political influence and particularly resent the extension of the quota system for government jobs and seats in universities to lower-caste groups (see pages 306–307). Meanwhile, politically mobilized lower castes are no longer willing to follow the dictates of the dominant castes.

Political parties based on religious nationalism have gained popularity throughout South Asia. Although their members think of these parties, such as the Bharatiya Janata Parishad (BJP) in India, as forces that will purge their country of corruption and violence, they are usually no less corrupt or violent than other parties.

Terrorist Attacks in Mumbai in 2008 A 3-day terrorist attack in Mumbai in November 2008 left 160 dead in luxury hotels, a Jewish center, and a railroad station. To link this attack directly with religious nationalism as it has played out since 1947 would be incorrect, however. The attack appears, on the one hand, to be related to Islamic militancy centered in Afghanistan and Pakistan. But on the other hand, investigations have revealed that the terrorists were impoverished young men from Punjab recruited by Lashkar-e-Taiba, a Pakistan-based militant group, with origins based in the Kashmir dispute that is suspected of links with Al Qaeda. There is little doubt, however, that unchecked religious nationalism in the region helps create the antagonistic context for such violence.

THINGS TO REMEMBER

1. **Learning Goal 5: Democratization** Many conflicts have been worsened by an unwillingness on the part of governments and warring parties to recognize the results of elections, or even to allow people to vote. Some conflicts have been diffused, at least for the short term, by incorporating former combatants into the democratic process.

2. Most armed conflicts in South Asia were sparked by breakdowns in the democratic process and authoritarian behavior by governments. Supporters of political opposition groups have faced disenfranchisement, imprisonment, and sometimes execution. However, in some cases, democratization has paved the way for peaceful reconciliation between former combatants.

3. Increasingly, people frustrated by government inefficiency, corruption, and caste politics, and by the failure of governments to deliver on their promises of broad-based prosperity are joining religious nationalist movements.

Reflections on South Asia

Despite the number, scale, and complexity of problems facing South Asia, optimism regarding the future is not unwarranted. While this region's enormous population is straining resources, population growth is also slowing as women become educated and economically empowered. Changes in agriculture are addressing food shortages and low productivity, but are also creating environmental problems and forcing more rural people into the cities. At the same time, urbanization increases access to education, jobs, and health care, and these changes help slow population growth overall.

South Asian innovations such as microcredit have brought millions out of extreme poverty, and a recent boom in manufacturing promises to transform the incomes of India's vast population of urban poor (Reasons for Optimism A). Globalization has already boosted the ranks of India's middle class through jobs outsourced to India from abroad and through a flood of investment from South Asian emigrants that promises to go toward job creation. Meanwhile, some migrants are themselves returning to participate in development as entrepreneurs and experts.

South Asia's long history of democratic politics is threatened by rampant corruption, religious nationalism, terrorism, and communal conflict, but in general, democracy seems to be expanding (Reasons for Optimism B). Following years of civil war and authoritarian rule, both Afghanistan and Nepal are now nominal democracies, although political stability in both countries is a distant dream. Pakistan is experiencing an alarming increase in Islamic extremist activity and a collapse of civil society in several parts of the country, but it has at least passed from military dictatorship to elected governments. The tiny Himalayan kingdom

Reasons for Optimism in South Asia

Development: South Asian innovations such as microcredit have brought millions out of extreme poverty. **A** *Members of a women's cabinet-manufacturing cooperative in India funded by microcredit.* ▼

Thinking Geographically: Where in South Asia was microcredit first widely implemented?

Democratization: While threatened by rampant corruption, religious nationalism, terrorism, and communal conflict, democracy is expanding. **B** *A memorial to a Hindu, caste-based politician in India.* ▼

Thinking Geographically: Is there anything in this photo that suggests a religious influence on politics?

Climate Change: South Asia is pursuing many important solutions to the climate crisis, such as electric vehicles, alternative energy, and water conservation. **C** *An Indian-made Reva electric car charging in London.* ▼

Thinking Geographically: How do electric vehicles contribute to solving the problems of climate change?

of Bhutan granted its people the right to elect local representatives in 2002 and national leaders in 2008. In Bangladesh, after years of military dictatorship, democratic elections have been held with some regularity. However, government corruption is a recurring cause for public protest. In Sri Lanka, the recent end to a long-standing civil war was not the result of democratic processes, but rather of military force. Even so, democratization will be able to proceed faster with less violence. The greatest anchor of democratization in South Asia is its largest country, India, which is also the most democratized by far.

Of any world region, South Asia has the largest populations vulnerable to the effects of global climate change and the risks are only likely to increase. While the region is pursuing many important solutions to the climate crisis, such as electric vehicles, alternative energy, and water conservation, it is also on the verge of an explosion of motor vehicle ownership that will very likely overwhelm any energy savings and pollution control achievements (see Reasons for Optimism C). India's surging middle class now has the disposable income to afford Indian-produced economy cars. In this time of tumultuous change and expanding global connections, South Asia's problems are daunting but no more so than those faced by Mohandas Gandhi, the leader of India's independence movement, when he said, "We must be the change we wish to see." The emerging solutions to pressing global problems today suggest that many in the region have taken these words to heart.

Learning Goals Review

1. Climate Change and Water: How is South Asia vulnerable to water-related problems associated with climate change?

Why is the issue of melting glaciers in the Himalayas so important in this region? Compare the short- and long-term effects of this change. What additional threat do coastal areas face?

2. Gender and Population: How have issues related to gender transformed South Asian populations in recent decades?

What has created the strong preference for sons among many South Asian families? Why are sons preferred over daughters? Where in South Asia is this preference the weakest?

3. Food and Urbanization: In what ways have recent changes in South Asian food production systems contributed to urbanization?

What has been the impact on food production of the introduction of new seeds, fertilizers, pesticides, and equipment? How has this shift resulted in many rural people moving to cities? What kind of jobs and housing do these people usually find in urban areas?

4. Globalization and Development: How has globalization benefited some Indian workers in recent years?

What kinds of workers have gained from the recent boom in foreign investment? Why are they attractive to foreign investors? How are these jobs linked to the global economy? What kinds of workers are benefiting much less, or not at all, from these new foreign investments?

5. Democratization: What role has the democratic process played in South Asia's many violent conflicts?

How have many conflicts been made worse by an unwillingness on the part of governments and warring parties to allow free and fair elections to occur and be recognized? What democratic strategy has been used to diffuse some conflicts, at least for the short term?

Geographic Themes about South Asia

Look back at the Thematic Overview photos at the beginning of the chapter. Thinking geographically, answer the following questions about them:

(A) Climate Change: What does the presence of water buffalo (used as draft animals) suggest about the area being flooded?

(B) Food: How might this farmer be exposing himself to harmful chemicals?

(C) Water: What does this photo suggest about the water supply in Mumbai, India?

(D) Globalization: These troops might be a part of which military?

(E) Gender: Are there any clues in the photo that suggest seclusion?

(F) Population: How does this photo suggest that Afghanistan has a high population growth rate?

(G) Urbanization: What about the structures at the bottom of the photo suggest that very poor people live in them?

(H) Development: What is noticeable about the gender of the participants of this workshop?

(I) Democratization: Is there anything notable about the gender of the participants of this rally?

Key Terms

agroecology, 317
Buddhism, 308
caste system, 306
civil disobedience, 303
communal conflict, 309
dowry, 310
Harappa culture, 300
Hinduism, 305

Indus Valley civilization, 300
Jainism, 308
jati, 306
microcredit, 317
monsoon, 292
Mughals, 300
offshore outsourcing, 318
Partition, 303

purdah, 309
regional conflict, 320
religious nationalism, 323
Sikhism, 308
subcontinent, 289
Taliban, 311
varna, 306

9 East Asia

Learning Goals

After you read this chapter, you will be able to answer the following questions:

1. Climate Change and Water: How might climate change result in changes in water availability in East Asia?

2. Food and Globalization: How has East Asia's ability to feed itself been transformed by globalization?

3. Urbanization and Development: How is China's spectacular urban growth, and the massive increase in pollution that has accompanied it, linked to processes of globalization that were set in motion three decades ago?

4. Democratization: In what ways has China embraced changes that might eventually lead to democratization?

5. Population and Gender: How has the interaction between the government's population policy and Chinese culture resulted in a shortage of women in China?

FIGURE 9.1 Political map of East Asia.

Thematic Overview of East Asia

Climate Change: Japan is now second only to Germany in its installed solar power–generating capacity. **A** *Solar panels being installed in Japan.* ▼

Water: Nearly every year, abnormally low rainfall or abnormally high temperatures result in a drought somewhere in China. **B** *A dry field in Inner Mongolia.* ▼

Food: East Asian countries have become more able to import food from other regions but less able to produce all the food they need at home. **C** *A Japanese fishing boat off the coast of New Zealand.* ▼

Development: After three decades of rapid economic growth, China has the third-largest economy in the world, following the EU and the United States. **D** *A knife factory in Guangzhou, China.* ▼

Globalization: Since enacting economic reforms in 1980, China has become the world's largest producer of manufactured goods, supplying stores and street vendors across the globe. **E** *A Chinese boat carrying automobiles for export.* ▼

Urbanization: China's urban population has grown by 450 million people since 1980, due mainly to massive migration from poor rural areas to booming and globalized coastal cities. **F** *Shanghai, China.* ▼

Democratization: While Japan, South Korea, and Taiwan are relatively democratized, China holds only a few competitive elections, and North Korea holds none. **G** *Parliamentary election posters in Japan.* ▼

Population: Population growth has slowed in response to government policies, urbanization, and changing gender roles. **H** *A father and daughter in Japan, where single-child families are increasingly the norm.* ▼

Gender: A cultural preference for sons over daughters has produced gender imbalances throughout East Asia, with males outnumbering females substantially in many age groups. **I** *High school students in Korea.* ▼

Global Patterns, Local Lives In 2000, at age 18, Li Xia (Li is her family name) left her farming village in China's Sichuan Province for the city of Dongguan in the Guangdong Province of southern China. She was accompanied by two friends. A few months earlier, the government had taken their families' farmland for an urban real estate project, paying compensation of only U.S.$2000 per family. The three young women accepted an offer to work in a Dongguan toy factory so they could send money back to their families, who now have to pay cash for food and housing.

When the young women arrived in the city, they joined 5 million other recent internal migrants. Sixty percent, like Li Xia, were illegal migrants without government-approved residency rights, and thus were dependent on their employers for housing. As did many others, Li Xia and her friends soon found that the labor recruiters had lied about their wages. Per month they would be paid not U.S.$45, as promised, but U.S.$30. In addition, they would work 12-hour shifts in 100°F heat, receive no overtime pay, and have only one day off a month. But there was no point in protesting. Their families in Sichuan needed the money they would send home, and the recruiters purposely brought in thousands of extra workers to replace any complainers.

Xia felt better when she saw that the toy factory was a clean, modern building known locally as the "Palace of Girls" (nearly all

3500 employees were female). Within a day, she had completed her training, signed a 3-year contract, and mastered her task of putting eyes on stuffed animals destined for toddlers in the United States and Europe. Her enthusiasm faded, however, when she learned that she and her friends would be spending much of their money on the expensive but low-quality food provided by the company, and would be sharing one small room and a tiny bath with eight other women and a rat or two.

In less than 6 months, Xia broke her contract and returned home to her village in Sichuan. There, using ideas and assertiveness she had gained from her time in Dongguan, she opened a snack stand. In just 1 month, she made 10 times her investment of U.S.$12. But Xia yearned to return to the southern coast to try again for a well-paid factory job. A few months later, she returned to Dongguan with her sister, leaving the stand to her sister's husband (who also cared for the couple's baby).

Since her first arrival, the city had grown by 20 percent and now had 1400 foreign companies trying to hire thousands of workers. Through connections, Xia and her sister found jobs requiring midlevel skills in a Taiwanese-owned fiber optics factory at twice the wages Xia had earned at the toy factory. One year after her first trip to Dongguan, Xia was making a bit more than U.S.$100 a month, enough to live relatively comfortably

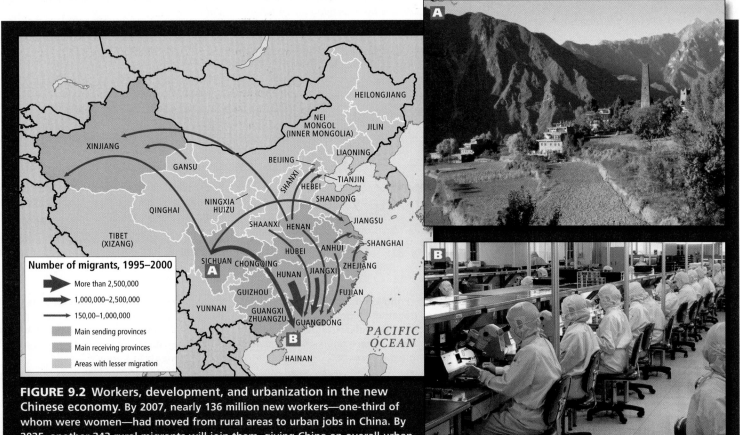

FIGURE 9.2 Workers, development, and urbanization in the new Chinese economy. By 2007, nearly 136 million new workers—one-third of whom were women—had moved from rural areas to urban jobs in China. By 2025, another 243 rural migrants will join them, giving China an overall urban population of over 1 billion. The map identifies the interior provinces from which most of the rural workers are coming and the primarily coastal areas where they find work. **(A)** A village in rural Sichuan province. **(B)** Workers in Dongguan, Guangdong Province, assemble and test fiber-optic systems.

with only three roommates. She had prospects for a raise, and she was sending money home and once again saving, this time to open a bar in her home village. *[Adapted from Peter S. Goodman, "In China's Cities, a Turn from Factories," Washington Post, September 25, 2004, at http://www.washingtonpost.com/wp-dyn/articles/A48818-2004Sep24.html; Louisa Lim, "The End of Agriculture in China," Reporter's Notebook, National Public Radio, May 19, 2006; with background information from Kathy Chen, "Boom-Town Bound," Wall Street Journal, October 29, 1996, p. A6; and "Life Lessons," Wall Street Journal (July 9, 1997)].* ∎

The experiences of Li Xia and her sister illustrate first how the needs of rural areas are being subverted to the needs of China's burgeoning cities. Developers are increasingly targeting rural land, and farmers are rarely given a fair price for their land. Meanwhile, urbanization, globalization, and changes in gender roles are transforming East Asia as millions of rural young adults flock to work in factories in East Asia's coastal cities, producing goods for sale on global markets, learning new skills, and gaining new confidence (Figure 9.2).

Despite its popularity, most rural-to-urban migration in China is illegal because the ***hukou*** (household registration) **system**, an ancient practice reinforced in the Maoist era (see page 344) and only now becoming less strict, effectively ties rural people to the place of their birth. In a desperate search for work, over 136 million people (Table 9.1) like Li Xia (nearly half of the urban workforce) are ignoring the *hukou* system; by doing so, they become part of what is called the **floating population**, a term used in China to indicate those who have no rights to housing, schools, or health care. These migrants generally work in menial, low-wage jobs and make agonizing sacrifices to send money home to children and spouses living in

> ***hukou* system** the system in China by which citizens' permanent residence is registered
>
> **floating population** the Chinese term for jobless or underemployed people who have left economically depressed rural areas for the cities and move from place to place looking for work

rural areas. Between 2000 and 2007, the number of people migrating from rural to urban areas in China in search of work increased 58 percent. This number is approaching half of the working people in China's cities.

China is part of the region of *East Asia*, home to nearly one-fourth of humanity. This vast territory stretches from the Taklimakan Desert in far western China to Japan's rainy Pacific coastline, and from the frigid mountains of Mongolia in the north to the subtropical forests of China's southeastern coastal provinces (see Figure 9.3 on pages 330–331). East Asia (see Figure 9.1 on page 326) includes the countries of China, Mongolia, North Korea, South Korea, Japan, and Taiwan (the last has operated as an independent country since World War II but is claimed by China as a province). These countries are grouped together because of their cultural and historical roots, many of which are in China. Because of China's great size, historical influence, enormous population, and huge economy, this chapter gives it particular emphasis. Japan, whose large and prosperous economy makes it a major player on the world stage, is also emphasized.

THINGS TO REMEMBER

1. The shift of workers from rural China to cities includes at least 136 million workers who have moved without legal papers and who now constitute nearly half (46.5 percent) of China's urban labor force.

2. Most rural workers who have moved to China's cities without official permission are in low-paying jobs with little or no access to social services or education.

TABLE 9.1	Migrant workers and urban employment, 2000–2007		
Year	Rural migrant workers (millions)	Urban employment (millions)	Percent of rural to urban employed
2000	78.49	212.74	36.9
2001	83.99	239.4	35.1
2002	104.7	247.8	42.3
2003	113.9	256.39	44.4
2004	118.23	264.76	44.7
2005	125.78	273.31	46.0
2006	132.12	283.1	46.7
2007	136.49	293.5	46.5

[Adapted from Cai Fang, Du Yang, and Wang Meiyan, *Human Development Research Paper 2009/09: Migration and Labor Mobility in China* (United Nations Development Programme, April 2009), p. 4, at http://hdr.undp.org/en/reports/global/hdr2009/papers/HDRP_2009_09.pdf.]

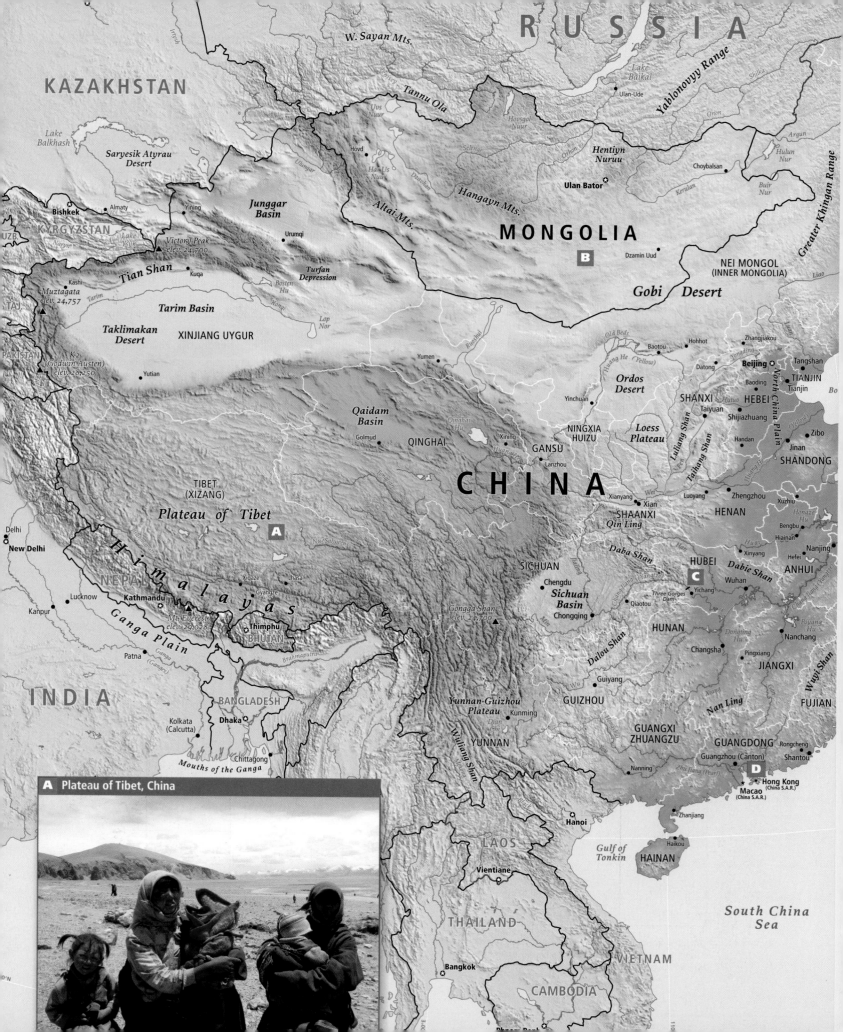

A Plateau of Tibet, China

B Mongolia

C Three Gorges Dam, Chang Jiang, Hubei Province, China

D Hong Kong Island and the South China Sea

E Mount Fugi, Japan

FIGURE 9.3 Regional map of East Asia.

I THE GEOGRAPHIC SETTING

Terms in This Chapter

East Asian place-names can be very confusing to English-speaking readers. We give place-names in English transliterations of the appropriate Asian language, taking care to avoid redundancies. For example, *he* and *jiang* are both Chinese words for *river*. Thus the Yellow River is the Huang He, and the Long River, also called the Yangtze in English, is the Chang Jiang in China; it is redundant to add the term *river* to either name. The word *shan* appears in many place-names and usually means "mountain."

Pinyin (a spelling system based on Chinese sounds) versions of Chinese place-names are now commonplace. For example, the city once called Peking in English is now Beijing, and Canton is Guangzhou.

The region popularly known as Manchuria is here referred to as China's Far Northeast to emphasize its geographical location. Although China refers to Tibet as Xizang, people around the world who support the idea of Tibetan self-government avoid using that name. This text uses Tibet for the region (with Xizang in parentheses), and Tibetans for the people who live there.

Physical Patterns

A quick look at the regional map of East Asia (see Figure 9.3) reveals that the topography here is perhaps the most rugged in the world. East Asia's varied climates result from a dynamic interaction between huge warm and cool air masses and the land and oceans. The region's rapidly expanding human populations have affected the variety of ecosystems that have evolved there over the millennia and that still contain many important and unique habitats.

Landforms

The complex topography of East Asia is partially the result of the slow-motion collision of the Indian subcontinent with the southern edge of Eurasia over the past 60 million years. This tremendous force created the Himalayas and lifted up the Plateau of Tibet (see Figure 9.3A, depicted in gray and gold in the map of Figure 9.3), which can be considered the highest of four descending steps that define the landforms of mainland East Asia, moving roughly west to east.

The second step down from the Himalayas is a broad arc of basins, plateaus, and low mountain ranges (depicted in yellowish tan in Figure 9.3). These landforms include the broad, rolling highland grasslands and deep, dry basins and deserts of western China (Taklimakan Desert, Junggar Basin) and Mongolia (see Figure 9.3B) and also include the Sichuan Basin and the rugged Yunnan–Guizhou Plateau to the south, which is dominated by a system of deeply folded mountains and valleys that bend south through the Southeast Asian peninsula.

The third step, directly east of this upland zone, consists mainly of broad coastal plains and the deltas of China's great rivers (shown in shades of green in the map in Figure 9.3). Starting

from the south, this step is defined by three large lowland river basins: the Zhu Jiang (Pearl River) basin, the massive Chang Jiang basin (see Figure 9.3C), and the lowland basin of the Huang He on the North China Plain. Each of these rivers has a large delta. Despite the deltas being subject to periodic flooding, they have long been used for agriculture; but now coastal cities have spread into these deltas, filling in wetlands. Each is now a zone of dense population and industrialization (see Figure 9.15 on page 362). Low mountains and hills (shown in light brown) separate these river basins. China's far northeast and the Korean Peninsula are also part of this third step.

The fourth step consists of the continental shelf, covered by the waters of the Yellow Sea, the East China Sea, and the South China Sea. Numerous islands—including Hong Kong, Hainan, and Taiwan—are anchored on this continental shelf; all are part of the Asian landmass (see Figure 9.3D).

The islands of Japan have a different geological origin: they are volcanic, not part of the continental shelf. They rise out of the waters of the northwestern Pacific in the highly unstable zone where the Pacific, Philippine, and Eurasian plates grind against one another. Lying along a portion of the Pacific Ring of Fire (see Figure 1.22 on page 46), the entire Japanese island chain is particularly vulnerable to disastrous eruptions, earthquakes, and **tsunamis** (seismic sea waves). The volcanic Mount Fuji, perhaps Japan's most recognizable symbol (see Figure 9.3E), last erupted in 1707. However, the mountain is still classed as active and deep internal rumblings have been detected since 2001. In March of 2011, the largest earthquake in recorded Japanese history (9.2 on the Richter scale) hit off the coast of Honshu, Japan, near the city of Sendai. The quake and subsequent tsunami killed tens of thousands of people and damaged several nuclear reactors located on the coast, resulting in the second-worst nuclear accident ever (discussed further on page 336).

The East Asian landmass has few flat portions, and most flat land is either very dry or very cold. Consequently, the large numbers of people who occupy the region have had to be particularly inventive in creating spaces for agriculture. They have cleared and terraced entire mountain ranges, until recently using only simple hand tools (see Photo Essay 9.2A on page 335). They have irrigated drylands with water from melted snow, drained wetlands using elaborate levees and dams, and applied their complex knowledge of horticulture and animal husbandry to help plants and animals flourish in difficult conditions.

Climate

East Asia has two principal contrasting climate zones (Photo Essay 9.1): the dry interior west and the wet (monsoon) east. Recall from Chapter 8 that the term *monsoon* refers to the seasonal reversal of surface winds that flow from the Eurasian continent to the surrounding oceans during winter and from the oceans inland during summer.

> **tsunami** a large sea wave caused by an earthquake

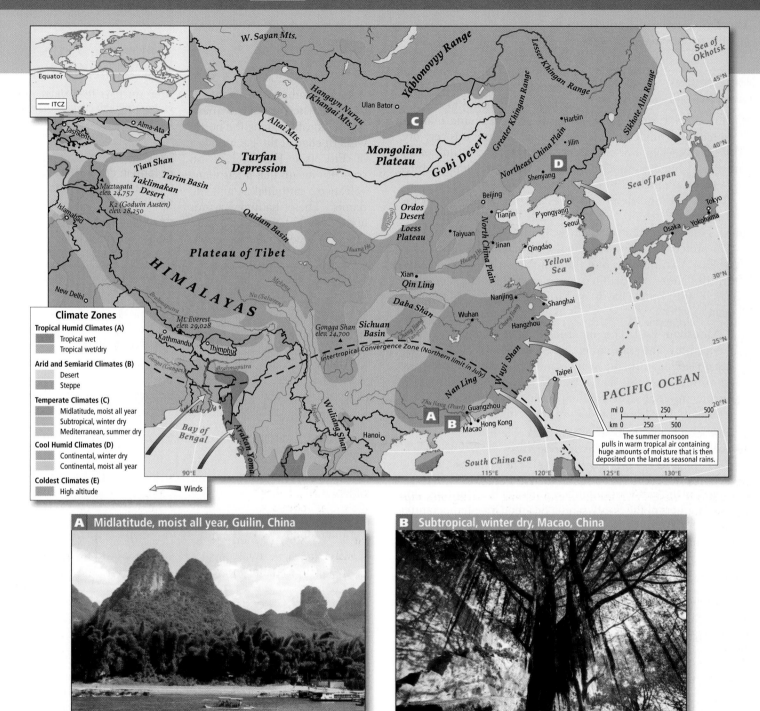

Equator

— ITCZ

W. Sayan Mts.

Yablonovyy Range

Hangayn Nuruu (Khangai Mts.)

Ulan Bator

C

Altai Mts.

Mongolian Plateau

Gobi Desert

Greater Khingan Range

Lesser Khingan Range

Sikhote Alin Range

Sea of Okhotsk

Harbin

Jilin

Northeast China Plain

D

Shenyang

Sea of Japan

45°N

40°N

Turfan Depression

Tian Shan

Taklimakan Desert

Tarim Basin

Muztagata elev. 24,757

K2 (Godwin Austen) elev. 28,250

Islamabad

Jashkent

Alma-Ata

Qaidam Basin

Ordos Desert

Loess Plateau

Huang He (Yellow)

Beijing

Tianjin

P'yongyang

Seoul

Taiyuan

Jinan

Qingdao

North China Plain

Yellow Sea

Tokyo

Osaka

Yokohama

Plateau of Tibet

New Delhi

HIMALAYAS

Mt. Everest elev. 29,028

Kathmandu

Thimphu

Brahmaputra

Nu (Salween)

Mekong

Gongga Shan elev. 24,700

Sichuan Basin

Xian

Qin Ling

Daba Shan

Wuhan

Chang Jiang

Nanjing

Shanghai

Hangzhou

30°N

Intertropical Convergence Zone (Northern limit in July)

Ganges (Ganga)

Brahmaputra

Bay of Bengal

Arakan Yoma

Wuliang Shan

Mekong

Hanoi

Nan Ling

Wuyi Shan

Taipei

PACIFIC OCEAN

25°N

20°N

Zhu Jiang (Pearl)

Guangzhou

A

B

Macao

Hong Kong

South China Sea

mi 0   250   500

km 0   250   500

The summer monsoon pulls in warm tropical air containing huge amounts of moisture that is then deposited on the land as seasonal rains.

90°E   115°E   120°E   125°E   130°E

Climate Zones

Tropical Humid Climates (A)
- Tropical wet
- Tropical wet/dry

Arid and Semiarid Climates (B)
- Desert
- Steppe

Temperate Climates (C)
- Midlatitude, moist all year
- Subtropical, winter dry
- Mediterranean, summer dry

Cool Humid Climates (D)
- Continental, winter dry
- Continental, moist all year

Coldest Climates (E)
- High altitude

← Winds

A Midlatitude, moist all year, Guilin, China

B Subtropical, winter dry, Macao, China

C Steppe, Mongolia

D Continental, winter dry, Liaoning, China

The Dry Interior Because land heats up and cools off more rapidly than water, the interiors of large landmasses in the midlatitudes tend to be intensely cold in winter and extremely hot in summer. Western East Asia, roughly corresponding to the first two topographic steps described above, is an extreme example of such a midlatitude continental climate because it is very dry. With little vegetation or cloud cover to retain the warmth of the sun after nightfall, summer daytime and nighttime temperatures may vary by as much as 100°F (55°C).

> **typhoon** a tropical cyclone or hurricane

Grasslands and deserts of several types cover most of the land in this dry region (see Photo Essay 9.1C). Only scattered forests grow on the few relatively well-watered mountain slopes and in protected valleys supplied with water by snowmelt. In all of East Asia, humans and their impacts are least conspicuous in the large, uninhabited portions of the deserts of Tibet (Xizang), the Tarim Basin in Xinjiang, and the Mongolian Plateau.

The Monsoon East The monsoon climates of the east are influenced by the extremely cold conditions of the huge Eurasian landmass in the winter and the warm temperatures of the surrounding seas and oceans in the summer. During the dry winter monsoon, descending frigid air sweeps south and east through East Asia, producing long, bitter winters on the Mongolian Plateau, on the North China Plain, and in China's Far Northeast (see Photo Essay 9.1D). While occasional freezes may reach as far as southern China, winters there are shorter and less severe. The cold air of the dry winter monsoon is partially deflected by the east-west mountain ranges of the Qin Ling, and the warm waters of the South China Sea moderate temperatures on land.

During the summer monsoon, as the continent warms, the air above it rises, pulling in wet tropical air from the adjacent seas. The warm, wet air from the ocean deposits moisture on the land as seasonal rains. As the summer monsoon moves northwest, it must cross numerous mountain ranges and displace cooler air. Consequently, its effect is weakened toward the northwest. Thus the Zhu Jiang basin in the far southeast is drenched with rain and enjoys warm weather for most of the year (see Photo Essay 9.1A), whereas the Chang Jiang basin, which lies in central China to the north of the Nan Ling range, receives only about 5 months of summer monsoon weather. The North China Plain, north of the Qin Ling and Dabie Shan ranges, receives only about 3 months of monsoon rain. Very little monsoon rain reaches the dry interior.

Korea and Japan have wet climates year-round, similar to those found along the Atlantic coast of the United States, because of their proximity to the sea. They still have hot summers and cold winters because of their northerly location and exposure to the continental effects of the huge Eurasian landmass. Japan and Taiwan actually receive monsoon rains twice: once in spring, when the main monsoon moves toward the land, and again in autumn, as the winter monsoon forces warm air off the continent. This retreating warm air picks up moisture over the coastal seas, which is then deposited on the islands. Much of Japan's autumn precipitation falls as snow.

Natural Hazards The entire coastal zone of East Asia is intermittently subject to **typhoons** (tropical storms, hurricanes). Japan's location along the northwestern edge of the Pacific Ring of Fire (see Figure 1.22 on page 46) means that it has many volcanoes, earthquakes, and tsunamis. These natural hazards are a constant threat in Japan; the heavily populated zone from Tokyo southwest through the Inland Sea (between Shikoku and southern Honshu) is particularly endangered. Earthquakes are also a serious natural hazard in Taiwan and in China's mountainous interior.

Thinking Geographically

After you have read about vulnerability to climate change in East Asia, you will be able to answer the following questions:

A When people begin to live or farm in dry environments, what is one process that can result?

B What is causing the water table to fall each year on the North China Plain?

C Flooding is especially likely in what part of China?

D What causes of desertification are visible in this photograph?

THINGS TO REMEMBER

1. There are four main topographical zones, or "steps," that form the East Asian continent.

2. Japan was created by volcanic activity along the Pacific Ring of Fire.

3. East Asia has two principal contrasting climates: the dry continental interior (west) and the monsoon east.

4. East Asia faces a wide range of natural hazards, including earthquakes, tsunamis, volcanic eruptions, and tropical storms (typhoons).

Environmental Issues

Today, East Asia's worst environmental problems result from high population density combined with rapid urbanization and environmentally insensitive economic development (Photo Essay 9.2).

Climate Change: Emissions and Vulnerability in East Asia

> **Learning Goal 1**
> **Climate Change and Water:** How might climate change result in changes in water availability in East Asia?

Climate change, especially global warming, is a growing concern in East Asia. China now leads the United States as the world's largest overall producer of greenhouse gases (but not on a per capita basis; see page 22 in Chapter 1), and Japan is also a major emitter. China has recently given attention

China is particularly vulnerable to drought, desertification, flooding, and other hazards that climate change may intensify.

A The terraced hillsides of China's eastern Gansu Province are in a dry upland zone, where agriculture depends on rainfall or irrigation water taken from rivers or underground aquifers. Rainfall is already unpredictable and could become more so with climate change. Melting glaciers on the Tibetan Plateau threaten to reduce river flows, and overuse of groundwater for irrigation is depleting aquifers. Large poor, rural populations are increasingly being forced to relocate to other parts of China.

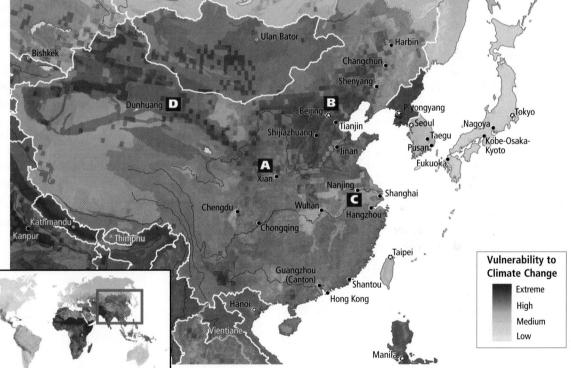

Vulnerability to Climate Change

Extreme
High
Medium
Low

B A resident of Beijing takes water from a canal. Beijing faces severe water shortages, and projected major population increases are prompting the government to consider massive canal projects to divert water from distant rivers in southern and western China.

C Flooding hits the Chang Jiang at Nanjing. Climate change could result in changes in rainfall patterns that would increase flooding. This is especially likely in eastern China, where large amounts of rain fall during the summer monsoon.

D Sand dunes loom over Dunhuang, China, where ongoing desertification could intensify if climate change brings more droughts. Past attempts to farm dry areas in western China have led to desertification, especially when droughts are accompanied by strong winds that blow sand in from nearby deserts.

to the need to limit its emissions, perhaps out of recognition of its responsibility as an increasingly influential global economic power. However, even with large increases in efficiency through new technologies, China's greenhouse gas emissions will double by 2030 as its urban households use more cars, air conditioning, appliances, and computers.

China's Vulnerability to Glacial Melting Glaciers on the Tibetan Plateau capture monsoon moisture and store it frozen for slow release. Two of China's largest rivers, the Huang He and the Chang Jiang, are partially fed by these glaciers, which are now melting so rapidly that scientists predict they will eventually disappear (see Photo Essay 9.2A). One result could be significantly lower flows in these two great rivers during the winter when little rain falls. Both have already begun to run low during winter, and trade is suffering as river boats and barges become stranded on sandbars. Irrigation for dry-season farming could also suffer.

Water Shortages Nearly every year, abnormally low rainfall or abnormally high temperatures result in a drought somewhere in China (see Thematic Overview B on page 327). These droughts often cause more suffering and damage than any other natural hazard.

Droughts can be worsened by human activity on a local or regional scale. When people begin to live or farm in dry environments, as many millions have done in China and Mongolia during the twentieth century, *desertification* (see pages 218–220 in Chapter 6) can result. Natural vegetation is cleared to grow crops, which have higher water requirements than the natural vegetation, so the crops must be irrigated with water pumped to the surface from underground aquifers. Many dry areas in China are subject to strong winds that can blow away topsoil once natural vegetation is removed. Further, irrigated crops are often less able to hold the soil than the natural vegetation. As a result, huge dust storms have plagued China in recent years. High dunes of dirt and sand have appeared almost overnight in some areas that border the desert, threatening crops, roads, and homes (see Photo Essay 9.2D). Particulate matter from these dust storms circles the globe in the upper atmosphere.

Vignette In Ningxia Huizu Autonomous Region on the Loess Plateau in northwest China, Wang Youde squints out at what are now sand-colored low hills barren of vegetation. He explains how these vast tracts of former farmland have been transformed into deserts by a combination of human error and climate change.

The area was already prone to drought and wind (*loess* means "wind-deposited soil"), and then agricultural expansion into this fragile, wind-deposited soil led to the removal of thick, natural, deep-rooted grasses. One can still see the agricultural terraces on the arid slopes where now not even grass grows. Wang Youde's family and 30,000 others fled the area when Wang was 10 years old because one day a sand dune covered their village. From his early childhood, he remembers flowers, bird song, and occasional snowfalls; all have vanished. Now Wang is back, heading up a project to revegetate thousands of hectares with drought-resistant plants that will hold the soil. By hand, squares of braided straw or stones are laid down to keep water from running off the land and to protect planted seedlings. The seedling survival rate is only 20 to

30 percent. Yet, Wang says, "Every time we see an oasis that we have created we are very satisfied . . . [b]ecause we have poured sweat and blood into our work." His adult children are helping him, hoping to remain with the family and escape the hardships of migrating to find urban factory work. [Adapted from Maria Siow, "Desertification: One of the Challenges Faced by China," Asia Pacific News, October 29, 2009, at http://www.channelnewsasia.com/stories/eastasia/view/1014326/1/.html.] ∎

Water shortages are particularly intense in the North China Plain, which produces half of China's wheat and a third of its corn. Here the water table is falling more than 10 feet a year because of increased use of groundwater for irrigation and urban needs. Meanwhile, withdrawals of water from the Huang He often make its lower sections run completely dry during the winter and spring.

Japan, the Koreas, and Taiwan have monsoon rainfall patterns, which give them a generally wetter climate and make them less vulnerable to drought than China and Mongolia.

Flooding in Central China The same shifting patterns of rainfall that may worsen droughts can also worsen flooding. Under usual conditions, the huge amounts of rain deposited on eastern China during the summer monsoon periodically can cause catastrophic floods along the major rivers. If global warming leads to even slight changes in rainfall patterns, flooding could be much more severe (see Photo Essay 9.2C). Engineers have constructed elaborate systems of dikes, dams, reservoirs, and artificial lakes to help control flooding. However, these systems failed in 2004, when heavy rains in the Chang Jiang basin caused some of the worst flooding in two centuries. Two hundred and forty million people were affected, with 3656 drowned and 14 million left homeless.

Responses to the Climate Crisis East Asia is increasingly responding to the warming aspect of global climate change. Japan has led such efforts for decades. In 1997, the Japanese city of Kyoto hosted the meeting in which countries first committed to reduce their greenhouse gas emissions. Japanese automakers such as Toyota were among the first to develop and sell hybrid gas-electric vehicles, and Japan is now second only to Germany in its installed solar power–generating capacity (see Thematic Overview A). However, Japan has made only small actual reductions in its greenhouse gas emissions.

Until the Sendai earthquake and tsunami of 2011, Japan, along with many other countries, planned to increase the use of nuclear power as a way of reducing greenhouse gas emissions. However, the severity of the nuclear disaster that followed in the wake of the tsunami halted or curtailed many plans for expanding nuclear power across the globe. The accident forced the evacuation of over 200,000 people; temporarily contaminated the water supply of Tokyo; contaminated the ground, food, and livestock for miles around the reactor; released thousands of gallons of water containing 10,000 times the normal level of iodine 131, a radioactive substance, into the Pacific Ocean; and sent enough radiation into the atmosphere that scientists detected measurable radiation increases in the United States and Europe. The accident is expected to boost reliance on both renewable resources and coal in the near future.

With regard to its water-related vulnerabilities, China is focusing on water conservation to stretch existing supplies. Already, 30 percent of China's urban water is recycled, and many cities are trying to raise this percentage. A major new national effort to remove pollutants from wastewater discharged by industry and farming is indicative of official support for these conservationist policies.

Food Security and Sustainability in East Asia

Learning Goal 2
Food and Globalization: How has East Asia's ability to feed itself been transformed by globalization?

> **food security** the ability of a state to consistently supply a sufficient amount of basic food to its entire population

East Asia's **food security**—the capacity of the people in a geographic area to consistently provide themselves with adequate food—is increasingly linked to the global economy. The wealthiest countries in the region are already very dependent on food imports from around the globe. About three-quarters of the food consumed in Japan, South Korea, and Taiwan is imported. In China, on the other hand, self-sufficiency in grain production is important to national identity because devastating famines were a recurring problem before and during the Communist Revolution. China is now nearly self-sufficient with regard to basic necessities, but it relies on imports of commodities to supply growing demands for luxury food items, including grain and soybeans for animal feed (see below).

From one perspective, high levels of imported food are a sign of increasing food security in East Asia, because countries with competitive globalized economies can afford to bring in food from elsewhere. However, recent dramatic increases in global food prices illustrate the perils of dependence on food imports, especially for the poor. In 2007, for example, the world market price for corn shot up because of a variety of proposals, among them proposals to use corn to produce ethanol fuel as a replacement for gasoline in the Americas and elsewhere.

▐▌▶ 217. FOOD PRICES SKYROCKETING IN CHINA

Food Production Only a relatively small portion of East Asia's vast territory can support agriculture (**Figure 9.4**). In much of this area, food production has been pushed well beyond what can be sustained over the long term. As a result, East Asia's fertile zones are shrinking. In China, roughly one-fifth of the agricultural land has been lost since the Communist Revolution of 1949, largely because of urban and industrial expansion and agricultural mismanagement that created soil erosion and desertification (see the vignette on page 335 and Photo Essay 9.2D).

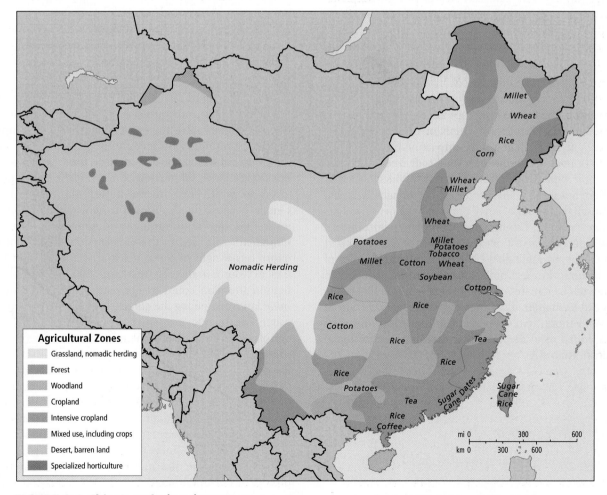

FIGURE 9.4 China's agricultural zones. As part of the economic reforms instituted in recent years in China, there is more regional specialization in agricultural products.

As urban populations grow more affluent, they consume more meat and other animal products that require more land and resources than did the plant-based national diet of the past. Soybeans, which used to be important as an export for China, are one example of this change in diet. More than 70 percent of the soybeans used in China are now imported, primarily from the United States, Brazil, and Argentina. Soybeans are no longer used mainly for human food but for animal feed, for farmed fish food, and especially for high-grade cooking oil.

Rice Cultivation The region's most important grain is rice, and over the millennia its cultivation has transformed landscapes throughout central and southern China, Japan, Korea, and Taiwan. In these areas, rainfall is sufficient to sustain **wet rice cultivation**, which can be highly productive.

> **wet rice cultivation** a prolific type of rice production that requires the submersion of the plant roots in water for part of the growing season

Wet rice cultivation requires elaborate systems of water management, as the roots of the plants must be submerged in water early in the growing season. Centuries of painstaking human effort have channeled rivers into intricate irrigation systems, and whole mountainsides have been transformed into descending terraces that evenly distribute the water. Writing about wet rice cultivation in Sichuan Province, geographer Chiao-Min Hsieh describes how "[e]verywhere one can hear water gurgling like music as it brings life and growth to the farms." However, these same cultivation techniques, combined with extensive forestry and mining, have led to the loss of most natural habitats in all but the most mountainous, dry, or remote areas. With so little suitable agricultural land left, further expansion of the area under wet rice cultivation is unlikely.

Fisheries and Globalization Many East Asians depend heavily on ocean-caught fish for protein (see Thematic Overview C). The Japanese in particular have had a huge impact on the seas of not only this region but of the entire world. With only 2 percent of the world's population, Japan consumes 15 percent of the global wild fish catch. There are some 4000 coastal fishing villages in Japan, sending out tens of thousands of small crafts to work nearby waters each day. Large Japanese fishing ships, complete with onboard canneries and freezers, harvest oceans around the world. In part because of this, environmentalists have been longtime critics of Japan for overfishing. For example, Japanese fishing off western Africa has reduced the catches of local fishers so much that many have been forced to migrate to Europe for work (see page 255 in Chapter 7). With more than 75 percent of the world's fisheries either fully exploited or in decline, there is little room for Japan to expand its fish imports.

THINGS TO REMEMBER

1. Learning Goal 1: Climate Change and Water East Asia has long suffered from lengthy searing droughts and devastating floods. These disruptions are likely to increase with global climate change.

2. China's main river systems are being affected by the melting of its glaciers.

3. The region's most important grain is rice, and over the millennia, rice cultivation has transformed landscapes throughout central and southern China, and Japan, Korea, and Taiwan.

4. Learning Goal 2: Food and Globalization About three-quarters of the food consumed in Japan, South Korea, and Taiwan is imported. China is currently self-sufficient with regard to basic necessities such as rice, but it is highly dependent on imports for other foods.

Three Gorges Dam: The Power of Water

The Chinese environmental activist, Dai Qing, notes that China has 22,000 large dams, all of which have displaced people—perhaps as many as 60 million—without any attention to their rights as stakeholders in the projects. The Three Gorges Dam (Photo Essay 9.3A) is at this time the largest dam in the world at 600 feet (183 meters) high and 1.4 miles (2.3 kilometers) wide, and is the second-largest engineering project in China's history (the largest is the South–North Transfer Project, not scheduled to be completed until 2050). It is designed to improve navigation on the Chang Jiang and control flooding, but it is most lauded for its role in generating hydroelectricity for this energy-hungry country that longs to improve its greenhouse gas emissions record.

Many experts involved with the Three Gorges project see serious design flaws. Of greatest concern is the dam's position above a seismic fault. The dam was built at the east end of the Three Gorges because the deep canyons provided a prodigious reservoir for water to power turbines, providing electricity for all of central China from the sea to the Tibetan Plateau. Unfortunately, the enormous weight of the water and its percolation through geological fissures in the 370-mile-long (600-kilometer-long) reservoir behind the dam could trigger earthquakes. Already at issue are huge landslides along the gorges, lubricated by the rising reservoir water. Meanwhile, as many as 80 visible cracks in the dam raise doubts about its structural integrity. Similar defects led to the failure of China's much smaller Banqiao Dam in 1975, which was responsible for 171,000 deaths from the flooding and an ensuing famine. Even if the dam holds, its potential to generate power will probably be reduced by the buildup of eroded silt behind the dam.

Any failure of the dam would be a financial as well as human disaster. Official construction costs are $25 billion, but the real costs may be three times this figure, due in part to unforeseen negative environmental and social impacts and to theft from the project by corrupt officials.

Also of concern are the incalculable costs associated with relocating the 1.2 million people who once lived where the

Thinking Geographically

After you have read about human impact on the Biosphere in East Asia, you will be able to answer the following questions:

A What design flaw is the greatest concern to experts involved with the Three Gorges Dam project?

B Where is the worst air pollution located in Japan?

C Of all the coal burned on earth each year, how much is burned in China?

D What allows particulates from China's coal burning to be transported globally?

Economic development in East Asia has often proceeded without adequate environmental protection or safeguards for human health. Here we focus on the Three Gorges Dam, air pollution, and energy issues.

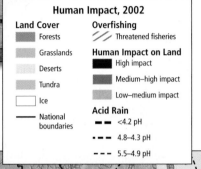

Human Impact, 2002

Land Cover
- Forests
- Grasslands
- Deserts
- Tundra
- Ice
- ⎯ National boundaries

Overfishing
- ⫽⫽ Threatened fisheries

Human Impact on Land
- High impact
- Medium–high impact
- Low–medium impact

Acid Rain
- ▪ ▪ <4.2 pH
- ▪ ▪ ▪ 4.8–4.3 pH
- ▪ ▪ ▪ 5.5–4.9 pH

A China's Three Gorges Dam is the largest dam in the world. It is capable of supplying about 3 percent of China's electricity needs, resulting in significantly less air pollution than the several large coal-fired power plants that it replaces. However, the dam has also resulted in significant environmental damage due to its 370-mile-long (600-km-long) reservoir, which has submerged the habitats of endangered species and the dwellings of 1.2 million people. Structural and design flaws in the dam threaten further environmental damage.

B A commuter in Tokyo wears a mask for protection against the city's air pollution. Recent restrictions on the use of diesel vehicles in Tokyo have resulted in improved local air quality.

mi 0 300 600
km 0 300 600

C An open-pit coal mine in China's province of Inner Mongolia. China is already the world's largest producer and consumer of coal, and its demand for coal is projected to double by 2030, which will make it by far the largest contributor of greenhouse gases.

D A toxic haze of air pollution engulfs Beijing and the North China Plain, often blowing eastward to reach Korea, Japan, and even the U.S. and Europe. China's coal-fired power plants, industries, and vehicles are emerging as major global environmental concerns.

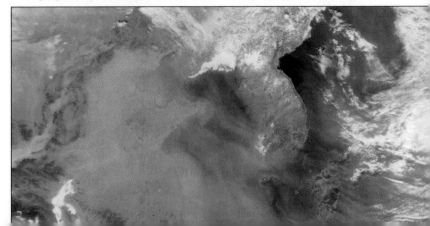

dam now forms a reservoir. Thirteen major cities have been submerged, along with 140 large towns, hundreds of small villages, 1600 factories, and 62,000 acres (25,000 hectares) of farmland. The reservoir has destroyed important archaeological sites, as well as some of China's most spectacular natural scenery. There are significant environmental costs as well. The giant sturgeon, for example, a fish that can weigh as much as three-quarters of a ton and is as rare as China's giant panda, may become extinct. Sturgeon used to swim more than 1000 miles (1600 kilometers) up the Chang Jiang past the location of the dam to spawn. Now the sturgeon's reproductive process has been irretrievably interrupted.

‖▶ 209. THREE GORGES DAM LEAVES SOME CHINESE SWAMPED

The plan to build the Three Gorges Dam came just as UN development specialists were deciding that the benefits dams could bring did not sufficiently outweigh the many problems they caused. Decades ago, international funding sources such as the World Bank withdrew their support for the dam because of concerns over the social and environmental costs and other shortcomings of the project. However, Chinese industrialists who need the energy, construction companies that have prospered from building the dam and its many ancillary projects, and government officials eager to impress the world and leave their mark on China continue not only to support the *Da Ba* (the Big Dam), but to seek other locations around the world where China can gain influence and profit by building dams.

Air Pollution: Choking on Success

Air pollution is often severe throughout East Asia, but the air quality in China's cities is the worst in the world. Coal burning is the primary cause. China is the world's largest consumer and producer of coal, accounting for 40 percent of all the coal burned on the planet each year. Between 1975 and 2005, China's coal consumption more than quadrupled (see Photo Essay 9.3C). By 2006, China was bringing a new coal-fired power plant online every week.

Air Pollution in China The combustion of coal releases high levels of two pollutants—suspended particulates and sulfur dioxide, both of which can cause respiratory ailments. In Chinese cities, these emissions can be 10 times higher than World Health Organization guidelines. The worst pollution is in northeastern China, where homes are in close proximity to industries that also depend heavily on coal for fuel.

Vignette Every winter, the elderly couple got through the biting Beijing winters by feeding 1200 one-kilogram coal bricks into a small iron stove. The ashes and coal dust blackened their belongings, and they worried about carbon monoxide poisoning. Then, in 2009 they were given a new electric space heater by the city government. In a move to clean up Beijing's air and reduce the city's contribution to China's greenhouse gas emissions (**Figure 9.5**), the city replaced nearly 100,000 old coal stoves with electric heaters and cut electric nighttime rates to just 3 cents per kilowatt-hour. The couple was astonished with the change the new cheap heater made in their lives. But of course, 100,000 more homes were now

dependent on electrical power, most of it generated by coal-fired power plants. *[Source: Michael Wines, Beijing's Air Is Cleaner, but Far from Clean, New York Times/International Herald Tribune Global Edition, October 16, 2009, at http://www.nytimes.com/2009/10/17/world/asia/17beijing. html?partner=rss&emc=rss.?]* ∎

Sulfur dioxide from coal burning also contributes to acid rain, which is displaced to the northeast by prevailing winds, reaching Korea, Japan, Taiwan, and beyond. Photo Essay 9.3D and the map above it show the zones of heavy pollution. Japan and Taiwan are particularly afflicted (see Photo Essay 9.3B). Particulates from China's coal burning are transported globally by high-altitude west-to-east flowing jet streams, thus affecting air quality in North America and Europe.

Air pollution from vehicles is also severe, even though the use of cars for personal transportation in China is just beginning. For years, China's vehicles have had very high rates of lead and carbon dioxide emissions. Quite recently, the government has taken action. As of 2008, new Chinese cars had to meet EU standards for mileage and emissions—which are much stricter than those for U.S. cars. Even so, to provide decent air quality during the 2008 Summer Olympics, China removed half of Beijing's cars from the road.

‖▶ 203. WORLDWATCH INSTITUTE: 16 OF 20 OF WORLD'S MOST POLLUTED CITIES IN CHINA

Air Pollution Elsewhere in East Asia Public health risks related to air pollution are serious in the largest cities of Japan (Tokyo and Osaka), Taiwan (Taipei), Mongolia (Ulan Bator), and South Korea (Seoul), and in adjacent industrial zones. Even with anti-pollution legislation and increased enforcement, high population densities and rising expectations for better living standards make it difficult to improve environmental quality. Taiwan is a case in point.

Taiwan has some of the dirtiest air on earth (see the map in Photo Essay 9.3). Some of the main causes are the island's extreme population density of 1600 people per square mile (615 per square kilometer), its high rate of industrialization, and its close proximity to industrialized south China. In Taiwan there are now 4 motor vehicles (cars or motorcycles) for every 5 residents—more than 16.5 million exhaust-producing vehicles on this small island. In addition, there are nearly eight registered factories for every square mile (three per square kilometer), all emitting waste gases. The government of Taiwan acknowledges that the air is 6 times dirtier than that of the United States or Europe.

Both North and South Korea depend on hydro- or nuclear-generated electricity to run factories and heat buildings, unlike elsewhere in East Asia. North Korea has relatively few industries and uses very few cars, so its air pollution levels are thought to be low for the region. South Korea uses fossil fuels in its many industries and has many gasoline-powered cars. These are the sources of most of its internally generated air pollution. Mongolia generally has the region's cleanest air; but in Ulan Bator, pollution from coal heaters has inspired some imaginative projects. For example, whole sections of Ulan Bator are now heated by centrally located boilers, which supply hot water to apartment buildings and individual dwellings.

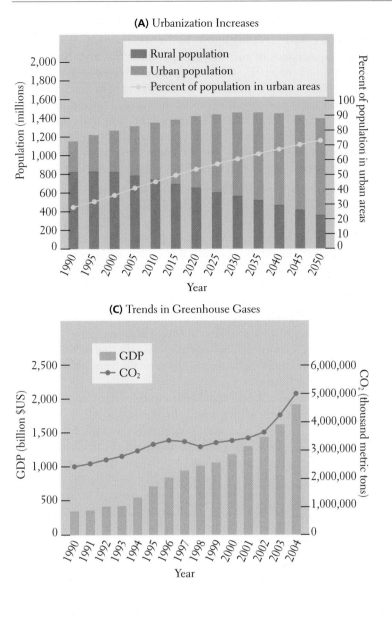

(A) Urbanization Increases

(B) Energy Consumption

(C) Trends in Greenhouse Gases

FIGURE 9.5 Urbanization, economic growth, and CO₂ emissions in China. As urban areas grow larger, GDP is increasing and the cities are putting increased pressure on available resources and on clean air. As China's GDP increases—especially in urban areas—demands for energy for households, industry, and transportation also go up. **(A)** By 2050, projections indicate that 73 percent of China's population will live in cities. **(B)** Between 1990 and 2005, GDP rose about 385 percent and energy consumption rose 149 percent, and is projected to rise another 131 percent by 2030. **(C)** As GDP increases, CO_2 emissions also rise. Between 1990 and 2004, CO_2 emissions rose 109 percent. Even if China improves its environmental regulation and enforcement, CO_2 emissions are projected to increase another 10 percent by 2020.

THINGS TO REMEMBER

1. In an effort to address droughts, floods, and the burgeoning need for electricity, China has built many dams, the largest of which is the Three Gorges Dam, which itself now poses many potential problems.

2. Rapid urban economic development in East Asia, combined with weak environmental protection, has resulted in some of the most polluted cities on the planet.

Human Patterns over Time

East Asia is home to some of the most ancient civilizations on earth. Settled agricultural societies have flourished in China for more than 7000 years.

Chinese civilization evolved from several hearths, including the North China Plain, the Sichuan Basin, and the lands of interior Asia that were inhabited by Mongolian nomadic pastoralists. On East Asia's eastern fringe, the Korean Peninsula and the islands of Japan and Taiwan were profoundly influenced by the culture of China, but they were isolated enough that each developed a distinctive culture and maintained political independence most of the time. In the early twentieth century, Japan industrialized rapidly by integrating European influences that China disdained.

Bureaucracy and Imperial China

Although humans have lived in East Asia for hundreds of thousands of years, the region's earliest complex civilizations appeared in various parts of what is now China about 4000 years ago. Written records exist only from the civilization that was located in north-central China. There, a small, militarized, feudal aristocracy controlled vast estates on which the majority of the population lived and worked as semi-enslaved farmers and laborers. The landowners usually owed allegiance to one of the petty kingdoms that dotted northern China. These kingdoms were relatively self-sufficient and well defended with private armies.

An important move away from feudalism came with the Qin empire (beginning in 221 B.C.E.), which instituted a trained and salaried bureaucracy in combination with a strong military to extend the monarch's authority into the countryside (Timeline B).

The Qin system proved more efficient than the old feudal allegiance system it replaced. The estates of the aristocracy were divided into small units and sold to the previously semi-enslaved farmers. The empire's agricultural output increased because the people worked harder to farm land they now owned. In addition, the salaried bureaucrats were more responsible than the aristocrats they replaced, especially about building and maintaining levees, reservoirs, and other tax-supported public works that reduced the threat of flood, drought, and other natural disasters. Although the Qin empire was short-lived, subsequent empires maintained Qin bureaucratic ruling methods, which have proved essential in governing a united China.

Confucianism Molds East Asia's Cultural Attitudes

Closely related to China's bureaucratic ruling tradition is the philosophy of **Confucianism**. Confucius, who lived from 551 to 479 B.C.E., was an idealist who was interested in reforming government and eliminating violence from society. He thought human relationships should involve a set of defined roles and mutual obligations. Confucian values include courtesy, knowledge, integrity, and respect for and loyalty to parents and government officials. These values diffused across the region and are still widely shared throughout East Asia (see Timeline A).

> **Confucianism** a Chinese philosophy that teaches that the best organizational model for the state and society is a hierarchy based on the patriarchal family

The Confucian Bias Toward Males The model for Confucian philosophy was the patriarchal extended family. The oldest male held the seat of authority and was responsible for the well-being of everyone in the family. All other family members were aligned under the patriarch according to age and gender. Beyond the family, the Confucian patriarchal order held that the emperor was the grand patriarch of all China, charged with ensuring the welfare of society. Imperial bureaucrats were to do his bidding and commoners were to obey the bureaucrats.

Over the centuries, Confucian philosophy penetrated all aspects of East Asian society. Concerning the ideal woman, for example, a student of Confucius wrote: "A woman's duties are to cook the five grains, heat the wine, look after her parents-in-law, make clothes, and that is all! When she is young, she must submit to her parents. After her marriage, she must submit to her husband. When she is widowed, she must submit to her son." These concepts about limited roles for women affected society at large, where the idea developed that sons were the more valuable offspring, with public roles; while daughters were primarily servants within the home.

The Bias Against Merchants For thousands of years, Confucian ideals were used to maintain the power and position of emperors and their bureaucratic administrators at the expense of merchants.

In parable and folklore, merchants were characterized as a necessary evil, greedy and disruptive to the social order. At the same time, the services of merchants were sorely needed. Such conflicting ideas meant the status of merchants waxed and waned. At times, high taxes left merchants unable to invest in new industries or trade networks. At other times, however, the anti-merchant aspect of Confucianism was less influential, and trade and entrepreneurship flourished. Under communism, merchants again acquired a negative image; then when marketization was encouraged, the social status of merchants once again rose.

Cycles of Expansion, Decline, and Recovery Although the Confucian bureaucracy at times facilitated the expansion of Imperial China (Figure 9.6), its resistance to change also led to periods of decline. Heavy taxes were periodically levied on farmers, bringing about farmer revolts that weakened imperial control. Threats from outside, particularly invasions by nomadic people from what are today Mongolia and western China, inspired the creation of massive defenses, such as the Great Wall (see Timeline C), built along China's northern border. Nevertheless, the Confucian bureaucracy always recovered, and China's cultural and economic sophistication usually overwhelmed the invaders. After a few generations, the nomads were indistinguishable from the Chinese.

One nomadic invasion did result in important links between China and the rest of the world. In the 1200s, the Mongolian military leader Genghis Khan and his descendants were able to conquer all of China. They then pushed west across Asia as far as Hungary and Poland (see page 184 in Chapter 5). It was during the time of this Mongol empire (also known as the Yuan empire) that traders such as the Venetian Marco Polo made the first direct contacts between China and Europe. These connections proved much more significant for Europe, which was dazzled by China's wealth and technologies, than for China, which saw Europe as backward and barbaric.

Indeed, from 1100 to 1600, China remained the world's most developed region, despite enduring several cycles of imperial expansion, decline, and recovery. It had the largest economy, the highest living standards, and the most magnificent cities. Improved strains of rice allowed dense farming populations to expand throughout southern China and supported large urban industrial populations. Nor was innovation lacking: Chinese inventions included paper making, printing, paper currency, gunpowder, and improved shipbuilding techniques.

Why Did China Not Colonize an Overseas Empire?

During the well-organized Ming dynasty, 1368–1644, Zheng He, a Chinese Muslim admiral in the emperor's navy, directed an expedition that could have led to China conquering a vast overseas empire. From 1405 to 1433, Zheng He sailed 250 ships—the biggest and most technologically advanced fleet that the world had ever seen. Zheng He took his fleet throughout Southeast Asia, across the Indian Ocean, and along the coasts of India, continuing into the Persian Gulf and then down the east coast of Africa.

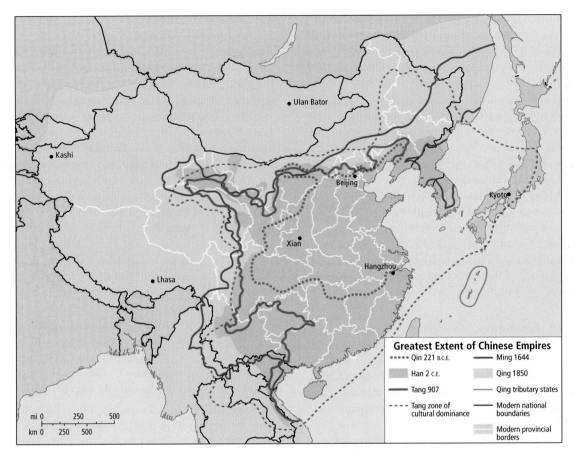

FIGURE 9.6 The extent of Chinese empires, 221 B.C.E.–1850 C.E. The Chinese state has expanded and contracted throughout its history.

The lavish voyages of Zheng He were funded for almost 30 years, but they never resulted in an overseas empire like those established by European countries a century or two later. These newly explored regions simply lacked trade goods that China wanted. Moreover, back home the empire was continually threatened by the armies of nomads from Mongolia, so any surplus resources were needed for upgrading the Great Wall. Eventually the emperor decided that Zheng He's explorations were not worth the effort, and China as a whole turned inward. In the years following Zheng He's voyages, contacts with the rest of the world were minimized as the emperor focused on repelling invaders from Mongolia. As a result, the pace of technological change slowed, leaving China ill prepared to respond to growing challenges from Europe after 1600.

European and Japanese Imperialism

By the mid-1500s, during Europe's Age of Exploration, Spanish and Portuguese traders interested in acquiring China's silks, spices, and ceramics found their way to East Asian ports. To exchange, they brought a number of new food crops from the Americas, such as corn, peppers, peanuts, and potatoes. These new sources of nourishment contributed to a spurt of economic expansion and population growth during the Qing, or Manchurian, dynasty (1644–1912), and by the mid-1800s, China's population was more than 400 million.

By the nineteenth century, European merchants gained access to Chinese markets and European influence increased markedly. In exchange for Chinese silks and ceramics, British merchants supplied opium from India, which was one of the few things that Chinese merchants would trade for. The emperor attempted to crack down on this drug trade because of its debilitating effects on Chinese society. The result was the Opium Wars (1839–1860), in which Britain badly defeated China. Hong Kong became a British possession, and British trade, including opium, expanded throughout China (see Timeline D).

The final blow to China's long preeminence in East Asia came in 1895, when a rapidly modernizing Japan won a spectacular naval victory over China in the Sino-Japanese War. After this first defeat by the Japanese, the Qing dynasty made only halfhearted attempts at modernization, and in 1912, it was overthrown by an internal revolt and collapsed. Beginning with the decline of the Qing empire (1895) until China's Communist Party took control in 1949, much of the country was governed by provincial rulers in rural areas and by a mixture of Chinese, Japanese, and European administrative agencies in the major cities.

China's Turbulent Twentieth Century

Two rival reformist groups arose in China in the early twentieth century. The Nationalist Party, known as the Kuomintang (KMT), was an urban-based movement that appealed to workers as well as the middle and upper classes. The Chinese Communist Party (CCP), on the other hand, found its base in rural areas among peasants. At first the KMT gained the upper hand, uniting the country in 1924. However, Japan's invasion of China in 1931 changed the dynamic.

By 1937, Japan had control of most major Chinese cities. The KMT did not resist the Japanese effectively and were confined to the few deep interior cities not in Japanese control. The CCP, however, waged a constant guerilla war against the Japanese throughout rural China. This resistance gained the CCP heroic status. Japan's brutal occupation caused 10 million Chinese deaths. When Japan finally withdrew in 1945, defeated at the end of World War II by the United States, Russia, and other Allied forces, the vastly more popular CCP pushed the KMT out of the country and into exile in Taiwan. In 1949, the CCP, led by Mao Zedong, proclaimed the country the "People's Republic of China," with Mao as president.

Mao's Communist Revolution Mao Zedong's revolutionary government became the most powerful China ever had. It dominated all the outlying areas of China—the Far Northeast, Inner Mongolia, and western China (Xinjiang Uygur)—and launched a brutal occupation of Tibet (Xizang). The People's Republic of China was in many ways similar to past Chinese empires. The Chinese Communist Party replaced the Confucian bureaucracy and Mao Zedong became a sort of emperor with unquestioned authority (see Timeline E).

Among the early beneficiaries of the revolution were the masses of Chinese farmers and landless laborers. On the eve of the revolution, huge numbers lived in abject poverty. Famines were frequent, infant mortality was high, and life expectancy was low. The vast majority of women and girls held low social status and spent their lives in unrelenting servitude.

The revolution drastically changed this picture. All aspects of economic and social life became subject to central planning by the Communist Party. Land and wealth were reallocated, often resulting in an improved standard of living for those who needed it most. Heroic efforts were made to improve agricultural production and to reduce the severity of floods and droughts. The masses, regardless of age, class, or gender, were mobilized to construct almost entirely by hand huge public works projects—roads, dams, canals, whole mountains terraced into fields for rice and other crops. "Barefoot doctors" with rudimentary medical training dispensed basic medical care, midwife services, and nutritional advice to people in the remotest locations. Schools were built in the smallest of villages. Opportunities for women became available, and some of the worst abuses against them were stopped—such as the crippling binding of women's feet to make their feet small and childlike. Most Chinese people who are old enough to have witnessed these changes say that the revolution did a great deal to improve overall living standards for the majority.

Mao's Missteps Nonetheless, progress came at enormous human and environmental costs. During the **Great Leap Forward** (a government-sponsored program of massive economic reform initiated in the 1950s), 30 million people died from fam-

> **Great Leap Forward** an economic reform program under Mao Zedong intended to quickly raise China to the industrial level of Britain and the United States
>
> **Cultural Revolution** a series of highly politicized and destructive mass campaigns launched in 1966 to force the entire population of China to support the continuing revolution

ine brought on by poorly planned development objectives. Meanwhile, deforestation, soil degradation, and agricultural mismanagement became widespread. In the aftermath of the Great Leap Forward, some Communist Party leaders tried to correct the inefficiencies of the centrally planned economy only to be demoted or jailed as Mao Zedong remained in power.

In 1966, partially in response to the failures of the Great Leap Forward, a series of highly politicized and destructive mass campaigns, known as the **Cultural Revolution**, enforced support for Mao and punished dissenters. Everyone was required to study the "Little Red Book" of Mao's sayings. Educated people and intellectuals were a main target of the Cultural Revolution because they were thought to instigate dangerously critical evaluations of Mao and Communist Party central planning. Tens of millions of Chinese scientists, scholars, and students were sent out of the cities to labor in mines and industries or to jail, where as many as 1 million died. Children were encouraged to turn in their parents. Petty traders were punished for being capitalists, as were those who adhered to any type of organized religion. The Cultural Revolution so disrupted Chinese society that by Mao's death in 1976, the Communists had been seriously discredited.

Changes After Mao Two years after Mao's death, a new leadership formed around Deng Xiaoping. Limited market reforms were instituted, but the Communist Party retained tight political control. In 2009, after more than 30 years of reform and remarkable levels of economic growth, China's economy became the third largest in the world behind the European Union and the United States. It passed Japan when China managed to weather the world recession beginning in 2008 better than expected. However, the disparity of wealth in China has been increasing for some years, human rights are still often abused, and political activity remains tightly controlled even as discontent boils over into open protests against government foibles.

Japan Becomes a World Leader

Although China's influence was dominant in East Asia for thousands of years, for much of the twentieth century, Japan, with only one-tenth the population and 5 percent of the land area of China, controlled East Asia economically and politically. Japan's rise as a modern global power results largely from its response to challenges from Europe and North America.

Beginning in the mid-sixteenth century, active trade with Portuguese colonists brought new ideas and technology that strengthened Japan's wealthier feudal lords (*shoguns*), allowing them to unify the country under a military bureaucracy. However, the shoguns monopolized contact with the European outsiders, allowing no Japanese people to leave the islands, on penalty of death.

A second period of radical change came with the arrival in Tokyo Bay in 1853 of a small fleet of U.S. naval vessels. The foreigners, carrying military technology far more advanced than Japan's, forced the Japanese government to open the economy to international trade and political relations. In response, a group of reformers (the Meiji) seized control of the Japanese government, setting the country on a crash course of modernization and industrial development. They sent Japanese students abroad and recruited experts from around the world, especially from Western nations, to teach everything from foreign languages to modern military technology. The innovations that resulted enabled Japan's economy to grow rapidly, surpassing China's size in the early twentieth century.

Between 1895 and 1945, Japan fueled its economy with resources from a vast colonial empire. Equipped with imported European and North American military technology, its armies occupied first Korea, then Taiwan, then coastal and eastern China, and eventually Indonesia and much of Southeast Asia, as shown in Figure 9.7. Many of the people of these areas still harbor resentment about the brutality they suffered at Japanese hands. Japan's imperial ambitions ended with its defeat in World War II and its subsequent occupation by U.S. military forces until 1952.

▶ 204. JAPANESE STILL RESOLVING FEELINGS ABOUT THE WAR

In the immediate post-war era, the U.S. government imposed many social and economic reforms. Japan was required to create a democratic constitution and reduce the emperor to symbolic status. Its military was reduced dramatically, forcing it to rely on U.S. forces to protect it from attack. With U.S. support, Japan rebuilt rapidly after World War II, and it eventually became a giant in industry and global business, exporting automobiles, electronic goods, and many other products (see Timeline F). In September 2010, Japan's economy was still among the world's largest and most technologically advanced.

Chinese and Japanese Influences on Korea, Taiwan, and Mongolia

The history of the entire East Asia region is largely grounded in what transpired in China and Japan.

The Korean War and Its Aftermath Korea was a unified but poverty-stricken country until 1945. At the end of World War II, in an effort to extend its global influence, the Soviet Union declared war against Japan and invaded Manchuria (northeast China) and the northern part of Korea; it was prevented from taking the whole peninsula only by U.S. military intervention. The United States took control of the southern half of the Korean peninsula, where it instituted reforms similar to those in Japan. After the

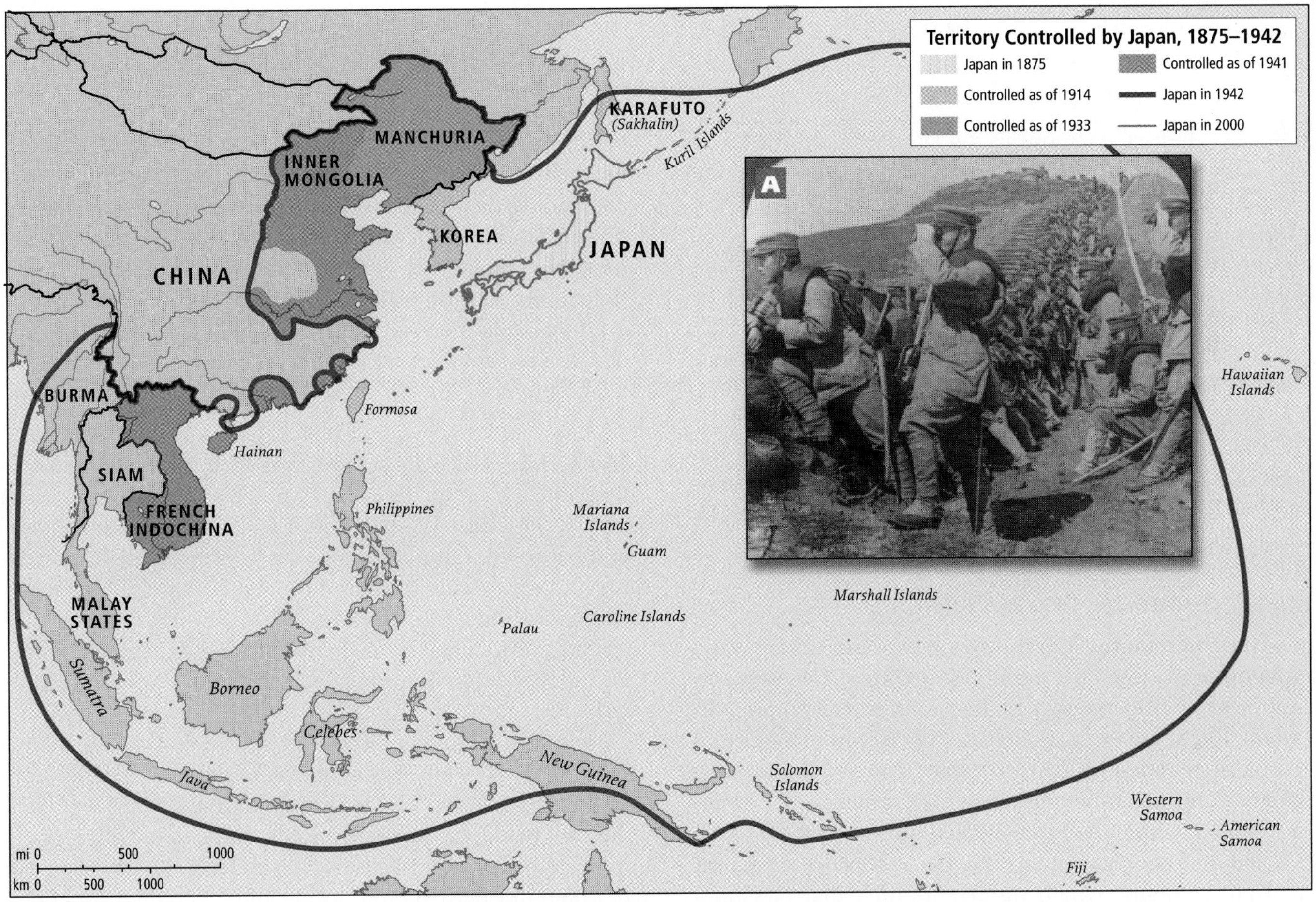

FIGURE 9.7 Japan's expansions, 1875–1942. Japan colonized Korea, Taiwan (then known as Formosa), Manchuria, China, parts of Southeast Asia, and several Pacific islands to further its program of economic modernization and to fend off European imperialism in the early twentieth century. **(A)** Japanese soldiers await an attack by the Russian army in Manchuria. Russia and Japan competed for dominance in Manchuria in the early twentieth century.

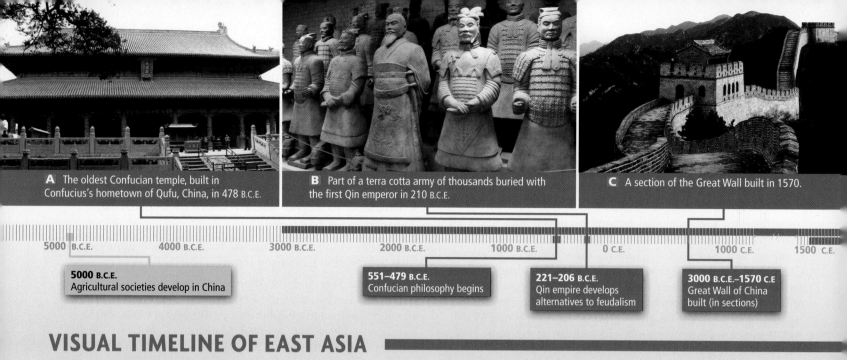

A The oldest Confucian temple, built in Confucius's hometown of Qufu, China, in 478 B.C.E.

B Part of a terra cotta army of thousands buried with the first Qin emperor in 210 B.C.E.

C A section of the Great Wall built in 1570.

5000 B.C.E.   4000 B.C.E.   3000 B.C.E.   2000 B.C.E.   1000 B.C.E.   0 C.E.   1000 C.E.   1500 C.E.

5000 B.C.E.
Agricultural societies develop in China

551–479 B.C.E.
Confucian philosophy begins

221–206 B.C.E.
Qin empire develops alternatives to feudalism

3000 B.C.E.–1570 C.E
Great Wall of China built (in sections)

VISUAL TIMELINE OF EAST ASIA

Thinking Geographically

After you have read about the human history of East Asia, you will be able to answer the following questions:

A The model for Confucian philosophy is what?

B Although the Qin empire was short-lived, what did subsequent empires do to survive?

United States withdrew its troops in the late 1940s, North Korea attacked South Korea. The United States returned to defend the south, leading a 3-year war against North Korea and its allies, the Soviet Union and Communist China.

After great loss of life on both sides and devastation of the peninsula's infrastructure, the Korean War ended in 1953 in a truce and the establishment of a demilitarized zone (DMZ) at the 38th parallel. North Korea closed itself off from the rest of the world, and to this day it remains isolated, impoverished, and defensive, occasionally gaining international attention by hinting at its nuclear potential. Meanwhile, South Korea has developed into a prosperous and technologically advanced market economy. Relations between the two countries remain tense, with occasional skirmishes breaking out along the border.

II ▶ 201. INTER-KOREAN COOPERATION GROWS
II ▶ 210. HIGH-TECH KOREA EYES 'UBIQUITOUS' FUTURE

Taiwan's Uncertain Status For thousands of years, Taiwan was a poor agricultural island on the periphery of China; then between 1895 and 1945 it became part of Japan's regional empire. In 1949, when the Chinese nationalists (the Kuomintang) were pushed out of mainland China by the Chinese Communist Party, they set up an anti-communist government in Taiwan, naming it the Republic of China (ROC). For the next 50 years, with U.S. aid and encouragement, the ROC became a modern industrialized economy, which quickly dwarfed that of China and remained dominant until the 1990s, serving as a prosperous icon of capitalism right next door to massive Communist China.

Today, Taiwan remains an economic powerhouse. Taiwanese investors have been especially active in Shanghai and the cities of China's southeast coast. Yet mainland China has never relinquished claim to Taiwan and occasionally threatens to invade and militarily occupy it. As China's economic and military power has increased, the United Nations, the World Bank, and most other countries, agencies, and institutions have judiciously tiptoed around the issue of whether Taiwan should continue as an independent country or be downgraded to simply a province of China. Taiwan itself remains divided over just how strongly it should hold on to its sovereignty (see Photo Essay 9.5C on page 356).

Mongolia Seeks Its Own Way For millennia, Mongolia's nomadic horsemen periodically posed a threat to China, so much so that the Great Wall was built and reinforced and extended to combat them. China has long been obsessed with both deflecting and controlling its northern neighbor; and China did control Mongolia from 1691 until the 1920s. Revolutionary communism spread to Mongolia soon thereafter, and Mongolia continued as an independent communist country under Soviet, not Chinese, guidance until the breakup of the Soviet Union in 1989. Communism brought education and basic services. Literacy for both men and women rose above 95 percent. Deeply suspicious of both Russia and China, Mongolia has since 1989 been on a difficult road to a market economy. In need of cash to participate in the modern world, families have elected to abandon nomadic herding and permanently locate their portable *ger* homes near Ulan Bator and search for paid employment. This rapid and drastic change in lifestyle has resulted in broken homes, increasing personal debt, and poverty, in a society that formerly took pride in an egalitarian if not prosperous standard of living.

D A British warship in Shanghai's harbor in 1931.

E A 1950 stamp celebrating a treaty between China and the Soviet Union, both communist countries.

F A "bullet train" in Kyoto, Japan, in 2006.

1600 C.E.   1700 C.E.   1800 C.E.   1900 C.E.   2000 C.E.

1500s–1949
European and Japanese imperialism throughout East Asia

1949–Present
Communist China

1945–Present
Japan rebuilds after WWII

1980s–Present
Economic reforms in China

2006

C The Great Wall was built in response to what?

D What major Chinese city and port was a British possession from the time of the Opium Wars until recently?

E After the Chinese Communist Revolution, all aspects of economic and social life became subject to _____.

F After World War II, Japan eventually became _____.

THINGS TO REMEMBER

1. The histories of East Asian countries are deeply intertwined, and Confucian thought still permeates the entire region.

2. For six centuries, from 1100 to 1600, China was the world's most developed region, despite cycles of imperial expansion, decline, and recovery.

3. In recent times, the countries of East Asia have followed very different philosophical paths toward development.

4. Japan has long-term cultural ties to the mainland of East Asia and has also exercised influence over the region, primarily through conquest. For most of the twentieth century, it has been the dominant economy in the region, only recently losing that status to China in 2010.

5. Taiwan and South Korea emerged after World War II as rapidly industrializing countries, while Mongolia until recently retained communist connections to the USSR. North Korea remains a communist country, and is extremely defensive and cut off from the wider world.

II CURRENT GEOGRAPHIC ISSUES

Although the countries of East Asia adopted new economic systems only after World War II, most of them are making progress toward creating a better life for their citizens. In fact, Japan, South Korea, and China's Hong Kong Special Administrative Unit have among the highest standards of living in the world.

Economic and Political Issues

After World War II, the countries of East Asia established two basic types of economic systems. The communist regimes of China, Mongolia, and North Korea relied on central planning by the government to set production goals and to distribute goods among their citizens. In contrast, first

> **state-aided market economy** an economic system based on market principles such as private enterprise, profit incentives, and supply and demand, but with strong government guidance; in contrast to the free market (limited government) economic system of the United States and Europe
>
> **export-led growth** an economic development strategy that relies heavily on the production of manufactured goods destined for sale abroad

Japan, and then Taiwan and South Korea, established **state-aided market economies** with the assistance and support of the United States and Europe. In this type of economic system, market forces, such as supply and demand and competition for customers, determine many economic decisions. However, the government intervenes strategically, especially in the financial sector, to make sure that certain economic sectors develop in a healthy fashion. Investment in the country by foreigners is also limited so that the government can retain greater control over the direction of the economy. In the cases of Japan, South Korea, and Taiwan, government intervention was designed to enable **export-led growth**. This economic development strategy relies heavily on the production

347

of manufactured goods destined for sale abroad, primarily to the large economies of North America and Europe, while limiting imports for local consumers.

More recently, the differences among East Asian countries have diminished as China and Mongolia have set aside strict central planning and adopted reforms that rely more on market forces. China now also relies heavily on exports of its manufactured goods to North America and Europe. Politically, however, contrasts remain stark. While Japan, South Korea, Taiwan, and Mongolia have become democracies, unelected governments in China and North Korea maintain a tight grip on politics and the media. In China, there is some experimentation with democracy at the local level, but the central government is primarily a force against widespread democratic participation.

The Japanese Miracle

Throughout the nineteenth century, the economies of Japan, Korea, and Taiwan were minuscule compared with China's. Then, during the twentieth century, all three grew tremendously due in part to ideas originating in Japan.

Japan Rises from Ashes Japan's recovery after its crippling defeat at the end of World War II is one of the most remarkable tales in modern history. Except for Kyoto, which was spared because of its historical and architectural significance, all of Japan's major cities were destroyed by the United States. Most notably, the United States bombed Hiroshima and Nagasaki with nuclear weapons, and leveled Tokyo with incendiary bombs.

▐▐ ▶ 205. SURVIVORS RECALL THE NUCLEAR BOMBING OF HIROSHIMA

Key to Japan's rapid recovery was its state-aided market economy, in which government guided private investors in creating new manufacturing industries. The overall strategy was one of export-led economic growth, with the Japanese government negotiating trade agreements with the United States and Europe. These and other deals ensured that large and wealthy foreign markets would be willing to import Japanese manufactured goods. These two aspects of Japan's economic recovery were imitated in Korea, Taiwan, countries in Southeast Asia, and eventually in China and many other parts of the developing world. Also central to Japan's recovery, but less imitated abroad, were arrangements between the government, major corporations, and labor unions that guaranteed lifetime employment by a single company for most workers in return for relatively modest pay.

The government-engineered trade and labor arrangements produced explosive economic growth of 10 percent or more annually between 1950 and the 1970s. The leading sectors were export-oriented automobile and electronics manufacturing. Japanese brand names such as Sony, Panasonic, Nikon, and Toyota became household words in North America and Europe. Products made by these companies sold at much higher volumes than would have been possible in the then relatively small Japanese and nearby Asian economies. Although growth slowed considerably both during the 1990s and again

beginning in 2007, Japan's postwar "economic miracle" continues to have an immense worldwide impact as a model, and Japan remains a significant actor in the world economy. Japan purchases resources from all parts of the world for its industries and domestic use. These purchases and investments in various local economies continue to create jobs for millions of people around the globe.

Productivity Innovations in Japan Over the years, Japan has made two major innovations in manufacturing that have boosted its productivity and been diffused to other industrial economies, changing the spatial arrangements of industries. The **kanban**, or "just in time," **system** clusters together companies that are part of the same production process so that they can deliver parts to each other when they are needed (Figure 9.8). For example, factories that make automobile parts are clustered around the final assembly plant, delivering parts literally minutes before they will be used. This saves money by making production more efficient and reducing the need for warehouses.

A related innovation is the **kaizen system**, or "continuous improvement system." This system ensures that fewer defective parts are produced because production lines are constantly surveyed for errors. Production lines are also constantly adjusted and improved to save time and energy. Both kanban and kaizen systems have been imitated by companies around the world and have been taken overseas by Japanese companies that invest abroad. For example, Toyota uses kaizen and kanban in all of its U.S. plants (see page 85 in Chapter 2).

> **kanban system** the "just-in-time" system pioneered in Japanese manufacturing that clusters companies that are part of the same production system close together so that they can deliver parts to each other precisely when they are needed
>
> **kaizen system** the "continuous improvement system" pioneered in Japanese manufacturing which ensures that fewer defective parts are produced because production lines are constantly surveyed for errors

FIGURE 9.8 Japan's kanban system. In this example of the kanban system, related industries are clustered together on a human-made island in Yokohama, Japan. They supply each other with various chemical inputs needed for their production processes.

Mainland Economies: Communists in Command

After World War II, economic development on East Asia's mainland proceeded on a dramatically different course than Japan's. Communist economic systems transformed poverty-ridden China, Mongolia, and North Korea. Private property was abolished, and the state took full control of the economy, loosely following the example of the Soviet *command*, or *centrally planned*, economy (see Chapter 5 on page 186). These sweeping changes transformed life for the poor majority, but ultimately proved less resilient and successful than was hoped.

By design, most people in the communist economies were not allowed to consume more than the bare necessities. On the other hand, the "iron rice bowl" policy guaranteed nearly everyone a job for life, sufficient food, basic health care, and housing that was better than what they had before. One drawback was that overall productivity remained low.

The Commune System When the Communist Party first came to power in China in 1949, its top priority was to make monumental improvements in both agricultural and industrial production. Similar early goals were held by the communist governments in North Korea and Mongolia, though they had much smaller populations and resource bases to work with.

In the years following World War II, an aggressive agricultural reform program joined small landholders together into cooperatives so that they could pool their labor and resources to increase production. In time, the cooperatives became full-scale communes, with an average of 1600 households each. The communes, at least in theory, took care of all aspects of life. They provided health care and education, and built rural industries to supply such items as simple clothing, fertilizers, small machinery, and eventually even tractors. The rural communes also had to fulfill the ambitious expectations that the leaders in Beijing had of better flood control, expanded irrigation systems, and especially, increased food production.

The Chinese commune system met with several difficulties. Rural food shortages developed because farmers had too little time to farm. They were required to spend much of their time building roads, levees, terraces, and drainage ditches, or working in the new rural industries. Local Communist Party administrators often compounded the problem by overstating harvests in their communes to impress their superiors in Beijing. The leaders in Beijing responded by requiring larger food shipments to the cities, which created yet greater food shortages in the countryside.

Although the Chinese agricultural communes were inefficient and resulted in devastating food scarcities during the Great Leap Forward, they did eventually result in a stable food supply that kept Chinese people well fed.

In North Korea, the Communists have pushed military strength at the expense of broader economic development. Meanwhile, agriculture has been so neglected that production is precarious, with food aid from its immediate neighbors and the United States a yearly necessity, and famine a constant threat. In recent years, North Korea has used rocket launchings and nuclear tests to intimidate its neighbors, and is selling its military technology abroad to pay for food imports.

In Mongolia, communist policy followed the Soviet model of collectivization, but was tailored to Mongolia's economy, which at the time was largely one of herding and agriculture There were minimal changes to the nomadic pastoralist way of life, but the role of mining and industry was emphasized, and the contribution to GDP by herding and forestry declined to less than 20 percent.

Focus on Heavy Industry The communist leadership in China, North Korea, and Mongolia believed that investment in heavy industry would dramatically raise living standards. Massive investments were made in the mining of coal and other minerals and in the production of iron and steel. Especially in China, heavy machinery produced equipment to build roads, railways, dams, and other infrastructure improvements that leaders hoped would increase overall economic productivity. However, much as in India (see pages 318–320 in Chapter 8), the vast majority of the population remained poor agricultural laborers who received little benefit from industries that created jobs mainly in urban areas. Not enough attention was paid to producing consumer goods (such as cheap pots, pans, hand tools, and other household items) that would have driven modest internal economic growth and improved living standards for the rural poor. Even in the urban areas, growth remained sluggish because, as also was the case in the Soviet command economy, small miscalculations by bureaucrats resulted in massive shortages and production bottlenecks that constrained economic growth.

regional self-sufficiency an economic policy in Communist China that encouraged each region to develop independently in the hope of evening out the wide disparities in the national distribution of production and income

responsibility system in the 1980s, a decentralization of economic decision making in China that returned agricultural decision making to the farm household level, subject to the approval of the commune

Struggles with Regional Disparity For centuries, China's interior west has been poorer and more rural than its coastal east. The interior west has been locked into agricultural and herding economies, while even in the most restricted times, the economies of the east have benefited from trade and industry. The first effort to address regional disparities, right after the revolution, was an economic policy focused on **regional self-sufficiency**. Each region was encouraged to develop as an independent entity with both agricultural and industrial sectors that would create jobs and produce food and basic necessities. Despite these efforts, spatially uneven development continues even today, and can also be seen on the provincial level where rural–urban disparities are sometimes extreme (see Figure 9.9 on page 352). Indeed, the floating population discussed at the beginning of this chapter is a direct consequence of this spatially uneven development.

Market Reforms in China

In the 1980s, China's leaders enacted market reforms that changed the country's economy in four ways. First, economic decision making was decentralized and given the name **responsibility system**, which meant that agricultural decision making was returned to the farm household level, subject to the approval of the commune. Second, market reforms allowed existing and newly established businesses to sell their produce and goods in competitive markets. Third, **regional specialization**, rather than regional self-sufficiency, was encouraged in order to take advantage of regional variations in climate, natural resources, and location, and thereby encourage national economic integration. Finally, the government allowed **foreign direct investment** in Chinese export-oriented enterprises and the sale of foreign products in China. This four-part shift to a more market-based economy dramatically improved the efficiency with which food and goods were produced and distributed, and to some extent, addressed problems related to regional disparities.

China's market reforms have transformed not only China's economy but also the economies of wider East Asia and indeed the whole world. Today China is the world's largest producer of manufactured goods, supplying consumers across the globe (see Thematic Overview D and E). Of the other Communist-led countries, Mongolia has participated in this revolution only modestly, and North Korea has not participated at all.

Urbanization, Development, and Globalization in East Asia

> **Learning Goal 3**
> **Urbanization and Development**: How is China's spectacular urban growth, and the massive increase in pollution that has accompanied it, linked to processes of globalization that were set in motion three decades ago?

Across East Asia, cities have grown rapidly over the last several decades as urban industrial development has accelerated, focused on production for the global market (see the map in Photo Essay 9.4). China has undergone the most massive urbanization in the world's history. Since 1980, when market reforms were initiated, the urban population has more than quadrupled, and is now estimated at nearly 600 million. This explosive economic development is fueled largely by the highly globalized economies of China's big coastal cities, whose urban factories now supply consumers throughout the world. Despite the rapid growth, urbanites still represent only 46 percent of China's total population, so more growth is likely, whereas in every other country in the region—even North Korea—the majority has been urban for at least two decades. Most East Asian cities are extremely crowded, so future growth is a challenge for planners and residents (see Photo Essay 9.4 A, B).

Spatial Disparities The patterns of public and private investment that accompany urbanization have led to rural areas lagging behind urban areas in access to jobs and income, education, and medical care. The map in Figure 9.9 depicts the problem. GDP per capita is significantly lower in China's interior provinces than it is in coastal provinces, and within each province there is a disparity between rural and urban places, here shown in pie diagrams of rural versus urban incomes. Notice that rural–urban GDP disparities, while still significant, are less extreme in northeastern and coastal provinces than in interior and western provinces. Similar rural-to-urban disparity patterns are found

in all other countries in East Asia, including North Korea.

⏸▶ 206. OLD BEIJING MAKING WAY FOR MODERN DEVELOPMENT

⏸▶ 219. SOME CHINESE FEAR PRIVATE PROPERTY LAW WILL COST THEM THEIR HOMES

International Trade and Special Economic Zones One way China has tried to address regional disparities is to increase economic growth in underdeveloped areas by opening them up to foreign investment and trade. Wary of the disruption that could result from abruptly opening the economy to international trade, China first selected five coastal cities to function as free trade zones, or **special economic zones (SEZs)** (Figure 9.10 on page 353). In this sense, SEZs were similar to the EPZs found in other world regions (see Chapter 3 on page 116). In the late 1990s, the program was expanded to 32 other cities, many of them in the interior. These new locations were designated *economic and technology development zones* (ETDZs in Figure 9.10). Like SEZs, the ETDZs provide footholds for international investors and multinational companies eager to establish operations in the country.

This program was successful. Today the SEZs and ETDZs are China's greatest **growth poles**, meaning that their development, like a magnet, is drawing yet more investment and migration. The first coastal SEZs were spectacularly successful. In only 25 years, many coastal cities, including Dongguan (the city Li Xia migrated to, in the vignette that opens this chapter) grew from medium-sized towns or even villages into some of the largest urban areas in the world. SEZs and ETDZs in the interior have had higher rates of growth lately than those on the coast, but foreign direct investment remains concentrated on the eastern coast, which accounts for 94 percent of China's exports. As shown in Figure 9.9, GDP per capita (PPP) rates remain noticeably lower in China's interior than on the eastern coast.

Hong Kong's Unique Role Hong Kong is one of the most densely populated cities on earth, and its residents have China's highest per capita income. In 2007, its annual GDP per capita (adjusted for PPP) was more than U.S.$42,000. Hong Kong was a British Crown colony until July 1997, when Britain's 99-year lease ran out and Hong Kong became a special administrative region (SAR) of China. Many wealthy citizens fled, worried that

After you have read about urbanization in East Asia, you will be able to answer the following questions:

A What happened to many of the former residents of Pudong, the new financial center of Shanghai?

C Migrants to urban areas who ignore the *hukou* system are considered part of what population?

D The urbanized region that stretches from the cities of Tokyo and Yokohama on Honshu south through the coastal zones of the inland sea to the islands of Shikoku and Kyushu is home to what percent of Japan's population?

> **regional specialization** the encouragement of specialization rather than self-sufficiency in order to take advantage of regional variations in climate, natural resources, and location
>
> **foreign direct investment** investment in a country's businesses by citizens, corporations, or governments of other countries
>
> **special economic zones (SEZs)** free trade zones within China
>
> **growth poles** zones of development whose success draws more investment and migration to a region

Some of the fastest growing, largest, and wealthiest urban areas in the world are in East Asia. China's cities are undergoing particularly rapid growth.

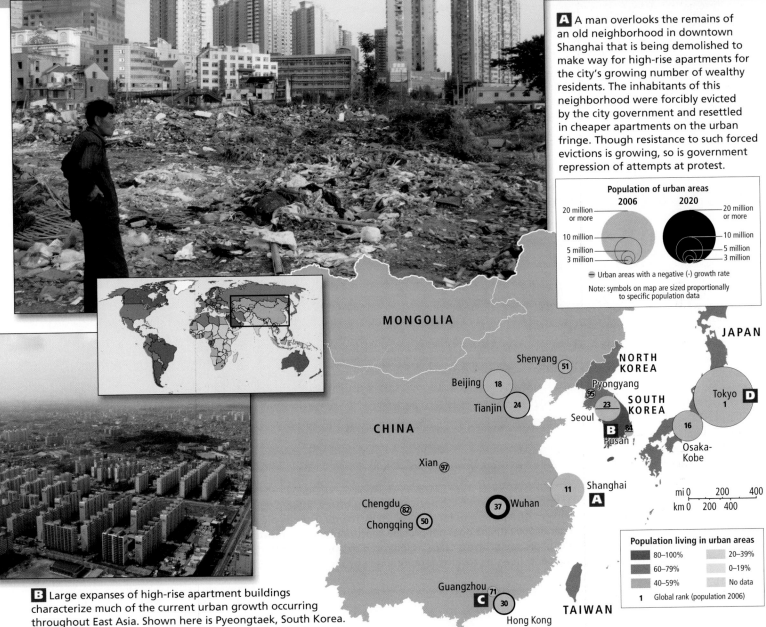

A A man overlooks the remains of an old neighborhood in downtown Shanghai that is being demolished to make way for high-rise apartments for the city's growing number of wealthy residents. The inhabitants of this neighborhood were forcibly evicted by the city government and resettled in cheaper apartments on the urban fringe. Though resistance to such forced evictions is growing, so is government repression of attempts at protest.

Population of urban areas

2006   2020

20 million or more — 20 million or more
10 million — 10 million
5 million — 5 million
3 million — 3 million

⊖ Urban areas with a negative (-) growth rate

Note: symbols on map are sized proportionally to specific population data

MONGOLIA

JAPAN

NORTH KOREA

Shenyang (51)

Beijing (18)   Pyongyang (95)   SOUTH KOREA   Tokyo **D** 1

Tianjin (24)   (23)

Seoul   (16)

B (84)   Osaka-Kobe

CHINA   Pusan

Xian (97)

Shanghai **A**

Chengdu (82)   (11)

Chongqing (50)   (37) Wuhan

mi 0   200   400
km 0   200   400

Population living in urban areas

80–100%	20–39%
60–79%	0–19%
40–59%	No data

1   Global rank (population 2006)

Guangzhou (71)

C (30)

TAIWAN

Hong Kong

B Large expanses of high-rise apartment buildings characterize much of the current urban growth occurring throughout East Asia. Shown here is Pyeongtaek, South Korea.

C Migrants in Guangzhou, China, head home for the spring holidays. China's urban growth is based on migration from rural areas.

D Ginza is the most lavish shopping district in Tokyo, the world's wealthiest and most populous city.

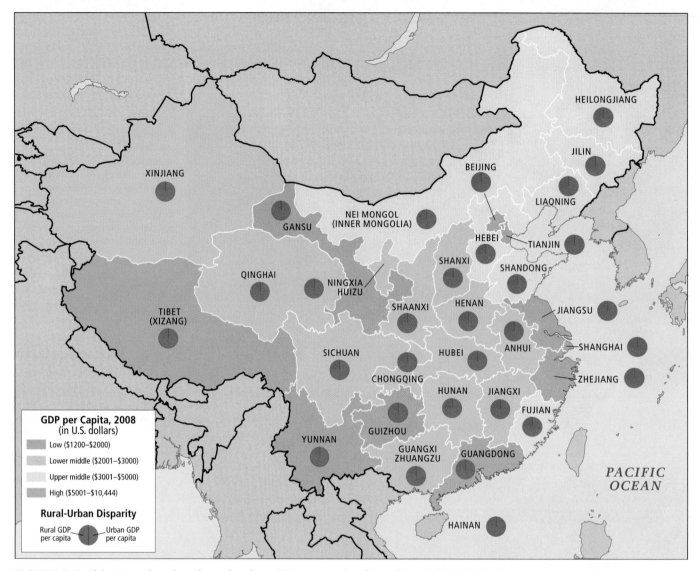

FIGURE 9.9 China's regional and rural–urban GDP per capita disparities, 2008–2009. The province colors represent the range of GDP per capita for the year 2008. The rural–urban pie diagrams are from data for the third quarter of 2009 (China does not provide concurrent data by province for 2008 or 2009). Notice the disparity in GDP per capita across China as indicated by the colors of the provinces as well as the rural–urban income disparity in each province as represented by the pie diagrams.

China would absorb Hong Kong and no longer allow it economic and political freedom.

While Hong Kong's democracy has been curtailed, political freedom is somewhat more evident there than elsewhere in China, and its role as China's unofficial link to the global economy has continued. Before 1997, some 60 percent of foreign investment in China was funneled through Hong Kong, and since then Hong Kong has remained the financial hub for China's booming southeastern coast (Figure 9.11). Because so much of this investment comes from Japan, Korea, and Taiwan, Hong Kong is an important regional financial hub as well.

II▶ 220. AS HONG KONG ENTERS SECOND DECADE UNDER CHINA, CITY PONDERS PLACE IN ECONOMIC GIANT

Shanghai's Latest Transformation Shanghai has a long history as a trendsetter. Its opening to Western trade in the early nineteenth century spawned a period of phenomenal economic growth and cultural development that led it to be called the "Paris of the East." As a result of China's recent reentry into the global economy, Shanghai is undergoing another boom, which has enriched some people and dislocated others (see Photo Essay 9.4A). Shanghai is today the world's busiest cargo port, and the region around Shanghai is now responsible for as much as a quarter of China's GDP.

In less than a decade, the city's urban landscape has been remade by the construction of more than a thousand business and residential skyscrapers; subway lines and stations; highway overpasses; bridges; and tunnels (see Thematic Overview F).

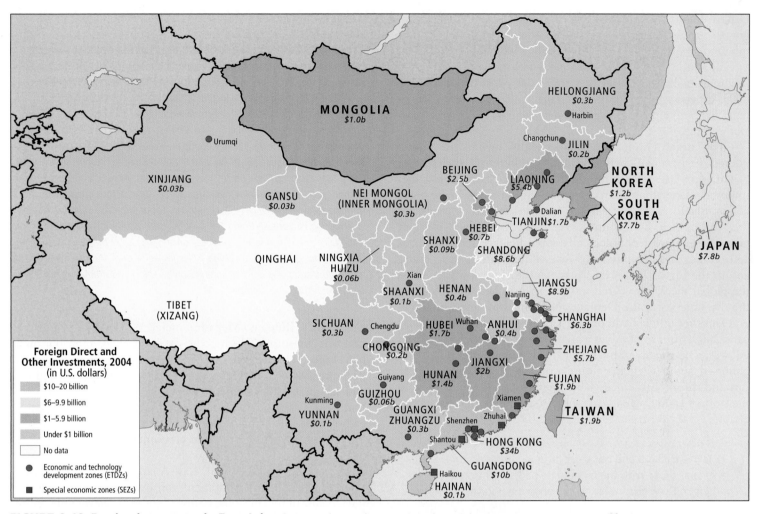

FIGURE 9.10 Foreign investment in East Asia. The map shows China's original special economic zones (SEZs) and more recently designated economic and technology development zones (ETDZs). The colors on the map reflect levels of foreign investment (direct and otherwise) in each of the countries of the region and in each of China's provinces.

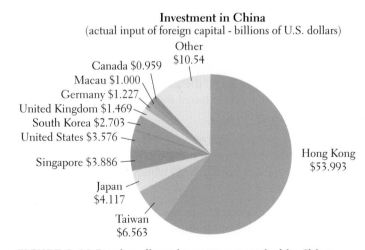

Investment in China
(actual input of foreign capital - billions of U.S. dollars)

Other $10.54
Canada $0.959
Macau $1.000
Germany $1.227
United Kingdom $1.469
South Korea $2.703
United States $3.576
Singapore $3.886
Japan $4.117
Taiwan $6.563
Hong Kong $53.993

FIGURE 9.11 Foreign direct investment capital in China, 2009. Foreign investment in China reached more than $90 billion in 2009. The graph depicts the actual amount invested by the top 10 investors—whose investments comprised 88.3 percent of the total FDI—and several other countries whose investments make up the remaining amount.

For hundreds of miles into the countryside, suburban development linked to Shanghai's economic boom is gobbling up farmland, and displaced farmers have rioted.

Pudong, the new financial center for the city, sits across the Huangpu River from the Bund—Shanghai's famous, elegant row of big brownstone buildings that served as the financial capital of China until half a century ago. Previously, Pudong was a maze of dirt paths and sprawling neighborhoods of simple tile-roofed houses, but its former residents were pushed out to make way for soaring high-rises.

Shanghai's long-standing role as a window on the outside world has meant that it often is host to quirky behavior that is less common elsewhere in the country. For example, for years people in Shanghai have relaxed in public in their pajamas—light, loose cotton tops and bottoms stamped with images of puppies or butterflies. The city government has tried to squelch the custom, but the citizens have proved recalcitrant. The police seem to understand that images of them arresting pudgy grandmothers for wearing pajamas would be ludicrous, so for now the issue is unresolved.

Urban Labor Surpluses and Shortages Spectacular urban growth in East Asia was based on hundreds of millions of new urban migrants who were willing to put up with almost any abuse to earn a little cash. However, so many factories, businesses, and shopping malls were built or were under construction that by 2005, experienced and skilled workers of all types were increasingly in short supply. Those with management skills were especially scarce.

To attract employees, some factory owners in China offered higher pay, better working conditions, and shorter workdays or increased time off. The extra costs these changes imposed meant that China was no longer the cheapest place to manufacture products. Some factories moved to Vietnam, the Philippines, and countries in Africa with even cheaper labor. The more technologically sophisticated items, such as cell phones and computers, which sell for more money, could absorb the higher wages for workers, at least for a while. Then the global recession, which was felt in Asia by 2008, changed the dynamic yet again. Demand for China's products dropped, and factories quickly laid off workers or shut down entirely. By early 2009, throngs of urban workers returned to their farms and villages, dejected about not being able to help their families and confused about what the future would hold for them. These ups and downs in China's labor market added a new twist to the story of Li Xia, whom we met in the vignette that opens this chapter.

Vignette Li Xia and her sister returned home to rural Sichuan a second time. With the money she had saved from her second job in Dongguan, Xia tried to open a bar in the front room of her parents' house. She hoped to introduce the popular custom of karaoke singing she had enjoyed in the city, but people in her village could not stand the noise, and family tensions rose. Xia's sister, delighted to be home with her husband and baby, quickly found a job in a small new fruit-processing factory—a rural enterprise. Her husband began farming again to fill the new demand for organic vegetables among the middle class in the Sichuan city of Chongqing.

In early 2007, amid all this success in her family and the failure of her own bar, news that training was now available in Dongguan for skilled electronics assemblers convinced Xia to try again. Past experience with the bureaucracy and her knowledge of Dongguan helped Xia to sign up for the electronics training. The cost of tuition was to come out of her wages, which were only U.S.$350 a month rather than the U.S.$400 she had thought she would be paid. Also, the waiting list for an apartment was long. She would need several years to repay her tuition, so she went back to shared quarters with her friends.

In November of 2008, Xia heard rumors that a global recession was causing orders for electronics to be cut and that she might soon be laid off. But her factory limped along into 2009 with a reduced staff. Then, miraculously, in June of 2009, orders picked up. Consumers in America had continued to buy electronics even as they downsized their homes and cars. Electronic inventories in America were down, so orders for products from Xia's factory went up significantly. And with so many laid-off

workers returning home, the apartment waiting list had shortened drastically. Now the apartment developers were only too happy to give her a firm move-in date. *[Sources: Paul Wiseman, "Chinese Factories Struggle to Hire," USA Today, April 11, 2005, at http://www.usatoday.com/money/world/2005-04-11-china-labor_x.htm; Louisa Lim, "The End of Agriculture in China," Reporter's Notebook, National Public Radio, May 19, 2006, at http://www.npr.org/templates/story/story.php?storyId=5411325; Mei Fong, "A Chinese Puzzle: Surprising Shortage of Workers Forces Factories to Add Perks; Pressures on Pay—and Prices," Wall Street Journal, August 16, 2004, p. B1; David Barboza, "Labor Shortage in China May Lead to Trade Shift," New York Times, April 3, 2006, Business Section, p. 1, at http://www.nytimes.com/2006/04/03/business/03labor.html?pagewanted=1&_r=1; Qiu Quanlin, "Labor Shortage Hinders Guangdong Factories," China Daily, August 25, 2009, at http://www.chinadaily.com.cn/china/2009-08/25/content_8612599.htm; Nina Ying Sun, "Labor Shortage Returns to China," Plastics News.Com, August 10, 2009, at http://plasticsnews.com/china/english/chinablog/2009/08/labor_shortage_returns_to_chin.html.]* ■

East Asia's Role as Mega-Financier When Americans were rampant consumers during the 2000s decade, they borrowed both from themselves and others. By 2009, foreign banks owned 25 percent of U.S. public debt—and of that 25 percent, Chinese and Japanese banks owned 44 percent, or about U.S.$1.5 trillion (**Figure 9.12**). China and Japan both had a great deal of cash because of their successful export economies, and they lent the money partly to ensure that consumers in the United States would continue to buy their goods.

Elsewhere in the world, East Asians are able to use their reserves to finance development and thereby secure privileged trade deals. In recent years, China's lending to developing countries has outpaced even that of the World Bank. Often these loans are made in return for guarantees that the recipient country will supply China with needed industrial materials, especially fossil fuels.

THINGS TO REMEMBER

1. Key to Japan's rapid economic recovery after World War II was close cooperation between the government and private investors to create new export-based industries, primarily automotive and electronic.

2. Japanese management innovations known as the kanban and kaizen systems also were important to economic recovery and were so successful that they diffused globally.

3. In the 1980s, China's leaders enacted market reforms that changed the country's economy in significant ways, and subsequently those market reforms transformed the region and indeed the whole world.

4. Learning Goal 3: Urbanization and Development China's cities are growing fast, with explosive economic development fueled by foreign direct investment and access to overseas markets.

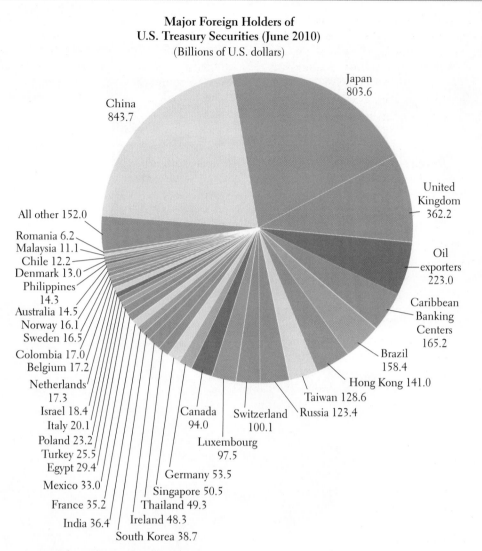

Major Foreign Holders of
U.S. Treasury Securities (June 2010)
(Billions of U.S. dollars)

China 843.7
Japan 803.6
United Kingdom 362.2
Oil exporters 223.0
Caribbean Banking Centers 165.2
Brazil 158.4
Hong Kong 141.0
Taiwan 128.6
Russia 123.4
Switzerland 100.1
Luxembourg 97.5
Canada 94.0
Germany 53.5
Singapore 50.5
Thailand 49.3
Ireland 48.3
South Korea 38.7
India 36.4
France 35.2
Mexico 33.0
Egypt 29.4
Turkey 25.5
Poland 23.2
Italy 20.1
Israel 18.4
Netherlands 17.3
Belgium 17.2
Colombia 17.0
Sweden 16.5
Norway 16.1
Australia 14.5
Philippines 14.3
Denmark 13.0
Chile 12.2
Malaysia 11.1
Romania 6.2
All other 152.0

FIGURE 9.12 Major foreign holders of U.S. Treasury securities, June 2010.

Democratization in East Asia

The demand for greater democracy is growing throughout East Asia. Japan's democracy was established after World War II, South Korea's in the late 1980s, and Taiwan's in the mid-1990s. All have steadily expanded democratization since their inception (see Thematic Overview G). Mongolia has had a dramatic expansion of democracy since abandoning socialism in 1992. North Korea and China, however, remain under the tight control of undemocratic regimes. In the Photo Essay 9.5 map on page 356, the colors of the countries roughly indicate the level of democracy enjoyed by the populace. The red starbursts show where civil unrest has broken out.

Learning Goal 4
Democratization: In what ways has China embraced changes that might eventually lead to democratization?

With China now a globalized economy, many wonder how much longer the Communist Party can remain in control without instituting democracy throughout the entire country. The Communist Party officially claims that China is a democracy, and indeed some

elections have long been held at the village level and within the Communist Party. However, representatives to the *National People's Congress*, the country's highest legislative body, are appointed by the Communist Party elite, who maintain tight control throughout all levels of government.

While radical change is unlikely in the near future, most experts on China agree that a steady shift toward greater democracy is underway. This change might be inevitable as the population becomes more prosperous, educated, and widely traveled, thereby becoming exposed to places that have greater freedoms and more open government. Demands for political change built to a crescendo less than a decade after market reforms began, culminating in a series of pro-democracy protests that drew hundreds of thousands to Beijing's Tiananmen Square in 1989 (see Photo Essay 9.5A). These protests were brutally repressed, with thousands (the precise number is uncertain) of students and labor leaders massacred by the military. Since then, pressure for change has continued to mount both within China and internationally, but it seems the link between economic development and democratization is only tenuous.

A The annual vigil commemorating those who died in the 1989 Tiananmen Square massacre in Beijing has become a focus for China's pro-democracy movement. As many as 150,000 people have shown up for the vigil in recent years. Hong Kong is the only part of China where the right to political protest is protected by law.

MONGOLIA

NORTH KOREA

A

CHINA

JAPAN
B

D

SOUTH KOREA

C

TAIWAN

Democratization and Conflict
Democratization Index

- Full democracy
- Flawed democracy
- Hybrid regime
- Authoritarian regime
- No data

B Tokyo's governor, Shintar Ishihara, campaigning in 2009, when his Liberal Democratic Party lost control of Japan's legislature for only the second time since 1955. The defeat marked a more "Chinacentric" era in Japanese politics.

C A protest in Taipei, Taiwan, organized by a political opposition party against the government's moves to strengthen ties with China. The issue of Taiwan's continued independence from China—including Taiwan's ability to maintain democratic political freedoms—plays a central role in the island's electoral politics.

Armed Conflicts and Genocides with High Death Tolls Since 1945

- Ongoing conflict
- 1000-10,000 deaths
- 10,000-100,000 deaths
- 100,000-1,000,000 deaths
- 1,000,000-3,000,000 deaths
- 30,000,000 deaths

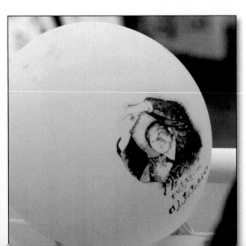

D A memorial balloon bearing the image of the late Kim Dae Jung, the president of South Korea from 1998 to 2003. Jung received the Nobel Peace Prize in 2000 for his work in supporting democracy and human rights during several periods of military dictatorship in South Korea during the 1970s and 1980s. He was also recognized for his efforts to reconcile with North Korea's authoritarian government.

Thinking Geographically

After you have read about democratization and conflict in East Asia, you will be able to answer the following questions:

A What happened in Tiananmen Square in 1989?

B Many Japanese people support closer relations with China in part because they think that China will likely be _____.

C When was democracy in Taiwan established?

D When was democracy in South Korea established?

International Pressures China's admission to the World Trade Organization in 2001 brought pressure for political change. Informed consumers and environmentalists in developed countries have long criticized China for its "no holds barred" pursuit of economic growth. Much of China's growth has been built on environmentally destructive activities and the abuse of workers, both of which effectively lower production costs so that goods can be sold at lower prices.

The 2008 Olympics in Beijing highlighted a number of issues that have also created an impetus for increasing democracy. Before the games, the international media shined its spotlight on human rights abuses of workers, protesters, prisoners, ethnic minorities, and spiritual and mystical groups such as Falun Gong. The media paid particular attention to protests in Tibet and elsewhere around the world against China's ongoing repression of Tibetan Buddhist monasteries and the mass importation of non-Tibetans into the province. More recently, China's repressive treatment of the Uygur people in Xinjiang Uygur Autonomous Region (discussed on pages 362–363) has received international attention. The extravagance of the 2008 Olympics became a subject of criticism by the global media, with many journalists pointing out that no democracy would ever be able to devote so many tax revenue (estimated at $40 billion) to such an event, especially not in a country with as many pressing human and environmental problems as China has.

▐▌▶ 211. OLYMPIC RELAY CUT SHORT BY PARIS PROTESTS

▐▌▶ 213. BEIJING OLYMPICS: POLITICAL BATTLEGROUND?

▐▌▶ 221. REPORTS OF SALE OF EXECUTED FALUN GONG PRISONERS' ORGANS IN CHINA CALLED "SHOCKING"

Information Technology and Democracy The spread of information via the Internet has increased the push for democratization from within China. Since the revolution in 1949, China's central government has controlled the news media in the country. By the late 1990s, however, the expanding use of electronic communication devices was loosening central control over information. By 2001, twenty-three million people were connected to the Internet in China, and by 2009, more than 338 million (or 25 percent of the population) were connected (see Figure 9.13 on page 358). Just a few years ago, telephones were very rare and required permission, but now anyone can buy temporary cell phone access without showing identification. As a result, millions of Chinese people have anonymous access to an international network of information. It is now much more difficult for the government to give inaccurate explanations for problems caused by inefficiency and corruption. Reporters can now check the accuracy of government explanations by phoning witnesses or the principal actors directly and then posting the explanations on the Internet. Analysts both inside and outside China see the availability of the Internet to ordinary Chinese citizens as a watershed event that supports democracy.

Nevertheless, the Internet in China is not the open forum that it is in most Western countries. For example, if people writing blogs in China use the words "democracy," "freedom," or "human rights," they may receive the following reminder: "The title must not contain prohibited language, such as profanity. Please type a different title." Such censorship has been aided by U.S. technology firms such as Yahoo and Microsoft, which allow the Chinese government to use software that blocks access to certain Web sites for users in China. Google, now in dispute with the Chinese government over state-sponsored censorship, invasion of human rights activists' email accounts, and cyber attacks, has moved operations to Hong Kong, where somewhat more freedom is allowed (see pages 350–352). The final chapter of Google's Internet ideology "war" with the Chinese government over freedom of Internet access and privacy has yet to be written.

Urbanization, Protests, and Technology May Enhance Democracy Protests by workers for better pay and living conditions, or riots by farmers and urban dwellers displaced by new real estate developments, have exposed average citizens to the depth of dissatisfaction felt by their compatriots. In 2003, China's government reported that there were 58,000 public protests in the country. This number rose to 74,000 in 2004, and 87,000 in 2005. Since then, the government has stopped issuing complete statistics on protests.

Once the people involved in this collective discontent gained access to communications technology, they were able to form *interest groups*, which are associations of individuals with particular complaints that require government action and perhaps compensation. These groups communicate face-to-face or via cell phones (or, to a lesser extent, on the Internet), formulate a position, and then coalesce into larger organizations with increasing power and momentum to demand governmental action. One of the most publicized of such protests was by parents whose children died during the earthquake (which registered 7.8 on the Richter scale) in Sichuan in May 2008. They demanded redress for poorly constructed schools that collapsed, killing thousands.

Groups that work toward gaining compensation are actually less threatening to the central government than those that work for political reform. In December 2008, people seeking more political freedoms signed a manifesto called the *08 Charter*, calling for a decentralized federal system of government (that is, more power to the provinces), democratic elections, and the end of the Communist Party's political monopoly. Liu Xiaobo, coauthor of the 08 Charter, was arrested, and in December 2009 was sentenced to 11 years in prison, even as prominent international human rights activists lobbied on his behalf. Three hundred other signatories were arrested, interrogated, and released. In 2010, Liu Xiaobo was awarded the Nobel Peace Prize, though

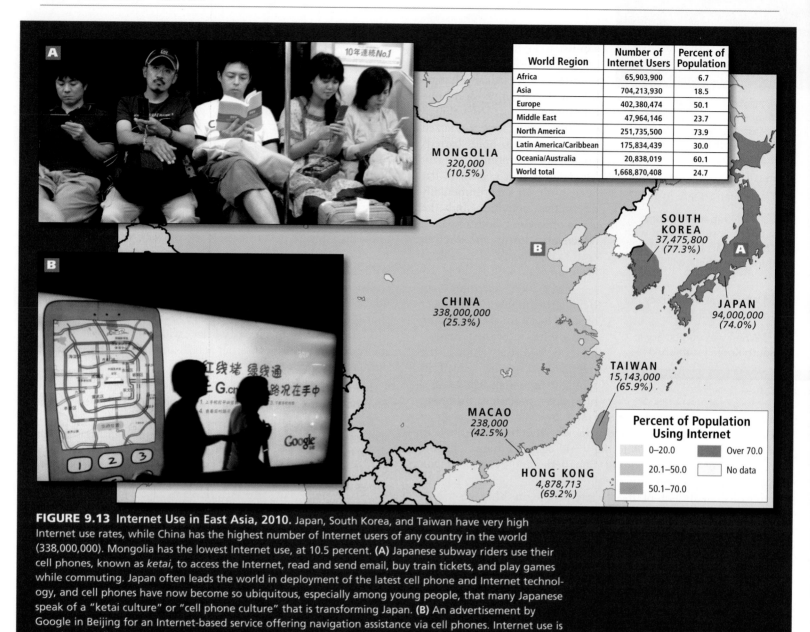

World Region	Number of Internet Users	Percent of Population
Africa	65,903,900	6.7
Asia	704,213,930	18.5
Europe	402,380,474	50.1
Middle East	47,964,146	23.7
North America	251,735,500	73.9
Latin America/Caribbean	175,834,439	30.0
Oceania/Australia	20,838,019	60.1
World total	1,668,870,408	24.7

MONGOLIA
320,000
(10.5%)

SOUTH KOREA
37,475,800
(77.3%)

JAPAN
94,000,000
(74.0%)

CHINA
338,000,000
(25.3%)

TAIWAN
15,143,000
(65.9%)

MACAO
238,000
(42.5%)

HONG KONG
4,878,713
(69.2%)

Percent of Population Using Internet

0–20.0
20.1–50.0
50.1–70.0
Over 70.0
No data

FIGURE 9.13 Internet Use in East Asia, 2010. Japan, South Korea, and Taiwan have very high Internet use rates, while China has the highest number of Internet users of any country in the world (338,000,000). Mongolia has the lowest Internet use, at 10.5 percent. **(A)** Japanese subway riders use their cell phones, known as *ketai*, to access the Internet, read and send email, buy train tickets, and play games while commuting. Japan often leads the world in deployment of the latest cell phone and Internet technology, and cell phones have now become so ubiquitous, especially among young people, that many Japanese speak of a "ketai culture" or "cell phone culture" that is transforming Japan. **(B)** An advertisement by Google in Beijing for an Internet-based service offering navigation assistance via cell phones. Internet use is growing rapidly in China.

neither he nor his family were allowed to attend the ceremony in Norway.

Political protests by minorities such as the Tibetans and the Uygurs are seen by the Chinese government as a very serious threat to the country. The government has used extreme repression against Tibetans and Uygurs. The effect that the protests by minorities will ultimately have on government policies is not yet clear.

In an effort to maintain control over China's increasingly articulate protestors, the government has allowed elections to be held for "urban residents' committees." The idea seems to be to provide a peaceful outlet for voicing frustrations and achieving limited change at the local level. Many of these elections are hardly democratic, with candidates selected by the Communist Party. However, in cities where unemployment is high and where

protests have been particularly intense, elections tend to be more free, open, and truly democratic. It could be that interest group protests are an important first step toward actual participatory democracy, even when some, such as those who signed the 08 Charter, are harshly stifled.

Japan's Recent Political Shifts Significantly, the government that played such a central role during the post–World War II rise of the Japanese economy was controlled from 1955 to 2009 by one political party, the Liberal Democratic Party (LDP) (see Photo Essay 9.5B). This has long led to criticism that Japan is lacking a meaningful democracy. However, in late August 2009, for the first time in 54 years, the Japanese people elected a government led by the opposition party, the Democratic Party of Japan (DPJ). With government debt at 200 percent of GDP, the new government faced the prospect of cutting many social services that had supported Japan's high standard of living. The DPJ-headed government appears to favor closer relations with China, recognizing that it will likely be the market of the future for Japanese exports, but it is also trying to reduce Japan's dependence on exports for income and to spur consumption of domestic products by lowering household taxes.

THINGS TO REMEMBER

1. **Learning Goal 4: Democratization** While radical change is unlikely in the near future, most experts on China agree that a steady shift toward greater democracy is underway. This change might be inevitable as the population becomes more prosperous, educated, and widely traveled, thereby becoming exposed to places that have greater freedoms and more open government.

2. By 2009, China had 338 million people using the Internet, but they represent only 25 percent of the country's population. China has strict controls on Internet access and use.

3. Chinese citizens who share a grievance against the government are increasingly cooperating with one another in their protests. In December 2008, a manifesto for change, the 08 Charter, was signed and submitted to the government by 350 protestors.

4. In Japan, the government was controlled from 1955 to 2009 by one political party, leading to criticism that Japan lacked a meaningful democracy. However, in August 2009, the Japanese people elected a government led by the opposition party.

Sociocultural Issues

East Asia's economic progress has led to social and cultural change throughout the region. Population growth, long a dominant issue, has slowed, while other demographic issues are receiving greater attention. Modernized economies, bearing some features of capitalism, are changing work patterns and family structures.

Population Patterns

Although East Asia remains the most populous world region, families there are having many fewer children than in the past. Of all world regions, only Europe has a lower rate of natural increase (<0.0 percent increase per year, compared to East Asia's 0.5 percent). In China this is partially due to government policies that harshly penalize families for having more than one child. But in China as well as elsewhere, urbanization and changing gender roles are also encouraging smaller families, regardless of official policy. Women in Mongolia (with 2.7 million people) and North Korea (with 22.7 million) are still averaging two or more children each, but even in these two countries family size is shrinking.

Responding to an Aging Population

Low birth rates mean that fewer young people are being added to the population, and improved living conditions mean that people are living longer across East Asia. The overall effect is that the average age of the populations is rising. Put another way, populations are aging. For Mongolia, South Korea, North Korea, and Taiwan, it will be several decades before the financial and social costs of supporting numerous elderly people will have to be addressed. China faces especially serious future problems with elder care because the one-child policy and urbanization have so drastically reduced family kin-groups. Japan, on the other hand, has already confronted the problem of having a large elderly population that requires support and a reduced number of young people to do the job.

Japan's Options Japan's population is growing slowly and aging rapidly, raising concerns about economic productivity and humane ways to care for dependent people. The demographic transition (see page 18 in Chapter 1) is well underway in this highly developed country where 86 percent of the population live in cities. Japan's rate of natural increase is in the negative range (<0.0), the lowest in East Asia, and on par with Europe's (see Thematic Overview H). If this trend continues, Japan's population will plummet from the current 127.6 million to 95.2 million by 2050.

At the same time, the Japanese have the world's longest life expectancy, at 83 years. As a result, Japan also has the world's oldest population, 23 percent of which is over the age of 65. By 2055, this age group will account for 40 percent of the population. By 2050, Japan's labor pool could be reduced by more than a third, but it would still need to produce enough to take care of more than twice as many retirees as it does now. Clearly, these demographic changes will have a momentous effect on Japan's economy, and the search for solutions is underway. One possibility is increasing immigration to bring in younger workers who will fill jobs and contribute to the tax rolls, as the United States and Europe have done. Another is to keep the elderly fit so they can work if necessary and take care of themselves.

In Japan, recruiting immigrant workers from other countries is a very unpopular solution to the aging crisis. Many Japanese people consider foreigners a source of "pollution," and the few small minority populations with cultural connections to China or Korea have long faced discrimination. The children of foreigners born in Japan are not granted citizenship, and some communities that have been in the country for generations are still thought of as foreign. Today, immigrants are fingerprinted and photographed upon entering Japan and must carry an "alien registration card" at all times.

Nonetheless, foreign workers are dribbling into Japan in a multitude of legal and illegal guises. Many are "guest workers" from South Asia brought in to fill the most dangerous and low-paying jobs with the understanding that they will eventually leave. Others are the descendants of Japanese people who once migrated to South America (Brazil and Peru). Regardless, Japan's foreign population remains tiny, making up only 1.2 percent of the total population (approximately 1.5 million people). A recent

United Nations report estimates that Japan would have to import over 640,000 immigrants per year just to maintain its present workforce and avoid a 6.7 percent annual drop in its GDP.

In a novel approach to Japan's demographic changes, the government has invested enormous sums of money in robotics over the past decade. Robots are already widespread in Japanese industries such as auto manufacturing, and their industrial use is growing. Now they are also being developed to care for the elderly, to guide patients through hospitals, to look after children, and even to make sushi. By 2025, the government plans to replace up to 15 percent of Japan's workforce with robots.

China's Options The proportion of China's population over 65 is now only 8 percent, but this will change rapidly as conditions improve and life expectancies increase by 10 or more years, as they have done elsewhere. However, a crisis in elder care is already upon China for two other reasons: the high rate of rural-to-urban migration and the shrinkage of family support systems because of the one-child policy (discussed below).

When hundreds of millions of young Chinese were lured into cities to work, most thought rural areas would benefit from remittances, and this has happened. However, few anticipated that the one-child family would mean that for every migrant two aging parents would be left to fend for themselves, often in rural, underdeveloped areas. The Chinese are being proactive: when throngs of elderly Chinese are seen exercising in public parks, few outsiders realize that this movement to keep the elderly fit is linked to far-sighted policies aimed at lowering the costs of supporting the aged.

China's One-Child Policy

> **Learning Goal 5**
> **Population and Gender:** How has the interaction between the government's population policy and Chinese culture resulted in a shortage of women in China?

In response to fears of overpopulation and environmental stress, China has had a one-child-per-family policy since 1979. The policy is enforced with rewards for complying and with large fines. As a result, China's rate of natural increase (0.5 percent) is slightly lower than that of the United States (0.6 percent), and roughly half the world average (1.2 percent). If the one-child family pattern continues, China's population will start to shrink sometime between 2025 and 2050 (Figure 9.14B), creating many of the same economic and social problems that Japan is now facing.

The one-child policy transformed Chinese families and Chinese society at large. For example, an only child has no siblings, so within two generations, the kinship categories of brother, sister, cousin, aunt, and uncle have disappeared from most families, meaning that any individual has very few if any related age peers with whom to share family responsibilities. The effect on society of the one-child family is that most children are doted on by several adults, and children are not taught by siblings to share. Conscious efforts must be made to instill self-sufficiency in only children. The one-child policy has sometimes been enforced brutally. In various times and places, the government has waged a campaign of forced sterilizations and forced abortions for mothers who already have one child.

Cultural Preference for Sons: Missing Females and Lonely Males The prospect of a couple's only child being a daughter, without the possibility of having a son in the future, is the aspect of the one-child policy that has caused families the most despair. For years, the makers of Chinese social policy have sought to eliminate the old Confucian-based preference for males by empowering women economically and socially. In many ways the policymakers are succeeding in empowering young women, but the preference for sons persists.

The population pyramid in Figure 9.14A illustrates the extent to which the preference for sons has created a gender imbalance in China's population. The normal sex ratio at birth is 105 boys to 100 girls, which evens out by the age of 5. In China, however, for nearly every category until age 70, males outnumber females. The census data show that there are already at least 22 million more men than women, and that this could rise to 24 million by 2020. What happened to the missing girls?

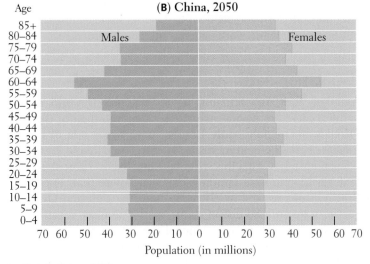

FIGURE 9.14 Population pyramids for China, 2009 and 2050 (projected). (A) China, 2009
(B) China, 2050

There are several possible answers. Given the preference for male children, the births of these girls may simply have gone unreported as families hoped to conceal their daughter and try again for a son. There are many anecdotes of girls being raised secretly or even disguised as boys. Also, adoption records indicate that girls are given up for adoption much more often than boys. Or the girls may have died in early infancy, either through neglect or infanticide. Finally, some parents have access to medical tests that can identify the sex of a fetus. There is evidence that in China, as elsewhere around the world, some of these parents choose to abort a female fetus.

The cultural preference for sons persists elsewhere in East Asia as well. A deficit of girls appears on the 2000 population pyramids for Japan, the Koreas, Mongolia, and Taiwan (see Thematic Overview I). Nonetheless, evidence shows that attitudes may be changing. In Japan, South Korea, and Mongolia, the percentage of women receiving secondary education equals or exceeds that of men.

A Shortage of Brides A major side effect of the preference for sons is that there is now a growing shortage of women of marriageable age throughout East Asia. China alone had an estimated deficit of 10 million women aged 20 to 35 in 2000. Females are also effectively "missing" from the marriage rolls because many educated young women are too busy with career success to meet eligible young men.

Research suggests that at least 10 percent of young Chinese men will fail to find a mate; and poor, rural, uneducated men will have the greatest difficulty. The shortage of women will lower the birth rate yet further, causing a more rapid shrinkage of the population over the next century. When single men age, without spouses, children, or even siblings, there will be no one to care for them. Furthermore, China's growing millions of single young men are emerging as a threat to civil order as they may be more prone to drug abuse, violent crime, HIV infection, and sex crimes. Cases of kidnapping and forced prostitution of young girls and women are already increasing.

Recently, some officials in China have supported abolishing the one-child policy, or adopting a two-child policy. Enforcement of the one-child policy is already lax in rural areas, and many ethnic minorities have long been exempt. Given China's growing urban populations, which face numerous pressures for small families, it is unlikely that abandoning the one-child policy would produce a population boom.

Population Distribution

In East Asia, people are not evenly distributed on the land (see Figure 9.15 on page 362). China, with 1.34 billion people, has more than one-fifth of the world's population. However, 90 percent of these people are clustered on the approximately one-sixth of the total land area that is suitable for agriculture, and roughly half of these live in urban areas. People are concentrated especially densely in the eastern third of China in the North China Plain, the coastal zone from Tianjin to Hong Kong that includes the delta of the Zhu Jiang (Pearl River) in the southeast, the Sichuan Basin, and the middle and lower Chang Jiang (Yangtze) basin.

The west and south of the Korean Peninsula are also densely settled, as are northern and western Taiwan. In Japan, settlement is concentrated in a band that stretches from the cities of Tokyo and Yokohama on the island of Honshu, south through the coastal zones of the Inland Sea to the islands of Shikoku and Kyushu. This urbanized region is one of the most extensive and heavily populated metropolitan zones in the world, accommodating 86 percent of Japan's total population. The rest of Japan is mountainous and more lightly settled. Mongolia is only lightly settled, with one modest urban area.

THINGS TO REMEMBER

1. More people live in East Asia than in any world region, but population growth is now slowing as family size decreases. Most of the region now faces the challenge of caring for large elderly populations.

2. The Japanese have the world's longest life expectancy, at 83 years, and the world's oldest population—23 percent of Japanese people are over the age of 65. By 2055, this age group will account for 40 percent of the population.

3. Learning Goal 5: Population and Gender China is grappling with an unexpected consequence of its one-child-per-family policy: a shortage of women. It is also struggling with the demise of sibling relationships and the extended family.

Cultural Diversity in East Asia

Most countries in East Asia have one dominant ethnic group, but all countries have considerable cultural diversity. In China, for example, 93 percent of Chinese citizens call themselves "people of the Han." The name harks back about 2000 years to the Han empire, but it gained currency only in the early twentieth century, when nationalist leaders were trying to create a mass Chinese identity. The term *Han* simply connotes people who share a general way of life, pride in Chinese culture, and a sense of superiority to ethnic minorities and outsiders. The main language spoken by the Han is Mandarin, although it is only one of many Chinese dialects.

China's non-Han minorities number about 117 million people in more than 55 different ethnic or culture groups scattered across the country. Most live outside the Han heartland of eastern China (see **Figure 9.16** on page 363). Some of these areas have been designated autonomous regions, where minorities theoretically manage their own affairs. In practice, however, the Han-dominated Communist Party in Beijing controls the fate of the minorities, especially those considered to be security risks or who have resources of economic value. We profile a few of China's ethnic groups here.

Western China's Muslims Muslims of various ethnic origins have long been prominent minorities in China. All are

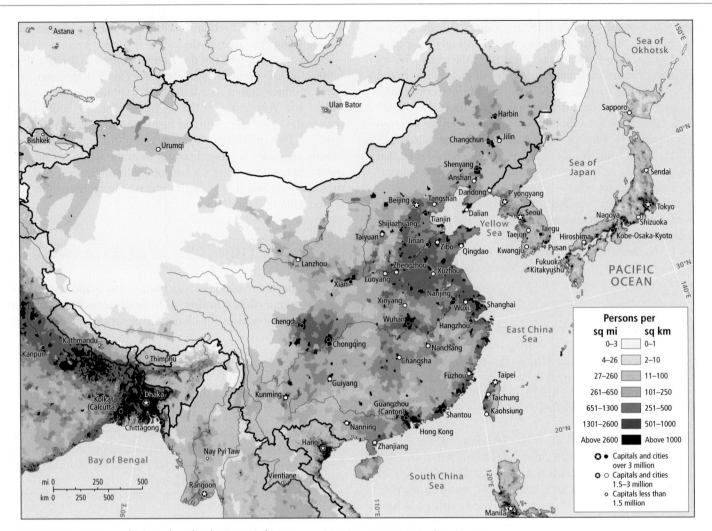

FIGURE 9.15 Population density in East Asia. More people live in East Asia than in any other world region, but population growth is now slowing as family size decreases. Most of the region now faces the challenge of caring for large elderly populations. Meanwhile, China is grappling with an unexpected consequence of its one-child-per-family policy: a shortage of women.

originally of Central Asian origins; most are Turkic people who historically have been seminomadic herders. Others specialized in trading. They tend to be concentrated in China's northwest and to think of themselves as quite separate from mainstream China.

In the autonomous region of Xinjiang Uygur in the far northwest (see Figure 9.3), live the Uygurs and Kazakhs, who are Turkic-speaking Muslims. Historically, these peoples were nomadic herders, and some still are. Contact with culturally related peoples in Central Asia has been revived since China's market reforms began (Figure 9.17). The Beijing government wishes to claim this area's oil, other mineral resources, and irrigable agricultural land for national development. Accordingly, it has sent troops and by now many millions of Han settlers to Xinjiang. The Han settlers fill most managerial jobs in the bureaucracy, in mineral extraction, the military, and power generation. An important secondary role of the Han is to dilute the power of Uygurs and Kazakhs within their own lands. In

Xinjiang Uygur, there are now as many Han as Uygurs (9 million each), plus small numbers of other minorities.

The Beijing government has rushed to develop Xinjiang and its capital Urumqi with special development zones (ETDZs), but the Uygurs have been left out of most policy-making roles and indeed have been excluded from participating in the economic boom. For example, one young Uygur man in Urumqi writes of his discontent: "I am a strong man, and well-educated. But [Han] Chinese firms won't give me a job. Yet go down to the railroad station and you can see all the [Han] Chinese who've just arrived. They'll get jobs. It's a policy to swamp us."

Until recently, the Uygur people of Xinjiang expressed their resistance to Han dominance merely by reinvigorating their Islamic culture. Islamic prayers were increasingly heard publicly, more Muslim women were wearing Islamic dress, Uygur was spoken rather than Chinese, and Islamic architectural traditions were being revived. Then more active resistance groups were formed. The Beijing government responded by harshly punishing

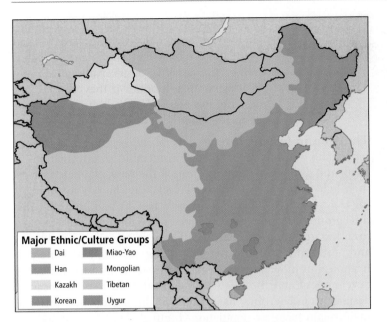

FIGURE 9.16 Major ethnic groups of China. This map shows the areas traditionally occupied by the Han and the ethnic minorities. It does not show the recent resettling of Han in Xinjiang and Tibet, nor does it show the Hui, people from many ethnic groups whose ancestors converted to Islam and who are found in disparate locations along the old Silk Road and in coastal southeastern China.

the resisters and broadcasting the accusation that they are Islamic fundamentalists bent on terrorism. In July 2009, violence erupted in the streets between Uygurs and the Han and about 200 people were killed.

Distinct from the Uygurs and Kazakhs are the *Hui*, who altogether number about 8 million. The original Hui people were descended from ancient Turkic Muslim traders who traveled

FIGURE 9.17 The Silk Road market in Kashi. Uygur men inspect currency before selling their goats at the Sunday market in Kashi, Xinjiang Uygur Autonomous Region, China. Kashi has been an important trading center along the Silk Road for at least 2000 years.

the Silk Road from Europe across Central Asia to Kashgar (now Kashi) and on to Xian in the east (see the map of the Silk Road in Figure 5.5 on page 183). Subgroups of Hui live in the Ningxia Huizu Autonomous Region and throughout northern, western, and southwestern China. There they continue as traders, farmers, and artisans. The long tradition of commercial activity among the Hui has facilitated their success in China. Many are active in the new free market economies of southeastern China as businesspeople, technicians, and financial managers, using their money not only to buy luxury goods, but also to revive religious instruction and to fund their mosques, which are now more obvious in the landscape.

The Tibetans In contrast to the prosperous assimilated Hui are the Tibetans, an impoverished ethnic minority of nearly 5 million individuals scattered thinly over a huge, high, mountainous region in western China. The history of Tibet's political status vis-à-vis China is long and complex, characterized by both cordial relations and conflict. During China's imperial era (prior to the twentieth century), Tibet maintained its own government but faced the constant threat of invasion and of Chinese meddling in its affairs. In the early 1900s, Tibet declared itself separate and free from China and conducted its affairs as an independent country. It was able to maintain this status until 1949–1950, when the Chinese Communist army "liberated" Tibet, promising a "one-country, two-systems" structure. A Tibetan uprising in 1959 led China to abolish the Tibetan government and violently reorder Tibetan society. Since the 1950s, the Chinese government has referred to Tibet as the Xizang Autonomous Region. The Chinese government suppressed the Tibetan Buddhist religion by destroying thousands of temples and monasteries and massacring many thousands of monks and nuns. In 1959, the spiritual and political leader of Tibet, the Dalai Lama, was forced into exile in India along with thousands of his followers.

‖ ▶ 216. DALAI LAMA CALLS FOR AUTONOMY BUT NOT INDEPENDENCE

By the 1990s, the Beijing government's strategy was to overwhelm the Tibetans with secular social and economic modernization and with Han Chinese settlers rather than outright military force (though China maintains a military presence in Tibet). To attract trade and quell foreign criticism of its treatment of Tibetans, China is spending hundreds of millions of dollars on housing and on roads, railroads, and a tourism infrastructure that capitalizes on European and American interest in Tibetan culture. China presents its actions in Tibet as part of its overall strategy to integrate the entire country economically and socially. Schools are being built and jobs opened up to young Tibetans. A new railway link connecting Tibet more conveniently to the rest of China was completed in 2006. The Han in Tibet see the railway as a public service that will promote Tibetan development, but Tibetan activists see it as a conveyor belt for more Han dominance over the Tibetan economy and culture.

‖ ▶ 214. TIBETAN BUDDHIST NUNS' PROTEST SONGS BRING PUNISHMENT FROM CHINA

Within Tibetan culture, women have held a somewhat higher position than in other cultures of East Asia. Buddhism did introduce many patriarchal attitudes to Tibet, but these did not curtail other more equitable traditions. Among nomadic herders, women could have more than one husband, just as men were free to have more than one wife; at marriage, a husband often joins the wife's family. By comparison, Han Chinese culture has typically regarded the women of Tibet and other western minorities as barbaric precisely because their roles were not circumscribed: they were not secluded, they rode horses, they worked alongside the men in herding and agriculture, and they were assertive.

Indigenous Diversity in Southern China In Yunnan Province in southern China, more than 20 groups of ancient native peoples live in remote areas of the deeply folded mountains that stretch into Southeast Asia. These groups speak many different languages, and many have cultural and language connections to the indigenous people of Tibet, Burma, Thailand, or Cambodia. Women and men are treated more equally in this area than among the Han. A crucial difference may be that among several groups, most notably the Dai, the husband moves in with the wife's family at marriage and provides her family with labor or income. A husband inherits from his wife's family rather than from his birth family, and female children are valued just as highly as males.

Taiwan's Many Minorities In Taiwan and the adjacent islands, the Han account for 95 percent of the population, but Taiwan is also home to 60 indigenous minorities. Some have cultural characteristics—languages, crafts, and agricultural and hunting customs—that indicate a strong connection to ancient cultures in

> **Ainu** an indigenous cultural minority group in Japan characterized by their light skin, heavy beards, and thick, wavy hair who are thought to have migrated thousands of years ago from the northern Asian steppes

far Southeast Asia and the Pacific. The mountain dwellers among these groups have resisted assimilation more than the plains peoples. Both groups may live on reservations set aside for indigenous minorities if they choose, but most are now being absorbed into mainstream urbanized Han-influenced Taiwanese life.

The Ainu in Japan There are several indigenous minorities in Japan and most have suffered considerable discrimination. A small and distinctive minority group is the **Ainu**, characterized by their light skin, heavy beards, and thick, wavy hair. Now numbering only about 16,000, the Ainu are a racially and culturally distinct group thought to have migrated many thousands of years ago from the northern Asian steppes. They once occupied Hokkaido and northern Honshu, living by hunting, fishing, and some cultivation, but they are now being displaced by forestry and other development activities. Few full-blooded Ainu remain because, despite prejudice, they have been steadily assimilated into the mainstream Japanese population (Figure 9.18).

East Asia's Largest Cultural Export: The Overseas Chinese

China has had an impact on the rest of the world not only through its global trade, but also through the migration of its people to nearly all corners of the world. The first recorded emigration by the Chinese took place over 2200 years ago. Since then, China's contacts spread eastward to Korea and Japan, westward into Central and Southwest Asia via the Silk Road, and by the fifteenth century, to Southeast Asia, coastal India, the Arabian Peninsula, and even Africa.

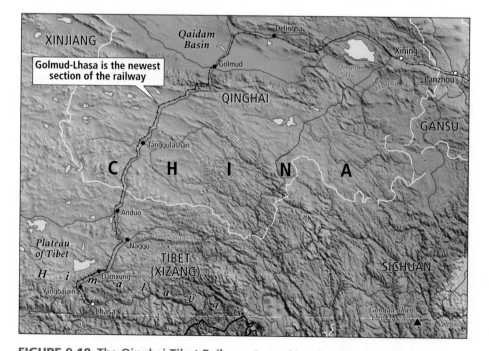

FIGURE 9.18 The Qinghai-Tibet Railway. Opened in July 2006, the world's highest railway travels nearly 596 miles (1,000 kilometers) at an altitude above 13,123 feet (4,000 meters). It now connects Beijing to Lhasa in less than 48 hours, transporting goods and people at a much lower cost than the existing highway. However, many Tibetans worry that it will become a conduit for more Han migration to Tibet.

Trade was probably the first impetus for Chinese emigration. The early merchants, artisans, sailors, and laborers came mainly from China's southeastern coastal provinces. Taking their families with them, some settled permanently on the peninsulas and islands of what are now Indonesia, Thailand, Malaysia, and the Philippines. Today they form a prosperous urban commercial class known as the "Overseas Chinese."

In the nineteenth century, economic hardship in China and a growing international demand for labor spawned the migration of as many as 10 million Chinese people to countries all over the world. By the middle of the twentieth century, they were joined by many others fleeing the repression of China's Communist Revolution. As a result, "Chinatowns" are found in most major world cities. The term *Overseas Chinese* has been extended to apply to Chinese emigrants and their descendants in all such locations.

THINGS TO REMEMBER

1. There are several distinct minority groups in East Asia: the Uygurs, Kazakhs, and Tibetans in western China; the Hui, scattered throughout central China; smaller indigenous groups in southern China and Taiwan; and the Ainu in Japan.

2. There is resentment and sometimes open resistance to Han Chinese domination in the various minority homelands.

3. Throughout East Asia, minorities have experienced discrimination.

4. Millions of "Overseas Chinese" live in cities and towns around the world. Most have settled permanently in their foreign locations, and many (perhaps most) maintain relationships with China.

Reflections on East Asia

Just as the astounding economic rise of Japan, South Korea, and Taiwan became the model for developing countries over the past 50 years, the opening of China to the global economy will continue to transform East Asia and the world for the next 50. The changes underway in this region today are immense. Incomes have risen and cities have boomed, but at the cost of some of the worst air and water pollution in the world. Greenhouse gas emissions have increased, but so has the ability to use new technologies to create cleaner energy economies (Reasons for Optimism A). The ability of countries to feed their own people with domestic agricultural resources is declining, but the ability to pay for food imported from abroad is increasing.

Wealth disparities in China have increased dramatically, with the income of millions hurt by the reformulation of the communist command economy and the loss of the safety nets it provided, while the income of others has been helped by new economic opportunities created by the forces of free market

Reasons for Optimism in East Asia

Climate Change: While greenhouse gas emissions have increased, so has the ability to use new technologies to create cleaner energy economies. **A** *A wind farm in China's Changshan islands.* ▼

Globalization and Development: Economic development related to globalization has boosted incomes throughout the region. **B** *A worker in Wuxi, China, assembles computer hardware.* ▼

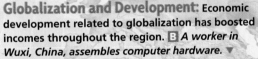

Democratization: Tensions created by China's recent economic and social transformations may eventually lead to democratization. **C** *A pro-democracy demonstration in Hong Kong in 2010.* ▼

Thinking Geographically: For what main reason will China's greenhouse gas emissions double, even if there are large increases in efficiency through new technologies?

Thinking Geographically: The extra costs imposed by offering Chinese workers higher pay, better working conditions, and shorter work days or increased time off means that China is _____.

Thinking Geographically: What factors bolster the shift toward greater democracy in China?

globalization (see Reasons for Optimism B). Shortages of skilled labor are bringing better wages and working conditions, and may also hasten a shift toward the production of more technologically sophisticated and higher-quality goods.

A major question remains regarding China's growing and future prosperity. Where will China find the natural resources to supply increasing living standards for so many people? China's massive importation of food, minerals, and oil is already affecting many regions of the world (see page 105 in Chapter 3, page 271 in Chapter 7, and page 389 in Chapter 10). These areas may not be willing or able to sustain such massive resource exports for long.

Throughout East Asia, as birth rates decline and populations age, some economies may suffer if they fail to either import labor or increase mechanization. This shift is already occurring in Japan, South Korea, and Taiwan, whose existing broad prosperity has given them more breathing room. China's economy may take a bigger hit because the one-child-per-family policy has reduced population growth drastically before the country has achieved broader prosperity. This, combined with China's gender imbalance, may produce considerable social instability. And as more women throughout East Asia pursue careers instead of family life, birth rates will continue to fall, bringing further social and economic changes.

Social and political tensions created by these transformations may eventually force China to adopt the democratic systems that already prevail among East Asia's wealthier countries (see Reasons for Optimism C). Still, many of China's problems will remain even if and when strong democratic institutions are put in place. Democracy has not solved the problems of an aging population in Japan, Taiwan, and South Korea, or of environmental degradation. In this ever-more-globalized world, the ways in which China and East Asia as a whole solve these problems will influence all of our lives.

Learning Goal Review Questions

1. Climate Change and Water: How might climate change result in changes in water availability in East Asia?

East Asia has long suffered from enormously destructive droughts and devastating floods. Now China's highest glaciers are melting. How might this effect the country's rivers? In what ways is China more vulnerable to global warming than Japan, the Koreas, or Taiwan?

2. Food and Globalization: How has East Asia's ability to feed itself been transformed by globalization?

East Asian countries have become more able to buy food on the global market and less able to produce the food they need at home. What kinds of problems might this create both for East Asia and the rest of the world?

3. Urbanization and Development: How is China's spectacular urban growth, and the massive increase in pollution that has accompanied it, linked to processes of globalization that were set in motion three decades ago?

China's cities have grown by 450 million people since 1980, when China's economy was opened to the global economy. How did globalization and China's response to it lead the country's cities to where they are today?

4. Democratization: In what ways has China embraced changes that might eventually lead to democratization?

As China's economy continues to develop and contacts with the rest of the world continue to multiply, its population is becoming more prosperous, educated, and well traveled. How might these changes help bring about greater democratization?

5. Population and Gender: How has the interaction between the government's population policy and Chinese culture resulted in a shortage of women in China?

China's one-child policy has left most couples with only one chance to give their families an heir. Why do most families want a male child? What problems have been created by the resulting shortage of women?

Geographic Themes about East Asia

Look back at the Thematic Overview photos on page 327. Thinking geographically, answer the following questions about them:

(A) Climate Change: What has been happening to the levels of greenhouse gas emitted by Japan in the last several decades?

(C) Food: How much of the global fish catch does Japan consume in a year?

(D) Development: China is currently the largest producer of what?

(E) Globalization: China's industries supply consumers where?

(F) Urbanization: The region around Shanghai is responsible for how much of China's GDP?

(G) Democratization: When was Japan's democracy established?

(H) Population: Compared with the rest of East Asia, what is Japan's rate of natural increase?

(I) Gender: What is the major side effect of the preference for sons in East Asia?

Key Terms

Ainu 364

Confucianism 342

Cultural Revolution 344

export-led growth 347

floating population 329

food security 337

foreign direct investment 350

Great Leap Forward 344

growth poles 350

hukou system 329

kaizen system 348

kanban system 348

regional self-sufficiency 349

regional specialization 350

responsibility system 349

special economic zones (SEZs) 350

state-aided market economy 347

tsunami 332

typhoon 334

wet rice cultivation 338

10 Southeast Asia

Learning Goals

After you read this chapter, you will be able to answer the following questions:

1. Climate Change, Food, and Water: How is global warming linked to deforestation, food production, and water resources in Southeast Asia?

2. Globalization and Development: How has globalization produced both spectacular successes and tragic failures in the economies of Southeast Asia?

3. Democratization: Is democratization influencing levels of violence and corruption in Southeast Asia?

4. Population, Development, and Gender: How is economic development affecting gender roles in ways that contribute to slower population growth in Southeast Asia?

5. Urbanization: What changes are driving urbanization in this region?

FIGURE 10.1 Political map of Southeast Asia.

Thematic Overview of Southeast Asia

Climate Change: Deforestation is a major contributor to global climate change, and Southeast Asia has accounted for 25 percent of deforestation globally over the past 10 years. **A** *Forest cleared for oil palms in Indonesia.* ▼

Food: Southeast Asia's most productive form of agriculture is wet rice cultivation, which has transformed landscapes throughout the region. **B** *A farmer applies pesticide to his field in the Philippines.* ▼

Water: Many parts of Southeast Asia have already adapted to dramatic seasonal floods, which have occurred in this region for most of human history. **C** *A houseboat in Cambodia.* ▼

Globalization: Export-led economic development has focused investment on industries that manufacture products for export, primarily to developed countries. **D** *Cranes being used to load goods for export onto cargo ships in Bangkok, Thailand.* ▼

Development: After decades of economic development based on manufacturing industries, the service sector is emerging as a powerful force. **E** *A shopping mall in Singapore, where the service sector dominates the economy.* ▼

Democratization: Movement toward democracy has been uneven. **F** *Demonstrators in 2010 in Bangkok, Thailand, protest a government that came to power after a military coup in 2006.* ▼

Population: Economic development is changing gender roles, and government policies are influencing population growth. **G** *A father and daughter in Singapore, where single-child families are increasingly common.* ▼

Urbanization: Southeast Asia as a whole is only 43 percent urban, but the rural–urban balance is shifting in response to declining agricultural employment and booming urban industries. **H** *Manila, Philippines.* ▼

Gender: Southeast Asia has become one of several global centers for the sex industry, supported in large part by international visitors willing to pay for sex. **I** *Patpong, Bangkok, Thailand's red-light district.* ▼

Global Patterns, Local Lives In December 2005, a group of indigenous people in the Malaysian state of Sarawak, on the island of Borneo (see **Figure 10.4** on pages 372–373), attended a public meeting wearing orangutan masks. They carried signs informing onlookers that, although the government protects orangutan territories, it ignores the basic right of indigenous people to live on their own ancestral lands.

Over the years, Sarawak forest dwellers have tried many tactics to save their lands from deforestation by logging companies and expanding oil palm plantations. In the mid-twentieth century, the state government began licensing logging companies to cut down forests occupied by indigenous peoples. By the 1980s, 90 percent of Sarawak's lowland forests had been degraded and 30 percent had been clear-cut, with much of the cleared area replaced by oil palm plantations. As a consequence, indigenous people found that even on uncleared land their hunts declined. Meanwhile, streams and rivers became polluted with eroded sediment and by fertilizers and the pesticides applied to the palm trees. Both the cut timber and the oil from the palm plantations are exported to distant markets in China, Japan, Europe, and North America, and few of the profits return to the communities affected by the conversion of forests to plantations.

In the 1990s, a group of citizens in Berkeley, California, who were concerned about news reports of the deforestation in Sarawak, organized the Borneo Project. They offered to become a "sister city" to one indigenous group in Sarawak, the Uma Bawang, giving help wherever it was needed. With the help of the Borneo Project, the Uma Bawang began a community-based mapping project in 1995. Using rudimentary compass and tape techniques, they began mapping both the extent and content of their forest home (**Figure 10.2**). Since then, indigenous people from across Sarawak have learned how to use global positioning systems (GPS), geographic information systems (GIS), and satellite imagery to make more sophisticated maps. The power of these maps was demonstrated in 2001, when they helped win a precedent-setting court case that protected indigenous lands from an encroaching oil palm plantation.

This favorable court ruling was challenged in 2005, when the government appealed, but in 2009, the Malaysian federal court ruled in favor of the Uma Bawang, stating that native customary land rights had been protected since 1939, when British colonial officials had directed the district lands and survey departments to map the boundaries of native lands. Now indigenous people have the right to sue the government for past illegal leases to logging companies, and some 203 such cases are in litigation. [Adapted from the Borneo Wire (newsletter of the Borneo Project), Jessica Lawrence, ed., Spring 2006, at http://borneoproject.org/article.php?id=623; "Community Stops Illegal Logging and Bulldozing Towards Protected Rainforests," Borneo Wire, Jessica Lawrence, ed., Fall 2006, at http://borneoproject. org/article.php?list=type&type=39; Mark Bujang, "A Community Initiative: Mapping Dayak's Customary Lands in Sarawak," presented at the Regional Community Mapping Network Workshop, November 8–10, 2004, Diliman, Quezon City, Philippines; "Malaysia Highest Court Affirms Tribes' Land Rights," Julia Zappei, Associated Press, May 10, 2009, available at http://borneoproject.org/article.php?id=762; and "Actualization of the UN Declaration on the Rights of Indigenous Peoples (UNDRIP)," International Forum on Globalization: Indigenous Rights Programs, at http://www.ifg.org/ programs/indigenousrights.htm.] ∎

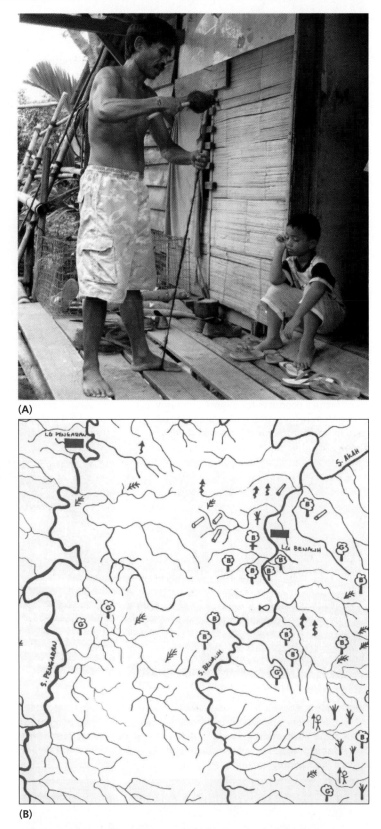

(A)

(B)

FIGURE 10.2 Indigenous people in Sarawak, Borneo. **(A)** An indigenous man in Sarawak keeps his traditional rope-making skills alive by demonstrating them to his son. **(B)** Villagers drew the boundaries of their lands in Penan, Sarawak. They included important features such as longhouses (communal dwellings for extended kin-groups), sago palm trees, hunting areas, and graveyards, as well as physical features such as rivers.

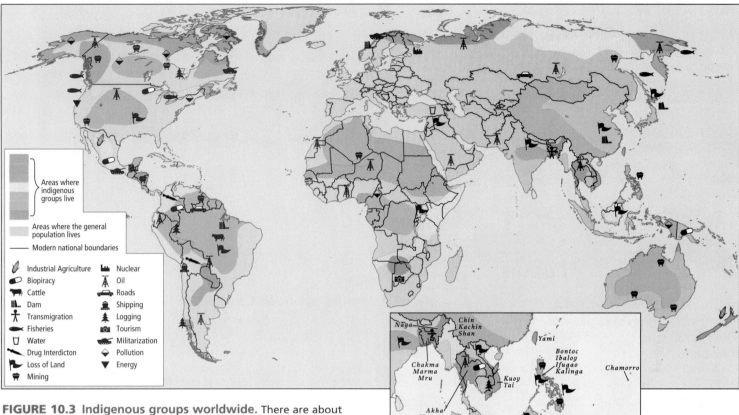

FIGURE 10.3 Indigenous groups worldwide. There are about 5000 distinct indigenous groups in the world. On this map, each color represents one or more of these groups that are related by language, culture, or an affinity to a geographic location. Many of these peoples have participated in community mapping projects, similar to those of the Sarawak forest dwellers, in order to identify and protect their rights to their traditional lands. The symbols reflect global issues that affect a particular group.

This account of the tactics that indigenous groups are using to secure their rights to ancestral lands highlights the extent to which local or national issues are becoming global issues. As problems like climate change become better understood, once-local issues are becoming international or even global in nature. This presents both challenges and opportunities for the world's indigenous people (**Figure 10.3**). While distant global markets provide the demand for timber and palm oil, international support from the Borneo Project also enables communities like the Uma Bawang to have their land claims validated. Moreover, mapping as a tactic for helping indigenous groups secure their legal rights to ancestral lands has now become a global phenomenon. Hundreds of indigenous groups throughout the world are collaborating with mapping specialists, many of them geographers, often with the result that their land rights are formally recognized by governments. The 2009 United Nations *State of the World's Indigenous Peoples* report documents several of these efforts, and is available online at http://www.un.org/esa/socdev/unpfii/documents/SOWIP_web.pdf.

THINGS TO REMEMBER

1. Indigenous land claims are no longer just local issues. Indigenous groups have shifted their efforts to secure their rights to ancestral lands to the global scale, highlighting a trend in which many issues that were once local or national are now becoming global concerns.

I THE GEOGRAPHIC SETTING

Terms in This Chapter

Many governments in Southeast Asia (see **Figure 10.1** on page 368) choose to dispense with place-names that originated in their colonial past. However, when the governments that make these changes earn broad disrespect in the international community by violating the human rights of their citizens, their chosen name may not be acknowledged. Such is the case with Burma, where a military government seized control in a coup d'état in 1990, changing the country's name to Myanmar. In this text, we use the country's traditional name of Burma instead of Myanmar to

A Gorges, Burma (Myanmar)

B Mekong River Plains, Laos

C Philippine Archipelago

D Volcanoes, Java, Indonesia

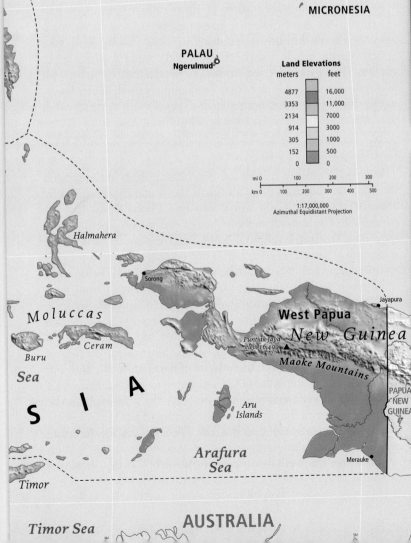

E Post-Earthquake/Tsunami, Sumatra, Indonesia

FIGURE 10.4 Regional map of Southeast Asia.

(Japan)

Okinawa Islands

PACIFIC OCEAN

MICRONESIA

PALAU
Ngerulmud

Land Elevations

meters	feet
4877	16,000
3353	11,000
2134	7000
914	3000
305	1000
152	500
0	0

mi 0   100   200   300

km 0   100   200   300   400   500

1:17,000,000
Azimuthal Equidistant Projection

Halmahera

Sorong

Jayapura

West Papua

New Guinea

Puntjak Jaya
elev 16,499

Maoke Mountains

Moluccas

Ceram

Buru

Sea

PAPUA NEW GUINEA

S I A

S

Aru Islands

Arafura Sea

Merauke

Timor

Timor Sea

AUSTRALIA

acknowledge the repression of the people of that country by a government they had no role in choosing.

Another potential point of confusion is Borneo, a large island that is shared by three countries. The part of the island known as Kalimantan is part of Indonesia; Sarawak and Sabah are part of Malaysia; and Brunei is a very small, independent, oil-rich country.

Physical Patterns

The physical patterns of Southeast Asia have a continuity that is not immediately obvious on a map. A map of the region shows a unified mainland region that is part of the Eurasian continent and a vast and complex series of islands arranged in chains and groups. These landforms are actually related in origin. Climate is another source of continuity, with most of the region tropical or subtropical.

Landforms

Southeast Asia is a region of peninsulas and islands (see Figure 10.4). Although the region stretches over an area larger than the continental United States, most of that space is ocean; the area of all the region's land amounts to less than half that of the contiguous United States. The large Indochina peninsula that extends to the south of China is occupied by Burma, Thailand, Laos, Cambodia, and Vietnam. This peninsula itself sprouts the long, thin Malay peninsular that is shared by outlying parts of Burma and Thailand, a part of Malaysia, and the city-state of Singapore, which is built on a series of islands at the southern tip. The **archipelago** (a series of large and small islands) that fans out to the south and east of the mainland is grouped into the countries of Indonesia, Malaysia, Brunei, Timor-Leste (East Timor), and the Philippines. Indonesia alone has some 17,000 islands, and the Philippines has 7000.

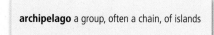

archipelago a group, often a chain, of islands

The irregular shapes and landforms of the Southeast Asian mainland and archipelago are the result of the same tectonic forces that were unleashed when India split off from the African Plate and gradually collided with Eurasia (see Figure 1.21 on page 45). As a result of this collision, which is still under way, the mountainous folds of the Plateau of Tibet reach heights of almost 20,000 feet (6100 meters). These tectonic folds, which bend out of the high plateau, turn south into Southeast Asia. There, they descend rapidly and then fan out to become the Indochina peninsula. Deep gorges widen out into valleys that stretch toward the sea, each containing hills of 2000 to 3000 feet (600 to 900 meters) and a river or two flowing from the mountains of the Yunnan–Guizhou Plateau of China to the Andaman Sea, the Gulf of Thailand, and the South China Sea (see Figure 10.4A). The major rivers of the peninsula are the Irrawaddy and the Salween in Burma; the Chao Phraya in Thailand; the Mekong, which flows through Laos, Cambodia, and Vietnam; and the Black and Red rivers of northern Vietnam. Several of these rivers have major delta formations—especially the Irrawaddy, Chao Phraya, and the Mekong—that are densely used for agriculture and increasingly for settlement (see the discussion on pages 371 and 392; see also Figure 10.4B).

The curve formed by Sumatra, Java, the Lesser Sunda Islands (from Bali to Timor), and New Guinea conforms approximately to the shape of the Eurasian Plate's leading edge (see Figure 1.21 on page 45). As the Indian-Australian Plate plunges beneath the Eurasian Plate along this curve, hundreds of earthquakes and volcanoes occur, especially on the islands of Sumatra and Java. Volcanoes and earthquakes also occur in the Philippines, where the Philippine Plate is pushing against the eastern edge of the Eurasian Plate. The volcanoes of the Philippines are part of the Pacific Ring of Fire (see Figure 1.22 on page 46; see also Figure 10.4C).

Volcanic eruptions, and the mudflows and landslides that occur in their aftermath, endanger and complicate the lives of many Southeast Asians (see Figure 10.4D). Over the long run, though, the volcanic material creates new land and provides minerals that enrich the soil for farmers. Earthquakes are especially problematic because of the tsunamis they can set off. The tsunami of December 2004, triggered by a giant earthquake (9.3 in magnitude) just north of Sumatra, swept east and west across the Indian Ocean, taking the lives of 230,000 people and injuring many more (see Figure 10.4E). It was one of the deadliest natural disasters in recorded history. A series of strong earthquakes have occurred since, with several in August and September of 2009 and April and May 2010 along coastal Sumatra, but they did not generate notable tsunamis.

The now-submerged shelf of the Eurasian continent that extends under the Southeast Asian peninsulas and islands was above sea level during the recurring ice ages of the Pleistocene epoch, during which much of the world's water was frozen in glaciers. The exposed shelf, known as Sundaland (Figure 10.5), allowed ancient people and Asian land animals (such as elephants, tigers, rhinoceroses, and orangutans) to travel south to what, when sea levels rose, became the islands of Southeast Asia.

Climate and Vegetation

The largely tropical climate of Southeast Asia is distinguished by continuous warm temperatures in the lowlands—consistently above 65°F (18°C)—and heavy rain (see Photo Essay 10.1 on page 376). The rainfall is the result of two major processes: the monsoons (seasonally shifting winds) and the movement of the intertropical convergence zone (ITCZ), an area centered roughly around the equator where surface winds converge and rise upward, resulting in rainfall. The wet summer season extends from May to October, when the warming of the Eurasian landmass sucks in moist air from the surrounding seas and pulls the ITCZ northward (see Photo Essay 10.1). Between November and April, there is a long dry season on the mainland, when the seasonal cooling of Eurasia causes dry air from the interior of the continent to flow out toward the sea, pushing the ITCZ southward (see Photo Essay 11.1 on page 413). On the many islands, however, the winter can also be wet because the air that flows from the continent picks up moisture as it passes south and east over the seas. The air releases its moisture as rain after ascending high enough over landforms to cool. With rains coming from both the monsoon and the ITCZ, the island part of Southeast Asia is one of the wettest areas of the world.

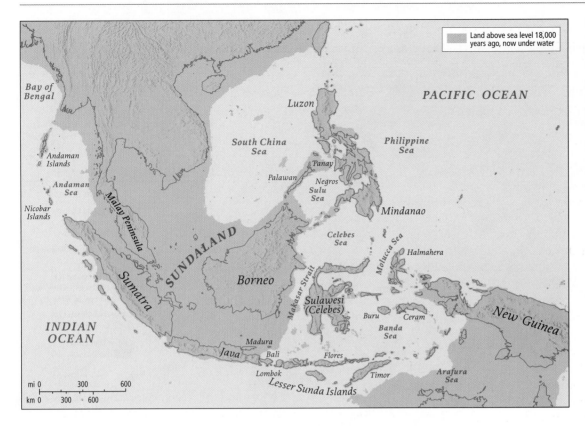

FIGURE 10.5 Sundaland 18,000 years ago, at the height of the last ice age. The now-submerged shelf of the Eurasian continent that extends under Southeast Asia's peninsulas and islands was exposed during the last ice age and remained above sea level until about 16,000 years ago, when that ice age was ending.

Irregularly every 2 to 7 years, the normal patterns of rainfall are interrupted, especially in the islands, by the **El Niño** phenomenon (see Figure 11.5 on page 414). In an El Niño event, the usual patterns of air and water circulation in the Pacific are reversed. Ocean temperatures are cooler than usual in the western Pacific near Southeast Asia. Instead of warm, wet air rising and condensing as rainfall, cool, dry air sits on the ocean surface. The result is severe drought, with often catastrophic results for farmers and tinder-dry forests that regularly catch fire. The cool El Niño air can trap smoke and other pollutants at the earth's surface, creating unusually toxic smog and low visibility.

The soils in Southeast Asia are typical of the tropics. Although not particularly fertile, they will support dense and prolific vegetation when left undisturbed for long periods. The warm temperatures and damp conditions promote the rapid decay of **detritus** (dead organic material) and the quick release of useful minerals. These minerals are taken up directly by the living forest rather than enriching the soil. Because rainfall is usually abundant during the summer wet season (when drought is only episodic), this region has some of the world's most impressive forests in both tropical and subtropical zones (see Photo Essay 10.1 B, C). On the mainland, human interference, coupled with the long winter dry season can negatively affect forest cover (see Photo Essay 10.1A). Human–environment relations that affect the forests are discussed further below.

> **El Niño** a recurring phenomenon during which the usual patterns of air and water circulation in the Pacific are reversed. Ocean temperatures are cooler than usual in the western Pacific near Southeast Asia and warmer than usual along the southern Pacific coast of South America
>
> **detritus** dead organic material (such as plants and insects) that collects on the ground

THINGS TO REMEMBER

1. The irregular shapes and landforms of the Southeast Asian mainland and archipelago are the result of the same tectonic forces that were unleashed when India split off from the African Plate and crashed into Eurasia.

2. The tropical climate of Southeast Asia is distinguished by continuous warm temperatures in the lowlands and heavy rain, except during those irregular years of the El Niño phenomenon.

Environmental Issues

Many of the environmental issues in Southeast Asia are in some way related to climate change. However, as we saw in the chapter introduction, deforestation has multiple effects (see **Figure 10.6** on page 377; see also **Photo Essay 10.2** on page 378). One is the loss of living space and resources for indigenous people of the forest. Another is the loss of habitat for orangutans, the Sumatran tiger, the Sumatran rhinoceros, and tens of thousands of other less spectacular species that are lost when rain forests are converted to agricultural uses. Finally, the emissions of greenhouse gases that accompany deforestation processes foster climate change.

II ▶ 222. INDONESIA'S POLLUTED ENVIRONMENT THREATENS HAWKSBILL TURTLE

The presence of so much wet tropical air, combined with the movement of the Intertropical Convergence Zone (ITCZ) across the region twice a year, makes Southeast Asia one of the wettest regions in the world.

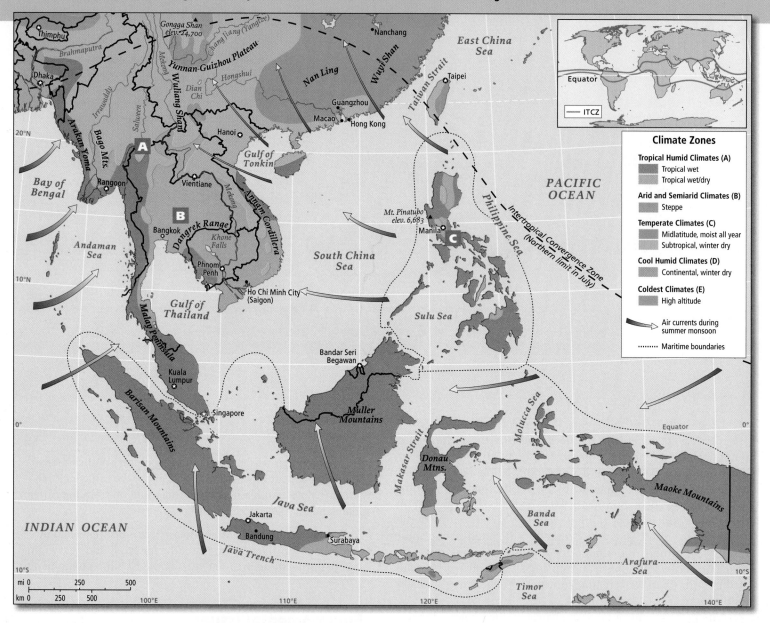

Climate Zones

Tropical Humid Climates (A)
Tropical wet
Tropical wet/dry

Arid and Semiarid Climates (B)
Steppe

Temperate Climates (C)
Midlatitude, moist all year
Subtropical, winter dry

Cool Humid Climates (D)
Continental, winter dry

Coldest Climates (E)
High altitude

Air currents during summer monsoon

Maritime boundaries

A Subtropical, winter dry, Pai, Thailand

B Tropical, wet/dry, Khao Yai, Thailand

C Tropical wet, Calauan, Philippines

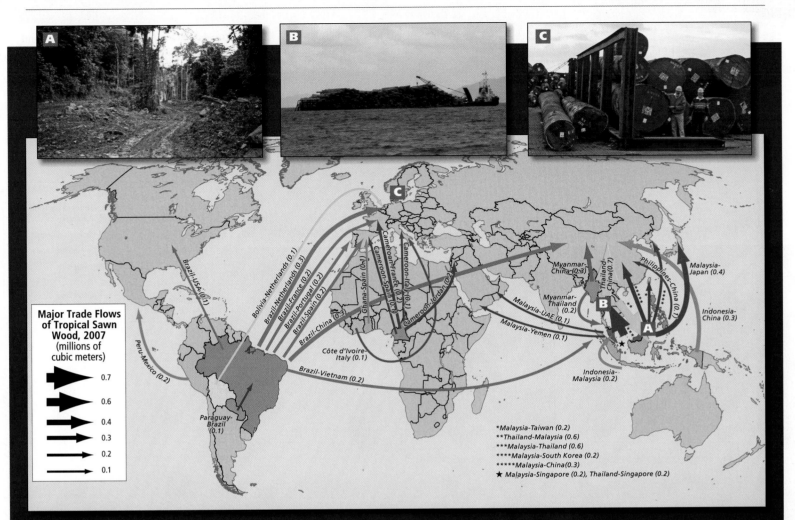

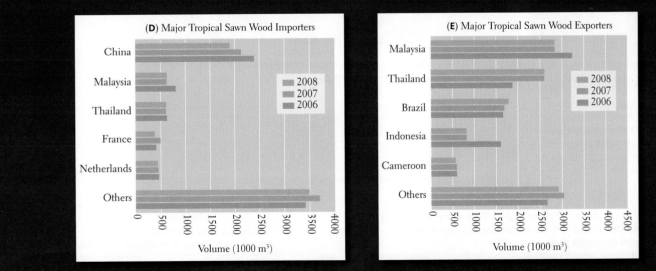

FIGURE 10.6 Trade in tropical timber. Most of the consumer demand for tropical timber is in North America, Europe, and Japan. Increasingly large quantities of tropical timber sold to China are made into furniture, plywood, and flooring, which are then sold to consumers in the developed world. The trade shown on the map is mostly legal, but much of the wood fueling China's wood-processing industries is harvested illegally in Southeast Asia. **(A)** A logging road in Kalimantan, Indonesia. **(B)** A logging ship in the Gulf of Thailand. **(C)** Tropical hardwoods being inspected at the port of Hamburg, Germany.

Thinking Geographically: What happens to the atmosphere when there are fewer trees due to legal or illegal logging?

Deforestation and the conversion of land to agricultural uses in Southeast Asia are having a profound impact on both regional ecosystems and on the entire biosphere.

A Land recently cleared of forest to make way for an oil palm plantation in Malaysia. Deforestation in much of Malaysia and Indonesia is now driven by the expansion of oil palm plantations.

Human Impact, 2002

Land Cover	Human Impact on Land
Forests	High impact
Grasslands	Medium–high impact
Deserts	Low–medium impact
Tundra	**Overfishing**
Ice	Threatened fisheries
Acid Rain	National boundaries
5.5–4.9 pH	Maritime boundaries

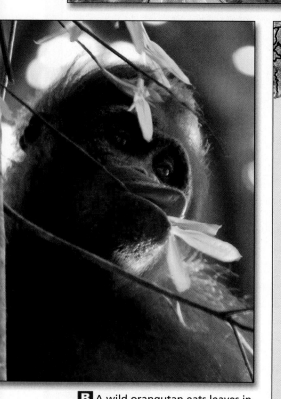

B A wild orangutan eats leaves in Kutai National Park, Kalimantan, Indonesia. The deforestation and fragmentation of habitats caused by roads have made orangutans an endangered species.

C A satellite image of fires used to clear forests for agriculture in Indonesia and Malaysia. The CO_2 produced contributes to climate change and degrades local air quality.

D Rice terraces in the Philippines. Agricultural expansion is a major force behind deforestation in the region, and wet rice agriculture is a significant contributor of greenhouse gas emissions through the release of methane.

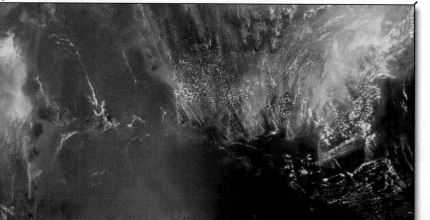

Climate Change and Deforestation

Deforestation is a major contributor to global climate change. Enormous amounts of CO_2 are released when forests are logged and the underbrush is burned (Photo Essay 10.2C). Fewer trees mean that less carbon dioxide is absorbed from the atmosphere. According to 2010 UN Food and Agriculture Organization (UNFAO) reports, although Southeast Asia contains only 5 percent of the world's forests, in the last 10 years it accounted for nearly 25 percent of deforestation globally. The region has the world's second-highest rate of deforestation, after sub-Saharan Africa. Every day, 13 to 19 square miles (34 to 50 square kilometers) of Southeast Asia's rain forests are destroyed, much of it done illegally. A great deal of this takes place in Malaysia and Indonesia, where the logging of tropical hardwoods for sale on the global market is a major activity (see Figure 10.6). Additionally, when the forests are converted to oil palm plantations, especially when the land for the plantations is cleared by burning the forest, significant amounts of CO_2 are released (Photo Essay 10.2A). For decades, these activities have made Indonesia among the world's largest contributors to global climate change. Figure 10.7 shows the activities that result in legal deforestation across the region. Illegal logging may be even more widespread but is difficult to measure.

Logging, Legal and Illegal Efforts to limit logging through government regulation have long been constrained by rampant corruption, but renewed efforts may be yielding better results. With a vision of sustainable logging as a long-term source of income for his country, Indonesia president Susilo Bambang Yudhoyono has lead the fight to reduce deforestation. Yudhoyono has worked

on enforcing existing logging regulations, shifting the focus from efforts of low-level workers (such as laborers and truck drivers) to provincial forestry officials and military personnel, who have been running illegal logging operations for years. Deforestation has decreased somewhat, though this is difficult to measure accurately. Since 1950, tropical forest cover on the Indonesian island of Borneo alone has decreased by about 60 percent. Oil palm plantations now constitute the predominant use of the land.

Oil Palm Plantations Palm oil, which is used across the globe as a cooking oil, in food products, in soaps and cosmetics, and as a machine oil, is at the center of debates over global climate change in Southeast Asia. Indonesia and Malaysia are the world's largest palm oil producers, and the expansion of their oil palm plantations has resulted in extensive deforestation. Much of the deforestation is in lowland peat swamps, which in a natural state are capable of storing huge amounts of carbon in their soils. To create space for oil palm plantations, the swamps are drained and most of their vegetation is burned (see Thematic Overview A on page 369). Often the fires spread underground to the peat beneath the forests, smoldering for years after the surface fires have been extinguished. These subsurface fires release enormous amounts of carbon into the atmosphere. In recent years, smoke from burning forests has periodically covered much of the region, dramatically reducing air quality. During El Niño events, the smoke is so bad that airplanes have difficulty landing in the cities, and even in rural areas, people are urged to wear masks.

There is some support for palm oil as a potential solution to global climate change because it can be converted into fuel for automobiles. The claim is that because the palm trees absorb carbon dioxide from the atmosphere when they are growing, palm oil could be considered a "carbon neutral" fuel. However, critics point out that palm trees do not absorb carbon at a rate equivalent to natural rain forest and if forests are burned to clear land for plantations, it can take decades to counteract the initial carbon emissions of deforestation. If a peat swamp is cleared for an oil palm plantation, more carbon is released through the burning of subsurface peat than centuries of carbon dioxide absorption by the growing oil palms can counteract.

Climate Change and Food Production

Food production contributes to global climate change in two ways. First, it is a major contributor to deforestation, and second, some types of cultivation actually produce significant greenhouse gases.

Shifting Cultivation Shifting cultivation, also known as *slash-and-burn* or *swidden* cultivation, has been practiced for thousands of years in the hills and uplands of mainland Southeast Asia and in many parts of the islands (see Figure 10.8C on page 380). To maintain soil fertility in these warm, wet environments where nutrients are quickly lost to decay, farmers move their fields every 3 years or so, letting old plots lie fallow for 15 years or more. The regrowth of forest on once-cleared fields not only regenerates the soil, it also absorbs significant amounts of carbon dioxide from

Sources of Legal Deforestation

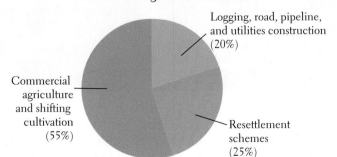

Logging, road, pipeline, and utilities construction (20%)

Commercial agriculture and shifting cultivation (55%)

Resettlement schemes (25%)

FIGURE 10.7 Legal deforestation in Southeast Asia, 2004. The diagram shows an estimate of various legal deforestation activities in Southeast Asia. Much deforestation activity is the result of illegal logging. The World Wildlife Fund estimated that 83 percent of timber production in Indonesia in 2004 stemmed from illegal logging.

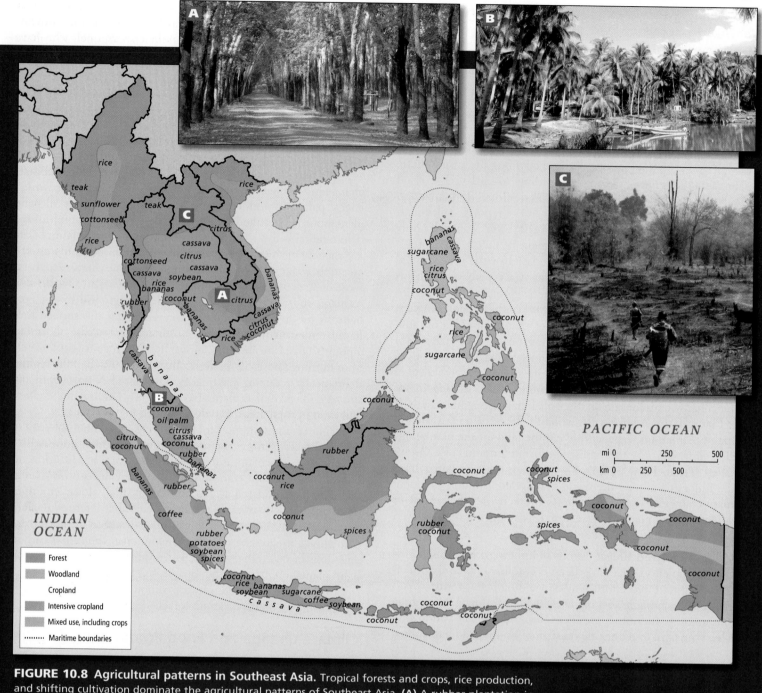

FIGURE 10.8 Agricultural patterns in Southeast Asia. Tropical forests and crops, rice production, and shifting cultivation dominate the agricultural patterns of Southeast Asia. **(A)** A rubber plantation in Cambodia. **(B)** A coconut plantation in Malaysia. **(C)** Slash-and-burn agriculture in Thailand.

Thinking Geographically: How does increased population density in Southeast Asia affect traditional farming methods of shifting cultivation and providing for fallow periods?

the atmosphere. However, if fallow periods are shortened, or are disrupted by logging, soil fertility can collapse, making cultivation impossible. In some cases, fertility can be restored through the use of chemical or organic fertilizers, but these can be too expensive for most farmers and chemical fertilizers may be ineffective after a year or be too unhealthy for food cultivation. Tropical soils left bare of forest for too long will eventually turn into hard, infertile, sun-baked clay.

Where population densities are relatively low, subsistence farmers can practice sustainable shifting cultivation indefinitely, but to allow for long fallow periods and still support human populations, this system requires larger areas than other types of agriculture. As population density increases, farmers may be forced to shorten fallow periods, thus inhibiting forest regrowth. Even though plots are small, because shifting cultivation requires clearing forest at each move, it accounts for a significant portion of the region's deforestation. Moreover, because the forests are usually cleared by burning, shifting cultivation can result in wildfires. This is especially true during an El Niño period, when rainfall is low. These wildfires have increased in recent years, particularly in Indonesia, further contributing to deforestation there.

Wet Rice Cultivation Another major contributor to global climate change is Southeast Asia's most productive form of agriculture. *Wet rice cultivation* (sometimes called *paddy rice*) entails planting rice seedlings by hand in flooded and often terraced fields that are first cultivated with hand-guided plows pulled by water buffalo or tractors (see Thematic Overview B). Wet rice cultivation has transformed landscapes throughout Southeast Asia. It is practiced throughout this generally well-watered region, especially on rich volcanic soils and in places where rivers and streams bring a yearly supply of silt (see Photo Essay 10.2D; see also Photo Essay 10.3D on page 382).

The flooding of rice fields also results in the production of methane, a powerful greenhouse gas responsible for about 20 percent of global climate change. It is estimated that up to one-third of the world's methane is released from flooded rice fields where organic matter in soil undergoes fermentation as oxygen supplies are cut off. Wet rice has been cultivated for thousands of years, but the increase of the human population in the last 25 years has driven a 17 percent expansion of the area devoted to wet rice. Figure 10.8 shows patterns of field and forest crops across the region.

Climate Change and Water

Many of Southeast Asia's potential vulnerabilities to global climate change are related to water resources. Four areas of vulnerability may affect the region's economy and food supply: glacial melting, increased evaporation, coral reef bleaching, and storm surge flooding (see Photo Essay 10.3).

Melting Glaciers and Increased Evaporation Like much of Asia, mainland Southeast Asia's largest rivers (the Irrawaddy, Salween, Mekong, and Red rivers) are fed during the dry season (November–March) by glaciers high in the Himalayas. These glaciers are now melting so rapidly that they may eventually disappear.

As glacial melting accelerates, the immediate risk is flooding, though many areas are already adapted to dramatic seasonal floods, which have occurred in this region for most of human history (see Thematic Overview C). The longer-term concern is reduced dry-season flows in the rivers. As much as 15 percent of this region's rice harvest depends on the dry-season flows of the major rivers. The loss of these harvests would strain many farmers' incomes. In addition, it could possibly lead to food shortages in cities. Coming on top of a global rise in food prices in recent years, this would place further strain on the incomes of poor people throughout the region.

The higher temperatures associated with present trends in global climate change mean evaporation rates will rise, result-

ing in drier conditions in fields, lower lake levels, and lower fish catches because of changing habitats for aquatic animals. Evaporation resulting in reduced river and groundwater flows can also cause saltwater intrusions into estuaries and fresh water aquifers.

Coral Reef Bleaching Global climate change is expected to increase sea temperatures in Southeast Asia, threatening the coral reefs that sustain much of the region's fishing and tourism. A coral reef is an intricate structure composed of the calcium-rich skeletons of millions of tiny living creatures called coral polyps.

The polyps are subject to **coral bleaching** (see Photo Essay 10.3C), or color loss, which results when photosynthetic algae that live in the corals are expelled by a variety of human-instigated changes (rising water temperature related to climate change, sedimentation).

> **coral bleaching** color loss that results when photosynthetic algae that live in the corals are expelled

Bleaching may also occur when corals become polluted from a nearby city or industrial activity, or when they are overfished. Under normal conditions, the coral will recover within weeks or months. However, severe or repeated bleaching can cause corals to die. Unprecedented global coral bleaching events occurred in 1998 and again in 2002. Roughly half of the world's coral reefs were affected, causing significant coral die-offs in some areas. Scientists generally agree that these events are the result of global climate change.

II ▶ 224. NEW SPECIES OF UNDERSEA LIFE FOUND NEAR INDONESIA

Many of the fish caught in Southeast Asia's seas are dependent on healthy coral reefs for their survival. The thousands of rural communities throughout coastal Southeast Asia that depend on these fish for food are thus also threatened by coral bleaching. So far, however, the greatest observable impacts on people have been in the tourism industry. In the Philippines, the coral bleaching event of 1998 brought a dramatic decline in tourists who come to dive the country's usually spectacular reefs, resulting in a loss of about U.S.$30 million to the economy.

Storm Surges and Flooding Although the relationship is not yet entirely understood, violent tropical storms seem to be increasing as the climate warms. Normally the Philippines can expect six typhoons (known in the Atlantic as hurricanes) in a year, but in 2009, in October alone, four typhoons struck these islands. Climatologists studying data on tropical storms predict that the entire Southeast Asian region will have a somewhat higher number of typhoons over the next few years, and that, in particular, the duration of peak winds along coastal zones will increase. This is significant because many poor, urban migrants have crowded into precarious dwellings in low-lying coastal cities such as Manila, Bangkok, Rangoon, and Jakarta, where coping with high winds, flooding, and the aftermath of storms will be a common experience (see Photo Essay 10.3A).

Responses to Climate Change

The reduction of fossil fuel consumption is a goal of all governments in this region, and some are delivering on those goals despite often high start-up costs. Both the Philippines and Indonesia have significant potential for generating electricity from *geothermal*

Much of Southeast Asia's vulnerability to climate change relates to water. Here we explore vulerabilities related to tropical storms, flooding, and coral reefs.

A Residents of Manila, Philippines, caught in floods created by typhoon Ketsana. Typhoons (known in the Atlantic as hurricanes) are projected to increase in frequency as a result of the warmer temperatures created by climate change.

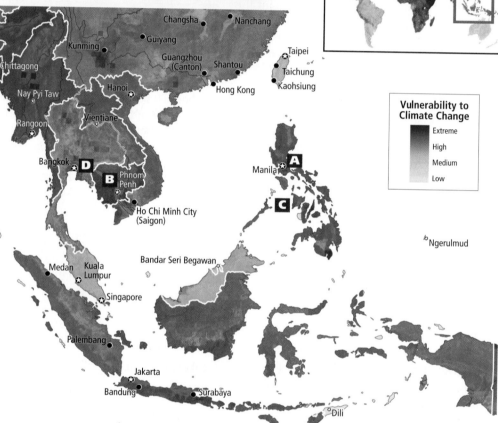

Vulnerability to Climate Change

- Extreme
- High
- Medium
- Low

B A girl in Cambodia's Tonlé Sap Lake helps her family catch snakes. Sixty percent of Cambodia's protein intake comes from this inland freshwater lake. Higher temperatures related to climate change may be increasing rates of evaporation from the lake, causing water levels to drop and fish catches to decline.

C A partially bleached coral on Tubbataha Reef, Philippines. Climate change is raising ocean temperatures, resulting in more coral bleaching events that severely degrade the ability of reefs to provide fish with habitat.

D A researcher in Thailand inspects a variety of rice that can withstand severe flooding. In part because of the work of the International Rice Research Institute in the Philippines, Southeast Asia has become a hub for research about the ways in which tropical agricultural systems can adapt to climate change.

Thinking Geographically

After you have read about Vulnerability to Climate Change in Southeast Asia, you will be able to answer the following questions:

A The increase in the frequency of typhoons is particularly significant given the many people living where?

C The global coral bleaching events that occurred in 1998 and again in 2002 affected how much of the world's coral reefs?

D Why might flood-resistant rice help communities adapt to climate change?

energy (heat stored in the earth's crust). This energy is particularly accessible near active volcanoes, which both countries have in abundance. Already, the Philippines generates 27 percent of its electricity from geothermal energy, and is second only to the United States in the amount of geothermal power it generates. By some estimates, geothermal energy could eventually provide a majority of Indonesia's energy needs. Solar energy is another attractive source, given that the entire region lies near the equator, the part of the earth that receives the most solar energy. For most countries, however, wind is the most cost-effective option, especially in Laos and Vietnam, where many population centers are in high-wind areas.

THINGS TO REMEMBER

1. Learning Goal 1: Climate Change, Food, and Water Southeast Asia has the world's second-highest rate of deforestation, after sub-Saharan Africa, and because of its widespread deforestation, is a major contributor of greenhouse gas emissions.

2. Much of this region's rice harvest depends on the dry-season flows of the major rivers of mainland Southeast Asia, which are threatened by glacial melting.

3. Climate change poses special challenges for people in low-lying cities or in typhoon zones.

4. Alternative energy sources in this region are numerous but have high start-up costs.

Human Patterns over Time

First settled in prehistory by migrants from the Eurasian continent, Southeast Asia was later influenced by Chinese, Indian, and Arab traders. Later still, it was colonized by Europe (from the 1500s to the early 1900s) and the United States (from 1898 to 1946 in the Philippines), and was occupied by Japan during World War II. By the late twentieth century, domination by outsiders had ended and the region was profiting from the sale of manufactured goods to its former colonizers.

The Peopling of Southeast Asia

The modern indigenous populations of Southeast Asia arose from two migrations widely separated in time. In the first migration,

about 40,000 to 60,000 years ago, **Australo-Melanesians**, a group of hunters and gatherers from what is now northern India and Burma, moved to the exposed landmass of Sundaland and into present-day Australia. Their descendants still live in Indonesia's easternmost islands and in small, usually remote pockets on other islands and the Malay Peninsula, and in Australia and parts of Oceania.

In the second migration (about 10,000 years ago, at the end of the last ice age), people from southern China began moving into Southeast Asia. Their migration gained momentum about 5000 years ago, when a culture of skilled farmers and seafarers from southern China, the **Austronesians**, migrated first to present-day Taiwan, then to the Philippines, and then into the islands of Southeast Asia and the Malay Peninsula (see Timeline A on page 386). Some of these sea travelers eventually moved westward to southern India and to Madagascar (off the east coast of Africa), and eastward to the far reaches of the Pacific islands (see Chapter 11).

> **Australo-Melanesians** a group of hunters and gatherers who moved from the present northern Indian and Burman parts of southern Eurasia to the exposed landmass of Sundaland about 60,000 to 40,000 years ago
>
> **Austronesians** a Mongoloid group of skilled farmers and seafarers from southern China who migrated south to various parts of Southeast Asia between 10,000 and 5000 years ago

Diverse Cultural Influences

Southeast Asia has been and continues to be shaped by a steady circulation of cultural influences, both internal and external. Overland trade routes and the surrounding seas brought traders, religious teachers, and sometimes even invading armies from China and India, as well as Arab armies from southwest Asia. These newcomers brought religions, trade goods (such as cotton textiles), and food plants (such as mangoes and tamarinds) deep into the Indonesian and Philippine archipelagos and throughout the mainland. The monsoon winds, which blow from the west in the spring and summer, facilitated access for merchant ships from South Asia and the Persian Gulf. The ships sailed home on winds blowing from the east in the autumn and winter. These winds carried people, spices, bananas, sugarcane, silks, and other Southeast Asian items to the wider world.

Religious Legacies Spatial patterns of religion in Southeast Asia reveal an island–mainland division that reflects the history of influences from India, China, Southwest Asia, and Europe. Both Hinduism and Buddhism arrived thousands of years ago via monks and traders traveling by sea and along overland trade routes that connected India and China through Burma (see Timeline B). Many early Southeast Asian kingdoms and empires switched back and forth between Hinduism and Buddhism as their official religion. Spectacular ruins of these Hindu–Buddhist empires are scattered across the region; the most famous is the city of Angkor in present-day Cambodia. At its zenith in the 1100s, Angkor was among the largest cities in the world, and its ruins are now a World Heritage Site (see Figure 10.13A on page 392). Today, Buddhism dominates mainland Southeast Asia, while Hinduism is dominant only on the Indonesian islands of Bali and Lombok.

In Vietnam, people practice a mix of Buddhist, Confucian, and Taoist beliefs that reflect the thousand years (ending in 938 C.E.) when Vietnam was part of various Chinese empires. China's traders and laborers also brought cultural influences to scattered coastal zones throughout Southeast Asia.

Islam is now dominant in the islands of Southeast Asia. Islam came mainly through South Asia after India fell to Muslim (Mughal) conquerors in the fifteenth century. Muslim spiritual leaders and traders converted many formerly Hindu–Buddhist kingdoms in Indonesia, Malaysia, and parts of the southern Philippines, where Islam is still dominant. Roman Catholicism is the predominant religion in Timor-Leste, which was colonized by Portugal, and in most of the Philippines, which was colonized by Spain.

European Colonization

Over the last five centuries, several European countries established colonies or quasi-colonies in Southeast Asia (Figure 10.9). Drawn by the region's fabled spice trade, the Portuguese established the first permanent European settlement in Southeast Asia at the port of Malacca, Malaysia, in 1511. Although better ships and weapons gave the Portuguese an advantage, their anti-Islamic and pro-Catholic policies provoked strong resistance in Southeast Asia. Only in Timor-Leste did the Portuguese establish Catholicism as the dominant religion.

Arriving first in 1521, the Spanish had established trade links across the Pacific between the Philippines and their colonies in the Americas by 1540 (see Timeline C). Like the Portuguese, they practiced a style of colonial domination grounded in Catholicism, but they met less resistance because of their greater tolerance of non-Christians. The Spanish ruled the Philippines for more than 350 years, and as a result, the Philippines is the most deeply Westernized and certainly the most Catholic part of Southeast Asia.

The Dutch were the most economically successful of the European colonial powers in Southeast Asia. From the sixteenth to the nineteenth centuries, they extended their control of trade over most of what is today called Indonesia, known previously as the Dutch East Indies. The Dutch became interested in growing cash crops for export. Between 1830 and 1870, they forced indigenous farmers to leave their own fields and work part time without pay in Dutch coffee, sugar, and indigo plantations. The resulting disruption of local food production systems caused severe famines and provoked resistance that often took the form of Islamic religious movements. Such movements hastened the spread of Islam throughout Indonesia, where the Dutch had made little effort to spread their Protestant version of Christianity.

Beginning in the late eighteenth century, the British established colonies at key ports on the Malay Peninsula. They held these ports both for their trade value and to protect the Strait of Malacca, the passage for sea trade between China and Britain's empire in India. In the nineteenth century, Britain extended its rule over the rest of modern Malaysia to benefit from Malaysia's tin mines and plantations. Britain also added Burma to its empire, which provided access to forest resources and overland trade routes into southwest China.

The French first entered Southeast Asia as Catholic missionaries in the early seventeenth century. They worked mostly in the eastern mainland area in the modern states of Vietnam, Cambodia, and Laos. In the late nineteenth century, spurred by rivalry with Britain and other European powers for greater access to the markets of nearby China, the French formally colonized the area, which became known as French Indochina.

In all of Southeast Asia, the only country not to be colonized was Thailand (then known as Siam). Like Japan, it protected its sovereignty through both diplomacy and a vigorous drive toward European-style modernization.

Struggles for Independence

Agitation against colonial rule began in the late nineteenth century when Filipinos fought first against Spain in 1896. They then fought against the United States, which took control of the Philippines in 1898 after the Spanish-American War (see Figure 10.9C). However, the Philippines and the rest of Southeast Asia did not win independence until after World War II. By then, Europe's ability to administer its colonies had been weakened by the devastation of the war, during which Japan had conquered most of European-controlled Southeast Asia. Japan held these lands until it was defeated by the United States at the end of the war (see Timeline D; see also Figure 9.7 on page 345). By the mid-1950s, the colonial powers had granted self-government to most of the region, and all of Southeast Asia was independent by 1975.

The Vietnam War The most bitter battle for independence took place in French Indochina (the territories of Vietnam, Laos, and Cambodia). Although all three became nominally independent in 1949, France retained political and economic power over them. Various nationalist leaders, most notably Vietnam's Ho Chi Minh, headed resistance movements against continued French domination. The resistance leaders accepted military assistance from Communist China and the Soviet Union, even though they were not doctrinaire communists and despite ancient antipathies toward China for its previous millennia of domination. In this way, the Cold War was brought to mainland Southeast Asia.

In 1954, the French were defeated by Ho Chi Minh at Dien Bien Phu, in northern Vietnam. Although the United States was against aiding the French continuation of quasi-colonial powers, it stepped in because it was increasingly worried about the spread of international communism should the resisters, now supported by Communists, succeed. The **domino theory**—the geographically based idea that if one country fell to communism other nearby countries would follow—was a major influence in this decision, because both North Korea and China had recently become communist. The Vietnamese resistance, which controlled the northern half of the country, attempted to wrest control of the southern half from

> **domino theory** a foreign policy theory that used the idea of the domino effect to suggest that if one country "fell" to communism, others in the neighboring region would also fall

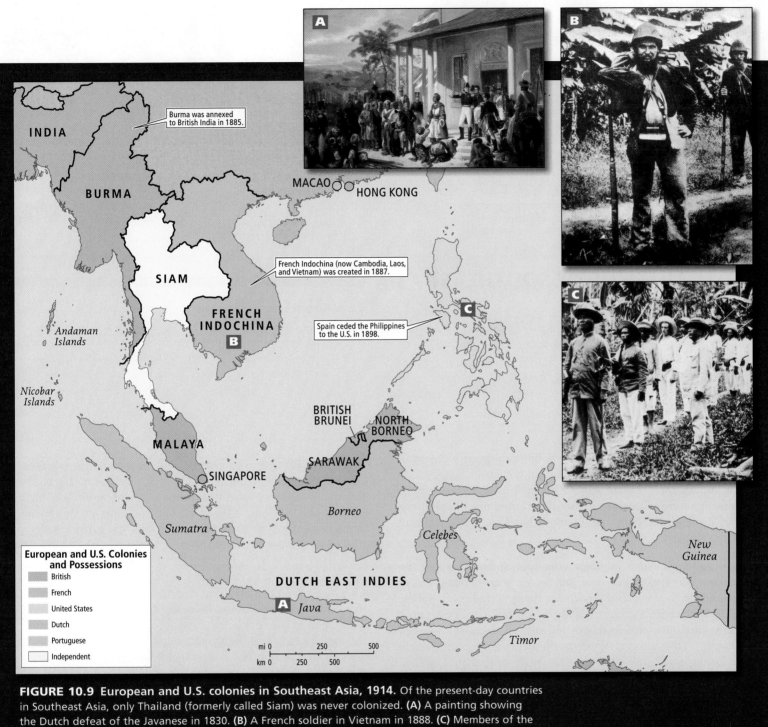

FIGURE 10.9 European and U.S. colonies in Southeast Asia, 1914. Of the present-day countries in Southeast Asia, only Thailand (formerly called Siam) was never colonized. **(A)** A painting showing the Dutch defeat of the Javanese in 1830. **(B)** A French soldier in Vietnam in 1888. **(C)** Members of the Philippine army fighting the U.S. takeover of the Philippines in 1898.

the United States and a U.S.-supported and quite corrupt South Vietnamese government. The pace of the war accelerated in the mid-1960s. After many years of brutal conflict, public opinion in the United States forced U.S. withdrawal from the conflict in 1973. The civil war continued in Vietnam, finally ending in 1975, when the North defeated the South and established a new national government.

More than 4.5 million people died during the Vietnam War, including more than 58,000 U.S. soldiers. Another 4.5 million on both sides were wounded, and bombs, napalm, and defoliants ruined much of the Vietnamese environment. Land mines continue to be a hazard to this day, and the effects of the highly toxic defoliant known as "Agent Orange" are still producing debilitating birth defects among many rural Vietnamese and

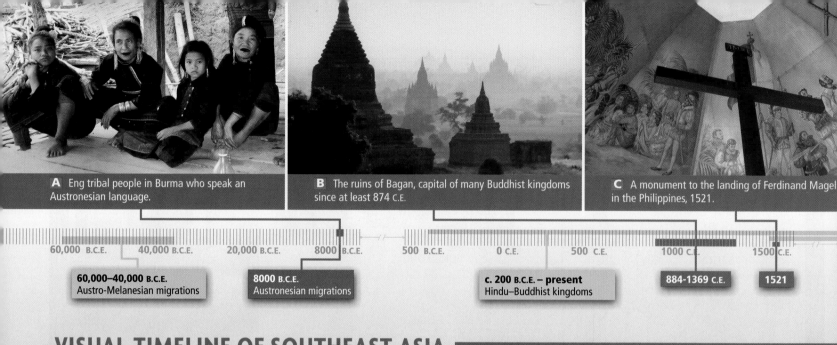

A Eng tribal people in Burma who speak an Austronesian language.

B The ruins of Bagan, capital of many Buddhist kingdoms since at least 874 C.E.

C A monument to the landing of Ferdinand Magell in the Philippines, 1521.

60,000 B.C.E.   40,000 B.C.E.   20,000 B.C.E.   8000 B.C.E.   500 B.C.E.   0 C.E.   500 C.E.   1000 C.E.   1500 C.E.

60,000–40,000 B.C.E.
Austro-Melanesian migrations

8000 B.C.E.
Austronesian migrations

c. 200 B.C.E. – present
Hindu–Buddhist kingdoms

884-1369 C.E.

1521

VISUAL TIMELINE OF SOUTHEAST ASIA

Thinking Geographically

After you have read about the human history of Southeast Asia, you will be able to answer the following questions:

A Who were the Austronesians?

B How did Hinduism and Buddhism arrive in Southeast Asia?

Laotian people (Timeline E). The withdrawal from Vietnam in 1973 ranks as one of the most profound defeats in U.S. history. After the war, the United States crippled Vietnam's recovery by imposing severe economic sanctions that lasted until 1993. Since then, the United States and Vietnam have become significant trading partners.

The "Killing Fields" in Cambodia In Cambodia, where the Vietnam War had spilled over the border, a particularly violent revolutionary faction called the Khmer Rouge seized control of the government in 1975. Inspired by the vision of a rural communist society, they attempted to destroy virtually all traces of European influence. They targeted Western-educated urbanites in particular, forcing them into labor camps, where more than 2 million Cambodians—one-quarter of the population—starved or were executed in what became known as the "killing fields" (see Timeline F).

In 1979, Vietnam deposed the Khmer Rouge and ruled Cambodia through a puppet government until 1989. A 2-year civil war then ensued. Despite a major UN effort to establish a multiparty democracy in Cambodia throughout the 1990s, the country remains plagued by political tensions between rival factions and by government corruption. In March 2009, Kang Kek Iew, the first of the Khmer Rouge leaders to be tried for war crimes and genocide, was forced to listen to and watch lengthy accounts of the torture that he supervised of men, women, and children. He was convicted in 2010 and sentenced to thirty-five years in prison. Others will be tried, but most operatives in the killing fields will never be prosecuted.

‖▶ 223. CAMBODIAN HIP-HOP ARTIST TELLS STORY THROUGH RAP

Vignette One night in 1977, soldiers came to the home of Samrith Phum in Cambodia and took away her husband. She thought he was just going to a meeting, but he never came home. Samrith was then only 20 years old and had three young children, one a newborn. With her infant in her arms, she went to talk to Choch, the Khmer Rouge village chief, and asked him, "Brother, do you know where my husband is?" The village chief told her not to worry about other people's business. A short while later, Choch appeared at Samrith's door and said he would take her to see her husband. Instead, he drove to the nearby prison and locked her up with her baby. She was released a year later, after the Vietnamese drove the Khmer Rouge from power. Today, Samrith still lives just down the street from Choch. For his part, Choch denies any involvement in the killings or even ever being at the prison. *[Adapted from the Frontline/World video, "Cambodia: Pol Pot's Shadow" and Amanda Pike "Reporter's Diary: In Search of Justice: The Killer Next Door," at http://www.pbs.org/frontlineworld/stories/cambodia/diary03.html.]* ∎

THINGS TO REMEMBER

1. Over the last five centuries, several European countries, and later the United States and Japan, established colonies or quasi-colonies that covered almost all of Southeast Asia.

2. Colonial rule in Southeast Asia came to a violent end in Vietnam and Cambodia, where war took the lives of millions, including more than 58,000 U.S. soldiers.

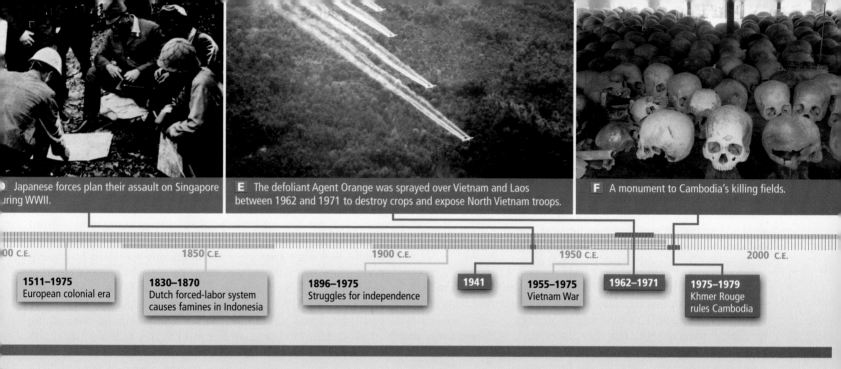

Japanese forces plan their assault on Singapore uring WWII.

E The defoliant Agent Orange was sprayed over Vietnam and Laos between 1962 and 1971 to destroy crops and expose North Vietnam troops.

F A monument to Cambodia's killing fields.

| 1850 C.E. | | 1900 C.E. | | 1950 C.E. | | 2000 C.E. |

1511–1975
European colonial era

1830–1870
Dutch forced-labor system causes famines in Indonesia

1896–1975
Struggles for independence

1941

1955–1975
Vietnam War

1962–1971

1975–1979
Khmer Rouge rules Cambodia

At what point in time did the Spanish establish trade links across he Pacific between the Philippines and the Spanish colonies in the mericas?

During World War II, Japan conquered which parts of Southeast sia?

E What has been the consequence of exposure to Agent Orange for many rural Vietnamese and Laotian people?

F How many Cambodians were executed or starved in the killing fields?

II CURRENT GEOGRAPHIC ISSUES

Like Middle and South America, Africa, and South Asia, Southeast Asia is expanding its links to the global economy. However, it has had greater success than other areas in achieving widespread prosperity, largely by following the example of some of East Asia's most successful countries.

Economic and Political Issues

Initially, trade among countries within the region was inhibited by the fact that they all exported similar goods—primarily food and raw materials—and traditionally imposed tariffs against one another. They imported consumer products, industrial materials, machinery, and fossil fuels mostly from the developed world. Several decades ago, the economic and political situation in the region changed dramatically. Southeast Asian countries had some of the highest economic growth rates in the world. The growth was based on an economic strategy that emphasized the export of manufactured goods—initially clothing and then more sophisticated technical products. In the late 1990s, though, economic growth stagnated and as people's financial expectations were dashed, political instability increased. The high level of corruption revealed by the crisis strengthened calls for democratization, which has expanded only unevenly across the region.

Strategic Globalization: State Aid and Export-Led Economic Development

Learning Goal 2

Globalization and Development: How has globalization produced both spectacular successes and tragic failures in the economies of Southeast Asia?

From the 1960s to the 1990s, some national governments in Southeast Asia achieved strong and sustained economic expansion by emulating two strategies for economic growth pioneered earlier by Japan, Taiwan, and South Korea (see pages 347–348 in Chapter 9). One was the formation of state-aided market economies. National governments in Indonesia, Malaysia, Thailand, and to some extent, the Philippines, intervened strategically in the financial sector to make sure that certain economic sectors developed. Investment by foreigners was limited so that the governments could have more control over the direction of the economy. The other strategy was export-led economic development, which focused investment on industries that manufactured products for export, primarily to developed countries (see Thematic Overview D). These strategies amounted to a limited and selective embrace of globalization in that global markets for the region's products were sought but foreign sources of capital were not.

387

These approaches were a dramatic departure from those used in other developing areas. In Middle and South America and parts of Africa, post-colonial governments relied on import substitution industries that produced manufactured goods mainly for local use. By contrast, export-led growth allowed Southeast Asia's industries to earn much more money in the vastly larger markets of the developed world. Standards of living increased markedly, especially in Malaysia, Singapore, Indonesia, and Thailand. Other important results of Southeast Asia's success were a decrease in wealth disparities and improvements in vital statistics—lower population growth rate, infant mortality, and maternal mortality, and longer life expectancy.

Export Processing Zones   In the 1970s, some governments in the region began adopting an additional strategy for encouraging economic development. This time foreign sources of capital were sought, but the places they could invest were limited to specially designated free trade areas. Such Export Processing Zones (EPZs) (see the discussion in Chapter 3) are places in which foreign companies can set up their facilities using inexpensive, locally available labor to produce items only for export (maquiladoras in Mexico are an example of companies set up by foreign firms). Taxes are eliminated or greatly reduced. Since the 1970s, EPZs have expanded economic development in Malaysia, Indonesia, Vietnam, and the Philippines, and are now used in China and Middle and South America.

The Feminization of Labor   Between 80 and 90 percent of the workers in the EPZs are women, not only in Southeast Asia but in other world regions as well (**Figure 10.10**). The **feminization of labor** has been a distinct characteristic of globalization over the past three decades (see pages 428–429 in Chapter 9). Employers prefer to hire young, single women because they are perceived as the cheapest, least troublesome employees. Statistics do show that, generally, women will work for lower

> **feminization of labor** the increasing representation of women in both the formal and the informal labor force

wages than men, will not complain about poor and unsafe working conditions, will accept being restricted to certain jobs on the basis of sex, and are not as likely as men to agitate for promotions.

Working Conditions   In general, the benefits of Southeast Asia's "economic miracle" have been unequally apportioned. In the region's new factories and other enterprises, it is not unusual for assembly-line employees to work 10 to 12 hours per day, 7 days per week, for less than the legal minimum wage, and without benefits. Labor unions that typically would address working conditions and wage grievances are frequently repressed by governments, and international consumer pressure to improve working conditions at U.S. companies like Nike has been only partially effective. By 2010, for example, Nike had sidestepped the entire issue of customer complaints by outsourcing the manufacturing to non-U.S. contractors in the region. U.S. protestors now have a more difficult time tracing abuse of workers back to Nike.

A much more powerful force that is driving up pay and improving working conditions is the service sector, which is growing throughout the region. In Singapore, the Philippines, and Timor-Leste, the service sector already dominates the economy (see Thematic Overview E). In Malaysia and Cambodia, the service sector contributes more to the country's GDP than the rest of the sectors. In other countries in the region, the service sector is approaching parity with the industrial or agricultural sectors. Only Brunei is still dominated by its industrial sector, which is entirely based on oil production. Many service sector jobs require at least a high school education, and competition for the smaller number of educated workers means that wages and working conditions are already better than in manufacturing and are likely to improve faster.

Economic Crisis and Recovery: The Perils of Globalization

The economic crisis that swept through Southeast Asia in the late 1990s forced millions of people into poverty and changed the political order in some countries. A major cause of the crisis was the lifting of controls on Southeast Asia's once highly regulated financial sector. There were some geographic elements to the crisis: the banks involved were located primarily in Singapore and the main cities of Thailand, Malaysia, and Indonesia. These were also the countries to feel the immediate effects of the crisis. Eventually the effects filtered into the hinterland, and workers in remote areas lost jobs and access to essentials.

Deregulating Investment   As part of a general push by the IMF to open national economies to the free market, Southeast Asian governments relaxed controls on the financial sector in the 1990s. Soon, Southeast Asian banks were flooded with money from investors in the rich countries of the world who hoped to profit from the region's growing economies. Flush with cash and newfound freedoms, the banks often made reckless decisions. For example, bankers made risky loans to real estate developers,

FIGURE 10.10 Feminization of labor. Women in Vietnam work at a small garment factor in Hoi An, Vietnam.

Thinking Geographically: From Southeast Asia and other world regions, what percentage of workers in EPZs are women?

often for high-rise office building construction. As a result, many Southeast Asian cities soon had far too much office space, with millions of investment dollars tied up in projects that contributed little to economic development.

One of the forces that led bankers to make such bad decisions was a kind of corruption known as **crony capitalism**. In most Southeast Asian countries, as elsewhere, corruption is encouraged by the close personal and family relationships between high-level politicians, bankers, and wealthy business owners. In Indonesia, for example, the most lucrative government contracts and business opportunities were reserved for the children of former president Suharto, who ruled the country from 1967 to 1997. His children became some of the wealthiest people in Southeast Asia. This kind of corruption expanded considerably with the new foreign investment money, much of which was diverted to bribery or unnecessary projects that brought prestige to political leaders.

> **crony capitalism** a type of corruption in which politicians, bankers, and entrepreneurs, sometimes members of the same family, have close personal as well as business relationships
>
> **Association of Southeast Asian Nations (ASEAN)** an organization of Southeast Asian governments established to further economic growth and political cooperation

The cumulative effect of crony capitalism and the lifting of controls on banks was that many ventures failed to produce any profits at all. In response, foreign investors panicked, withdrawing their money on a massive scale. In 1996, before the crisis, there was a net inflow of U.S.$94 billion to Southeast Asia's leading economies. In 1997, inflows had ceased and there was a net outflow of U.S.$12 billion.

The IMF Bailout and Its Aftermath The International Monetary Fund (IMF) made a major effort to keep the region from sliding deeper into recession by instituting reforms designed to make banks more responsible in their lending practices. The IMF also required structural adjustment policies (see pages 116–118 in Chapter 3), which required countries to cut government spending (especially on social services) and abandon policies intended to protect domestic industries.

After several years of economic chaos, and much debate over whether the IMF bailout helped or hurt a majority of Southeast Asians, economies began to recover. In the largest of the region's economies, growth resumed by 1999, and by 2006 the crisis, while still serving as an ominous reminder of the risks of globalization, had been more or less overcome.

The Global Recession of 2008 The region was hit again in 2008 by the effects of the global recession; growth slowed markedly because high oil and food prices restricted disposable cash worldwide and consumers in wealthy countries, faced with crippling debt, seriously curtailed their spending. However, while the recession dragged on in North America and Europe, by late 2009, East and Southeast Asia appeared to be recovering. In part, the recovery was based on the pent-up demand for goods and services within the domestic economies of China and Southeast Asia. Southeast Asia's ability to respond to this demand was facilitated by policies that opened up intraregional trade and access to China. The recovery was carefully modulated by tight trade and monetary regulations aimed at controlling inflation. Older strategies, such as the establishment of EPZs, were expanded to attract additional multinational corporations to Southeast Asia.

II ▶ 230. BANKERS, ANALYSTS SEE RESURGENT ASIA 10 YEARS AFTER ECONOMIC CRISIS

The Impact of China's Growth During the Southeast Asian financial crisis of the 1990s, Singapore, Malaysia, and Thailand lost out to China in attracting new industries and foreign investors. By the early 2000s, China attracted more than twice as much foreign direct investment (FDI) as Southeast Asia. However, China's growth also became an opportunity for Southeast Asia. Singapore, Malaysia, Indonesia, Thailand, and the Philippines began to "piggy-back" on China's growth by winning large contracts to upgrade China's infrastructure in areas such as wastewater treatment, gas distribution, and shopping mall development.

Southeast Asia has also used its comparative advantages over China to attract investment. In some poorer countries, such as Vietnam, wages have remained considerably lower than in China. This has attracted some investment in low-skill manufacturing that had been going to China. Meanwhile, the region's wealthier countries, such as Singapore, Malaysia, and Thailand, began positioning themselves as locations offering more highly skilled labor and greater high-tech infrastructure than China.

Regional Trade and ASEAN

During the 1980s and 1990s, Southeast Asian countries traded more with China and the rich countries of the world than they did with each other. This issue of insufficient reciprocal trade within Southeast Asia was the reason behind the creation of the **Association of Southeast Asian Nations (ASEAN)**, an increasingly strong organization of all ten Southeast Asian nations (Brunei, Burma, Cambodia, Indonesia, Laos, Malaysia, the Philippines, Singapore, Thailand, and Vietnam).

Regional Integration Although started in 1967 as an anticommunist, anti-China association, ASEAN now focuses on agreements that strengthen regional cooperation, including agreements with China. One example is the Southeast Asian Nuclear Weapons–Free Zone Treaty signed in December 1995. Another is the ASEAN Economic Community, or AEC, a trade bloc patterned after the North American Free Trade Agreement and the European Union. Tariffs between countries are being reduced in order to lower production costs and make ASEAN's manufacturing industries more efficient. And increasingly, ASEAN is focused on increasing trade between all ten members of the association. By 2008, the total intraregional trade between ASEAN countries was more than trade with any one outside country or region. Exports between ASEAN countries represented 25.9 percent of total imports and 27.6 percent of total exports (see Figure 10.11 on page 390).

Tourism International tourism is an important and rapidly growing economic activity in most Southeast Asian countries. Between 1991 and 2001, the number of international visitors to the region doubled to more than 40 million, and by 2008 international visitors numbered more than 65 million. As in other trade

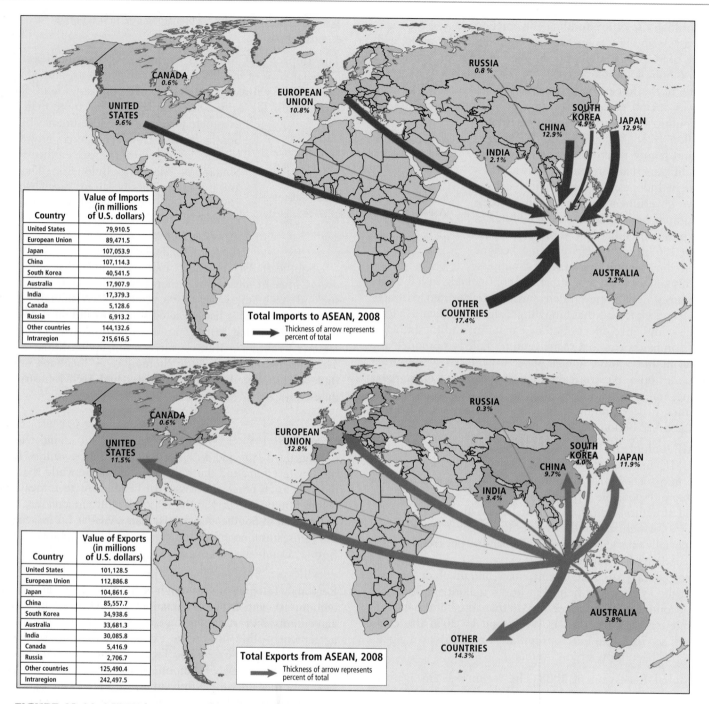

Country	Value of Imports (in millions of U.S. dollars)
United States	79,910.5
European Union	89,471.5
Japan	107,053.9
China	107,114.3
South Korea	40,541.5
Australia	17,907.9
India	17,379.3
Canada	5,128.6
Russia	6,913.2
Other countries	144,132.6
Intraregion	215,616.5

Total Imports to ASEAN, 2008
→ Thickness of arrow represents percent of total

Country	Value of Exports (in millions of U.S. dollars)
United States	101,128.5
European Union	112,886.8
Japan	104,861.6
China	85,557.7
South Korea	34,938.6
Australia	33,681.3
India	30,085.8
Canada	5,416.9
Russia	2,706.7
Other countries	125,490.4
Intraregion	242,497.5

Total Exports from ASEAN, 2008
→ Thickness of arrow represents percent of total

FIGURE 10.11 ASEAN imports and exports, 2008. The ASEAN countries are very active in world trade as importers and exporters. However, the largest share of trade is among the ASEAN countries themselves, with imports at 25.9 percent and exports at 27.6 percent.

matters, Southeast Asians are themselves increasingly touring neighboring countries. By 2008, close to 50 percent of tourists in ASEAN countries were from within the region (Figure 10.12). This is a positive trend because familiarity between neighbors lays the groundwork for various forms of regional cooperation, such as infrastructure improvements.

In response to its popularity with global and regional tourists, ASEAN members have been working to improve the region's transportation infrastructure. One such project is the Asian Highway, a web of standardized roads that loop through the main-

land and connect it with Malaysia, Singapore, and Indonesia (the latter via ferry) (see Figure 10.13 on page 392). Eventually, the Asian Highway will facilitate ground travel through 32 Eurasian countries from Moscow to Indonesia and from Turkey to Japan.

The surge of tourism in Southeast Asia has also raised concerns about too great a dependency on an industry that leaves economies vulnerable to events that precipitously stop the flow of visitors, or that leave locals vulnerable to the sometimes destructive demands of tourists (see the discussion of sex tourism on pages 402–403).

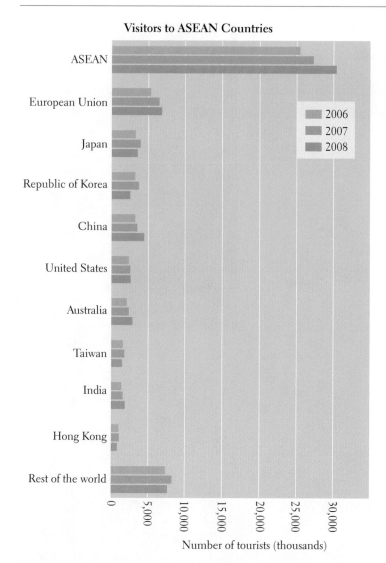

Visitors to ASEAN Countries

Number of tourists (thousands)

FIGURE 10.12 Top country/regional sources of visitors to Southeast Asian countries, 2006–2008. Like trade, the largest share of visitors to Southeast Asia comes from countries within the region, accounting for 46.5 percent of all visitors in 2008.

Pressures For and Against Democracy

Learning Goal 3

Democratization: Is democratization influencing levels of violence and corruption in Southeast Asia?

Movements toward democracy have been uneven in Southeast Asia. Across the region, there are significant barriers to democratic participation (see Photo Essay 10.4 map on page 394), but the types of barrier vary. Several countries are plagued with violent conflict, but the experience of Indonesia in recent years shows how democracy can reduce the tensions that produce violent conflict over the long term. However, democratic governments are not always durable. For example, Thailand seemed to have constructed a democratic system that allowed for protest and provided mechanisms for constitutional adjustments and smooth transitions after regular elections. Nevertheless, Thailand's democracy has

recently weakened, due largely to a 2006 coup d'état against an apparently corrupt but popular prime minister, Thaksin Shinawatra (see Thematic Overview F). As argued below, it appears that democratization does influence levels of violence and corruption, but it often works in combination with other factors, especially economic ones. The many barriers to democratization found in this region also highlight the fact that there are other viable systems of government.

▐▐▶ 322. THAILAND'S PROTESTERS HIGHLIGHT RIFTS, POLITICAL PARTICIPATION

Can Democracy Work in Indonesia? The greatest recent shift toward democracy occurred in Indonesia in the wake of the economic crisis of the late 1990s. After three decades of semidictatorial rule by President Suharto, the economic crisis spurred massive demonstrations that forced Suharto to resign. Since then, democratic parliamentary and presidential elections have initiated a new political era in the country (see Reasons for Optimism B on page 404).

Indonesia is the largest country in Southeast Asia, and the most fragmented—physically, culturally, and politically. It comprises more than 17,000 islands (3000 of which are inhabited), stretching over 3000 miles (8000 kilometers) of ocean. It is also the most culturally diverse, with dozens of ethnic groups and multiple religions. But periodic instability in Indonesia has many people wondering whether this multi-island country of 243 million might be headed for disintegration.

Until the end of World War II, Indonesia was not a nation at all but rather a loose assemblage of distinct island cultures, which Dutch colonists managed to hold together as the "Dutch East Indies." When Indonesia became an independent country in 1945, its first president, Sukarno, hoped to forge a new nation out of these many parts. To that end, he articulated a national philosophy known as *Pancasila*, which was aimed at holding the disparate nation together primarily through religious tolerance.

Encouraging Cohesiveness with Government Policies *Pancasila* embraces five precepts: belief in God, and the observance of *conformity*, *corporatism* (often defined as organic social solidarity with the state), *consensus*, and *harmony*. These last four precepts could be interpreted as discouraging dissent or even loyal opposition, and they seem to require a perpetual stance of boosterism. For some people, the strength of *Pancasila* is the emphasis on harmony and consensus that seems to ensure there will never be a religiously based government. Others note that conformity and corporatism counteract the extreme ethnic diversity and geographic dispersion of the country. But conformity and corporatism have also had a chilling effect on participatory democracy and on criticism of the government, president, and the army. Indeed, the first orderly democratic change of government did not take place until national elections in 2004. Since then, there have been several peaceful elections, and the government is stable enough to allow citizens to publicly protest policies (see Photo Essay 10.4B).

In recent years, separatist movements have sprouted in four distinct areas. The only one to succeed was in Timor-Leste, which became an independent country in 2002. However, its case is unique in that this area was under Portuguese control until 1975, when it was forcibly integrated into Indonesia. Two

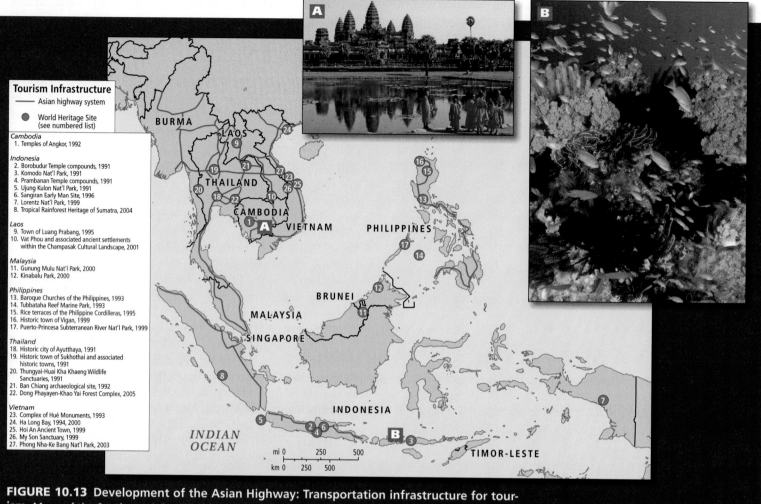

FIGURE 10.13 Development of the Asian Highway: Transportation infrastructure for tourism. Many of the Southeast Asian World Heritage Sites are located along the partially completed Asian Highway, which will eventually connect Europe to Indonesia. The UNESCAP report, "Development of the Asian Highway" (http://www.unescap.org/ttdw/common/TIS/AH/ files/AH_2008.pdf) describes the project and includes a map of the entire system. **(A)** Buddhist monks visiting an Angkor Wat temple complex in Cambodia. **(B)** A coral reef in Komodo National Park, Indonesia.

Thinking Geographically: How many international visitors came to Southeast Asia in 2008? Of those, how many were from ASEAN countries?

other separatist movements have grown largely in response to Indonesia's **resettlement schemes**. Also known as *transmigration schemes*, between 1965 and 1995 these programs relocated approximately 8 million people from crowded islands such as Java to less densely settled islands. The policies were originally initiated under the Dutch in 1905 to relieve crowding and provide agricultural labor for plantations in thinly populated areas.

After independence, Indonesia used resettlement schemes for the same purposes and also to bring outlying areas under closer control of the central government in Jakarta. However, the government ultimately lost rather than gained control as the newcomers inspired separatist movements in two main areas of resettlement: in the Indonesian half of the island of

> **resettlement schemes** government plans to move large numbers of people from one part of a country to another to relieve urban congestion, disperse political dissidents, or accomplish other social purposes; also called *transmigration*

New Guinea (the Indonesian provinces of Papua and West Papua) and in the Malucca islands. In 2001, three years after the expansion of Indonesian democracy following Suharto's resignation in 1998, the resettlement schemes were cancelled. This was in response to violence in the resettlement areas and to the expense of the programs. both of which made the schemes politically unpopular. The separatist movements were further diffused by the extension of greater local autonomy to the Maluccas, Papua, and West Papua. Violence has declined as local and provincial governments have become more democratic, making them more responsive to their own people and less subject to the at times authoritarian leanings of national leaders in Jakarta. Nevertheless, sporadic violence is still common in both of these areas.

The most dramatic turnaround was in the far western province of Aceh in Sumatra, where in 2005, separatists who had battled Indonesian security forces for decades laid down their arms. Conflicts had originally developed because most of the wealth yielded by Aceh's resources, especially oil, was going to the central government in Jakarta. The conflict began to decline after the democratization that followed Suharto's resignation. A final boost came from the tsunami of 2004. Recovery efforts following the disaster created a powerful incentive for separatists and the government to cooperate in order to receive outside aid. A peace accord signed in 2005 brought many former combatants into the political process as democratically elected local leaders. Violence has decreased dramatically since then.

Southeast Asia's Authoritarian Tendencies Despite the democratization that has occurred in Indonesia, sporadically in Thailand, and to a lesser extent in Malaysia, authoritarianism (see pages 273–275 in Chapter 7) is still a powerful force in Southeast Asia. Undemocratic socialist regimes still control Laos and Vietnam (see Photo Essay 10.4D), and a military dictatorship runs Burma. Cambodia's democracy is precarious and violence there is common. Brunei is an authoritarian sultanate, and from 2006 to 2008, Thailand's government was taken over by the military—with subsequent elections not having quieted protesters who took over part of Bangkok in 2010 (see Thematic Overview F). Powerful and corrupt leaders have subverted the democratic process even in the region's oldest democracy, the Philippines, as well as occasionally in the wealthier countries of Malaysia and Singapore.

II ▶ 230. BANKERS, ANALYSTS SEE RESURGENT ASIA 10 YEARS AFTER ECONOMIC CRISIS

Some Southeast Asian leaders, such as Singapore's former prime minister, Lee Kuan Yew, have argued that Asian values are not compatible with Western ideas of democracy. Yew and other leaders assert that Asian values are grounded in the Confucian view that individuals should be submissive to authority, so Asian countries should avoid the highly contentious public debate of electoral politics. Nevertheless, when confronted with governments that abuse their power, people throughout Southeast Asia have repeatedly rebelled, often in the form of pro-democracy movements (see Photo Essay 10.4 A, B).

II ▶ 319. AUNG SAN SUU KYI TRIAL UNDER WAY IN BURMA

Some pro-democracy movements have resulted in real change, as in Thailand (before 2007) and Indonesia, but others have not. For more than two decades, people in Burma have been futilely protesting the rule of a corrupt and undemocratic military regime. The regime refused to step aside when the people elected Aung San Suu Kyi to lead a civilian reformist government in 1990. Suu Kyi was under house arrest for nearly 15 years between 1989 and 2010, despite having won the Nobel Peace Prize in 1991. Widespread pro-democracy protests throughout the country in 2007 were brutally repressed (see Photo Essay 10.4A).

II ▶ 234. NEW TECHNOLOGY BEAMS BURMA PROTESTS ACROSS GLOBE

II ▶ 235. JIMMY CARTER CALLS FOR MORE INTERNATIONAL PRESSURE ON BURMA

Terrorism, Politics, and Economic Issues Like authoritarianism, terrorism has long loomed as a counterforce to democracy in this region. Terrorist violence short-circuits the public debate that is at the heart of the democratic processes. However, it is important to recognize how economic factors like poverty also contribute to terrorism.

Terrorist movements often thrive in the context of both economic deprivation and political repression, two factors that are often highly interrelated. Terrorist attacks may be carried out by militant groups pursuing local or national political agendas and grievances—such as revenge for government campaigns against Muslim separatists (see Photo Essay 10.4C). But from a different perspective, terrorism seems to draw on the resentment of people who feel shut out from opportunities for economic advancement. Increasingly, the solution seems to be to address both the political and economic dimensions of terrorism. This could lead to solutions similar to what occurred in Aceh, where conflict was resolved in part by giving local communities more political control over their own economic resources.

Solutions to terrorism are further complicated by the growing international cooperation among terrorist groups. During the late 1990s and through the 2000s, a series of bombs exploded in the Philippines and across Indonesia. Small terrorist cells were discovered in Malaysia, the Philippines, Indonesia, southern Thailand, and Singapore. In October 2002, the Sari Hotel in Bali, Indonesia, was bombed, killing nearly 200 foreign tourists, and luxury hotels in Jakarta were bombed in 2003 and again in 2009. Investigations into these acts suggest growing links between local groups and international Islamist terrorist networks. Hence, efforts to combat terrorism are increasingly requiring greater political and economic cooperation on an international level. While this is often very difficult to achieve, especially among countries experiencing political repression and economic deprivation, greater international cooperation can help countries learn from each other's experiences. Indeed, Indonesia's recent success in Aceh is based in part on strategies that have also been successful in North Africa and Southwest Asia (see pages 237–238 in Chapter 6).

II ▶ 225. TERROR AND ISLAMIC STRUGGLE IN INDONESIA

II ▶ 232. VIOLENCE IN THAILAND'S MUSLIM SOUTH INTENSIFIES

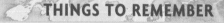

THINGS TO REMEMBER

1. Globalization has produced long periods of strong economic growth in Southeast Asia punctuated by brief but dramatic periods of decline.

2. Urban incomes have increased significantly during times of growth, but many poor people remain highly vulnerable to periods of global economic decline that threaten their access to jobs, food, and adequate living quarters.

3. Learning Goal 3: Democratization Several countries are plagued with violent conflict, but the experience of Indonesia in recent years shows how democracy can reduce some of the tensions that contribute to violent conflict over the long term.

There are significant barriers to democratization in this region, such as high levels of political violence and authoritarian political cultures. However, demands for greater public participation in governance are growing.

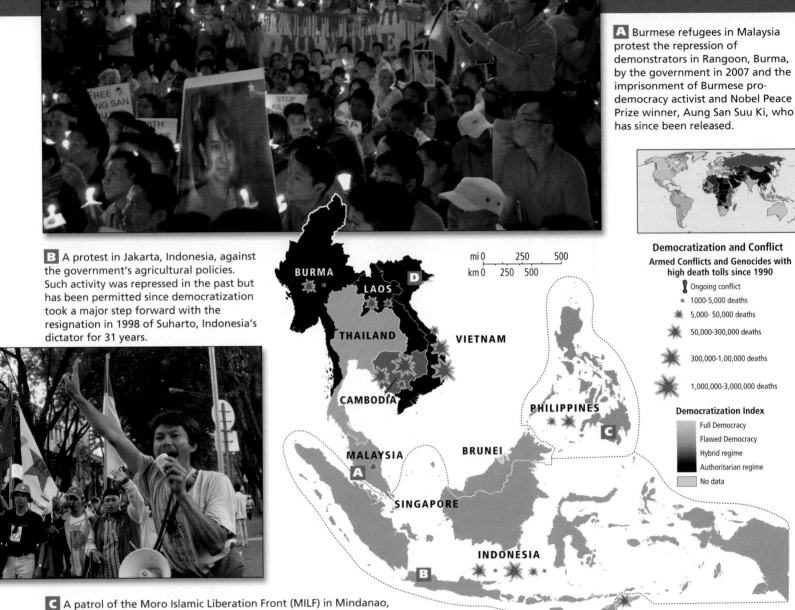

A Burmese refugees in Malaysia protest the repression of demonstrators in Rangoon, Burma, by the government in 2007 and the imprisonment of Burmese pro-democracy activist and Nobel Peace Prize winner, Aung San Suu Ki, who has since been released.

B A protest in Jakarta, Indonesia, against the government's agricultural policies. Such activity was repressed in the past but has been permitted since democratization took a major step forward with the resignation in 1998 of Suharto, Indonesia's dictator for 31 years.

Democratization and Conflict

Armed Conflicts and Genocides with high death tolls since 1990

- Ongoing conflict
- ✳ 1000-5,000 deaths
- ✳ 5,000- 50,000 deaths
- ✳ 50,000-300,000 deaths
- ✳ 300,000-1,00,000 deaths
- ✳ 1,000,000-3,000,000 deaths

Democratization Index
- Full Democracy
- Flawed Democracy
- Hybrid regime
- Authoritarian regime
- No data

C A patrol of the Moro Islamic Liberation Front (MILF) in Mindanao, Philippines. In response to what they see as discrimination against Muslims by the Catholic-dominated government, the MILF demands a separate Islamic state in Mindanao. Violence subsided recently after the Muslim-dominated areas of Mindanao were given greater political autonomy by the central government.

D A sign in Hanoi, Vietnam, features a portrait of Ho Chi Minh along with Communist symbols and icons. Vietnam's government remains Communist and authoritarian, allowing few political freedoms.

Thinking Geographically

After you have read about Democratization and Conflict in Southeast Asia, you will be able to answer the following questions:

A When was Aung San Suu Kyi elected to lead a civilian reformist government in Burma?

B When did the greatest recent shift toward democracy in Indonesia occur?

C What was the level of terrorist activity before the October 2002 bombing of the Sari Hotel in Bali, Indonesia, which killed nearly 200 foreign tourists?

Learning Goal 4
Population, Development, and Gender: How is economic development affecting gender roles in ways that contribute to slower population growth in Southeast Asia?

Sociocultural Issues

Southeast Asia is home to 586 million people (almost double the U.S. population) who occupy a land area that is about one-half the size of the United States. Due to a long and complex history, these people have a great diversity of cultures and religious traditions.

Population Patterns

Southeast Asia's population is large and growing, but economic development, changing gender roles, and population-control efforts are slowing the growth rate. At present rates, Southeast Asia's population is projected to reach 826 million by 2050, by which time much of this population will live in cities (less than half do now) (see Figure 10.14 on page 396). However, population projections could be inaccurate, both because rates of natural increase are slowing markedly and because many Southeast Asians are migrating to find employment outside the region.

Population Dynamics There is considerable variety in the population dynamics of this region, due in part to differences in economic development, gender roles, government policy, and broader religious and cultural practices. Most countries are nearing the last stage of the demographic transition, where births and deaths are low and growth is minuscule or slightly negative. In the last several decades, overall fertility rates in Southeast Asia have dropped by more than half (see Figure 10.15 on page 397). Whereas women formerly had 5 to 7 children, they now have 2 or 3. The only major exception is Timor-Leste, where fertility rates are still over 7, with rates in Laos (3.5) and the Philippines (3.3) also relatively high for the region. Nonetheless, in most countries populations are young; between one-quarter and one-third of the people are aged 15 years or younger. Steady population growth is ensured for several decades because so many are just coming into their fertile years.

On the other hand, Brunei, Singapore, and Thailand have reduced their fertility rates so steeply—below replacement levels—that they will soon need to cope with aging and shrinking populations (see Figure 10.15). The Singapore government is now so concerned about the low fertility rate that it offers young couples various incentives for marrying and procreating (see Thematic Overview G). A greater source of population growth for Singapore is the steady stream of highly skilled immigrants that its vibrant

economy attracts. Thailand's low fertility rate of 1.8 children per adult woman was achieved in part via a government-sponsored condom campaign, and in part by rapid economic development, which made many couples feel that smaller families would be best (see Reasons for Optimism A). As women have gained more opportunities to work and study outside the home, they have decided to have fewer children. High literacy rates for both men and women and Buddhist attitudes that accept the use of contraception have also been credited for the decline in Thailand's fertility rate.

The poorest and most rural countries in the region show the usual correlation between poverty, high fertility, and infant mortality. In Cambodia and Laos, fertility rates average 3.0 and 3.5 children, respectively. Infant mortality rates are 62 per 1000 births for Cambodia and 64 per 1000 births for Laos. In Vietnam, where people are only slightly more prosperous and urbanized, the fertility rate (2.1 children per adult woman) and infant mortality rate (15 per 1000 births) are much lower. Vietnam's lower rates are explained by the fact that this socialist state provides basic education and health care to all of its people, regardless of income. In Vietnam, literacy rates are more than 94 percent for men and 87 percent for women, whereas only 68 percent of women in Cambodia and only 63 percent in Laos can read. In addition, Vietnam's rapidly developing economy is attracting foreign investment, which provides more employment for women, and careers are replacing child rearing as the central focus of many women's lives. Out of concern that these changes weren't happening rapidly enough, and that a rapidly growing population would jeopardize Vietnam's upswing in economic development, the government of Vietnam recently introduced a two-child family policy.

Vignette In Roman Catholic Philippines, Gina Judilla, who works outside the home, has had six children with her unemployed husband. They wanted only two, but because of the strong role of the Catholic Church and the political pressure it exerts, birth control was not available to the poor and abortion is illegal, so with every succeeding pregnancy she tried folk methods of inducing an abortion. None worked. Now she can afford to send only two of her six children to school.

A move by family planners to provide national reproductive health services and sex education is underway. A recent survey showed that 54 percent of all pregnancies in the Philippines in 2008 were unintended, and only one-third of Philippine women have access to modern birth control methods. *[Source: Carlos H. Conde, "Bill to Increase Access to Contraception Is Dividing Filipinos," New York Times, October 25, 2009, at http://www.nytimes.com/2009/10/26/world/asia/26iht-phils.html?hpw.]* ∎

Population Pyramids The youth and gender features of Southeast Asian populations are best appreciated by looking at the population pyramids for Indonesia, 2009 and projected to 2050 (see Figure 10.16 on page 397). The wide bottom of the 2009 pyramid shows that most people are under age 30, but the projections to 2050 show that eventually, with declining birth rates, Indonesia will accumulate ever larger numbers in the upper age groups and the pyramid will eventually be more box shaped, as those

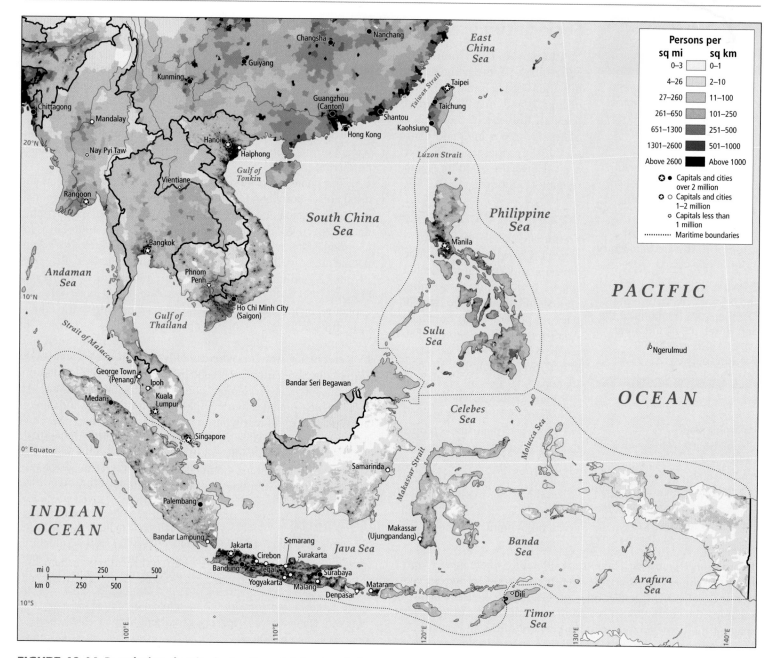

FIGURE 10.14 Population density in Southeast Asia. Population growth in Southeast Asia is slowing, due largely to economic development, urbanization, changing gender roles, and government policies. Fertility rates have declined sharply in all countries since the 1960s, but are still high in the poorest areas.

for Europe are now. The gender disparities (more males than females) can be seen by carefully examining the sizes of each age group on the male and female sides of the pyramid. The difference is slight but significant, because there is a rather consistent deficit of females, apparently selected out before birth.

Urbanization

Learning Goal 5
Urbanization: What changes are driving urbanization in this region?

Southeast Asia as a whole is only 43 percent urban, but the rural–urban balance is shifting steadily in response to declining agricultural

employment and booming urban industries. The forces driving farmers into the cities are called the *push factors* in rural-to-urban migration. They include the rising cost of farming caused by the use of new technologies. *Pull factors*, in contrast, are those that attract people to the city, such as abundant manufacturing jobs and education opportunities. In Southeast Asia, as in all other regions, these factors have come together to create steadily increasing urbanization. Malaysia is already 68 percent urban; the Philippines, 63 percent; Brunei, 72 percent; and Singapore, 100 percent.

Throughout Southeast Asia, employment in agriculture has been declining since new production methods were introduced

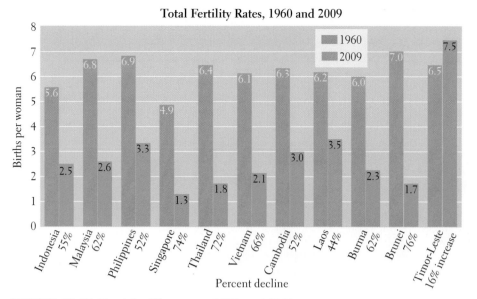

FIGURE 10.15 Total fertility rates, 1960 and 2009. Total fertility rates declined during this period for all Southeast Asian countries except Timor-Leste.

that increase the need for labor-saving equipment and reduce the need for human labor. Meanwhile, the use of chemical pesticides and fertilizers has also spread. While such additives can increase harvests dramatically, they also drive the cost of production higher than what most farmers can afford. Many family farmers have sold their land to more prosperous farmers or to corporations and moved to the cities. These people, skilled at traditional farming but with little formal education, often end up in the most menial of urban jobs (see Photo Essay 10.5C on page 398).

Labor-intensive manufacturing industries (garment and shoe making, for instance), are expanding in the cities and towns of the poorer countries, such as Cambodia, Vietnam, and parts of Indonesia and Timor-Leste. In the urban and suburban areas of the wealthier countries—Singapore, Malaysia, Thailand, and parts of Indonesia and the northern Philippines—technologically sophisticated manufacturing industries are also growing. These

include automobile assembly, chemical and petroleum refining, and assembly of computers and other electronic equipment. Riding on this growth in manufacturing are innumerable construction projects that often provide employment to recent migrants (see Photo Essay 10.5D).

Push and pull factors have sent rural migrants streaming into cities like Jakarta, Manila, and Bangkok, which are among the most rapidly growing metropolitan areas on earth. Such cities are *primate cities*—cities that, with their suburbs, are vastly larger than all others in a country. Bangkok is over 20 times larger than Thailand's next-largest metropolitan area, Udon Thani, and Manila is over 9 times larger than Davao, the second-largest city in the Philippines (see Thematic Overview H). Thanks to their strong industrial base, political power, and the massive immigration they attract, primate cities can dominate whole countries.

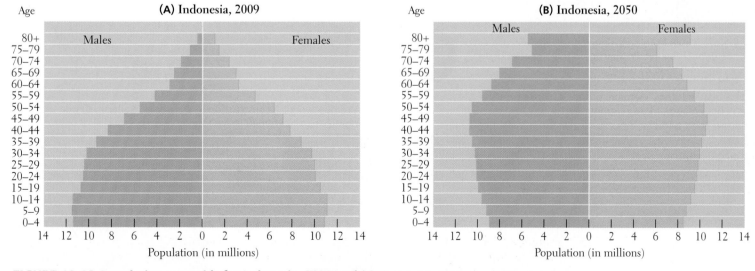

FIGURE 10.16 Population pyramids for Indonesia, 2009 and 2050. Total fertility rates declined during this period for all Southeast Asian countries except Timor-Leste. Since 2009, Timor-Leste has experienced dramatic declines in total fertility rates due to the same factors at work throughout the rest of the region.

Southeast Asia is rapidly urbanizing as growing manufacturing and service sector industries pull in people from rural areas and as changes in agriculture push farmers to the cities. Many cities are struggling to cope with rapid growth.

A A slum area in Jakarta, Indonesia, where 62 percent of the population lives in slums. Jakarta will grow by almost 50 percent by 2020.

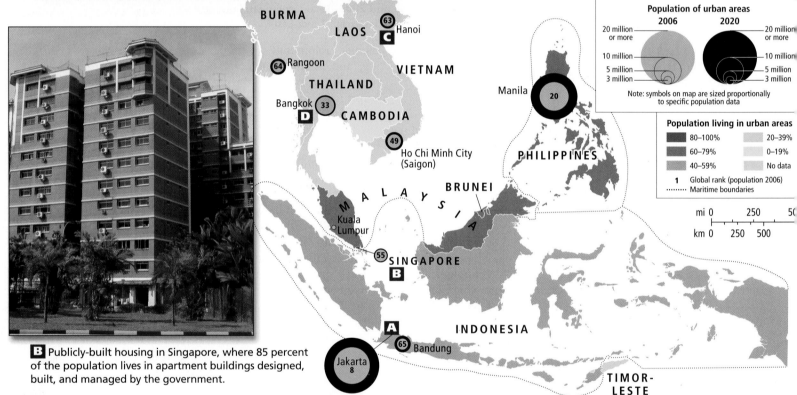

BURMA

LAOS

63 Hanoi **C**

64 Rangoon

VIETNAM

THAILAND

Bangkok **33**
D

CAMBODIA

49

Ho Chi Minh City (Saigon)

Manila **20**

PHILIPPINES

BRUNEI

M A L A Y S I A

Kuala Lumpur

55 SINGAPORE
B

INDONESIA

65 Bandung

Jakarta **8**
A

TIMOR-LESTE

Population of urban areas

2006 | 2020

20 million or more
10 million
5 million
3 million

Note: symbols on map are sized proportionally to specific population data

Population living in urban areas

80–100%	20–39%
60–79%	0–19%
40–59%	No data

1 Global rank (population 2006)
.... Maritime boundaries

mi 0 — 250 — 50
km 0 — 250 — 500

B Publicly-built housing in Singapore, where 85 percent of the population lives in apartment buildings designed, built, and managed by the government.

C A man in Hanoi, Vietnam, pushes a bicycle converted to carry heavy loads. Many migrants start out in low-wage, urban manual-labor jobs like this.

D Women working construction in Bangkok, Thailand. Gender roles are changing as more women move to the cities.

Rarely can such cities provide sufficient housing, water, sanitation, or even decent jobs for all the new arrivals. Of all the cities in Southeast Asia, only Singapore provides well for nearly all of its citizens (Photo Essay 10.5B). Even there, however, a significant illegal, noncitizen population lives in poverty on islands surrounding the city. More typical is the experience of rural-to-urban migrants who go to Bangkok or Jakarta. Living conditions in Jakarta can be especially difficult (Photo Essay 10.5A). The city is expected to grow by 50 percent in the next 10 years.

Migration Related to Globalization

The same push and pull factors driving urbanization are also driving millions to migrate out of Southeast Asia. These migrants are a major force of globalization, as they supply much of the world's growing demand for low- and middle-wage workers who are willing to travel or live temporarily in foreign countries. They are also a globalizing force within their home countries because their remittances (monies sent home) boost family incomes and supply governments with badly needed **foreign exchange** (foreign currency) that countries need to purchase imports. For example, Filipinos working abroad are that country's largest source of foreign exchange, sending home over U.S.$6 billion annually, and increasing household annual income by an average of 40 percent.

foreign exchange foreign currency that countries need to purchase imports

The Maid Trade Recently, women have constituted well over 50 percent of the more than 8 million migrants from Southeast Asia. Many skilled nurses and technicians from the Philippines work in European, North American, and Southwest Asian cities. About 3 million participate in the global "maid trade" (Figure 10.17 map). Most are educated women from the Philippines and Indonesia who work under 2- to 4-year contracts in wealthy homes throughout Asia. An estimated 1 to 3 million Indonesian maids now work outside of the country, mostly in the Persian Gulf.

The maid trade has become notorious for abusive working conditions and employers who often do not pay what they promise. In Saudi Arabia, the NGO Human Rights Watch is

FIGURE 10.17 Globalization: The "maid trade." In the 1990s, between 1 million and 1.5 million Southeast Asian women were working elsewhere in Asia (including the Arab states) as domestic servants. The number is estimated to have more than doubled by 2009, with the majority of the workers coming from the Philippines, Indonesia, and Sri Lanka. **(A)** Filipina housemaids in Hong Kong relax on their day off in temporary enclosures made of cardboard boxes. The government of the Philippines requires that Filipinas working abroad be given Sunday off, so many public areas in central Hong Kong are occupied on Sundays by small groups of Filipina housemaids talking, trading goods, playing games, and packing up items to send home.

Thinking Geographically: For what practices has the maid trade become notorious?

monitoring the cases of Muslim Indonesian women who were brutally abused—two of whom were killed—by members of a privileged Saudi family. The Philippines went so far as to ban the maid trade in 1988, but reestablished it in 1995 after better pay and working conditions were negotiated with countries receiving the workers.

Vignette Every Sunday is amah (nanny) day in Hong Kong. Gloria Cebu and her fellow Filipina maids and nannies stake out temporary geographic territory on the sidewalks and public spaces of the central business district. Informally arranging themselves according to the different dialects of Tagalog (the official language of the Philippines) they speak, they create room-like enclosures of cardboard boxes and straw mats where they share food, play cards, give massages, and do each other's hair and nails (Figure 10.17A). Gloria says it is the happiest time of her week, because for the other 6 days she works alone caring for the children of two bankers.

Gloria, who is a trained law clerk, has a husband and two children back home in Manila. Because the economy of the Philippines has stagnated, she can earn more in Hong Kong as an amah than in Manila in the legal profession. Every Sunday she sends most of her income (U.S.$125 a week) home to her family. *[Sources: Adapted from notes by Kirsty Vincin, an Australian teacher in Hong Kong, November 17, 2009, and "The Filipina Sisterhood: An Anthropology of Happiness,"* Economist, *December 20, 2001, at http://www.economist.com/world/asia/displaystory.cfm?story_id=E1_RRPJDJ.]* ∎

THINGS TO REMEMBER

1. **Learning Goal 4: Population, Development, and Gender** Southeast Asia's population is large and growing, but economic development, changing gender roles, and population control efforts are slowing the rate of growth.

2. **Learning Goal 5: Urbanization** Southeast Asia as a whole is only 43 percent urban, but the rural–urban balance is shifting steadily in response to declining agricultural employment and booming urban industries.

3. Often the focus of migration is the capital of a country, which may become a primate city—a city that, with its suburbs, is vastly larger than all others in a country.

4. Thousands of people emigrate from the region each year to seek temporary or long-term employment in the Middle East, Europe, North America, or Japan.

Cultural and Religious Pluralism

Southeast Asia is a place of **cultural pluralism** in that it is inhabited by groups of people from many different backgrounds. Over the past 40,000 years, migrants have come to the region from India, the Tibetan plateau, the Himalayas, China, Southwest Asia, Japan,

cultural pluralism the cultural identity characteristic of a region where groups of people from many different backgrounds have lived together for a long time but have remained distinct

Korea, and the Pacific. Many of these groups have remained distinct, partly because they lived in isolated pockets separated by rugged topography or seas. However, the religious practices and traditions of many groups show diverse cultural influences.

The major religious traditions of Southeast Asia include Hinduism, Buddhism, Confucianism and Taoism, Islam, Christianity, and animism (Figure 10.18). The patterns of religious practice are complex. All originated outside the region, with the exception of the animist belief systems of the indigenous peoples. Animism takes many different forms in this region. In general, in animism such natural features as trees, rivers, crop plants, and the rains all carry spiritual meaning. These natural phenomena are the focus of festivals and rituals to give thanks for bounty and to mark the passing of the seasons, and these ideas have permeated all the imported religious traditions of the region.

Vignette Although arranged marriages were long the tradition across Southeast Asia, now in most urban and rural areas marriages are love matches. Such is the case for Harum and Adinda, who live in Tegal, on the island of Java in Indonesia. They met in high school and some 10 years later, after saving a considerable sum of money, decided to formally ask both sets of parents if they could marry. Harum, an accountant, would normally be expected to pay the wedding costs, which could run to many thousands of dollars; however, Adinda was able to contribute from her salary as a teacher. Like nearly all Javanese, both are Muslims but because Islam does not have elaborate marriage ceremonies, colorful rituals from Christianity, Buddhism, and indigenous animism will enhance the elaborate and festive occasion.

In preparation, both bride and groom undergo unique Javanese rituals that remain from the days when marriages were arranged and the bride and groom did not know each other. A *pemaes*, a woman who prepares a bride for her wedding and whose role is to inject mystery and romance into the marriage relationship, bathes and perfumes the bride. She also puts on the bride's makeup and dresses her, all the while making offerings to the spirits of the bride's ancestors and counseling her about how to behave as a wife and how to avoid being dominated by her husband (see **Figure 10.19** on page 402). The groom also undergoes ceremonies meant to prepare him for marriage. Both are counseled that their relationship is bound to change over the course of the decades as they mature and as their family grows older.

Despite the elaborate preparations for marriage, divorce in Indonesia (and also in Malaysia) is fairly common among Muslims, who often go through one or two marriages early in life before they settle into a stable relationship. Although the prevalence of divorce is lamented by society, it is not considered outrageous. Apparently, ancient indigenous customs predating Islam allowed for mating flexibility early in life, and this attitude is still tacitly accepted. *[Sources: Adapted from personal communications with the anthropologist Jennifer W. Nourse, University of Richmond, a specialist in Southeast Asia, 2010; and Walter Williams,* Javanese Lives *(Piscataway, NJ: Rutgers University Press, 1991), pp. 128–134.]* ∎

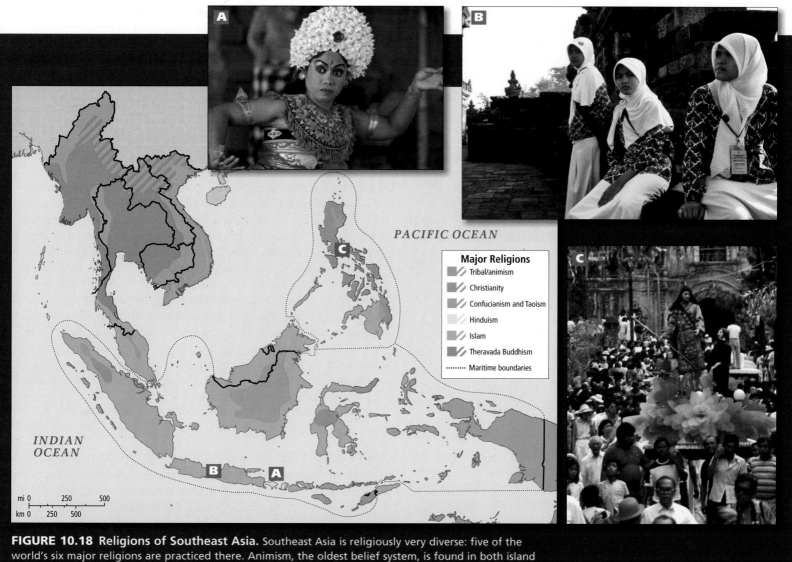

FIGURE 10.18 Religions of Southeast Asia. Southeast Asia is religiously very diverse: five of the world's six major religions are practiced there. Animism, the oldest belief system, is found in both island and mainland locations and has many subtle influences. **(A)** A dancer in Bali, Indonesia, where traditional Hindu performances incorporate Buddhist and animist symbols and stories. **(B)** Muslim students visit Borobudur, an ancient Buddhist monument on the island of Java, Indonesia. Many Islamic traditions on Java incorporate Buddhist ideas and practices. **(C)** San Isidro Labrador, patron saint of farmers, is carried by Catholics during the Pahiyas Festival in Lucban, Quezon, Philippines. The Pahiyas Festival is a Catholic festival that was originally an animist harvest celebration.

Gender Patterns in Southeast Asia

Gender roles are being transformed throughout Southeast Asia by urbanization and the changes it brings to family organization and employment. Moving to the city shifts people away from extended families and toward the nuclear family. Women have made significant gains in political empowerment and in education, but they still lag well behind men.

Family Organization, Traditional and Modern Throughout the region, it has been common for a newly married couple to reside with, or close to, the wife's parents. Along with this custom is a range of behavioral rules that empower the woman in a marriage, despite some basic patriarchal attitudes. For example, a family is headed by the oldest living male, usually the wife's father. When he dies, he passes on his wealth and power to the husband of his oldest daughter, not to his own son. (A son goes to live with his wife's parents and inherits from them.) Hence, a husband may live for many years as a subordinate in his father-in-law's home. Instead of the wife being the outsider, subject to the demands of her mother-in-law—as is the case, for example, in South Asia—it is the husband who must show deference. The inevitable tension between the wife's father and the son-in-law is resolved by the custom of ritual avoidance—in daily life they simply arrange to not encounter each other much. The wife manages communication between the two men by passing messages and even money back and forth. Consequently, she has access to a wealth

FIGURE 10.19 A Javanese wedding. In this phase of a traditional wedding, the bridegroom makes three small balls of the food and feeds them to the bride; she then does the same for him. The ritual reminds them that they should share joyfully whatever they have.

of information crucial to the family and has the opportunity to influence each of the two men.

Urbanization has brought a shift to the nuclear family that has transformed these traditional relationships. Young couples now frequently live apart from the extended family, an arrangement that takes the pressure off the husband in daily life. Because this nuclear family unit is often dependent entirely on itself for support, wives usually work for wages outside the home. Although married women lose the power they would have if they lived among their close kin, they are empowered by the opportunity to have a career and an income. The main drawback of this compact family structure, as many young families have discovered in Europe and the United States, is that there is no pool of relatives available to help working parents with child care and housework. Further, no one is left to help elderly parents maintain the rural family home.

Political and Economic Empowerment of Women Women have made some impressive gains in politics in Southeast Asia. Economically, they still earn less money than men and work less outside the home, but this will likely change if their level of education in relation to that of men continues to increase (see Reasons for Optimism C).

Southeast Asia has had several prominent female leaders over the years, most of whom have risen to power in times of crisis as the leaders of movements opposing corrupt or undemocratic regimes. In the Philippines, Corazon Aquino, a member of a large and powerful family, became president in 1986 after leading the opposition to Ferdinand Marcos, whose 21-year presidency was infamous for its corruption and authoritarianism. Gloria Macapagal-Arroyo, from another powerful family, became president in 2001 after opposing a similarly corrupt president, and then she was accused of corruption herself.

> **sex tourism** the sexual entertainment industry that services primarily men who travel for the purpose of living out their fantasies during a few weeks of vacation

In Indonesia, Megawati Sukarnoputri became president in 2001 after decades of leading the opposition to Suharto's notoriously corrupt 31-year reign. And in Burma, for more than two decades, opposition to the military dictatorship has been led by a woman, Aung San Suu Kyi. All of these women were wives or daughters of powerful political leaders, which raises some questions of nepotism. However, family favoritism cannot account for the several countries where the percentage of female national legislators is well above the world average of 18 percent: Vietnam (26 percent), Laos (25 percent), Singapore (24 percent), and the Philippines (20 percent).

Despite their successes in politics and their acknowledged role in managing family money, women still lag well behind men in terms of economic empowerment. Throughout the region, men have a higher rate of employment outside the home than women, and are paid more for doing the same work (see Figure 1.12 on page 21). But changes may be on the way. In Brunei, Malaysia, the Philippines, and Thailand, significantly more women than men are completing training beyond secondary school. If training qualifications were the sole consideration for employment, women would appear to have an advantage over men. This advantage may be significant if service sector economies, which generally require more education, become dominant in more countries. The service economy already dominates in the Philippines and Singapore.

❚❚ ▶ 318. THAILAND'S "THIRD SEX" WANTS ACCEPTANCE, LEGAL SUPPORT

Globalization and Gender: The Sex Industry

Southeast Asia has become one of several global centers for the sex industry, supported in large part by international visitors willing to pay for sex. **Sex tourism** in Southeast Asia grew out of the sexual entertainment industry that served foreign military troops stationed in Asia during World War II, the Korean War, and the Vietnam War. Now, primarily civilian men arrive from around the globe to live out their fantasies during a few weeks of vacation. The industry is found throughout the region but is most prominent in Thailand (see Thematic Overview I). In 2010, sixteen million tourists visited Thailand alone, up from 250,000 in 1965, and some observers estimate that as many as 70 percent were looking for sex. Even though the industry is officially illegal, some Thai government officials have even publicly praised sex tourism for its role in helping the country weather the economic crisis of 1997, because it created jobs. Some corrupt officials also support sex tourism because it provides them with a source of untaxed income from bribes.

One result of the "success" of sex tourism is a high demand for sex workers, and this demand has attracted organized crime. Estimates of the numbers of sex workers vary from 30,000 to more than a million in Thailand alone. Once they have been first forced into the trade, girls and women are often coerced by gangs into remaining in sex work. Demographers estimate that 20,000 to 30,000 Burmese girls taken against their will—some as young as 12—are working in Thai brothels. Their wages are too low to enable them to buy their own freedom. In the course of their work, they must service more than 10 clients per day, and they are routinely exposed to physical abuse and sexually transmitted diseases, especially HIV-AIDS.

Vignette Twenty-five-year-old Watsanah K. (not her real name) awakens at 11:00 every morning, attends afternoon classes in English and secretarial skills, and then goes to work at 4:00 P.M. in a bar in Patpong, Bangkok's red light district. There she will meet men from Europe, North America, Japan, Taiwan, Australia, Saudi Arabia, and elsewhere, who will pay to have sex with her. She leaves work at about 2:00 A.M., studies for a while, and then goes to sleep.

Watsanah was born in northern Thailand to an ethnic minority group who are poor subsistence farmers. She married at 15 and had two children shortly thereafter. Several years later, her husband developed an opium addiction. She divorced him and left for Bangkok with her children. There, she found work at a factory that produced seat belts for a nearby automobile plant. In 1997, Watsanah lost her job as the result of the economic crisis that ripped through Southeast Asia. To feed her children, she became a sex worker.

Although the pay, between U.S.$400 and U.S.$800 a month, is much better than the U.S.$100 a month she earned in the factory, the work is dangerous and demeaning. Sex work, though widely practiced and generally accepted in Thailand, is illegal, and the women who do it are looked down on. As a result, Watsanah must live in constant fear of going to jail and losing her children. Moreover, she cannot always make her clients use condoms, which puts her at high risk of contracting AIDS and other sexually transmitted diseases. "I don't want my children to grow up and learn that their mother is a prostitute," says Watsanah. "That's why I am studying. Maybe by the time they are old enough to know, I will have a respectable job." [Sources: Adapted from the field notes of Alex Pulsipher and Debbi Hempel, 2000; coverage of the HIV-AIDS conference in Thailand, July 11–16, 2004, by the Kaiser Family Foundation; "Life as a Thai Sex Worker," BBC News, February 22, 2007, at http://news.bbc.co.uk/2/hi/asia-pacific/6360603.stm.] ■

THINGS TO REMEMBER

1. The major religious traditions of Southeast Asia include Hinduism, Buddhism, Confucianism and Taoism, Islam, Christianity, and animism. All originated outside the region, with the exception of the animist belief systems.

2. Gender roles are shifting in this region as more women work outside the home. In some areas, women's economic and political empowerment builds on cultural traditions that give special responsibilities to women.

3. Some countries have become centers for the global sex industry, which puts many women at risk of kidnapping and disease.

Reflections on Southeast Asia

Southeast Asia is a geographically complex region that is often held up as a model for other developing regions for its "miraculous" economic development. While the economic success of this region is indeed remarkable, so are the environmental and social crises that have accompanied this success. The enormous amounts of carbon released into the atmosphere by deforestation have put this region at the center of efforts to deal with global climate change. Indigenous peoples, threatened by rapacious deforestation, are using new technologies to chart effective strategies for challenging corporations that are trying to appropriate their resources. Many cities are also facing crises as changes in rural food production systems, combined with the growth of employment in urban-based, export-oriented manufacturing, have created a massive tide of rural-to-urban migration. The infrastuctures of the region's largest cities have been swamped by the ever-increasing need for water, housing, and sanitation.

Yet, in the face of these challenges, there have been many positive changes. Population growth is slowing due to urbanization, development, and the success of some government policies to reduce birth rates (see Reasons for Optimism A). Despite the continuation of military dictatorships or undemocratic rule in a number of countries, democracy is strengthening throughout the region (see Reasons for Optimism B). The bright side of the economic crisis of the 1990s was that it exposed corruption and brought reform to both Thailand and Indonesia. The global recession that began in 2008 hit the region later than elsewhere and seems to be waning more quickly. Even on issues of gender, there are signs of change. Women's political and economic empowerment is increasing, as is their access to education. Meanwhile, greater attention is also being paid to the abuse of women and girls in the region's huge sex industry (see Reasons for Optimism C).

Looking toward the future, China's role as a trading partner with Southeast Asia will continue to grow. As the China–ASEAN free trade agreement continues to come into effect, the pressure is increasing on Southeast Asia to strengthen the areas where it can have an edge over China—high-value, technologically sophisticated industries. If it doesn't, Southeast Asia will be stuck playing second fiddle to its much larger neighbor, able to attract investment only when wages and other production costs in China rise beyond those in Southeast Asia. And yet, trade with China also offers many opportunities, and Southeast Asian companies are already profiting from some of them. Whether the majority of Southeast Asians can similarly benefit from China's growth is a key question for this region's future.

Learning Goal Review

1. Climate Change, Food, and Water: How is global warming linked to deforestation, food production, and water resources in Southeast Asia?

Reasons for Optimism in Southeast Asia

Population: Urbanization, development, changing gender roles, and government policies have come together to slow population growth. **A** *A poster supporting Thailand's population-control program.* ▼

Democratization: As corruption is increasingly exposed throughout the region, democracy is strengthening. **B** *Megawati Sukarnoputri, who became president of Indonesia in 2001.* ▼

Gender: Women's political and economic empowerment is increasing, as is their access to education. Meanwhile, greater attention is being paid to the abuse of women and girls in the sex industry. **C** *A Cambodian girl learning to write.* ▼

Thinking Geographically: What government-sponsored campaign has been partially responsible for Thailand's low fertility rate of 1.8 children per adult woman?

Thinking Geographically: Democratization in Indonesia received a major boost after the economic crisis spurred what event?

Thinking Geographically: Women would appear to have an advantage over men in gaining employment if _____.

This region is a major contributor to greenhouse gas emissions via widespread deforestation. To what extent is agriculture, both for export and local consumption, driving this process? How do deforestation and agriculture contribute to global warming? How is global warming likely to impact this region's water resources?

2. Globalization and Development: How has globalization produced both spectacular successes and tragic failures in the economies of Southeast Asia?

The Southeast Asian economy grew rapidly from the mid-1980s to the late 1990s. Beginning in 1997, however, a disastrous financial crisis shook the region for several years. How were both the rapid growth and the financial crisis related to globalization?

3. Democratization: Is democratization influencing levels of violence and corruption in Southeast Asia?

While significant barriers to democracy remain, some countries have had dramatic expansions of democracy in recent years. How has the expansion of democracy affected corruption in this region? What is the evidence that democracy can reduce political violence associated with separatist movements?

4. Population, Development and Gender: How is economic development affecting gender roles in ways that contribute to slower population growth in Southeast Asia?

How have new employment opportunities for women reduced incentives for large families? How has urbanization influenced gender roles?

5. Urbanization: What changes are driving urbanization in this region?

What are the main push factors that are driving rural people into the cities? What are the main pull factors that attract people to the cities?

Geographic Themes about Southeast Asia

Look back at the Thematic Overview photos on page 369. Thinking geographically, answer the following questions about them:

(A) Climate Change: What is one reason that the replacement of forests by oil palm plantations contributes to climate change?

(B) Food: How does wet rice cultivation contribute to climate change?

(C) Water: What phenomena related to climate change is increasing the (short-term) risk of flooding in this region?

(D) Globalization: Export-led development was a dramatic departure from what strategy used in other developing areas in Middle and South America and part of Africa?

(E) Development: In what countries (other than Singapore) does the service sector dominate the economy?

(F) Democratization: What prime minister was deposed by a coup in 2006?

(G) Population: What is Singapore's chief population concern?

(H) Urbanization: How much larger is Manila, relative to the next largest city in the Philippines?

(I) Gender: Why have some Thai government officials publicly praised sex tourism?

Key Terms

archipelago 374
Association of Southeast Asian
  Nations (ASEAN) 389
Australo-Melanesians 383
Austronesians 383

coral bleaching 381
crony capitalism 389
cultural pluralism 400
detritus 375
domino theory 384

El Niño 375
feminization of labor 388
foreign exchange 399
resettlement schemes 392
sex tourism 402

11 Oceania: Australia, New Zealand, and the Pacific

Learning Goals

After you have read this chapter, you will be able to answer the following questions:

1. Climate Change and Water: What impact will climate change have on Oceania, particularly in relation to water?

2. Food and the Environment: In what ways have European systems for producing food and fiber changed environments in Oceania?

3. Globalization and Development: How has globalization transformed patterns of economic development in Oceania?

4. Population and Urbanization: What are the two main patterns of population and urbanization developing across Oceania?

5. Gender and Democratization: How are issues of gender, democracy, and economic empowerment developing differently across Oceania?

FIGURE 11.1 Political map of Oceania. Thirteen of these entities are independent countries; the others are territories of, or are otherwise affiliated with, other nations. Not depicted on the map are the Pacific Island Wildlife Refuges, a widely scattered, essentially uninhabited group of northern Pacific islands that constitute a U.S. territory.

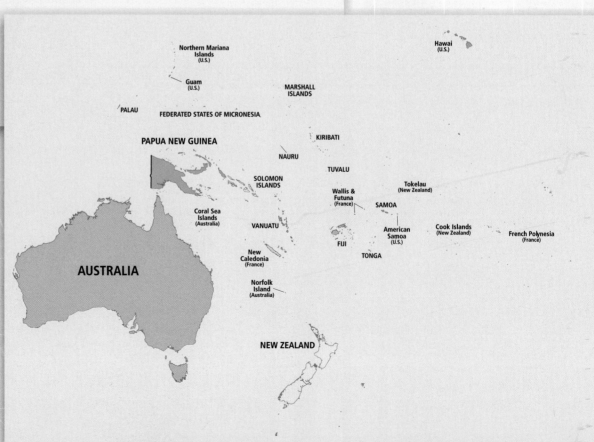

Thematic Overview of Oceania

Climate Change: As the oceans warm, coral reefs are threatened by coral bleaching. This jeopardizes fishing resources for many Pacific islands. **A** *A healthy part of the Great Barrier Reef.* ▼

Water: Parts of Australia and some low, dry Pacific islands have prolonged droughts and freshwater shortages. **B** *A drought-stricken wheat field in Australia.* ▼

Food: Throughout Oceania, the introduction of food production systems from elsewhere has transformed landscapes and resulted in the spread of ecologically damaging invasive species. **C** *A herd of sheep in New Zealand.* ▼

Globalization: Oceania is reorienting away from Europe and toward Asia, which is now Oceania's largest trading partner. **D** *Japanese investors visit an Australian winery.* ▼

Development: Australia and New Zealand are somewhat unusual in having achieved broad prosperity largely through raw material exports. **E** *A statue near Melbourne honoring Australia's gold miners.* ▼

Democratization: While parts of Oceania are highly democratized, many Pacific islands are ruled by chiefs and royalty. **F** *A gathering of Tonga's aristocracy (Tonga is ruled by a king and allied aristocracy).* ▼

Population: Two patterns are emerging in Oceania. Relative to Australia, New Zealand, and Hawaii, most Pacific islands are rural, poor, and have younger and more rapidly growing populations. **G** *Children in Fiji.* ▼

Urbanization: Seventy-three percent of Oceania's population live in urban areas, with most employed by growing service economies. **H** *Hotels, restaurants, marinas, and other tourism infrastructure in Honolulu, Hawaii.* ▼

Gender: While women are gaining political and economic empowerment throughout Oceania, this shift is slowest in Papua New Guinea and other Pacific islands. **I** *A young mother sells tobacco in Papua New Guinea.* ▼

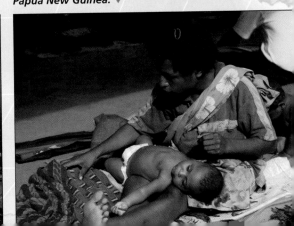

Global Patterns, Local Lives In October 2003 in Brisbane, Australia, a crowd of 47,000 waited eagerly for the Haka as New Zealand's national rugby team, the All Blacks, took the field opposite the team from Tonga, a Polynesian archipelago in the South Pacific (see Figure 11.4 on pages 410–411).

The Haka is a highly emotional and physical dance traditionally performed by the **Maori**, the indigenous (Polynesian) people of New Zealand, to motivate fellow warriors and intimidate opponents before entering battle. Dances like this have long been a part of many cultures in the islands of Oceania (**Figure 11.2A**), but the Haka has now become an integral part of rugby, the region's most popular sport. Before almost every international match for the past century, the All Blacks have performed the Haka in unison: chanting, screaming, jumping, stomping their feet, poking out their tongues, widening their eyes to show the whites, and beating their thighs, arms, and chests.

> **Maori** Polynesian people indigenous to New Zealand

Until 2003, only the New Zealand team performed the Haka, but that night it was a different story. Halfway through New Zealand's Haka, the Tongan team responded with its own Haka. The crowd roared its approval of this scene, a revival of the traditional prelude to battle throughout Oceania's long history.

The Haka has now spread in the world of rugby and beyond. Outside of Oceania, those who perform the Haka include the rugby teams at Jefferson High in Portland, Oregon, and Middlebury College in Vermont (Figure 11.2B), and the football teams at Brigham Young University and the University of Hawaii. Many of these teams have players who are of Polynesian heritage. Most practitioners speak of the Haka as filling them with the necessary exuberance, aggression, and spirituality to play a vigorous and successful game.

To see videos of a Haka, go to http://www.YouTube.com and type in "haka." [Adapted from articles by Phil Wilkins, "Tonga Can Only

Match the Kiwis in the Haka," Brisbane, Australia, October 25, 2003; and Mark Falcous, "The Decolonizing National Imaginary: Promotional Media Constructions During the 2005 Lions Tour of Aotearoa," New Zealand, Journal of Sport and Social Issues, November 2007, Vol. 31, No. 4, pp. 374–393.] ■

The Haka is an example of how, in the postcolonial modern era, indigenous cultures in Oceania are being revived, celebrated, and appropriated—in this case by those who wish to project a multicultural national image for New Zealand. And yet the Haka itself seems to be a globalizing phenomenon. The fact that a Maori war dance is being performed before a game of rugby, which was brought to Oceania by the British (Figure 11.3), shows how deeply globalization has penetrated this region. Oceania was dominated politically and economically by people of European descent for more than 200 years. Now, Oceania finds itself subject to a new wave of globalization emanating from Asia—a situation that has far-reaching economic implications.

THINGS TO REMEMBER

1. For more than two centuries, European peoples dominated Oceania economically, politically, and to some extent, culturally.

2. Today, European influence is challenged by reinvigorated native traditions and by economic globalization.

3. The Haka today is an example of globalization—the merging of a Maori war dance with the European game rugby—that is spreading beyond Oceania.

(A)

(B)

FIGURE 11.2 The Haka, a Maori tradition. (A) A Haka performed by Maori in Christchurch, New Zealand. **(B)** The Middlebury College men's rugby team performing their Haka before a match in Middlebury, Vermont.

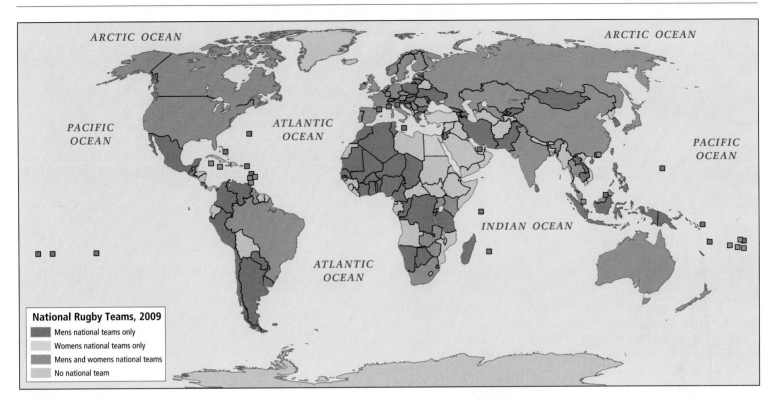

FIGURE 11.3 Rugby around the world. In over 136 countries, women, men, boys, and girls play rugby: 136 countries have men's national rugby teams, and 58 countries have women's national teams. The men's World Cup rugby competition began in 1987 and the women's World Cup competition began in 1991. In April 2010, New Zealand had the top team for both men and women, but the ranking of the men's teams can change weekly.

I THE GEOGRAPHIC SETTING

Oceania is made up of Australia, New Zealand, Papua New Guinea, and the many small islands scattered across the Pacific Ocean (see Figure 11.1 on page 406). It is a unique world region in that it is composed primarily of ocean and covers the largest portion of the earth's surface of any region, yet has only 37 million people, the vast majority of whom live in Australia (21.9 million), Papua New Guinea (6.6 million), and New Zealand (4.3 million).

Terms in This Chapter

Some maps in this chapter show different place names in similar locations. This reflects the political evolution of this region, where some islands previously were grouped under one name that remains in use but are now regrouped into country units with another name. For example, the Caroline Islands, located north of New Guinea, still go by that name on maps and charts, but they have been divided into two countries: Palau (a small group of islands at the western end of the Caroline Islands) and the Federated States of Micronesia, which extends over 2000 miles west to east, from Yap to Kosrae. In addition, there are three commonly used names for island groupings—Micronesia, Melanesia, and Polynesia—that

are not political units, but are based on ancient ethnic and cultural links (see Figure 11.10 on page 426).

Physical Patterns

With the exception of the continent of Australia, this region consists of islands, some quite large, some very small, separated by the huge expanse of the Pacific Ocean. The ocean moderates climates and simultaneously links and inhibits movement across this vast region.

Continent Formation

The largest landmass in Oceania is the ancient continent of Australia at the southwestern perimeter of the region (see the map in Figure 11.4 on pages 410–411). The Australian continent is partially composed of some of the oldest rock on earth and has been relatively stable for more than 200 million years, with very little volcanic activity and only an occasional mild earthquake. Australia was once a part of the great landmass, called **Gondwana**, which formed the southern part of the ancient supercontinent Pangaea (see Figure 1.21 on page 45). What became present-day Australia broke free from Gondwana and drifted until

> **Gondwana** the great landmass that formed the southern part of the ancient supercontinent Pangaea

CHINA
Zhengzhou
Nanjing
Fukuoka
Osaka
Shanghai
Hangzhou
JAPAN
Changsha
Nanchang
Guangzhou
(Canton)
Taipei
Taichung
TAIWAN
Hong Kong
Kaohsiung

A Australia's Eastern Highlands, Blue Mountains

South
China
Sea
Manila
PHILIPPINES
Philippine
Sea

Northern
Mariana
Islands
(U.S.)

Saipan
Tinian
Agana
Guam
(U.S.)

Sulu
Sea
BRUNEI
DARUSSALAM
Bandar Seri Begawan
MALAYSIA
Celebes
Sea
Borneo
Sulawesi
(Celebes)

Yap

Caroline
Islands

FEDERATED STATES
OF MICRONESIA

PALAU
Ngerulmud

Pohnpei
Truk
Palikir

Kosrae

MARSHALL
ISLANDS
Majuro

M i c r o n e s i a

M e l a n e s i a

NAURU
Yaren

Tarawa
KIRIBATI
(GILBERT ISLANDS)

Phoenix
Islands

INDONESIA
Surabaya
Java
Bali

Banda
Sea

West
Papua
Jayapura
Wewak
New Guinea
PAPUA
NEW GUINEA
Salamaua
Daru

Admiralty
Islands
Kavieng
Bismarck
Archipelago
Rabaul
New Ireland
New
Britain
Buka
Bougainville
SOLOMON
ISLANDS
Malaita
Honiara

TUVALU
(ELLICE ISLANDS)
Funafuti

Tokelau
(New Zealand)

Dili
TIMOR-LESTE

Arafura
Sea

Thursday
Island
Torres
Strait
Port
Moresby

Wallis & Futuna
(France)
Mata Utu
Apia
Pago Pago

SAMOA
(WESTERN
SAMOA)
American
Samoa
(U.S.)
Alofi
Niue
(New Zealand)

Timor
Sea

INDIAN
OCEAN

Darwin
Jabiru
Katherine
Daly Waters

Gulf of
Carpentaria

Cape
York
Penninsula
Cooktown
Cairns

Great Barrier Reef

Coral Sea
Islands
(Australia)

Coral
Sea

VANUATU
(NEW HEBRIDES)
Port-Vila

FIJI
Vanua Levu
Viti Levu
Suva

King
Leopold
Ranges
Derby

Hamersley
Range

Great Sandy Desert
NORTHERN
TERRITORY
Tennant Creek

Selwyn
Range

Macdonnell
Ranges
Alice Springs

QUEENSLAND

Great
Dividing
Range

Mackay

E

New
Caledonia
(France)
Noumea

Carnarvon

Gibson Desert

Musgrave
Ranges

Uluru
(Ayers Rock)

Great
Artesian
Basin

AUSTRALIA

WESTERN
AUSTRALIA
B
Great
Victoria
Desert

Geraldton

SOUTH
AUSTRALIA

Perth

Nullarbor Plain

Flinders
Range
Broken
Hill

Darling

Brisbane

Norfolk
(Australia)

Kingston

Esperance
Albany

Elliston
Adelaide

Great Australian Bight

Murray

NEW
SOUTH
WALES

Great Dividing Range

Sydney
A

Kermadec Islands
(New Zealand)

INDIAN OCEAN

VICTORIA
Canberra
Australian Alps

Tasman
Sea

TONGA
Nuku'alofa

Melbourne

B Australian Desert

Launceston
TASMANIA
Hobart

NEW
ZEALAND

Auckland
North
Island
Gisborne

Westport
Wellington
Christchurch

South
Island
Southern Alps
D
Dunedin
Invercargill

Chatham Islands
(New Zealand)

Bounty Islands
(New Zealand)

Auckland Islands
(New Zealand)

Antipode Islands
(New Zealand)

Campbell Island
(New Zealand)

NORTH PACIFIC OCEAN

Land Elevations

meters	feet
4877	16,000
3353	11,000
2134	7000
914	3000
305	1000
152	500
0	0

Ocean Depths

meters	feet
0	0
300	984
3500	11,483
5000	16,404

mi 0   200   400   600   800
km 0 200 400 600 800 1000 1200
1:42,000,000
Mercator Projection

30°N

Kauai
Oahu   Molokai
Honolulu   Maui
Hawaii
(U.S.)   Hawaii

20°N

10°N

Line Islands
Kiritimati

0° Equator

Polynesia

Marquesas Islands

10°S

F

Cook Islands
(New Zealand)   Papeete   Tahiti
French Polynesia
(France)

C

20°S

Avarua
Rarotonga

SOUTH PACIFIC OCEAN

Pitcairn Islands
(U.K.)

30°S

C Atoll de Réao, Tuamotu Archipelago

E Great Barrier Reef

D Milford Sound, South Island, New Zealand

F The high island of Tahiti

FIGURE 11.4 Regional map of Oceania.

it eventually collided with the part of the Eurasian Plate on which Southeast Asia sits. That impact created the mountainous island of New Guinea to the north of Australia.

Australia is shaped roughly like a dinner plate with a lumpy, irregular rim and two bites taken out of it: one in the north (the Gulf of Carpentaria) and one in the south (the Great Australian Bight). The center of the plate is the great lowland Australian desert, with only two hilly zones and rocky outcroppings (Figure 11.4B). The lumpy rim of Australia is composed of uplands; the highest and most complex of these are the long, curving Eastern Highlands (labeled "Great Dividing Range" on the map in Figure 11.4, and shown in Photo A). Over millennia, the forces of erosion—both wind and water—have worn most of Australia's landforms into low, rounded formations, some of which are quite spectacular.

Off the northeastern coast of the continent lies the **Great Barrier Reef**, the largest coral reef in the world and a World Heritage Site since 1981 (Figure 11.4E). It stretches in an irregular arc for more than 1250 miles (2000 kilometers) along the coast of Queensland, covering 135,000 square miles (350,000 square kilometers). The Great Barrier Reef is so large that it influences Australia's climate. It interrupts the westward-flowing ocean currents in the mid–South Pacific circulation pattern, shunting warm water to the south, where it warms the southeastern coast of Australia. Threats to the health of the Great Barrier Reef are discussed on page 416.

Island Formation

The islands of the Pacific were created (and are being created still) by a variety of processes related to the movement of tectonic plates. The islands found in the western reaches of Oceania—including New Guinea, New Caledonia, and the main islands of Fiji—are remnants of the Gondwana landmass; they are large, mountainous, and geologically complex. Other islands in the region are volcanic in origin and form part of the Ring of Fire (see Figure 1.22 on page 46). Many of this latter group are situated in boundary zones where tectonic plates are either colliding or pulling apart. For example, the Mariana Islands east of the Philippines are volcanoes that were formed when the Pacific Plate plunged beneath the Philippine Plate. The two much-larger islands of New Zealand were created when the eastern edge of the Indian-Australian Plate was thrust upward by its convergence with the Pacific Plate.

The Hawaiian Islands were produced through another form of volcanic activity associated with **hot spots**, places where particularly hot magma moving upward from Earth's core breaches the crust in tall plumes. Over the past 80 million years, the Pacific Plate has moved across one of these hot spots, creating a string of volcanic formations 3600 miles (5800 kilometers) long. The youngest volcanoes, only a few of which are active, are on or near the islands known as Hawaii.

Volcanic islands exist in three forms: volcanic high islands, low coral atolls, and coral platforms raised or uplifted by volca-

nism known as *makatea*. High islands are usually volcanoes that rise above the sea into mountainous, rocky formations that contain a rich variety of environments. New Zealand, the Hawaiian Islands, Tahiti, and Easter Island are examples of high islands (Figure 11.4 D, F). An **atoll** is a low-lying island, or chain of islets, formed of coral reefs that have built up on the circular or oval rim of a submerged volcano (Figure 11.4C). These reefs are arranged around a central lagoon that was once the volcano's crater. As a consequence of their low elevation, atoll islands tend to have only a small range of environments and very limited supplies of fresh water.

Climate

Although the Pacific Ocean stretches nearly from pole to pole, most of Oceania is situated within the tropical and subtropical latitudes of that ocean. The tepid water temperatures of the central Pacific bring mild climates year-round to nearly all the inhabited parts of the region (Photo Essay 11.1). The seasonal variation in temperature is greatest in the southernmost reaches of Australia and New Zealand.

Moisture and Rainfall The vast interior of Australia is mostly desert, due to the belt of dry air that circles the globe between roughly 20 and 30 degrees latitude (see page 211 in Chapter 6). Much of the rest of Oceania is warm and humid nearly all the time. New Zealand and the high islands of the Pacific receive copious rainfall and once supported dense forest vegetation, although much of that forest is gone after 1000 years of human impact (see Photo Essay 11.2D on page 417).

Travelers approaching New Zealand, either by air or by sea, sometimes notice a distinctive long, white cloud that stretches above the North Island. A thousand years ago, the Maori settlers also noticed this phenomenon, and they named that place *Aotearoa*, "land of the long white cloud," a name that is now the Maori name for the whole country.

The distinctive mass of moisture that is represented by Aotearoa is brought in by the legendary **roaring forties** (named for the 40th parallel south), which are powerful air and ocean currents that speed around the far Southern Hemisphere virtually unimpeded by landmasses. These westerly winds (blowing west to east) deposit a drenching 130 inches (330 centimeters) of rain per year in the New Zealand highlands and more than 30 inches (76 centimeters) per year on the coastal lowlands. At the southern tip of New Zealand's North Island, the wind averages more than 40 miles per hour (64 kilometers per hour) about 118 days a year. Farmers in the area stake their cabbages to the ground so they will not blow away.

By contrast, two-thirds of the continent of Australia is overwhelmingly dry. The dominant winds affecting Australia (see the map in Photo Essay 11.1) are the north and south easterlies (blowing east to west) that converge east of the continent. The Great Dividing Range blocks the movement of moist, westward-moving air, so that rain does not reach the interior (an orographic

---

Great Barrier Reef the largest coral reef in the world, located off the northeastern coast of Australia

hot spots individual sites of upwelling material (magma) that originate deep in the mantle of the earth and surface in a tall plume; hot spots tend to remain fixed relative to migrating tectonic plates

atoll a low-lying island, formed of coral reefs that have built up on the circular or oval rims of a submerged volcano

roaring forties powerful air and ocean currents at about 40° S latitude that speed around the far Southern Hemisphere virtually unimpeded by landmasses

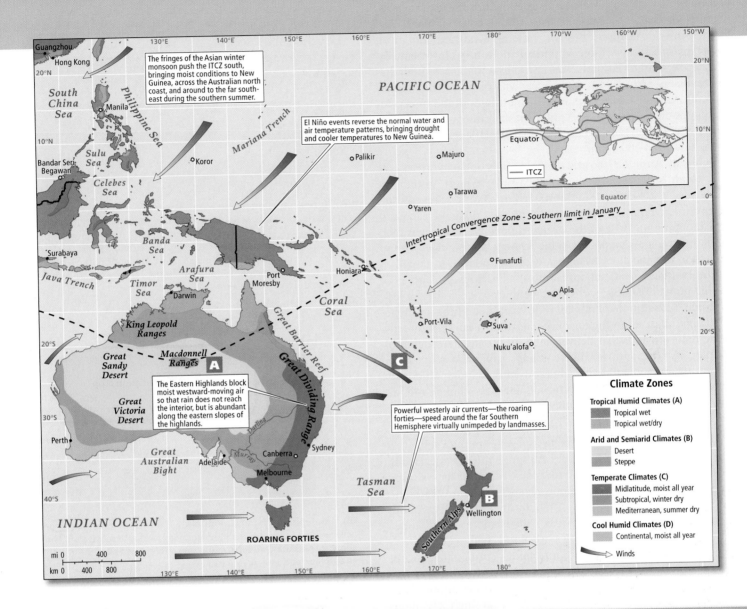

The fringes of the Asian winter monsoon push the ITCZ south, bringing moist conditions to New Guinea, across the Australian north coast, and around to the far southeast during the southern summer.

El Niño events reverse the normal water and air temperature patterns, bringing drought and cooler temperatures to New Guinea.

Intertropical Convergence Zone - Southern limit in January

The Eastern Highlands block moist westward-moving air so that rain does not reach the interior, but is abundant along the eastern slopes of the highlands.

Powerful westerly air currents—the roaring forties—speed around the far Southern Hemisphere virtually unimpeded by landmasses.

ROARING FORTIES

PACIFIC OCEAN

Equator

ITCZ

Climate Zones

Tropical Humid Climates (A)
- Tropical wet
- Tropical wet/dry

Arid and Semiarid Climates (B)
- Desert
- Steppe

Temperate Climates (C)
- Midlatitude, moist all year
- Subtropical, winter dry
- Mediterranean, summer dry

Cool Humid Climates (D)
- Continental, moist all year
- Winds

A Desert, Alice Springs, Australia

B Midlatitude, moist all year, New Zealand

C Tropical wet/dry, New Caledonia

pattern; see also Figure 1.23 on page 47). As a result, a large portion of Australia receives less than 20 inches (50 centimeters) of rain per year, and humans have found rather limited uses for this interior territory. But the eastern (windward) slopes of the highlands receive more abundant moisture. This relatively moist eastern rim of Australia was favored as a habitat by both the indigenous people and the Europeans who displaced them after 1800. During the southern summer, the fringes of the monsoon that passes over Southeast Asia and Eurasia bring moisture across Australia's northern coast. There, annual rainfall varies from 20 to 80 inches (50 to 200 centimeters).

Overall, Australia is so arid that it has only one major river system, which is in the temperate southeast where most Australians live. There, the Darling and Murray rivers drain one-seventh of the continent, flowing west and south into the Indian Ocean near Adelaide. One measure of the overall dryness of Australia is that the entire average *annual* flow of the Murray-Darling river system is equal to just *one day's* average flow of the Amazon in Brazil.

In the island Pacific, mountainous high islands also have orographic rainfall patterns, with a wet windward side and a dry leeward side. The amount of rainfall varies considerably between the low-lying islands across the region. Some of the islands lie directly in the path of winds that deliver between 60 and 120 inches of rain per year on average. These islands support a remarkable variety of plants and animals. Other low-lying islands, particularly those on the equator, receive considerably less rainfall and are dominated by grasslands that support little animal life.

El Niño Recall from Chapter 3 the El Niño phenomenon, a pattern of shifts in the circulation of air and water in the Pacific that occurs irregularly every 2 to 7 years. Although these cyclical shifts, or oscillations, are not yet well understood, scientists have worked out a model of how the oscillations may occur (**Figure 11.5**).

The El Niño event of 1997–1998 illustrates the effects of this phenomenon. By December of 1997, the island of New Guinea (north of Australia; see also Figure 11.4) had received very little rainfall for almost a year. Crops failed, springs and streams dried up, and fires broke out in tinder-dry forests. The cloudless sky allowed heat to radiate up and away from elevations above 7200 feet (2200 meters), so temperatures at high elevations dipped below freezing at night for stretches of a week or more. Tropical plants died, and people unaccustomed to chilly weather sickened. Meanwhile, along the Pacific coasts of North, Central, and South America, the warmer-than-usual weather brought unusually strong storms, high ocean surges, and damaging wind and rainfall (Figure 11.5C). Recently, an opposite pattern, in which normal conditions become unusually strong, has been identified and named *La Niña*.

Fauna and Flora

The fact that Oceania comprises an isolated continent and numerous islands has affected its animal life (*fauna*) and plant life (*flora*). Many of its species are **endemic**, meaning they exist in a particular place and nowhere else on earth. This is especially true of Australia, but many Pacific islands also have endemic species.

endemic belonging or restricted to a particular place

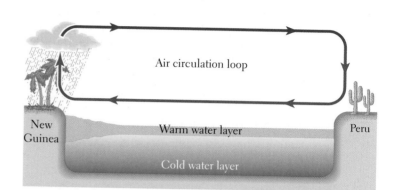

(A) Normal equatorial conditions. Water in the equatorial western Pacific (the New Guinea/Australia side) is warmer than water in the eastern Pacific (the Peru side). Due to prevailing wind patterns, the warm water piles up in the west. Warm air rises above this warm-water bulge in the western Pacific and forms rain clouds. The rising air cools and, once in the higher atmosphere, moves in an easterly direction. In the east, the dry cool air descends, bringing little rainfall to Peru.

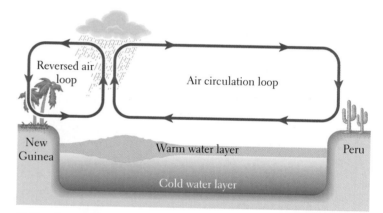

(B) Developing El Niño conditions. As an El Niño event develops, the ocean surface's warm-water bulge (orange) begins to move east. The air rising above it splits into two formations, one circulating east to west in the upper atmosphere and one west to east.

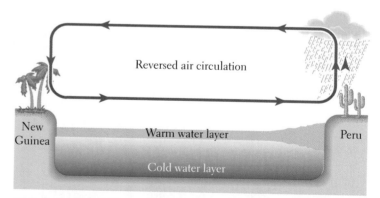

(C) Fully developed El Niño. Slowly, as the bulge of warm water at the surface of the ocean moves east, it forces the whole system into the fully developed El Niño with air at the water surface and in the upper atmosphere flowing in reverse of normal (A). Instead of warm, wet air rising over the mountains of New Guinea and condensing as rainfall, cool, dry, cloudless air descends to sit at the earth's surface. Meanwhile, in the east, the normally dry, clear coast of Peru experiences clouds and rainfall.

FIGURE 11.5 A model of the El Niño phenomenon.

Animal and Plant Life in Australia The uniqueness of Australia's animal and plant life is the result of the continent's long physical isolation, large size, relatively homogeneous landforms, and arid climate. Since Australia broke away from Gondwana more than 65 million years ago, its animal and plant species have evolved in isolation. One spectacular result of this long isolation is the presence of more than 144 living species of endemic marsupial animals. **Marsupials** are mammals that give birth to their young at a very immature stage and then nurture them in a pouch equipped with nipples. The best-known marsupials are the kangaroos; other species include wombats, koalas, and bandi-coots. The various marsupials fill ecological niches that in other regions of the world are occupied by rats, badgers, moles, cats, wolves, ungulates (grazers), and bears. The **monotremes**, egg-laying mammals that include the duck-billed platypus and the spiny anteater, are endemic to Australia and New Guinea. Some of the 750 species of birds known in Australia migrate in and out, but more than 325 species are endemic.

> **marsupials** mammals that give birth to their young at a very immature stage and nurture them in a pouch equipped with nipples
>
> **monotremes** egg-laying mammals, such as the duck-billed platypus and the spiny anteater

Most of Australia's endemic plant species are adapted to dry conditions. Many of the plants have deep taproots to draw moisture from groundwater and small, hard, pale green or shiny leaves to reflect heat and to hold moisture. Much of the continent is grassland and scrubland with bits of open woodland; there are only a few true forests, found in pockets along the Eastern Highlands, the southwestern tip, and in Tasmania (Figure 11.6). Two plant genera account for nearly all the forest and woodland plants: *Eucalyptus* (450 species, often called "gum trees") and *Acacia* (900 species, often called "wattles").

Plant and Animal Life in New Zealand and the Pacific Islands Naturalists and evolutionary biologists have long been interested in the species that inhabited the Pacific islands before humans arrived. Charles Darwin formulated many of his ideas about evolution after visiting the Galápagos Islands of the eastern Pacific (see Figure 3.3 on pages 102–103) and the islands of Oceania.

Islands gain plant and animal populations from the sea and air around them as organisms are carried from larger islands and continents by birds, storms, or ocean currents. Once these organisms "colonize" their new home, they may evolve over time into new species that are unique to one island. High, wet islands generally contain more varied species because their more complex environments provide niches for a wider range of wayfarers and thus greater opportunities for evolutionary change.

The flora and fauna of islands are also modified by human inhabitants once they arrive. In prehistoric times, Asian explorers in oceangoing sailing canoes brought plants such as bananas and breadfruit, and animals such as pigs, chickens, and dogs. Today, human activities from tourism to military exercises to urbanization continue to change the flora and fauna of Oceania.

II ▶ 240. BREADFRUIT ADVOCATES SAY IT COULD SOLVE HUNGER IN TROPICAL REGIONS

Generally, the diversity of land animals and plants is richest in the western Pacific, near the larger landmasses. It thins out to the east, where the islands are smaller and farther apart. The natural rain forest flora is rich and abundant on New Zealand, New Guinea, and also on the high islands of the Pacific. However, the natural fauna is much more limited on these islands. While New Guinea has fauna comparable to Australia, to which it was once connected via Sundaland (see Figure 10.5 on page 375), New Zealand and the Pacific islands have no indigenous land mammals, almost no indigenous reptiles, and only a few indigenous species of frogs. New Zealand and the islands were never connected to Australia and New Guinea by a land bridge that land animals could cross. On the other hand, indigenous birds have been numerous and varied. Two examples are New Zealand's kiwi and the huge moa—a bird that grew up to 12 feet [3.7 meters] tall and was a major source of food for the Maori people until they hunted it to extinction before Europeans arrived. Today, New Zealand may well be the country with the most introduced species of mammals, fish, and fowl, nearly all brought in by European settlers.

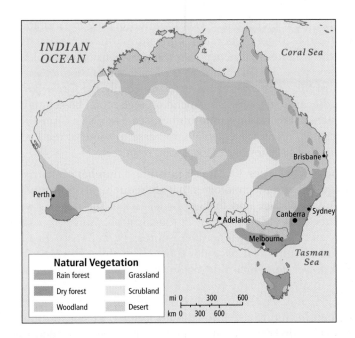

FIGURE 11.6 Australia's natural vegetation. Much of Australia is grassland and scrubland; a few forests can be found in the Eastern Highlands, in the far southwest, and in Tasmania.

THINGS TO REMEMBER

1. This largest region of the world is primarily water and has a very small population of just 37 million.

2. The largest land areas are the continent of Australia and the two island nations of New Zealand and the eastern half of New Guinea, known as Papua New Guinea.

3. The thousands of islands of the Pacific are made up of three main types of volcanic formations.

4. The climate of the region is mostly tropical and subtropical.

5. The fauna and flora of the region are unique in many distinctive ways that have provided information about evolution. Partly because they have been isolated for lengthy periods, many species are endemic to a specific locale.

Environmental Issues

Oceania faces a host of environmental problems, and despite its relatively small population, public awareness of environmental issues is keen. Global climate change, primarily warming, has brought a broad array of threats to the region's ecology, as has human introduction of many nonnative species. The expansion of herding, agriculture, fishing fleets, and the human population itself have also impacted much of the region (Photo Essay 11.2).

Global Climate Change

Oceania is a minor contributor of greenhouse gases, except for Australia, which has some of the world's highest greenhouse gas emissions on a per capita basis (see Figure 1.20 on page 40). Much like the United States, its emissions result from the use of automobile-based transportation systems to connect a widely dispersed network of cities and towns. Also, as in the United States, heavy dependence on coal for electricity generation leads to high emissions. However, because Australia has a relatively small population (22 million people in 2010), it accounts for only slightly more than 1 percent of global emissions. Despite negligible contributions of greenhouse gases by the islands of the Pacific, they, as well as Australia, are highly vulnerable to the effects of global climate change (see Photo Essay 11.3 on page 418).

> **Learning Goal 1**
> **Climate Change and Water:** What impact will climate change have on Oceania, particularly in relation to water?

Sea Level Rise As we have seen, global warming may raise sea levels by melting glaciers and ice caps. Obviously, this issue is of great concern to residents of islands that already barely rise above the waves (see Photo Essay 11.3A). If sea levels rise the 4 inches (10 centimeters) per decade predicted by the International Panel on Climate Change, many of the lowest-lying Pacific atolls, such as Tuvalu, will disappear under water within 50 years. Other islands, some with already very crowded coastal zones, will be severely reduced in area and will become more vulnerable to storm surges and cyclones.

Vignette Nicholas Hakata is a community leader and youth worker on the Carteret Islands that are administered by Papua New Guinea. The islands rise less than 2 meters above the sea. Since salty seawater began washing over their cultivation plots, he and his fellow citizens have often gone hungry, subsisting on fish and coconuts. As the sea rises, standing water encourages mosquito infestations, and now malaria is spreading among the children already weakened by hunger. The children used to do well in school, but now, sick and hungry, they have lost interest. Hakata organizes community meetings to study the situation and

the people are reluctantly reaching a consensus to try to migrate to the New Guinea mainland; but New Guinea, already dealing with others wishing to migrate, is not willing to take them. *[Source: Neil MacFarquhar, "Refugees Join List of Climate-Change Issues," International Herald Tribune, May 28, 2009; UNUChannel, "Local Solutions on a Sinking Paradise, Carterets Islands, Papua New Guinea," at http://www.vimeo.com/4177527.]* ∎

Other Water-Related Vulnerabilities Much of Oceania is vulnerable to changes in the region's climate that could result from global warming. Parts of Australia and some low, dry Pacific islands are already undergoing prolonged droughts and freshwater shortages that are requiring major changes in daily life and livelihoods (see Thematic Overview B). It is possible that recent severe droughts are not merely periodic dry spells but may represent permanent alterations in rainfall patterns that could worsen wildfires. Such fires emerged as a major issue in Australia in February 2009, when 173 people in the state of Victoria died in a rural firestorm near Melbourne.

Global warming is also causing the oceans to warm, which can bring stronger tropical storms (see Photo Essay 11.3B), such as those that brought catastrophic flooding to Australia in 2010–2011. Warmer ocean temperatures threaten coral reefs and the fisheries that depend on them by causing coral bleaching (see page 381 in Chapter 10), a phenomenon affecting all reefs in this region in recent years, especially the Great Barrier Reef. Because so many fish depend on reefs, coral bleaching also threatens many fishing communities (see Thematic Overview A on page 407). This is especially the case in some of the Pacific islands that have few other local food resources.

Responses to Crises Oceania is pursuing a number of alternative energy and water technologies that are reducing its contributions to climate change and making better use of the region's water resources. New Zealand has set a goal of obtaining 90 percent of its electricity from renewable sources by 2025 (see Reasons for Optimism C on page 437). Much of this will come from wind power, for which this region has excellent potential, especially in areas near the roaring forties. Geothermal energy (see page 383 in Chapter 10) is either already being used or is being planned for future use throughout Oceania, except in some of the low, nonvolcanic Pacific islands. Solar energy is widely used in some remote Pacific islands where the cost of importing fuel is prohibitive.

Thinking Geographically

After you have read about the Human Impact on the Biosphere in Oceania, you will be able to answer the following questions:

A Who has conducted the most nuclear tests in Oceania?

B For what reason were many plants and animals brought to Oceania from Europe?

C What treaty allows islands to claim rights to ocean resources 200 miles (320 kilometers) out from the shore?

D Before the arrival of humans, midlatitude rain forest covered what percent of New Zealand?

Despite its vast size and relatively small population, Oceania has been severely impacted by human activity. Much damage has been wreaked by people from distant countries, as well as by countries in Oceania that export resources outside the region.

A One of the first underwater tests of a nuclear weapon and its effects on naval vessels took place on Bikini Atoll in the Marshall Islands (then a U.S. territory) in 1946. Over 300 nuclear tests have been conducted in Oceania since then.

B Once hunted to near extinction to protect sheep herds, the Tasmanian Devil is now threatened by low genetic diversity, which leaves it vulnerable to disease.

C Part of a Ukrainian fishing fleet at port for repairs in Lyttleton, New Zealand. Fleets from around the world come to Oceania to take advantage of its fisheries, many of which are now over-exploited. Nevertheless, many Pacific islands are still selling fishing rights to foreign fleets because they need the money.

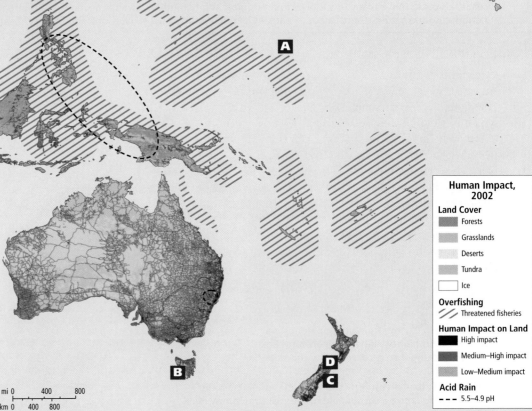

Human Impact, 2002

Land Cover
- Forests
- Grasslands
- Deserts
- Tundra
- Ice

Overfishing
- Threatened fisheries

Human Impact on Land
- High impact
- Medium–High impact
- Low–Medium impact

Acid Rain
- 5.5–4.9 pH

mi 0   400   800
km 0 400 800

D Logging in Marlborough, New Zealand. Despite increasing moves toward conservation of remaining forests and wildlife habitats, log and wood products are still a major export for New Zealand.

Oceania is vulnerable to a wide variety of hazards related to climate change, including sea level rise, increased tropical storm intensity, and less certain water availability. Fortunately, several countries in this region are already implementing solutions that are increasing resilience to climate hazards.

A A low-lying atoll in Tuvalu. With its highest point only 14.7 feet (4.5 meters) above sea level, Tuvalu is highly vulnerable to sea level rise. Combined with increased flooding during tropical storms, higher sea levels could make many low islands in this region uninhabitable. Tuvalu's government is already negotiating the future resettlement of parts of its population in nearby nations, such as New Zealand.

B A child in Guam after Typhoon Pongsona destroyed his home. Higher temperatures are bringing stronger tropical storms to this region.

C A rainwater-harvesting system provides drinking water to a house in Ceres, a sustainable residential community near Melbourne, Australia. Such systems provide resilience in the face of drought and reduce potential for flooding.

Vulnerability to Climate Change

Extreme
High
Medium
Low

D A vineyard fitted with a drip irrigation system in Marlborough, New Zealand. Drip irrigation uses substantially less water than other irrigation methods.

Thinking Geographically

After you have read about the Vulnerability to Climate Change in Oceania, you will be able to answer the following questions:

A If sea levels rise the 4 inches (10 centimeters) per decade predicted by the International Panel on Climate Change, many of the lowest-lying Pacific atolls, such as Tuvalu, will disappear under water within ___ years.

C Where in Oceania is rainwater harvesting most common?

Oceania is a world leader in implementing water technologies. Some of these technologies are age-old methods that are simple but effective, such as harvesting rainwater from roofs and the ground itself for household use (Photo Essay 11.3C). Most buildings in rural Australia, New Zealand, and many Pacific islands get at least part of their water this way, relieving surface and groundwater resources. Australia and New Zealand are now stretching water resources further with extensive use of highly efficient drip irrigation technologies in agriculture (see pages 216–218 in Chapter 6; see also Photo Essay 11.3D).

Invasive Species and Food Production

Learning Goal 2
Food and the Environment: In what ways have European systems for producing food and fiber changed environments in Oceania?

The many unique endemic plants and animals of Oceania have been displaced by **invasive species**, organisms that spread into regions outside of their native range, adversely affecting economies or environments. Many exotic plants and animals were brought to this region by Europeans to support their food production systems. Ironically, many of these same species are now major threats to food production.

Australia When Europeans first settled the continent, they brought many new animals and plants with them, sometimes intentionally, sometimes unintentionally. European rabbits are among the most destructive of the introduced species. Rabbits were brought to Australia by early British settlers who enjoyed eating them. Many were released for hunting, and with no natural predators, they multiplied quickly, consuming so much of the native vegetation that many indigenous animal species starved. Moreover, rabbits became a major source of agricultural crop loss and reduced the capacity of many grasslands to support herds of introduced sheep and cattle. Attempts to control the rabbit population by introducing European foxes and cats backfired as these animals became major invasive species themselves. Foxes and cats have driven several native Australian predator species to extinction without having much effect on the rabbit population. Intentionally introduced diseases have proven more effective at controlling the rabbit population, though rabbits have repeatedly developed resistances to them.

Herding has also had a huge impact on Australian ecosystems. Because the climate is arid and soils in many areas are relatively infertile, agriculture is limited and the dominant land use in Australia is the grazing of introduced domesticated animals—primarily sheep, but also cattle. More than 15 percent of the land has been given over to grazing, and Australia leads the world in exports of sheep and cattle products.

Dingoes, the indigenous wild dogs of Australia, prey on the introduced sheep and young cattle. To separate the wild dogs from the herds, the Dingo Fence—the world's longest fence—was built, extending 3200 miles (5800 kilometers). The fence is also a major ecological barrier to other wild species. However, kangaroos, the natural prey of dingoes, have been able to cross over to the sheep side of the fence, where their populations have boomed in the absence of dingoes.

invasive species organisms that spread into regions outside of their native range, adversely affecting economies or environments

New Zealand The environment of New Zealand has been transformed by introduced species and food production systems even more extensively than has Australia's environment. No humans lived in New Zealand until about 700 years ago, when the Polynesian Maori people settled there. When they arrived, dense midlatitude rain forest covered 85 percent of the land. The Maori were cultivators who brought in yams and taro as well as other nonnative plants, pacific rats, and birds. By the time of European contact (1642), forest clearing and overhunting by the Maori had already degraded many environments and driven several species of birds to extinction.

European settlement in New Zealand dramatically intensified environmental degradation associated with food production (Figure 11.7) and invasive species. Attempts to recreate European farming and herding systems in New Zealand resulted in environments that are actually hostile to many native species, a growing number of which are now extinct. Today, only 23 percent of the

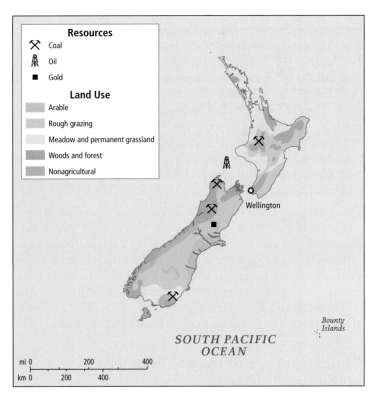

FIGURE 11.7 Land uses and natural resources of New Zealand. As a result of European settlement and the clearing of land for farming, only 23 percent of New Zealand remains forested.

(A) **(B)** **(C)** **(D)**

FIGURE 11.8 The Ok Tedi Mine. **(A)** The Ok Tedi open-pit copper and gold mine. A large hole now exists where Mount Fubila once stood. **(B)** An aerial view of the Ok Tedi River, which flows downstream of the mine. Each year since its opening in 1984, the mine has discharged 80 million tons of contaminated mine tailings and mine-induced erosion, resulting in a once deep and slow-moving river becoming a shallow, wide, and sometimes fast-moving river. **(C)** A close-up of the Ok Tedi riverbed, which is now contaminated with copper and other chemicals that kill fish and produce illness in the local human population. Sediment from the polluted river has contaminated 500 square miles (1300 square kilometers) of farmland. **(D)** A meeting about the mine and potential court settlements in the village of Serki. The Ok Tedi Mine has affected 50,000 people in 120 villages. Litigation against the mine owners is ongoing.

Thinking Geographically: **(A)** How many indigenous subsistence cultivators were forced off their land and into new mining market towns as a result of the huge amounts of mine waste resulting from the Ok Tedi Mine on Papua New Guinea? **(D)** How many indigenous people affected by the OK Tedi Mine have sued the Australian parent mining company?

country remains forested, with ranches, farms, roads, and urban areas claiming more than 90 percent of the lowland area.

Most of the cleared land is used for export-oriented farming and ranching. Grazing has become so widespread that today there are 15 times as many sheep as people, and 3 times as many cattle. Both of these activities have severely degraded the environment (see Thematic Overview C). Soils exposed by the clearing of forests have proven infertile, forcing farmers and ranchers to augment them with agricultural chemicals. The chemicals, along with feces from sheep and cattle, have severely polluted many waterways, causing the extinction of some aquatic species.

Pacific Islands In the Pacific islands, many unique species of plants and animals have been driven to extinction as islands have been deforested and converted to agriculture. In Hawaii, for example, extensive conversion of forests to land used for agriculture caused the extinction of numerous plant, bird, and snail species. Hawaii is home to more threatened or endangered species than any other U.S. state, despite having less than 1 percent of the U.S. landmass.

❚❚▶ 239. HAWAII CONSIDERED AMERICA'S ENDANGERED SPECIES CAPITAL

Globalization and the Environment in the Pacific Islands

As the Pacific islands have become more connected to the global economy over the years, flows of both resources and pollutants have increased dramatically. Mining, nuclear pollution, and

tourism are all examples of how globalization has transformed environments in the Pacific islands.

Mining in Papua New Guinea and Nauru Two mines have had particularly devastating effects on the environment. Both were operated by foreign-owned mining companies that took advantage of poorly enforced or nonexistent environmental laws. In the case of the Ok Tedi Mine on Papua New Guinea, huge amounts of mine waste devastated river systems (Figure 11.8). Tens of thousands of indigenous subsistence cultivators were forced into new mining market towns, where their skills were of little use and where they needed cash to buy food and pay rent.

In 2007, thirty thousand of these people sued the Australian parent mining company, which was then BHP Billiton, for U.S.$4 billion. Two villagers, Rex Dagi and Alex Maun, traveled to Europe and the United States to explain their cause and meet with international environmental groups. They and their supporters convinced U.S. and German partners in the Ok Tedi mine to divest their shares. The case is still awaiting a decision by a court in Papua New Guinea.

The most extreme case of environmental disaster due to mining took place on the once densely forested Melanesian island of Nauru (one-third the size of Manhattan, located northeast of the Solomon Islands (see Figure 11.4). During most of the last half of the twentieth century, the people of Nauru lived in prosperity. In fact, for a few years, the country had the highest per capita income in the world, thus attracting many immigrants. Nauru's wealth was based on the proceeds from the strip-mining of high-grade phosphates used in the manufacture of fertilizer, derived from eons of bird droppings (guano). The mining companies were owned first by Germany, then by Japan, and finally by Australia.

The phosphate reserves are now depleted, the proceeds ill-spent, and the environment destroyed. Junked mining equipment sits on miles of bleached white sand where forest once stood. To avoid financial disaster, the government has been exploring controversial and possibly illegal ways of making money, including money laundering for the international crime syndicates.

Nuclear Pollution The geopolitical aspects of globalization have hit this region especially hard. From the 1940s to the 1960s, the United States exploded 106 nuclear bombs in tests, primarily over the Marshall Islands (see Photo Essay 11.2 A), which were a U.S. territory until 1986. Similarly, Mururoa Atoll in French Polynesia has been the site of 180 nuclear weapons tests and the recipient of numerous shipments of nuclear waste from France. The weapons tests and imported waste have become major environmental issues for the Pacific islands.

In July 1985, the ship *Rainbow Warrior*, owned by the environmental group Greenpeace, was blown up in New Zealand by the French secret service. In response, the 1986 Treaty of Rarotonga established the South Pacific Nuclear Free Zone. Most independent countries in Oceania have signed this treaty, which bans nuclear weapons testing and nuclear waste dumping on their lands. Because of political pressure from France and the United States, however, French Polynesia and the U.S. territories have not signed the treaty.

Tourism Even tourism, which until recently was considered a "clean" industry, has created environmental problems. Foreign-owned tourism enterprises have often accelerated the loss of wetlands and worsened beach erosion by clearing coastal vegetation for construction, golf courses, and waterfront-related entertainment.

Tourism has also strained island water resources. Tourists increase the population dependent on this scarce resource, and they inevitably consume more water per capita than they do at home, because of increased showering, laundering, and other services that consume fresh water. Furthermore, inadequate methods of disposing of sewage and trash from resorts have polluted many once-pristine areas. Ecotourism (see page 111 in Chapter 3), aimed at reducing these impacts, is now a common element of development throughout the Pacific, but environmental impacts from tourism are still generally high.

The United Nations Convention on the Law of the Sea Implementation of the 1994 UN Convention on the Law of the Sea (UNCLOS) has revealed how the globalization of Pacific island economies has thwarted environmental protection. UNCLOS is based on the idea that all problems of the world's oceans are interrelated and need to be addressed as a whole. It establishes rules governing all uses of the world's oceans and seas and has been ratified by 157 countries (although not the United States).

The treaty allows islands to claim rights to ocean resources 200 miles (320 kilometers) out from the shore. Island countries can now make money by licensing privately owned fleets from Japan, South Korea, Russia, the United States, and elsewhere to fish within these offshore limits. However, there is no overarching enforcement agency, and protecting the fisheries from overfishing by these rich and powerful licensees has turned out to be an enforcement nightmare for tiny island governments with few resources. Similarly, it has proved difficult to monitor and control the exploitation of seafloor mineral deposits by foreign companies.

II ▶ 241. ENDANGERED HAWAIIAN MONK SEAL POPULATION CONTINUES TO DECLINE

II ▶ 242. NEW SPECIES OF UNDERSEA LIFE FOUND NEAR INDONESIA

II ▶ 243. SCIENTISTS WARN OF DEPLETION OF OCEAN FISH IN 40 YEARS

THINGS TO REMEMBER

1. Learning Goal 1: Climate Change and Water Sea level rise threatens some Pacific islands with submersion, and water supplies throughout the region may be strained as rainfall patterns change. If sea levels rise the predicted 4 inches per decade, many of the lowest-lying atolls will disappear under water, leaving a multitude of refugees.

2. As part of its efforts to combat global warming, New Zealand has set a goal of obtaining 95 percent of its energy from renewable sources by 2025.

3. Learning Goal 2: Food and the Environment Throughout Oceania, the introduction of food production systems from elsewhere has resulted in the spread of ecologically and economically damaging invasive species.

4. Globalization and the patterns of consumption by people who live far from the Pacific are seriously affecting life in this region.

Human Patterns over Time

Vignette *"With courage, you can travel anywhere in the world and never be lost. Because I have faith in the words of my ancestors, I'm a navigator."*

—MAU PIAILUG

In 1976, Mau Piailug made history by sailing a traditional Pacific island voyaging canoe across the 2400 miles (3860 kilometers) of deep ocean between Hawaii and Tahiti. He did so without a compass, charts, or other modern instruments, using only methods passed down through his family. He relied mainly on observations of the stars, the sun, and the moon to find his way. When clouds covered the sky, he used the patterns of ocean waves and swells, as well as the presence of seabirds, to tell him of distant islands over the horizon.

Piailug reached Tahiti 33 days after leaving Hawaii and made the return trip in 22 days. His voyage resolved a major scholarly debate over how people settled the many remote islands of the Pacific without navigational instruments, thousands of years before the arrival of Europeans. Some thought that navigation without instruments was impossible and argued that would-be settlers simply drifted about on their canoes at the mercy of the winds, most of them starving to death on the seas, with a few happening upon new islands by chance. It was hard to refute this argument because local navigational methods had died out almost everywhere. However, in isolated Micronesia, where Piailug lived until his death in 2010, indigenous navigational traditions survive.

After the successful 1976 voyage, Piailug trained several students in traditional navigational techniques, which have become a symbol of cultural rebirth and a source of pride throughout the Pacific. In 2007, these protégés of Mau Piailug sailed from Hawaii through the Marshall Islands to Yokohama, Japan (see **Timeline B** on page 424), as a gesture to celebrate peace and the human need to stay connected with nature. *[Source: Richard Nile and Christian Clerk, Cultural Atlas of Australia, New Zealand, and the South Pacific (New York: Facts on File, 1996), pp. 63–65; Emma Brown, "Mau Piailug, Micronesian Who Sailed by Navigating Sun and Stars, Dies at 78," Washington Post, July 10, 2010, at http://www.washingtonpost.com/wp-dyn/content/article/2010/07/20/AR2010072002941_pf.html.]* ■

The Peopling of Oceania

The longest-surviving inhabitants of Oceania are Australia's **Aborigines**, who migrated from Southeast Asia 50,000 to 70,000 years ago (see Timeline A; see also Figure 11.9C). Amazingly, some memory of this ancient journey may be preserved in Aboriginal oral traditions, which recall mountains and other geographic features that are now submerged under water. At about the same time that the Aborigines were settling Australia, related groups were settling nearby areas.

Melanesians, so named for their relatively dark skin tones, a result of high levels of the protective pigment *melanin*, migrated throughout New Guinea and other nearby islands, giving this area its name, **Melanesia** (Figure 11.9A). Archeological evidence indicates that they first arrived more than 50,000 to 60,000 years ago from Sundaland (see Figure 10.5 on page 375), a shelf exposed during the Pleistocene epoch. They lived in isolated pockets, which resulted in the evolution of hundreds of distinct yet related languages. Like the Aborigines, the Melanesians survived mostly by hunting, gathering, and fishing, although some groups—especially those inhabiting the New Guinea highlands—also practiced agriculture.

Much later, between 5000 to 6000 years ago and as recently as 1000 years ago, linguistically related *Austronesians* settled **Micronesia** and **Polynesia**, sometimes mixing with the Melanesian peoples they encountered. Micronesia consists of small islands that lie east of the Philippines and north of the equator. Polynesia is made up of numerous islands situated inside a large irregular triangle formed by New Zealand, Hawaii, and Easter Island (Figure 11.9 B, D). (Easter Island is a tiny speck of land in the far eastern Pacific, at 27° S 109° W, not shown in the figures in this chapter.) Although Europeans long questioned the skills of Pacific navigators, experiments run by Mau Piailug (see the vignette on page 421) proved

that ancient sailors could navigate over vast distances, using seasonal winds, astronomical calculations, avian and aquatic life, and wave patterns to reach the most far-flung islands of the Pacific (see Timeline B). They were fishers, hunter-gatherers, and cultivators who developed complex cultures and maintained trading relationships among their widely spaced islands.

Many Austronesians coexisted in a state of moderate antagonism. Although warfare occurred, hostilities were often settled ritualistically and by means of annual tribute-paying ceremonies, rather than by resorting to combat. Individual rulers rarely amassed large territories or controlled them for long. On these islands, many societies were and still are hierarchical, with layers of ruling elites at the top and undifferentiated commoners at the bottom.

Arrival of the Europeans

The earliest recorded contact between Pacific peoples and Europeans took place in 1521, when the Portuguese navigator Ferdinand Magellan (exploring for Spain), landed on the island of what is now called Guam in Micronesia. The encounter ended badly. The islanders, intrigued by European vessels, tried to take a small skiff. For this crime, Magellan had his men kill the offenders and burn their village to the ground. A few months later, Magellan was himself killed by islanders in what became the Philippines, which he had claimed for Spain. Nevertheless, by the 1560s, the Spanish had set up a lucrative Pacific trade route between Manila in the Philippines and Acapulco on the west coast of Mexico. Explorers from other European states followed, first taking an interest mainly in the region's valuable spices. The British and French explored extensively in the eighteenth century (see Timeline C).

The Pacific was not formally divided among the colonial powers until the nineteenth century, and by that time, the United States, Germany, and Japan had joined France and Britain in taking control of various island groups. As in other regions, European colonization of Oceania emphasized extractive agriculture and mining. Because native people were often displaced from their lands or exposed to exotic diseases to which they had no immunity, their populations declined sharply.

The Colonization of Australia and New Zealand

Although all of Oceania has been under European or American rule at some point, the most Westernized parts of the region are Australia and New Zealand. The colonization of these two countries by the British has resulted in many parallels with North America. In fact, the American Revolution was a major impetus for "settling" Australia because once the North American colonies became independent, the British needed somewhere else to send their convicts. In early nineteenth-century Britain, a relatively

---

Aborigines the longest-surviving inhabitants of Oceania, whose ancestors, the Australoids, migrated from Southeast Asia as early as 50,000 years ago over the Sundaland landmass that was exposed during the ice ages

Melanesians a group of Australoids named for their relatively dark skin tones, a result of high levels of the protective pigment melanin; they settled throughout New Guinea and other nearby islands

Melanesia New Guinea and the islands south of the equator and west of Tonga (the Solomon Islands, New Caledonia, Fiji, and Vanuatu)

Micronesia the small islands that lie east of the Philippines and north of the equator

Polynesia the numerous islands situated inside an irregular triangle formed by New Zealand, Hawaii, and Easter Island

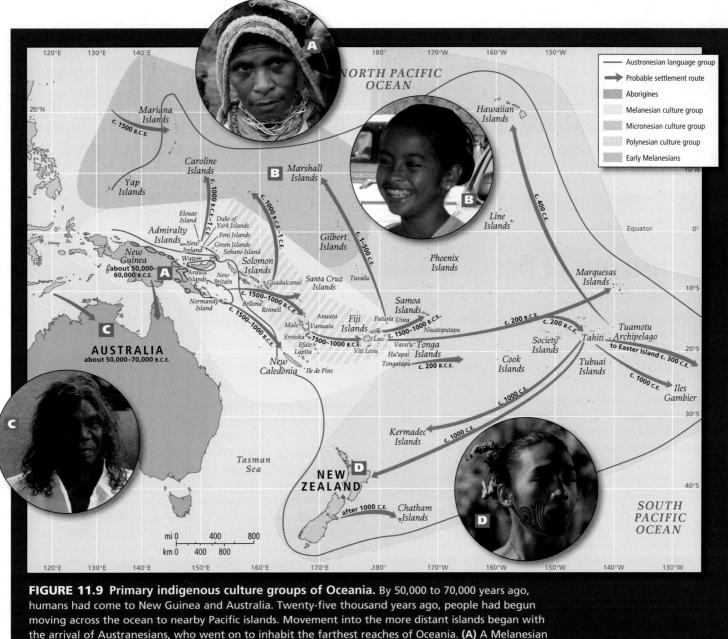

FIGURE 11.9 Primary indigenous culture groups of Oceania. By 50,000 to 70,000 years ago, humans had come to New Guinea and Australia. Twenty-five thousand years ago, people had begun moving across the ocean to nearby Pacific islands. Movement into the more distant islands began with the arrival of Austranesians, who went on to inhabit the farthest reaches of Oceania. **(A)** A Melanesian woman in New Guinea. **(B)** A Micronesian girl on Lelu Island, Micronesia. **(C)** An Australian Aborigine from Northern Australia. **(D)** A young Maori (Polynesian culture group) man in Rotorua, New Zealand.

minor theft—for example, of a piglet—might be punished with 7 years of hard labor in Australia (see Timeline D).

A steady flow of English and Irish convicts arrived in Australia until 1868, at which point the population of Australia was approximately 1.7 million. Most of the convicts chose to stay in the colony after their sentences were served. They are given credit for Australia's rustic self-image and egalitarian spirit. They were joined by a much larger group of voluntary immigrants from the British Isles who were attracted by the availability of inexpensive farmland. Waves of these immigrants arrived until World War II. New Zealand was settled somewhat later than Australia,

in the mid-1800s. Although its population also derives primarily from British immigrants, New Zealand was never a penal colony.

Another similarity among Australia, New Zealand, and North America was the treatment of indigenous peoples by European settlers. In both Australia and New Zealand, native peoples (numbering approximately 750,000 in Australia, and 200,000 in New Zealand at the time of the British arrival) were killed outright, annihilated by infectious diseases, or shifted to the margins of society. The few who lived on territory the Europeans thought undesirable were able to maintain their traditional way of life. However, the vast majority of the survivors

A Aborigines demonstrating hunting tools.

B Hōkūleʻa 2, a functioning replica of a traditional Hawaiian voyaging canoe.

C A painting of early British contact with Pacific Islanders in 1783.

75,000 B.C.E.   50,000 B.C.E.   25,000 B.C.E.   1500 C.E.   1550 C.E.   1600 C.E.   1650 C.E.

70,000–50,000 B.C.E.
Australian Aborigines migrate from Southeast Asia

4000–3000 B.C.E.
Settlement of Micronesia and Polynesia

1500s C.E.
Spanish explorations

VISUAL TIMELINE OF OCEANIA

Thinking Geographically

After you have read about the human history of Oceania, you will be able to answer the following questions:

A Aboriginese migrated to Oceania from_____.

B How did Polynesians use boats like the one in this painting?

lived and worked in grinding poverty, either in urban slums or on cattle and sheep ranches. Today, native peoples still suffer from pervasive discrimination and maladies such as alcoholism and malnutrition. Even so, some progress is being made toward improving their lives (as described on pages 431–432). In 2008, the then newly elected prime minister of Australia, Kevin Rudd, officially apologized to Aboriginal people for the treatment they had received since the land had first been colonized.

Closely related to attitudes toward indigenous people were attitudes toward any people of color as immigrants. Although Pacific island people (indigenous) were brought in as laborers during the nineteenth century, by 1901 a "whites-only" policy governed Australian immigration. The favored migrants were those from the British Isles; then after World War II, southern Europeans were encouraged. This discrimination persisted until the mid-1970s, when the whites-only policy on immigration was ended. In New Zealand, while similar racist attitudes prevailed, there was not an official whites-only policy, and by the 1970s, students and immigrants were arriving from Asia and the Pacific islands.

Oceania's Shifting Global Relationships

During the twentieth century, Oceania's relationship with the rest of the world went through three phases: from a predominantly European focus to identification with the United States and Canada to the currently emerging linkage with Asia. Until roughly World War II, the colonial system gave the region a European orientation. In most places, the economy depended largely on the export of raw materials to Europe (see Timeline E). Thus, even when a colony gained independence from Britain, as Australia did in 1901 and New Zealand did in 1907, people

remained strongly tied to their mother countries. Even today, the Queen of England remains the titular head of state in both countries. During World War II, however, the European powers provided only token resistance to Japan's invasion of much of the Pacific and its bombing of northern Australia. Such European impotence began a change in political and economic orientation.

II ▶ 244. VETERANS REMEMBER TRAGEDY OF WAR IN PACIFIC

After the war, the United States, with an already strong foothold in the Philippines, became the dominant power in the Pacific, and U.S. investment became increasingly important to the economies of Oceania. Australia and New Zealand joined the United States in a Cold War military alliance, and both fought alongside the United States in Korea and Vietnam, suffering considerable casualties during a time of significant antiwar activity at home. U.S. cultural influences were strong, too, as North American products, technologies, movies, and pop music penetrated much of Oceania.

By the 1970s, another shift was taking place as many of the island groups were granted self-rule by their European colonizers, and Oceania became steadily drawn into the growing economies of Asia. Since the 1960s, Australia's thriving mineral export sector has become increasingly geared toward supplying Asian manufacturing industries (first Japan in the 1960s, and increasingly China since the 1990s). Similarly, since the 1970s, New Zealand's wool and dairy exports have gone mostly to Asian markets. Despite occasional backlashes against "Asianization," Australia, New Zealand, and the rest of Oceania are becoming increasingly transformed by Asian influences (see Timeline F). Many Pacific islands have significant Chinese, Japanese, Filipino, and Indian minorities, and the small Asian minorities of Australia and New Zealand are increasing. On some Pacific islands, such

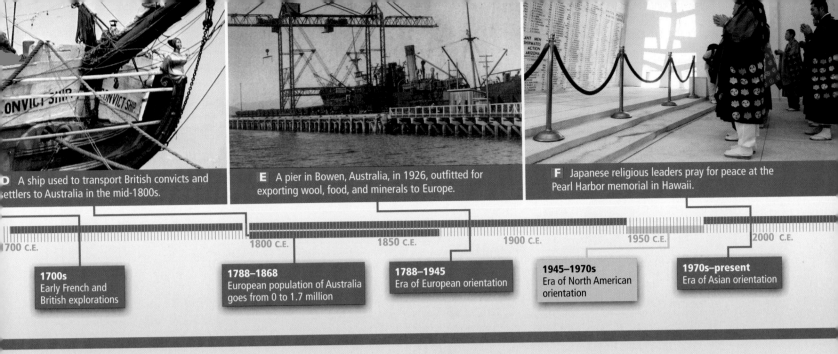

D A ship used to transport British convicts and settlers to Australia in the mid-1800s.

E A pier in Bowen, Australia, in 1926, outfitted for exporting wool, food, and minerals to Europe.

F Japanese religious leaders pray for peace at the Pearl Harbor memorial in Hawaii.

1700 C.E.   1800 C.E.   1850 C.E.   1900 C.E.   1950 C.E.   2000 C.E.

1700s
Early French and British explorations

1788–1868
European population of Australia goes from 0 to 1.7 million

1788–1945
Era of European orientation

1945–1970s
Era of North American orientation

1970s–present
Era of Asian orientation

C What were the early Europeans interested in when they came to Oceania in the sixteenth, seventeenth, and eighteenth centuries?

D By the time that the flow of convicts to Australia from England and Ireland ended in 1868, what was the population of Australia?

E Until roughly World War II, on what did the economy in most parts of Oceania depend?

F Asians make up what percentage of Hawaii's population?

---

as Hawaii, Asians now constitute the largest portion (42 percent) of the population.

THINGS TO REMEMBER

1. Oceania can be divided into four distinct cultural regions: Australia and Tasmania, settled originally by Aborigines; Melanesia, settled by Melanesians, so identified because of their skin tone derived from melanin; Micronesia, whose original inhabitants came from coastal Asia and from the more northern Melanesian islands; and Polynesia, settled by a variety of Austronesian people.

2. Through colonization, Europeans dominated Oceania from the early sixteenth century until the end of World War II. During the 50 years after the war, the United States was the principal power in the region. Beginning in the 1970s, Asian countries—China and Japan in particular—have had increasing influence throughout Oceania.

II CURRENT GEOGRAPHIC ISSUES

Many current geographic issues in Oceania are related to the transition under way from European to Asian and inter-Pacific cultural influences. Oceania's old relationships were built on historical factors, such as the settlement of Australia and New Zealand by Europeans and the depth of ancient Pacific cultural affiliations. By contrast, its new relationships are influenced by economic and geographic considerations, particularly the physical proximity to Asia.

Economic and Political Issues

Oceania has been powerfully transformed by global relationships over the past 200 years. Now, new forces of globalization, driven largely by Asia's growing affluence and enormous demand for resources, are shifting trade, migration, and tourism within Oceania.

Globalization, Development, and Oceania's New Asian Orientation

Learning Goal 3
Globalization and Development: How has globalization transformed patterns of economic development in Oceania?

One could say that globalization began in Oceania when the first European explorers came into the region, beginning the trend of settlement and cultural and economic influence by outsiders, especially Europeans. More recently, the United States has exerted a powerful influence on trade and politics. For the past several decades, however, globalization has reoriented this region toward Asia, which buys more than 76 percent of Australia's exports (mainly coal, iron ore, and other minerals), nearly 35 percent of New Zealand's exports (mainly meat, wool, and dairy products (see Figure 11.10 on page 426), and many

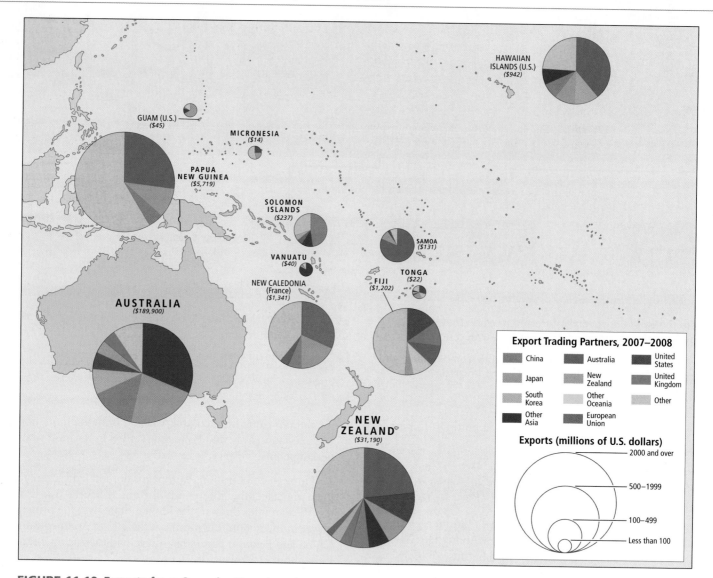

FIGURE 11.10 Exports from Oceania. The colors of each pie chart indicate a country's export trading partners. The "other" sections can include trade with Canada, Mexico, the Caribbean, non-EU Europe, sub-Saharan Africa, and other locales, some of them new trading partners. (Figures for Hawaii do not include exports to other parts of the United States.)

other products and services from across island Oceania (see Thematic Overview D).

Asia is also increasingly a source of the region's imports. Because there is little manufacturing in Oceania, most manufactured goods are imported from China, Japan, South Korea, Singapore, and Thailand—Oceania's leading trading partners. Both Australia and New Zealand have free trade relationships either completed or in negotiation with Asia's two largest economies, China and Japan.

The Pacific islands are even further along in their reorientation toward Asia than either Australia or New Zealand. Not only are coconut, forest, and fish products from the Pacific islands sold to Asian markets, but Asian companies increasingly own these industries. Fishing fleets from Asia regularly ply the offshore waters of Pacific island nations. Asians also dominate the Pacific island tourist trade, both as tourists and as investors in tourism infrastructure. And increasing numbers of Asians are taking up residence in the Pacific islands, exerting widespread economic influence.

The Stresses of Asia's Economic Development "Miracle" on Australia and New Zealand

For Australia and New Zealand, Asia's global economic rise has meant not only increased trade but also increased competition with Asian economies in foreign markets. Throughout the region, local industries used to enjoy protected or "preferential" trade with Europe. They have lost that advantage because new European Union regulations stemming from the EU's membership in the World Trade Organization prohibit such arrangements. In their trade with Europe, these industries now face stiff competition from larger companies in Asia that benefit from much cheaper labor.

Australia and New Zealand are somewhat unusual in having achieved broad prosperity largely on the basis of raw materials exports (see Thematic Overview E). Preferential trade with Europe allowed higher profits for many export industries.

Meanwhile, strong labor movements in Australia and New Zealand meant that the profits from these industries were more equitably distributed throughout society than were those in other raw materials–based economies in Middle and South America or in sub-Saharan Africa. Australian coal miners' unions successfully agitated for the world's first 35-hour workweek. Other labor unions won a minimum wage, pensions, and aid to families with children long before such programs were enacted in many other industrialized countries. For decades, these arrangements were highly successful. Both Australia and New Zealand enjoyed living standards comparable to those in North America but with a more egalitarian distribution of income. However, since the 1970s, competition from Asian companies has meant that increasing numbers of workers in Australia and New Zealand have lost jobs and seen their hard-won benefits scaled back or eliminated.

Competition from Asian companies led to lower corporate profits. As corporate profits fell, so did government tax revenues, which necessitated cuts to previously high rates of social spending on welfare, health care, and education. The loss of social support, especially for those who have lost jobs, has contributed to rising poverty in recent years. Australia now has the second-highest poverty rate in the industrialized world, after that of the United States.

Maintaining Raw Materials Exports as Service Economies Develop The shift toward trading more with Asia than with Europe or North America has had little effect on Oceania's dependence on exporting unprocessed raw materials, because most Asian economies have a much greater need to import raw materials than manufactured goods. Nevertheless, although their dollar contribution to national economies remains high, raw materials export industries are of decreasing prominence in the economies of Australia and New Zealand, in that they now employ fewer people because of mechanization. This shift to lower labor requirements has been essential for these industries to stay globally competitive with other countries that have much cheaper labor.

Today, the economies of both countries are dominated by diverse and growing service sectors, which have links to the countries' export sectors. Extracting minerals and managing herds and agriculture have become technologically sophisticated enterprises that depend on many supporting services and an educated workforce. Australia is now a world leader in providing technical and other services to mining companies, sheep farms, and winemakers. Meanwhile, New Zealand's well-educated workforce and well-developed marketing infrastructure have helped it break into luxury markets for dairy products, meats, and fruits. Perhaps the most visible success has been New Zealand's global marketing of the indigenous "Kiwi" fruit, which is now also a slang term for anyone from New Zealand.

Economic Change in the Pacific Islands In general, the Pacific islands are also making a shift away from extractive industries, such as mining and fishing, and toward such service sector industries as tourism and government. On many islands, the stress of economic change is cushioned by self-sufficiency and resources from abroad. Many households still rely on fishing and subsistence

> **MIRAB economy** an economy based on migration, remittance, aid, and bureaucracy
>
> **subsistence affluence** a lifestyle whereby self-sufficiency is achieved for most necessities, while some opportunities to earn cash allow for travel and occasional purchases of manufactured goods

cultivation for much of their food supply. On the islands of Fiji, for example, part-time subsistence agriculture engages more than 60 percent of the population, although it accounts for just under 17 percent of the economy. Remittances sent home from those working abroad are essential to many Pacific Island economies, and account for more than half of all income on some islands.

In the relatively poor and undereducated nations (the Solomon Islands, Tuvalu, and parts of Papua New Guinea, for example), conditions typify what has been termed a **MIRAB economy**—one based on migration, remittances, aid, and bureaucracy. Foreign aid from former or present colonial powers supports government bureaucracies that supply employment for the educated and semiskilled. This type of economy has little potential for growth. Nevertheless, islanders who can be self-sufficient in food and shelter while saving extra cash for travel and occasional purchases of manufactured goods are sometimes said to have achieved **subsistence affluence**. Where there is poverty, it is often related to geographic isolation, which means a lack of access to information and economic opportunity. Although computers and the new global communication networks are not yet widely available in the Pacific islands, they have the potential to alleviate some of this isolation.

The Advantages and Stresses of Tourism Development

Tourism is a growing part of the economy throughout Oceania, with tourists coming largely from Japan, Korea, Taiwan, Southeast Asia, the Americas, and Europe. In some Pacific island groups, the number of tourists far exceeds the island population. Guam, for example, annually receives tourists equivalent to 5 times its population. Palau and the Northern Marianas Islands annually receive more than 4 times their populations. Such large numbers of visitors, expecting to be entertained and graciously accommodated, can place a special stress on local inhabitants. And although they bring money to the islands' economies, these visitors create problems for island ecology, place extra burdens on water and sewer systems, and require a standard of living that may be far out of reach for local people. Perhaps nowhere in the region are the issues raised by tourism clearer than in Hawaii.

A Case Study of Civil Conflict: Hawaii

Since the 1950s, travel and tourism has been the largest industry in Hawaii, producing nearly 18 percent of the gross state product in 2008. Tourism is related in one way or another to nearly 75 percent of all jobs in the state. (By comparison, travel and tourism accounts for 9 percent of GDP worldwide.) In 2008, tourism employed one out of every six Hawaiians and accounted for 25 percent of state tax revenues.

Dramatic drops in tourist visits, often driven by forces far removed from Oceania, can wreak havoc on local economies. Decreases in tourism impact not just tourist facilities, but supporting industries, too. For example, construction thrives by building condominiums, hotels, resorts, and retirement facilities. The Asian recession of the late 1990s, the terrorist attacks of September 11, 2001, and the global recession of 2008–2009 all impacted Hawaii's economy by creating dramatic slumps in tourist visits.

Sometimes the mass tourism can seem like an invading force to ordinary citizens. For example, an important segment of the Honolulu tourist infrastructure—hotels, golf courses, specialty shopping centers, import shops, and nightclubs—is geared to visitors from Japan, and many such facilities are owned by Japanese investors. Hawaiian citizens and other non-Japanese shoppers and vacationers can be made to feel out of place.

Another example of the impositions of mass tourism is the demand by tourists for golf courses on many Hawaiian islands, which has resulted in what Native (indigenous) Hawaiians view as desecration of sacred sites. Land that in precolonial times was communally owned, cultivated, and used for sacred rituals was confiscated by the colonial government and more recently sold by the state to Asian golf course developers. Now the only people with access to the sacred sites are fee-paying tourist golfers. By 2007, there were 83 golf courses in Hawaii, and the golf industry alone contributed $1.6 billion to the state's economy—more than twice the amount from agriculture.

Sustainable Tourism Some islands have attempted to deal with the pressures of tourism by adopting the principle of *sustainable tourism*, which is designed to decrease the imprint of tourism and minimize disparities between hosts and visitors. Samoa, for example, has created the Samoan Tourism Authority in conjunction with the South Pacific Regional Environment Programme. (The name *Samoa* refers to the independent country that was formerly known as Western Samoa; that country is politically distinct from American Samoa, a U.S. territory.) With financial aid from New Zealand, the Authority develops and monitors sustainable tourism components: beaches, wetlands, and forested island environments, and provides knowledge-based tourism experiences for visitors (information-rich explanations of island political, social, and environmental issues).

The Future: Diverse Orientations?

Despite the powerful forces pushing Oceania toward Asia, important factors still favor strong ties with Europe and North America. In spite of increasing trade links and a recent effort by China to expand diplomatic and cultural relations with Australia, both Australia and New Zealand remain staunch military allies of the United States. Over the years, both have participated in U.S.-led wars in Korea, Vietnam, Afghanistan, and Iraq.

In parts of the Pacific islands, strong links to Europe and North America are also upheld by continuing political domination. In Micronesia, the United States controls Guam and the Northern Mariana Islands, and in Polynesia, the United States controls American Samoa. Just as the Hawaiian Islands are a state of the United States, similarly, the 120 islands of French

Polynesia—including Tahiti and the rest of the Society Islands, the Marquesas Islands, and the Tuamotu Archipelago—are considered Overseas Lands of France. Any desire for independence in these possessions has not been sufficient to override the financial benefits of aid, subsidies, and investment money provided by France and the United States.

THINGS TO REMEMBER

1. **Learning Goal 3: Globalization and Development** Globalization is reorienting Oceania (especially Australia and New Zealand) toward Asia as a major destination for exports and an increasing source of Oceania's imports.

2. Service industries are becoming the dominant income source for most of the region's economies, although extractive industries remain important.

3. Tourism is a significant and growing part of the economies in Oceania.

Population Patterns

Although Oceania occupies a huge portion of the planet, its population is only 37 million people, close to that of the state of California (37.3 million) (Figure 11.11). They live on a total land area slightly larger than the contiguous United States but spread out in bits and pieces across an ocean larger than the Eurasian landmass. The Pacific islands, including Hawaii, have nearly 4.5 million people; Australia has 21.9 million; Papua New Guinea, 6.6 million; and New Zealand, 4.3 million.

Disparate Population Patterns in Oceania

Learning Goal 4
Population and Urbanization:
What are the two main patterns of population and urbanization developing across Oceania?

Different population patterns exist in Oceania. Australia, New Zealand, and Hawaii, like many wealthy countries, have highly urbanized, relatively older, and more slowly growing populations, with life expectancies of about 80 years. The Pacific islands and Papua New Guinea, like many developing countries, have much more rural, younger, and rapidly growing populations, with life expectancies in the 60s or low 70s. The overall trend throughout the region, however, is toward smaller families, aging populations, and urbanization (see Thematic Overview G, and see Reasons for Optimism A).

Population densities remain low in Australia, at 7.8 people per square mile (3 per square kilometer), and New Zealand, at 41.4 per square mile (16 per square kilometer). Densities vary widely in the Pacific islands; some islands are sparsely settled or uninhabited, while others—including some of the smallest and poorest, such as the Marshall Islands and Funafuti, the capital of Tuvalu, which have 772 and 4847 people respectively per square mile (298 and 1871 respectively per square kilometer)—are densely populated.

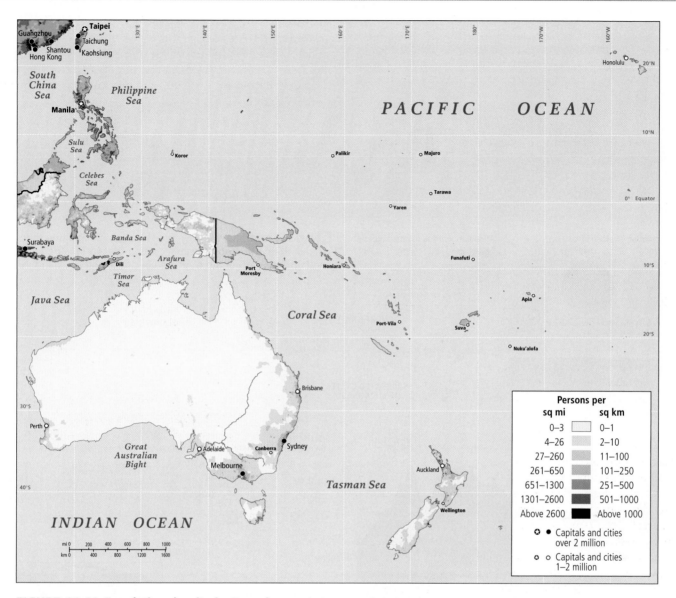

FIGURE 11.11 Population density in Oceania. Population growth is slowing throughout Oceania as more people move to the cities, where health care is better and women are more likely to pursue careers—factors that make large families less likely. Australia, New Zealand, and Hawaii are furthest along in this process and now are highly urbanized, with smaller families and increasingly old populations.

Urbanization in Oceania

The global trend of migration from the countryside to cities is highly visible in Oceania, where 73 percent of the population live in urban areas (see Thematic Overview H). The shift from agricultural and resource-based economies toward service economies is a major driver of urbanization, especially in Australia, New Zealand, and some of the wealthier Pacific islands such as Hawaii and Guam. These trends are weakest in Papua New Guinea and many smaller Pacific Islands.

Australia and New Zealand Australia and New Zealand have among the highest percentages of city dwellers outside Europe. More than 83 percent of Australians live in cities, mostly along the country's well-watered and relatively fertile eastern and southeastern coasts. Similarly, 86 percent of

New Zealanders live in urban areas. The vast majority of the people in these two countries live in modern comfort, work in a range of occupations typical of highly industrialized societies, and have access to tax-supported leisure facilities (see Photo Essay 11.4 map, A, B on page 430). Vibrant, urban-based service economies have expanded to employ about three-quarters of the population in both countries. Declining employment in mining and agriculture, where mechanization has dramatically reduced the number of workers needed, has also contributed to urbanization.

Pacific Islands Throughout the Pacific, urban centers have transformed natural landscapes, and in some small countries, such as Guam, Palau, and the Marshall Islands, they have become the dominant landscape. Although cities are places of opportunity, they

There are two patterns of urbanization in Oceania. While Australia, New Zealand, and Hawaii are already highly urbanized places with high standards of living, Papua New Guinuea and many Pacific islands are much more rural and have lower standards of living.

Population of urban areas

2006   2020

20 million or more — 20 million or more
10 million — 10 million
5 million — 5 million
3 million — 3 million

Note: symbols on map are sized proportionally to specific population data

Population living in urban areas

- 80–100%
- 60–79%
- 40–59%
- 20–39%
- 0–19%
- No data

1 Global rank (population 2006)

A Sydney, Australia, the largest city in Oceania and one that is consistently ranked among the most liveable cities in the world, along with Melbourne and Perth (two other large Australian cities), and Auckland, New Zealand.

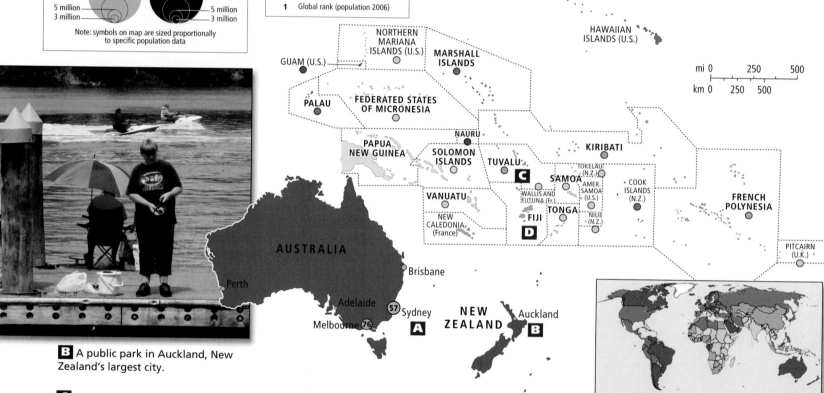

mi 0   250   500
km 0 250 500

B A public park in Auckland, New Zealand's largest city.

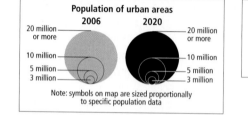

C Part of Funafuti atoll, capital of Tuvalu. Funafuti is one of the most densely populated islands in the world, with 4847 people per square mile (1871 per square kilometer). Tuvalu is among the poorest nations in Oceania, with GDP (PPP) per capita of about U.S.$1600.

D Children in Ravi Ravi, a village in rural Fiji where incomes are relatively low, as is access to health care and education. The home behind the children was constructed by a nonprofit funded by foreign aid.

can also be sites of cultural change, conflict, and environmental hazards (see Photo Essay 11.4C).

The great majority of Pacific island towns and all the capital cities are located in ecologically fragile coastal settings. Many of these waterfront towns were established during the colonial era as ports or docking facilities and were situated in places suitable for only limited numbers of people. Consequently, little land is available for development, and access to housing is limited. Unplanned "squatter" settlements have been a visible feature of Pacific island urban areas for several decades (see Chapter 3 on pages 131–134; see also Chapter 7 on gage 278).

THINGS TO REMEMBER

1. **Learning Goal 4: Population and Urbanization** Australia, New Zealand, and Hawaii, like many wealthy countries, have highly urbanized, relatively older, and more slowly growing populations, with life expectancies of about 80 years. The Pacific islands and Papua New Guinea, like many developing countries, have much more rural, younger, and rapidly growing populations, with life expectancies in the 60s or low 70s.

2. Urbanization is the dominant settlement pattern in Oceania.

3. The great majority of Pacific island towns and all of the capital cities are located in ecologically fragile coastal settings.

Sociocultural Issues

The shift in Oceania away from Europe and toward Asia and the Pacific has been accompanied by new respect for indigenous peoples. In addition, a growing sense of common economic ground with Asia has heightened awareness of Asian culture.

Ethnic Roots Reexamined

Until very recently, most people of European descent in Australia and New Zealand thought of themselves as Europeans in exile. Many considered their lives incomplete until they had made a pilgrimage to the British Isles or the European continent. In her book *An Australian Girl in London* (1902), Louise Mack wrote: "[We] Australians [are] packed away there at the other end of the world, shut off from all that is great in art and music, but born with a passionate craving to see, and hear and come close to these [European] great things and their home[land]s."

These longings for Europe were accompanied by racist attitudes toward both indigenous peoples and Asians. Most histories of Australia written in the early twentieth century failed to even mention the Aborigines, and later writings described them as amoral. From the 1920s to the 1960s, whites-only immigration policies barred Asians, Africans, and Pacific Islanders from migrating to Australia and discouraged them from entering New Zealand. And, as we have seen, trading patterns further reinforced connections to Europe.

Weakening of the European Connection When migration from the British Isles slowed after World War II, both Australia and New Zealand began to lure immigrants from southern and eastern Europe, many of whom had been displaced by the war. Hundreds of thousands came from Greece, what was then Yugoslavia, and Italy. The arrival of these non-English-speaking people began a shift toward a more multicultural society. With the demise of the whites-only immigration policy, there was an influx of Vietnamese refugees in the early 1970s, following the United States' withdrawal from Vietnam. More recently, growing demand for information technology specialists throughout the service sector has been met by recruiting skilled workers from India.

People of Asian birth or ancestry remain a minor percentage of the total population in both Australia and New Zealand, but new immigration policies are increasing the numbers of Asian immigrants, especially from China, Vietnam, and India. In 2006, the latest year for which complete statistics are available, 42 percent of Australia's foreign-born residents were from Europe; 15 percent were from Asia (see Figure 11.12 on page 432). There are similar patterns in New Zealand, where people of European descent are declining as a proportion of the population, although they are expected to remain the most numerous segment throughout the twenty-first century. Auckland now has the largest Polynesian population (including Native Maori) of any city in the world.

The Social Repositioning of Indigenous Peoples in Australia and New Zealand The makeup of the populations of Australia and New Zealand is also changing in another respect. In Australia, for the first time in 200 years, the number of people who claim indigenous origins is increasing. Between 1991 and 1996, the number of Australians claiming Aboriginal origins rose by 33 percent. By 2009, the indigenous population was estimated at 528,600 (2.3 percent of the total), 39 percent of whom were 15 years of age or younger. In New Zealand, the number claiming Maori background rose by 20 percent between 1991 and 2009. In 2009, there were 652,900 Maori; they comprise about 15 percent of New Zealand's population.

These increases are not the result of a population boom, but rather more positive attitudes toward indigenous peoples, which encourage more people to acknowledge their Aborigine or Maori ancestry. It is broadly recognized now that discrimination has been the main reason for the low social standing and impoverished state of indigenous peoples. Marriages between European and indigenous peoples are also more common and more open, and so the number of people with mixed heritage is increasing.

In recent decades, respect for Aboriginal and Maori culture has also increased. Aborigines base their way of life on the idea that the spiritual and physical worlds are intricately related

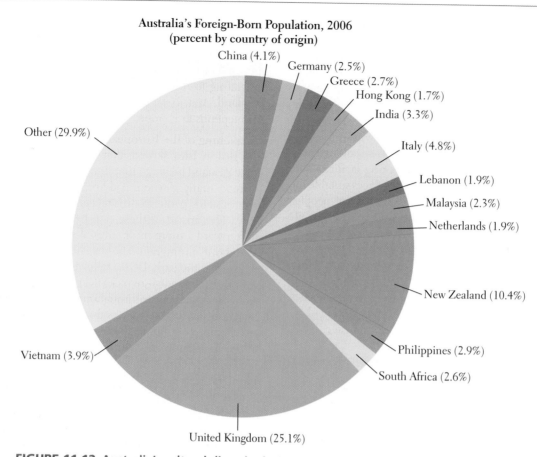

Australia's Foreign-Born Population, 2006
(percent by country of origin)

China (4.1%)
Germany (2.5%)
Greece (2.7%)
Hong Kong (1.7%)
India (3.3%)
Italy (4.8%)
Lebanon (1.9%)
Malaysia (2.3%)
Netherlands (1.9%)
New Zealand (10.4%)
Philippines (2.9%)
South Africa (2.6%)
United Kingdom (25.1%)
Vietnam (3.9%)
Other (29.9%)

FIGURE 11.12 Australia's cultural diversity in 2006. Over 24 percent (4.95 million) of Australia's people were born in other places, making Australia one of the world's most ethnically diverse nations.

(Figure 11.13). The dead are everywhere present in spirit, and they guide the living in how to relate to the physical environment. Much Aboriginal spirituality refers to the *Dreamtime*, the time of creation when the human spiritual connections to rocks, rivers, deserts, plants, and animals were made clear. Unfortunately, the social repositioning of Aboriginal and Maori people remains incomplete. Very few Aboriginal people practice their own cultural traditions or live close to ancient homelands. Instead, many live in impoverished urban conditions.

Aboriginal Land Claims In 1988, during a bicentennial celebration of the founding of white Australia, a contingent of some 15,000 Aborigines protested that they had little reason to celebrate. During the same 200 years, they had been excluded from their ancestral lands, had lost basic civil rights, and had effectively been erased from Australian national consciousness. Into the 1960s, Aborigines had only limited rights of citizenship, and it was even illegal for them to drink alcohol.

For over two centuries, Aborigines were assumed to have no prior claim to any land in Australia, as their cultures were largely nomadic and nonhierarchical, resulting in few concepts related to land ownership. The Australian High Court declared this position void in 1993. After that, Aboriginal groups began to win some land claims, mostly for land in the arid interior previously controlled by the Australian government. Court cases to restore Aboriginal rights and lands continue. Figure 11.14 shows the Aboriginal tent embassy, a permanent installation in Canberra that advocates for Aborigines.

Maori Land Claims In New Zealand, relations between the majority European-derived population and the indigenous Maori have proceeded only somewhat more amicably. In 1840, the Maori signed the Waitangi Treaty with the British, assuming they were granting only rights of land usage, not ownership. The Maori did not regard land as a tradable commodity, but rather as an asset of the people as a whole, used by families and larger kin groups to fulfill their needs. The geographer Eric Pawson, in *Inventing Places-Studies in Cultural Geography*, writes: "To the Maori the land was sacred . . . [and] the features of land and water bodies were woven through with spiritual meaning and the Maori creation myth." The British, on the other hand, assumed that the treaty had transferred Maori lands to them, giving them *exclusive* rights to settle the land with British migrants and to extract wealth through farming, mining, and forestry.

By 1950, the Maori had lost all but 6.6 percent of their former lands to European settlers and the government. Maori numbers shrank from a probable 120,000 in the early 1800s to 42,000 in 1900, and they came to occupy the lowest and most impoverished rung of New Zealand society. In the 1990s, however, the Maori began to reclaim their culture, and they established a tribunal that forcefully advances Maori interests and land claims through the courts. Since then, nearly half a million acres

FIGURE 11.14 The Aboriginal tent embassy in Canberra, Australia. Intermittently since 1972, and continuously since 1992, Aboriginal activists have camped out on the grounds of Australia's parliament in Canberra, Australia. Considered by many to be the most effective political action ever taken by Australian Aborigines, the first "tent embassy" was a response to the Australian government's denial of land ownership and other "land rights" by Aborigines regarding territories they had continuously occupied for thousands of years. With increasing recognition of Aboriginal land rights, other causes have been championed by the tent embassy, including opposition to mining activity that threatens Aboriginal communities and cultural sites, and the plight of Aboriginal slums in major cities, such as the community of Redfern in Sydney, Australia. The tent embassy remains controversial; in 2003 anti-Aboriginal arsonists burned down parts of it. The Australian government plans a more permanent structure and a ban on camping at the embassy.

FIGURE 11.13 Aboriginal rock art. A "Mimi spirit" painted on a rock at Kakadu National Park, Australia. To the Aborigines, Mimi spirits are teachers who pass between this world and another dimension via crevices in rocks. They are responsible for many teachings on hunting, food preparation, use of fire, dance, and sexuality.

of land and several major fisheries have been transferred back to Maori control. Nonetheless, the Maori still have notably higher unemployment, lower education levels, and poorer health than the New Zealand population as a whole.

Politics and Culture: Different Definitions of Democracy

In recent decades, stark divisions have emerged in Oceania over definitions of democracy—the system of government that dominates in New Zealand and Australia—and the political and cultural philosophy that influences governments throughout the rest of the region, known as the **Pacific Way** (see Photo Essay 11.5 on page 434).

Democracy as it is practiced in Australia and New Zealand is a parliamentary system based on universal suffrage, debate, and majority rule. The Pacific Way is based on traditional notions of power and problem

> **Pacific Way** the idea that Pacific Islanders have a regional identity and a way of handling conflicts peacefully that grows out of their particular social experience

solving and refers to a way of settling issues based in the traditional culture of many Pacific Islanders. It favors consensus and mutual understanding over open confrontation, and respect for traditional leadership (especially the usually patriarchal leadership of families and villages) over free speech, personal freedom, and democracy.

The Pacific Way is a political and cultural philosophy developed in Fiji around the time of its independence from the United Kingdom in 1970. It subsequently grew in popularity in many Pacific islands, most of which gained independence in the 1970s and 1980s. The Pacific Way was born, in part, out of resistance to Europeanization; it has often been invoked to uphold the notion of a regional identity shared by Pacific islands that grows out of their own unique history and social experience. It was particularly important to the academics who were given the task of writing new textbooks to replace those used by their former colonial masters. The new texts focused students' attention on their own cultures and places before they studied Britain, France, or the United States. Appeals to the Pacific Way have also been used to uphold attempts by Pacific island governments to control their own economic development and solve their own political and social problems, even when their methods are criticized by outsiders, foreign governments, and their own citizens.

Levels of democratization vary significantly between Australia, New Zealand, and Hawaii (which are among the world's more democratized places) and New Guinea and many Pacific islands (which are much less democratized).

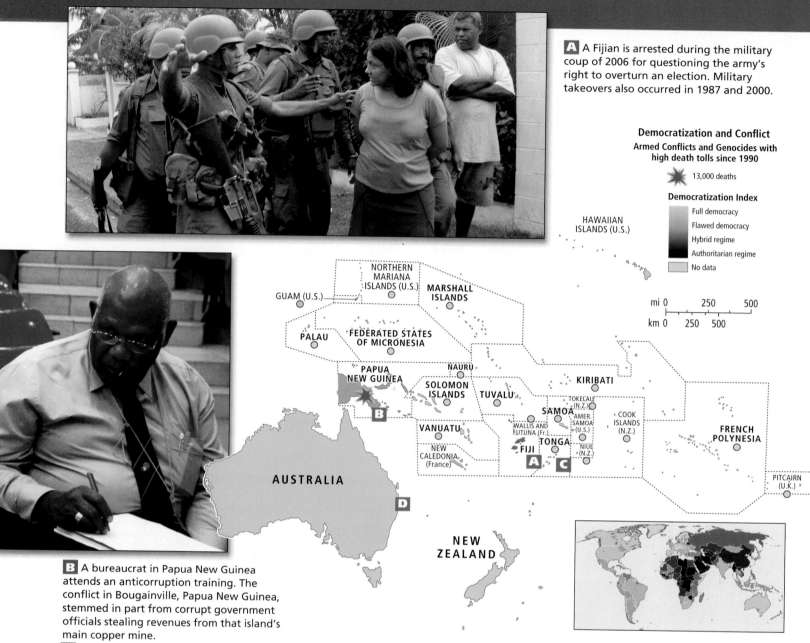

A A Fijian is arrested during the military coup of 2006 for questioning the army's right to overturn an election. Military takeovers also occurred in 1987 and 2000.

Democratization and Conflict
Armed Conflicts and Genocides with high death tolls since 1990

✳ 13,000 deaths

Democratization Index

Full democracy
Flawed democracy
Hybrid regime
Authoritarian regime
No data

mi 0   250   500

km 0   250   500

B A bureaucrat in Papua New Guinea attends an anticorruption training. The conflict in Bougainville, Papua New Guinea, stemmed in part from corrupt government officials stealing revenues from that island's main copper mine.

C The streets of Nuku'alofa, capital of the Kingdom of Tonga, after riots that broke out when the legislature refused to enact democratic reforms that would have reduced the role of the royal family in politics.

D A march in Brisbane, Australia, protesting present and past discrimination against Aborigines, who were first allowed to vote throughout Australia only in 1967.

In politics, the Pacific Way has occasionally been invoked as a philosophical basis for overriding democratic elections that challenge the power of indigenous Pacific Islanders. In 1987, 2000, and 2006, indigenous Fijians used the Pacific Way to justify coups d'état against legally elected governments (Photo Essay 11.5A). All three of the overthrown governments were dominated by Indian Fijians, the descendants of people from India who were brought to Fiji more than a century ago by the British to work on sugar plantations.

Fiji's population is now about evenly divided between indigenous Fijians and Indian Fijians. Indigenous Fijians are generally less prosperous and tend to live in rural areas where community affairs are still governed by traditional chiefs. By contrast, Indian Fijians hold significant economic and political power, especially in the urban centers and in areas of tourism and sugar cultivation. In response to the coups, many Indian Fijians left the islands, resulting in a loss of badly needed skilled labor, which has slowed economic development.

Political responses around Oceania to the coups have been divided. Australia, New Zealand, and the United States (via the state government of Hawaii) have demanded that the election results stand and the Indian Fijians be returned to office. The rest of Oceania has generally supported the coup leaders with appeals to the Pacific Way. Like Fiji, many islands are governed by leaders of indigenous descent, including a number of indigenous royal families and associated aristocracies who have not always respected democracy, especially when it could threaten their hold on power (see Thematic Overview F).

On the global stage, Fiji has been made to suffer officially for subverting majority rule democracy by being suspended from the Commonwealth of Nations (a union of former British colonies). As a result, it will be ineligible for aid from the Commonwealth.

Regardless of its global political status, the Pacific Way is likely to endure, especially as a concept that upholds regional identity and traditional culture. Further, some organizations now use the Pacific Way as the basis of an integrated approach to economic development and environmental issues. For example, the South Pacific Regional Environmental Programme builds on traditional Pacific island economic activities — such as fishing and local traditions of environmental knowledge and awareness — to promote grassroots economic development and environmental sustainability.

Gender Roles in Oceania

Perceptions of Oceania are colored by many myths about how men and women are and should be. As always, the realities are more complex.

Gender Myths and Realities Because of Oceania's cultural diversity, many different roles for men and women exist. In the Pacific islands, men traditionally were cultivators, deepwater fishers, and masters of seafaring. In Polynesia, they also were responsible for many aspects of food preparation, including cooking. In the modern world, men fill many positions, but idealized male images continue to be associated with vigorous activities.

In Australia and New Zealand, the supermasculine, white, working-class settler has long been prominent in national mythologies (see **Figure 11.15A** on page 436). In New Zealand, he was a farmer and herdsman. In Australia, he was more often a many-skilled laborer — a stockman, sheep shearer, cane cutter, or digger (miner) — who possessed a laconic, laid-back sense of humor. Labeled a "swagman" for the pack he carried, he went from station (large farm) to station or mine to mine, working hard but sporadically, gambling, and then working again until he had enough money or experience to make it in the city. There, he often felt ill at ease and chafed to return to the wilds. Now immortalized in songs, novels, and films, these men are portrayed as a rough and nomadic tribe whose social life was dominated by male camaraderie and frequent brawls. No small part of this characterization of males derived from the fact that many of Australia's first immigrants were convicts.

Today, as part of larger efforts to recognize the diversity of Australian society, new ways of life for men are emerging and are breaking down the national image of the tough heterosexual male loner. For example, the Australian city of Sydney annually hosts one of the world's largest gay pride parades (see Figure 11.15B). Nonetheless, the old model persists and remains prominent in the public images of Australian businessmen, politicians, and movie stars.

Perhaps the most enduring myth Europeans created regarding Oceania was their characterization of the women of the Pacific islands as gentle, simple, compliant love objects. (Tourist brochures still promote this notion.) There is ample evidence to suggest that Pacific Islanders did have more sexual partners in a lifetime than Europeans did. However, the reports of unrestrained sexuality related by European sailors were no doubt influenced by the exaggerated fantasies one might expect from all-male crews living at sea for months at a time. The notes of Captain James Cook are typical: "No women I ever met were less reserved. Indeed, it appeared to me, that they visited us with no other view, than to make a surrender of their persons." Over the years, such notions about Pacific island women have been encouraged by the paintings and prints of Paul Gauguin, the writings of novelist Herman Melville (*Typee*), and the studies of anthropologist Margaret Mead (*Coming of Age in Samoa*), as well as by movies and musicals such as *Mutiny on the Bounty* and *South Pacific*.

In reality, women's roles in the Pacific islands varied considerably from those in Europe, but not in the ways European explorers imagined. Women often exercised a good bit of power in family and clan, and their power increased with motherhood and advancing age. In Polynesia, a woman could achieve the rank of ruling chief in her own right, not just as the consort of a male chief. Women were primarily craftspeople, but they also

(A)

(B)

FIGURE 11.15 Australia's growing diversity in gender identity. **(A)** A "swagman," or itinerant worker, still a powerful icon of Australian maleness, in 1901. **(B)** Sydney's annual Mardi Gras Parade in 2008, one of the world's largest lesbian, gay, bisexual, and transgender (LGBT) events.

contributed to subsistence by gathering fruits and nuts and by fishing. And in some places—Micronesia, for example—lineage was established through women, not men.

Gender, Democracy, and Economic Empowerment

Learning Goal 5
Gender and Democratization: How are issues of gender, democracy, and economic empowerment developing differently across Oceania?

Today, there is a trend toward equality across gender lines throughout Oceania, but there is persistent inequality as well. A striking disparity is emerging between Australia and New Zealand (where women and other once-marginalized groups are gaining political and economic empowerment) and Papua New Guinea and the Pacific islands (where change is much slower) (see Thematic Overview I; see also Reasons for Optimism B on page 437).

In Australia and New Zealand, women's access to jobs and policy-making positions in government has improved, particularly over the last few decades. New Zealand and the Australian province of South Australia were among the first places in the world to grant European women full voting rights (1893 and 1895, respectively). New Zealand has elected two female prime ministers, and Australia, one female prime minister and one deputy prime minister (similar to the vice president in the United States). In both countries the proportion of females in national legislatures (34 percent in New Zealand, 30 percent in Australia) is well above the global average of 18 percent. In Papua New Guinea and the Pacific islands, women are generally less empowered politically and economically. No woman has yet been elected to a top-level national office, and women are a tiny minority in national legislatures, if present at all.

In both Australia and New Zealand, young women are pursuing higher education and professional careers and postponing marriage and childbearing until their thirties. (This is also a trend in Hawaii, Guam, and those islands with French affiliation.) Nonetheless, both societies continue to reinforce the housewife role for women in a variety of ways. For example, the expectation is that women, not men, will interrupt their careers to stay home to care for young or elderly family members. Women in Australia receive on average only about 70 percent of the pay (69 percent in New Zealand) that men receive for equivalent work. This is, however, a smaller gender pay gap than in many developed countries.

Throughout Papua New Guinea and the Pacific islands, gender roles and relationships vary greatly over the course of a lifetime. Today, many young women fulfill traditional roles as mates and mothers and practice a wide range of domestic crafts, such as weaving and basketry. Then in middle age, they may return to school and take up careers. Some Pacific Island women, with the aid of government scholarships, pursue higher education or job training that takes them far from the villages where they raised their children. Accumulating age and experience may boost women into positions of considerable power in their communities.

Throughout their lives, women in Papua New Guinea and the Pacific islands contribute significantly to family assets through the formal and informal economies. Most traders in marketplaces are women, and the items they sell are usually made and transported by women. Yet like women everywhere, they have trouble obtaining credit to expand their businesses. The World Bank in 2007–2008 ranked Papua New Guinea 131st out of 181 nations in terms of the access that female entrepreneurs have to bank loans. Only women employed in the formal economy qualify for bank loans; other women need a husband with a salary to act as a guarantor. Recently, however, the Asian Development Bank notes that microcredit loans to women are increasing significantly. The repayment rate is very high.

THINGS TO REMEMBER

1. The number of Asian immigrants into Australia and New Zealand has been increasing over the past two decades, while the number of Europeans has been declining. Asians, however, still compose less than 20 percent of Australia's population and less than 10 percent of New Zealand's.

2. The indigenous populations of both Australia (Aborigines) and New Zealand (Maori) are relatively small, but represent an increasing percentage of each country: 2.4 percent in Australia and 14.9 percent in New Zealand. People in Oceania are now more willing to claim native roots than they once were.

3. **Learning Goal 5: Gender and Democratization** Gender roles and relationships vary considerably in Oceania. In Australia and New Zealand, compared with elsewhere in the region, women generally are better off in terms of political power and pay equity with men. In some traditional societies, however, women inherit or accrue considerable power in their own communities over the course of a lifetime.

Reflections on Oceania

In this region that has been so powerfully shaped by the globalizing influence of outsiders, several ongoing waves of transformation will define much of what happens in the near future. The economic reorientation away from Europe and toward Asia is likely to continue, with a boom in Asian tourism in the Pacific islands that will strengthen this shift. As service economies expand, average living standards will rise and urbanization will continue. All of this will make population growth likely to slow, even in the Pacific islands (see Reasons for Optimism A). However, disparities in income and well-being may also increase.

Positive ongoing transformations set Oceania on a course toward weathering new challenges. Democracy is well established and expanding in Australia, New Zealand, and Hawaii, as is the political and economic empowerment of women and other groups (Reasons for Optimism B). While both of these trends are less well established in the Pacific islands, especially Papua New Guinea, the islands also have traditions of cooperation and consensus that may help societies navigate future stresses.

It is difficult to make detailed predictions about the future of Oceania because of the uncertainties brought by climate change. While some low-lying islands are clearly threatened with submersion, the precise nature of the impacts facing Australia and New Zealand is less certain. Water stress could constrain agricultural expansion, and fire could become a more pressing concern. Changes in Oceania's regional climate could further threaten the many unique ecosystems of this region, much as the importing of European food systems did. Advances in water technologies, though, could reduce the vulnerability of many areas to water shortage, drought, and food shortage. Further, New Zealand's ambitious goals for renewable energy are among the most advanced in the world (Reasons for Optimism C). The future of Oceania, much like its past, will likely remain powerfully linked to events in distant places. Perhaps more than any other world region, Oceania is not the master of its own destiny.

Reasons for Optimism in Oceania

Population: As service economies expand, average living standards will rise and urbanization will continue. All of this will likely slow population growth. **A** *Children in rural Fiji.* ▼

Democratization: Democracy is well established and expanding in Australia, New Zealand, and Hawaii, as is the political and economic empowerment of women. **B** *A political demonstration in Australia.* ▼

Climate Change: New Zealand plans to use renewable energy to supply 90 percent of its electricity by 2025. **C** *A wind farm in New Zealand.* ▼

Thinking Geographically: The economic reorientation away from Europe and toward Asia is likely to continue, bolstered in the Pacific islands by _____.

Thinking Geographically: What percentages of the national legislatures of Australia and New Zealand are female?

Thinking Geographically: Where is the there excellent potential for wind power in this region?

Learning Goals Review Questions

1. Climate Change and Water: What impact will climate change have on Oceania, particularly in relation to water?

Sea level rise threatens some Pacific islands with submersion. Which ones are most threatened? What other major changes are likely to occur with climate change?

2. Food and the Environment: In what ways have European systems for producing food and fiber changed environments in Oceania?

Oceania has been transformed by plants and animals brought to the region as part of European farming and herding systems. How have these new systems impacted the native species and landscapes of Oceania?

3. Globalization and Development: How has globalization transformed patterns of economic development in Oceania?

Europe and the United States have both exerted a powerful influence on Oceania. Now, trade and social interaction with Asia is increasing. What changes are resulting from Oceania's reorientation toward Asia?

4. Population and Urbanization: What are the two main patterns of population and urbanization developing across Oceania?

In some parts of this region the shift to the cities has been accompanied by broad-based affluence, as expanding urban service economies bring decent jobs to more people. What are other urbanization patterns in the region?

5. Gender and Democratization: How are issues of gender, democracy, and economic empowerment developing differently across Oceania?

Women's access to jobs and elected office varies dramatically across this region. Where have women gained the most political and economic empowerment?

Geographic Themes about Oceania

Look back at the Thematic Overview photos on page 407 of this chapter. Thinking geographically, answer the following questions about them:

(A) Climate Change: What reefs in this region have been affected by coral reef bleaching in recent years?

(B) Water: Recent severe droughts might not merely be periodic dry spells, but may represent permanent alterations in rainfall patterns that could worsen what phenomena?

(C) Food: Following deforestation, what have farmers and ranchers in New Zealand used to augment infertile soils?

(D) Globalization: Asia buys what percent of Australia's exports?

(E) Development: What two factors came together to help Australia and New Zealand achieve broad prosperity on the basis of raw materials exports?

(F) Democratization: Who governs Fiji and many other Pacific islands?

(G) Population: What is the life expectancy (in decades) of most Pacific Islanders?

(H) Urbanization: Where in Oceania is the shift from rural to urban livelihoods furthest along?

(I) Gender: In Papua New Guinea and the Pacific islands, women have been elected to _____

Key Terms

Aborigines 422
atoll 412
endemic 414
Gondwana 409
Great Barrier Reef 412
hot spots 412

invasive species 419
Maori 408
marsupials 415
Melanesia 422
Melanesians 422
Micronesia 422

MIRAB economy 427
monotremes 415
Pacific Way 433
Polynesia 422
roaring forties 412
subsistence affluence 427

Despite its remote location and few human inhabitants, environmental change is accelerating in Antarctica as temperatures rise and surrounding fisheries are exploited.

A Tourists on a Southern Ocean cruise observe the break up of one of Antarctica's many coastal ice shelves. While most sea level–rise predicted for the near future is due to thermal expansion of oceans, sea levels would rise 200 feet (60 meters) if Antarctica's massive glaciers were all to melt.

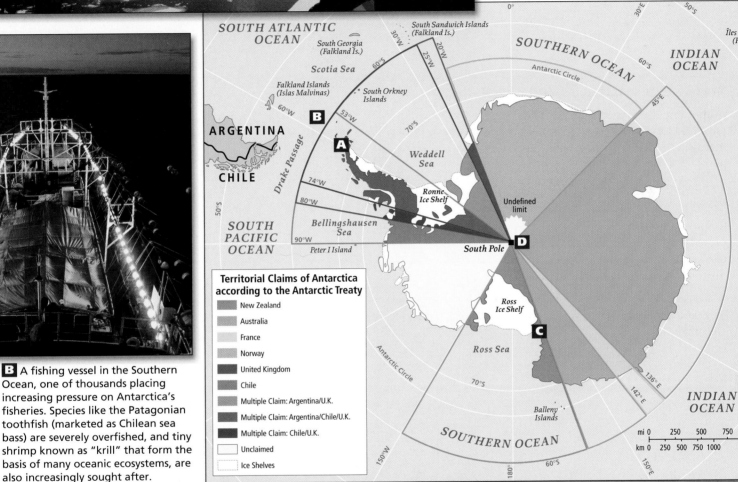

Territorial Claims of Antarctica according to the Antarctic Treaty

- New Zealand
- Australia
- France
- Norway
- United Kingdom
- Chile
- Multiple Claim: Argentina/U.K.
- Multiple Claim: Argentina/Chile/U.K.
- Multiple Claim: Chile/U.K.
- Unclaimed
- Ice Shelves

B A fishing vessel in the Southern Ocean, one of thousands placing increasing pressure on Antarctica's fisheries. Species like the Patagonian toothfish (marketed as Chilean sea bass) are severely overfished, and tiny shrimp known as "krill" that form the basis of many oceanic ecosystems, are also increasingly sought after.

C An emperor penguin dives beneath a hole in some sea ice on a fishing expedition. Emperor penguins are found only in Antarctica and are proving extremely sensitive to climate change. In warmer years declines in sea ice, their ideal hunting habitat, have resulted in widespread starvation. Meanwhile, in colder years fewer penguin chicks hatch.

D Amundsen-Scott South Pole Station, the dome-shaped structure, is a U.S. research station. Outside fly the flags of the first countries to sign the Antarctic Treaty in 1959, which bans all military or resource extraction-related activity, making the continent a scientific and nature reserve.

GLOSSARY

Aborigines (p. 422) the longest-surviving inhabitants of Oceania, whose ancestors, the Australoids, migrated from Southeast Asia as early as 50,000 years ago over the Sundaland landmass that was exposed during the ice ages

acculturation (p. 128) adaptation of a minority culture to the host culture enough to function effectively and be self-supporting; cultural borrowing

acid rain (p. 66) falling precipitation that has formed through the interaction of rainwater or moisture in the air with sulfur dioxide and nitrogen oxides emitted during the burning of fossil fuels, making it acidic

agribusiness (p. 80) the business of farming conducted by large-scale operations that purchase, produce, finance, package, and distribute agricultural products

agriculture (p. 23) the practice of producing food through animal husbandry, or the raising of animals, and the cultivation of plants

agroecology (p. 317) the practice of traditional, nonchemical methods of crop fertilization and the use of natural predators to control pests

agroforestry (p. 254) growing economically useful crops of trees on farms, along with the usual plants and crops, to reduce dependence on trees from nonfarmed forests and to provide income to the farmer

Ainu (p. 364) an indigenous cultural minority group in Japan characterized by their light skin, heavy beards, and thick, wavy hair who are thought to have migrated thousands of years ago from the northern Asian steppes

animism (p. 281) a belief system in which natural features carry spiritual meaning

apartheid (p. 262) a system of laws mandating racial segregation. South Africa was an apartheid state from 1948 until 1994.

aquifers (p. 59) natural underground reservoirs of water

archipelago (p. 374) a group, often a chain, of islands

assimilation (pp. 128, 164) the loss of old ways of life and the adoption of the lifestyle of another culture or country

Association of Southeast Asian Nations (ASEAN) (p. 389) an organization of Southeast Asian governments established to further economic growth and political cooperation

atoll (p. 412) a low-lying island, formed of coral reefs that have built up on the circular or oval rims of a submerged volcano

Australo-Melanesians (p. 383) a group of hunters and gatherers who moved from the present northern Indian and Burman parts of southern Eurasia to the exposed landmass of Sundaland about 60,000 to 40,000 years ago

Austronesians (p. 383) a Mongoloid group of skilled farmers and seafarers from southern China who migrated south to various parts of Southeast Asia between 10,000 and 5000 years ago

authoritarianism (p. 33) a political system based on the power of the state or of elitist regional and local leaders

Aztecs (p. 111) indigenous people of high-central Mexico noted for their advanced civilization before the Spanish conquest

biodiversity (p. 104) the variety of life forms to be found in a given area

biosphere (pp. 22, 142) the global ecological system that integrates all living things and their relationships

birth rate (p. 14) the number of births per 1000 people in a given population, per unit of time, usually per year

Bolsheviks (p. 186) a faction of communists who came to power during the Russian Revolution

brain drain (p. 129) the migration of educated and ambitious young adults to cities or foreign countries, depriving the communities from which the young people come of talented youth in whom they have invested years of nurturing and education

brownfields (p. 88) old industrial sites whose degraded conditions pose obstacles to redevelopment

Buddhism (p. 308) a religion of Asia that originated in India in the sixth century B.C.E. as a reinterpretation of Hinduism; it emphasizes modest living and peaceful self-reflection leading to enlightenment

capitalism (pp. 34, 152) an economic system based on the private ownership of the means of production and distribution of goods, driven by the profit motive and characterized by a competitive marketplace

capitalists (p. 186) usually a wealthy minority that owns the majority of factories, farms, businesses, and other means of production

carbon sequestration (p. 252) the removal and storage of carbon taken from the atmosphere

carrying capacity (p. 24) the maximum number of people that a given territory can support sustainably with food, water, and other essential resources

cartel (p. 232) a group of producers strong enough to control production and set prices for its products

cash economy (p. 17) an economic system in which the necessities of life are purchased with monetary currency

caste system (p. 306) an ancient Hindu system for dividing society into hereditary hierarchical classes

Caucasia (p. 198) the mountainous region between the Black Sea and the Caspian Sea

central planning (p. 152) a communist economic model in which a central bureaucracy dictates prices and output with the stated aim of allocating goods equitably across society according to need

centrally planned, or socialist, economy (p. 186) an economic system in which the state owns all real estate and means of production, while government bureaucrats direct all economic activity, including the locating of factories, residences, and transportation infrastructure

Christianity (p. 220) a monotheistic religion based on the belief in the teachings of Jesus of Nazareth, a Jew, who described God's relationship to humans as primarily one of love and support, as exemplified by the Ten Commandments

civil disobedience (p. 303) the breaking of discriminatory laws by peaceful means

civil society (pp. 33, 263) the totality of voluntary civic and social organizations and institutions that form the basis of a functioning society, as opposed to the structures of a state that are backed by force (regardless of that state's political system) and commercial institutions of the market

clear-cutting (p. 65) the cutting down of all trees on a given plot of land, regardless of age, health, or species

climate (p. 45) the long-term balance of temperature and precipitation that characteristically prevails in a particular region

climate change (p. 39) a slow shifting of climate patterns due to the general cooling or warming of the atmosphere

Cold War (pp. 152, 186) the contest that pitted the United States and western Europe, espousing free market capitalism and democracy, against the USSR and its allies, promoting a centrally planned economy and a socialist state

commercial agriculture (p. 254) farming in which crops are grown deliberately for cash rather than solely as food for the family

commodities (p. 264) raw materials that are traded, often to other countries, for processing or manufacturing into more valuable goods

Common Agricultural Program (CAP) (p. 159) a European Union program, meant to guarantee secure and safe food supplies at affordable prices, that places tariffs on imported agricultural goods and gives subsidies to EU farmers

communal conflict (p. 309) a euphemism for religiously based violence in South Asia

communism (pp. 34, 152, 186) an ideology, based largely on the writings of the German revolutionary Karl Marx, that calls on workers to unite to overthrow capitalism and establish an egalitarian society in which workers share what they produce; on behalf of the people, the state owns all farms, industry, land, and buildings (a version of socialism)

Communist Party (p. 186) the political organization that ruled the USSR from 1917 to 1991; other communist countries, such as China, Mongolia, North Korea, and Cuba, also have communist parties

Confucianism (p. 342) a Chinese philosophy that teaches that the best organizational model for the state and society is a hierarchy based on the patriarchal family

contested space (p. 120) any area that several groups claim or want to use in different and often conflicting ways, such as the Amazon or Palestine

coral bleaching (p. 381) color loss that results when photosynthetic algae that live in the corals are expelled

coup d'état (p. 121) a military- or civilian-led forceful takeover of a government

Creoles (p. 112) people mostly of European descent born in the Caribbean

crony capitalism (p. 389) a type of corruption in which politicians, bankers, and entrepreneurs, sometimes members of the same family, have close personal as well as business relationships

cultural homogenization (p. 139) the tendency toward uniformity of ideas, values, technologies, and institutions among associated culture groups

cultural pluralism (p. 400) the cultural identity characteristic of a region where groups of people from many different backgrounds have lived together for a long time but have remained distinct

Cultural Revolution (p. 344) a series of highly politicized and destructive mass campaigns launched in 1966 to force the entire population of China to support the continuing revolution

culture (p. 50) all the ideas, materials, and institutions that people have invented to use to live on earth that are not directly part of our biological inheritance

currency devaluation (p. 266) the lowering of a currency's value relative to the U.S. dollar, the Japanese yen, the European euro, or another currency of global trade

czar (p. 185) title of the ruler of the Russian empire; derived from the word "caesar," the title of the Roman emperors

death rate (p. 14) the ratio of total deaths to total population in a specified community, usually expressed in numbers per 1000 or in percentages

delta (p. 44) the triangular-shaped plain of sediment that forms where a river meets the sea

democratization (p. 33) the transition toward political systems guided by competitive elections

demographic transition (p. 16) the change from high birth and death rates to low birth and death rates that usually accompanies a cluster of other changes, such as change from a subsistence to a cash economy, increased education rates, and urbanization

desertification (pp. 216, 258) a set of ecological changes that converts nondesert lands into deserts; the process by which arid conditions spread to areas that were previously moist

detritus (p. 375) dead organic material (such as plants and insects) that collects on the ground

development (p. 19) usually used to describe economic changes like greater productivity of agriculture and industry that lead to better standards of living or simply to increased mass consumption

diaspora (p. 220) the dispersion of Jews around the globe after they were expelled from the eastern Mediterranean by the Roman Empire beginning in 73 C.E.; the term can now refer to other dispersed culture groups

dictator (p. 121) a ruler who claims absolute authority, governing with little respect for the law or the rights of citizens

digital divide (p. 83) the discrepancy in access to information technology between small, rural, and poor areas and large, wealthy cities that contain major government research laboratories and universities

divide and rule (p. 271) the deliberate intensification of divisions and conflicts by potential rulers; in the case of sub-Saharan Africa, by European colonial powers

domestication (p. 23) the process of developing plants and animals through selective breeding to live with and be of use to humans

domino theory (p. 384) a foreign policy theory that used the idea of the domino effect to suggest that if one country "fell" to communism, others in the neighboring region would also fall

double day (p. 165) the longer workday of women with jobs outside the home who also work as caretakers, housekeepers, and/or cooks for their families

dowry (p. 310) a price paid by the family of the bride to the groom (the opposite of *bride price*); formerly a custom practiced only by the rich

dual economy (p. 265) an economy in which the population is divided by economic disparities into two groups, one prosperous and the other near or below the poverty level

early extractive phase (p. 114) a phase in Central and South American history, beginning with the Spanish conquest and lasting until the early twentieth century, characterized by a dependence on the export of raw materials

ecological footprint (p. 22) the biologically productive area per person needed to sustain a particular standard of living

economic core (p. 72) the dominant economic region within a larger region

economic diversification (p. 234) the expansion of an economy to include a wider array of activities

economies of scale (p. 156) reductions in the unit cost of production that occur when goods or services are efficiently mass produced, resulting in increased profits per unit

ecotourism (p. 109) nature-oriented vacations, often taken in endangered and remote landscapes, usually by travelers from affluent nations

El Niño (pp. 106, 375) a recurring phenomenon during which the usual patterns of air and water circulation in the Pacific are reversed. Ocean temperatures are cooler than usual in the western Pacific near Southeast Asia and warmer than usual along the southern Pacific coast of South America

endemic (p. 414) belonging or restricted to a particular place

erosion (p. 24) the process by which fragmented rock and soil are moved over a distance, primarily by wind and water

ethnic cleansing (p. 36) the deliberate removal of an ethnic group from a particular area by forced migration

ethnic group (p. 50) a group of people who share a set of beliefs, a way of life, a technology, and usually a geographic location

ethnicity (p. 90) the quality of belonging to a particular culture group

euro (p. 156) the official (but not required) currency of the European Union as of January 1999

European Union (EU) (p. 138) a supranational organization that unites most of the countries of West, South, North, and Central Europe

evangelical Protestantism (p. 133) a Christian movement that focuses on personal salvation and empowerment of the individual through miraculous healing and transformation; some practitioners preach the "gospel of success"—that a life dedicated to Christ will result in prosperity for the believer—to the poor

Export Processing Zones (EPZs) (p. 116) specially created legal spaces or industrial parks within a country where, to attract foreign-owned factories, duties and taxes are not charged

export-led growth (p. 347) an economic development strategy that relies heavily on the production of manufactured goods destined for sale abroad

extended family (p. 128) a family that consists of related individuals beyond the nuclear family of parents and children

fair trade (p. 32) trade that values equity throughout the international trade system; now proposed as an alternative to free trade

favelas (p. 131) Brazilian urban slums and shantytowns built by the poor; called colonias, barrios, or barriadas in other countries

female earned income as a percent of male earned income (F/MEI) (p. 20) a measure of pay equity that shows average female earned income as a percent of average male earned income

female genital mutilation (FGM) (p. 281) removing the labia and the clitoris and sometimes the stitching the vulva nearly shut

female seclusion (p. 225) the requirement that women stay out of public view

feminization of labor (p. 388) the increasing representation of women in both the formal and the informal labor force

Fertile Crescent (p. 218) an arc of lush, fertile land formed by the uplands of the Tigris and Euphrates river systems and the Zagros Mountains, where nomadic peoples began the earliest known agricultural communities

floating population (p. 329) the Chinese term for jobless or underemployed people who have left economically depressed rural areas for the cities and move from place to place looking for work

floodplain (p. 44) that flat land around a river where sediment is deposited during flooding

food security (pp. 23, 337) the ability of a state to consistently supply a sufficient amount of basic food to the entire population

foreign direct investment (p. 350) investment in a country's businesses by citizens, corporations, or governments of other countries

foreign exchange (p. 399) foreign currency that countries need to purchase imports

formal economy (p. 20) all aspects of the economy that take place in official channels

fossil fuel (p. 208) a source of energy formed from the remains of dead plants and animals

free trade (p. 31) the movement of goods and capital without government restrictions

Gazprom (p. 188) in Russia, the state-owned energy company; it is the tenth-largest oil and gas entity in the world

gender (p. 17) the sexual category of a person

gender roles (p. 18) the socially assigned roles of males and females

genetic modification (GM) (p. 24) in agriculture, the practice of splicing together the genes from widely divergent species to achieve particular desirable characteristics

genocide (pp. 36, 272) the deliberate destruction of an ethnic, racial, or political group

gentrification (p. 88) the renovation of old urban districts by middle-class investment, a process that often displaces poorer residents

Geographic Information Science (GISc) (p. 7) the body of science that underwrites multiple spatial analysis technologies and keeps them at the cutting edge

geopolitics (p. 34) the use of strategies by countries to ensure that their best interests are served

glasnost (p. 190) literally, "openness"; the policies instituted in the late 1980s under Mikhail Gorbachev that encouraged greater transparency, openness, and publicity in the workings of all levels of Soviet government

global economy (p. 28) the worldwide system in which goods, services, and labor are exchanged

globalization (p. 27) the growth of interregional and worldwide linkages and the changes these linkages are bringing about

global scale (p. 7) the level of geography that encompasses the entire world as a single unified area

global warming (p. 39) the warming of the earth's climate as atmospheric levels of greenhouse gases increase

Gondwana (p. 409) the great landmass that formed the southern part of the ancient supercontinent Pangaea

grassroots economic development (p. 270) economic development projects designed to help individuals and their families achieve sustainable livelihoods

Great Barrier Reef (p. 412) the largest coral reef in the world, located off the northeastern coast of Australia

Great Leap Forward (p. 344) an economic reform program under Mao Zedong intended to quickly raise China to the industrial level of Britain and the United States

Green (p. 146) environmentally conscious

greenhouse gases (p. 39) harmful gases, such as carbon dioxide and methane, released into the atmosphere by human activities

green revolution (p. 23) increases in food production brought about through the use of new seeds, fertilizers, mechanized equipment, irrigation, pesticides, and herbicides

gross domestic product (GDP) (p. 16) the market value of all goods and services produced by workers and capital within a particular country's borders and within a given year

gross domestic product (GDP) per capita (p. 19) the market value of all goods and services produced by workers and capital within a particular country's borders and within a given year, divided by the number of people in the country

gross domestic product (GDP) per capita PPP (p. 20) the market value of all goods and services produced by modern workers and capital within a particular country's borders and within a given year, divided by the number of people in the country and adjusted for purchasing power parity

Group of Eight (G8) (p. 188) an organization of eight countries with large economies: France, the United States, Britain, Germany, Japan, Italy, Canada, and Russia

growth poles (p. 350) zones of development whose success draws more investment and migration to a region

guest workers (p. 163) legal workers from outside a country who help fulfill the need for temporary workers but who are expected to return home when they are no longer needed

Gulf states (p. 225) Saudi Arabia, Kuwait, Bahrain, Oman, Qatar, and the United Arab Emirates

haciendas (p. 114) a large agricultural estate in Middle or South America, more common in the past; usually not specialized by crop and not focused on market production

hajj (p. 223) the pilgrimage to the city of Makkah (Mecca) that all Muslims are encouraged to undertake at least once in a lifetime

Harappa culture (p. 300) see Indus Valley civilization

Hinduism (p. 305) a major world religion practiced by approximately 900 million people, 800 million of whom live in India; a complex belief system, with roots both in localized folk traditions (known as the Little Tradition) as well as in a broader system based on literary texts (known as the Great Tradition)

Hispanic (p. 58) term used to refer to all Spanish-speaking people from Middle and South America, although their ancestors may have been black, white, Asian, or Native American

Holocaust (p. 152) during World War II, a massive execution by the Nazis of 6 million Jews and 5 million gentiles (non-Jews), including ethnic Poles and other Slavs, Roma (Gypsies), disabled and mentally ill people, gays, lesbians, transgendered people, and political dissidents.

Horn of Africa (p. 252) the triangular peninsula that juts out from northeastern Africa below the Red Sea and wraps around the Arabian Peninsula

hot spots (p. 412) individual sites of upwelling material (magma) that originate deep in the mantle of the earth and surface in a tall plume; hot spots tend to remain fixed relative to migrating tectonic plates

Hukou **system** (p. 329) the system in China by which citizens' permanent residence is registered

human geography (p. 44) the study of patterns and processes that have shaped human understanding, use, and alteration of the earth's surface

humanism (p. 149) a philosophy and value system that emphasizes the dignity and worth of the individual

human well-being (p. 20) various measures of the extent to which people are able to obtain a healthy life in a community of their choosing

humid continental climate (p. 142) a midlatitude climate pattern in which summers are fairly hot and moist, and winters become longer and colder the deeper into the interior of the continent one goes

import substitution industrialization (ISI) (p. 115) policies that encouraged local production of machinery and other items that previously had been imported at great expense from abroad

Incas (p. 111) indigenous people who ruled the largest pre-Columbian state in the Americas, with a domain stretching from southern Colombia to northern Chile and Argentina

income disparity (p. 113) the gap in income between rich and poor

indigenous (p. 101) native to a particular place or region

Indus Valley civilization (p. 300) the first substantial settled agricultural communities, which appeared about 4500 years ago along the Indus River in modern-day Pakistan and northwest India

informal economy (pp. 20, 266) all aspects of the economy that take place outside official channels

information technology (IT) (p. 83) the part of the service sector that deals with the design and distribution of computer software and hardware and the management of digital data

infrastructure (p. 70) road, rail, and communication networks and other facilities necessary for economic activity

interregional linkages (p. 27) economic, political, or social connections between regions, whether contiguous or widely separated

intertropical convergence zone (ITCZ) (p. 252) a band of atmospheric currents that circle the globe roughly around the equator; warm winds from both north and south converge at the ITCZ, pushing air upward and causing copious rainfall

intifada (p. 240) a prolonged Palestinian uprising against Israel

invasive species (p. 419) organisms that spread into regions outside of their native range, adversely affecting economies or environments

iron curtain (p. 152) a long, fortified border zone that separated western Europe from (then) eastern Europe during the Cold War

Islam (p. 208) a monotheistic religion that emerged in the seventh century C.E. when, according to tradition, the archangel Gabriel revealed the tenets of the religion to the Prophet Muhammad

Islamism (p. 208) a grassroots religious revival in Islam that seeks political power to curb what are seen as dangerous secular influences; also seeks to replace secular governments and civil laws with governments and laws guided by Islamic principles

Jainism (p. 308) originally a reformist movement within Hinduism, Jainism is a faith tradition that is more than 2000 years old; found mainly in western India and in large urban centers throughout the region, Jains are known for their educational achievements, nonviolence, and strict vegetarianism

jati (p. 306) in Hindu India, the subcaste into which a person is born, which largely defines the individual's experience for a lifetime

Judaism (p. 220) a monotheistic religion characterized by the belief in one God, Yahweh, a strong ethical code summarized in the Ten Commandments, and an enduring ethnic identity

kaizen system (p. 348) the "continuous improvement system" pioneered in Japanese manufacturing which ensures that fewer defective parts are produced because production lines are constantly surveyed for errors

kanban system (p. 348) the "just-in-time" system pioneered in Japanese manufacturing that clusters companies that are part of the same production system close together so that they can deliver parts to each other precisely when they are needed

Kyoto Protocol (p. 42) an amendment to a United Nations treaty on global warming, the Protocol is an international agreement, adopted in 1997 and in force in 2005, that sets binding targets for industrialized countries for the reduction of emissions of greenhouse gases

landforms (p. 44) physical features of the earth's surface, such as mountain ranges, river valleys, basins, and cliffs

latitude (p. 3) the distance in degrees north or south of the equator; lines of latitude run parallel to the equator, and are also called parallels

liberation theology (p. 133) a movement within the Roman Catholic Church that uses the teachings of Jesus to encourage the poor to organize to change their own lives and the rich to promote social and economic equity

living wages (p. 32) minimum wages high enough to support a healthy life

local scale (p. 7) the level of geography that describes the space where an individual lives or works; a city, town, or rural area

longitude (p. 3) the distance in degrees east and west of Greenwich, England; lines of longitude, also called meridians, run from pole to pole (the line of longitude at Greenwich is 0° and is known as the prime meridian)

machismo (p. 129) a set of values that defines manliness in Middle and South America

Maori (p. 408) Polynesian people indigenous to New Zealand

map projections (p. 6) the various ways of showing the spherical earth on a flat surface

maquiladoras (p. 116) foreign-owned, tax-exempt factories, often located in Mexican towns just across the U.S. border from U.S. towns, that hire workers at low wages to assemble manufactured goods which are then exported for sale

marianismo (p. 128) a set of values based on the life of the Virgin Mary, the mother of Jesus, that defines the proper social roles for women in Middle and South America

marketization (p. 116) the development of a free market economy in support of *free trade*

marsupials (p. 415) mammals that give birth to their young at a very immature stage and nurture them in a pouch equipped with nipples

Mediterranean climate (p. 142) a climate pattern of warm, dry summers and mild, rainy winters

megalopolis (p. 86) an area formed when several cities expand so that their edges meet and coalesce

Melanesia (p. 422) New Guinea and the islands south of the equator and west of Tonga (the Solomon Islands, New Caledonia, Fiji, and Vanuatu)

Melanesians (p. 422) a group of Australoids named for their relatively dark skin tones, a result of high levels of the protective pigment melanin; they settled throughout New Guinea and other nearby islands

mercantilism (pp. 112, 149) the policy by which European rulers sought to increase the power and wealth of their realms by managing all aspects of production, transport, and commerce in their colonies

Mercosur (p. 119) a free trade zone created in 1991 that links the economies of Brazil, Argentina, Uruguay, and Paraguay to create a common market

mestizos (p. 112) people of mixed European, African, and indigenous descent

metropolitan areas (p. 86) cities of 50,000 or more and their surrounding suburbs and towns

microcredit (p. 317) a program based on peer support that makes very small loans available to very low-income entrepreneurs

Micronesia (p. 422) the small islands that lie east of the Philippines and north of the equator

Middle America (p. 101) in this book, a region that includes Mexico, Central America, and the islands of the Caribbean

migration (p. 15) movement of people from a place or country to another, often for safety or economic reasons

MIRAB economy (p. 427) an economy based on migration, remittance, aid, and bureaucracy

mixed agriculture (p. 254) the raising of a variety of crops and animals on a single farm, often to take advantage of several environmental riches

Mongols (p. 184) a loose confederation of nomadic pastoral people centered in East and Central Asia, who by the thirteenth century had established by conquest an empire stretching from Europe to the Pacific

monotheistic (p. 220) pertaining to the belief that there is only one god

monotremes (p. 415) egg-laying mammals, such as the duck-billed platypus and the spiny anteater

monsoon (pp. 47, 292) a wind pattern in which, in summer months, warm, wet air coming from the ocean brings copious rainfall, and in winter months, cool, dry air moves from the continental interior toward the ocean

Mughals (p. 300) a dynasty of Central Asian origin that ruled India from the sixteenth to the nineteenth century

multinational corporation (p. 30) a business organization that operates extraction, production, and/or distribution facilities in multiple countries

Muslims (p. 220) followers of Islam

nationalism (p. 151) devotion to the interests or culture of a particular country, nation, or cultural group; the idea that a group of people living in a specific territory and sharing cultural traits should be united in a single country to which they are loyal and obedient

nationalize (p. 118) to seize private property and place under government ownership, with some compensation

nomadic pastoralists (p. 183) people whose way of life and economy are centered on tending grazing animals who are moved seasonally to gain access to the best pasture

nongovernmental organization (NGO) (p. 36) an association outside the formal institutions of government in which individuals, often from widely differing backgrounds and locations, share views and activism on political, social, economic, or environmental issues

nonpoint sources of pollution (p. 178) diffuse sources of environmental contamination, such as untreated automobile exhaust, raw sewage, or agricultural chemicals that drain from fields into water supplies

North American Free Trade Agreement (NAFTA) (pp. 82, 119) a free trade agreement made in 1994 that added Mexico to the 1989 economic arrangement between the United States and Canada

North Atlantic Drift (p. 142) the easternmost end of the Gulf Stream, a broad warm-water current that brings large amounts of warm water to Europe

North Atlantic Treaty Organization (NATO) (p. 158) a military alliance between European and North American countries that was developed during the Cold War to counter the influence of the Soviet Union; since the breakup of the Soviet Union, NATO has expanded membership to include much of Eastern Europe and Turkey, and is now focused mainly on providing the international security and cooperation needed to expand the European Union

nuclear family (p. 92) a family consisting of a married father and mother and their children

occupied Palestinian Territories (OPT) (p. 209) Palestinian lands occupied by Israel in 1967

offshore outsourcing (p. 318) the contracting of certain business functions or production functions to providers where labor and other costs are lower

oligarchs (p. 188) in Russia, those who acquired great wealth during the privatization of Russia's resources and who use that wealth to exercise power

OPEC (Organization of Petroleum Exporting Countries) (p. 232) a cartel of oil-producing countries—including Algeria, Angola, Iran, Iraq, Kuwait, Libya, Nigeria, Qatar, Saudi Arabia, the United Arab Emirates, Ecuador, Venezuela, and Nigeria—that was established to regulate the production, and hence the price, of oil and natural gas

orographic rainfall (p. 46) rainfall produced when a moving moist air mass encounters a mountain range, rises, cools, and releases condensed moisture that falls as rain

Pacific Rim (p. 95) a term referring to all the countries that border the Pacific Ocean

Pacific Way (p. 433) the idea that Pacific Islanders have a regional identity and a way of handling conflicts peacefully that grows out of their particular social experience

Partition (p. 303) the breakup following Indian independence that resulted in the establishment of Hindu India and Muslim Pakistan

pastoralism (p. 258) a way of life based on herding; practiced primarily on savannas, on desert margins, or in the mixture of grass and shrubs called open bush

patriarchal (p. 225) relating to a social organization in which the father is supreme in the clan or family

perestroika (p. 190) literally, "restructuring"; the restructuring of the Soviet economic system in the late 1980s in an attempt to revitalize the economy

permafrost (p. 176) permanently frozen soil just a few feet beneath the surface

physical geography (p. 44) the study of the earth's physical processes: how they work, how they affect humans, and how they are affected by humans

plantation (p. 115) a large factory farm that grows and partially processes a single cash crop

plate tectonics (p. 44) the scientific theory that the earth's surface is composed of large plates that float on top of an underlying layer of molten rock; the movement and interaction of the plates create many of the large features of the earth's surface, particularly mountains

political ecologist (p. 22) a geographer who studies power allocations in the interactions among development, human well-being, and the environment

polygyny (p. 227) the taking by a man of more than one wife at a time

Polynesia (p. 422) the numerous islands situated inside an irregular triangle formed by New Zealand, Hawaii, and Easter Island

population pyramid (p. 15) a graph that depicts the age and gender structures of a country populist movements (p. 132) popularly based efforts, often seeking relief for the poor

primate city (p. 129) a city, plus its suburbs, that is vastly larger than all others in a country and in which economic and political activity is centered

privatization (pp. 116, 190) the selling of formerly government-owned industries and firms to private companies or individuals

purchasing power parity (PPP) (p. 20) the amount that the local currency equivalent of U.S.$1 will purchase in a given country

purdah (p. 309) the practice of concealing women from the eyes of nonfamily men

push/pull phenomenon of urbanization (p. 27) conditions, such as political instability, that encourage (push) people to leave rural areas, and urban factors, such as job opportunities, that encourage (pull) people to move to the urban area

Qur'an (or Koran) (p. 213) the holy book of Islam, believed by Muslims to contain the words Allah revealed to Muhammad through the archangel Gabriel

race (p. 53) a social or political construct that is based on apparent characteristics such as skin color, hair texture, and face and body shape, but that is of no biological significance

rate of natural increase (RNI) (p. 14) the rate of population growth measured as the excess of births over deaths per 1000 individuals per year without regard for the effects of migration

region (p. 7) a unit of the earth's surface that contains distinct patterns of physical features and/or of human development

regional conflict (p. 320) especially in South Asia, a conflict created by the resistance of a regional ethnic or religious minority to the authority of a national or state government

regional self-sufficiency (p. 349) an economic policy in Communist China that encouraged each region to develop independently in the hope of evening out the wide disparities in the national distribution of production and income

regional specialization (p. 350) the encouragement of specialization rather than self-sufficiency in order to take advantage of regional variations in climate, natural resources, and location

religious nationalism (p. 323) the association of a particular religion with a political unit

resettlement schemes (p. 392) government plans to move large numbers of people from one part of a country to another to relieve urban congestion, disperse political dissidents, or accomplish other social purposes; also called *transmigration*

responsibility system (p. 349) in the 1980s, a decentralization of economic decision making in China that returned agricultural decision making to the farm household level, subject to the approval of the commune

Ring of Fire (p. 44) the tectonic plate junctures around the edges of the Pacific Ocean; characterized by volcanoes and earthquakes

roaring forties (p. 412) powerful air and ocean currents at about 40° S latitude that speed around the far Southern Hemisphere virtually unimpeded by landmasses

Roma (p. 152) the now-preferred term in Europe for Gypsies

Russian Federation (p. 173) Russia and its political subunits, which include 30 internal republics and more than 10 so-called autonomous regions

Russification (p. 196) the assimilation of all minorities to Russian (Slavic) ways

Sahel (p. 252) a band of arid grassland, where steppe and savanna grasses grow, that runs east-west along the southern edge of the Sahara

salinization (p. 214) a process that occurs when large quantities of water are used to irrigate areas where evaporation rates are high, leaving behind dissolved salts and other minerals

scale (of a map) (p. 2) the proportion that relates the dimensions of the map to the dimensions of the area it represents; also, variable-sized units of geographical analysis from the local scale to the regional scale to the global scale

Schengen Accord (p. 163) an agreement signed in the 1990s by the European Union and many of its neighbors that called for free movement across common borders

seawater desalination (p. 216) the removal of salt from seawater—usually accomplished through the use of expensive and energy-intensive technologies—to make the water suitable for drinking or irrigating

secular states (p. 235) countries that have no state religion and in which religion has no direct influence on affairs of state or civil law

self-reliant development (p. 270) small-scale development schemes in rural areas that focus on developing local skills, creating local jobs, producing products or services for local consumption, and maintaining local control so that participants retain a sense of ownership

service sector (p. 82) economic activity that involves the sale of services

sex (p. 17) the biological category of male or female

sex tourism (p. 402) the sexual entertainment industry that services primarily men who travel for the purpose of living out their fantasies during a few weeks of vacation

shari'a (p. 223) literally, "the correct path"; Islamic religious law that guides daily life according to the interpretations of the Qur'an

Shi'ite (or Shi'a) (p. 223) the smaller of two major groups of Muslims, with different interpretations of shari'a; Shi'ites are found primarily in Iran and southern Iraq

shifting cultivation (pp. 111, 257) a productive system of agriculture in which small plots are cleared in forestlands, the dried brush is burned to release nutrients, and the clearings are planted with multiple species; each plot is used for only 2 or 3 years and then abandoned for many years of regrowth

Sikhism (p. 308) a religion of South Asia that combines beliefs of Islam and Hinduism

silt (p. 104) fine soil particles

Slavs (p. 183) a group of farmers who originated between the Dnieper and Vistula Rivers in modern-day Poland, Ukraine, and Belarus

slum (p. 27) densely populated area characterized by crowding, run-down housing, and poverty

social safety net (p. 79) the services provided by the government—such as welfare, unemployment benefits, and health care—that prevent people from falling into extreme poverty

social welfare (in the European Union, social protection) (p. 167) in Europe, tax-supported systems that provide citizens with benefits such as health care, pensions, and child care

South America (p. 101) the continent south of Central America

Soviet Union (p. 172) see **Union of Soviet Socialist Republics**

spatial distribution (p. 3) the arrangement of a phenomenon across the Earth's surface

spatial interaction (p. 3) the flow of goods, people, services, or information across space and among places

special economic zones (SEZs) (p. 350) free trade zones within China

state-aided market economy (p. 347) an economic system based on market principles such as private enterprise, profit incentives, and supply and demand, but with strong government guidance; in contrast to the free market (limited government) economic system of the United States and Europe

steppes (p. 173) semiarid, grass-covered plains

structural adjustment policies (SAPs) (p. 116) policies that require economic reorganization toward less govenment involvement in industry, agriculture, and social services; sometimes imposed by the World Bank and the International Monetary Fund as conditions for receiving loans

subcontinent (p. 289) a term often used to refer to the entire Indian peninsula, including Nepal, Bhutan, India, Pakistan, and Bangladesh

subduction zone (p. 101) a zone where one tectonic plate slides under another

subsidies (p. 159) monetary assistance granted by a government to an individual or group in support of an activity, such as farming, that is viewed as being in the public interest

subsistence affluence (p. 427) a lifestyle whereby self-sufficiency is achieved for most necessities, while some opportunities to earn cash allow for travel and occasional purchases of manufactured goods

subsistence agriculture (p. 254) farming that provides food for only the farmer's family and is usually done on small farms

subsistence economy (p. 16) an economy in which families produce most of their own food, clothing, and shelter

suburbs (p. 86) populated areas along the peripheries of cities

Sunni (p. 223) the larger of two major groups of Muslims, with different interpretations of shari'a

sustainable agriculture (p. 26) farming that meets human needs without poisoning the environment or using up water and soil resources

sustainable development (p. 22) improvement of standards of living in ways that will not jeopardize those of future generations

taiga (p. 176) subarctic forests

Taliban (p. 311) an archconservative Islamist movement that gained control of the government of Afghanistan in the mid-1990s

temperate midlatitude climate (p. 142) as in south-central North America, China, and Europe, a climate that is moist all year with relatively mild winters and long, mild to hot summers

temperature-altitude zones (p. 104) regions of the same latitude that vary in climate according to altitude

theocratic states (p. 235) countries that require all government leaders to subscribe to a state religion and all citizens to follow rules decreed by that religion

total fertility rate (TFR) (p. 15) the average number of children that women in a country are likely to have at the present rate of natural increase

trade deficit (p. 83) the extent to which the money earned by exports is exceeded by the money spent on imports

trade winds (p. 104) winds that blow from the northeast and the southeast toward the equator

tsunami (p. 332) a large sea wave caused by an earthquake

tundra (p. 176) a treeless area, between the ice cap and the tree line of arctic regions, where the subsoil is permanently frozen

typhoon (p. 334) a tropical cyclone or hurricane

underemployment (p. 190) the condition in which people are working too few hours to make a decent living or are highly trained but working at menial jobs

Union of Soviet Socialist Republics (USSR) (p. 172) the multinational union formed from the Russian empire in 1922 and dissolved in 1991

United Nations (UN) (p. 36) an assembly of 192 member states that sponsors programs and agencies that focus on scientific research, humanitarian aid, planning for development, fostering general health, and peacekeeping assistance

United Nations Human Development Index (HDI) (p. 20) the ranking of countries based on three indicators of well-being: life expectancy at birth, educational attainment, and income adjusted to purchasing power parity

urban sprawl (p. 65) the encroachment of suburbs on agricultural land

urbanization (p. 26) the movement of people from rural areas to cities

varna (p. 306) the four hierarchically ordered divisions of society in Hindu India underlying the caste system: *Brahmins* (priests), *Kshatriyas* (warriors/kings), *Vaishyas* (merchants/landowners), and *Sudras* (laborers/artisans)

veil (p. 226) the custom of covering the body with a loose dress and/or of covering the head—and in some places the face—with a scarf

virtual water (p. 37) the volume of water used to produce all that a person consumes in a year

water footprint (p. 37) the water used to meet a person's basic needs for a year, added to the person's annual virtual water

weathering (p. 44) the physical or chemical decomposition of rocks by sun, rain, snow, ice, and the effects of life-forms

welfare state (p. 151) a government that accepts responsibility for the well-being of its people, guaranteeing basic necessities such as education, employment, and health care for all citizens

West Bank barrier (p. 243) a 25-foot-high concrete wall in some places and a fence in others that now surrounds much of the West Bank and encompasses many of the Jewish settlements there

wet rice cultivation (p. 338) a prolific type of rice production that requires the submersion of the plant roots in water for part of the growing season

world region (p. 7) a part of the globe delineated according to criteria selected to facilitate the study of patterns particular to the area

World Trade Organization (WTO) (p. 32) a global institution made up of member countries whose stated mission is the lowering of trade barriers and the establishment of ground rules for international trade

Zionists (p. 240) those who have worked, and continue to work, to create a Jewish homeland (Zion) in Palestine

PHOTO CREDITS

Shankbone. 210 A: Bachmont/Flickr, B: Ben Boulaïd/Wikimedia, C: Ryckaert/Wikimedia, D: NASA/GSFC/METI/ERSDAC/JAROS, and U.S./Japan ASTER Science Team. 211 E: Arkady Chubykin/Fotolia.com, F: Bionet/Wikimedia, G: Bachmont/Flickr. 212 A: Pedronet/Flickr, B: Rouvin/Flickr, C: Clochard/Fotolia.com, D: Dagan/Fotolia.com. 215 A: SPC Timothy Kingston/DOD, B: Sgt. Paul L. Anstine II, USMC/DOD, C: Barber/USAID, D: Font de Matas/IRIN. 217 A: TekinT/Fotolia.com, B: Claire Mc Evoy/IRIN, C: Timothy McKulka/UNMIS/IRIN. 219 A: Patrick Syder/Lonely Planet Images/Getty Images. 222 A: Ellaine Dickinson/Flickr, B: Jeffery Beggerly/Wikimedia, C: Amedeo Preziosi/Wikimedia. 223 D: Detroit Photographic Company/Library of Congress, E: Wikipedia, F: R. D. Ward/DOD. 226 A: Barber/USAID, B: Kitkatcrazy/Wikimedia, C: Keller/Flickr, D: Jay Galvin/Flickr. 230 A: Haider Yousuf/Fotolia.com, B: Neil Alejandro/Flickr, C: Andrew Currie/Flickr. 236 A: Jim Gordon/DOD, E: Ronald Shaw/U.S. Army. 237 B: Paul Keller/Flickr, C: Timothy McKulka/UNMIS /IRIN, D: Jim Gordon/Wikipedia. 242 A: Briony Balsom/IRIN, B: Apollo Images/IRIN, C: James Emery/Flickr, D: Amos Gil/Wikimedia. 244 A: Hsing Wei/Flickr, B: Morello/USAID, C: Tanya Habjouga/USAID.

CHAPTER 7

247 A: Ridley/IRIN, B: Derek Sciba/World Concern, C: Jean-François Dontaine/FAO, D: Anna Jefferys/IRIN, E: Richard/Flickr, F: Ben Parker/IRIN, G: Siegfried Modola/IRIN, H: David Gough/IRIN, I: Save the Children UK/IRIN. 248 IRIN. 250 A: Stefanie van der Vinden/Fotolia.com, B: Paul Hampton/Fotolia.com. 251 C: Robert Hardholt/Fotolia.com, D: Terry Whalebone/Flickr, E: Andres de Wet/Wikimedia. 253 A: Axel Rouvin/Flickr, B: Genvessel/Flickr, C: Mistress_f/Flickr, D: NASA. 255 A: WRI/Flickr, B: WRI/Flickr, C: Marc Allart/Fotolia.com, D: CPWF Basin Focal Project/Flickr. 256 A: Manoocher Deghati/IRIN, B: Federico Rizzato/Fotolia.com, C: Curt Reynolds/USDA, D: Shared Interest/Flickr. 260 A: Amédée Forestier/Wikimedia, B: Ctsnow/Flickr, C: Heinrich Barth/Wikimedia. 261 D: William Rednbacher/Wikimedia, E: Library of Congress, F: Paul Birnie/Flickr, Bottom: G. Cranston/IRIN. 266 Sarah Simpson/IRIN. 268 Julius Mwelu/IRIN. 269 Lillian Liang/IRIN. 270 Ken Wiegand/USAID. 272 AP Photo/Khalil Senosi. 274 A: Julius Mwelu/IRIN, B: R. D. Ward/DOD, C: Tiggy Ridley/IRIN, D: Antony Kaminju/IRIN. 278 A: Nicholas Reader/IRIN, B: Manoocher Deghati/IRIN, C: Killingsworth/Flickr, D: Julius Mwelu/IRIN. 284 A: Tiggy Ridley/IRIN, B: USAID, C: Zahur Ramji/AKDN.

CHAPTER 8

287 A: Abdul Majeed Goraya/IRIN, B: IRIN, C: Jon Hurd/Flickr, D: Library of Congress, E: Manoocher Deghati/IRIN, F: Kate Holt/IRIN, G: Manoocher Deghati/IRIN, H: Niyam Bhushan/Flickr, I: Soman/Wikimedia. 288 A: TMAX/Fotolia.com, B: AceFighter19/Wikipedia, C: Mira Murphy/Flickr. 290 A: Wildxplorer/Flickr, B: NASA. 291 C: Arnim Schulz/Fotolia.com, D: K.C. Bimal/Flickr, E: Kppethe/Wikipedia, F: Mark Bold/Flickr. 294 A: Subharnab Majumdar/Flickr, B: Prakhar Amba/Flickr, C: Honza Soukup/Flickr, D: NASA. 297 A: IRRI, B: Ben Barber/USAID, C: Mohit Gupta/Flickr, D: Robin Murphy, World Resources Institute/Flickr. 299 A: Kamila Hyat/IRIN, B: Dylan Walters/Flickr, C: Ville Miettinen/Flickr, D: Chubby Chandru/Flickr. 304 A: Comrogues/Flickr, B: Getty Images. 305 C: Library of Congress, D: Wikimedia, E: Steve Evans/Flickr. 307 A: Ted Ollikkala/Flickr, B: Judith/Flickr, C: Whitney Lauren/Flickr. 314 A: WL/Flickr, B: Manoocher Deghati/IRIN, C: Niyam Bhushan/Flickr, D: Manoocher Deghati/IRIN. 321 A: Spc. Jean-Paul G. Li/DOD, B: Amantha Perera/IRIN, C: David Swanson/IRIN, D: Cyril Bèle/Flickr. 324 A: McKay Savage/Flickr, B: McKay Savage/Wikimedia, C: Frank Hebbert/Flickr.

CHAPTER 9

327 A: Richard Doolin/USN B: Gustavo Madico/Flickr, C: Phillip Capper/Flickr, D: Taro Taylor/Flickr, E: Britrob/Flickr, F: Jakob Montrasio/Flickr, G: Kimubert/Flickr, H: Ajari/Flickr, I: UNC-CFC-USFK/Flickr. 328 A: Rduta/Flickr, B: Steve Jurvetson/Flickr. 330 A: McKay Savage/Flickr. 331 B: John Pannell/Flickr, C. Staudamm/Fotolia.com, D: J Aaron Farr/Flickr, E: Kazuhiko Teramoto/Flickr. 333 A: Bernt Rostad/Flickr, B: Philip Poon/Flickr, C: Tiarescott/Flickr,

D: Sarcozona/Flickr. 335 A: Robert Thomson/Flickr, B: Dan Zen/Flickr, C: Paul Mannix/Flickr, D: Ian A. Inman/Wikipedia. 339 A: Staudamm/Fotolia.com, B: Jorge Hernández Valiñani/Flickr, C: Herry Lawford/Flickr, D: NASA-GSFC. 345 Library of Congress. 346 A: Gisling/Wikimedia, B: Bernt Rostad/Flickr, C: Marianna/Flickr. 347 D: Library of Congress, E: Wikimedia, F: Flowizm/Flickr. 348 Fotorin/Fotolia.com. 351 A: Augapfel/Flickr, B: SSgt Stacy Pearsall, USAF/DOD, C: Tom Booth/Flickr, D: Yasuhiro/Flickr. 356 A: Ryanne/Flickr, B: MCSN Adam York, USN/DOD, C: CT Snow/Flickr, D: Mark Zastrow/Flickr. 358 A: Isabelle Plante/Flickr, B: Scott Chang/Flickr. 363 Gustavo Jeronimo/Flickr. 365 A: Lycoo/Wikimedia, B: Robert Scoble/Flickr, C: David Yan/Flickr.

CHAPTER 10

369 A: H Dragon/Flickr, B: Chris Quintana/IRRI, C: Steve Jurvetson/Flickr, D: Bruce Tuten/Flickr, E: Jakrapong Kongmalai/Flickr, F: Pittaya Sroilong/Flickr, G: Chong Eileen/Flickr, H: IRRI, I: Ryan Lackey/Flickr. 370 A: Tajai/Flickr, B: Bruno Manser Fonds. 372 A: David/Flickr. 373 B: Mat Honan/Flickr, C: Angelo Juan Ramos/Flickr, D: Vberger/Wikimedia, E: PM2 Philip A. McDaniel, USN/DOD. 376 A: Mari.francille/Flickr, B: Laughlin Elkind/Flickr, C: Eric Molina/Flickr. 377 A: Manfred Mielke/USDA Forest Service, B: John J. Mosesso/National Biological Information Infrastructure, C: Manfred Mielke/USDA Forest Service. 378 A: Shariff Che'Lah/Fotolia.com, B: Neil Liddle/Flickr, C: Jesse Allen/NASA, D: Shubert Ciencia/Flickr. 380 A: J.F. Perigois/Fotolia.com, B: How Sia/Fotolia.com, C: Mattmangum/Flickr, D: IRRI. 382 A: IRRI, B: Allie Caulfield/Flickr, C: NOAA, D: IRRI. 385 A: Nicolaas Pieneman/Wikipedia, B: Wikipedia, C: Library of Congress. 386 A: Dan Bennett/Flickr, B: Nicholas Kenrick/Wikimedia, C: Magalhães/Wikimedia. 387 D: Wikimedia, E: USAF/Wikimedia, F: Adam Fletcher/Flickr. 388 Emilio Labrador/Flickr. 392 A: Sam Garza/Flickr, B: Fotolia.com. 394 A: Mohd Hafiz Noor Sham/Wikimedia, B: McIntosh/Wikimedia, C: Jason Gutierrez/IRIN, D: Alexis Lê-Quôc/Flickr. 398 A: Jonathan McIntosh/Wikimedia, B: Terence Ong/Flickr, C: Emilio Labrador/Flickr, D: Cubby Chandru/Flickr. 399 Paul Keller/Flickr. 401 A: Wid Sulistio/Flickr, B: Simon Monk/Flickr, C: IRRI. 402 Amrufm/Flickr. 404 A: Christian Haugen/Flickr, B: National Information Agency Republic of Indonesia/Wikimedia, C: Beth Kanter/Cambodia4Kids.org.

CHAPTER 11

407 A: Steve Evans/Flickr, B & C: Phillip Capper/Flickr, D: Mark Smith/Flickr, E: Tim in Sydney/Flickr, F: FearlessRich/Flickr, G: Alex Kehr/Flickr, H: Forest & Kim Starr/Wikimedia, I: Taro Taylor/Flickr. 408 A: Jessica Rabbit/Flickr, B: Erik Hersman/Flickr. 410 A: Safaris/Flickr, B: Mapu/Flickr, C: Mermoz/Fotolia.com, D: Henrickson/Wikipedia, E: Dropu/Fotolia.com, F: Alexander Edmonds/Fotolia.com. 413 A: Roger Blake/Flickr, B: Austin/Flickr, C: Kiwi Flickr/Flickr. 417 A: Library of Congress, B: Chen Wu/Flickr, C & D: Phillip Capper/Flickr. 418 A: Stefan Lins/Flickr, B: Andrea Booher/FEMA News Photo, C: Peter Halasz/Wikipedia, D: Tim Parkinson/Flickr. 420 A–D: Ok Tedi Mine CMCA Review. 423 A: Panvorax/Wikimedia, B: MCS2 Nathaniel J. Karl/USN, C: Stephen Michael Barnett/Flickr, D: Andrew Turner/Flickr. 424 A: Steve Evans/Wikimedia, B: HongKongHuey/Wikimedia, C: George Carter/Wikimedia. 425 D: Library of Congress, E: Wikimedia, F: Michael A. Lantron/USN. 430 A: Bkamprath/Wikimedia, B: Sandy Austin/Flickr, C: Stefan Lins/Flickr, D: Alex Kehr/Flickr. 433 Left: Jon Connell/Flickr, Right: Feral Arts/Flickr. 434 A: Michaelfield/Flickr, B: Global Integrity/Flickr, C: Photo by the Salvation Army from www.salvationist.org, D: David Jackmanson/Flickr. 436 A: New South Wales Government Printer, B: Mark Scott Johnson/Flickr. 437 A: Alex Kehr/Flickr, B: Cascade_of_rant/Flickr, C: Karl Baron/Flickr.

EPILOGUE

439 Jason Auch/Flickr
439 Pablo Fernando Filippo/Forum for the Conservation of the Patagonian Sea and Areas of Influence
439 Emily Stone/U.S. Antarctic Program/National Science Foundation
439 PO2 John K. Sokolowski, USN/DOD.

FIGURE CREDITS

CHAPTER 1

2 Fig 1.2: Courtesy of Julia Stump. **6** Fig 1.4: Adapted from *The New Comparative World Atlas* (Maplewood, NJ: Hammond, 1997), p. 6. **6** Fig 1.5: Mercator and Robinson projections adapted from *The New Comparative World Atlas* (Maplewood, NJ: Hammond, 1997), pp. 6–7. Goode's interrupted homolosine projection adapted from *Goode's World Atlas*, 19th ed. (Chicago: Rand McNally, 1995), p. 13. **9** Fig 1.7: Adapted from G. Tyler Miller, Jr., *Living in the Environment*, 8th ed. (Belmont, CA: Wadsworth, 1994), p. 4. **15** Fig 1.9: Adapted from U.S. Census Bureau, Population Division, "Population Pyramids for Austria," and "Population Pyramids for Jordan," International Data Base, May 2009, at http://www.census.gov/ipc/www/idb/country.php. **16** Fig 1.10: Adapted from Joni Seager, *The State of Women in the World Atlas* (New York: Penguin, 1997), p. 35; and updated from the Department of Economic and Social Affairs of the United Nations Secretariat, Population Division, *World Population Prospects: The 2008 Revision*, at http://esa.un.org/unpp/p2k0data.asp. **17** Fig 1.11: Adapted from G. Tyler Miller, Jr., *Living in the Environment*, 8th ed. (Belmont, CA.: Wadsworth, 1994), p. 218. **18** Table 1.1: Source: *United Nations Human Development Report*, 2007–2008 (New York: United Nations Development Programme), Table 28, "Gender-Related Development Index," pp. 326–329, and Table 29, "Gender Inequality in Education," pp. 330–333. **21** Fig 1.12: Maps adapted from "Human development indices," (Maps A and B): Table 2, pp. 28–32; (Map C): Table 5, pages 41–44, at http://hdr.undp.org/en/media/HDI_2008_EN_Tables.pdf. **24** Photo Essay 1.1: Adapted from United Nations Environment Programme, "Human impact, year 1700 (approximately)," and "Human impact, year 2002," (New York: United Nations Development Program), 2002, 2003, 2004, 2005, 2006, at http://maps.grida.no/go/graphic/ human_impact_year_1700_approximately and http://maps.grida.no/go/graphic/human_impact_year_2002. **26** Fig 1.13: Adapted from the Food and Agriculture Organization of the United Nations, FAO Statistics Division, Rome 2009: Map 14, Year 2003–2005, at http://www.fao.org/economic/ess/publications-studies/statistical-yearbook/fao-statistical-yearbook-2007-2008/g-human-welfare/en/. **37** Table 1.2: Source: Arjen Y. Hoekstra and Ashok K. Chapagain, *Globalization of Water—Sharing the Planet's Freshwater Resources* (Malden, MA: Blackwell, 2008), p. 15, Table 2.2. **38** Fig 1.18: *United Nations World Water Development Report 2: Water—A Shared Responsibility*. Published jointly in 2006 by the UN Educational, Scientific and Cultural Organization (UNESCO), Paris; and Berghahn Books, New York, pp. 391–392. **40** Fig 1.20: Adapted from the UN Department of Economic and Social Affairs, Statistics Division, *Environmental Indicators: Greenhouse Gas Emissions*, 2009, at http://unstats.un.org/unsd/environment/air_greenhouse_emissions.htm. **45** Fig 1.21: Adapted from Frank Press, Raymond Siever, John Grotzinger, and Thomas H. Jordan, *Understanding Earth*, 4th ed. (New York: W. H. Freeman, 2004), pp. 42–43. **46** Fig 1.22: Adapted from United States Geological Survey, *Active Volcanoes and Plate Tectonics*, "Hot Spots" and the "Ring of Fire," at http://vulcan.wr.usgs.gov/Glossary/PlateTectonics/Maps/map_plate_tectonics_world.html; and Frank Press, Raymond Siever, John Grotzinger, and Thomas H. Jordan, *Understanding Earth*, 4th ed. (New York: W. H. Freeman, 2004), p. 27. **47** Fig 1.23: (A) Adapted from Frank Press, Raymond Siever, John Grotzinger, and Thomas H. Jordan, *Understanding Earth*, 4th ed. (New York: W. H. Freeman, 2004), p. 281. **53** Fig 1.25: Map courtesy of UNEP/GRID-Arendal at http://maps.grida.no/go/graphic/skin-colour-map-indigenous-people, Emmanuelle Bournay, cartographer. Data source: G. Chaplin, "Geographic Distribution of Environmental Factors Influencing Human Skin Coloration," *American Journal of Physical Anthropology*, 125, 292–302, 2004; map updated in 2007.

CHAPTER 2

62 Fig 2.4: Map: USGS/National Wetlands Research Center. **69** Fig 2.6: Graphic adapted from *National Geographic* (March 1993): 84–85, with supplemental information from High Plains Underground Water Conservation District 1, Lubbock, Texas, at http://www.hpwd.com; and Erin O'Brian, Biological and Agricultural Engineering, National Science Foundation Research Experience for Undergraduates, Kansas State University, 2001. **71** Fig 2.7: U.S. Census Bureau, "Percent of the Total Population Who Are Black or African-American Alone: 2007," *American Community Survey 2005–2007*, M0202. **72** Fig 2.8: Adapted from James A. Henretta, W. Elliot Brownlee, David Brody, and Susan Ware,

America's History, 2nd ed. (New York: Worth, 1993), pp. 400–401; and James L. Roark, Michael P. Johnson, Patricia Cline Cohen, Sarah Stage, Alan Lawson, and Suan M. Hartmann, *The American Promise: A History of the United States*, 3rd ed. (Boston: Bedford/St. Martin's, 2005), p. 601. **75** Fig 2.9: Data source: U.S. Energy Information Administration, "Crude Oil and Total Petroleum Imports Top 15 Countries," at http://www.eia.doe.gov/pub/oil_gas/petroleum/data_publications/company_level_imports/current/import.html. **79** Fig 2.10: Adapted from *National Geographic* (February 1990): 106–107, and augmented with data from the Office of Travel and Tourism Industries, at http://tinet.ita.doc.gov/outreachpages/inbound.country_in_north_america.canada.html; http://internationaltrade.suite101.com/article.cfm/canadas_top_exports_imports; http://import-export.suite101.com/article.cfm/canadian_imports_exports_2008; and http://www.bea.gov/newsreleases/international/fdi/2009/pdf/fdi08.pdf; http://www.thaindian.com/newsportal/business/canada-gets-record-fdi-since-tech-bubble_10045881.html; http://www.thaindian.com/newsportal/business/canadian-companies-on-buying-spree-in-us_100281211.html; http://canadaonline.about.com/gi/o.htm?zi=1/XJ/Ya&zTi=1&sdn=canadaonline&cdn=newsissues&tm=20&f=10&tt=14&bt=1&bts=1&zu=http%3A//www.ec.gc.ca/acidrain/acidfact.html; http://en.wikipedia.org/wiki/Immigration_to_the_United_States#Origin; http://www.answerbag.com/q_view/958850; Table 2.1: *Sources: United Nations Human Development Report 2007–2008* (New York: United Nations Development Programme), Table 6 and Table 10; "Crude Death Rate (per 1,000 Population)," *United Nations World Population Prospects: The 2006 Revision*, at http://data.un.org/Data.aspx?q=world+population&d=PopDiv&f=variableID%3A65%3BcrID%3A900; and *Income, Poverty, and Health Insurance Coverage in the United States: 2007*, U.S. Census Bureau, p. 19, at http://www.census.gov/prod/2008pubs/p60-235.pdf. **81** Fig 2.11: Adapted from Arthur Getis and Judith Getis, eds., *The United States and Canada: The Land and the People* (Dubuque, IA: William C. Brown, 1995), p. 165. **84** Fig 2.13: Adapted from Walmart Corporate, *International Operations Data Sheet—July 2009*, at http://walmartstores.com/FactsNews/NewsRoom/9350.aspx. **89** Fig 2.15: U.S. Census Bureau, "Percent of People Who Are Foreign Born: 2007," *American Community Survey, 2005–2007*, M0501. **91** Fig 2.16: Adapted from Jorge del Pinal and Audrey Singer, "Generations of Diversity: Latinos in the United States," *Population Bulletin*, 52 (October 1997): 14; and U.S. Census Bureau, "Race by Sex, for the United States, Urban and Rural, 1950, and for the United States, 1850 to 1940," *Census of Population: 1950*, Vol. 2, Pt. 1, United States Summary, Table 36, 1953, at http://www2.census.gov/prod2/decennial/documents/21983999v2p1ch3.pdf; Fig 2.17: U.S. Census Bureau, Income, Earnings, and Poverty Data, *2007 American Community Survey*, 2008, at http://www.census.gov/prod/2008pubs/acs-09.pdf. **92** Fig 2.18: Adapted from Jerome Fellmann, Arthur Getis, and Judith Getis, *Human Geography* (Dubuque, IA: Brown & Benchmark, 1997), p. 164. **93** Fig 2.19: Data source: U.S. Census Bureau, "Households and Persons per Household by Type of Household: 1990 to 2008," Population: Households, Families, Group Quarters, Table 61, 2010, at http://www.census.gov/compendia/statab/cats/population/households_families_group_quarters.html; Fig 2.20: Data sources: (A) U.S. Census Bureau, "Money Income of Households—Distribution by Income Level and Selected Characteristics: 2007," Income, Expenditures, Poverty, & Wealth: Household Income, Table 676, 2010, at http://www.census.gov/compendia/statab/cats/income_expenditures_poverty_wealth/household_income.html; (B) U.S. Census Bureau, "Mean Earnings by Highest Degree Earned: 2007," Education: Educational Attainment, Table 227, 2010, at http://www.census.gov/compendia/statab/cats/education/educational_attainment.html. **96** Fig 2.22: Adapted from U.S. Census Bureau, Population Division, "Population Pyramids of the United States" and "Population Pyramids of Canada," International Data Base, 2009, at http://www.census.gov/ipc/www/idb/pyramids.html.

CHAPTER 3

106 Fig 3.4: Illustration by Tomo Narashima, based on fieldwork and a drawing by Lydia Pulsipher. **112** Fig 3.5: Adapted from *Hammond Times Concise Atlas of World History* (Maplewood, NJ: Hammond, 1994), pp. 66–67. **113** Fig 3.6: Adapted from *Hammond Times Concise Atlas of World History* (Maplewood, NJ: Hammond, 1994), p. 69. **117** Fig 3.7: Adapted from *United Nations Human Development Report 2005* and *2007–2008* (New York: United Nations

Development Programme), Tables 20 and 18, respectively. **120** Fig 3.9: Adapted from *Goode's World Atlas*, 21st ed. (Chicago: Rand McNally, 2005), p. 137, Minerals and Economic map. **123** Fig 3.10: Adapted from *2008 World Drug Report* (Vienna: United Nations Office on Drugs and Crime, 2005), pp. 13, 15, at http://www.unodc.org/documents/wdr/WDR_2008/WDR_2008_eng_web.pdf. **125** Fig 3.11: Sources: "Internet Usage Statistics—The Big Picture" and "Internet Usage Statistics for the Americas," at http://www.internetworldstats.com/stats.htm, http://www.internetworldstats.com/stats10.htm, and http://www.internetworldstats.com/stats11.htm. **127** Fig 3.13: Adapted from *United Nations Human Development Report 2009* (New York: United Nations Development Programme), Table 5. **129** Fig 3.15: From *Yearbook of the Association of Pacific Coast Geographers* 57(28): 1995; printed with permission.

CHAPTER 4

146 Fig 4.4: Adapted from a NASA image created by Jesse Allen, Earth Observatory, using data provided courtesy of the SeaWiFS Project, NASA/Goddard Space Flight Center and ORBIMAGE, available at http://earthobservatory.nasa.gov/Newsroom/NewImages/images.php3?img_id517332. **150** Fig 4.5: Adapted from Alan Thomas, *Third World Atlas* (Washington, D.C.: Taylor & Francis, 1994), p. 29. **157** Fig 4.6: Source: *Europe in Figures: Eurostat Yearbook 2009* (Eurostat Statistical Books, Luxembourg: Office for Official Publications of the European Commission, 2009), pp. 388 and 392, Figures 10.5, 10.6, 10.9, and 10.10, at http://epp.eurostat.ec.europa.eu/cache/ITY_OFFPUB/KS-CD-09-001/EN/KS-CD-09-001-EN.PDF. **158** Fig 4.7: Source: *United Nations Human Development Report 2009* (New York: United Nations Development Programme), Table H, at http://hdr.undp.org/en/reports/global/hdr2009/. **164** Fig 4.10: Adapted from "Population Pyramids of Germany" and "Population Pyramids of Sweden," U.S. Census Bureau International Data Base, 2009, at http://www.census.gov/ipc/www/idb/country.php; and from *Europe in Figures: Eurostat Yearbook 2009* (Luxembourg: Eurostat, 2009), p. 139, Figure 3.5, at http://epp.eurostat.ec.europa.eu/cache/ITY_OFFPUB/KS-CD-09-001/EN/KS-CD-09-001-EN.PDF. **167** Fig 4.13: Source: *Europe in Figures: Eurostat Yearbook 2009* (Luxembourg: Eurostat, 2009), p. 27, Figure 7.3: Employment Rate by Gender, 2007, at http://epp.eurostat.ec.europa.eu/cache/ITY_OFFPUB/KS-CD-09-001-03/EN/KS-CD-09-001-03-EN.PDF.

CHAPTER 5

180 Fig 5.4: Adapted from *National Geographic*, February 1990, pp. 72, 80–81; and the NASA Earth Observatory, at http://earthobservatory.nasa.gov/IOTD/view.php?id=9036. **183** Fig 5.5: Adapted from "The Silk Road and Related Trade Routes" map, the Silk Road Project and the Stanford Program on International and Cross-Cultural Education, at http://www.silkroadproject.org/tabid/177/default.aspx. **185** Fig 5.7: Adapted from Robin Milner-Gulland with Nikolai Dejevsky, *Cultural Atlas of Russia and the Former Soviet Union*, rev. ed. (New York: Checkmark Books, 1998), pp. 56, 74, 128–129, 177. **187** Fig 5.8: Adapted from Clevelander, at http://en.wikipedia.org/wiki/Image:New_Cold_War_Map_1980.png. **189** Fig 5.9: Map adapted from U.S. Department of Energy, Energy Information Administration, "Russia Country Analysis Brief," May 2008, at http://www.eia.doe.gov/emeu/cabs/Russia/images/Russian%20Energy%20at%20a%20Glance%202007.pdf. **191** Fig 5.11: Adapted from Robin Milner-Gulland with Nikolai Dejevsky, *Cultural Atlas of Russia and the Former Soviet Union*, rev. ed. (New York: Checkmark Books, 1998), pp. 186–187, 198–199, 204–205, 216–217; and http://www.travelcenter.com.au/russia/images/trans-sib-map-v3.jpg. **193** Fig 5.13: Adapted from Robin Milner-Gulland with Nikolai Dejevsky, *Cultural Atlas of Russia and the Former Soviet Union*, rev. ed. (New York: Checkmark Books, 1998), pp. 186–187, 198–199, 204–205, 216–217. **194** Fig 5.15: Source: Olga Kryshtanovskaya, "The Making of a Neo-KGB State," The *Economist*, August 23, 2007, at http://www.economist.com/world/displaystory.cfm?story_id=E1_JGRNGNT. **197** Fig 5.16: Adapted from James H. Bater, *Russia and the Post-Soviet Scene* (London: Arnold, 1996), pp. 280–281; Graham Smith, *The Post Soviet States* (London: Arnold, 2000), p. 75; and *The World Factbook 2009*, at https://www.cia.gov/library/publications/the-world-factbook/. **201** Fig 5.18: Adapted from U.S. Census Bureau, Population Division, "Population Pyramids of Russia," "Population Pyramids of Belarus," "Population Pyramids of Kyrgyzstan," and "Population Pyramids of Kazakhstan," International Data Base, 2009, at http://www.census.gov/ipc/www/idb/pyramids.html. **202** Fig 5.20: Adapted from *United Nations Human Development Report 2009* (New York: United Nations Development Programme), Gender Empowerment Measure, Table K.

CHAPTER 6

213 Fig 6.4: Adapted from United Nations Environmental Programme, *Vital Water Graphics—An Overview of the State of the World's Fresh and Marine Waters*, 2nd ed., 2008, available at http://www.unep.org/dewa/vitalwater/article69.html. **218** Fig 6.6: Adapted from United Nations Environment Programme, *Vital Water Graphics: Problems Related to Freshwater Resources*, "Turning the Tides" map, at http://www.unep.org/dewa/assessments/ecosystems/water/vitalwater/22.htm. **219** Fig 6.7: Map adapted from Bruce Smith, *The Emergence of Agriculture* (New York: Scientific American Library, 1995), p. 50. **221** Fig 6.8: Adapted from Richard Overy, ed., *The Times History of the World* (London: Times Books, 1999), pp. 98–99. **224** Fig 6.9: (A) Adapted from *Hammond Times Concise Atlas of World History* (Maplewood, N.J.: Hammond, 1994), pp. 100–101. (B) Adapted from *Rand McNally Historical Atlas of the World* (Chicago: Rand McNally, 1965), pp. 36–37; and *Cultural Atlas of Africa* (New York: Checkmark Books, 1988), p. 59. **225** Fig 6.10: Adapted from Joni Seager, *The Penguin Atlas of Women in the World*, 3rd ed. (New York: Viking Penguin, 2003), pp. 14–15. **228** Fig 6.13: Adapted from Joni Seager, *Women in the World: An International Atlas* (New York: Viking Penguin, 1997), pp. 66–67; and *United Nations Human Development Report 2007–2008* (New York: United Nations Development Programme), Table 31. **229** Fig 6.14: Adapted from U.S. Census Bureau, Population Division, "Population Pyramids for Iran," "Population Pyramids for Qatar," and "Population Pyramids for Israel," International Data Base, May 2009, at http://www.census.gov/ipc/www/idb/country.php. **233** Fig 6.15: Adapted from U.S. Department of Energy, Country Analysis Briefs, at http://www.eia.doe.gov/emeu/cabs/Region_me.html; U.S. Department of Energy, "Selected Oil and Gas Pipeline Infrastructure in the Middle East," at http://www.eia.doe.gov/cabs/Saudi_Arabia/images/Oil%20and%20Gas%20Infrastructue%20Persian%20Gulf%20(large)%20(2).gif; and "Who Has the Oil," available at http://gcaptain.com/maritime/blog/who-has-the-oil-a-map-of-world-oil-reserves/. **234** Fig 6.16: Adapted from Organization of the Petroleum Exporting Countries, *Annual Statistical Bulletin 2008*, 2009. **235** Fig 6.17: Adapted from *United Nations Human Development Report 2007–2008* (New York: United Nations Development Programme), Table 19, at http://hdr.undp.org/en/media/HDR_20072008_EN_Complete.pdf. **240** Table 6.1: Source: *United Nations Human Development Report 2005* (New York: United Nations Development Programme). **241** Fig 6.18: Adapted from Colbert C. Held, *Middle East Patterns—Places, Peoples, and Politics* (Boulder, CO: Westview Press, 1994), p. 184; Fig 6.19: Adapted from Colbert C. Held, *Middle East Patterns—Places, Peoples, and Politics* (Boulder, CO: Westview Press, 1994), p. 184; and the Israeli Settlements in the Occupied Territories, 2002, Foundation for Middle East Peace, March 2002, http://www.firstpr.com.au/nations.

CHAPTER 7

248 Fig 7.3: Export data is from GlobalTimber.org.uk, "Exports by, and Imports from, Africa," at http://www.globaltimber.org.uk/africa.htm; Ikea standards are available at http://www.ikea.com/ms/en_CA/about_ikea/pdf/IWAY_marketing_standard.pdf. **260** Fig 7.5: Adapted from the work of Joseph H. Harris, in Monica B. Visona et al., *A History of Art in Africa* (New York: Harry N. Abrams, 2001), pp. 502–503. **262** Fig 7.7: Adapted from Alan Thomas, *Third World Atlas* (Washington, D.C.: Taylor & Francis, 1994), p. 43. **265** Fig 7.8: Adapted from George Kurian, ed., *Atlas of the Third World* (New York: Facts on File, 1992), and the Central Intelligence Agency, The *World Factbook*, at https://www.cia.gov/library/publications/the-world-factbook/fields/2090.html?countryName=Angola&countryCode=ao&regionCode=af&#ao. **267** Fig 7.10: Debt and trade data from the Central Intelligence Agency, The *World Factbook 2010*, at https://www.cia.gov/cia/publications/factbook/index.html. **268** Fig 7.11: Source: Quote from IRIN, Lucy Mugure: "*I Do Not Want My Children to Bring Up Their Children Here*," July 2007, at http://www.irinnews.org/HOVReport.aspx?ReportId=73094.]; Fig 7.12: Source: UN Conference on Trade and Development, *Economic Development in Africa Report 2009: Strengthening Regional Economic Integration for Africa's Development* (New York and Geneva: United Nations, 2009), pp. 41–42, 76, at http://www.unctad.org/en/docs/aldcafrica2009_en.pdf. **271** Fig 7.15: Adapted from James M. Rubenstein, *An Introduction to Human Geography* (Upper Saddle River, N.J.: Prentice Hall, 1999), p. 246. **275** Fig 7.17: Data courtesy of Deborah Balk, Gregory Yetman, et al., Center for International Earth Science Information Network, Columbia University, at http://www.ciesin.columbia.edu. **276** Fig 7.18: Adapted from

U.S. Census Bureau, Population Division, "Population Pyramids of Nigeria" and "Population Pyramids of South Africa," International Data Base, 2009, at http://www.census.gov/ipc/www/idb/pyramids.html, and Population Reference Bureau, *2009 World Population Data Sheet*. **277** Fig 7.19: Adapted from the World Health Organization, *The Global Burden of Disease 2004 Update* (Geneva: World Health Organization Press, 2008), available at http://www.who.int/healthinfo/global_burden_disease/2004_report_update/en/index.html. **279** Fig 7.20: Adapted from UNAIDS, *2008 Report on the Global AIDS Epidemic*, at http://www.searo.who.int/LinkFiles/Facts_and_Figures_global_HIV_epidemiology.pdf.] **281** Fig 7.21: Adapted from the World Health Organization, *Eliminating Female Genital Mutilation—An Interagency Statement* (Geneva: World Health Organization Press, 2008), p. 5, at http://whqlibdoc.who.int/publications/2008/9789241596442_eng.pdf. **282** Fig 7.22: Maps adapted from Matthew White, "Religions in Africa," in *Historical Atlas of the Twentieth Century* (October 1998), at http://users.erols.com/mwhite28/afrorelg.htm; revised with new data from the CIA, *The World Factbook*, 2009, at https://www.cia.gov/cia/publications/factbook/index.html.

CHAPTER 8

295 Fig 8.5: Adapted from *National Geographic*, June 1993, p. 125. **301** Fig 8.6: Adapted from Gordon Johnson, *Cultural Atlas of India* (New York: Facts on File, 1996), p. 158. **302** Fig 8.7: Data from the Ministry of External Affairs, Non Resident Indians & Persons of Indian Origin Division, *Report of the High Level Committee on The Indian Diaspora*, at http://indiandiaspora.nic.in/contents.htm; specifically, "Estimated Size of Overseas Community: Country-Wise," at http://indiandiaspora.nic.in/diasporapdf/part1-est.pdf. **303** Fig 8.8: Adapted from *National Geographic*, May 1997, p. 18. **306** Fig 8.9: Adapted from Alisdair Rogers, ed., *Peoples and Cultures* (New York: Oxford University Press, 1992), p. 204. **308** Fig 8.11: Adapted from Gordon Johnson, *Cultural Atlas of India* (New York: Facts on File, 1996), p. 56. **310** Fig 8.12: Adapted from "District Wise Female Literacy Rate of India," Maps of India, at http://www.mapsofindia.com/census2001/femaleliteracydistrictwise.htm; and *United Nations Human Development Report 2009* (New York: United Nations Development Programme), Table J, "Gender-Related Development Index and Its Components," at http://hdr.undp.org/en/reports/global/hdr2009/. **313** Fig 8.14: Adapted from U.S. Census Bureau, Population Division, "Population Pyramids for Pakistan" and "Population Pyramids for Sri Lanka" (Washington, D.C.: U.S. Census Bureau, International Data Base), at http://www.census.gov/ipc/www/idb/country.php. **316** Fig 8.16: Adapted from John Dixon and Aidan Gulliver with David Gibbon, *Farming Systems and Poverty: Improving Farmers' Livelihoods in a Changing World* (Rome and Washington, DC: FAO and World Bank, 2001), at http://www.fao.org/farmingsystems/FarmingMaps/SAS/01/FS/index.html. **319** Fig 8.17: Data from the Directorate of Economics & Statistics of each of the respective state governments, 2006; GDP data from the Reserve Bank of India, Table 8, "Per Capita Net State Domestic Product at Factor Cost—State-Wise," at http://rbidocs.rbi.org.in/rdocs/Publications/PDFs/8T_HB150909.pdf.

CHAPTER 9

329 Table 9.1: Adapted from Cai Fang, Du Yang, and Wang Meiyan, *Human Development Research Paper 2009/09: Migration and Labor Mobility in China* (United Nations Development Programme, April 2009), p. 4, at http://hdr.undp.org/en/reports/global/hdr2009/papers/HDRP_2009_09.pdf. **337** Fig 9.4: Adapted from "World Agriculture," *National Geographic Atlas of the World*, 8th ed. (Washington, D.C.: National Geographic Society, 2005), p. 19; and "China: Economic, Minerals" map, *Goode's World Atlas*, 21st ed. (New York: Rand McNally, 2005), pp. 39 and 207. **341** Fig 9.5: Adapted from Country Network China, CAI-Asia Center China Program, "Clean Air in the People's Republic of China: Summary of Progress on Improving Air Quality," November 2008, at http://www.cleanairnet.org/caiasia/1412/articles-70822_PRC.pdf. **343** Fig 9.6: Adapted from *Hammond Times Concise Atlas of World History* (Maplewood, NJ: Hammond, 1994). **345** Fig 9.7: Adapted from *Hammond Times Concise Atlas of World History* (Maplewood, NJ: Hammond, 1994). **352** Fig 9.9: Adapted from Invest in China, "Per Capita Cash Income of Rural Households by Region (Third Quarter, 2009)," at http://www.fdi.gov.cn/pub/FDI_EN/Economy/Investment%20Environment/Macro-economic%20Indices/Population%20

&%20GDP/t20091103_113868.htm, and "Income of Urban Households by Region (Third Quarter, 2009)," at http://www.fdi.gov.cn/pub/FDI_EN/Economy/Investment%20Environment/Macro-economic%20Indices/Population%20&%20GDP/t20091120_114779.htm. **353** Fig 9.10: FDI data from Invest in China, at http://www.fdi.gov.cn/common/info.jsp?id5ABC00000000000022787. Specific Web site no longer available without registering at http://www.fdi.gov.cn/pub/FDI_EN/Statistics/default.htm; Fig 9.11: Adapted from Invest in China, "News Release of National Assimilation of FDI from January to December 2009," at http://fdi.gov.cn/pub/FDI_EN/Statistics/FDIStatistics/ExpressofForeignInvestment/default.htm. **355** Fig 9.12: Adapted from U.S. Treasury, "Major Foreign Holders Of Treasury Securities," at http://www.treasury.gov/resource-center/data-chart-center/tic/Documents/mfh.txt. **360** Fig 9.14: Adapted from U.S. Census Bureau, Population Division, "Population Pyramids of China," International Data Base, 2009, at http://www.census.gov/ipc/www/idb/country.php.] **363** Fig 9.16: Adapted from Chiao-min Hsieh and Jean Kan Hsieh, *China: A Provincial Atlas* (New York: Macmillan, 1995), p. 12. The Web site http://www.index-china.com/minority/minority-english.htm includes a comprehensive survey of minorities in China.

CHAPTER 10

371 Fig 10.3: Adapted from "Struggling Cultures," *National Geographic Atlas of the World*, 8th ed. (Washington, DC: National Geographic Society, 2005), p. 15; and from the "Globalization: Effects on Indigenous Peoples" map, International Forum on Globalization, at http://www.ifg.org/programs/indig.htm. **375** Fig 10.5: Adapted from Victor T. King, *The Peoples of Borneo* (Oxford: Blackwell, 1993), p. 63. **377** Fig 10.6: Map adapted from "Annual Review and Assessment of the World Timber Situations 2008," by the International Tropical Timber Organization, Figure 7, p. 13, at http://www.itto.int/en/annual_review/. **378** Photo Essay 10.2: Map adapted from *United Nations Environment Programme, 2002, 2003, 2004, 2005, 2006* (New York: United Nations Development Programme), at http://maps.grida.no/go/collection/globio-geo-3. **379** Fig 10.7: Data from "Forestry Issues—Deforestation: Tropical Forests in Decline," at http://www.canadian-forests.com/Deforestation_Tropical_Forests_in_Decline.pdf; http://www.panda.org/about_our_earth/about_forests/deforestation/forestdegradation/forest_illegal_logging/. **380** Fig 10.8: Adapted from *Hammond Citation World Atlas* (Maplewood, N.J.: Hammond, 1996), pp. 74, 83, 84. **385** Fig 10.9: Adapted from *Hammond Times Concise Atlas of World History* (Maplewood, N.J.: Hammond, 1994), p. 101. **390** Fig 10.11: Adapted from http://www.aseansec.org/22122.htm and ASEAN trade statistics, 2008, Table 19, at http://www.aseansec.org/Stat/Table19.pdf. **391** Fig 10.12: Adapted from ASEAN tourism statistics, Table 30, at http://www.aseansec.org/stat/Table30.pdf. **392** Fig 10.13: Adapted from "World Heritage List," UN World Heritage Convention, at http://whc.unesco.org/en/list/; "Tourism Attractions Along the Asian Highway," UN Economic and Social Commission for Asia and the Pacific, 2004, at http://www.unescap.org/ttdw/common/tis/ah/tourism%20attractions.asp; and *Asia Times Online*, at http://www.atimes.com/atimes/Asian_Economy/images/highways.html. **397** Fig 10.15: Data from Sasha Loffredo, *A Demographic Portrait of South and Southeast Asia* (Washington, D.C.: Population Reference Bureau, 1994), p. 9; Globalis, at http://globalis.gvu.unu.edu/; and *2009 World Population Data Sheet* (Washington, DC: Population Reference Bureau, 2009); Fig 10.16: Adapted from U.S. Census Bureau, Population Division, "Population Pyramids of Malaysia" and "Population Pyramids of Indonesia," International Data Base, 2009, at http://www.census.gov/ipc/www/idb/pyramids.html. **399** Fig 10.17 Map adapted from Joni Seager, *The Penguin Atlas of Women in the World* (New York: Penguin Books, 2003), p. 73; with updated information from the Migration Policy Institute, at http://www.migrationinformation.org/Profiles/display.cfm?ID5364. **401** Fig 10.18: Map adapted from *Oxford Atlas of the World* (New York: Oxford University Press, 1996), p. 27.

CHAPTER 11

409 Fig 11.3: For additional information, see http://www.irb.com/EN/IRB1Organisation/. **414** Fig 11.5: Adapted from Environmental Dynamics Research, Inc., 1998; Ivan Cheung, George Washington University, Geography 137, Lecture 16, October 29, 2001. **415** Fig 11.6: Adapted from Tom L. McKnight, *Oceania* (Englewood Cliffs, N.J.: Prentice Hall, 1995), p. 28. **419** Fig 11.7: Adapted from Richard Nile and Christian Clerk, *Cultural Atlas of Australia, New Zealand, and the South Pacific* (New York: Facts on File, 1996),

p. 194. **423** Fig 11.9 Adapted from Richard Nile and Christian Clerk, *Cultural Atlas of Australia, New Zealand, and the South Pacifie* (New York: Facts on File, 1996), pp. 58–59. **426** Fig 11.10: Sources: Central Intelligence Agency, *The World Factbook,* at https://www.cia.gov/library/publications/the-world-factbook/; U.S. Department of State, "Background Note: Kiribati," at http://www.state.gov/r/pa/ei/bgn/1836.htm; Australian Government, Department of Foreign Affairs and Trade, "Guam Country Brief," at http://www.dfat.gov.au/geo/guam/guam_brief.html; Samoa Bureau of Statistics, "Economic Statistics," at http://www.sbs.gov.ws/Portals/138/PDF/Trade/Export%20trading%20partners.pdf; World Bank, "World Trade Indicators 2008," at http://info.worldbank.org/etools/wti/docs/ wti2008/mainpaper.pdf; Encyclopedia of the Nations, "Tuvalu—International Trade," at http://www.nationsencyclopedia.com/economies/Asia-and-the-Pacific/Tuvalu-INTERNATIONAL-TRADE.html#ixzz0XctzTLD8; U.S. Census Bureau, "Foreign Trade: Total U.S. Exports (Origin of Movement) via Hawaii," at http://www.census.gov/foreign-trade/statistics/state/data/hi.html#ctry. **432** Fig 11.12: Data from Australian Bureau of Statistics, "Main Countries of Birth," *Year Book Australia,* 2008, Table 7.39, at http://www.abs.gov.au/ausstats/abs@.nsf/bb8db737e2af84b8ca2571780015701e/F1C38FAE9E5F2B82CA2573D200110333?opendocument.

Boldface indicates a definition; *italics* indicate a figure.

Abkhazia, 196, 197

Aboriginal Americans. *See* Native Americans

Aborigines (Australia), **422**, *423*, *424*, 424, 425, 431–432, *433*, *434*, 435, 437

Abyssinia, 261

Acculturation, **128**

Aceh province, 393

Acid rain, *66*, 68, 78, *108*, 144, *144*, 145, *179*, *215*, 255, 299, 300, 339, 340, 378, *417*

Acquired immunodeficiency syndrome (AIDS), 279. *See also* HIV/AIDS

Adivasis, 306, 307

Afghanistan. *See also* Kabul
  Arab traders and, 300
  British colonialism and, 301, 303
  climate change vulnerability, *43*, 297
  deforestation and, 296
  democratization and, 77, 324
  family compound in, 287
  immigrants from, 163
  Islamic fundamentalism and, 324
  Kabul Valley, 293, 295
  Muslims in, 308
  NATO and, 74, 158
  population growth, 325
  poverty in, 296, 297
  purdah and, 309
  Al Qaeda and, 74, 77, 322, 323, 324
  refugees, in Iran, 231
  South Asia region and, 289
  Soviet-Afghanistan war, 187, 194, *195*
  Taliban in, 203, 311, 321, 322
  U.S.-lead war in, 74, 76, 239, 321, 428
  war/reconstruction in, 322–323
  water shortage, 296, 297
  women and, 309, 311

Africa. *See also* North Africa and Southwest Asia; Sub-Saharan Africa
  HIV/AIDS and, 14
  Horn of, 252, 272
  naming of countries, 249
  peopling of, 259
  as plateau continent, 44
  term usage, 249

African National Congress (ANC), 263

African Plate, *45*, 59, 139, 209, 374, 375

African slave trade (European slave trade), 259–260, 261, 264

African-Americans
  culture of poverty and, 91, 96
  infant mortality rates, 90
  median household income, *91*
  population, *71*, *91*
  southern settlements (Europeans) and, 70
  unemployment, 90

Afrikaners, 263, 265

Age structures, 15–16. *See also* Population pyramids

Agent Orange, 385, 387

Aging populations, *10*, 17
  Central Europe, 14
  East Asia, 359–360
  Europe, *137*, 161, 163
  North America, 57, 95

Agnostics, 92

Agra, *304*

Agribusiness, **80**
  Bangladesh and, 316
  Del Monte, 120
  India and, 316
  Mexican job losses and, 84
  Ogallala aquifer and, 66, 69
  Pakistan and, 316

Agriculture. *See also* Food production; Green revolution agriculture; Sustainability
  animal husbandry and, 23, 148, 152, 332
  commercial, *11*, 22, 26, 83, 176, 209, 254, 257, 259, 261, 284, 379
  consequences of, 23
  corporate, 119, 159
  defined, 23
  domestication and, 23
  GM and, 24, 26, 81
  mixed, 254, 256, 259
  in North America, *81*
  organic, 26, 54, 81, 97, 159
  origins of, 218–219
  plant cultivation and, 23
  in Russia and post-Soviet states, *193*
  shifting cultivation, 25, 111, 113, 120, 135, 257, 379–381
  slash and burn, 379, *380*
  subsistence, *11*, 16, 22, 71, 120, 121, 247, 254, 255, 256, 257, 259, 284, 380, 403, 420, 427
  sustainable, 26

Agricultural subsidies, 66, 83, 115, *137*, 138, **159**, 169, 192, 428

Agricultural zones
  Arabian Peninsula, *214*
  China, 337
  Middle and South America, *120*
  South Asia, 293, 316, *316*, 317

Agroecology, **317**, 320

Agroforestry, **254**, 258

Aguilar, Javier, 58

AIDS (acquired immunodeficiency syndrome), 279. *See also* HIV/AIDS

Ainu, **364**, 365

Air pollution (industrial air pollution)
  acid rain, 66, 68, 78, *108*, 144, *144*, 145, *179*, *215*, 255, 299, 300, 339, 340, 378, *417*
  Beijing, 339
  Canada, 78
  China, 340
  East Asia, 340–341
  Europe, 145
  New Delhi, 298, 299
  North America, 68
  smog, 66, 68, 375
  South Asia, 299, 300
  Tokyo, 339
  U.S., 78

Air pressure, 46

Air transportation (North America), 81–82

Al Jazeera, 208, 209, 238, 239, 243

Alaska
  Barrow, 67
  Bering land bridge and, 69
  Cook Inlet, *47*
  Exxon Valdez disaster in, *215*, 272
  Natives, 62, *91*
  pipeline, 62, *64*, 65
  selling of, *185*

Albania, *147*

Alcohol abuse
  Aborigines, 332
  Brahmins, 306
  Central Valley, 58
  Oceania, 424
  Russia, 199, 201
  Secoya people, 100

Alexander the Great, 300

Alexandria, 216, *217*

Algeria, 148, *152*, 153, 209, 221, 223, 232, 233, 236

Alhambra, *152*

Alice Springs, *410*, *413*

Allamakee County, 82

Alps, 44, 139, 141, 142

Altruism, 53

Amazon Basin, *103*
  biodiversity in, 104, 107
  deforestation, 107
  oil spill, 100
  shifting cultivation, 111
  soybeans and, 121

Amazon River, 99, 104

American Samoa, 428

Amsterdam, *202*

Amu Darya, 180

Amundsen-Scott South Pole Station, *439*

An Australian Girl in London (Mack), 431

Anatolia Project, 216, 218

ANC (African National Congress), 263

Ancient Greece/Rome, 148

Andaman Sea, 374

Andes, 48, 101, *102*, 104, *105*, *110*, 111, 115

Angelou, Maya, 91

Angkor Wat temple, 392

Angola
  China and, 269
  Portugal and, *154*

Animal husbandry, 23, 148, 152, 332

Animals/plants. *See* Fauna/flora

Animism, 271, 281, 308, 400, *401*, 403

Antarctic Plate, *45*, 46

Antarctic Treaty, *439*

Antarctica, *439*
  Benguela Current and, 252
  Middle and South America (world region) and, 101

Anteaters, 415

Anti-foreigner sentiment (Europe), 168

Aotearoa, 408, 412

Apartheid, 238, *261*, **262**, 263, 265, 272, 273, 276

Appalachian Mountains, 58, 59, 61, 70, 139, 173

Aquifers, **59**, 66, 67, 69, 97, 182, 209, 213, 216, 289, 297, 335, 336, 381

Aquino, Corazon, 402

Arab traders, 300

Arab world, 209. *See also* North Africa and Southwest Asia

Arabian Peninsula, *214*, 220, 221, 238, 252, 364

Arabian Plate, 45, 209, 249
Aral Sea, *180*, 180–181, 182, 183
Archipelagos, **374**, 375, 383, 408, 428
Arctic climate, 48, *48*, *63*, *177*
Arctic Ocean, 145, 146, 173, 176, 178
Argentina
  independence from Spain, *115*
  pampas, *103*, 104, 105, 121
Arid/semiarid climates, 48, *48*
Arizona
  Arizona–Mexico border, 90
  illegal immigrants and, 90
  irrigation and, *81*
  Phoenix, 88
Arranged marriages, 227, 400
ASEAN (Association of Southeast Asian Nations), 32, **389–391**, 390, 392, 403
Asia. *See* East Asia; North Africa and Southwest Asia; South Asia; Southeast Asia
Asian economy, New Zealand/Australia and, 425–427
Asian Highway, 392
Asian snakehead fish, 65
Asian-Americans, 90, 91
Assimilation, **128**, 163–**164**, *165*, 196, 363, 364
Association of Southeast Asian Nations. *See* ASEAN
Astana, Kazakhstan, 198, *200*
Aswan Dams, 214, 216
Asymmetries (U.S./Canada relationships), 75
Atacama, Chile, 104, *105*
Atheism, 92, 203, 204
Atlanta, Georgia, 66, 87
Atlas Mountains, 209, 212
Atoll de Réao, *411*
Atolls, **412**
Auckland, *430*
Aung San Suu Ki, 393, *394*, 395, 402
Australia
  Aborigines, *422*, 423, *424*, *424*, 425, 431–432, *433*, *434*, 435, *437*
  Asian economy and, 425–427
  Blue Mountains, *410*
  Bowen, *425*
  climate, *412*, 413, *414*
  colonization of, 422–424
  continent formation, 409, 412
  convict heritage of, 422–423, 425, 435
  cultural diversity, *432*
  development, *407*
  drought-stricken wheat field in, *407*
  Eastern Highlands, *410*
  fauna/flora, 414–415
  formation, 409
  invasive species, 419
  political demonstration in, *437*
  Sydney, *430*
Australian Desert, 412
Australo-Melanesians, **383**
Austria
  GDP per capita, 16
  population pyramid for, *15*, 15–16
  TFR for, 15
Austronesians, **383**, 386, 422, 423, 425
Authoritarianism, 12, 33, 393. *See also* Dictators
Automobiles
  Albanians and, *147*
  dependency, 87
  greenhouse gas emissions and, 163
  Indian auto industry, 318
Autonomous regions
  Ningxia Huizu Autonomous Region, 336, 363
  Russian Federation and, 173
  Xinjiang Uygur Autonomous Region, 188, 334, 357, 362, 363
Average population density, 14. *See also* Population densities
Aztecs, **111**, 114, *115*

Baby boomers, 95, 96, 163
Badaga ethnic group, 298
Bagan, 386
Baghdad, 35, 220, 236, 239
Bahrain, 40, 41, 225, 238
Bali, 374, 383, 393, 395, *401*
Banana worker (Costa Rica), 120
Bangkok, 369, 381, 393, 397, 398, 399, 403
Bangladesh
  agribusiness and, 316
  climate change vulnerability, *43*
  dictators and, 325
  East Pakistan and, 321
  global warming and, *12*
  hurricanes and, *43*
  pesticide application in, 287
  sea level rise, *43*, 295
  war of independence (1971), 304, *321*
  winter monsoon and, 294
Banqiao Dam, 338
Baptists, 92, 203
Barcelona, *162*
Barrios, 27, 131
Barrow, Alaska, 67
Bauxite, 115, 120, 265, 266
Bazaars, 192, *192*, 267
Bechtel, 30, 37, 109
Beijing
  air pollution, 299, 339
  Han-dominated Communist Party in, 361
  Olympics, 340, 357

Tiananmen Square, 355, 356, 357
  water shortage in, 335
Belarus, *201*
Belgian town festival, *152*
Belgium, *165*
Belief systems, 51. *See also* Religions
Bemba, Jean-Pierre, 35
Benares. *See* Varanasi
Bengal Famine, 304
Bengali language, 309
Benguela Current, 252
Bering land bridge, 69, 70, *114*
Berlin
  Berlin Conference, 261
  old urban core, *162*
Bhopal gas tragedy, 300
Bhutan, 289, 301, 303, 308, 309, 316, 320, 325
Bhutto, Benazir, 309
Bible Belt, 92
Bicycle rickshaw driver, *11*
Bihari refugees, 320, *321*
Bin Laden, Osama, 74, 322
Biodiversity, 22, 24, 25, **104**, 107, 133, *134*, 142
Biopiracy, 371
Biosphere, **22**, 24, *142*. *See also* Human impact on biosphere
Birth control, 14, 227, 231, 276, 277, 395
Birth rates, **14**
  demographic transition and, *17*
  East Asia, 359, 361, 366
  Europe, 163, 164, 166
  Middle and South America, 126, 127, 133
  North America, 95, 96
  Pakistan, *313*
  Russia and post-Soviet states, 198–199, 201, 204
  Southeast Asia, 395, 403
  Sri Lanka, *313*
  Sub-Saharan Africa, 275, 276, 277, 278, 283
Black market, 196, 197–198
Blue Mountains (Australia), *410*
Boer War, 263
Boers, 262, 263, 265
Bolivia
  Cochabamba water utility in, 37, 109
  drought-prone highlands in, *110*
  malnutrition, 26
Bollywood, 315. *See also* Mumbai
Bolsheviks, 185, **186**, 187
Bombay. *See* Mumbai
Borneo, 370, 374, 379
Borneo Project, 370, 371
Bosnia civil war, *155*
Boston, oil pollution and, 57
Botswana, 180, 250, 252, 273, 276, 279

Bottled water, 38
Bougainville, *434*
Bowen, Australia, *425*
Boys. *See* Men
BP (British Petroleum) oil spill, 62
Brahmaputra river basin, 293, 295, *295*. *See also* Ganga-Brahmaputra delta
Brahmins, 298, 306, 307
Brain drain, **129**
Brazil. *See also* Amazon Basin
  cultural diversity, *128*
  deforestation in, *108*
  ecotourism lodge in, *134*
  Fortaleza, 131–132
  João Pessoa, 99
  landless movement, 121
  Pará, *108*
  Rio de Janeiro, 27, *103*, *131*
BRIC, 189
Brides
  price, 310
  shortage of, 13, 361
British colonialism, 301, 303
British East India Company, 302, 303, 305
British Indian Empire, *301*
British Petroleum (BP) oil spill, 62
Brownfields, 88
Brundtland, Gro Harlem, 166
Brunei, 374, 388, 393, 395, 396, 402
Buddhism, 51, 52, 91, 92, 176, 203, 221, 305, 308, **308**, 323, 357, 363, 383, 384, 386, 392, 395, 400, 401, 403
Bulgaria, 152, 154
Burma (Myanmar), 371, 372, 394
Bush, George W., 74, 194, 239
Bush administration, 239
Bushmeat, 258, 267

Cahokia, 70
Cairo, 27, 207, 229, 230, 231, 237
Calauan, 376
Calcutta. *See* Kolkata
Call centers, 318
Cambodia
  girl, *404*
  houseboat, 369
  Khmer Rouge, 386, 387
  killing fields, 386, 387
  malnutrition, 26
Campo de Dalía, *144*
Canada
  air pollution, 78
  democratic system, 57, 76, 78–79
  economic issues, 80–85
  health-care system, 57, 79
  immigration and, 78, 88–90
  indigenous groups and, 96
  infant mortality rates, *79t*

population pyramid, 96
social safety net, 79
U.S./Canada relationships, 75–97
Canton. *See* Guangzhou
CAP (Common Agricultural Program), **159**
Cape Town, 27, 49, 250, 251
Capitalism, **34**, **152**, **154**, **186**
  Chile and, 101
  Cold War and, 152
  communism *v.*, 154
  crony, 389
Carbon dioxide, 3, 12, 39, 42, 99, 107, 109, 134, 340, 379
Carbon footprint, 22
Carbon sequestration, **252**
Caribbean Plate, 45, 101
Caribbean sea, 5
Caroline Islands, 409, 410
Carpathian Mountains, 173
Carrying capacity, **24**
Cartels, 124, **232**
Carteret Islands, 416
Cartography, 3, 6, 7
Casablanca, 230
Cash crops, 70, 71, 90, 115, 120, 124, 193, 264, 266, 280, 384
Cash economies, **17**
Caste system, **306**–308, 309, 310, 323, 324
Castro, Fidel, 124
Çatalhöyük, 219
Catalonia, 147
Catholic Church. *See* Roman Catholicism
Caucasia, 185, 190, 193, 195, **198**, 199, 205
Caucusus Mountains, 174
Cell phones
  China, 354, 357, 358
  Europe, 160
  GISc and, 7
  Iran, 238
  Japan, 358
  Masai herder with, *11*
  microcredit loans and, 317
  windmill and, 270
Central African Republic, 247, 255
Central America, 101. *See also* Middle America
Central China
  flooding in, 336
  Sichuan Province, 328, 338, 354, 357, 361
Central Europe, 136, 138
  coal power plants, *144*
  development and, *137*
  population aging/decline, 14
Central lowlands, 59, 60, 61
Central mountain zones, 139
Central planning, **152**, 171, 186, 193, 197, 198, 202, 344, 347, 348

Central Siberian Plateau, 173, *175*, 176, 183
Central Valley unemployment, 58
Centrally planned/socialist economy (command economy), 152, 154, 171, **186**, 187, 188, 189–190, 193, 197, 198, 202, 344, 347, 348, 349, 365
Ceylon, 301, 303. *See also* Sri Lanka
Chad, 247
Chang Jiang (Yangtze River), 332, 334, 335, 336, 338, 340, 361
Chang'aa, 268
Changshan Islands, 365
Chávez, Hugo, 118, 119, 121, 123
Chechnya, *195*, 196
Chennai (Madras), 289, 302
Chernobyl nuclear power disaster, 178
Cherokee, 73
Chevron, 100
Chiapas, Mexico, *108*
Children. *See also* Infant mortality rates
  child soldiers, 248
  childhood landscape map, 2, 2–3
  cultural preference, for boys, 13, 15, 18, 227, 312, 313, 327, 360–361
  one-child policy, 10, 13, 311, 359, 360–361, 362, 366
  two-child policy, 361, 395
Chile
  Andes, 101, *102*, 104, *105*, *110*, 111, 115
  Atacama, 104, *105*
  capitalism of, 101
  Frei and, 99
  independence from Spain, 115
Chimpanzees, 258
China. *See also* Beijing
  agricultural zones, 337
  air pollution, 340
  Angola and, 269
  Chinese empires, 343, *343*, 344, 384
  culture groups, 363
  FDI in, 353
  feudalism and, 341, 342, 344, 346
  GDP per capita PPP, 350, 352
  greenhouse gas emissions, *40*, 339, 341, 365
  Imperial, 341–342
  income disparities, 365
  international pressures, 357
  IT and, 357
  Japanese imperialism and, 343
  market reforms in, 349–350
  Mexico and, 118

  one-child policy, *10*, *13*, 311, 359, 360–361, 362, 366
  population pyramids, 360, *360*
  rural-to-urban migration, 329
  Soviet Union treaty and, 347
  Tibet and, 344, 357, 358, 363
  twentieth century and, 343–344
  women population, *13*
  Wuxi, 365
China's Far Northeast (Manchuria), 332, 334, 344, 345
Chinese Communist Party, 344, 346
Chipko (tree hugging), 296, 300
Chivu, Grigore, 138
Chloropleth maps, 4, *4*
Cholera, 38, 277
Christianity, 51, 52, **220**
  evangelical, 92, 203, 282, 283
  North America and, 91–92, *92*
  South Asia and, 298, 300, 305, 308, 309
  sub-Saharan Africa and, 282
Cienfuegos, Camillo, *122*
Cisco, 30
Cities. *See also* Slums; Urbanization
  growing, 26–27
  primate, 129, *130*, 198, 200, 277, 397, 399, 400
  technological innovation and, 28, 95, 96
Civil conflict
  Hawaii, 427–428
  Kashmir, 320, 322
  sub-Saharan Africa and, 273
Civil disobedience, **303**
Civil rights
  Aborigines and, 432
  ANC and, 263
  European notions of, 149
Civil society, 32, **33**, 36, 54, **263**, 264, 273, 324
Civil wars
  Afghanistan, *43*, 322, 324
  Bangladesh and, 303
  Bolsheviks and, 185, 186
  Bosnia, 155
  Civil War (1861-1865), 70, 71, 72
  Colombia, 123, 124
  Congo, 35
  democratization and, 33
  Liberia, 248
  Nepal, 321, 324
  Portugal, 154
  South Africa, 263
  Sri Lanka, 321, 323, 325
  sub-Saharan Africa, 271, 273, 274
  Sudan, *43*, 237, 263
  Vietnam, 385, 386
Class-based explanation (poverty), 91

Clear-cutting, 64, **65**, 252, 254, 370
Climates. *See also specific climates*
  Arctic, 48, *48*, 63, 177
  defined, 45
  East Asia, 332–334
  Europe, *143*
  Middle and South America, 104–107
  North Africa and Southwest Asia, 209
  North America, 57, 59, 63
  Oceania, 412–414
  Russia and post-Soviet states, 176, *177*
  South Asia, 292–295
  Southeast Asia, 374–375
  sub-Saharan Africa, 252–254
  weather *v.*, 45, 47
Climate change, **39**–43
  Antarctica, 439
  coral reef bleaching and, *12*, 381, 382, 383, 407, 416, 438
  defined, 39
  deforestation and, 99
  drivers of, 39–41
  hurricanes and, 41, 42, 43, *110*
  impacts, 41
  as thematic concept, *12*, 39–43
  urbanization and, 56, 65, 68, 96
  water and, 41–42
Climate change vulnerability, 41–43
  Afghanistan, 43, 297
  Antarctica, 439
  Bangladesh, *43*
  East Asia, 327, 334, 335, 336, 365
  Europe, 145–146, *147*, 169
  Mexico, 67
  Middle and South America, 107, 109, *110*, 134
  Morocco, 43, 207
  North Africa and Southwest Asia, 207, 216, 217, 218, 243, 244
  North America, 57, 65–66, 67, 69, 97
  Oceania, 407, 418, 419, 437, *437*
  Pacific Islands, 416
  photo essay, 42–43
  Russia and post-Soviet states, 181, *182*, 183
  South Asia, *13*, 287, 295–296, 297, 324
  Southeast Asia, 369, 379–381, 382, 383
  Spain, *43*
  sub-Saharan Africa, 247, 254, 256, 257, 283
  Sudan, *43*, 216
  U.S., *43*

Climate zones (climate regions), 47, 48–49
East Asia, 332, 333
Europe, 143
Middle and South America, 105
North Africa and Southwest Asia, 212
North America, 63
Oceania, 413
Russia and post-Soviet states, 177
South Asia, 293, 294
Southeast Asia, 376
sub-Saharan Africa, 253
world, 48–49
Clinton, Bill, 80
Clinton, Hillary, 80
Clouds, 46, 47, 292
Coal ash spill (Tennessee), 64, 65
Coal mining, 65
Coal power plants, Central Europe, 144
Coastal lowlands, 59, 61, 127, 139, 249, 290, 412
Cocaine seizures, 123, 123
Cochabamba water utility, 37, 109
Cocos Plate, 45, 46, 101
Coffee, 32, 104, 106, 115, 254, 265
Cold War, 152, 154, 185, 186, 187
alliances, 187
Cuba and, 124
geopolitics and, 34, 272
NATO and, 158
Oceania and, 424
Southeast Asia and, 384
Soviet Union and, 152
U.S. military presence in Europe and, 77
World War II and, 186–187
Collapse, of Soviet Union, 74, 154, 178, 181, 186, 188, 199, 200, 204. See also Union of Soviet Socialist Republics
Colombia
conflict in, 122
democratization and, 34
Colonialism (European colonialism)
Australia, 422–424
decolonization and, 154
globalization and, 28, 149–150
Middle and South America, 111–113, 126
New Zealand, 422–424
North Africa and Southwest Asia, 224
South Asia, 301–304
Southeast Asia, 384, 385
sub-Saharan Africa, 261–263, 264–265
sugar plantations and, 30
wealth transfers (colonies to Europe), 150
Colonias, 131

Color Revolutions, 194
Colorado River, 66
Columbus, Christopher, 101, 111
Command economy. See Centrally planned/socialist economy
Commercial agriculture, 11, 22, 26, 83, 176, 209, 254, 257, 259, 261, 284, 379
Commodities, 23, 28, 37, 231, 233, 234, 264, 265, 266, 268, 269, 337, 432
Common Agricultural Program (CAP), 159
Communal conflict, 309
Commune system, 349
Communism. See also Russia and post-Soviet states; Union of Soviet Socialist Republics
capitalism v., 154
central planning and, 152, 154, 171, 186, 193, 197, 198, 202, 344, 347, 348
defined, 34, 154
impact on Europe, 151
Marx and, 34, 145, 151, 152, 186, 189
post-communist welfare systems, 168
Communist Party, 186
Communist Revolution, 185–186, 337, 344, 347, 365
The Communist Manifesto (Marx), 151
Competitiveness, South Asia and, 318–320
Computer conference, in Mexico, 134
Condom use, 280, 395, 403
Condominiums, 229, 315, 428
Conflicts. See Democratization/conflict issues; Regional conflicts
Confucianism, 50, 52, 342, 344, 346, 347, 360, 384, 393, 400, 401, 403
Congo (Brazzaville/Republic of Congo), 210, 249, 267, 271, 272, 275
Congo (Kinshasa/Democratic Republic of Congo), 10, 35, 249, 252, 270, 273
Contested space
Jammu and Kashmir region, 303
Middle and South America, 120–121
Continent formation (Australia/Oceania), 409, 412
Continental climates, 48, 49, 105, 177, 333
Continental plates, 44
Contraception, 227, 276, 395
Convict heritage, Australia, 422–423, 425, 435
Cook, James, 435

Cook Inlet, 47
Cool humid climates, 48, 48
Copenhagen Accord, 42, 181
Coral reef bleaching, 12, 381, 382, 383, 407, 416, 438
Coral reefs, 392, 412, 416
Corazon Aquino, 402
Corporate agriculture, 119, 159
Corruption/organized crime (Russia and post-Soviet states), 197–198
Corsica, 136, 143
Costa Rica
banana worker in, 120
Internet use, 125
Côte d'Ivoire, 277, 377
Council of the European Union, 156, 160
Coup d'état, 121, 123, 125, 134, 135, 371, 391
Creoles, 112
Crony capitalism, 389
Crowley, William, 129
Crowley's model of urban land use, 129
Cuba
Cold War era and, 122, 124
infant mortality rates, 124
socialism, 101, 122, 124
Cultural blending (U.S.-Mexico border), 7
Cultural change, 203, 431
Cultural diversity, 50
Australia, 432
Brazil, 128
East Asia, 361–364
Middle and South America, 127–128
New Urbanism and, 88
North America, 58, 95
Oceania, 435
Russia and post-Soviet states, 196–197, 199
values and, 50
Cultural geography, 50–54
Cultural homogenization, 139
Cultural influences, Southeast Asia and, 383–384
Cultural issues. See Sociocultural issues
Cultural pluralism (Southeast Asia), 400
Cultural preference, for boys, 13, 15, 18, 227, 312, 313, 327, 360–361
Cultural Revolution, 344
Culture
defined, 50
European, sources for, 148–149
material, 8
Culture groups. See also Ethnic groups; Indigenous groups
China, 363
ethnic groups v., 50

ethnicity and, 90
non-Han minorities (China), 361
Oceania, 423
Culture of poverty, 91, 96
Cumulonimbus clouds, 292
Currency devaluation, 266
Cyclones, 43, 292, 334, 416
Czar, 185
Czech Republic, 162

Dadaab Refugee Camp, 272
Dai Qing, 338
Dalai Lama, 363
Dalal Street, 319
Dalits, 306, 307
Dam projects
Aswan Dam, 216
Grand Coulee Dam, 288
Haditha Dam, 215, 218
Niger River and, 257
Sardar Sarovar Dam, 288–289
Three Gorges Dam, 288, 331, 338, 339, 340, 341
on Tigris/Euphrates drainage basins, 218
Darfur region, 35, 217
Darjeeling, 293
Darwin, Charles, 415
Dead zones, 68
Death rates, 14. See also Infant mortality rates
Middle and South America, 127, 133
Russia and post-Soviet states, 198–199
Sub-Saharan Africa, 277
Debt crises, euro and, 156–158
Deccan Plateau, 290, 291, 292, 294, 308
Deepwater Horizon oil rig, 62, 64, 65, 75
Deforestation. See also Logging
Afghanistan, 296
Amazon Basin, 24
Brazil, 108
Brazilian Amazon, 24
carbon dioxide and, 39
climate change and, 99
food and, 25
global warming and, 107
Laos, 25
Middle and South America, 107–108
in Sarawak, 370, 372, 374
South Asia, 296, 298
Southeast Asia, 378, 379
sub-Saharan Africa, 252, 254
Deindustrialization
Europe and, 157
South Asia, 301–302
Del Monte, 120
Delhi metropolitan area, 29. See also New Delhi
Delta, 44, 45
Delta works, 147

Democratic Republic of Congo.
See Congo
Democratic Society Party, 51
Democratic system, U.S./Canada, 57, 76, 78–79
Democratization, 33–36. See also Democratization/conflict issues
    Afghanistan and, 77, 324
    authoritarianism v., 12, 33
    defined, 33
    expansion of democracy and, 33
    factors for, 33
    geopolitics and, 34, 36
    lower levels of, 34, 354
    as thematic concept, 12, 33–36
    urbanization and, 136, 150–151, 153, 169
Democratization/conflict issues, 34–35
    East Asia, 327, 355, 356, 357–359, 365
    Europe, 137, 150–154, 154–155, 169
    Georgia (country), 172, 172
    Indonesia, 391
    Iraq War, 238–239
    Middle and South America, 121–125, 122, 134
    North Africa and Southwest Asia, 207, 236–237, 238, 243, 244
    North America, 76–77
    Oceania, 407, 433, 434, 435–436, 437
    pro-democracy demonstrations, 169, 207, 365
    Russia and post-Soviet states, 194, 195, 196–197, 198
    South Asia, 310, 320, 321, 322–323, 324
    Southeast Asia, 369, 391–393, 394, 395, 404
    sub-Saharan Africa, 77, 247, 271–273, 274, 275, 284
    U.S., 76–77
Demographic transition, 16–17, 17, 127, 198, 275–276, 280, 284, 312, 359, 395
Deng Xiaoping, 344
Denmark, 80, 113, 137, 145, 154, 166
Derry, Northern Ireland, 155
Desalination, 144, 216
Desertification, 207, 215, 216, 218, 244, 255, 258, 335, 336, 337
Detritus, 375
Devaluation, currency, 266
Development, 10, 19–22. See also F/MEI maps; GDP per capita PPP; HDI rank maps; Sustainability
    Australia, 407

Central Europe, 137
    defined, 19
    deforestation and, 24
    East Asia, 327, 350–354, 365
    GDP per capita and, 19–20
    globalization and, 27–28, 33
    grassroots, 269, 270, 435
    indigenous peoples and, 100, 100
    in Madagascar, 13
    measures of, 19–20
    Middle and South America, 99, 107, 109, 112–116, 134
    mining and, 25
    New Zealand, 407
    North Africa and Southwest Asia, 207, 232–233, 244
    North America and, 57
    Oceania, 407
    PPP and, 20
    self-reliant, 270
    South Asia and, 324
    Southeast Asia, 369
    Sub-Saharan Africa and, 247, 269–270, 284
    sustainable, 22
    as thematic concept, 10, 19–22
    urbanization and, 170, 178, 181, 204, 326, 328, 350–354, 366
"Devil's Highway," 90
Dhaka
    Bihari refugees and, 321
    migrants (sleeping on sidewalks) and, 29
    slums in, 287, 314
    urbanization and, 287
    workers, 314
Dharavi, 315
Dhobi Ghat slum, 314
Dialects, 53
Diamond mining, 247, 249, 263, 265, 284
Diaspora, 220, 269, 282, 302
Dictators, 121
    Aung San Suu Kyi and, 402
    Bangladesh and, 325
    Cold War and, 34
    Franco, 153
    Gyanendra, 323
    Hussein, 236, 238, 239
    Iraq and, 231, 236
    Middle and South America and, 99, 121, 122, 124, 134
    North Africa and Southwest Asia and, 235
    Pakistan, 324
    Portugal and, 153, 154
    South Korea and, 356
    Southeast Asia and, 393, 403
    Suharto, 389, 391, 392, 393, 394, 402
Digital divide, 83
Disease, Native Americans and, 70

Distribution, spatial, 3
Diversity. See Biodiversity; Cultural diversity
Divide-and-rule tactics, 271, 272, 273
Djenné, 259
Doctors Without Borders, 36
Dogrib territory, 73
Domestication, 23
Dominica, 5
Dominican Republic, 26, 111, 127
Domino theory, 384
Donetsk basin, 179
Double day, 18, 165, 166, 201
Dowry, 310
Drip irrigation, 214, 243, 296, 418, 419
Drought-stricken wheat field, 407
Drug trade
    Mexico, 122
    Middle and South America, 122, 123, 123–124
    Opium Wars (1839–1860) and, 343, 347
Dual economies, 265
Dubayy (Dubai), 11, 207, 229, 230, 231, 232, 244
Duck-billed platypus, 415, 415
Dunhuang, 335
Dysentery, 277, 299
Dzerzhinsk, 178

Early extractive phase (Middle and South America), 114–115
Early humans
    South Africa, 260
    South Asia, 304
Early Middle Ages, 148
Earthquakes
    Haiti, 36, 101
    Middle and South America, 101
    North Africa and Southwest Asia, 209
    Ring of Fire and, 44, 46
    South Asia, 292, 295
    in Sumatra, 292, 373
    tectonic plates and, 44, 58
    tsunamis and, 36, 292, 332, 334, 336, 373, 374, 393
East Asia (world region), 326–367. See also China; Japan; Mongolia; North Korea; South Korea; Taiwan
    aging population, 359–360
    air pollution, 340–341
    birth rates, 359, 361, 366
    climate, 332–334
    climate change vulnerability, 327, 334, 335, 336, 365
    climate zones, 332, 333
    Confucianism and, 342, 344, 346, 347, 360
    cultural diversity, 361–364
    democratization/conflict issues, 327, 355, 356, 357–359, 365

development, 327, 350–354, 365
    economic issues, 347–359
    environmental issues, 334–341
    FDI, 350, 353
    food production, 327, 337–338
    gender inequalities, 327
    geographic issues, 347–365
    geographic setting, 332–347
    globalization and, 327, 350–354, 365
    Hong Kong and, 350, 352
    human impact on biosphere, 338, 339, 340
    human patterns over time, 341–347
    indigenous groups, 364, 365
    Internet usage, 358, 358
    landforms, 332
    as mega-financier, 354
    natural hazards, 334
    optimism for, 365, 366
    physical patterns, 332–334
    political issues, 347–359
    political map, 326
    population density, 334, 340, 362
    population distribution, 361
    population growth, 13
    population patterns, 359–360
    reflections on, 365–366
    regional map, 330–331, 332
    rural-to-urban migration, 350
    sociocultural issues, 359–366
    sustainability in, 337–338
    terms usage, 332
    thematic concepts overview, 327, 366
    urbanization, 326, 327, 328, 329, 334, 341, 350–354, 351, 357, 359, 366
    visual timeline, 346–347
    water issues, 327
    as world region, 1, 3, 7
East Europe, 136, 138, 139
East Pakistan, 303, 321. See also Bangladesh; Pakistan
Eastern Hemisphere, 6
Eastern Highlands (Australia), 410
Eastern Orthodox churches, 52
Ecological footprint, 22
Ecology
    agroecology, 317, 320
    political, 22
Economies. See also Economic issues; Global economic downturn; Informal economies; Service economies
    cash, 17
    centrally planned/socialist, 152, 154, 171, 186, 187, 188, 189–190, 193, 197, 198, 202, 344, 347, 348, 349, 365

Economies (*Continued*)
  dual, 265
  EU, *158*
  formal, 20
  free market, 116, 152, 204, 363
  global, 8, 28, 30–31
  Internet, 83
  knowledge, 82, 83
  MIRAB, 427
  of scale, 156
  state-aided market, 347,
    348, 387
  subsistence, 16
Economic and technology
  development zones
  (ETDZs), 350, 353, 362
Economic core, **72**
Economic development.
  See Development
Economic diversification, 216,
  234, *244*
Economic issues. *See also*
  Global economic downturn;
  Political issues
  Canada, 80–85
  East Asia, 347–359
  Europe, 154–160
  Middle and South America,
    113–120
  North Africa and Southwest
    Asia, 231–243
  North America, 80–85
  Oceania, 425–428
  Russia and post-Soviet states,
    188–198, 204
  South Asia, 315–320
  Southeast Asia, 387–394
  sub-Saharan Africa,
    264–273, 267
  U.S., 80–85
Economic reforms, Russia and
  post-Soviet states, 189–192
The Economist magazine,
  33, 34, 86. *See also*
  Democratization/conflict
  issues
Ecotourism, 100, **109**, *134*,
  258, 421
Ecuador
  OPEC and, 232
  Secoya people and, 100
  Texaco oil development
    and, 100
Edmonton, Alberta, *57*
EEC (European Economic
  Community), 154
Egypt
  Cairo, 27, 207, 229, 230,
    231, 237
  New York City *v.*, 227
El Niño, *105*, **106**, **375**, 379, 381,
  413, 414, *414*
Electric cars, 296, 324
Emir, 208, 226, 238
Emperor penguin, *439*

Empires
  British Indian Empire, *301*
  Chinese, 343, *343*, 344, 384
  Great Zimbabwe Empire,
    259, *260*
  Mughal Empire, 220, 221, *304*
  Ottoman Empire, 148,
    220–221, 222, 223, 224,
    240, 262
  Qin, 342, 343, 346
  Russian Empire, 183–185, *185*
Endemic, **414**, 415, 416, 419
Energy resources
  Europe, 142–143
  North Africa and Southwest
    Asia, 243
Eng tribal people, 386
Environmental issues. *See also* Air
  pollution; Climate change
  vulnerability; Deforestation;
  Development; Globalization;
  Human impact on
  biosphere; Mining; Oil
  spills; Urbanization; Water
  pollution
  East Asia, 334–341
  EU and, 142
  Europe, 142–148
  indigenous peoples and,
    100, *100*
  Middle and South America,
    107–109
  North Africa and Southwest
    Asia, 213–218
  North America, 59–69
  Oceania, 416–421
  Russia and post-Soviet states,
    176–182
  South Asia, 295–300
  Southeast Asia, 375–383
  sub-Saharan Africa, 252–259
  urbanization and, 170, 178,
    181, 204
EPZs. *See* Export Processing
  Zones
Erosion, **24**
Escarpments, 249, 251
ETDZs. *See* Economic and
  technology development
  zones
Ethiopia
  Abyssinia and, 261
  early agriculture, 259, 264
  population growth, 276
  Somalia and, 272
Ethnic cleansing, **36**
Ethnic groups. *See also* Culture
  groups
  Badaga, 298
  China, 361–362, 363
  cleansing of, 36
  culture groups *v.*, 50
  defined, 50
  East Asia, 361
  Indonesia, 391

Iraq, 239
  Russia and, 183, *197*
  South Asia, 305, 309
  sub-Saharan Africa, 263, 271
Ethnicity, **90**
  culture groups and, 90
  in North America, 90–91,
    103–105
Ethnolinguistic groups. *See*
  Ethnic groups
EU. *See* European Union
EU-27, 142, 145, *157*, *164*, *166*
Euphrates River, 148, 216, 218
Eurasian Plate, 45, 46, 139, 176,
  209, 292, 332, 374, 412
Euro (€), **156**–158
Europe (world region), 136–169.
    See also Colonialism
  aging population, *137*, 161, 163
  air pollution, 145
  birth rates, 163, 164, 166
  climate change vulnerability,
    145–146, *147*, 169
  climate zones, *143*
  communism's impact on, 151
  culture, sources for, 148–149
  deindustrialization and, 157
  democratization/conflict issues,
    137, 150–154, *154–155*, 169
  economic issues, 154–160
  energy resources, 142, 145
  environmental issues, 142–148
  feudalism and, 149, 153
  food production, *137*
  gender roles, 164–167
  geographic issues, 153–168
  geographic setting, 139–153
  globalization and, *169*
  green policies, 146, 148
  human impact on biosphere,
    142, *144*, 145–146
  human patterns over time,
    148–153
  immigration and, 163–164
  Industrial Revolution, 28, 30,
    101, 149, 150, 153, 169,
    221, 259
  Islamic civilization and, 148
  landforms, 139–142
  migration into, 163–164, *165*
  multimodal transport, 148
  Muslims in, 164, *165*
  optimism for, 168, *169*
  physical patterns, 139–142
  political issues, 154–160
  political map, *136*
  population density, 146, 160, *161*
  population distribution, 160–161
  population patterns,
    160–161, 163
  reflections on, 168
  refugees in, 163
  regional map, *140–141*
  seawater pollution, *146*
  service economies and, 160

social welfare/protection
    systems, *167*, 167–168
  sociocultural issues, 160–168
  thematic concepts overview,
    137, 169
  tourism, 159
  urbanization, 136, *137*, 149,
    150–151, 153, 160, *162*,
    163, 169
  vegetation, 142
  visual timeline, *152–153*
  water pollution, 136, *137*, 142,
    145, *146*, 148, 168
  wealth transfers (colonies to
    Europe), *150*
  western, 139, *144*, 166
  women at work, *164*
  women's roles in, 164–167
  as world region, *1*, *3*, 7
  World War II and, 151–154, 161
European arrival
  North America, 70, 73
  Oceania, 422
European conquest (Middle and
  South America), 111–113
European Economic Community
  (EEC), 154
European Parliament, 156, 160
European settlements (North
  America)
  expansion (west of Mississippi
    River), 72–73
  Native Americans and, 70, 73
  transformation and, 70–72
European slave trade (African
  slave trade), 259–260,
  261, 264
European Union (EU), **138**–139
  agricultural subsidies and, *137*
  CAP and, 159
  creation of, 154, 156
  economy of, *158*
  environmental issues and, 142
  EU-27 and, 142, 145, *157*,
    164, *166*
  farms and, 138–139
  food production and, 159
  Gazprom and, 188, *205*
  GDP per capita PPP, 156, 158
  as global peacemaker, 158
  global warming and, *137*
  globalization and, 157–158
  governing institutions of,
    156, 160
  greenhouse gas emissions
    and, 148
  infant mortality rates, 167
  Internet economy, 83
  NATO and, 158–159, 160
  population growth, migration
    and, 15
  population pyramid, *164*
  as regional trade bloc, 32, 269
  renewable energy and, 145
  as rising superpower, 154–156

Romania and, 138
trading partners, *157*
women and, *137*, 166
world trade and, *157*
Evangelical Christianity, 92, 203, 282, 283
Evangelical Protestantism, **133**
Export Processing Zones (EPZs), **116**, *117*, 118, 350, 388, 389
Export trading partners (Oceania), *426*
Export-led growth, **347–348**
Exposure, 41
Extended family, 92, **128**, 129, 133, 135, 310, 342, 361, 401, 402
Extinction
    invasive species and, 419, 420
    languages, 53
    moa, 415
    sturgeon fish, 340
    Tasmanian Devil, *417*
Extraction, 19, 22, 114–115
Exxon Valdez disaster, *215*, 272

Facebook, 83, 202
Factory girls, 71
Fair trade, **32**
Families
    extended, 92, 128, 129, 133, 135, 310, 342, 361, 401, 402
    Middle and South America, 128–129
    North Africa and Southwest Asia, 223–225
    North America, 92–94
    nuclear, 92, 93, 128, 401, 402
    Southeast Asia, 401–402
Famines
    agriculture and, 23
    Bengal Famine, 304
    China, 337, 338, 344
    Indonesia, 387
    Malawi, 270
    North Korea, 349
    OXFAM and, 36
    Southeast Asia, 384
    sub-Saharan Africa, 254
Farfan, Jefferson, 205
Farming. *See also* Agriculture; Agricultural zones
    EU and, 138–139
    farmland preservation, 86
    South Asian systems, *316*
Fauna/flora. *See also* Species; Vegetation
    Australia, 414–415
    habitat loss, 59, 62, 65, 86, 299
    Middle and South America, 112
    New Zealand, 414–415
    North America, 59, 62
    Oceania, 414–415
    Pacific Islands, 415
Favelas, 27, **131**–132. *See also* Slums

FDI. *See* Foreign direct investment
Female earned income as percent of male earned income. *See* F/MEI
Female genital mutilation (FGM), **281**, 283
Female infanticide, 13, 310, 312, 361
Female seclusion, **225–226**
Feminization of labor, **388**
Fertile Crescent, *218*, *219*, 222
Fertility rates
    North Africa and Southwest Asia, 227, 228
    Pakistan, 312
    South Asia, *313*
    Sri Lanka, 312, *313*
    sub-Saharan Africa, 276
    TFR, 15, 313, 397
    Thailand, 395, *404*
Feudalism, 149, 153, 341, 342, 344, *346*
FGM. *See* Female genital mutilation
Fiji, *407*
Finland, 80, 142, *143*, 154, 166, 180, 185
Fishbone pattern, *24*
Fisheries
    Antarctica, *439*
    East Asia, 338
    Gulf of Mexico, 62
    Oceania, 416, 417, 421, 433
    threatened, *64*, *108*, *144*, *215*, 255, 299, 378, *417*
Five Pillars of Islamic Practice, 222, 223, 227
Floating population, 329, **329**, 349
Flooding
    Central China, 336
    Mexico City, 29
    Southeast Asia, *369*
Floodplain, **44**, **45**
Flora. *See* Fauna/flora
F/MEI (female earned income as percent of male earned income), **20**
F/MEI maps, 20–21
    global, 20, *21*
Food production, *11*, **22–26**. *See also* Agriculture
    deforestation and, 25
    East Asia, 327, 337–338
    EU and, 159
    Europe, *137*
    Middle and South America, 99, 120–121
    North Africa and Southwest Asia, 207, 214, 216
    North America, 57, 80–81, 97
    Oceania, *407*
    South Asia, 287, 316–317
    Southeast Asia, *369*, 379–381
    sub-Saharan Africa, 254, 257–258

sustainability and, 81, 82, 97
technology and, 24
as thematic concept, *11*, 22–26
urbanization and, 54, 98, 121, 134, 286, 287, 317, 320, 325
Food scarcity. *See* Famines
Food security, **23**, 255, 269, **337**
Footprints
    carbon, 22
    ecological, 22
    water, 37, 38
Foreign direct investment (FDI), 268, **350**, 353, 389
Foreign exchange, **399**
Foreign holders, of Treasury securities, 355, *355*
Foreign investment, 84, 101, 114, 115, 119, 124, 172, 180, 268, 269, 325, 350, 352, 389, 395
Foreign involvement, in Middle and South America, 124
Formal economies, 20
Fortaleza, Brazil, 131–132
Fossil fuel, **208**
Fossil fuel exports, 232
Fossil water, 66, 216
France
    hijab and, 164
    immigration and, *165*
    riots of 2005, *165*
Franco, Francisco, 153
Free market economies, 116, 152, 204, 363
Free trade, **31**–33. *See also* NAFTA
Free Trade Area of the Americas (FTAA), 84, 119
Free trade zones, 116, 119, 350
Freedom of press, 208, 236
Freetown, Sierra Leone, *247*, 278
Frei, Eduardo, 99
Freshwater. *See* Water issues/scarcity
Fria, Guinea, 266
Frontal precipitation, 47
Fronts, 47
FTAA (Free Trade Area of the Americas), 84, 119
Fuji mountain, *331*, 332
Fujimori, Alberto, 128
Funafuti, *430*

G8 (Group of Eight), **188**
Galápagos Islands, 415
Gandhi, Mohandas, 303, 304, *305*, 307, 308, 309
Ganga River (Ganges River), 289, 298, 299, 300
Ganga-Brahmaputra delta, 291, 298
Ganges River. *See* Ganga River
Gansu Province, *335*
Garcia, Matias, 90
Garzweiler open pit coal mine, 144, *144*

Gaza Strip, 231, 237, 239–241, *241*. *See also* occupied Palestinian Territories
Gazprom, **188**, *205*
GDP (gross domestic product), **16**, 19
GDP per capita, **19–20**
GDP per capita PPP, **20**
    China, 350, 352
    EU, 156, *158*
    global map, *21*
    India, 318, *319*
    Kazakhstan, *190*
    maps, 19–20, *20*
    Pakistan, 312
    Sri Lanka, 312
Gender, 10, **17–19**. *See also* Men; Women
    families (North America) and, 92–94
    sex *v.*, 17, 19
    as thematic concept, *10*, 17–18
    U.S. politics and, 80
Gender inequalities, 18–19
    cultural preference for boys, 13, 15, 18, 227, 312, 313, 327, 360–361
    East Asia, 327
    EU, *137*
    North Africa and Southwest Asia, 207, 225, 225–226
    South Asia, 287, 310–311, 312–313
    sub-Saharan Africa, 280–281, 284
    UAE, 229
Gender patterns
    migrants to U.S. and, 90
    Southeast Asia, 401–402
Gender roles, **18–19**
    Europe, 164–167
    Middle and South America, 99, 128–129
    North Africa and Southwest Asia, 219, 225–226
    Oceania, 435–436, 437
    Papua New Guinea, 436
    Southeast Asia, 369, 398, *404*
Gendered spaces, 222, 225–226
General Motors, 85
Genetic modification (GM), 24, 26, 81
Genocide, **36**, *122*, *155*, *195*, 237, **272**, 273, 274, 321, 356, 386, 394, *434*. *See also* Refugees
Gentrification, **88**
Geodesy, 7
Geography, 2–3, 44–54
    cultural, 50–54
    defined, 3
    human, 50–54
    physical, 3, 44–50
    regional, 3
    where/why questions, 2–3

Geographic Information Science (GIS), **7**
Geographic issues
  East Asia, 347–365
  Europe, 153–168
  Middle and South America, 113–133
  North Africa and Southwest Asia, 222–243
  North America, 74–112
  Oceania, 425–437
  Russia and post-Soviet states, 188–204
  South Asia, 304–324
  Southeast Asia, 387–403
  sub-Saharan Africa, 264–283
Geographic setting
  East Asia, 332–347
  Europe, 139–153
  Middle and South America, 101–113
  North Africa and Southwest Asia, 209–222
  North America, 58–73
  Oceania, 409–425
  Russia and post-Soviet states, 173–187
  South Asia, 289–304
  Southeast Asia, 371–387
  sub-Saharan Africa, 249–264
Geographic thematic concepts. *See* Thematic concepts
Geoinformatics, 7
Geomatics, 7
Geomorphology, 44
Geopolitics, **34**, 36, 272
Georgia (country)
  agriculture and, 204
  cabbage-planting season, *193*
  democratization, 172, *172*
  EU and, 156
  Rose Revolution, 172, 194
  Russia and, 193, 196–197
  Saakashvili and, 172
  Stalin and, 196
  Tbilisi, *171*, 172
Georgia (U.S.)
  Atlanta, 66, 87
  cash crops, 70
  industrial pig farm in, *138*, *138*
Geothermal energy, 65, 381, 383, 416
Germany. *See also* World War II
  Hamburg, 377
  iron works in, *153*
  Nazis, 151, 152, 186, 240
  open pit coal mines, *25*, *144*
  population pyramid, *164*
  pro-democracy uprising in, *169*
  Ramstein Air Base, 77
  windfarms, 145
Gers (yurts), 183, 346
Ghana, 20, 252, 259, 263, 277
Ghettos, 27
Giant panda, 340

Ginza, *351*
Girls
  cultural preference for boys, 13, 15, 18, 227, 312, 313, 327, 360–361
  mortality rates, 15–16, *16*
The Girls of Riyadh (al-Sanea), 227
GIS. *See* Geographic Information Science
Glacial melting
  Andes, *110*
  Antarctica, *439*
  Bangladesh and, 297
  Bering land bridge and, 69
  China, 336, 338, 366
  global warming and, 41, 416
  Great Lakes and, 59
  Himalayan Mountains, *13*, 293, 296, 300, 325, 381
  ice ages and, 45, 59, 69
  Kyrgyzstan, 180
  Middle and South America, 107, 110
  Russia and post-Soviet states, 181, 182
  Southeast Asia, 381, 383
  Tibetan Plateau, 335, 336
Glasnost, **190**
Global climate change. *See* Climate change
Global economic downturn (2008-2010). *See also* Unemployment
  Chinese-Mexican relations and, 118
  globalization and, 27–28
  Las Vegas and, 87
  repercussions of, 85
  Southeast Asia and, 389
Global economy
  globalization and, 83–85
  photo interpretation and, 8
  Russia and, 188–189
  workers in, 30–31
Global patterns, local lives
  Central Valley unemployment, 58
  Georgia (country) democratization, 172, *172*
  Haka (dance), 408, *408*
  Li Xia (rural-to-urban migration), 328–329, 350, 354
  logging fraud (Liberia), 248, *248*, 249
  Oz/Khoury peace efforts, 208
  pig farms (Romania), 138, *138*
  Sarawak forest dwellers, 370, 371
  Sardar Sarovar Dam, 288–289
  Secoya people, 100
Global relationships, Oceania's, 424–425
Global scale, 7

Global warming, **39–40**. *See also* Glacial melting
  Antarctica, *439*
  Bangladesh and, *12*
  carbon dioxide and, 3, 12, 39, 42, 99, 107, 109, 134, 340, 379
  carbon footprint and, 22
  deforestation and, 107
  EU and, *137*
  oceans and, *12*
Globalization, 27–33
  colonialism and, 28, 149–150
  defined, 27–28
  development and, 27–28, 33
  East Asia and, 327, 350–354, 365
  economic downturn and, 27–28
  EU and, 157–158
  Europe and, *169*
  fossil fuel exports and, 232
  free trade *v.*, 31–33
  global economy and, 83–85
  Haka and, 408, *408*
  interregional linkages and, 27, 33
  Kentucky Fried Chicken and, 57, 84, 97
  Madagascar and, *13*
  maid trade and, 399–400
  marketization and, 116
  Middle and South America and, 99, 107
  North Africa and Southwest Asia and, 207
  North America and, 57, 83–85
  of nuclear pollution, 178
  Oceania and, *407*
  Pacific Islands and, 420–421
  of resource extraction, 178–180, *189*
  South Asia and, 301–304, 318–320
  Southeast Asia and, 369, 387–389, 399–400
  sub-Saharan Africa and, 247, 264–269
  as thematic concept, *11*, 27–33
  urbanization and, 206, 227–231, *230*, 244, 350–354
GM (genetic modification), **24**, 26, 81
GMOs (genetically modified organisms), 321
Golan Heights, 237, 240, *241*, 242
Gold mining, 115, 259, 263, *407*, 420
Gold Rush, 71, 73
Gondwana, 45, 409, **412**, 415. *See also* Pangaea
Goode's interrupted homolosine projection, 6, *6*
Google, 357, 358
Gorillas, 258
Gospel of success, 133, 282, 283

Grameen Bank (Village Bank), 317–318, 320
Grand Coulee Dam, 288
Grassroots economic development, 269, **270**, 435
Great Barrier Reef, *407*, *411*, **412**, 416
Great Depression, 58
Great Lakes, 59, 72
Great Leap Forward, **344**
Great Man-Made River, 216
Great Plains, 66, 69, 73, 81
Great Rift Valley, 249, 251
Great Tradition, 305
Great Wall, 342, 343, 346, 347
Great Zimbabwe Empire, 259, *260*. *See also* Zimbabwe
Greece
  ancient, 148
  debt crisis, 157
  Iberian Peninsula and, 139
Green Belt Movement, 254
Green energy, 42
Green living, 68
Green policies, *146*, 148
Green revolution agriculture, **23–24**, 54
  Middle and South America, 98, 120, 121, 132, 133, 134
  North Africa and Southwest Asia, 229
  North America, 80–81, 96
  South Asia, 287, 304, 305, 316–317
Greenhouse gas emissions, 39–41
  automobiles and, 163
  China, *40*, 339, 341, 365
  India, *40*
  Japan, *40*
  North America, 57
  Russia, *40*, 181
  UAE, *40*
  U.S., *40*
  world, *40*
Greenhouse gases, **39**
  carbon dioxide, 3, 12, 39, 42, 99, 107, 109, 134, 340, 379
  methane, 12, 39, 40, 42, 378, 381
  nitrous oxide, 39
Gross domestic product. *See* GDP
Gross national happiness (Bhutan), 320
Groundwater pumping, 216
Groundwater recharge, 289
Group of Eight (G8), **188**
Growth poles, **350**
Guadeloupe, 5
Guangdong Province, *11*
Guangzhou (Canton), 332, *351*
Guatemala
  family in, 99
  liberation theology and, 133
  voters in, *134*

Guest workers, **163**, 227, 229, 231, 244, 359
Guevara, Che, *122*
Guinea, 266
Gujarat, 288, 289, 292, 303
Gulf of Carpentaria, 412
Gulf of Mexico, 62
Gulf of Thailand, 374, 377
Gulf States, 208, **225–226**, 229, 231, 232, 234, 235, 238, 243
Gulf War, *215*, 231
Gyanendra (King), 3*21*, 323
Gypsies. *See* Roma

Habitat loss, 59, 62, 65, 86, 299
Haciendas, **114–115**, 120
Haditha Dam, 215, *218*
Haiti
    earthquake, 36, 101
    foreign aid for, *110*
    malnutrition, *26*
    refugees from, 76
Haka, 408, *408*
Hamburg, 377
Han settlers, 362
Han-dominated Communist Party, 361
Hanoi, 394, 398
Harappa culture. *See* Indus Valley civilization
Hawaii
    conflict in, 427–428
    creation of, 412
    Pearl Harbor memorial in, *425*
    tourism and, *407*
    tropical wet climate and, *48*
HDI (United Nations Human Development Index), **20**
HDI rank maps, 20–21
Health-care systems (Canada/U.S.), 57, 79
Hemispheres, 6
Herding. *See* Pastoralism
High-altitude climates, 48, *48*
Highlands (Middle and South America), 101
Hijab, 164
Hijras, 307
Himalayan highlands, 293
Himalayan Mountains, *13*, 44
Hindi, 51, 303, 305, 309
Hinduism, 52, **305**
    caste system and, 306–308, 309, 310, 323, 324
    South Asia and, 305–308
Hindu-Muslim relationship, 308–309
Hindutva, 323
Hispanics, **58**
    culture of poverty and, 91, 96
    Latinos *v.*, 91
    median household income, *91*
    religion and, 92
    U.S. population, *91*

History. *See* Human patterns over time
HIV (Human immunodeficiency virus), 279
HIV/AIDS
    in Africa, 14
    population patterns and, 14
    in Southeast Asia, 403
    in sub-Saharan Africa, 267, 275, 276, 277, *277*, 279, 279–280
Ho Chi Minh, 384, *394*
Ho Chi Minh City, *372*, 382, 396, 398
Hodder, Ian, 219
Hoi An, 388
Hokole'a canoe, *424*
Holocaust, **152**, 221, 231
Homo erectus, 259
Homo sapiens, 259, 260, 263
Homo sapiens sapiens, 53
Hong Kong
    Amah day in, 400
    East Asia and, 350, 352
    pro-democracy demonstration, 365
Hong Kong Island, *331*
Horn of Africa, **252**, 272
Hot spots, **412**
Houseboat, Cambodia, 369
Households (North America), 91, 93, *93*
Huang He (Yellow River), 291, 332, 333, 336
Hub-and-spoke network, 82
Hui people, 363, 365
hukou system, **329**
Humans
    early, *260*, *304*
    well-being, *20*
Human Development Index. *See* HDI
Human geography, 50–54
    components of, 50
    defined, 3, 44
    physical geography *v.*, 3
Human immunodeficiency virus (HIV), 279. *See also* HIV/AIDS
Human impact on biosphere, 22, *24–25*
    East Asia, 338, 339, 340
    Europe, 142, *144*, 145–146
    Middle and South America, 107, *108*, 109
    North Africa and Southwest Asia, 213–214, *215*, 216
    North America, 59, 62, *64*, 65–66, 68
    Oceania, 412, 416, *417*, 419–421
    Russia and post-Soviet states, 176, 178, *179*, 180–181
    South Asia, 295–296, 298, 299, 300

Southeast Asia, 375, 378, 379–381
    sub-Saharan Africa, 252, 254, *255*, 257
    urbanization and, 24
    Western Europe, *144*
Human patterns over time (history). *See also* Visual timeline
    East Asia, 341–347
    Europe, 148–153
    Middle and South America, 109–113
    North Africa and Southwest Asia, 218–222
    North America, 69–73
    Oceania, 421–425
    Russia and post-Soviet states, 183–187
    South Asia, 300–304
    Southeast Asia, 383–387
    sub-Saharan Africa, 259–264
Human Rights Watch, 203, 399
Humanism, **149**
Humid continental climates, **142**
Hunger strikes, 288–289, 303. *See also* Malnutrition
Hunting and gathering, 9, 22, 23, 70, 219, 254, 422
Hurricanes
    Bangladesh and, *43*
    climate change and, 41, *42*, 43, *110*
    frontal precipitation and, *47*
    Ike, *43*
    Katrina, 57, 67
    Middle and South America, 106–107
    Mitch, *42*, *110*
    typhoons, 334, 381, 382, 383, *418*
Hussein, Saddam, 236, 238, 239
Hutments, 27
Hydroelectric industries
    Central Valley, 58
    Three Gorges Dam and, 288, *331*, 338, 339, 340, 341

Iberian Peninsula, 139
Ice ages, 45, 59, 69, 374, 375, 383, 422
Ike (Hurricane Ike), *43*
IKEA (furniture manufacturer), 248
Illegal immigration. *See* Undocumented immigration
Imbalance, sex, 15
IMF. *See* International Monetary Fund
Immigration. *See also* Migration; U.S.-Mexico border
    Afghanistan and, 163
    Canada and, 78, 88–90
    Europe and, 163–164
    France and, *165*
    Mexico to U.S., 84

Oceania and, 437
    undocumented, 30, 76, 84, 89, 90, 95, 119, 163
    U.S. and, 78, 88–90
    U.S./Canada, 78
Imperial China, 341–342
Imperial expansion, Russian, *185*
Import quotas, 31
Import substitution industrialization (ISI), 114, **115**–116, 119, 234, 235, 318, 388
Incas, **111**, 114, 115
Income (median U.S. income)
    household/education and, 93, *93*
    race and, *91*
Income disparities (wealth disparities), **113**. *See also* F/MEI maps
    China, 365
    female *v.* male, 18*t*, 57
    Middle and South America, 113–114
    Moscow, *192*
    SAPs and, 32
    South Asia, *314*
India
    agribusiness and, 316
    Gandhi and, 303, 304, *305*, 307, 308, 309
    GDP per capita PPP, 318, *319*
    greenhouse gas emissions, *40*
    independence, 302–303, *303*
    IT sector and, 304, *319*, 320, 431
    literacy rates, 309, *310*
    nuclear weapons and, 322
Indian Ocean tsunami (2004), 36, 292, 393
Indian-Australian Plate, 44, 45, 46, 292, 374, 412
Indigenous African belief systems, 281–282
Indigenous groups, **101**, *371*. *See also* Native Americans
    Aborigines, 422, 423, 424, *424*, 425, 431–432, 433, *434*, 435, 437
    Alaska Natives, 62, *91*
    Aztecs, 111, 114, *115*
    Canada and, 96
    East Asia, 364, 365
    environmental/development issues and, 100, *100*
    Incas, 111, 114, 115
    Maori, 408, 412, 415, 419, *423*, 431–432, 433, 437
    Middle and South America, 101, 104, 109, 111–112, 113, 121, 123, 126, 127, 132, 133
    Nigeria and, 271
    Nilgiris, 298
    North Africa and Southwest Asia, 240

Indigenous groups (*Continued*)
  Occidental Petroleum and, 100
  Oceania, 408, 414, 420, *423*,
    431–432, 437
  religions, 52
  Russia and post-Soviet states,
    176, *179*, 184, 196, 203
  Sarawak, 370, 371
  skin color of, 53
  South Asia, 300, 308
  Southeast Asia, 370, 371, 375,
    383, 384, 400, 403
  Sub-Saharan Africa belief
    systems, 281–282
  U.S. and, 96
  worldwide, *371*
  Zapatistas and, 120–121, 124
Indo-Gangetic Plain, *291*,
  292, 309
Indonesia
  Bali, 374, 383, 393, 395, *401*
  deforestation, *378*
  democratization, 391
  ethnic groups, 391
  famines, 387
  Java, 373, 400, *401*, *402*
  Kalimantan, 374, 377, 378
  Megawati and, 402, *404*
  population pyramids,
    395–396, *397*
  Suharto and, 389, 391, 392,
    393, 394, 402
  Sumatra, 292, 373, 374,
    375, 393
Indus Valley, *291*
Indus Valley civilization (Harappa
  culture), **300**
Industrial areas/land transport
  routes (Russia and post-
  Soviet states), 191, *191*
Industrial hog farm, *138*
Industrial pollution. *See also*
  Air pollution
  Dzerzhinsk and, *178*
  Europe, 151
  Ganga River and, 298
  Norilsk Nickel, 178, *179*, 180
  Poland, 145
  Russia and post-Soviet states,
    178, 181
Industrial Revolution, 28, 30, 101,
  149, 150, 153, 169, 221, 259
Industrial self-sufficiency, South
  Asia and, 318
Infant mortality rates
  African-Americans, 90
  Canada, *79t*
  Cuba, 124
  EU, 167
  Israel, *240t*
  Middle and South America, 129
  Pakistan, 312
  Palestine, *240t*
  Southeast Asia, 388, 395
  Sri Lanka, 312

sub-Saharan Africa, 275, 276
  U.S., *79t*
  young girls, 15–16, *16*
Infanticide, *13*, 310, 312, 361
Informal economies, **20**, **266**
  Middle and South America,
    119, 129, 131, 132, 133
  Russia and post-Soviet states,
    191–192
  sub-Saharan Africa, 266–267,
    268, 271
Informal religions. *See* Religions
Information technology (IT)
    industries, **83**
  China and, 357
  Europe and, 160
  India and, 304, *319*, 320, 431
  Middle and South America, 124
  outsourcing of, 85
Infrastructure, **70**
Inner Mongolia, 339
Interaction, spatial, **3**
Interdependencies (U.S./Canada
    relationships), 76
International Atomic Energy
    Agency, 198
International Monetary Fund
    (IMF), 32, 36, 116, 118, 156,
    265, 267, 268, 318, 388, 389
International Panel on Climate
    Change, 258
International Rice Research
    Institute, 382
Internet economy, 83
Internet usage. *See also* Information
    technology industries
  East Asia, 358, *358*
  Middle and South America,
    124, *125*, *134*
  social networking, 57, 83
Interregional linkages, **27**, 33
Interstate Highway System, 81, 85
Intertropical convergence zone
    (ITCZ), **252**, 253, 258,
    292, 293, *293*, 333, 374,
    376, *413*
Intifada, **240**, 242
Introduced species, 415, 419
Invasions, into South Asia,
    300–301
Invasive species, 254, 407,
    419–420, 421
Iowa, 80, 82, 295
Iran, 229
  pro-democracy demonstrations
    in, *207*
  refugees in, 231
Iraq
  democratization and, 35
  dictators, 231, 236
  ethnic groups, 239
  mother and child in, *207*
  oil facility/rocket fire
    and, *25*
  refugees, in Iran, 231

Iraq War (2003-2010), 238–239
  Hussein and, *236*, 238, 239
  Obama administration and, 74
  U.S. strategic and economic
    interests and, 76, 77
Ireland (Republic of Ireland),
    155. *See also* Northern
    Ireland
Iron curtain, **152**
Iron works, German, *153*
Irrigation, *12*, *13*
  California, 73, 81
  drip, 214, 243, 296, 418, 419
  East Asia, 335, 336, 338, 349
  Incas, 111
  North Africa and Southwest
    Asia, 209, 213, 214, 215,
    216, 218, 219, 243
  North America, 66, 67
  Oceania, 418, 419
  Ogallala aquifer and, 66, 69
  Russia and post-Soviet states,
    180, 181, *182*, 205
  South Asia, 288, 289, 296, 297,
    300, 317
  sub-Saharan Africa, 247, 254,
    255, 257, 279, 284
ISI. *See* Import substitution
    industrialization
Islam, 52, **208**. *See also* Muslims
  Europe and, 148
  Pillars of Islamic Practice, 222,
    223, 227
  Qur'an, 213, 220, 223, 227, 235
  shari'a and, 223, 322
  spread of, 220–221, *221*
  sub-Saharan Africa and, 282
  in U.S., 91, 92
  women's rights and, 226–227
Islamic fundamentalism, 324
Islamism, **208**, 235, 236, 322
Isorhythmic maps, 4, *4*
Israel
  creation of, 240
  infant mortality rates, *240t*
  population pyramid, 229
Israeli-Palestinian conflict,
    239–243
  conflict/cooperation in, 242
  Gaza Strip and, 231, 237,
    239–241, *241*
  Golan Heights and, 237, 240,
    241, 242
  Israel and Palestine
    (1923-1949), *241*
  Israel and Palestine Territory
    (after 1949), *241*
  West Bank and, 208, 209, 231,
    237, 239, 240, *241*, *242*,
    243, *244*
IT industries. *See* Information
    technology industries
ITCZ. *See* Intertropical
    convergence zone
Ivan IV, 184

J curve, 9, 14
Jainism, 305, **308**
Jaisalmer, 294
Jakarta, 394
jalee, 309
Jamaica, 40, 41, 127
Japan
  Ainu in, 364, 365
  creation of, 332, 334
  early history, 344–345
  expansionism, 345, *345*
  greenhouse gas emissions, *40*
  imperialism, China and, 343
  political shifts, 358–359
  population growth,
    359–360
  rise of, 348
  Yokohama, 348, 350,
    361, 422
jatis, **306**, 307
Java, 373, 400, *401*, *402*
Al Jazeera, 208, 209, 238,
    239, 243
Jesus statue, *203*
Jews. *See also* Israeli-Palestinian
    conflict
  Holocaust and, 152, 221, 231
  Judaism and, 52, 132, 176,
    218, 220, 222
  Russia and post-Soviet states
    and, 203
  A Tale of Love and Darkness
    (Oz) and, 208
  yarmulkes for, 164
João Pessoa, 99
Johnson-Sirleaf, Ellen, 248, 273,
    274, 275
Jordan
  demographic transition of, 16
  GDP per capita, 16
  population pyramid for,
    15, *15*
  TFR for, 15
Judaism, 52, 91, 92, 132, 176,
    218, **220**, 222

Kabila, Joseph, 274
Kabul (Afghanistan)
  climate change vulnerability, *43*
  water access and, 39
Kabul Valley, 293, 295
Kaizen system, **348**
Kalahari Desert, 252, 253
Kalimantan, 374, 377, 378
Kamkwamba, William, 270
Kanban system, 348, **348**
Karaoke, *137*, 354
Kasanka National Park, 258
Kashmir
  conflict in, 320, 322
  high altitude climate
    of, *294*
Kassem, Hisham, 238
Kasur, 299
Katrina hurricane, 57, 67

Kazakhstan
  Astana, 198, *200*
  climate/vegetation, 176, *177*
  GDP per capita PPP, *190*
  nuclear testing in, 178
  pipeline construction in, *190*
  population pyramid, *201*
  Russia and post-Soviet states
      and, 173
  Stalin and, 196
  water issues, 180, 181
Kentucky Fried Chicken, 57,
      84, 97
Kenya
  Nairobi, *11*, 36
  presidential elections (2008)
      and, 77
  riots, 77
  slum, 39
Ketsana (typhoon), 382
Khan, Genghis, 184, 342
Khao Yai, 376
Kharkov, *200*
Khmer Rouge, 386, 387
Khoury, Elias, 208
Kibbutzim, 240
Kibera slums, *11*, 278
Kidnapping, 124, 202, 248, 259,
      323, 361, 403
Kilimanjaro mountain, 249,
      250, 251
Killing fields, 386, 387
Kim Dae-jung, 356
King Louis XIV, *153*
Kinshasa. *See* Congo
Kirkuk, 215
Knowledge economy, 82, 83
Koli village, 315
Kolkata (Calcutta), 298, 302
Konya, 219
Köppen classification system, 47
Koran (Qur'an), **213**, 220, 223,
      227, 235
Korean Peninsula, 332, 341,
      345, 361
Korean War, 345–346
Kosovo Liberation Army, *155*
Krill, *439*
Kroo Bay slum, 278
Kryshtanovskaya, Olga, 194
Kshatriyas, 306, 307
Kuala Lumpur, 208
Kurds, 50, *51*, 208, 209, 237, 239
Kutai National Park, 378
Kuzmic, Vera, 159, *159*
Kyoto Protocol, **42**, 181, 336
Kyrgyzstan, 176, 180, 194, 195,
      201, *201*, 202, 203

La Rambla, *162*
Laccadives Islands, 289
Ladakh, 296, 297, 322
Lake Chad, 258
Land claims, 371, 432
Land reform, 121

Land transport routes/industrial
      areas (Russia and post-Soviet
      states), 191, *191*
Landforms, **44–45**. *See also* Plate
      tectonics
  East Asia, 332
  Europe, 139–142
  Middle and South America,
      101–104
  North Africa and Southwest
      Asia, 209–213, 218
  North America, 58–59
  photo interpretation and, 8
  physical geography and, 44–45
  Russia and post-Soviet States,
      173–176
  South Asia, 292
  Southeast Asia, 374
  sub-Saharan Africa, 249–251
Land-for-peace formula, 242–243
Landless movement, 121,
      302, 344
Languages, 51, 53. *See also
      specific languages*
  dialects and, 53
  South Asia, 305, 306
Laos
  Agent Orange, 385, 387
  deforestation, 25
  wind power and, 383
Las Vegas, 87
Lashkar-e-Taiba, 324
Latin America, 101. *See also*
      Middle and South America
Latinos, *91*. *See also* Hispanics
Latitude lines (parallels), 6, **6**
Lattice screens, 225, 309
Leather tanneries, 299, 308
Lebanese oil spill, *215*, 240
Legends (maps), 4, *4*
Lenin, Vladimir, *185*, 186,
      187, 189
Lenin statue, *203*
Leonhardt, David, 58
Li Xia (rural-to-urban migration),
      328–329, 350, 354
Liberation theology, **133**
Liberia, logging fraud in, 248,
      248, 249
Lima, Peru, *115*, *130*
Limpopo River, 257
Lines of latitude (parallels), 6, **6**
Lines of longitude (meridians),
      3, 6
Linkages, interregional, 27, 33
Lions Club, 33
Literacy rates
  Delhi, 315
  India, 309, *310*
  Kerala, 313
  Mumbai, 315
  Sri Lanka, 309
  sub-Saharan Africa, 276
  Thailand, 395
  Vietnam, 395

Little Tradition, 305
Liu Xiaobo, 357
Livability, 86, 87, 88
Living wages, **32**
Local scale, **7**
Location (photo interpretation), 8
Loess Plateau, 330, 333, 336
Logging. *See also* Deforestation
  Brazilian Amazon, *24*
  in Central African Republic, 255
  clear-cutting, 64, 65, 252,
      254, 370
  logging fraud (Liberia), 248,
      248, 249
  North America, 62, 65
  Sarawak forest dwellers and,
      370, 371
  tropical timber trade, 377
London
  bombing (2005), 164
  multinational corporations
      and, 160
  service economies and, 159
Longitude lines (meridians), **3**, 6
Louis XIV, King, *153*
Louisiana wetlands, 62
Lower levels, of democratization,
      34, 154
Lowlands
  central, 59, *60*, *61*
  coastal, 59, 61, 127, 139, 249,
      290, 412
  East Asia, 332
  Europe, 139
  India, 290, 292, 293
  Mali, 257
  Middle and South America,
      101, *103*, 104, 111, 127
  Oceania, 412, 420
  Sarawak, 370
  Southeast Asia, 374, 375
  sub-Saharan Africa, 249, 258
  West Siberian Plain, 173, 174,
      176, 198

Maathai, Wangari, 254
Macapagal-Arroyo, Gloria, 402
Machismo, **129**, 133
Machu Picchu, *114*
Mack, Louise, 431
Madagascar, *13*, 255, 383
Al Madinah (Medina), *211*,
      220, 223
Madras. *See* Chennai
Madrid bombing, 164
Mafia, Russian, 197, 202
Magellan, Ferdinand, *386*
Magma, 44
Maharashtra, 297
Maid trade, 399–400
Makatea, 412
Makkah (Mecca), 211, 220,
      223, 259
Malaria, 38, 257, 277, 279,
      280, 416

Malawi, 267, 270
Malaysia
  deforestation, 378
  foreign workers and, 30
  Kuala Lumpur, 208
  oil palm plantations, 378, 379
  Overseas Chinese in, 364, 365
  unemployment in, 30
Mali, 253, 257
Malnutrition
      (undernourishment), 26
  Congo (Kinshasa), 35
  hunger strikes, 288–289, 303
  Middle and South America, 132
  North Africa and Southwest
      Asia, 227
  Oceania, 424
  South Asia, 26, 317
  unemployment and, 58
  world, 26
  Zimbabwe, 274
Manchuria, 332, 334, 344, 345
Mandela, Nelson, 263
Manila, 369, 381, 382, 397, 400
Mao Zedong, 323, 344
Maoists, 320, 321, 323, 329
Maori people, **408**, 412, 415,
      419, 423, 431–432, 433, 437
Maps, 2, 2–7, *4–5*. *See also* F/
      MEI maps; GDP per capita
      PPP; HDI rank maps;
      Regional maps; *specific maps*
  projections, 6–7
  scales, 3, 5, *5*
Maquiladoras, **116**, 118, 119, 388
Mara River, 258
Marcos, Ferdinand, 402
Marianismo, **128**, 133
Market reforms, in China,
      349–350
Marketization, 114, **116**, 202,
      342. *See also* Sex industry
Marriages, arranged, 227, 400
Marshall Plan, 154
Marsupials, **415**
Marx, Karl, 34, 145, 151, 152,
      154, 186, 189
Masai, *11*, 258
Massachusetts, 57, 66, 70, 72
Material culture, 8
Mathare slum, 268, 278
Mato Grosso, *108*
Mau Piailug, 421–422
Mauritania, 259
Mauritius, 247, 276
Mayan temple, *103*
McKinley, Jesse, 58
Mead, Margaret, 435
Mecca (Makkah), 211, 220,
      223, 259
Medieval period, 153, 220
Medina (Al Madinah), *211*,
      220, 223
Mediterranean climates, 48, 49,
      73, **142**

Mediterranean Sea, 136, *137*, *142*, 145, *146*, 148, 168
Medvedev, Dimitri, 194
Mega-financier, East Asia as, 354
Megalopolis, **86**
Megawati Sukarnoputri, 402, *404*
Meghalaya, *294*
Mekong River Plains, *373*
Melanesia, **422**
Melanesians, **422**
Melbourne, 407, 416, 430
Melting glaciers. *See* Glacial melting
Melville, Herman, 435
Men. *See also* Gender
  cultural preference for boys, 13, 15, 18, 227, 312, 313, 327, 360–361
  machismo and, 129, 133
Mercantilism, 112, 114, **149**, 150, 264
Mercator projection, 6, *6*
Mercer Quality of Living Survey, 86
Mercosur (Southern Common Market), 32, 117, **119**
Meridians. *See* Lines of longitude
Mestizos, 101, **112**, 123, 128
Methane, 12, 39, 40, 42, 378, 381
Metropolitan areas, **86**, 94, 315, 361, 397
Mexicali workers (Thompson Electronics), 117–118
Mexico. *See also* NAFTA; U.S.-Mexico border; *specific cities*
  China and, 118
  climate change vulnerability, *67*
  computer conference in, *134*
  drug trade, *122*
  immigration to U.S., 84
  maquiladoras and, 116, 118, 119, 388
  Middle America and, 101
  Zapatistas and, 120–121, 124
Mexico City
  Catholics/swine flu and, *132*
  flooding in, *29*
  public dance in, *99*
Mickey Mouse statue, *203*
Microcredit, **317**–318, *324*
Micronesia, **422**
Mid-Atlantic economic core, **72**
Middle Ages, Early, 148
Middle America, **101**. *See also* Central America
Middle and South America (world region), 98–135
  agricultural zones, *120*
  birth rates, 126, 127, 133
  climate, 104–107
  climate change vulnerability, 107, 109, 110, *134*
  climate zones, *105*
  colonialism (European colonialism), 111–113, 126

Crowley's model of urban land use, *129*
  cultural diversity, 127–128
  deforestation, 107–108
  democratization/conflict issues, 121–125, *122*, *134*
  development, 99, 107, 109, 112–116, *134*
  dictators and, 121, *122*, 124, 134
  drug trade, *122*, *123*, 123–124
  economic development phases, 114–116
  economic issues, 113–120
  environmental issues, 107–109
  European conquest of, 111–113
  families, 128–129
  fauna/flora exchange, 112
  food production, 99, 120–121
  foreign involvement (politics) and, 124
  gender roles, 99, 128–129
  geographic issues, 113–133
  geographic setting, 101–113
  globalization and, 99, 107
  green revolution agriculture, 98, 120, 121, 132, 133, 134
  human impact on biosphere, 107, *108*, 109
  human patterns over time, 109–113
  hurricanes, 106–107
  income disparities, 113–114
  indigenous groups in, 101, 104, 109, 111–112, 113, 121, 123, 126, 127, 132, 133
  informal economy, 119, 129, 131, 132, 133
  Internet usage, 124, *125*, *134*
  IT and, 124
  landforms, 101–104
  map scale and, 5
  migration and, 126, 129
  mineral zones, *120*
  optimism for, 109, 121, 124, 133, *134*
  peopling of, 111
  physical patterns, 101–107
  political issues, 113–120, 121–125
  political map, *98*
  population density, 111, *126*
  population distribution, 126–127
  population growth, 99, 127, *127*
  population patterns, 126–127
  race and, 128
  reflections on, 133–134
  regional map, *102–103*
  religions, 132–133
  rural-to-urban migration, 126, 129, 132, 133
  SAPs and, 116, 118
  skin color and, 128
  sociocultural issues, 126–133

temperature-altitude zones, 104–106, *106*
  thematic concepts overview, 99, 134–135
  underdevelopment, 112–113
  urbanization, 98, 99, 109, 121, 126, 129, *130*, 131–132, *133*, 134, 135
  visual timeline, *114–115*
  water crisis, 99, 109
  as world region, *1*, 3, 7
Middle East, 209. *See also* North Africa and Southwest Asia
Middle West, 95
Midlatitude climates, 48, 49, 333, *413*
Migiro, Asha-Rose, *247*
Migration. *See also* Immigration; Rural-to-urban migration
  defined, 15
  diaspora and, 220, 269, 282, *302*
  into Europe, 163–164, *165*
  maid trade and, 399–400
  Middle and South America and, 126, 129
  North Africa and Southwest Asia and, 229
  return-migration (South Asia), 319
  sub-Saharan Africa and, 277
  urbanization and, 129, 135, 227–231
  to U.S., 90
Milford Sound, *411*
Mindanao, *394*
Mineral zones, Middle and South America, *120*
Mining
  Alaska, 64
  coal, 65
  development and, *25*
  diamonds, 249, 263, 265, 284
  East Asia, 338, 349
  Europe, 148, 150
  extraction and, 19, 22
  Garzweiler open pit coal mine, 144, *144*
  gold, 115, 259, 263, *407*, *420*
  Middle and South America, 115, *120*
  Norilsk Nickel and, 180
  Oceania, 420, 421, 422, 427, 429, 432, 433
  Ok Tedi Mine, *420*
  in Papua New Guinea, 420–421
  Russia and post-Soviet states, *191*, 198
  steel industry and, 72
  strip, 65, 421
  sub-Saharan Africa, 261, 265
  Zimbabwe, 259
MIRAB economy, **427**

Mississippi River, 62, 68, 72
  delta, 59
  European settlements and, 72–73
Mitch (Hurricane Mitch), 42, *110*
Mixed agriculture, **254**, 256, 259
Miya, Maurício, *128*
Moa, 415
Mobility, in North America, 95
Modi, Narendra, 288
Mohenjo-daro, *304*
Mongolia
  history, 346
  Inner Mongolia v., 339
  malnutrition, 26
  steppes in, 49, 333
Mongols, **184**
Monotheistic belief systems, 208, **220**, 222
Monotremes, **415**
Monsoons, 47, **292**
  East Asia, 334
  rain belt and, 46–47
  South Asia, 292, 293, 294, 295, *295*, 296, 297, 298, 300
Montserrat, 101, *103*
Mormon religion, 52, 91, 92
Moro Islamic Liberation Front, *394*
Morocco, 221, 225, 232, 234, 236
  Casablanca, *230*
  climate change vulnerability, *43*, 207
Mortality rates. *See* Death rates; Infant mortality rates
Moscow
  income disparities, *192*
  as primate city, *200*
  School of Management, 188
Mosques, 51, 88, 91, 164, 223, 236, 239, 363
Mount Fuji, *331*, 332
Mount Kilimanjaro, 249, 250, 251
Mountain zones
  central, 139
  Pacific Mountain Zone, *175*, 176
  Rocky Mountain Zone, 58
  tropical, 104
Mountaintop removal, 65
Mt. Ararat, 211
Mudumali Wildlife Sanctuary, 298
Mugabe, Robert, 269, 273
Mughal Empire, 220, 221, *304*
Mughals, **300**
Muhammad (Prophet), 208, 213, 220, 221, 222, 223, 226, 282
Mujahedeen, 187, 322
Mullahs, 223
Mulley, Phillip, 298
Multimodal transport (Europe), 148
Multinational corporations, **30**
  Bechtel, 30, 37, 109
  Cisco, 30

Del Monte, 120
ETDZs and, 350
living wages and, 32
London and, 160
privatization and, 116
Russia and, 188
Shell, 8, 30, 272
Southeast Asia and, 389
Walmart, 30, 82, 83, 84
water ownership and, 37
Multiplier effect, 71
Mumbai (Bombay)
lunch delivery, 305
monsoons, 297
sea level rise, 41
terrorist attacks (2008), 323–324
Murray-Darling river system, 414
Muslims, **220**. *See also* Islam
in Afghanistan, 308
in Europe, 164, *165*
Hindu-Muslim relationship, 308–309
Mutiny on the Bounty, 435
Myanmar. *See* Burma
MySpace, 83

NAFTA (North American Free
Trade Agreement), **82**,
83–84, 86, 90, 117, *117*,
119, 120, 269
Nairobi, *11*, 36, 274, 278
Namibia, *49*
Narmada River, 288, 289
National People's Congress, 355
Nationalism
defined, 151
religious, 309, 323–324
Nationalize, **118**
Native Americans. *See also*
Indigenous groups
culture of poverty and,
91, 96
disease/technology and, 70
European arrival/settlements
and, 70, 73
median household income, *91*
population growth, 73
NATO (North Atlantic Treaty
Organization), **158**
Afghanistan and, 74, 158
Cold War and, 158
EU and, 158–159, 160
Kosovo Liberation Army
and, *155*
Taliban and, 322
Ukraine and, 196
Natural gas resources. *See* Oil/
natural gas resources
Natural hazards, East Asia, 334
Nauru, 420–421
Nazca Plate, 45, 46, 101
Nazi death camps, 240
Nazis, 151, 152, 186, 240
Nepal, rebels, 323

Netherlands
climate change and, 139
delta works, *147*
European colonialism and,
149, 150, 153
Rotterdam, 142, 148, *169*
sea level rise, 139, *147*
New Delhi, 288, 289, 298, 299.
See also Delhi metropolitan
area
New England, 70, 71, 72, 73
New Guinea, 52. *See also* Papua
New Guinea
New Mexico, 60, 90
New Testament, 220
New Urbanism, 88
New York City
Egypt *v.*, 227
nineteenth-century
transportation times, 72, 72
population density, 87
wealth, 87
New Zealand
Asian economy and, 425–427
climate, 412, *413*, 414
colonization of, 422–424
democratization and, 35
development, *407*
fauna/flora, 414–415
invasive species, 419–420
Maori people, 408, 412, 415,
419, 423, 431–432, 433, 437
sheep in, *407*
wind farm in, *437*
NGOs (nongovernmental
organizations), **36**
Human Rights Watch, 203, 399
Open Society Institute, 172
OXFAM International and, 36
Saakashvili and, 172
SOS Racisme, 165
Ukraine and, 172, *195*
Nicaragua, 26, 114, 118, 122,
124, 129, 187
Nicholas II, Czar, 185
Nickel, *120*, 178, *179*, 180
Nicobar Islands, 289
Niger, 247, 261
Niger River, 255, 257, 259, 272
Niger River Delta, 251, 272, 288
Nigeria
conflict in, 271–272
indigenous languages and, 271
population pyramid, *276*
Rukpokwu state (oil pipeline
burst), 8
Nile Delta, 216, 217
Nile River, 216, 249
Nilgiri Hills, 298
Nilgiris people, 298
9/11 attacks. *See* September 11th
attacks
Ningxia Huizu Autonomous
Region, 336, 363
Nitrous oxide, 39

Nobel laureates
Aung San Suu Ki, *394*
Kim Dae-jung, 356
Liu Xiaobo, 357
Maathai, Wangari, 254
Mandela, Nelson, 263
Sen, Amartya, 20
Yunus, Muhammad, 317
Nogales, Mexico, *110*, *111*, *118*
Nomadic pastoralists, 183, **184**,
341, 349
Nongovernmental organizations.
See NGOs
non-Han minorities, 361
Nonnative species, 65, 254,
416, 419
Nonpoint sources of pollution, 178
Norilsk Nickel, 178, *179*, 180
Norms, 50–51
North Africa and Southwest Asia
(world region), 206–245
Arab world *v.*, 209
climate, 209
climate change vulnerability,
207, 216, 217, 218, 243, 244
climate zones, *212*
colonial regimes in, 224
democratization/conflict issues,
207, 236–237, 238, 243, 244
development, 207, 232–233, 244
earthquakes, 209
economic diversification and,
216, 234, 244
economic issues, 231–243
environmental issues, 213–218
family values, 223–225
fertility rates, 227, 228
food production, 207, 214, 216
gender inequalities, 207, 225,
225–226
gender roles, 219, 225–226
gendered spaces, 222, 225–226
geographic issues, 222–243
geographic setting, 209–222
globalization and, 207
green revolution agriculture, 229
human impact on biosphere,
213–214, 215, 216
human patterns over time,
218–222
landforms, 209–213, 218
Middle East *v.*, 209
migration and, 229
optimism for, 243, 244
physical patterns, 209–213
political issues, 231–243
political map, 206
population densities, 227
population density, 228
population growth, 207, 227,
243, 244
population patterns, 227
reflections on, 243
regional map, *210–211*
religions, 222–223

rural-to-urban migration, 207,
225, 229
sociocultural issues, 222–231
thematic concepts overview,
207, 244
unsolvable problems and, 243
urbanization, 206, 207, 218,
222, 227–231, 230, 243
vegetation, 209, 213
visual timeline, 222–223
water issues, 207, 213–216
as world region, *1*, *3*, *7*
North America (world region),
56–97. *See also* Canada;
United States
aging population, 57, 95
agriculture in, *81*
air pollution, 68
air transportation, 81–82
birth rates, 95, 96
changing regional composition
of, 73
Christianity in, 91–92, *92*
climate change vulnerability,
57, 65–66, 67, 69, 97
climate zones, 63
climates, 57, 59, 63
cultural diversity, 58, 95
culture of poverty in, 91, 96
democratization/conflict issues,
76–77
development and, 57
economic issues, 80–85
environmental issues, 59–69
ethnicity in, 90–91
European arrival (1600s), 70
families, 92–94
fauna/flora, loss of habitat,
59, 62
food production systems, 57,
80–81, 97
geographic issues, 74–112
geographic setting, 58–73
globalization and, 57, 83–85
green revolution agriculture,
80–81, 96
greenhouse gas emissions, 57
households, *91*, 93, *93*
human impact on biosphere,
59, 62, *64*, 65–66, 68
human patterns over time,
69–73
landforms, 58–59
mobility in, 95
optimism for, 81, 86, 96, 97
peopling of, 69–70
physical patterns, 58–59
political issues, 74–75
political map, 56
population density, 94
population patterns, 94–96
race in, 90–91
reflections on, 96
regional map, *60–61*
religion, 91–92, *92*

North America (*Continued*)
  sociocultural issues, 86–94
  thematic concepts overview,
    57, 97
  transportation networks, 81–82
  urbanization, 56, 57, 65, 68,
    86, 87, 88, 95, 97
  visual timeline, 70–71
  water pollution, 57, 66, 68, 68
  as world region, 1, 3, 7
North American Free Trade
    Agreement. *See* NAFTA
North American Plate, 45, 58,
    59, 101
North Atlantic Drift, **142**, *143*
North Atlantic Treaty
    Organization. *See* NATO
North China Plain, 330, 332,
    333, 334, 336, 339, 341, 361
North Europe, *136*, 138, 139
North European Plain, 139, 141,
    142, 143, 145, 149, 173, *174*
North Korea
  malnutrition, *26*
  nuclear weapons, 349
North Sea, 143, 145
Northern Hemisphere, 6
Northern Ireland, *155. See also*
    Ireland
Northern settlements (North
    America), 71–72
Norway, *48*
Nuclear family, **92**, 93, 128,
    401, 402
Nuclear pollution, *100*, 178, 332,
    336, 420, 421
Nuclear power, 65, 142, 145,
    178, 336, 340
Nuclear power disaster
    (Chernobyl), 178
Nuclear weapons
  black market materials for, 196,
    197–198
  on Hiroshima/Nagasaki, 348
  India and, 322
  North Korea, 349
  Pakistan and, 322
  Russia and post-Soviet states, 178
  USSR/Soviet Union, *185*
Nuclear Weapons–Free Zone
    Treaty, 389
Nunavut territory, 73, 80

Obama, Barack, 80, 90, 239, 317
Obama administration, 74, 75
Occidental Petroleum, 100
occupied Palestinian Territories
    (OPT), **209**, 242. *See also*
    Gaza Strip; Israeli-Palestinian
    conflict; West Bank
Oceania (world region), 406–438.
    See also Australia; New
    Zealand; Pacific Islands;
    Papua New Guinea
  climate, 412–414

climate change vulnerability,
    407, 418, 419, 437, *437*
climate zones, *413*
components of, 409
continent formation, 409, 412
cultural diversity, 435
democratization/conflict issues,
    407, 433, 434, 435–436, 437
development, 407
economic issues, 425–428
environmental issues, 416–421
European arrival, 422
export trading partners, *426*
fauna/flora, 414–415
food production, 407
future of, 428, 437
gender roles, 435–436, 437
geographic issues, 425–437
geographic setting, 409–425
global relationships, 424–425
globalization and, *407*
human impact on biosphere,
    412, 416, 417, 419–421
human patterns over time,
    421–425
immigration into, 437
indigenous groups, 408, 414,
    420, 423, 431–432, 437
island formation, 412
mining and, 420, 421, 422,
    427, 429, 432, 433
nuclear pollution, 420, 421
nuclear testing, *417*
optimism for, 416, 428, 436,
    437, *437*
peopling of, 422
physical patterns, 409–415
political issues, 425–428
political map, *406*
population density, 428, *429*
population growth, *407*, 437
population patterns, 428–429
reflections on, 437
regional map, *410–411*
service economies, 427, 429,
    437, 438
sociocultural issues, 431–437
terms usage, 409
thematic concepts overview,
    407, 438
tourism and, 427–428
urbanization, 406, *407, 415*,
    429–431, *430*, 437, 438
visual timeline, 424–425
as world region, 1, 3, 7
Oceanic plates, 44, 101
Oceans, global warming and, *12*
Offshore outsourcing, **318**
Ogallala aquifer, 66, 69
Oil contamination, 100
Oil dependency (U.S.), 74–75
Oil development (environmental
    damage)
  Ecuador, 100
  Rukpokwu state, 8

Oil drilling, 62, 272
Oil palm plantations, 378, 379
Oil spills
  Alaska pipeline, 62
  Amazon Basin, 100
  Boston, *57*
  Deepwater Horizon oil rig, 62,
    64, 65, 75
  Exxon Valdez, 215, 272
  Gulf War (1991), *215*
  Lebanese, *215*, 240
  Rukpokwu state, 8
  Russian arctic, *179*
Oil/natural gas resources
  North America, 57
  OPEC and, 8, 232, 233, 234,
    234, 236
  Russia and post-Soviet states,
    188, *189*
Ok Tedi Mine, *420*
Old Testament, 220
Oligarchs, **188**, 189, 197
Olympic National Park, *64*
Olympics
  Beijing, 340, 357
  South Africa and, 263
  Winter Games (2014), 181, 205
One-child family, 18, 163
One-child policy (China), 10,
    13, 311, 359, 360–361,
    362, 366
One-commodity countries, 265
OPEC (Organization of
    Petroleum Exporting
    Countries), 8, **232**, 233, 234,
    234, 236
Open Society Institute, *172*
Opium Wars (1839-1860),
    343, 347
OPT. *See* occupied Palestinian
    Territories
Optimism. *See* Reasons for
    optimism
Orange Free State, 263
Orange Revolution, 194, 195,
    195, 196, 204, 205
Orangutans, 378
Organic agriculture, 26, 54, 81,
    97, 159
Organization of Petroleum
    Exporting Countries.
    See OPEC
Organized crime/corruption
    (Russia and post-Soviet
    states), 197–198
Orographic rainfall, **46**, *47*
Orthodox churches, 52
Ottoman Empire, 148, 220–221,
    222, 223, 224, 240, 262
Overseas Chinese, 364–365
Ownership of water, 37–38, 39
OXFAM International, 36
Oz, Amos, 208
Oz, Nily, 208
Oz/Khoury peace efforts, 208

Pacific coasts, 46, 58, 59, 73, 81, 95,
    121, 176, 183, 191, 329, 414
Pacific Islanders (in U.S.), *91*
Pacific Islands
  Austronesians and, 383
  British contact with, 424
  climate change vulnerability, 416
  economic change in, 427
  El Niño and, *105*, 106, 375,
    379, 381, 413, 414, *414*
  endemic species, 414
  fishing rights, *417*
  flora/fauna, 415
  globalization and, 420–421
  Great Barrier Reef and, *407*,
    411, 412, 416
  invasive species, 420
  Japan's expansions and, *345*
  Oceania and, 409
  population patterns, *407*
  water issues, *407*, 416
Pacific Mountain Zone, 175, 176
Pacific Plate, 45, 46, 58, 176, 412
Pacific Rim, **95**
Pacific Ring of Fire, **44**, *46*
Pacific Way, **433**
Paddy rice, 381. *See also* Wet rice
    cultivation
Pahiyas Festival, *401*
Pakistan
  agribusiness and, 316
  dictators and, 324
  East Pakistan and, 321
  East Pakistan *v.*, 303, 321
  fertility rates, 312
  GDP per capita PPP, 312
  leather tanning in, 299
  Mohenjo-daro, 304
  nuclear weapons and, 322
  population pyramid, 312, *313*
  West Pakistan and, 321
  West Pakistan *v.*, 303, 321
Palestine, 120, 220, 239–241,
    240t, *241*, 243
Palestinian Territories. *See*
    Israeli-Palestinian conflict;
    occupied Palestinian
    Territories
Palm Jumeirah, 229, 230, 232
Palm oil, 22, 262, 371, 379
Pamirs, 176, 181, 182
Pampas (Argentina), *103*, 104,
    105, 121
Pancasila, 391
Panda, giant, 340
Pangaea, 44
  Africa and, 249
  breakup of, 44, *45*, 58
  Gondwana and, *45*, 409,
    412, 415
  hypothesis, 44, *47*
Panipur communal conflict, *309*
Papua New Guinea
  Bougainville, *434*
  Carteret Islands and, 416

gender roles, 436
Gondwana and, 412
mining in, 420–421
MIRAB economy and, 427
mother selling tobacco, 407
Oceania and, 409
Ok Tedi Mine, 420
population patterns, 409,
428, 431
positive transformations, 437
urbanization, 429, 430
Pará, Brazil, 108
Parachuting, 28
Paraguay, 40, 41, 99, 114, 119, 121
Parallels. See Lines of latitude
Paris
riots, 164, 165
tram in, 162
Parliaments. See Women in
parliaments
Partition, **303**
Pastoralism (herding), 183, 184,
257, **258**, 341, 349
Patriarchal, 220, **225**, 227, 238,
342, 364, 401, 433
Pearl Harbor memorial, 425
Pelican, 64
pemaes, 400
Penguin (Emperor penguin), 439
People (photo interpretation), 8
People's Republic of China, 323,
344. See also China
Peopling
of Africa, 259
of Middle and South
America, 111
of North America, 69–70
of Oceania, 422
of Southeast Asia, 383
of sub-Saharan Africa, 259
Perestroika, **190**
Permafrost, **176**
Persian Gulf, 74, 75, 208, 209,
213, 216, 229, 232, 322, 342,
383, 399
Personal vignettes. See Vignettes
Peru, 99, 101, 102, 104, 107, 111,
112, 114, 119, 121, 123, 127,
128, 130
Peru Current, 104, 105, 106, 252
Pesticides, 11, 24, 57, 68, 80, 81,
85, 120, 144, 160, 231, 266,
287, 300, 317, 320, 325, 369,
370, 397
Philippine Archipelago, 373
Philippine Plate, 45, 46, 374, 412
Philippines
birth control in, 395
Calauan, 376
Magellan and, 386
Manila, 369, 381, 382, 397,
400
Quezon, 401
rice terraces, 378
Phoenix, Arizona, 88

Photogrammetry, 7
Photograph interpretation, 7, 8, 8
Physical geography, 44–50
defined, 3, 44
human geography v., 3
Physical patterns. See also
Climates; Climate zones;
Landforms; Vegetation
East Asia, 332–334
Europe, 139–142
Middle and South America,
101–107
North Africa and Southwest
Asia, 209–213
North America, 58–59
Oceania, 409–415
Russia and post-Soviet states,
173–176
South Asia, 289–295
Southeast Asia, 374–375
sub-Saharan Africa, 249–252
Piaguaje, Colon, 100
Pillars of Islamic Practice, 222,
223, 227
Pipeline, Trans-Alaska, 62, 64, 65
Piracy, 371
Pizarro, Francisco, 115
Plains
floodplain, 44, 45
Great Plains, 66, 69, 73, 81
Indo-Gangetic Plain, 291, 292,
309
Mekong River Plains, 373
North China Plain, 330, 332,
333, 334, 336, 339, 341, 361
North European Plain, 139,
141, 142, 143, 145, 149,
173, 174
West Siberian Plain, 173, 174,
176, 198
Plant cultivation, 23
Plantations, **115**
oil palm, 378
sugar, 30
Plants/animals. See Fauna/flora
Plate tectonics, **44**. See also
Landforms
African Plate, 45, 59, 139, 209,
374, 375
Antarctic Plate, 45, 46
Arabian Plate, 45, 209, 249
Caribbean Plate, 45, 101
Cocos Plate, 45, 46, 101
continental plates, 44
Eurasian Plate, 45, 46, 139,
176, 209, 292, 332, 374, 412
Indian-Australian Plate, 44, 45,
46, 292, 374, 412
Nazca Plate, 45, 46, 101
North American Plate, 45, 58,
59, 101
oceanic plates, 44, 101
Pacific Plate, 45, 46, 58,
176, 412
Pangaea breakup and, 45

Philippine Plate, 45, 46, 374, 412
Ring of Fire and, 44, 46
South American Plate, 45,
46, 101
Plateau continent, 44
Platypus, 415, 415
Political ecology, **22**
Political issues. See also
Economic issues
East Asia, 347–359
Europe, 154–160
Middle and South America,
113–120, 121–125
North Africa and Southwest
Asia, 231–243
North America, 74–75
Oceania, 425–428
Russia and post-Soviet states,
188–198, 204
South Asia, 320–324
Southeast Asia, 387–394
sub-Saharan Africa, 264–273
Political maps
East Asia, 326
Europe, 136
Middle and South America, 98
North Africa and Southwest
Asia, 206
North America, 56
Oceania, 406
Russia and post-Soviet states, 170
South Asia, 286
sub-Saharan Africa, 246
Pollution. See also Air pollution;
Industrial pollution; Water
pollution
nonpoint sources of, 178
nuclear, 100, 178, 332, 336,
420, 421
urban, 178, 181, 205
Polo, Marco, 342
Polygyny, **227**
Polynesia, **422**
Population, 14–17. See also Aging
populations
African-Americans, 71, 91
floating, 329, 349
HIV/AIDS and, 14
as thematic concept, 9, 10,
14–17
U.S. (by race/ethnicity), 91
Population decline, 10. See also
Aging populations
Central Europe, 14
Russia and post-Soviet states,
198–199, 201
Population densities
average population density, 14
East Asia, 334, 340, 362
Europe, 146, 160, 161
Middle and South America,
111, 126
New York City, 87
North Africa and Southwest
Asia, 227, 228

North America, 94
Oceania, 428, 429
Russia and post-Soviet
states, 199
South Asia, 287, 295, 311, 312
Southeast Asia, 380, 396
sub-Saharan Africa, 257,
275, 280
world, 14
Population distribution
East Asia, 361
Europe, 160–161
Middle and South America,
126–127
Population growth
Afghanistan, 325
East Asia, 13
Ethiopia, 276
global patterns, 9, 10, 14
J curve, 9, 14
Japan, 359–360
Middle and South America,
99, 127, 127
migration and, 15
Native Americans, 73
North Africa and Southwest
Asia, 207, 227, 229, 243, 244
Oceania and, 407, 437
RNI and, 14
slowing of, 10
South Asia, 311–312
Southeast Asia, 369, 404
sub-Saharan Africa, 247,
276–277
urbanization and, 246, 275,
276–280, 284
wealth and, 16–17
Population patterns
East Asia, 359–360
Europe, 160–161, 163
Middle and South America,
126–127
North Africa and Southwest
Asia, 227
North America, 94–96
Oceania, 428–429
Pacific Islands, 407
Papua New Guinea, 409,
428, 431
Russia and post-Soviet states,
198–201
South Asia, 311–315
Southeast Asia, 395–396
sub-Saharan Africa, 275–280
urbanization and, 160, 198,
311, 324
Population pyramids, **15–16**
Austria, 15, 15–16
Belarus, 201
Canada, 96
China, 360, 360
EU-27, 164
Germany, 164
Indonesia, 395–396, 397
Iran, 229

Population pyramids (*Continued*)
Israel, 229
Jordan, 15, *15*
Kazakhstan, *201*
Kyrgyszstan, *201*
Nigeria, *276*
Pakistan, 312, *313*
Qatar, 229
Russia, *201*
South Africa, *276*
Southeast Asia, 395–396
Sri Lanka, 312, *313*
Sweden, *164*
U.S., *96*
Populist movements, 123, **132–133**
Portugal
Angola and, *154*
dictatorship, 153, *154*
Portugese/Spanish trade routes (circa 1600), *112*
Spain colonialism, 112–113
Post-communist welfare systems, 168
Postindependence period, South Asia, 303–304
Post-Soviet states. *See* Russia and post-Soviet states
Poverty. *See also* Slums
Afghanistan, 296, 297
class-based explanation for, 91
culture of, 91, 96
Lima, Peru, *130*
social safety net and, 79
South Asia, *314*
Poverty Reduction Strategy Papers. *See* PRSPs
Powell, Colin, 91
Poznan, Poland, *144*
PPP (purchasing power parity), **20**. *See also* GDP per capita PPP
Prague, Czech Republic, *162*
Precipitation, 45, 46–47
acid rain and, 66, 68, 78, *108*, *144*, 144, 145, 179, 215, 255, 299, 300, 339, 340, 378, *417*
Middle and South America and, 104, 106
Prejudice, 53, 90, 96, 163, 364
Press freedom, 208, 236
Primate cities, **129**, 130, 198, 200, 277, 397, 399, 400
Privatization, 37, 38, **116**, 180, 188, **190**, 192, 197, 204, 318
Pro-democracy demonstrations
Germany, *169*
Hong Kong, *365*
Iran, 207
Prophet Muhammad, 208, 213, 220, 221, 222, 223, 226, 282
Prostitutes, 202, *202*, 204, 261, 280, 307, 361, 402, 403
Protection money, 192
Protestant Reformation, 149, *152*

Protestantism
evangelical, 133
in world, *52*
PRSPs (Poverty Reduction Strategy Papers), 32, 267, 268
Puerto Misahualli, *109*
Puerto Rico, sugar plantation workers in, *115*
Punjab conflict, 324
Purchasing power parity (PPP), **20**. *See also* GDP per capita PPP
Purdah, **309**
Push/pull phenomenon of urbanization, **27**, 396
Putin, Vladimir, 159, 194, 197
Pyeongtaek, *351*
Pyramids, 111, 184. *See also* Population pyramids

Al Qaeda, 74, 77, 322, 323, 324
Qanats, 213
Qatar, 229
Qin empire, 342, 343, 346
Qin Ling, 334
Qing dynasty, 343
Qinghai–Tibet Railway, 364
Quagmire season, 176
Quetta Province, 292
Quezon, *401*
Qur'an (Koran), **213**, 220, 223, 227, 235

Race, **53**. *See also* Skin color
defined, 53
Middle and South America and, 128
North America and, 90–91
skin color map and, 53, *53*
Racism, 53, 96
Radio Sahar, 311
Railroads/rail transport
nineteenth-century transportation and, 72, *72*
Qinghai-Tibet Railway, 364
Rain, acid, **66**, 68, 78, 108, 144, *144*, 145, 179, 215, 255, 299, 300, 339, 340, 378, 417
Rain belt, 46, 47
Rain forests, 100, 104, 107, 109, 115, 252, 292, 375, 379, 415, 416, 419
Rain shadows, 46, 47
Rainbow Warrior, 421
Rajasthan, 288
Ramadan, 223
Rampell, Catherine, 58
Ramstein Air Base, 77
Rangoon, 394
Rasputitsa, 176
Rate of natural increase (RNI), **14**
Ravi Ravi, *430*
Reasons for optimism
East Asia, 365, 366
Europe, 168, *169*

Middle and South America, 109, 121, 124, 133, *134*
North Africa and Southwest Asia, 243, *244*
North America, 81, 86, 96, 97
Oceania, 416, 428, 436, 437, *437*
Russia and post-Soviet states, 181, 188, 196, 204, *205*
South Asia, 296, 324, 325
Southeast Asia, 391, 395, 402, 403, *404*
Sub-Saharan Africa, 269, 283, 284
Recession. *See* Global economic downturn
Reconstruction, in Afghanistan, 322–323
Red Crescent, 36
Red Cross, 36
Red light district, 369, 403
Red Sea, 209, 220, 249, 252
Reflections. *See also* Reasons for optimism
on East Asia, 365–366
on Europe, 168
on Middle and South America, 133–134
on North Africa and Southwest Asia, 243
on North America, 96
on Oceania, 437
on Russia and post Soviet States, 204
on South Asia, 324–325
on Southeast Asia, 403
on sub-Saharan Africa, 283
Refugee camps
Dadaab, 272
Kabul, *43*
Palestinian, 231, 239, 240
Sudan, 231
Refugees
Bihari community, 320, 321
Burma, 394
Chechnya, 196
Congo, 273
Darfur, 217
Europe, 163
Haiti, 76
Iran, 231
Kabul, *43*
Somalia, 217
sub-Saharan Africa, 272
Tamil, 298, 321, 323
Uganda, *43*
U.S. and, 89
Vietnam, 431
Regional conflicts, **325**
Regional development
Soviet, 190–191
Sub-Saharan Africa, 269–270
Regional geography, 3
Regional maps
East Asia, 330–331, 332
Europe, 140–141

Middle and South America, 102–103
North Africa and Southwest Asia, 210–211
North America, 60–61
Oceania, 410–411
Russia and post-Soviet states, 174–175
South Asia, 290–291
Southeast Asia, 372–373
sub-Saharan Africa, 250–251
Regional self-sufficiency, **349**
Regional specialization, **350**
Regional trade, 32, 119
Regions. *See also* World regions
climate regions, 47, 48–49
defined, 7, 9
world regional approach, 3
Reincarnation, 305, 308
Religions. *See also specific religions*
belief systems v., 51
distribution of, 52
Middle and South America, 132–133
monotheistic, 208, 220, 222
North Africa and Southwest Asia, 222–223
North America, 91–92, 92
Russia and post-Soviet states, 203
South Asia, 305–309, 308
Southeast Asia, 383–384, 400–401, *401*
sub-Saharan Africa, 281–282
Religious nationalism, 309, **323–324**
Remittances, 20, 119, 129, 232, 268, 269, 360, 399, 427
Republic of Congo. *See* Congo
Republic of Ireland. *See* Ireland
Resettlement schemes, 288, 363, 379, **392**, 418
Resilience, 41
Resource extraction
Antarctica and, *439*
Russia and post-Soviet states and, 178–180, *189*
Responsibility system, **349**
Return-migration (South Asia), 319
Reva company, 296
Reva electric car, 324
Revolutionary War, 72
Rhine River, 139
Rhodesia, 273. *See also* Zimbabwe
Rice cultivation (wet rice cultivation), 255, **338**, 369, 381, 404
Rice terraces, 378
Ring of Fire, **44**, 46
Rio de Janeiro, 27, *103*, *131*
Riyadh, 227
RNI. *See* Rate of natural increase
Roaring forties, **412**
Robinson projection, 6, *6*
Rocky Mountain Zone, 58
Rocky Mountains, 58, 59, 60, 73

Roma (Gypsies), **152**, 164
Roman Catholicism
in Mexico City, *132*
Middle and South America
and, 132–133
in U.S., 92
in world, 52
Romance languages, 148
Romania
pig farms in, 138, *138*
Smithfield firm in, 138
Rome (early civilization), 148, *152*
Rose Revolution, 172, 194
Rotary International, 36
Rotterdam, 142, 148, *169*
Roy, Beth, 309
Rub'al Khali, 209, *210*
Rudimentary welfare systems, 168
Rugby, 408, *409*
Rukpokwu state (oil pipeline
burst), 8
Rural-to-urban migration
China, 329
East Asia, 350
Li Xia and, 328–329, 350, 354
Middle and South America, 126,
129, 132, 133
North Africa and Southwest
Asia, 207, 225, 229
South Asia, 311
Southeast Asia, 396, 399, 403
Russia (Russian Federation)
ethnic groups and, 183, *197*
Georgia and, 193, 196–197
global economy and, 188–189
greenhouse gas emissions,
40, 181
multinational corporations
and, 188
population pyramid, *201*
Russian Empire, 183–185, *185*
as Russian Federation, 173
Russia and post-Soviet states
(world region), 170–205
agriculture in, *193*
birth rates, 198–199, 201, 204
climate change vulnerability,
181, *182*, 183
climate zones, *177*
climates, 176, *177*
corruption/organized crime,
197–198
cultural diversity, 196–197, *199*
death rates, 198–199
democratization/conflict issues,
194, *195*, 196–197, 198
economic issues, 188–198, 204
economic reforms, 189–192
environmental issues, 176–182
geographic issues, 188–204
geographic setting, 173–187
human impact on biosphere,
176, 178, *179*, 180–181
human patterns over time,
183–187

indigenous groups and, 176,
179, 184, 196, 203
industrial areas/land transport
routes, 191, *191*
informal economy, 191–192
landforms, 173–176
natural gas resource areas,
188, *189*
nuclear weapons, 178
optimism for, 181, 188, 196,
204, *205*
organized crime/corruption,
197–198
physical patterns, 173–176
political issues, 188–198, 204
political map, *170*
population decline,
198–199, 201
population density, *199*
population distribution, 198–199
population patterns, 198–201
reflections on, 204
regional map, *174–175*
religious revival, 203
resource extraction and,
178–180, *189*
Sochi, 181, *205*
sociocultural issues, 198–203
Soviet Union and, 172–173
term usage, 173
thematic concepts overview,
171, 205
unemployment and, 190,
202, 204
urbanization, 170, *171*, 178,
181, 198, 200, 201, 204, 205
vegetation, 176
visual timeline, *184–185*
water issues, 180–181
women legislators, 202
as world region, *1, 3, 7*
Russian Empire, 183–185, *185*
Russian Federation, **173.**
See also Russia
Russian imperial expansion, *185*
Russian Mafia, 197, 202
Russification, **196**
Rwanda, 272, 273, *274, 278*

Saakashvili, Rustaveli, 172, *172*
Sacramento river, 58
Sahara, 209, 249, 252, 258, 282
Sahel region, **252**, 254, 258, 259,
260, 264, 282
Salinization, **214**, 216, 317
Salvador, street vendor in, *130*
Samoa, 428
San Joaquin river, 58
San'a, 217, 238
Sanea, Rajaa al-, 227
Santiago, Chile, *130*
SAPs (structural adjustment
policies), 32, **116**
Middle and South America
and, 116, 118

PRSPs *v.*, 32
sub-Saharan Africa and,
265–268
Sarawak, 370, 372, 374
Sarawak forest dwellers, 370, 371
Sardar Sarovar Dam, 288–289
Sarkozy, Nicolas, 166
Saudi Arabia
desert in, 212
female/male incomes, 18*t*
Islam and, 223
Medina (Al Madinah) and,
211, 220, 223
oil in, 189, 221, 232
religious extremists, 203
seawater desalination and, 216
women and, 226, 227
Scales
economies of, 156
global, 7
local, 7
map scales, *3, 5, 5*
world region, 7
Schengen Accord, **163**
Scotland, 139, *143*
Sea level rise, 42
Bangladesh, *43*, 295
Mumbai, 41
Netherlands, 139, *147*
North Africa and Southwest
Asia, 217
North America, 67
South Asia, 295
Seawater desalination, 144, **216**
Seawater pollution, Europe, 145
Secoya people, 100
Secular states, **235**, 323, 363
Secularism, 320
Securocrats, 194
Self-reliant development, **270**
Sen, Amartya, 20
Senegal, *10, 257*, 280
Sensitivity, 41
September 11th attacks
oil dependency and, 74–75
al Qaeda and, 74, 77, 322,
323, 324
Taliban and, 203, 311,
321, 322
U.S.-lead war in Afghanistan
and, 74, 76, 239, 321, 428
War on Terror and, 74, 77,
235, 323
Sequestration, carbon, **252**
Serfs, 149, 185
Service economies, *10, 11*, 27
Europe, 160
North Africa and Southwest
Asia, 231, 234
Oceania, 427, 429, 437, 438
South Asia, 304
Southeast Asia, 402
Service sector, 65, **82**, 83, 85, 124,
125, 134, 263, 265, 316, 320,
369, 388, 398, 402, 427, 431

Settlements in North America.
See European settlements
Sex, **17**, 19. *See also* Gender
Sex imbalance, 15
Sex industry, 202, 280, 369,
402–403, 404
Sex structures, 15–16. *See also*
Population pyramids
Sex tourism, 390, **402**, 405
Sex workers, 202, 202, 204, 261,
280, 307, 361, 402, 403
SEZs. *See* Special economic
zones
Shanghai
transformation, 352–353
urbanization, 28, *351*
Shantytowns, 27, 131
Shari'a, **223**, 322
Sharki, Raufa Hassan al-, 238
Sheikh Hamad bin Khalifa, 208
Sheikh Moza, 226
Shell Oil Company, 8, 30, 272
Shifting cultivation, 25, **111**, 113,
120, 135, **257**, 379–381
Shi'ite Islam, 52
Shi'ite (Shi'a) Muslims, **223**,
236, 239
Shintar Ishihara, 356
Shintoism, 52
Siakor, Silas, 248, 249
Siam, 385. *See also* Thailand
Siberia, western, 184
Sichuan Province, 328, 338, 354,
357, 361
Sierra Leone, 247, 248, 273,
278, 284
Sierra Madre, 102, 104, 105
Sikhism, 52, **308**
Silk Road, *183*, 184, 187,
363, 364
Siloviki, 194
Silt, **104**
Similarities (U.S./Canada
relationships), 75–76
Singapore, 387
Singhalese nationalism, 323
Skin color
of indigenous people, 53
as intelligence/ability
marker, 90
Middle and South America
and, 128
race and, 53
Skin color map, *53, 53*
Slash and burn agriculture,
379, 380. *See also* Shifting
cultivation
Slavery, African slave trade,
259–260, 261, 264
Slavs, **183**
Slovenia, 156, 159–160
Slums
Dhaka, 287, *314*
Dhobi Ghat, *314*
favelas, 27, 131–132

Slums (*Continued*)
 Freetown, 247, 278
 Jakarta, 398
 João Pessoa, 99
 Kenya, 39
 Kibera, *11*, 278
 Kroo Bay, 278
 Mathare, 268, 278
 Nairobi, *11*, 278
 unplanned urbanization, 27,
  41, 99, 129, *130, 131*, 247,
  277, 284, 431
 wastewater and, 39
Smart growth, 86, 88, 96, 97
Smithfield firm, 138
Smog, 66, 68, 375
Sochi, Russia, *181, 205*
Social Forestry Movement,
 296, 300
Social networking, 57, 83
Social protection, **167**. *See also*
 Social welfare/protection
  systems
Social safety net, 79
Social Security, 95
Social welfare/protection systems,
 167, 167–168
Socialism, 101, *122*, 124
Socialist economy. *See* Centrally
 planned/socialist economy
Society, civil, 32, **33**, 36, 54, **263**,
 264, 273, 324
Sociocultural issues
 East Asia, 359–366
 Europe, 160–168
 Middle and South America,
  126–133
 North Africa and Southwest
  Asia, 222–231
 North America, 86–94
 Oceania, 431–437
 Russia and post-Soviet states,
  198–203
 South Asia, 304–315
 Southeast Asia, 395–403
 sub-Saharan Africa, 273–283
Solar thermal power plant, *169*
Somalia, 217, 256, 272
Somoza, Anastasio, *122*
Soros, George, *172*
SOS Racisme, *165*
Soter Mineral Springs Vineyard
 harvest, *58*
Soufrière, 30–31
South Africa. *See also* North
 Africa and Southwest Asia;
 Sub-Saharan Africa
 colonization of, 261–263
 early humans, *260*
 nurse in, *10*
 Olympics and, 263
 population pyramid, *276*
 World Cup soccer, *261*, 263
South America, **101**. *See also*
 Middle and South America

South American Plate, 45, *46*, 101
South Asia (world region), 286–325
 agricultural zones, 293, *316*,
  316, 317
 air pollution, 299, 300
 caste system, 306–308, 309,
  310, 323, 324
 Christianity, 298, 300, 305,
  308, 309
 climate, 292–295
 climate change vulnerability,
  13, 287, 295–296, 297, 324
 climate zones, 293, 294
 colonialism and, 301–304
 competitiveness and, 318–320
 deforestation, 296, 298
 deindustrialization, 301–302
 democratization/conflict issues,
  310, 320, 321, 322–323, 324
 development, 324
 diaspora of, *302*
 early humans enter, *304*
 economic issues, 315–320
 environmental issues, 295–300
 ethnic groups, 305, 309
 farming systems, *316*
 female infanticide and, 310, 312
 fertility rates, *313*
 food production, 287, 316–317
 gender imbalance, 287,
  310–311, 312–313
 geographic issues, 304–324
 geographic setting, 289–304
 globalization and, 301–304,
  318–320
 green revolution agriculture,
  287, 304, 305, 316–317
 Hinduism and, 305–308
 human impact on biosphere,
  295–296, 298, 299, 300
 human patterns over time,
  300–304
 indigenous groups, 300, 308
 industrial self-sufficiency, 318
 invasions into, 300–301
 landforms, 292
 languages, 305, *306*
 malnutrition, *26*
 microcredit and, 317–318, 324
 monsoons, 292, 293, 294, 295,
  295, 296, 297, 298, 300
 optimism for, 296, 324, 325
 physical patterns, 289–295
 political issues, 320–324
 political map, *286*
 population density, 287, 295,
  311, *312*
 population patterns, 311–315
 postindependence period,
  303–304
 reflections on, 324–325
 regional map, *290–291*
 religions, 305–309, *308*
 religious nationalism, 309,
  323–324

 return-migration and, 319
 rural-to-urban migration, 311
 sociocultural issues, 304–315
 subcontinent and, 289
 thematic concepts overview,
  287, 325
 urbanization, 286, 287, 311,
  314, 315, 317, 320, 324, 325
 vegetation, 292–295
 visual timeline, *304–305*
 water issues, 287, 296, 298,
  299, 300
 as world region, *1, 3, 7*
South China Sea, *331*
South Europe, *136*, 138
South Korea
 anti-WTO demonstration, *32*
 dictatorships, 356
South Ossetia, 196–197
South Pacific, *435*
South Pacific Nuclear Free
 Zone, 421
Southeast Asia (world region),
 368–405
 agricultural patterns, 380
 ASEAN and, 32, 389–391, *390*,
  392, 403
 authoritarian tendencies, 393
 birth rates, 395, 403
 climate, 374–375
 climate change vulnerability,
  369, 379–381, 382, 383
 climate zones, 376
 colonization, 384, 385
 cultural influences, 383–384
 deforestation, 378, 379
 democratization/conflict issues,
  369, 391–393, *394*, 395, *404*
 development, 369
 dictators and, 393, 403
 economic crisis, 388–389
 economic issues, 387–394
 environmental issues, 375–383
 family organization, 401–402
 fertility rates, 395, 396, 397
 flooding, *369*
 food production, 369, 379–381
 gender patterns in, 401–402
 gender roles, 369, 398, *404*
 geographic issues, 387–403
 geographic setting, 371–387
 globalization and, 369,
  387–389, 399–400
 HIV/AIDS in, 403
 human impact on biosphere,
  375, 378, 379–381
 human patterns over time,
  383–387
 independence and, 384–386
 indigenous groups, 370, 371,
  375, 383, 384, 400, 403
 landforms, 374
 maid trade and, 399–400
 optimism for, 391, 395, 402,
  403, *404*

 Overseas Chinese in, 364–365
 peopling of, 383
 physical patterns, 374–375
 political issues, 387–394
 population density, 380, 396
 population growth, *369, 404*
 population patterns, 395–396
 population pyramids, 395–396
 reflections on, 403
 regional map, *372–373*
 religions, 383–384, 400–401, *401*
 rural-to-urban migration, 396,
  399, 403
 sex industry, 402–403
 sociocultural issues, 395–403
 Sundaland and, 374, *375, 375*,
  383, 415, 422
 terms usage, 371
 terrorism and, 393, 395
 thematic concepts overview,
  369, 404–405
 tourism, 389–390
 tropical timber trade, 377
 urbanization, 368, 369,
  396–399, *398*, 400, 401, 402,
  403, *404*
 vegetation, 374–375
 visual timeline, *386–387*
 water issues, 381
 as world region, *1, 3, 7*
Southeastern Anatolia Project,
 216, 218
Southeastern Europe, 34
Southern Baptists, 92, 203
Southern Common Market.
 See Mercosur
Southern Hemisphere, 6
Southern Rockies, *60*
Southern settlements (North
 America), 70–71
Southwest Asia. *See* North Africa
 and Southwest Asia
Southwest settlements (North
 America), 73
Sovereignty, 36, 78, 346, 384
Soviet regional development
 schemes, 190–191
Soviet Union. *See* Russia and
 post-Soviet states; Union of
 Soviet Socialist Republics
Soviet-Afghanistan war, 187,
 194, *195*
Soweto, 278
Soybeans, 121
Spain
 Argentina/Chile independence
  from, *115*
 climate change vulnerability, 43
 Portugal colonialism, 112–113
 solar thermal power plant in, *169*
 Spanish/Portugese trade routes
  (circa 1600), *112*
Spatial analysis, 3, 7
Spatial distribution, **3**
Spatial interaction, **3**

Spatial scales. *See* Scales
Special economic zones (SEZs), **350**, 353
Species
  introduced, 415, 419
  invasive, 254, 407, 419–420, 421
  nonnative, 65, 254, 416, 419
Sprawl, urban, **65**, 76, 86, 88, 93, 94, 95, 96
"Squatter" settlement, *130*
Sri Lanka
  civil war, 321, 323, 325
  fertility rates, 312, *313*
  GDP per capita PPP, 312
  population pyramid, 312, *313*
  terrorism and, 323
St. Basil's Cathedral, *184*
Stalin, Joseph, 176, 185, 186, 187, 189, 196
Stalin Era, *185*, 186
State-aided market economies, **347**, 348, 387
Steel industry, 72
Steppes, 48, 49, **173**, 176, *177*, 209, *212*, 333, 364
Stop the Islamization of Europe (group), *165*
Street vendors, *130*, 207
Strip mining, 65, 421
Structural adjustment policies. *See* SAPs
Sturgeon fish, 340
Subcontinent, **289**. *See also* South Asia
Subduction, 44
Subduction zone, **101**
Sub-Saharan Africa (world region), 246–285
  African slave trade (European slave trade) and, 259–260, 261, 264
  birth rates, 275, 276, 277, 278, 283
  climate, 252–254
  climate change vulnerability, 247, 254, 256, 257, 283
  climate zones, 253
  colonialism, 261–263, 264–265
  commodities and, 264, *265*, 266, 268, 269
  death rates, 277
  deforestation, 252, 254
  democratization/conflict issues, 77, 247, 271–273, 274, 275, 284
  demographic transition, 275–276, 280, 284
  development and, 247, 269–270, 284
  early history (agriculture/industry/trade), 259
  economic issues, 264–273, 267
  environmental issues, 252–259
  ethnic groups, 263, *271*
  FDI inflows into, 268

fertility rates, 276
food production, 254, 257–258
gender issues, 280–281, 284
genocide and, 237, 272, 273, *274*
geographic issues, 264–283
geographic setting, 249–264
globalization and, 247, 264–269
HIV/AIDS in, 267, 275, 276, 277, *277*, 279, 279–280
human impact on biosphere, 252, 254, 255, 257
human patterns over time, 259–264
indigenous belief systems, 281–282
informal economies, 266–267, 268, 271
landforms, 249–251
local development, 269–270
malnutrition, 26
migration and, 277
naming of countries, 249
optimism for, 269, 283, 284
peopling of, 259
physical patterns, 249–252
political issues, 264–273
political map of, *246*
population density, 257, 275, 280
population growth, 247, 276–277
population patterns, 275–280
reflections on, 283
regional development, 269–270
regional map, *250–251*
religions, 281–282
Sahel region, 252, 254, 258, 259, 260, 264, 282
SAPs and, 265–268
sociocultural issues, 273–283
term usage, 249
thematic concepts overview, 247, 284
urbanization, 246, 247, 273, 275, 276–280, *278*, 284
vegetation, 252
visual timeline, *260–261*
water issues, 257–258
wildlife issues and, 258
as world region, *1*, *3*, *7*
Subsidence, 216
Subsidies, 66, 83, 115, *137*, 138, **159**, 169, 192, 428
Subsistence affluence, **427**
Subsistence agriculture, 11, 16, 22, 71, 120, 121, 247, **254**, 255, 256, 257, 259, 284, 380, 403, 420, 427
Subsistence economies, **16**
Subtropical climates, 48, 49, 253, 333, 376
Suburbanization, 86, 95. *See also* Urbanization
Suburbs, **86**

Sudan
  Aswan Dam and, 216
  climate change vulnerability, 43, 216
  conquest of, 221
  Darfur region, 35, *217*
  refugee camps, 231
  Sudan Liberation Movement Army, 35
Sudras, 306, 307
Sugar plantations, *30*, *115*
Suharto, 389, 391, 392, 393, 394, 402
Sumatra, 292, 373, 374, 375, 393
Sundaland, 374, 375, *375*, 383, 415, 422
Sunni Islam, 52
Sunni Muslims, 196, **223**, 239
Sustainability. *See also* Green revolution agriculture; Human impact on biosphere
  carrying capacity and, 24
  East Asia, 337–338
  green revolution, 23–24
  organic agriculture, 26, 54, 97
  sustainable agriculture, 26
  sustainable development, 22
  sustainable food production, 81, 82, 97
  sustainable tourism, 428
Swagman, *435*, 436
Sweden, 164, *164*
Swidden cultivation, 379
Swine flu, *132*
Sydney, *430*
Syr Darya, 180
Syria, 50, 213, 216, 218, 231, 232, 234, 237, 239, 240, 241

Tahiti, *411*
Taiga, **176**
Tailings, 65, 420
Taiwan
  minorities in, 364
  protests in, 356
  uncertain status of, 346
Taj Mahal, 300, 301, *304*
Tajikistan, 171, 173, 176, 181, 182, 202, 203
A Tale of Love and Darkness (Oz), 208
Taliban, 203, **311**, 321, 322
Tamil minority, 323
Tanzania, *10*, *11*, 247, *250*, *251*, 256, 266, 277, *284*
Tariffs
  ASEAN and, 389
  CAP and, 159
  defined, 31
  EEC and, 154
  SAPs and, 32, 116, 266
  sub-Saharan Africa and, 266, 269, 283
Tasmania, 415, 425
Tasmanian Devil, *417*

Tata Motors, 318
Taylor, Charles, 248, 269
Technology. *See also* Information technology industries
  food production and, 24
  innovation and, 28, 95, 96
  Native Americans and, 70
Tectonic plates. *See* Plate tectonics
Temperate climates, 48, *48*, 63
Temperate midlatitude climate, 63, **142**
Temperature
  air pressure and, 45–46
  defined, 46
Temperature-altitude zones, 48, **104–106**, *106*
Temporary workers, 163. *See also* Guest workers
Tennessee coal ash spill, 64, 65
Tenochtitlan, *114*
Tent villages, 27
Teotihuacan, *114*
Terrebonne Parish, 62
Territorial claims (Antarctica), 439
Terrorism. *See also* September 11th attacks
  Chechnya and, *195*, 196
  Iraq War and, 239
  London bombing, 164
  Madrid bombing, 164
  Mumbai, 2008, 323–324
  North Africa and Southwest Asia and, 236, 243
  Southeast Asia and, 393, 395
  Sri Lanka and, 323
  War on Terror, 74, 77, 235, 323
Texaco (in Ecuador), 100
TFR. *See* Total fertility rate
Thailand
  authoritarianism in, 393
  Bangkok, 369, 381, 393, 397, 398, 399, 403
  fertility rates, 395, 404, *404*
  Gulf of Thailand, 374
  Khao Yai, 376
  literacy rates, 395
  red light district, 369, 403
  Siam and, 385
Thaksin Shinawatra, 391
Thematic concepts, 9–43. *See also* Climate change; Democratization; Development; Food production; Gender; Globalization; Population; Reasons for optimism; Urbanization; Water
  defined, 9
  East Asia, 327, 366
  Europe, *137*, 169
  Middle and South America, 99, 134–135
  North Africa and Southwest Asia, 207, 244

Thematic concepts (*Continued*)
North America, *57, 97*
Oceania, *407, 438*
Russia and post-Soviet states, *171, 205*
South Asia, *287, 325*
Southeast Asia, *369, 404–405*
sub-Saharan Africa, *247, 284*
Theocratic states, 187, **235**, 236
Thompson Electronics, 117–118
Three Gorges Dam, 288, *331,* 338, 339, 340, 341
Tiananmen Square, 355, 356, 357
Tibet (Xizang), 332, 363–364
Buddhism, 52, 203, 357, 363
China and, 344, 357, 358, 363
Dalai Lama and, 363
high altitude climate and, *48*
Himalayan highlands, 293
Plateau of, 183, 292, 330, 364, 374
Qinghai–Tibet Railway, *364*
Tierra caliente, 104, 106, 127
Tierra del Fuego, 101, *102, 105*
Tierra fria, 104, 106
Tierra helada, 104, 106
Tierra templada, 104, 106, 127
Tigris River, 148, 207, 213, 218
Timbuktu (Tombouctou), 259, *260*
Timeline. *See* Visual timeline
Timor-Leste, 384, 388, 391, 395, 397
Tokyo
air pollution, *339*
Disneyland, 29
Ginza, *351*
wealth in, *351*
Tombouctou (Timbuktu), 259, 260
Tonle Sap Lake, 382
Torture, 124, 322, 386
Total fertility rate (TFR), **15,** 313, 397
Tourism
Antarctica, *439*
Europe, 159
Hawaii and, *407*
Oceania, 427–428
sex, 390, 402, 405
Southeast Asia, 389–390
sustainable, 428
U.S./Canada, 78
water pollution and, 38
Town charters, 149
Trade. *See also* Drug trade
ASEAN and, 389–391, *390*
BRIC and, 189
EU and, *157*
fair, 32
free, 31–33
maid, 399–400
SEZs and, 350, 353
Silk Road, *183, 184,* 187, 363, 364
Spanish/Portugese trade routes (circa 1600), *112*

tropical timber (Southeast Asia), *377*
U.S./Canada interdependencies and, *78*
Trade blocs. *See also* ASEAN; European Union; Mercosur; NAFTA
in Middle and South America, *117*
regional, 32
Trade deficit, **83,** 156, 157
Trade winds, **104**
Trail of Tears, 71, 73
Tram, *162*
Trans-Alaska pipeline, 62, *64, 65*
Trans-Amazon Highway, 107, 108
Transmigration schemes. *See* Resettlement schemes
Transportation
railroads/rail, 72, *72*
transportation networks (North America), 81–82
Transvaal, 263
Treasury securities, foreign holders of, 355, *355*
Tree hugging (*Chipko*), 296, 300
Tripura, 287
Tropical cyclones, 292, 334
Tropical mountain zones, 104
Tropical savanna. *See* Tropical wet/dry climates
Tropical timber, 377
Tropical wet/dry climates, *48, 48, 105, 253, 376, 413*
"the Troubles," 155
Tsunamis, 36, 292, **332,** 334, 336, 373, 374, 393
Tubbataha Reef, 382
Tulip Revolution, 194
Tundra, *24, 48, 64, 108, 144, 176, 179, 215, 299, 339, 378, 417*
Turkey, 50, *51,* 142, 145, 148, 156, 158, 196, 208, 209, 212, 236, 239, 243
Twitter, 83, 238
Two child policy, 361, 395
Tymoshenko, Yulia, 205
Typhoid, 277
Typhoon Pongsona, *418*
Typhoons, **334,** 381, 382, 383, *418.* *See also* Hurricanes

UAE (United Arab Emirates)
Dubai, *11*
gender imbalance, 229
greenhouse gas emissions, *40*
Palm Jumeirah, 229, 230, 232
Udon Thani, 397
Uganda, *10, 12, 43,* 280, 282
Ukraine
NATO and, 196
NGOs and, *172, 195*
Tymoshenko and, *205*
Ulan Bator, 340, 346

Uma Bawang, 370, 371
UN. *See* United Nations
UN Food and Agriculture Organization, 379
UNCLOS (United Nations Convention on the Law of the Sea), 421
Underdevelopment, Middle and South America, 112–113
Underemployment, **190,** 317, 329
Undernourishment. *See* Malnutrition
Undocumented immigration, 30, 76, 84, 89, 90, 95, 119, 163
Unemployment. *See also* Global economic downturn
Afghanistan, 296
African-Americans, 90
Canada, 79
Central Valley, 58
deindustrialization and, 157, 301–302
economic downturn and, 30
EPZs and, 116
EU and, 139, 157, 166
Europe and, 167, 168
Israel, 240*t*
Malaysia, 30
malnutrition and, 58
Maori and, 433
Mendota, 58
Muslims (in Europe), *165*
North Africa and Southwest Asia, 235
Palestine, 240*t*
Russia and post-Soviet states, 190, 202, 204
Sardar Sarovar controversy and, 288–289
social safety net and, 79
sub-Saharan Africa, 266
U.S., 79
Zimbabwe, 273
UNHDI. *See* HDI
Union Carbide Corporation, 300
Union of Soviet Socialist Republics (USSR/Soviet Union), **172–173.** *See also* Cold War; Russia and post-Soviet states
China treaty and, 347
collapse of, 74, 154, 178, 181, 186, 188, *199, 200,* 204
Russia and post-Soviet states and, 172–173
Soviet-Afghanistan war, 187, 194, *195*
Stalin and, 176, 185, 186, 187, 189, 196
United Arab Emirates. *See* UAE
United Nations (UN), **36**
United Nations Convention on the Law of the Sea (UNCLOS), 421

United Nations Human Development Index. *See* HDI
United States (U.S.). *See also* Global economic downturn; Iraq War; North America; Obama, Barack; September 11th attacks
abroad, 74
Afghanistan war and, 74, 76, 239, 321, 428
air pollution, 78
atheism in, 92
Canada/U.S. relationships, 75–97
climate change vulnerability, *43*
democratic system, 57, 76, 78–79
democratization/conflict issues, 76–77
economic issues, 80–85
greenhouse gas emissions, *40*
health-care system, 57, 79
immigration issues, 78, 88–90
income (race/household/education), *91, 93, 93*
indigenous groups and, 96
infant mortality rates, 79*t*
oil dependency, 74–75
population (by race/ethnicity), *91*
population pyramid, *96*
refugees and, 89
social safety net, 79
virtual water and, 37
women in parliament, 80, 202
Unplanned urbanization, 27, 41, 99, 129, *130,* 131, 247, 277, 284, 431. *See also* Slums
Ural Mountains, 139, 173, 183, 185
Urals, 173, *175,* 176
Urban pollution, 178, 181, 205. *See also* Air pollution; Water pollution
Urban sprawl, **65,** 76, 86, 88, 93, 94, 95, 96
Urbanization, **26–27.** *See also* Slums
climate change and, 56, 65, 68, 96
defined, 26–27
democratization and, 136, 150–151, 153, 169
demographic transition and, 16
development and, 170, 178, 181, 204, 326, 328, 350–354, 366
East Asia, 326, 327, 328, 329, 334, *341,* 350–354, *351,* 357, 359, 366
environmental issues and, 170, 178, 181, 204
Europe, 136, *137,* 149, 150–151, 153, 160, *162,* 163, 169

food production and, 54, 98, 121, 134, 286, 287, 317, 320, 325
globalization and, 206, 227–231, 230, 244, 350–354
habitat loss and, 65
human impact on biosphere and, 24
Middle and South America, 98, 99, 109, 121, 126, 129, 130, 131–132, 133, 134, 135
migration and, 129, 135, 227–231
mobility and, 95
North Africa and Southwest Asia, 206, 207, 218, 222, 227–231, 229, 230, 243
North America, 56, 57, 65, 68, 86, 87, 88, 95, 97
Oceania, 406, 407, 415, 429–431, 430, 437, 438
Papua New Guinea, 429, 430
photo essay, 28–29
population growth and, 246, 275, 276–280, 284
population patterns and, 160, 198
push/pull phenomenon of, 27, 396
Russia and post-Soviet states, 170, 171, 178, 181, 198, 200, 201, 204, 205
South Asia, 286, 287, 311, 314, 315, 317, 320, 324, 325
Southeast Asia, 368, 369, 396–399, 398, 400, 401, 402, 403, 404
sub-Saharan Africa, 246, 247, 273, 275, 276–280, 278, 284
suburbanization and, 86, 95
as thematic concept, 11, 26–27
unplanned, 27, 41, 99, 129, 130, 131, 247, 277, 284, 431
water pollution and, 38, 134
Uribe, Álvaro, 34
U.S. See United States
U.S.-lead war in Afghanistan, 74, 76, 239, 321, 428
U.S.-Mexico border
cultural blending along, 7
immigration issues, 90
living standards along, 118
maquiladoras and, 116, 118, 119, 388
Nogales, Mexico, 110, 111, 118
Utah, 63, 88, 91
Utah Valley, 81, 95
Uygurs, 358, 362, 363
Uzbekistan, 171, 173, 176, 180, 181, 182, 184, 188, 202, 203

Vaccinations, 36, 279
Vaishyas, 306, 307
Valdez oil spill, 215, 272
Valdez port, 62

Values, 50–51
Vancouver, 87
Varanasi (Benares), 289, 298, 299
varna, 306, 307
Vedas, 305
Vegetation. See also Fauna/flora
Europe, 142
North Africa and Southwest Asia, 209, 213
photo interpretation and, 8
Russia and post-Soviet states, 176
South Asia, 292–295
Southeast Asia, 374–375
sub-Saharan Africa, 252
Veils/veiling, 226, 226, 309
Venezuela, Chávez and, 118, 119, 121, 123
Vietnam
Agent Orange, 385, 387
feminization of labor, 388
Hanoi, 394, 398
Ho Chi Minh City, 372, 382, 396, 398
refugees from, 431
Vietnam War, 384, 385, 386, 387, 402
Vignettes (personal vignettes)
Amah day, 400
Arizona-Mexico border tragedy, 90
banana plantation (Costa Rica), 120
birth control (Philippines), 395
ecotourism (Ecuador), 109
favelas (Fortaleza, Brazil), 131–132
The Girls of Riyadh, 227
gospel of success, 282, 283
Grameen Bank, 317–318, 320
informal economy, Moscow, 203
Khmer Rouge, 386
Koli village, 315
Loess Plateau, 336
Mexicali workers (Thompson Electronics), 117–118
Panipur/communal conflict, 309
pemaes, 400
Piailug (sailing trip), 421–422
Radio Sahar, 311
red light district (Watsanah K.), 403
Soufrière, 30–31
street bazaar (Moscow), 192
windmill (Malawi), 270
Yemen (Raufa Hassan al-Sharki), 238
Village Bank (Grameen Bank), 317–318, 320
Virtual water, 37, 37t
U.S. and, 37
Visual timeline
East Asia, 346–347
Europe, 152–153

Middle and South America, 114–115
North Africa and Southwest Asia, 222–223
North America, 70–71
Oceania, 424–425
Russia and post-Soviet states, 184–185
South Asia, 304–305
Southeast Asia, 386–387
sub-Saharan Africa, 260–261
Volcanoes
in Java, 373
Montserrat, 103
Ring of Fire and, 44, 46
tectonic plates and, 44
Volga River, 173, 174, 183, 185
Voodoo, 132, 282
Voters, women, 19
Afghanistan, 77
Congo, 10, 35
Guatemala, 134
North Africa and Southwest Asia, 238, 239
North America, 80
Oceania, 436
Yemen, 12
Vulnerability to climate change. See Climate change vulnerability

Walls, Michael, 90
Walmart, 30, 82, 83, 84
War on Drugs, 124
War on Terror, 74, 77, 235, 323
Wars. See also Civil wars; Cold War; Democratization/conflict issues; Iraq War; Regional conflicts; World War I; World War II
Boer, 263
environment and, 25
Gulf War, 215, 231
Revolutionary War, 72
Soviet-Afghanistan war, 187, 194, 195
U.S.-lead war in Afghanistan, 74, 76, 239, 321, 428
Vietnam War, 384, 385, 386, 387, 402
Water. See also Dam projects; Glacial melting; Irrigation; Virtual water
access to, 38
bottled water, 38
climate change and, 41–42
fossil water, 66, 216
ownership of, 37–38, 39
privatization, 37
quality, 38
Sub-Saharan Africa and, 257–258
as thematic concept, 12, 37–38
urbanization and, 38

Water footprints, 37, 38. See also Virtual water
Water issues/scarcity, 39. See also Virtual water
Afghanistan, 296, 297
East Asia, 327
Kazakhstan, 180, 181
Middle and South America, 99, 109
North Africa and Southwest Asia, 207, 213–216
Pacific Islands, 407, 416
South Asia, 287, 296, 298, 300
Southeast Asia, 381
Water pollution, 39. See also Oil spills
Europe, 136, 137, 142, 145, 146, 148, 168
Mediterranean Sea, 136, 137, 142, 145, 146, 148, 168
North America, 57, 66, 68, 68
Russia and post-Soviet states, 180–181
South Asia, 299
urbanization and, 38, 134
Wealth. See also Income disparities; Poverty
New York City, 87
population growth rates and, 16–17
Tokyo, 351
wealth transfers (colonies to Europe), 150
Weather. See also Climates
climate v., 45, 47
defined, 45, 47
Weathering, 44
Wegener, Alfred, 44
Welfare states, 151
Welfare systems. See Social welfare/protection systems
West Bank, 208, 209, 231, 237, 239, 240, 241, 242, 243, 244. See also occupied Palestinian Territories
West Bank barrier, 242, 242, 243
West Europe, 136, 138, 139
West Pakistan, 303, 321. See also Pakistan
West Siberian Plain, 173, 174, 176, 198
Western Europe
countries of, 166
human impact on biosphere, 144
term usage, 139
Western Ghats, 291, 292, 298
Western Hemisphere, 6
Western Samoa. See Samoa
Western Siberia, 184
Wet rice cultivation, 255, 338, 369, 381, 404
Wetland loss (Louisiana), 62
Where/why questions (geography), 2–3

Wildebeest migration, 256, 257, 258
Wildlife issues, 258
Wind farms, 97, 145, 365, *437*
Wind power, 139, 296, 416, 437
Wind turbines, *137*, 169
Windmill (Malawi), 270
Women. *See also* F/MEI maps; Gender; Gender inequalities; Sex industry; Women's status
    Afghanistan and, 309, 311
    cultural preference for boys, 13, 15, 18, 227, 312, 313, 327, 360–361
    double day and, 18, 165, 166, 201
    EU and, 166
    factory girls, 71
    female infanticide, *13*, 310, 312, 361
    female seclusion and, 225–226
    FGM and, 281, 283
    F/MEI and, 20
    girl mortality rates, 15–16, *16*
    in government/business (North America), 80
    maid trade and, 399–400
    marianismo and, 128, 133
    population, China and, *13*
    Saudi Arabia and, 226, 227
    sex workers, 202, *202*, 204, 261, 280, 307, 361, 402, 403
    Taliban and, 311
    veils/veiling and, 226, *226*, 309
    water collection and, 39

women's roles (Europe), 164–167
working women (Europe), *164*
working women (North Africa and Southwest Asia), 228
Women in parliaments
    Russia and post-Soviet states, *202*
    Sub-Saharan Africa, 247
    U.S., 80, 202
Women voters, 19
    Afghanistan, 77
    Congo, *10*, 35
    Guatemala, *134*
    North Africa and Southwest Asia, 238, 239
    North America, 80
    Oceania, 436
    Yemen, *12*
Women's status, 17–18, 18*t*
    Islam and, 226–227
    male/female income, 18*t*, 57
    North America, 85
    Russia and post-Soviet states, 202–203
    sub-Saharan Africa, 247
Workers, guest, **163**, 227, 229, 231, 244, 359
World Bank, 32, 36
World Cup
    rugby competition, *409*
    soccer, *261*, 263
World regions, *1*, *7*. *See also* East Asia; Europe; Middle and South America; North Africa and Southwest Asia; North

America; Oceania; Russia and post-Soviet states; South Asia; Southeast Asia; Sub-Saharan Africa
    scale, 7
    world regional approach, 3
World Trade Organization (WTO), **32**, 36
World War I, 93, 151–152, 153, 154, 161, 185, 221, 224, 240, 241, 261
World War II, 151–152
    air transportation after, 82
    Bengal Famine and, 304
    Cold War and, 186–187
    Europe and, 151–154, 161
    female workers and, 85
    Holocaust and, 152, 221, 231
    Indonesia and, 391
    Japanese defeat and, 345
    Japanese officials (surrendering) and, 77
    Japan's economic recovery after, 348–349, 354
    Nazi death camps, 240
    North Africa and Southwest Asia after, 221
    nuclear families after, 93
    Oceania and, 424
    Russia and post-Soviet states after, *195*, 198
    Soviet Union and, 185, 186–187
    urbanization after, 86, 94
    U.S. presence in Europe after, 77, *77*

World Wildlife Fund, 36
WTO. *See* World Trade Organization
Wuxi, China, *365*

Xian, *13*, 363
Xinjiang Uygur Autonomous Region, 188, 334, 357, 362, 363
Xizang. *See* Tibet

Yachana Lodge, 109
Yangtze River. *See* Chang Jiang
Yarmouk River, *215*
Yarmulkes, 164
Yellow River. *See* Huang He
Yemen
    Sharki and, 238
    voters, *12*
Yokohama, 348, 350, 361, 422
YouTube, 83, 226, 408
Yucatán, Mexico, *48*
Yucatan Lowlands, *103*
Yunnan Province, 364
Yunus, Muhammad, 317
Yurts (gers), 183, 346

Zambezi River, 257
Zanzibar, *284*
Zapata, Emiliano, 120
Zapatistas, 120–121, 124
Zebras, 258
Zheng He, 342, 343
Zimbabwe, 259, 273, 274, 275
Zionists, **240**

ARCTIC OCEAN
Ellesmere Island
Beaufort Sea
Baffin Bay
Greenland
Greenland Sea
Mt. McKinley (Denali) 20,320 ft.
Yukon
Mackenzie
Great Bear Lake
Baffin Island
Davis Strait
Prime Meridian
Bering Sea
Great Slave Lake
Hudson Bay
CANADIAN SHIELD
Iceland
Gulf of Alaska
ROCKY MOUNTAINS
Peace
Saskatchewan
Lake Winnipeg
Labrador Sea
Ireland
British Isles
Aleutian Islands
Newfoundland
Columbia
Missouri
Great Lakes
St. Lawrence
NORTH PACIFIC OCEAN
Great Salt Lake
GREAT PLAINS
Ohio
Appalachian Mts.
NORTH ATLANTIC OCEAN
Iberian Peninsula
Death Valley -282 ft.
Colorado
Arkansas
Azores
Hawaiian Islands
Baja California
Rio Grande
Mississippi
Gulf of Mexico
Greater Antilles
Canary Islands
Atlas Mts.
SAH
Cuba
Lesser Antilles
Hispaniola
Caribbean Sea
S
Galápagos Is.
Orinoco
Gulf of Guinea
Equator
Marañón
Negro
Amazon
Niger
AMAZON BASIN
Madeira
São Francisco
BRAZILIAN HIGHLANDS
Polynesia
ANDES
Paraná
Easter Island
Pampas
SOUTH ATLANTIC OCEAN

Land Elevations
meters	feet
4877	16,000
3353	11,000
2134	7000
914	3000
305	1000
152	500
0	0

Ocean Depths
meters	feet
0	0
300	984
3500	11,483
5000	16,404

SOUTH PACIFIC OCEAN
Aconcagua 22,834 ft.
Rio de la Plata
Valdes Peninsula -131 ft.
Patagonia
Falkland Islands
Tierra del Fuego
Weddell Sea

Scale at Equator
mi 0   500   1000   1500   2000
km 0   500   1000 1500 2000

Vinson Massif 16,066 ft.